Basic Mathematics
You can count on yourself

SECOND EDITION

Basic Mathematics
You can count on yourself

LINDA D. FALSTEIN
Director of Mathematics Skills
University of Massachusetts/Boston

ADDISON-WESLEY PUBLISHING COMPANY
Reading, Massachusetts • Menlo Park, California
Don Mills, Ontario • Wokingham, England • Amsterdam
Sydney • Singapore • Tokyo • Mexico City
Bogotá • Santiago • San Juan

Sponsoring Editor	Jeffrey M. Pepper
Art Consultant	Loretta Bailey
Cover Design	Bob Barner
Text Design	Lauren Bellinger
Illustrator	Illustrated Arts
Manufacturing Supervison	Ann Delacey
Copy Editor	Jackie Dormitzer
Production Supervisor	Marion E. Howe

Library of Congress Cataloging in Publication Data

Falstein, Linda D.
    Basic mathematics.

    Bibliography: p.
    1. Mathematics—1961–   .   I. Title.
QA39.2.F37   1968   510   85-5969
ISBN 0-201-13363-6

Reprinted with corrections, October 1986

CDEFGHIJ-MU-8987

Dedication

 o all my former students who had trouble with math but eventually learned that by thinking through a math problem, **they could count on themselves**.

To the loving memory of my wonderful companion, Aurie.

Preface

Improvements in this edition. This is the second edition of *BASIC MATHEMATICS YOU CAN COUNT ON YOURSELF*. The organization of topics has been improved, the examples have been expanded to show more applications of concepts, and the text now includes the algebraic topics of graphing and solving inequalities, rational exponents, rational equations and their applications, graphing of parabolas, and word problems that can be set up and solved by quadratic equations. Three appendixes have been added on number bases, social science graphs, and the metric system. The discussion of concepts, the examples, and the exercises are still friendly and supportive but I've tried to make them more interesting than they were in the first edition.

Most revisions were stimulated from reactions to using the book for three years in the two fundamental skills math courses at the University of Massachusetts/Boston. I have responded to the feedback I received from students in those courses, from other instructors, from reviewers of the manuscript, and from my own experience of teaching from this text.

I developed our skills program in 1978 for students whose math skills were so weak or rusty that they were not prepared to succeed in the Math department's fast-paced basic algebra course. We are delighted with the results from using these materials to prepare students for more advanced work in algebra.

Overcoming math anxiety or fear of failing. Students who had difficulty understanding basic math when it was first taught to them often come back to math courses as adults with a sense of discouragement. They feel skeptical about succeeding. Some people call this "math anxiety"; many of my students have called it "math terror." I call it "fear of failing." These materials were developed to help overcome that problem. I believe these students cannot learn math until they get some help in believing that this time there is a reason why things will be different, and that they will now do well. I've tried to address those issues throughout the text.

I hope that people using it will actually *read* the text and go through all the examples before they attempt exercises. I have tried to write this book not only so that students will find it a useful resource to accompany a course but also so that people not in a class could use it independently to learn or review basic mathematics.

I have tried to clearly define all the concepts I use so that the readers will *understand what they are doing* instead of having to memorize rules for computing that don't make sense to them. There is a Glossary in each chapter.

There is a letter to students following the table of contents. Please encourage your students to read it before proceeding into the text.

Strategy for improving test performances.   There is an outline of strategies for improving test performances right after the letter to the students. Most of those suggestions were generated by students participating in a math anxiety workshop at U Mass. I encourage students to read it at the beginning of the course. It has helped me structure discussions about how to study.

Avoiding careless errors.   I have demonstrated in the examples how to work out problems by performing only *one computation* per step and by making each step easy, thus helping eliminate the careless errors that result when students do too much in their heads at one time. There are examples for every type of problem that appears in the exercise sets.

I have made suggestions throughout the text that will simplify the work, avoid confusion, and help eliminate careless errors. I have slowly built up the complexity of the problems I have both shown students how to do and have asked students to do, so they can develop a real mastery of each skill. The success from doing easy problems will help them do the more difficult problems in the exercises.

Checking every problem to find mistakes.   I have included the answer to *every* exercise, indicated at the end of each exercise set where to find the answers, and encouraged students to check all their answers so they will know immediately whether they are making mistakes.

I have tried to make the real-life problems as interesting and relevant as possible.

End-of-chapter reviews and tests.   There are some special features of each chapter that students will find very helpful.

At the end of each chapter is a Chapter Review with problems indentified from each section. Students should attempt all the problems after completing the chapter and check their answers. If they had any difficulty, they can easily identify which sections they need to go back and re-read and/or ask for help with.

Once students feel confident of all skills from the chapter, they can get additional practice on topics they found to be troublesome in the Supplementary Exercises that follow each Chapter Review.

Each chapter also has a Chapter Test where all the skills discussed in that chapter appear. The Chapter Test serves two functions: (1) Since problems are not identified by section in the test, the students get feedback on how well they *recognize how* to do a problem that appears out of the context of related examples. (2) The test also provides feedback on how well the student *retained* a mastery of each skill learned earlier.

Students have told me of their discouragement and frustration when at the end of a course they would forget concepts that they had mastered earlier. In response to that need, I have included at the end of each chapter a Cumulative Review of problems from all the earlier chapters. By continually getting practice using all those skills, students have less trouble retaining that mastery. I assign those exercises at the end of each chapter, and find they help me identify those students who need extra help.

Instructor's test manual.   There is a test manual available to instructors using this text that provides three more versions of each chapter test, and three tests from cumulative chapters with a prescription guide for each exercise that directs students to where in this text they can review that material.

Appendixes.   Appendix A is about Number Bases. I find students develop a better understanding of place values and also the ability to compute with decimals after a brief discussion of number bases. I use it just before I begin discussing Chapter 4.

Appendix B is about Social Science Graphs. Most students who are learning basic mathematics need to develop familiarity with the graphs they encounter in social science courses, newspapers, and magazines. Appendix B can be used at any time, but the circle graphs assume a mastery of percents (which are covered in Chapter 4).

Appendix C is about the Metric Measurement System. Chapters 3 and 4 and Sections 5.1 should be read before using it.

Text used for courses at U Mass/Boston—materials available.   We now have two math skill courses in our program at U Mass, and we use this text for both of them. We use Chapters 1–4 in our Pre-Algebra skills course and Chapters 5–9 in our Introduction to Algebra skills course. The material in Chapters 5–15 are covered in the math department's basic algebra course.

In our algebra course we do not teach the pre-algebra topics, but we administer a cumulative review on that material to each student and provide a prescription guide to this text to help them diagnose the areas where they might need more work and help them find that help in this text.

I make up day-by-day syllabi for these courses for a 14-week semester both on a twice-a-week and a thrice-a-week schedule, and for six-week summer-school schedules. I also have assignment sheets available for using this text in each of those courses. I hope you enjoy using this text. Thank you for ordering it. If there is anyway I can be of assistance to you, by sharing materials that we have developed in our program, please let me know. Write to me:

Linda D. Falstein, Director of Mathematics Skills
Academic Support and Advising
University of Massachusetts/Boston
Harbor Campus
Boston, Massachusetts 02125

Acknowledgments

would like to thank all the students who have participated in the Academic Skills Math program at the University of Massachusetts/Boston Campus. They taught me better ways to teach mathematics by sharing with me the creative methods they used for thinking through math problems, and by sharing their insights on how they have learned how to learn. Their feedback is the basis for many parts of this book.

I am also grateful to instructors who have used earlier versions of this material and given me invaluable suggestions that improved it. In particular I want to thank Bob Lee, Jerry Russell, Peter DelTredici, Mark Pawlak, Judy Clark, Kurt Jacobsen, Ed Hoyt, Joe Sheppeck, Dorothy Forest, Nancy Levy, and Alan Zaslavsky.

I want to also thank Alec Marshall and Phil Lewis—two of the best math teachers I have ever met—who taught me invaluable lessons about how to teach math effectively and how rewarding and satisfying it could be to teach mathematics.

I am also appreciative of the several people who read earlier versions of this book and made helpful suggestions. They include Bernice Auslander. Patricia Davidson, Lou Ferleger, and Colin Godfrey from U Mass/Boston; Laura Cameron, University of New Mexico; Sandra Pryor Clarkson, Hunter College; Terry Czerwinski, University of Illinois at Chicago Circle; Les Glaser, University of Utah; Lucy Horwitz, U Mass/Boston/CPCS; Janice B. Koop, University of San Diego; John Loughlin, Lane Community College; Alexis Mancini, College of Staten Island; Michael Marks, Orthstar Enterprises, Inc.; Josiah Meyer, Elmira College; John Spellman, Southwest Texas State University; Robert Tolar, University of Northern Colorado; and Claudia Zaslavsky.

I want to thank Constance Emmett, Anne Goodman, and Roberta Young for the support they gave me in the preparation of earlier versions of this material.

I owe a special thanks to Stanley Phillips, whose weekly golf dates with me during the summer provided the relief I needed to work most of the rest of the time on this book; golf was often so frustrating that by contrast writing this book seemed relaxing.

I am very grateful to all my friends and family whose support and encouragement made it possible for me to be who I am and to have persevered on completing this project. A special thank you is reserved for Julie Schneider who is particularly supportive.

Most of all, I want to thank Dick Cluster for his thoughtful, insightful, invaluable contribution. He frequently helped me find a better way to do what I wanted in this manuscript.

I'm very appreciative of the help and valuable contributions of the staff of Addison-Wesley, particularly Marion Howe, with whom I did not always agree but to whom I am grateful, for her capable supervision of my book's production.

Boston, Massachusetts   L.F.

Contents

4 Decimals and Percents   125

5 Polynomials

171

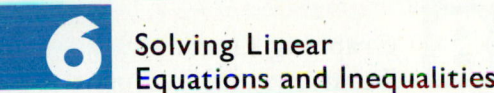

6 Solving Linear Equations and Inequalities 213

7 Applications of Linear Equations 257

13 Multiplying and Factoring Polynomials   549

14 Algebraic Fractions (Rational Expressions)   583

15 Quadratic Equations, Graphs, and Solutions

Appendixes   705

Answers to all Exercises   719
Index   765

Dear Student

elcome to this book! I hope this will be a successful and encouraging experience for you. I want to tell you a little about the ideas that lie behind *Basic Mathematics: You Can Count on Yourself* and about how I think you can make the best use of it.

You can succeed in math even if you never did before.   This book was written for people who thought they were not very good at math. Lots of people decide in their first few years of school that math is too hard for them. There are many reasons why that happens, but there is no one who can't learn math—not even you.

Math is a subject that builds on previously learned concepts. If you had trouble with some math concepts at school, you were at a real disadvantage when the new material was taught. But it is never too late to learn the early material and then go from there and learn the more advanced topics. That is why this book begins with basic math concepts. It doesn't matter what basic skills you did not understand in your early schooling. All the concepts are in this book, and you can learn them now.

Many people didn't learn math the first time because they weren't interested or weren't taught well. Some elementary school teachers have never felt very good about math themselves, and therefore they have a hard time teaching it. Elementary school teachers are unreasonably expected to be equally adept at instructing children in every major subject. It is not surprising that a teacher may be an exemplary instructor in social studies, reading, writing, grammar, and the like but be less talented at teaching math. Such teachers learned only one way to explain a topic. If an intelligent child does not understand an explanation and asks for another, the teacher may not be as receptive to such inquiries as he or she wants to or should be.

Most of us have seen children ridiculed by teachers or peers for asking such questions, and so we learn not to ask. That's the safe way to protect ourselves. In many situations it may be necessary, psychologically, but it carries a high price: Many of us go around for years believing we can't understand math. It is not necessary to remain in such a state. It doesn't matter why you didn't learn it the first time—you can learn it now.

Learning math is not a spectator sport.   Just as you suspected, there is a catch. You can learn math and learn it well, but there are two things you have to do. First, despite how discouraged you might have always felt, you must believe that it is possible to succeed now. Unless you believe you can do it (even in a small way), it is hard to succeed.

The second condition is even easier: You must do math problems in order to learn math. There may be some things that you can learn just by reading or by observing what other people do, but not math. I spent years as a teacher watching people fail to learn because they refused to do the work. If you understand how I do the problems in printed examples or how your teacher does them at the chalkboard, you will have taken the first step toward learning how to do such problems. But you won't really learn in a way that will last until you get some practice doing them correctly.

There is only one way to learn math and to retain the understanding. That way is to *do* math. Homework is not busywork! The process of practicing is where the learning takes place. Learning math is like learning to play a musical instrument or to master a sport. You learn by doing. Keep practicing each skill until you are so comfortable with it that you don't have to stop and think about how to do it—until you can do it automatically. That is what it means to really understand something.

Learn what that feels like to you. Practice until you are fluent.

Don't memorize rules—strive to **understand** what you are doing.   I do not assume you know how to do any of these math operations before reading the text, so if you have not studied these topics before, you are at no disadvantage. I have tried to develop these lessons to help you understand *why* you do the things I am asking you to do. If you do not understand something, ask someone for help.

The way many ingenious people try to learn math without understanding the basics is by trying to memorize everything. Memorizing concepts that you don't understand is much harder than learning to understand the concepts in this book. Such students make learning math much harder than it need be; they give math a bad name. Don't settle for not understanding something in this book. Get help. Ask your teacher or a friend. Don't give up and simply try to memorize it.

There isn't much math you do have to memorize. Most things will be "remembered" because you understand them and got enough practice doing the problems. There is a big difference between memorizing and remembering. The latter is better, easier, and more enjoyable.

You can learn a lot from making mistakes.   Making mistakes in math can be constructive. If you have the right attitude, you'll learn from your mistakes. We all make them. Learning how to solve problems is a process in which we constantly make mistakes. But by thinking about them, we figure out what is really happening or should be happening. A wrong decision helps narrow the strategy until we eventually make the right decisions.

Check your work.   You cannot benefit from an error unless you know it was wrong. The most important part of learning math is to check your answer to be sure you are right. The answer to every exercise is in the back of this book. The page number is at the end of each set of exercises. The pages are perforated—you can remove the answers and put them somewhere handy so you won't have to keep turning to the back, if you prefer.

If our answers disagree, try to figure out how I got my answer. That process will teach you a lot about the concepts and your means of working. Usually you will find the error quickly. By seeing where you went wrong, you can avoid making that same mistake in the future.

Using this book.   The book has 15 chapters altogether. The first 4 are about pre-algebra topics, which will provide you with a strong foundation to make learning algebra easier and more pleasant. The other 11 involve beginning algebra topics.

You may have covered some of these topics in other courses, and the teacher might have taught you a different method. If you know how to do the problems, that is the important thing. You do not have to use the methods I am suggesting, but I would encourage you to try them. I think that some of my methods, although different from those you may have learned, are easier. Some of my methods for computing numbers are required for computations in algebra. If you get used to these methods when working with numbers, it will be easier for you to use them when you work with letters in algebra. So try my methods, but remember: The important thing is that you know how to do these operations, and that you understand what you are doing.

As helpful as you find this book (and I hope that you find it very helpful), you can make it into a more valuable resource and reference for yourself than it is now. You can use the extra marginal space near the exercises to show all the computations and steps you perform. The answers are already here, so merely listing the answer won't have any effect. But if you show your work in your own (neat) handwriting now, you can refer to your computation in later years when a similar problem comes up.

Perhaps some teachers have preferred that you not show your work. When I was a student, everyone was impressed with those who did all the work in their heads. But that is not the impressive way to do math. Doing work in your head is an open invitation to make lots of careless errors. Now that you are finally learning these skills, why get answers wrong because of carelessly doing the work in your head? Intelligent students show their work. On exams a student who shows his or her work and makes a careless error loses only a few points in most situations, but the student who does not show the work and makes a careless error loses all possible points with the wrong answer. And don't do computations on scratch paper—scratch paper gets thrown out.

Do the review work at the end of each chapter after the chapter is completed. At the end of each chapter there is a Glossary of terms used in the chapter, a Chapter Review for reviewing the kinds of skills learned in each section, a set of Supplementary Exercises for each section, a Chapter Test, and (beginning with Chapter 2) a Cumulative Review involving many of the skills learned in the earlier chapters. After completing all the sections of a chapter, you should attempt all the problems in the Chapter Review. Check your answers to those problems. If you made any mistakes, go back and reread the section(s) where those problems came from. If you still have any questions about that material, ask someone for extra help. Once you feel confident that you understand, do all the Supplementary Exercises for each section where you made an error on the Chapter Review. Check all the answers to the Supplementary Exercises, and again ask for more help and more examples to do it there still is any confusion.

When you have learned how to do every kind of problem that appears in the Chapter Review, do all the problems in the Chapter Test. These problems are drawn from each section in the chapter. It is the first time you are seeing them all mixed together, out of the context of their sections. You have to recognize what skills are being asked for. Check your answers after completing the test.

After the Chapter Test, do all the problems in the Cumulative Review. Using your skills periodically helps to keep them strong. Even when you thoroughly understand a topic, you can forget the skill if you go a long time without practicing. By doing the Cumulative Review at the end of each chapter, you will keep your basic skills well practiced. That will result in less forgetting, and in easier relearning if you forget them later.

Don't do the reviews and tests at the end of a chapter until you have finished all the homework from that chapter. By waiting, you will get valuable feedback from those materials. Do you recognize how to do a problem if you haven't recently done ten others just like it? Do you still remember how to do the problems you learned a week ago? The only way to tell is by waiting to do those problems until you have finished the chapter.

The important goal is not only doing the problems—it is also being sure that you will be able to do them in the future too.

Avoid making careless errors.   You can tell what computation is being done in each step of the examples: It is highlighted by use of color in most cases and by underlining in others when that seemed the better way to clarify the computation for you.

EXAMPLE $3 + 4 \times 5 =$ (The 4 and 5 are being multiplied, so they appear in color.)
$3 + 20 = 23$

When 3 and 20 are being added, there is no highlighting, since that is the only possible computation to do. When numbers are plugged into expressions, they appear in color.

I hope that when you are working on the problems in the exercises, you will use the methods I demonstrate in the examples. Perform only *one* computation in each step, and completely rewrite the problem that results from each computation on the line beneath it. Make complicated problems into several easy ones.

Thank you for buying this book. I hope it helps you learn (or relearn) basic mathematics, so you will have a strong foundation and can go on to learn more mathematics, understand quantitative concepts in nonmath courses, and realize that of course you are capable of successfully learning anything!

Strategy for Improving Test Performances

**A** very important part of learning to become a successful student is to develop a strategy for being able to perform as well as possible while you are taking tests. Many intelligent, hard-working people fail to test well because anxiety, or fear of failing, keeps them from thinking clearly.

The key is to have an optimistic perspective about testing. Instead of dreading it as an experience in which you feel like a helpless victim, view the test as an opportunity to *demonstrate how well you have learned the concepts*.

If you cannot remember how to do a problem, or if you make any mistakes, don't feel discouraged—use the test as a tool to help you diagnose which topics are not as clear to you as you had thought. Tests help you recognize where you need either further clarification or more practice doing problems and checking that they were done correctly.

If you don't perform as well as you would like on tests, I hope this guide will help you develop a strategy for improvement. There are three stages:

Preparing for the test

Taking the test

Relating to the results

Preparing for the test

1. *Anticipate* that you will be tested on everything you are taught. Stay up to date and try not to get behind schedule. If you already *know* the material, you'll need only to review, not learn, it for the test.

2. Don't wait until the night before a test for your *heavy* studying. As much as possible, study two weekends in advance to allow time to act constructively on any weaknesses you unearth from your review. Try to get help on those areas from your teacher, a tutor, or someone else who understands the material.

3. Practice solving problems so that you become very adept and can do them quickly but accurately.

4. You know what chapters the test will cover. Make up your own practice test. If you find problems that you are unable to do when you close your book and simulate the test situation, you'll know where you need to do more work *before* taking the real test. Get help if necessary.

5. Close your eyes for a minute while you picture the exam room. Imagine taking the exam, staying relaxed, and doing well. Repeat this exercise as often as possible.

6. *Be able to think during the test.* Get a good night's sleep. Don't study late the night before. Don't study the last hour before trying to fall asleep. Be relaxed so that you *can* sleep. Eat a good breakfast—you need energy to think, and you don't want to be distracted by hunger.

Taking the test

1. Feel confident that you are going to do well. (Try to remember what it felt like to hit a home run when you were a kid or to do anything you were good at.)

2. Take a few deep breaths, or do any relaxation technique you are familiar with, as soon as you sit down.

3. Choose a comfortable place to sit.

4. If possible, plan to get to school earlier than usual to be sure you will be in class on time.

5. Have sharp pencils and an eraser (don't use pens), so you can turn in *accurate* and *readable* work.

6. Begin doing problems in the order they appear, and try to show as much work as you can (reflecting your good thinking) on the test paper.

7. If you come to something you don't remember, skip it and try not to worry about it. Use your time to show what you *do remember* rather than getting upset by trying to recall how to do that one problem.

    Worrying breaks down positive attitudes and discourages and exhausts you so much that it makes it harder to think clearly on the rest of the test.

8. Check all your work so *you* can find and correct any careless errors.

9. Don't change an answer that you are unsure of. Change an answer only if you know it is wrong and are sure what the right answer is.

10. Don't leave the test early if anything is unchecked or undone. Take a couple of deep breaths and try to relax before going back to try the hard ones. They'll seem easier once you know you have done all the other problems well.

11. Redo calculations, without looking at your original work, to check for careless errors. It's hard to *find careless mistakes*. You may not notice it as wrong when you write $3 + 2 = 6$ in your own handwriting, but you probably won't make that mistake again in a new computation. If you get different answers, it's a signal that there is a mistake somewhere. If you know there is an error, you'll find it.

Relating to the results

1. Use the results to help you focus on areas that were not clear to you. Get help on that material.

2. Ask for clarification if you don't understand why something was marked wrong.

3. Ask for an explanation if you think too few points were rewarded for partially correct work.

4. If you were surprised that on the test you couldn't do a problem you could do at home, think carefully about how it felt at home. Figure out the difference between how it felt when you practiced that kind of problem at home and how it felt when you practiced the ones that you remembered how to do *during the test.*

That feeling is the experience of *really understanding something*. It is your indication of when you have practiced a skill enough. Learn to recognize that feeling.

In the future, practice problems until you feel as confident about those skills as you did about the problems you remembered so well on your last test.

5. Figure out what you can learn about the teacher's testing techniques and your learning techniques to improve on the next test.

In general, always figure out *how you can do things differently so that your next experience will be more successful.*

Linda D. Falstein
Director of Math Skills
University of Massachusetts/Boston

Whole Numbers

⬜⬤⬜ Place Value

We can learn to understand the rules for computing with numbers because, for the most part, the rules make sense. We'll review the rules for computing with whole numbers, signed numbers, fractions, and decimals in the first four chapters of this book. By defining all the concepts we'll use, I hope you will easily *understand* the rules, agree that they make sense, and therefore readily be able to use and *remember* (not memorize) them.

This chapter reviews the arithmetic operations with **whole numbers**. Those are the numbers we are familiar with and use for counting. Examples of whole numbers are 0, 1, 2, 3, 4, 7, 28, and 315.

There is an infinite amount of whole numbers. But for representing these numbers, our number system makes use of only 10 different digits: 0, 1, 2, 3, 4, 5, 6, 7, 8, and 9. The prefix *deci* means ten, and that is why our number system is called the **decimal system**.

We use combinations of those 10 digits as symbols to represent whole numbers that are larger than 9.

EXAMPLE 1   437 is a symbol that represents

$$400 + 30 + 7.$$ ❑

Each digit is in a place value of 1, 10, 100, 1000, and so on and actually represents an instruction to multiply the digit by its place value.

EXAMPLE 2   In 437,

7 is in the *ones* place, so its value is $7 \times 1 = 7$.

3 is in the *tens* place, so its value is $3 \times 10 = 30$.

4 is in the *hundreds* place, so its value is $4 \times 100 = 400$.

That is why 437 represents $400 + 30 + 7$.   ❑

EXAMPLE 3   743 represents

$$700 + 40 + 3.$$

⇒ *Note*:   The *order* of the digits in a number determines its meaning. Even though 437 and 743 contain the *same digits*, they represent *different* numbers.

To figure out what the place values are, begin on the right with place value 1, and multiply by 10 each time you move one place to the left.

1

In order to make it easier to read numbers that contain more than three digits, we use commas to group together every three digits. Beginning at the right, the first group of three digits represents ones, tens, and hundreds. The second group is thousands, so its digits represent one thousands, ten thousands, and hundred thousands. The next group is millions, so its digits represent one millions, ten millions, and hundred millions—and so on.

EXAMPLE 4   The number

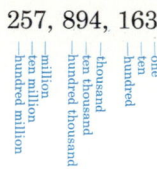

$$257, 894, 163$$

is read: "two hundred fifty-seven *million,* eight hundred ninety-four *thousand,* one hundred sixty-three."   ❏

When we get much past billions and trillions, saying the names of the groups becomes a lot of trouble, so we switch to the other way of expressing place values: exponents, to be discussed in Chapter 11.

EXERCISES 1.1

Write the name for each number symbolized.

1. 5,318   2. 27,294

3. 1,381,267   4. 57,309,173

5. 812,091,430   6. 4,308,194

Check your answers on page 719.   ✓

1.2 Adding and Subtracting Whole Numbers

Concept of Addition

When numbers are added, the result is called their **sum**. The sum of numbers is found by adding together digits representing the same place value. Calculations always begin in the ones place value and move to the left in the numbers.

Digits in the ones place value are added to digits in the ones place value.

Digits in the tens place value are added to digits in the tens place value, and so on.

The largest digit in the decimal system is 9, so if the sum is more than 9, part of that sum must be **carried** over to the higher place value to the left. The value of each place is 10 times the value of the place on its right. So 10 digits in one place are equal to 1 digit in the next place to the left. This concept is easiest seen in terms of money:

Thirty *one-dollar* bills are equal to 3 *ten-dollar* bills.

Ten *hundred-dollar* bills are equal to 1 *thousand-dollar* bill.

If the sum of digits in a place value is 10 or more, carry each 10 into the next place value as a 1.

EXAMPLE 1   Add 6,276 + 23,169 + 308.

$$
\begin{array}{r}
6^1 2^2 7\ 6 \\
2\ 3\ 1\ 6\ 9 \\
+\quad\ \ 3\ 0\ 8 \\
\hline
2\ 9\ 7\ 5\ 3
\end{array}
$$

The sum of the *ones* place digits (6 + 9 + 8) was 23 (2 tens + 3 ones). The 2 is carried to the tens place, and the 3 is put in the ones place.

The sum of the *tens* place digits (7 + 6 + 0 and the carried 2) was 15 (1 ten + 5 ones). The 1 is carried to the hundreds place, and the 5 is put in the tens place.

The sum of the *hundreds* place digits (2 + 1 + 3 and the carried 1) was 7, which is put in the hundreds place.

The sum of the *thousands* place digits (6 + 3) was 9, which is put in the thousands place.

The only *ten-thousands* place digit is 2, which is put in the ten-thousands place. ❏

Careless addition errors are avoided by writing the numbers below each other, as I did in Example 1.

To check an addition problem, add the digits once from top to bottom and then from bottom to top in each column to see if the sum is the same.

EXERCISES 1.2A

Write the numbers below each other and find the sums.

1. 307 + 512 + 713   **2.** 1,391 + 26 + 731

3. 23,061 + 438 + 51 + 3,009   **4.** 721,033 + 43 + 1,860

5. 403 + 1,304 + 197 + 43,017   **6.** 14,381 + 573 + 2,004

Check your answers on page 719.

Concept of Subtraction

Subtraction is the operation of taking one number away from another one. The answer to a subtraction problem is called the **difference**.

To subtract, you should also write the numbers so that digits representing the same place value are below each other, as you did in addition.

When we add numbers, it does not make any difference which number is written first and which numbers are written below, but that order is important in subtraction. The number we begin with is written first, and the number we are taking away from it is written below.

EXAMPLE 2   27 + 43 can be written as
$$
\begin{array}{r}
27 \\
+43 \\
\hline
\end{array}
\quad \text{or} \quad
\begin{array}{r}
43 \\
+27 \\
\hline
\end{array}
$$

57 − 23 must be written as
$$
\begin{array}{r}
57 \\
-23 \\
\hline
\end{array}
$$

57 is the number we start with, and 23 is the number we take away from it. ❏

We can add several numbers at one time. When subtracting we can deal with only two numbers at a time: the one we start with, and the one we take away.

In subtraction we sometimes must **borrow** digits from a place value. It is the reverse of when we *carry* in addition. If the digit that we are taking away is larger than the digit that we started with for any place value, we borrow from the next place value.

If we need a larger digit in the *ones place,* we can *borrow* a digit from the *tens place.* We know that 1 ten is the same as 10 ones. The digit 1 in any place value is the same as 10 in the place value to the right.

EXAMPLE 3

$$
\begin{array}{r}
473 \\
-258
\end{array}
\quad\text{is really}\quad
\begin{array}{r}
400 \\
-200
\end{array}
\begin{array}{c}
60+10 \\
\cancel{70} \rightarrow 3 \\
50 \qquad 8
\end{array}
\quad\text{or}
$$

	(hundreds)	(tens)	(ones)
	400	60	13
	-200	-50	-8
	200	+ 10	+ 5 or 215

If you wanted to take $258 away from $473, that is what you would be doing if the $473 was 4 hundred-dollar bills; 7 ten-dollar bills; and 3 one-dollar bills, or 3 singles.

To take away $8, you'd change 1 ten for 10 singles, giving you 13 singles, minus 8, leaving 5 singles ($5).

You then have only 6 ten-dollar bills ($60) minus $50, leaving 1 ten-dollar bill ($10).

Take 2 hundreds away from the 4 you started with, leaving 2 hundreds ($200). Thus

$$\$200 + \$10 + \$5 \text{ is } \$215.$$ ❏

EXAMPLE 4 $503 - 247 =$

$$
\begin{array}{r}
503 \\
-247
\end{array}
\rightarrow
\begin{array}{r}
{\overset{4}{\cancel{5}}}{}^{1}0\ 3 \\
-2\ 4\ 7
\end{array}
\rightarrow
\begin{array}{r}
4\,{\overset{9}{\cancel{\overset{1}{0}}}}\,{}^{1}3 \\
-2\ 4\ 7
\end{array}
\rightarrow
\begin{array}{r}
4\ {}^{9}\!1\!3 \\
-2\ 4\ 7 \\
\hline
2\ 5\ 6
\end{array}
$$

So $503 - 247 = 256.$ ❏

Addition and subtraction are the opposites of each other. (We say that they are **inverse** operations.) You can use addition to check a subtraction problem.

EXAMPLE 5   Since $503 - 247 = 256$, the sum of that difference and the number you took away should be the number you started with:

$$
\begin{array}{r}
256 \\
+247
\end{array}
\quad\text{is}\quad
\begin{array}{r}
{}^{1}2\,{}^{1}5\ 6 \\
+\ 2\ 4\ 7 \\
\hline
5\ 0\ 3
\end{array}
\quad\text{This check shows we were right.}\quad ❏
$$

When you add two numbers, you can check the answer by subtraction. Subtract either of the added numbers from the sum. That difference should be the other added number.

EXAMPLE 6

$$
\begin{array}{r}
147 \\
+379 \\
\hline
526
\end{array}
\qquad
\text{To check:}\quad
\begin{array}{r}
526 \\
-147 \\
\hline
379
\end{array}\checkmark
\quad\text{or}\quad
\begin{array}{r}
526 \\
-379 \\
\hline
147
\end{array}\checkmark
\qquad ❏
$$

EXERCISES 1.2B

Find the differences. Check by adding.

7. $291 - 57$ **8.** $435 - 284$

9. $704 - 61$ **10.** $6{,}281 - 5{,}749$

11. $326 - 209$ **12.** $43,053 - 21,407$

13. $673 - 495$ **14.** $18,047 - 5,253$

15. $764,071 - 635,849$ **16.** $730,428 - 617,749$

Check your answers on page 719.

Multiplication of Whole Numbers

Concept of Multiplication

For too many people, understanding how to multiply numbers means simply memorizing multiplication tables. But in fact, multiplication represents a computation that can be done without any memorizing.

For example, 3×2 represents a situation in which there are three groups of things, with two items in each group.

$$3 \times 2 =$$

$$(2) + (2) + (2) = 6$$

This situation could be illustrated as $(**)(**)(**) = 6$.

That same problem, 3×2, could also represent a situation in which there are two groups of things, with three items in each group.

$$3 \times 2 =$$

$$(3) + (3) = 6$$

This situation could be illustrated as $(***)(***) = 6$.

The answer to a multiplication problem is called the **product**. It is the same no matter which number tells how many groups and which number tells the amount in each group. Therefore *if you don't remember* what a product is, you can always represent it as an addition problem, which you can compute.

That computation is easier if you make the larger of the two numbers being multiplied tell the amount in each group, and the smaller number tell how many groups.

EXAMPLE 1 4×7 can be represented as four groups of seven:

$$(7) + (7) + (7) + (7) = 28$$

more easily than as seven groups of four:

$$(4) + (4) + (4) + (4) + (4) + (4) + (4) = 28.$$ ❏

EXERCISES 1.3A

Express each multiplication problem as an addition problem and compute the answer.

1. 17×3 **2.** 4×28

3. 3×53 **4.** 35×2

5. 5×23 **6.** 41×4

Check your answers on page 719.

Shortcuts   If both numbers in a multiplication problem are large, computing it as an addition problem is very awkward. There are some shortcuts that simplify the process.

Shortcut 1.   When you can't remember a product, recall the largest product you do know that involves one of the numbers in your problem.
    If you forget what 9×7 is, you might remember that $9 \times 5 = 45$.

$$7 \text{ nines} = 5 \text{ nines} + 2 \text{ nines}$$
$$5 \text{ nines} (9 \times 5) = 45, \qquad 2 \text{ nines} (9 \times 2) = 18$$
$$\text{So } 7 \text{ nines} (9 \times 7) =$$
$$45 + 18 = 63.$$

This situation can be illustrated with circles:

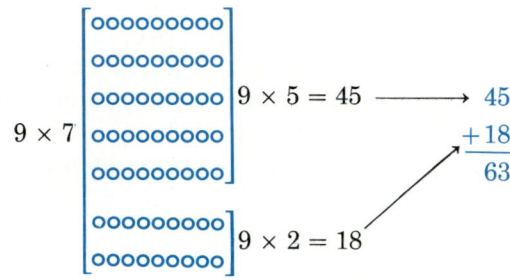

EXAMPLE 2   Use Shortcut 1 to find the product.

$$15 \times 7 =$$
$$(15 \times 4) + (15 \times 3) =$$
$$(60) + (45) = 105 \qquad \square$$

Shortcut 2.   This shortcut is slower but a little easier than Shortcut 1. Start by recalling the largest product you know involving one of the numbers in your problem, as you did in Shortcut 1.
    If you were trying to find the product of 9×7, you might again begin with $9 \times 5 = 45$.
    Now keep adding more nines, one at a time, to these 5 nines. Stop when you get to 7 nines.

$$5 \text{ nines} = 45$$
$$6 \text{ nines} = 45 + 9$$
$$7 \text{ nines} = 45 + 9 + 9$$
$$= 63$$

EXAMPLE 3   Use Shortcut 2 to find product. If you don't remember what 8×12 is, you might recall that $8 \times 10 = 80$.

$$10 \text{ eights} = 80$$
$$11 \text{ eights} = 80 + 8$$
$$12 \text{ eights} = 80 + 8 + 8$$
$$= 96 \quad \text{So } 8 \times 12 = 96. \qquad \square$$

You have probably used other clever shortcuts to figure out products when you forgot the multiplication tables for the numbers in the problem. Feel free to use those methods. You'll do all right as long as you understand a way of getting the right answer. And the more you use numbers, the more readily you will remember their products.

EXERCISES 1.3B

Find the products.

7. 13×5 **8.** 15×7

9. 15×8 **10.** 4×23

11. 23×9 **12.** 8×13

Check your answers on page 719.

Multiplication Notation

There are three ways to denote (write) a multiplication problem:

1. With an $\times$: 3×5

2. With a dot: $3 \cdot 5$

3. With parentheses: $3(5)$ or $(3)5$ or $(3)(5)$.

It is important for you to know all three notations so that you will always recognize when a problem's notation is instructing you to multiply. In algebra we will never use an x to represent multiplication, since x is often used to represent some unknown number.

The standard way to multiply large numbers is to write one beneath the other. Multiply the top number by the units digit in the bottom number and write that product below the line. Then multiply the top number by the tens digit in the botton number and write that product beneath the other product. Begin listing that product in the tens column and move left. Continue multiplying the top number by each digit in the bottom number. Always begin listing the product in the column of the digit from the bottom number that is being multiplied.

Multiplication always begins with the ones digit and moves to the left in the numbers. To find the total product, add the products from each line.

EXAMPLE 4

$$
\begin{array}{r}
34 \\
\times\, 41 \\
\hline
34 \\
+136 \\
\hline
1394
\end{array}
$$

$34 \times 1 = 34$
$34 \times 4 = 136$
So $34 \times 41 = 1394$.

EXERCISES 1.3C

Find the products. Use any method you understand.

13. 17×8 **14.** 32×5

15. 31×4 **16.** 12×17

17. 19×14 **18.** 21×11

19. 43×35 **20.** 52×23

21. 32×67 **22.** 27×83

23. 74×34 **24.** 61×25

Check your answers on page 719.

Division of Whole Numbers

Concept of Division

The number you are dividing is called the **dividend**. The number you are dividing it by is called the **divisor**. The answer is called the **quotient**.

Two notations can be used to indicate division:

1. When the symbol ÷ is written between two numbers, the first number is the dividend, and the second is the divisor.

EXAMPLE 1   12 ÷ 2 means 12 divided by 2.   ❑

2. When one number is written beneath a bar under another number, the number above the bar is the dividend, and the one below it is the divisor. (The bar notation is used more than the symbol ÷ in later chapters.)

EXAMPLE 2 $\dfrac{12}{2}$ means 12 divided by 2.   ❑

There are two ways to interpret a division problem, just as there are two ways to interpret a multiplication problem:

1. 12 ÷ 2 can represent the division of 12 items into 2 equal groups. The quotient indicates the number of items in each group.

(@ @ @ @ @ @) (@ @ @ @ @ @)

12 ÷ 2 = 6 means that if 12 items are split into 2 groups, 6 items will be left in each group. For example, if we divide 12 people into 2 volleyball teams, each team will include 6 people.

2. 12 ÷ 2 can represent the division of 12 items into groups, with 2 items in each group. The quotient indicates the number of groups.

(@ @) (@ @) (@ @) (@ @) (@ @) (@ @)

12 ÷ 2 = 6 means that if 12 items are split into groups of 2, there will be 6 groups. For example, if 12 people are paired up for dance teams, 6 teams will be formed.

⇒ *Note*:   We can never divide by *zero* because dividing items into *no* groups is meaningless.

The operation of division is the opposite of multiplication, just as the operation of subtraction is the opposite of addition. Another way to say this is that multiplication and division are **inverse** operations, just as addition and subtraction are.

In a division problem such as 12 ÷ 2 = 6, we notice that if the quotient is multiplied by the divisor, their product will be the same as the dividend.

$$6 \times 2 = 12$$

This observation leads us to see that multiplication and division can be used to check each other.

EXAMPLE 3   Divide 15 ÷ 5.

(@ @ @) (@ @ @) (@ @ @) (@ @ @) (@ @ @)

15 items divided into 5 groups leaves 3 items in each group. So 15 ÷ 5 = 3.

Check: Does 3 × 5 = 15? Yes.   ✓   So the answer is correct.   ❑

Division can check multiplication too. If a product is correct, then dividing it by either of the multiplying numbers should give the other multiplying number as a quotient.

EXAMPLE 4   Multiply 7×3.

$$7 \times 3 =$$
$$(7) + (7) + (7) = 21$$

Check: $21 \div 3 = 7$ ✓   or $21 \div 7 = 3$ ✓   ❏

These observations can also be helpful when you can't remember what the quotient of two numbers is. Just ask yourself what number multiplied by the divisor would produce the dividend. (Often it's easier to remember how to multiply than it is to remember how to divide.)

EXAMPLE 5   Divide $36 \div 4$.

If we can't remember how to divide 4 into 36 or can't picture it, we could turn the problem into the question: What number times 4 is equal to 36?

$$(?) \times 4 = 36$$

The answer is 9.   ✓   So $36 \div 4 = 9$.   ❏

EXERCISES 1.4A

Find each of the quotients. Check your answers by multiplying them by the divisors.

1. $10 \div 2$	**2.** $12 \div 4$
3. $16 \div 4$	**4.** $20 \div 5$
5. $\dfrac{28}{7}$	**6.** $\dfrac{18}{6}$
7. $18 \div 2$	**8.** $18 \div 3$

Check your answers on page 719.   ✓

Long Division

When large numbers are divided, we use a method of repeatedly dividing the divisor into the dividend. This operation is called **long division** and should not be done in your head. Write down each step as you proceed.

Starting at the leftmost digit, divide the divisor into as many digits in the dividend as necessary until they make a number bigger than the divisor.

Write the quotient above the rightmost digit used in the dividend.

Multiply that digit by the divisor, write their product below the digits that are being used in the dividend: subtract and write the difference below them. If the digit used for the quotient was correct, that difference is less than the divisor. (If the *difference is more than the divisor*, you chose too **small** a digit for that quotient; and if that *product was larger than the digits used from the dividend*, you choose too **large** a digit for that quotient.

Once you use the correct digit for that quotient, you will have that *difference* (a number smaller than the divisor) to work with. Bring down the next digit in the dividend that has not been used and write it to the right of the digits in the difference. That is the number you will be dividing into.

Write the quotient above the rightmost digit used in the dividend.

Continue until all digits in the dividend are used.

EXAMPLE 6 $1,656 \div 23$

$$
\begin{array}{r}
7 \\
\textbf{1. } 23\overline{)\ 1656}
\end{array}
$$

$$
\begin{array}{r}
7 \\
\textbf{2. } 23\overline{)\ 1656} \\
161 \quad (7 \times 23 = 161)
\end{array}
$$

$$
\begin{array}{r}
7 \\
\textbf{3. } 23\overline{)\ 1656} \\
-161 \\
\hline
4
\end{array}
$$

$$
\begin{array}{r}
7 \\
\textbf{4. } 23\overline{)\ 1656} \\
-161 \\
\hline
46
\end{array}
$$

$$
\begin{array}{r}
72 \\
\textbf{5. } 23\overline{)\ 1656} \\
-161 \\
\hline
46 \\
-46 \quad (2 \times 23 = 46) \\
\hline
0
\end{array}
$$

So $1,656 \div 23 = 72$.

Check: Does $72 \times 23 = 1,656$?

$$
\begin{array}{r}
72 \\
\times\ 23 \\
\hline
216 \\
144 \\
\hline
1656
\end{array}
$$
   Yes it does. So the answer is correct.

Long division is usually written in one step, as I'll do below in Example 7. I rewrote each step in Example 6 to help you see what was done in each step.

EXAMPLE 7 $2,717 \div 13 =$

$$
\begin{array}{r}
209 \\
13\overline{)\ 2717} \\
-26 \\
\hline
11 \\
-\ \ 0 \\
\hline
117 \\
-\ 117 \\
\hline
0
\end{array}
$$
   So $2,717 \div 13 = 209$.

Check: Does $209 \times 13 = 2{,}717$?

$$
\begin{array}{r}
209 \\
\times\ \ 13 \\
\hline
627 \\
209\ \ \ \\
\hline
2717
\end{array}
$$

✓   So the answer is correct.   ❏

We say that a number divides **exactly** into another number if their quotient is a whole number. Sometimes a whole number does not divide evenly into another one. In those cases there is a **remainder**.

For example, if an odd number of people try to pair off into dance partners, one person remains without a partner.

When we try to share 12 cookies equally among five people, we run into trouble. Either everyone gets 2 cookies (using only 10 of the cookies), or we have to break up the remaining 2 cookies into parts.

Since remainders are **parts of numbers** rather than whole numbers, it would be best to discuss them later in the chapters on parts of numbers ("Fractions": Chapter 3, and "Decimals": Chapter 4).

We will come back to long division of whole numbers that do not divide evenly into each other in those chapters.

EXERCISES 1.4B

Use long division to find the quotients. Check by multiplying the answers by the divisors.

9. $324 \div 4$ **10.** $1{,}284 \div 4$

11. $1{,}110 \div 6$ **12.** $4{,}530 \div 15$

13. $2{,}052 \div 19$ **14.** $3{,}216 \div 24$

15. $4{,}263 \div 21$ **16.** $3{,}644 \div 31$

17. $34{,}782 \div 34$ **18.** $90{,}343 \div 43$

Check your answers on page 719. ✓

1.5   Order of Operations

The Law of Order of Operations

In the problem $2 + 3 \times 7$, a sensible person's natural instinct might encourage him or her to perform the computations as they appear from left to right, just as one reads English. That would suggest adding 2 and 3, getting the answer 5, and multiplying that by 7 to get the answer 35.

As sensible and natural as that seems, it is *wrong*. In math problems involving more than one operation (such as addition and multiplication in this one), certain operations must be performed before others. *Multiplication* and *division* must always be done *before addition* and *subtraction*. This law is called the **order of operations**:

1. Calculate all **multiplications** and **divisions** as they appear from left to right.
2. Calculate all **additions** and **subtractions** as they appear from left to right.

Many errors result from not knowing that there is an *order of operations* or from forgetting about it during a calculation. Resist the temptation to add when it is the first calculation written in a problem. Your work will be definitely more accurate.

EXAMPLE 1 Perform the calculations following the order of operations: $2 + 3 \times 7$.

1. $2 + 3 \times 7 =$
2. $\quad 2 + 21 = 23$ ❏

It will be easier to calculate these problems if you take some time to think about what the notation in a problem is instructing you to do.

EXAMPLE 2 $3 + 5 \times 2$ is instructing you to:

1. Multiply 5 by 2.
2. Add the product to 3. ❏

EXAMPLE 3 $7 - 3 \times 2$ is instructing you to:

1. Multiply 3 by 2.
2. Subtract the product from 7. ❏

EXAMPLE 4 $5 \times 4 - 2 \times 3$ is instructing you to:

1. Multiply 5 by 4, and multiply 2 by 3.
2. Subtract the second product from the first. ❏

EXERCISES 1.5A _____

> Translate each problem into a set of instructions, as in Examples 2, 3, and 4.
>
> 1. $5 + 7 \times 2$ 2. $3 \times 4 + 8$
>
> 3. $4 + 3 \times 7$ 4. $1 + 3 \times 4 + 2$
>
> 5. $16 - 3 \times 4$ 6. $7 \times 3 - 2 \times 3$
>
> Check your answers on page 719. ✓

Calculating Order of Operation Problems

Many careless errors in a calculation result from trying to do more than one step at a time. You will find that math will be easier to do, and you will be more accurate, if you don't try to keep a lot of calculations going on in your head. In each example that follows, only the calculation for one step in the order of operations is performed in each line.

You can break up a complicated problem into several simple ones. Before calculating, underline the portion where you should work.

Remember, in a problem involving addition, subtraction, and multiplication, always perform the multiplication first.

EXAMPLE 5   Simplify $5 + 7 \times 2 + 6$.

1. $5 + 7 \times 2 + 6 =$
2. $5 + 14 + 6 = 25$ ❏

Remember, perform all *multiplication* and *division* operations as they appear from left to right. In Example 6 don't multiply *before* you divide.

EXAMPLE 6   Simplify $8 \div 4 \times 2$.
This notation instructs us first to divide 8 by 4, then to multiply the quotient by 2.

$$8 \div 4 \times 2 =$$
$$2 \times 2 = 4$$

The correct answer is 4.   ❏

If we had multiplied 4 by 2 first, and then divided 8 by the product, we would have come up with a different (and wrong) answer of 1.

EXAMPLE 7   Simplify $4 + 3 \times 6 + 5 \times 2 - 9$.

1. $4 + 3 \times 6 + 5 \times 2 - 9 =$
2. $4 + 18 + 10 - 9 = 23$ ❏

EXAMPLE 8   Simplify $15 - 3 \times 4 + 7 \times 2 - 5$.

1. $15 - 3 \times 4 + 7 \times 2 - 5 =$
2. $15 - 12 + 14 - 5 = 12$ ❏

EXERCISES 1.5B

Simplify.

7. $4 \times 3 + 7$

8. $8 + 2 \times 3$

9. $12 - 4 \times 2$

10. $15 - 4 \times 2 + 6$

11. $7 + 4 \times 5 + 3 \times 4$

12. $1 + 3 \times 6 + 9 - 5 \times 2$

13. $15 \div 3 + 2 - 4$

14. $2 + 18 \div 2 - 6$

Check your answers on page 719. ✓

Calculations Involving Parentheses

Parentheses are a mathematical punctuation used to group numbers that should be calculated first. Always simplify within the parentheses before doing anything else in the problem.

So the **order of operations** becomes:

1. Perform all calculations with **parentheses**.
2. Calculate all **multiplications** and **divisions** as they appear from left to right.
3. Calculate all **additions** and **subtractions** as they appear from left to right.

EXAMPLE 9 Simplify $4 + 3(8 - 1 - 5)$.

1. $4 + 3(8 - 1 - 5) =$ (Parentheses)
          $4 + 3(\ ) =$
2. $4 + 3(2) =$ (Multiply)
3. $4 + 6 = 10$ (Add)   ❑

EXAMPLE 10 Simplify $5(9 - 2) - 8 + 4(6 + 1 - 5)$.

1. $5(\ ) - 8 + 4(\ ) =$ (Parentheses)
2. $5(7) - 8 + 4(2) =$ (Multiply)
3. $35 - 8 + 8 = 35$ (Add and subtract)   ❑

EXAMPLE 11 Simplify $7 - 12 \div 3 + 2(8 - 5 + 1)$.

1. $7 - 12 \div 3 + 2(\ ) =$ (Parentheses)
2. $7 - 12 \div 3 + 2(4) =$ (Divide and multiply)
3. $7 - 4 + 8 = 11$ (Add and subtract)   ❑

EXERCISES 1.5C

Simplify the problems, as in the examples. Rewrite the problem in each step and perform only one computation in each step.

15. $3(4 + 1) + 6(1 + 2)$ **16.** $2 + 10 \div 5 - 3$

17. $8 + 3(2 + 4) + 6 \div 2$ **18.** $6 + 8(9 - 2) + 4 \times 7$

19. $1 + 5(9 - 1 - 4) \div 2$ **20.** $2(4 + 2) - 5 + 4(8 + 2 - 5)$

21. $12 \div 4 + 17 - 3(5 - 1 + 2)$ **22.** $19 - 6(8 + 1 - 7) + 4$

23. $7 + 2(12 - 5 - 4) - 8$ **24.** $3(18 - 10 - 3) - 5(3) + 1$

Check your answers on page 719. ✓

Calculations Involving the Multiplication of More than Two Numbers

Some problems ask you to multiply more than two numbers together. To do these problems, pair the numbers in any order you want and multiply the pairs. Continue to multiply until the final product is determined.

EXAMPLE 12 Find the product of $(2)(5)(7)$.

1. $(2)(5)(7) =$ or   1. $(2)(5)(7) =$
2. $(10)(7) = 70$ 2. $(2)(35) = 70$ ❑

⇒ *Note:* The result was the same even though we paired the numbers differently. When you multiply numbers, pair in any order.

EXAMPLE 13 Find the product of $(3)(2)(5)(4)(2)$.

1. $(3)(2)(5)(4)(2) =$
2. $(3)(10)(8) =$
3. $(3)(80) = 240$ ❑

Try to pair the numbers in this example differently. See if you still get the same answer.

EXAMPLE 14 Simplify $5 + 3(5)(6 - 1 + 2)$.

1. $5 + 3(5)(6 - 1 + 2) =$ (Parentheses)
2. $5 + 3(5)(7) =$ (Multiply)
3. $5 + 3(35) =$ (Multiply again)
4. $5 + 105 = 110$ (Add)   ❏

EXERCISES 1.5D

Simplify. In each step rewrite the problem and do only one computation, as in the examples.

25. $1 + 2(3)(5) - 9$

26. $6 + 3(2)(2) + 4 - 5$

27. $12 - 3(5 - 2) + 7(6 - 3)(2 + 2 - 1)$

28. $15 \div 3 + 4(3 - 1)(2 + 3 - 1) - 7$

29. $6 + 4(3 - 1 + 2)(5 - 2 + 1) - 6(7 - 5)$

30. $7 + (5 - 1)(3 + 2)(7 - 5)$

31. $3(4 - 2)(5 + 1) - 11 + 2$

32. $14 - 3(9 - 6) + 2(7 - 2 - 1)(8 - 3)$

33. $6 + 2(5)(8 - 4) + 12 \div 3$

34. $1 + 5 + 3(2 + 1 + 1)(6 + 2 - 5)$

Check your answers on page 719. ✓

Evaluating Formulas

Order of operations is particularly important in evaluating formulas. If there were no agreed-upon order, there would be no way to create and use meaningful formulas for calculating such things as averages or distance traveled or the odds of winning a bet. In formulas, letters are used to represent numbers. When we replace the letters with their number values, we are doing what is called **evaluating the formulas**.

The *easiest way to avoid* careless errors when evaluating is always to replace letters with their numerical values within a set of parentheses. Otherwise it is very easy to lose some of the notation necessary for doing calculations in the right order.

EXAMPLE 15 The formula $D = R \cdot T$ allows us to calculate distance (D) if we know rate (R) and time (T) traveled. Find the distance when we travel 50 mph for 3 hours.

$$D =$$
$$R \cdot T =$$
$$(50 \text{ mph}) (3 \text{ hours}) = 150 \text{ miles} \qquad ❏$$

EXAMPLE 16   The bowling average (A) is calculated by finding the total score from game one (g_1), game two (g_2), and game three (g_3), and dividing by the number of games (3).

$$\text{Average} = \frac{(g_1 + g_2 + g_3)}{3}$$

Find the average when the individual game scores are $g_1 = 106$, $g_2 = 90$, and $g_3 = 74$.

$$\text{Average} =$$
$$\frac{[(106) + (90) + (74)]}{3} =$$
$$\frac{270}{3} = 90$$

❑

⇒ **Note**:   I put the three scores into a bracket [ ] to clarify that their sum is divided by 3. I used a bracket instead of parentheses because the scores are within parentheses. Brackets make it easier to identify the beginning and end of the group.

EXAMPLE 17   Given the formula $F = 2a + 5b + 3c + 4d$, find the value of F if $a = 7$, $b = 3$, $c = 11$, and $d = 2$.

$$F =$$
$$2a + 5b + 3c + 4d =$$
$$2(7) + 5(3) + 3(11) + 4(2) =$$
$$14 + 15 + 33 + 8 = 70$$

So for those values of a, b, c, and d, $F = 70$.

❑

EXERCISES 1.5E

Evaluate each formula.

35. P is the perimeter of a rectangle. Find P if $w = 3$ and $l = 5$:
$P = 2w + 2l$.

36. D is distance traveled. Find D if $r = 40$ mph and $t = 2$ hours: $D = r \cdot t$.

37. A is the bowling average from three games (g). Find A if $g_1 = 80$, $g_2 = 105$, and $g_3 = 55$: $A = (g_1 + g_2 + g_3)/3$.

38. A is the average score from four tests (t). Find A if $t_1 = 78$, $t_2 = 84$, $t_3 = 68$, and $t_4 = 82$: $A = (t_1 + t_2 + t_3 + t_4)/4$.

39. Find K if $a = 3$, $b = 2$, $c = 1$, and $d = 5$: $K = a + 3b + 5c + 2d$.

40. Find M in $n = 5$, $p = 2$, $r = 1$, $k = 3$: $M = 3n + 4p + 7r + k$.

41. Find R if $a = 7$, $b = 3$, $c = 2$, $d = 4$, and $e = 5$:
$R = a + 2b + 5c + d + 2e$.

42. Find W if $x = 4$, $y = 3$, $m = 2$, $k = 7$: $W = x + 2y + 5m - k$.

43. Find H if $a = 3$, $b = 2$, and $c = 4$: $H = 3a + 5b - 2c$.

44. Find J if $w = 3$, $y = 2$, and $x = 5$: $J = 7w + 4y - 3x$.

Check your answers on page 719. ✓

 **Associative and Commutative Laws**

The **associative law** tells us we can group numbers any way without changing the result when we are adding or multiplying more than two numbers.

EXAMPLE 1 $(2 + 3) + 4 = 2 + (3 + 4)$

$$(2 + 3) + 4 = \qquad \text{or} \qquad 2 + (3 + 4) =$$
$$5 + 4 = 9 \quad \checkmark \qquad\qquad 2 + 7 = 9 \quad \checkmark \qquad \square$$

EXAMPLE 2 $5 \times (3 \times 7) = (5 \times 3) \times 7$

$$5 \times (3 \times 7) = \qquad \text{or} \qquad (5 \times 3) \times 7 =$$
$$5 \times 21 = 105 \quad \checkmark \qquad\qquad 15 \times 7 = 105 \quad \checkmark \qquad \square$$

We can use the associative law to regroup numbers when adding or multiplying if one grouping will simplify our calculation.

EXAMPLE 3   Use the associative law to simplify $2 \times (5 \times 13)$.

$$2 \times (5 \times 13) =$$
$$(2 \times 5) \times 13 =$$
$$10 \times 13 = 130$$

It is much easier to multiply 13 by 10 (we just add zero when we multiply by 10) than it is to multiply 13 by 5 and then multiply that product by 2. $\square$

The **commutative law** tells us we can change the order of numbers in addition or multiplication problems without changing the answer.

EXAMPLE 4 $7 + 15 = 15 + 7$

$$7 + 15 = 22 \quad \checkmark \qquad \text{or} \qquad 15 + 7 = 22 \quad \checkmark \qquad \square$$

EXAMPLE 5 $4 \times 9 = 9 \times 4$

$$4 \times 9 = 36 \quad \checkmark \qquad \text{or} \qquad 9 \times 4 = 36 \quad \checkmark \qquad \square$$

We can combine the associative and commutative laws to simplify our calculation.

EXAMPLE 6   Simplify and compute $18 + 43 + 32 + 57$.

First apply the commutative law: $18 + 43 + 32 + 57 =$
$$18 + 32 + 43 + 57 =$$

Now apply the associative law: $(18 + 32) + (43 + 57) =$
$$50 + 100 = 150 \qquad \square$$

EXAMPLE 7   Simplify and compute $25 \times 7 \times 4$.

First apply the commutative law: $25 \times 7 \times 4 =$
$$25 \times 4 \times 7 =$$

Then apply the associative law: $(25 \times 4) \times 7 =$
$$100 \times 7 = 700 \qquad \square$$

Use the associative and commutative laws to simplify each calculation.

1. $24 + 78 + 16 + 52 =$ **2.** $31 + 75 + 29 + 35 =$

3. $5 \times 83 \times 2 =$ **4.** $25 \times 17 \times 4 =$

5. $5 \times 47 \times 20 =$ **6.** $50 \times 470 \times 2 =$

7. $31 + 57 + 94 + 63 + 76 + 49 =$

8. $52 + 77 + 91 + 38 + 74 + 23 + 136 + 209 =$

9. $4 + 308 + 511 + 693 + 897 + 216 + 349 + 542 =$

10. $25 \times 37 \times 4 =$

Check your answers on page 719.

1.7 Geometric Formulas

Geometric Shapes

Letters are often used to represent numbers in formulas. Some familiar formulas have to do with geometric shapes. Let's review the definitions of some frequently used shapes.

A **polygon** is a many-sided closed figure that is formed when line segments join. In this section we will review some formulas for three- and four-sided polygons.

An angle that contains 90 degrees is called a **right angle**. It is an angle formed in the center of a circle cut into four equal parts.

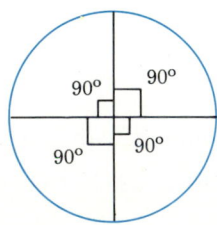

A **square** is a four-sided polygon with all sides of equal length and all right-angle corners.

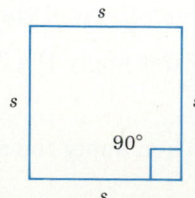

A **rectangle** is a four-sided polygon with opposite sides of equal length and all right-angle corners.

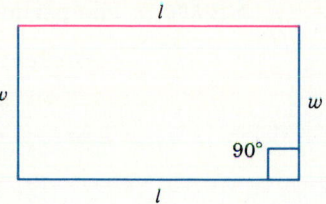

A **triangle** is a three-sided polygon.

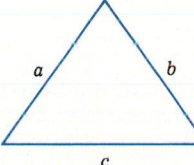 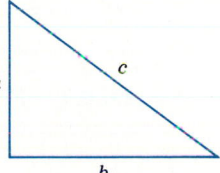

There are five different kinds of triangles that are useful to identify. A **right triangle** has one angle of 90 degrees.

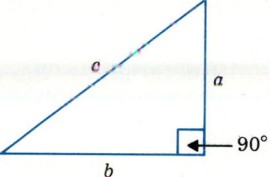

An **equilateral triangle** has all equal sides and equal angles.

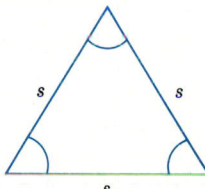

An **isosceles triangle** has two equal sides and two equal angles opposite those sides.

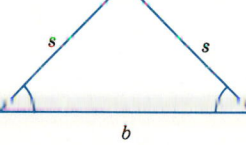

An **acute triangle** has all angles of less than 90 degrees. Angles of less than 90 degrees are called acute.

An **obtuse triangle** has one angle of more than 90 degrees. An angle of more than 90 degrees and less than 180 degrees is called obtuse.

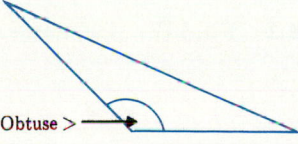

Perimeter   The **perimeter** of a polygon is the distance around it.

Square.   The perimeter of a square is the sum of its four sides.

Perimeter of square =

$$s + s + s + s = 4s$$

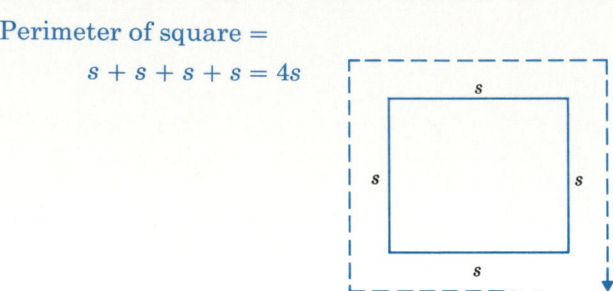

EXAMPLE 1   Find the perimeter of a square with sides of 3 inches.

Perimeter =

$$4s =$$

$$4(3 \text{ in.}) = 12 \text{ in.}$$ ❏

Rectangle.   The perimeter of a rectangle is the sum of its four sides.

Perimeter of rectangle =

$$w + l + w + l =$$

$$2 \text{ widths} + 2 \text{ lengths} = 2w + 2l$$

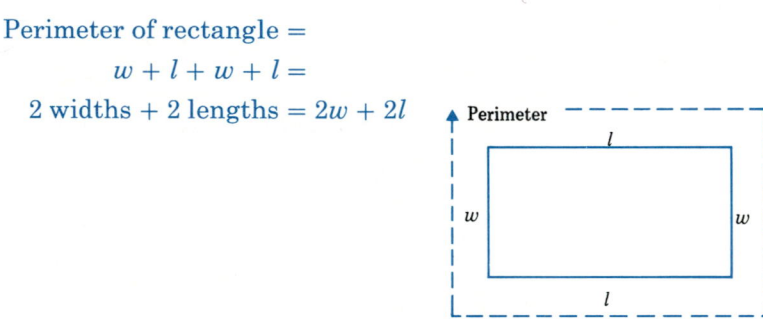

EXAMPLE 2   Find the perimeter of a rectangle with a width of 5 feet and a length of 9 feet.

Perimeter of rectangle =

$$2w + 2l =$$

$$2(5 \text{ ft}) + 2(9 \text{ ft}) =$$

$$10 \text{ ft} + 18 \text{ ft} = 28 \text{ ft}$$ ❏

Triangle.   The perimeter of a triangle is the sum of its three sides.

Perimeter of triangle = $a + b + c$

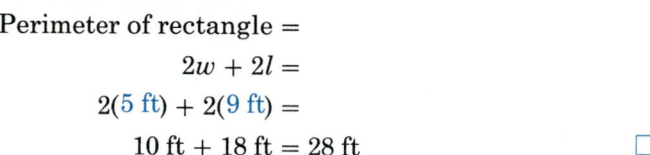

EXAMPLE 3   Find the perimeter of a triangle with sides of 5, 7, and 11 meters.

Perimeter =

$$5 \text{ m} + 7 \text{ m} + 11 \text{ m} = 23 \text{ m}$$ ❏

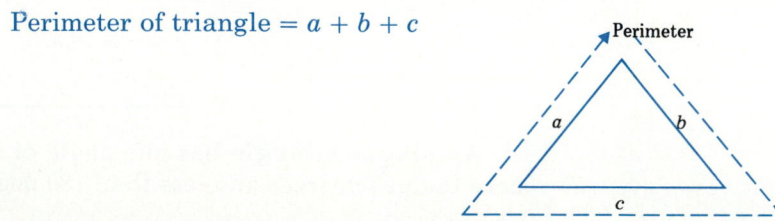

EXERCISES 1.7A

Find the perimeters of the shapes.

1. Rectangle with width 7 ft and length 10 ft.

2. Triangle with sides 4 cm, 10 cm, and 7 cm.

3. Square with sides 5 in.

4. Square with sides 8 ft.

5. Rectangle with length 12 m and width 8 m.

6. Triangle with sides 4 ft, 6 ft, and 8 ft.

7. Equilateral triangle with sides 11 cm.

8. Isosceles triangle with two sides 14 in. and one side 6 in.

Check your answers on page 719.

Area A **square unit** is a square that measures one unit for each of its sides. For example, a square measuring an inch on each side is called a square inch.

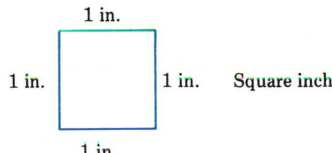

1 in.

1 in.   1 in.   Square inch

1 in.

A square foot is a square measuring one foot on each side.

The **area** of a surface is the number of square units it would take to cover that surface. When stating the area of an object, *the unit must be stated in terms of square units.*

Rectangle.   If w represents the width, then the rectangle could be cut into w columns from left to right, each being one unit wide.

If l represents the length, then the rectangle could be cut into l rows from top to bottom, each being one unit high.

Such divisions would cover the rectangle with unit squares. Since there are l rows of unit squares, with w in each row, the total number of unit squares could easily be found by multiplying the width by the length.

So the area of a rectangle with width w and length l is their product, lw.

Area of rectangle = (length)(width)

EXAMPLE 4 Find the area of a rectangle 7 meters wide and 3 meters long.

$$\text{Area} =$$
$$(w)(l) =$$
$$(7 \text{ m})(3 \text{ m}) = 21 \text{ sq m}$$

Count them.

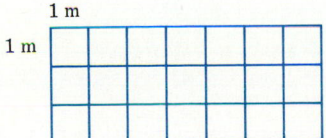

1 m

1 m

Square.   Since a square is a rectangle with the same width and length, the area of a square is the product of its side multiplied by itself.

$$\text{Area of square} = s \cdot s$$

EXAMPLE 5   Find the area of a square with sides of 5 feet.

$$\text{Area} =$$

$$(5 \text{ ft})(5 \text{ ft}) = 25 \text{ sq ft}$$

Count them

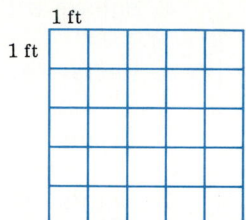

Triangle.   Every triangle is one-half of a rectangle that has the triangle's base as its width and the triangle's height as its length.

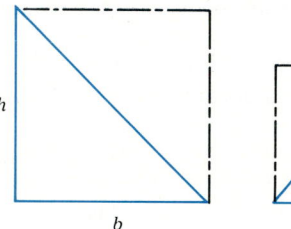 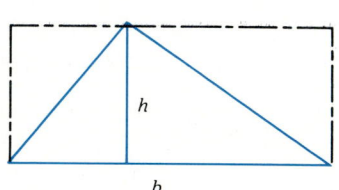

So the area of a triangle can be found by first finding the area of that rectangle, (base) (height), and then dividing it by two.

$$\text{Area of triangle} = \frac{\text{(base)(height)}}{2}$$

EXAMPLE 6   Find the area of a triangle with a base of 3 inches and a height of 10 inches.

$$\text{Area} =$$

$$\frac{(3 \text{ in.})(10 \text{ in.})}{2} =$$

$$\frac{30 \text{ sq. in.}}{2} = 15 \text{ sq. in.}$$

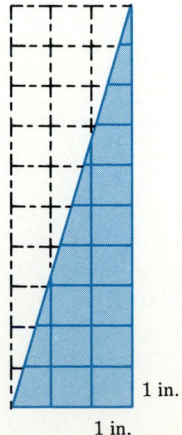

EXERCISES 17B

Find the areas of the shapes.

9. Square with sides 6 in.

10. Rectangle with length 8 cm and width 5 cm.

11. Triangle with base 4 in. and height 7 in.

12. Square with sides 3 m.

13. Rectangle with width 4 ft and length 9 ft.

14. Triangle with base 4 ft and height 9 ft.

15. Square with sides 9 yd.

16. Rectangle with width 2 mi and height 9 mi.

17. Triangle with base 10 cm and height 7 cm.

Check your answers on page 719.

Volume   A **unit cube** is a solid cube that measures one unit in its width, length, and height, and we call each a cubic unit. For example, a cube with a width, length, and height of 1 foot would be a cubic foot.

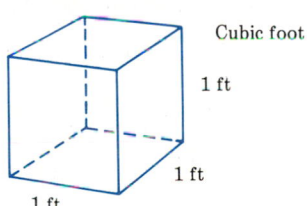

Cubic foot

1 ft

1 ft

1 ft

The **volume** of a solid is the number of unit cubes it would take to fill the solid.

In this section we will learn how to calculate the volume of solid figures whose *bases* (on the top and the bottom) are identical polygons that are parallel, and whose *height* is the perpendicular distance between those bases. These figures are called **cylindrical solids**, or cylinders.

To calculate the volume of a cylindrical solid, multiply the area of its base by its height. The area of its base is the number of *square units* that cover the surface of the polygon, and each one will be the base of a unit cube in the volume. There will be a layer of that number of unit cubes for every 1 unit in the solid's height. By multiplying the area by the height, you include all the layers of cubes needed.

Volume = (area of base) (height)

There are two kinds of cylindrical solids to consider:

1. A **rectangular cylindrical solid** has identical and parallel rectangular or square bases.

Volume = (area of base)(height of solid)

    = *(lw)(h)*   if the base is a rectangle with length *l* and width *w*

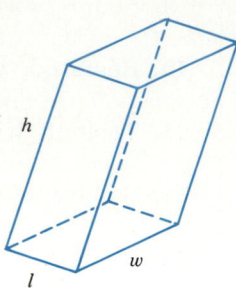

Volume = *(ss)(h)*   if the base is a square with side *s*

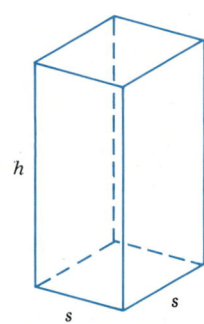

2. A triangular cylindrical solid has identical and parallel triangular bases.

Volume = $\dfrac{(bh)}{2}(H)$ if the base and height of the triangle are *b* and *h*, and the height of the solid is *H*

  = $\dfrac{(bhH)}{2}$ The answer is the same whether you divide by 2 before or after multiplying by *H*. The formula looks simplest in this form.

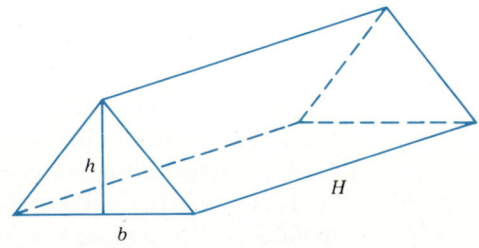

Volume must always be expressed in terms of *cubic* units.

EXAMPLE 7   Find the volume of a rectangular solid with the following dimensions: 3 millimeters length, 5 millimeters width, and 6 millimeters height. $[V = lwh]$

Volume = $(3 \text{ mm})(5 \text{ mm})(6 \text{ mm})$

    = $(15 \text{ sq mm})(6 \text{ mm})$

    = 90 cu mm

EXAMPLE 8   Find the volume of a triangular solid with the following dimensions: 4 foot base, 3 foot height of triangle, 10 foot height of solid. $\left[V = \dfrac{bhH}{2} \right]$

$$\text{Volume} = \frac{(4 \text{ ft})(3 \text{ ft})(10 \text{ ft})}{2}$$

$$= \frac{(4 \text{ ft})(30 \text{ sq ft})}{2}$$

$$= \frac{120}{2} \text{ cu ft} = 60 \text{ cu ft}$$

EXERCISES 17C

Find the volumes of the shapes.

18. Rectangular solid with width 5 ft, length 2 ft, and height 7 ft.

19. Square solid with sides 6 cm and height 10 cm.

20. Triangular solid with base 3 in., height of triangle 4 in., and height of solid 7 in.

21. Rectangular solid with length 8 in., width 3 in, and height 5 in.

22. Square solid with sides 12 cm and height 5 cm.

23. Triangular solid with base 9 yd, height of triangle 2 yd, and height of solid 6 yd.

24. Triangular solid with base 7 ft, height of triangle 5 ft, and height of solid 8 ft.

Check your answers on page 719.

Glossary

acute triangle   A triangle whose angles are all less than 90 degrees.

Example:

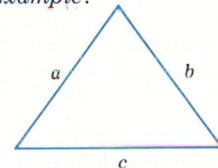 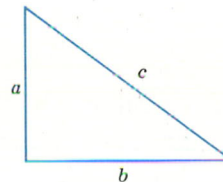

area   The number of square units it takes to cover a surface.

Example: The area of a rectangle with side 2 in. and length 6 in. is 12 sq in.

associative law   A law that tells us we can group numbers any way without changing the result when we are adding or multiplying more than two numbers.

Example: $29 + (59 + 87) = (29 + 59) + 87$

commutative law   A law that tells us we can change the order of numbers in addition or multiplication problems without changing the answer.

Example: $12 \times 8 = 8 \times 12$

cylindrical solid   A three-dimensional figure with identical polygons for its parallel bases. The distance between those bases is called its height.

rectangular cylindrical solid   A solid with rectangular bases.

Example:

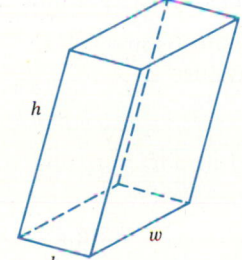

triangular cylindrical solid A solid with triangular bases.

Example:

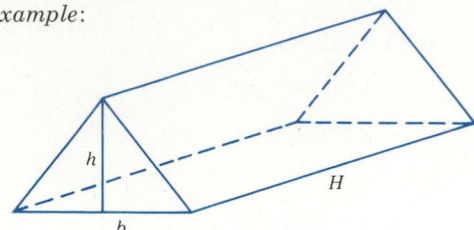

decimal system A number system in which combinations of only *ten* digits are used to represent the numbers. The place value of the digit furthest to the right in a whole number is *one,* and each place to the left has the value of *ten* times the place to its right.

difference The result of subtracting one number from another.

Example: The difference of 23 and 8 is 15.

dividend The number you are dividing in a division problem.

Example: 28 is the dividend in $28 \div 4$.

divisor The number you are dividing by in a division problem.

Example: 4 is the divisor in $28 \div 4$.

equilateral triangle A triangle whose sides are all equal. Each angle contains 60 degrees.

Example:

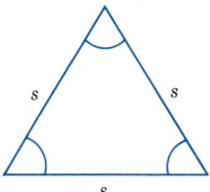

formula A way to represent relationships among data with letters.

Example: The perimeter of a rectangle is the sum of twice the width plus twice the length, and is represented by the formula:

$$P = 2w + 2l.$$

inverse operations Two operations that cancel each other out. When both are applied to a number, the result is that same original number.

Example: Addition and subtraction are inverse operations.

$$7 + 3 - 3 = 7$$

Multiplication and division are inverse operations.

$$12 \times 5 \div 5 = 12$$

isosceles triangle A triangle with two equal sides.

Example:

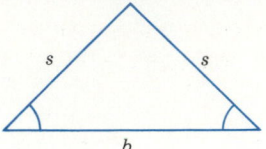

obtuse triangle A triangle with one angle larger than 90 degrees.

Example:

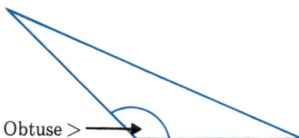

order of operations The law that clarifies, in problems involving more than one operation, which operations must be performed before the others.

1. Perform all calculations within *parentheses.*
2. Calculate all *multiplications* and *divisions* as they appear from left to right.
3. Calculate all *additions* and *subtractions* as they appear from left to right.

Example: $4 + 3(7 - 5) =$
$$4 + 3(2) =$$
$$4 + 6 = 10$$

perimeter The distance around a polygon.

Example: The perimeter of this triangle is 19 ft.

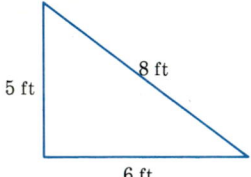

$$\text{Perimeter} = 8\,\text{ft} + 6\,\text{ft} + 5\,\text{ft}$$
$$= 19\,\text{ft}$$

place value The size $(1, 10, 100,$ and so on) that each digit in a number represents.

Example: 283 represents: $2(100) + 8(10) + 3(1)$.

So 2 is in place value **100,**

8 is in place value **10,**

and 3 is in place value **1.**

polygon A many-sided closed figure that is formed when line segments join.

product The result of multiplication.

Example: The product of 5 and 8 is 40.

quotient The result of division.

Example: 7 is the quotient of 28 divided by 4.

rectangle A four-sided polygon whose opposite sides are of equal length and whose corners are all right angles.

Example:

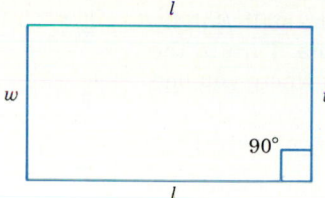

remainder The fractional part left over when one number does not divide exactly into another.

right angle The angle formed in the center when a circle is cut into four equal parts. It contains 90 degrees.

Example:

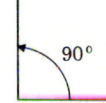

right triangle A triangle with one right angle.

Example:

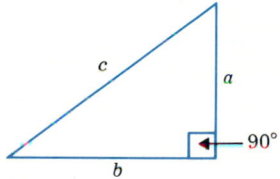

square A four-sided polygon whose sides are all equal and whose angles are all right angles.

Example:

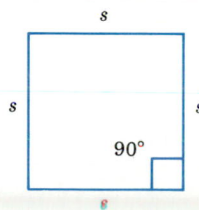

square unit A square that measures one unit for its side.

Example: A square foot is a square measuring 1 ft on each side.

sum The result of addition.

Example: The sum of 7 and 15 is 22.

triangle A three-sided polygon.

Example:

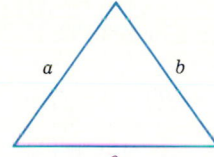 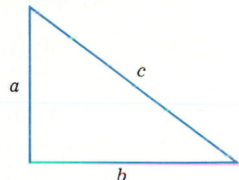

unit cube A solid cube that measures one unit in its width, length, and height.

Example: A cube with width, length, and height of 1 ft would be a cubic foot.

volume of a solid The number of unit cubes it would take to fill the solid. Volume must always be expressed in terms of cubic units.

Example: The volume of a solid might be 30 cu in.

whole numbers The numbers we are familiar with and use for counting: 0, 1, 2, 3, 4, 7, 28, 315, and so on.

Chapter I Review

Do each of the problems. After you are confident that you have done them as accurately as possible, compare your answers with those in the back of the book. If any are wrong, go back to the section where the problem came from (indicated to the left of the problem) and review the section.

Once you understand all the sections, you'll be sure to learn the skills solidly and remember how to do them if you practice them a little bit more. Turn to the Supplementary Exercises and do all the problems from any section where you had any difficulty on these review exercises.

1.1   1. Write out the name for 203,192.

1.2   2. Find the sum: $31 + 509 + 2 + 230 + 49$.

3. Find the difference: $2,381 - 427$.

4. Find the difference: $23,071 - 6,552$.

1.3   5. Determine the product: 53×37.

1.4   6. Determine the quotient: $3,991 \div 13$.

7. Determine the quotient: $54,864 \div 27$.

1.5   8. Simplify to one number: $7 + 3(6 - 2) + 18 \div 2 - 7$.

9. Simplify to one number: $3(5 - 2 + 1) + 7(3 - 1)(5 - 2) - 1$.

10. Find R if $m = 3$, $n = 2$, and $p = 5$.
$$R = 2m + 4n + 3p$$

Use the associative and commutative laws to simplify.

1.6   11. $67 + 81 + 42 + 39 + 13 + 58$

1.7   12. Find the perimeter of a rectangle with length 4 in. and width 7 in.

13. Find the area of a triangle with base 5 ft and height 10 ft.

14. Find the volume of a rectangular cylindrical solid with square bases and with sides of 3 in. and height 4 in.

Answers

1. _____

2. _____

3. _____

4. _____

5. _____

6. _____

7. _____

8. _____

9. _____

10. _____

11. _____

12. _____

13. _____

14. _____

Check your answers on page 720.

Supplementary Exercises

Do all the problems in every section that involves a skill in which you now lack complete mastery. With a little more practice you will achieve that sense of really understanding the topics, and you'll remember how to do these problems.

Write out the name for each number.

1.1

1. 421,084

2. 3,019,437

Perform these computations.

1.2

3. $28 + 53 + 190 + 3 + 115$

4. $11 + 92 + 304 + 103 + 357 + 9$

5. $8 + 239 + 51 + 104 + 17 + 6$

6. $371 - 183$

7. $2,347 - 1,792$

8. $70,348 - 23,972$

Determine the products.

1.3

9. 74×38

10. 29×76

11. 83×42

Determine the quotients.

1.4

12. $63,700 \div 14$

13. $46,782 \div 23$

14. $50,520 \div 24$

15. $130,231 \div 31$

Simplify to one number.

1.5

16. $7 + 3 \times 2 + 5 \times 3$

17. $9 + 3(7 - 3 + 1) + 5$

18. $4(3 + 5) - 6(7 - 2 - 1) + 7(2)(2)$

19. $3 + 4(5 - 2)(7 + 1 - 5) - 11$

20. Find T if $a = 3$, $b = 2$, $c = 5$, and $d = 4$.
$$T = a + 5b + 2c + d$$

21. Find M if $n = 7$, $m = 5$, $p = 3$, and $k = 2$.
$$M = n + 2m - p + 5k$$

1.6 Use the associative and commutative laws to simplify.

22. $55 + 72 + 80 + 35 + 98$

23. $237 + 584 + 696 + 323$

1.7

24. Find the perimeter of a triangle with sides 3 in., 11 in., and 7 in.

25. Find the area of a square with sides 6 m.

26. Find the volume of a rectangular cylindrical solid with length 2 ft, width 3 ft, and height 4 ft.

Check your answers on page 720.

Answers

1. _____

2. _____

3. _____

4. _____

5. _____

6. _____

7. _____

8. _____

9. _____

10. _____

11. _____

12. _____

13. _____

14. _____

15. _____

16. _____

17. _____

18. _____

19. _____

20. _____

21. _____

22. _____

23. _____

24. _____

25. _____

26. _____

Chapter 1 Test

1. Write out the name for 739,603.

2. Find the sum: $68 + 309 + 543 + 9 + 81$.

3. Find the difference: $6,037 - 3,862$.

4. Find the product: 57×87.

5. Find the quotient: $891,324 \div 27$.

6. Simplify to one number: $4 + 5(7 - 2 - 2)(3) - 6 \div 2 + 1$.

7. Use the associative and commutative laws to simplify: $13 + 68 + 104 + 52 + 66 + 627$.

8. What is the perimeter of a square with sides 9 in.?

9. What is the area of a triangle with base 6 in. and height 9 in.?

10. What is the volume of a rectangular solid with length 2 in., width 6 in., and height 4 in.?

Answers

1. _____

2. _____

3. _____

4. _____

5. _____

6. _____

7. _____

8. _____

9. _____

10. _____

Check your answers on page 720.

2 Signed Numbers

2.1 Concept of Signed Numbers

In Chapter 1 we reviewed calculations with whole numbers (0, 1, 2, 3, ...). Some situations involve numbers that are less than zero. Numbers smaller than zero are called **negative**. For example, on the coldest winter days or nights the temperature sometimes drops below 0 degrees. Those temperatures are described by negative numbers. As another example, if we write a check for more money than is in our checking account, we are left with a balance described by a negative number.

Numbers larger than zero are called **positive**. Once you recognize that some numbers are larger than zero and some are smaller, you must have a way to distinguish the negative from the positive numbers.

We make that distinction by placing a minus sign ($-$) in front of negative numbers. When a sign is placed before a positive number, that sign is a plus ($+$). We use the term **signed numbers** to describe positive and negative numbers. Before you knew about negative numbers, the only numbers you used were positive and so no sign was necessary. *When a number is written without a sign, it is always assumed to be positive.*

EXAMPLE 1   Negative 8 is written as -8; positive 14 is written as $+14$ or 14.   ❑

Integers is a term used to describe the following numbers; $\{\cdots -5, -4, -3, -2, -1, 0, 1, 2, 3, 4, 5, \ldots\}$. **Natural numbers** is the term we use when talking about just the positive integers.

In this chapter we will review the computations of addition, subtraction, multiplication, and division with *signed numbers*. The easiest way to get comfortable when computing with the negative numbers is to picture where those numbers are on a number line. Numbers increase from left to right on a number line.

EXAMPLE 2   7 represents the number seven units to the right of (or after) 0 on the number line. ($+7$ is seven units more than 0.)

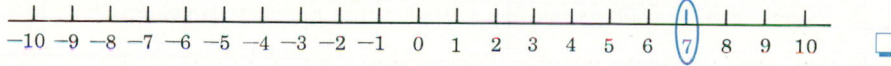
❑

EXAMPLE 3 -4 represents the number four units to the left of (or before) 0 on the number line. (-4 is four units less than 0.)

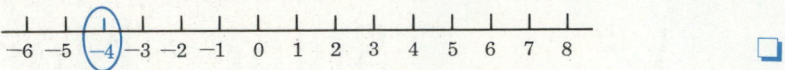

There are two properties about signed numbers that you must always consider in computations:

1. The **sign** of a number indicates whether the number is larger or smaller than zero.

    The sign is *positive* if the number is *larger* than zero.
    The sign is *negative* if the number is *smaller* than zero.

    On number lines, positive numbers are located to the right of zero, and negative numbers are located to the left of zero. So the sign of a number indicates the *direction* from zero where that number is located on a number line.

2. The **magnitude** (size of the number) indicates the distance (how many units away) it is from zero.

EXAMPLE 4   12 is a positive number with a magnitude of 12. It is 12 more than 0.

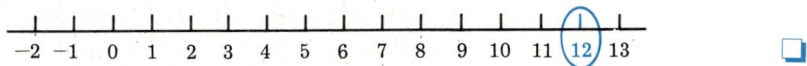

EXAMPLE 5 -12 is a negative number with a magnitude of 12. It is 12 less than 0.

When we refer only to a number's magnitude (that is, when we are interested only in how far it is from zero and not whether it is smaller or larger than zero), we are talking about the number's **absolute value**.

EXAMPLE 6   The absolute value of -12 is 12. The absolute value of $+8$ is 8.

To denote the *absolute value* of a number, we write the number between two vertical lines.

EXAMPLE 7   The absolute value of -6 is written as $|-6|$

EXAMPLE 8 $|-6| = 6$; $|3| = 3$; $|-38| = 38$; $|38| = 38$.

EXERCISES 2.1

State the sign and magnitude for each number.

1. 14

2. -8

3. -17

4. 23

5. 67

6. -84

Find the value of each number.

7. $|-28|$

8. $|74|$

9. $|308|$

10. $|-531|$

Check your answers on page 720.

2.2 Addition of Signed Numbers

Adding Numbers with the Same Sign

To add numbers is to find their sum or total. We can visualize what is actually happening in addition if we use a number line.

In *addition,* always begin at zero:

To add a *positive* number, move to the *right* as many units as the number's magnitude;

To add a *negative* number, move to the left as many units as the number's magnitude.

The *sum* is represented by *how far you are from zero* after you have made all the moves from computing.

EXAMPLE 1   Add $(+3) + (+5)$.

Begin at 0 and move 3 units to the right. That brings you to $+3$.

Then move 5 more units to the right. That leaves you at $+8$.

$$(+3) + (+5) = +8$$

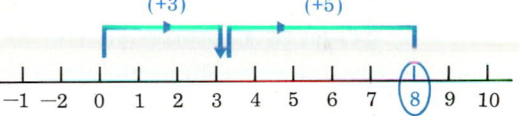

You can see that this is what you normally do, without thinking about signs, when you add the positive numbers 3 and 5. But when negative numbers are involved, you must think about the signs.

EXAMPLE 2   Add $(-2) + (-5)$.

Begin at 0 and move 2 units to the left. That brings you to -2.

Then move 5 more units to the left. That leaves you at -7.

$$(-2) + (-5) = (-7)$$

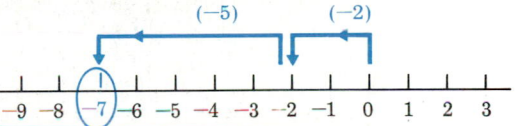

⇒ *Note:*   When we add numbers that have the same sign, all our movement on the number line is in the same direction. The sum must have the same sign as the numbers we added, since that is the only direction in which we moved. We could compute the answer without using a number line by observing that the sum of those numbers must have the *same sign as the sign of the numbers.*

The *sum* of *negative numbers* is always *negative.*

That observation makes it easy to add numbers that are too large to display on a number line.

> To add negative numbers together:
> *Add* their *magnitudes* and remember that their sum must be *negative.*

Their sum will be the total distance moved, in the negative direction, away from zero. In other words, the magnitude of the sum of negative numbers is the result of adding their absolute values.

EXAMPLE 3 Add $(-43) + (-58)$.

Since both numbers are negative, their sum will be negative.

The sum of $|-43| + |-58|$ is 101. So the magnitude of the sum is 101.

$$(-43) + (-58) = -101$$

EXERCISES 2.2A

Determine the sign and magnitude for each sum.

1. $(+7) + (+5)$ **2.** $(-3) + (-4)$ **3.** $(-1) + (-3)$

4. $(-3) + (-3)$ **5.** $(+6) + (+4)$ **6.** $(-6) + (-4)$

7. $(-21) + (-45)$ **8.** $(-38) + (-52)$ **9.** $(-184) + (-251)$

10. $(-357) + (-242)$

Check your answers on page 720.

Adding More Than Two Numbers

To add more than two *negative* numbers together, just add their magnitudes and remember that their sum must be *negative*.

EXAMPLE 4 Add $(-3) + (-5) + (-2) + (-1)$.

Since they are all negative, their sum will be negative.

$|-3| + |-5| + |-2| + |-1| = 11$ The magnitude of the sum is 11.

So $(-3) + (-5) + (-2) + (-1) = -11$.

EXERCISES 2.2B

Find the sums.

11. $(-1) + (-5) + (-3) + (-7) =$ **12.** $(3) + (4) + (7) + (1) =$

13. $(-5) + (-3) + (-3) + (-4) + (-1) =$

14. $(-2) + (-4) + (-3) + (-1) + (-1) =$

15. $(-12) + (-7) + (-33) + (-19) =$

16. $(-58) + (-24) + (-17) + (-31) =$

17. $(-18) + (-73) + (-14) + (-60) =$

18. $(-7) + (-3) + (-4) + (-1) + (-2) + (-1) =$

19. $(-2) + (-1) + (-4) + (-3) + (-2) =$

20. $(-6) + (-2) + (-5) + (-3) + (-4) =$

21. $(-7) + (-4) + (-3) + (-3) + (-2) =$

22. $(-31) + (-40) + (-53) + (-12) =$

23. $(-62) + (-38) + (-41) + (-27) =$

Check your answers on page 720.

Adding Numbers with Different Signs

Let's see what is involved in adding numbers with different signs by looking at this computation on a number line. Remember:

To add a positive number, move to the right.

To add a negative number, move to the left.

EXAMPLE 5 Add $(+5) + (-7)$.

Begin at 0 and move 5 to the right. That brings you to $+5$.

Then move 7 to the left. That leaves you at -2.

$(+5) + (-7) = -2$

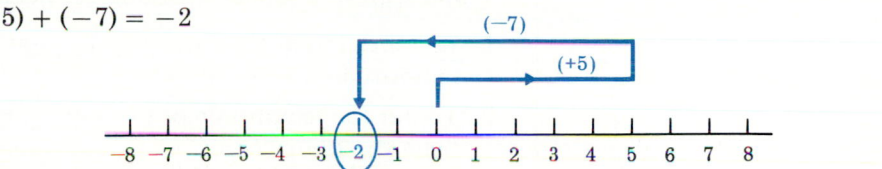

EXAMPLE 6 Add $(-3) + (+8)$.

Begin at 0 and move 3 to the left; then move 8 to the right.

$(-3) + (+8) = +5$

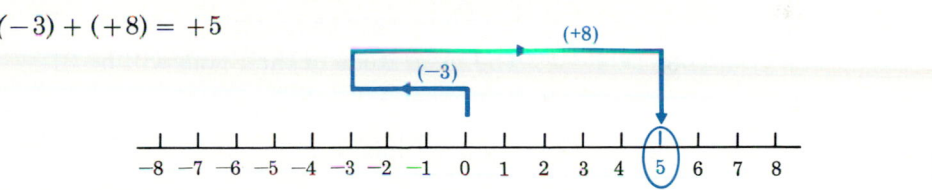

EXAMPLE 7 Add $(-6) + (+3)$.

Begin at 0 and move 6 to the left; then move 3 to the right.

$(-6) + (+3) = -3$

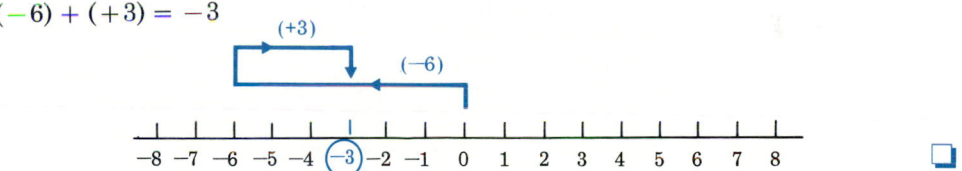

⇨ *Note*: When we add numbers with different signs, our movement on the number line is in both directions. After the first move from zero, the next move brings us back toward zero. Since the sum is represented by the distance from zero at the end of the computation, we observe that the sum is the difference between the total movement to the right and the total movement to the left.

We end up to the *right* of zero if we moved *more to the right* than to the left. This happens when the positive number has a larger magnitude than the negative number.

We end up to the *left* of zero if we move *more to the left* than to the right. This happens when the negative number has a larger magnitude than the positive number.

IN EXAMPLE 5 We added $(+5)$ and (-7).

The magnitude (7) of the negative number (-7) is larger than the magnitude (5) of the positive number $(+5)$. So the sum is negative.

The difference when we take 5 from 7 is 2.

So $(+5) + (-7) = -2$.

IN EXAMPLE 6 We added (-3) and $(+8)$.

The magnitude (8) of the positive number $(+8)$ is larger than the magnitude (3) of the negative number (-3). So the sum is positive.

The difference when we take 3 from 8 is 5.

So $(-3) + (+8) = +5$. ❑

These observations provide us with an easy way to add numbers with different signs that would be too large to display on a number line.

> To add signed numbers with different signs:
>
> **1.** For the sign of the sum, use the sign of the number with the larger magnitude.
>
> **2.** To get the magnitude of the sum, subtract the smaller magnitude from the larger one.

EXAMPLE 8 Add (-72) and $(+41)$.

1. The sign of their sum will be negative since the negative number (-72) has the larger magnitude.

2. The magnitude of their sum will be 31.

$$|-72| - |41| \text{ is } 72 - 41 = 31$$

So $(-72) + (+41) = -31$. ❑

EXAMPLE 9 Add (-45) and $(+83)$.

1. The sign of their sum will be positive since the positive number $(+83)$ has the larger magnitude.

2. The magnitude of their sum will be 38.

$$|83| - |-45| \text{ is } 83 - 45 = 38$$

So $(-45) + (+83) = +38$. ❑

EXERCISES 2.2C

Find the sign and magnitude of the sums.

24. $(-8) + (+2) =$ **25.** $(-3) + (+5) =$

26. $(+4) + (-1) =$ **27.** $(+7) + (-9) =$

28. $(-6) + (+9) =$ **29.** $(+3) + (-7) =$

30. $(+28) + (-16) =$ **31.** $(-47) + (+26) =$

32. $(-70) + (+48) =$ **33.** $(-37) + (+50) =$

34. $(+57) + (-34) =$ **35.** $(+39) + (-60) =$

Check your answers on page 720. ✓

Adding More Than Two Numbers with Different Signs

It is definitely easier to add numbers with the same sign than it is to add numbers with different signs. The easiest way to find the sum when you have to add more than two numbers with different signs is *first to combine all the numbers that have the same sign.* This method also cuts down sharply on careless errors.

Remember that a number written without a sign is always positive. When grouping numbers together within a parentheses, you don't need to write the sign of the first number *if it is positive.* But you must write a sign in front of all the other positive numbers to clarify that each of them is another number to be added.

EXAMPLE 10 Find the sum of $(-2) + (-3) + (6) + (-4) + (3) + (2)$. I'll group all the negative numbers in one pair of parentheses and all the positive numbers in another pair of parentheses.

$$(-2) + (-3) + (6) + (-4) + (3) + (2) =$$
$$(-2 - 3 - 4) + (6 + 3 + 2) =$$
$$(-9) + (+11) = +2$$

I had to put in the $(+)$ for (3) and (2) when I grouped all the positive numbers together. ❏

Sometimes a list of numbers to be added will be written without parentheses around each number, just as we did in Example 10 when we grouped the positive numbers together and the negative numbers together. You can always interpret the sign in front of a number as the sign (or direction) of that number.

So when a list of numbers is written without parentheses but with a sign before each number, you can read that list as the sum of those numbers.

EXAMPLE 11 $-6 + 8 + 3 - 4 - 1 + 7 - 2$ can be read as the sum of

negative 6, positive 8, positive 3, negative 4,

negative 1, positive 7, and negative 2.

The easiest way to get the right answer is first to add together all the numbers with the same sign.

$$-6 + 8 + 3 - 4 - 1 + 7 - 2 =$$
$$(-6 - 4 - 1 - 2) + (8 + 3 + 7) =$$
$$(-13) + (+18) = +5$$ ❏

EXERCISES 2.2D

Find the sums.

36. $(3) + (-2) + (-5) + (7) + (-1) =$

37. $(6) + (3) + (-4) + (-7) + (-5) =$

38. $(-8) + (7) + (-2) + (-3) + (5) =$

39. $(-4) + (-3) + (10) + (5) + (-6) =$

40. $(-11) + (-23) + (8) + (26) + (-10) =$

41. $(-23) + (56) + (31) + (-43) + (-15) =$

42. $(-73) + (-14) + (51) + (60) + (-40) =$

43. $(20) + (-32) + (-14) + (18) + (-10) =$

44. $-5 + 7 + 2 - 3 + 1 =$ **45.** $-1 - 4 - 3 + 7 + 5 - 2 + 6 =$

46. $6 + 1 - 4 + 5 - 7 - 1 =$

47. $11 - 15 - 4 + 20 + 17 - 3 =$

48. $-28 - 24 + 41 - 13 + 57 =$

49. $-14 - 7 - 3 + 2 + 8 - 3 =$

50. $34 - 58 - 24 - 7 + 18 + 9 =$

51. $-67 + 39 + 17 - 26 - 81 =$

Check your answers on page 720. ✓

Opposites

When numbers have the same magnitude but different signs, they are called **opposites**.

EXAMPLE 12 The numbers $(+4)$ and (-4) are opposites. ❑

Opposites are numbers that, when added, move the same distance from zero but in opposite directions. So their sum is always zero.

$$(-4) + (+4) = 0$$

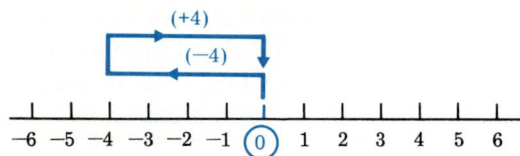

When adding a list of numbers, if you happen to notice that the list includes a pair of opposites, you can make the problem simpler.

Since the sum of any pair of opposites is zero, they can be crossed out.

EXAMPLE 13 Find the sum of $-4 + 7 + 3 - 5 - 7 + 1$.

This problem includes the pair of opposites: $+7$ and -7.

$$-4 + 7 + 3 - 5 - 7 + 1 =$$

We know that $(+7) + (-7) = 0$, so the problem becomes:

$$-4 + 3 - 5 + 1 + 7 - 7 =$$
$$-4 + 3 - 5 + 1 + 0 =$$

Since the 0 won't affect our sum, we can drop it from the next step:

$$(-4 - 5) + (+3 + 1) =$$
$$(-9) + (+4) = -5 \qquad ❑$$

Rather than putting in a zero and then dropping it, we could have simply crossed out $(+7)$ and (-7). But we must be careful to cross out only the pairs that have *opposite* signs.

EXAMPLE 14 Find the sum of $51 + 38 - 47 - 51 - 76 - 47 + 19$.

There is one pair of opposites: 51 and -51.

There are two -47s. Both are negative, so they are the same, not opposite. We cannot cross them out.

$$\cancel{51} + 38 - 47 \cancel{-51} - 76 - 47 + 19 =$$
$$38 - 47 - 76 - 47 + 19 =$$
$$(38 + 19) + (-47 - 76 - 47) =$$
$$(+57) + (-170) = -113 \qquad ❑$$

To add signed numbers:

1. Cross out any pairs of opposites in the list.
2. Group all the positive numbers in one pair of parentheses and group all the negative numbers in another pair of parentheses.
3. Add numbers with same sign by adding their magnitudes and keeping the sign of those numbers.
4. Add the resulting positive number and negative number by keeping the sign of the larger magnitude and subtracting the smaller magnitude from the larger one.

EXERCISES 2.2E

Find the sums.

52. $(-17) + (10) + (21) + (17) + (-11) =$

53. $(24) + (17) + (-15) + (-12) + (-31) + (15) =$

54. $(-43) + (31) + (-15) + (-12) + (31) + (43) =$

55. $(-74) + (-83) + (57) + (43) + (83) + (47) + (-21) =$

56. $11 + 7 - 9 + 5 - 4 + 9 - 11 - 13 =$

57. $6 + 15 - 38 - 26 - 6 - 38 + 15 =$

58. $-13 + 40 + 35 - 29 + 92 - 40 - 13 + 43 - 55 + 29 =$

59. $-20 - 30 + 40 + 70 - 20 - 30 - 40 + 70 - 40 - 10 =$

60. $74 - 28 + 36 - 36 - 28 + 74 - 17 + 40 - 68 - 74 =$

Check your answers on page 720. ✓

2.3 Subtraction of Signed Numbers

Interpreting Subtraction as Addition of Opposites

Now that you know how to add signed numbers, we can go back to the subtraction of natural numbers and make some useful observations. Notice that subtracting 3 from 10:

$$(10) - (3) = 7$$

and adding negative 3 to 10:

$$(10) + (-3) = 7$$

are two different ways to express the same problem.

Recall that (3) and (−3) are *opposites*. So we could say that *subtracting* 3 is the same as *adding* the *opposite* of 3. The same is true for all numbers:

Subtracting a number is the same as adding its opposite.

With that observation we don't need any new rules for subtracting signed numbers. We already know all there is to know about adding them, and now we see that every *subtraction problem can be interpreted as an addition problem.*

EXAMPLE 1   Subtract these numbers by interpreting the problem as one of addition: $(4) - (+8)$.

$$(4) - (+8) =$$
$$(4) + (-8) = -4$$ ❏

EXAMPLE 2   Subtract these numbers by interpreting the problem as one of addition: $(-9) - (-6)$.

$$(-9) - (-6) =$$
$$(-9) + (+6) = -3$$ ❏

To visualize better what we just did in subtraction, it helps to think in terms of money. Money that you have is *positive.* Money that you owe is *negative.*

In Example 1, if you take away something positive (like $8), you end up worse off. If you start with something positive ($4) and someone takes away something more positive ($8)), you end up in debt (−$4).

In Example 2, if you take away something negative (like −$6), you end up better off. If you start with something negative, like a debt of $9 (−$9), and someone takes away some (−$6) of that negative debt, you end up only $3 in debt (−$3).

In general:

Any time you take away something positive, you end up with less.

Any time you take away something negative, you end up with more.

EXAMPLE 3   Subtract these numbers by interpreting the problem as one of addition: $(-7) - (+3)$.

$$(-7) - (+3) =$$
$$(-7) + (-3) = -10$$ ❏

EXAMPLE 4   Subtract these numbers by interpreting the problem as one of addition: $(3) - (-5)$.

$$(3) - (-5) =$$
$$(3) + (+5) = +8$$ ❏

EXERCISES 2.3A

Change each subtraction problem to addition and combine the numbers.

1. $(5) - (7) =$ **2.** $(-3) - (4) =$

3. $(11) - (-3) =$ **4.** $(-4) - (11) =$

5. $(7) - (-15) =$ **6.** $(-13) - (12) =$

7. $(-13) - (-12) =$ **8.** $(13) - (12) =$

9. $(13) - (-12) =$ **10.** $(-23) - (-18) =$

11. $(-21) - (-38) =$ **12.** $(-16) - (30) =$

Check your answers on page 720.

Notation

How can you tell when the notation of a problem is instructing you to *subtract*? The number to subtract would have to be written within *parentheses* and be preceded by a *minus* (−) sign. Simply a minus (−) sign between numbers, with no parentheses, denotes that the number following the minus sign is *negative*.

EXAMPLE 5 The expression $8 - 14$ should be read as : Add negative 14 to 8.
The expression $3 - (7)$ should be read as: Subtract 7 from 3. ❏

The easiest way to compute a problem that contains a minus sign in front of a pair of parentheses is to *change the sign of the number in the parentheses* and then *add* that number to the other numbers.

EXAMPLE 6 Simplify and compute:

$$(-3) + (-5) - (7) - (-6) + (-4) - (-1) =$$
$$(-3) + (-5) + (-7) + (+6) + (-4) + (+1) =$$
$$(-3 - 5 - 7 - 4) + (6 + 1) =$$
$$(-19) + (+7) = -12$$ ❏

EXERCISES 2.3B

Change each problem to addition and then compute.

13. $(-2) - (-4) + (3) - (7) + (-2) =$

14. $5 + (-7) - (4) - (-3) + (-6) =$

15. $11 - (10) + (-7) - (-12) + (4) =$

16. $-(-7) + (-2) + (-8) - (10) =$

17. $(+6) + (-4) + (-7) - (+3) + (-5) =$

18. $(-2) + (-1) - (+6) + (+2) - (-6) + (5) =$

19. $(+5) - (+7) - (-2) - (+3) - (-5) - (+1) =$

20. $-(-12) + (-35) - (-46) - (+12) + (+35) =$

21. $(-4) - (-12) - (+20) + (-10) + (13) - (1) =$

22. $17 - (-8) + (-15) + (21) - (14) - (-20) =$

Check your answers on page 720. ✓

2.4 Multiplication and Division of Signed Numbers

Multiplication

To multiply signed numbers, multiply the magnitudes together to find the magnitude of the product. You need a way to determine what the sign will be.

In Section 1.3 we learned that a multiplication problem could be represented by an addition problem:

$$(3)(2) = (2) + (2) + (2) \quad \text{or} \quad (3) + (3)$$

Let's use that observation to visualize what is happening when a positive number is multiplied by a negative number.

EXAMPLE 1 Multiply (3) by (−2).

We'll use the positive number to represent how many groups we have: 3.

We'll use the negative number to represent the quantity in each group: −2.

So (3)(2) represents (2) + (2) + (2) = 6.

The product of these numbers of different signs is negative.

Say we have groups of missing buttons on shirts. (3)(−2) represents 3 shirts each missing 2 buttons, and so we're *missing* a total of 6 buttons. ❏

In general, every time a positive and a negative number are multiplied, the product can always be represented as a sum of negative numbers. The *sum of negative numbers is always negative.*

The *product* of two numbers with *different signs* will always be *negative.*

In Chapter 1 we worked only with positive numbers. When we multiplied them together, we always got another positive number. So we know that the product of two positive numbers is always positive. What about multiplying two negative numbers?

We can't visualize this as easily as we visualized a positive times a negative, but we can think of multiplying two negative numbers as something like the "double negative." In English we are discouraged from using a double-negative phrase, such as "I will not not go." It is preferable to use the more direct "positive" statement "I will go."

The *product* of *two negative* numbers is always *positive.*

We can visualize the double negative as using a mirror to read backwards writing. It's hard to read backwards writing (negative). It's hard to read normal writing in a mirror (negative). But to use a *mirror* to read *backwards* writing is a big help (positive).

If your mother always told you that two wrongs don't make a right, she wasn't talking about multiplication!

We can summarize this discussion as follows:

When we multiply two signed numbers, the sign of the product will be:

Positive if both signs are the same (both positive or both negative);

Negative if both signs are different (one positive and one negative).

EXAMPLE 2 Find the product of (−5)(−4).

Since both numbers are negative, the sign will be *positive.*

The product of their magnitudes is 5 times 4 = 20.

$$(-5)(-4) = +20$$ ❏

EXAMPLE 3 Find the product of (7)(−3).

Since the signs are different, the product will be *negative.*

The product of their magnitudes is 7 times 3 = 21.

$$(+7)(-3) = -21$$ ❏

Division   In Section 1.4 we learned that division and multiplication are very closely related. They are called the opposite, or *inverse,* operations of each other.

5 times 3 is 15, so 15 divided by 5 is 3, or 15 divided by 3 is 5.

$(5) \times (3) = 15$, so $(15) \div (5) = 3$, or $(15) \div (3) = 5$.

Remembering that relationship helps us see what the rules for dividing signed numbers must be.

The product of two negative numbers is positive. If we were to divide that positive product by either of those negative numbers, the quotient would have to be the other negative number.

EXAMPLE 4   Since $(-6)(-4) = +24$, $(+24) \div (-6) = -4$.

The *quotient* of numbers with *different signs,* must be *negative.*   ❏

The product of a negative number multiplied by a positive number is negative. If we were to divide that negative quotient by the negative number, the quotient would have to be the positive number.

EXAMPLE 5   Since $(-3)(+6) = -18$, $(-18) \div (-3) = +6$.

The *quotient* of *two negative* numbers must be *positive.*   ❏

And of course we already know that the quotient of two positive numbers will be positive. So we see that the rule for determining the sign of a quotient is the same as the one for determining the sign of the product of two numbers.

The sign of answers to a division problem or to a multiplication problem involving two numbers must be:

Positive if the numbers have the **same** sign;

Negative if the numbers have **different** signs.

EXAMPLE 6

$(+6)(+2) = +12 \qquad (-3)(+5) = -15$

$(-4)(-7) = +28 \qquad (+8)(-2) = -16$

$(+12) \div (+4) = +3 \qquad (+18) \div (-2) = -9$

$(-22) \div (-11) = +2 \qquad (-24) \div (+8) = -3$ ❏

EXERCISES 2.4A

Determine the sign for each problem and write the name of that sign in the first blank. Then write the answer (sign and magnitude) in the second blank. For example,

$(-4)(-2) =$ positive ; $+8$.

1. $(-6)(+3) =$ _____; ____

2. $(-7)(-4) =$ _____; ____

3. $(-10) \div (2) =$ _____; ____

4. $(14) \div (-2) =$ _____; ____

5. $(-20) \div (-5) =$ _____; ____

6. $(27) \div (9) =$ _____; ____

7. $(3)(-4) =$ _____; ____

8. $(-30) \div (6) =$ _____; ____

9. $(24) \div (-2) =$ _____; ____

10. $(-6)(+6) =$ _____; ____

11. $(-45) \div (-3) =$ _____; ____

12. $(2)(-8) =$ _____; ____

Check your answers on page 720. ✓

Multiplying More Than Two Signed Numbers

When there are more than two signed numbers to multiply, it's best to determine the *sign* of the entire product before multiplying the numbers.

Since every two negative numbers will produce a positive product, the easiest way to determine the sign is to count how many of the multiplying numbers are negative.

If an even number of negatives are to be multiplied, then all the negatives can be paired together to produce positives. The entire problem will then reduce to multiplying a series of positive numbers together, which will give us a positive answer.

EXAMPLE 7   Find the product of $(-2)(-3)(-1)(-5)$.

$$\frac{(-2)(-3)(-1)(-5)}{(+6)(+5)} = +30$$ ❑

If an odd number of negatives are to be multiplied, one will be left over after you try to pair them up. All the pairs become positive, and all the positive numbers will produce a positive product. When this product is multiplied by the one unpaired negative, the result will be negative.

EXAMPLE 8   Find the product of $(-1)(-3)(-4)(-1)(-2)$.

$$\frac{(-1)(-3)(-4)(-1)(-2)}{(+3)(+4)(-2)} =$$
$$(+12)(-2) = -24$$ ❑

⇒ *Note*:   The *positive* numbers don't have any effect on the *sign* of the final product. Just count the negative numbers.

The numbers with a magnitude of 1 don't have any effect on the magnitude of the product. Multiply the numbers (other than 1) together in pairs until you get the magnitude of the answer.

EXAMPLE 9   Find the product $(-1)(2)(-3)(-5)(4)(-1)$. Because four of the numbers are negative, the product is positive.

$$(-1)(2)(-3)(-5)(4)(-1) =$$
$$+ (2)(3)(5)(4) =$$
$$+ (6)(20) = +120$$ ❑

Find the sign and magnitude for each product.

13. $(-2)(-3)(+1) =$ **14.** $(-4)(+2)(-3)(+1) =$

15. $(-2)(-5)(2)(-1)(1) =$ **16.** $(-1)(3)(-3)(1)(-2) =$

17. $(-2)(-2)(3)(-1)(1) =$ **18.** $(-3)(2)(-5)(-1)(-1) =$

19. $(-1)(-4)(-2)(-1)(3) =$ **20.** $(-2)(-7)(-1)(1)(2) =$

21. $(5)(-1)(2)(-2)(-1)(-1) =$ **22.** $(4)(1)(-3)(-2)(1)(-1) =$

Check your answers on page 720. ✓

2.5   Order of Operations

Order of Operations Computations

Now that you know how to do each kind of computation with signed numbers, we can look at problems involving more than one kind of operation.

Remember, when there is more than one kind of operation in a problem, you must follow the law of order of operations, as we did in Chapter 1, Section 1.5.

The law of **order of operations** is:

1. Perform all calculations within parentheses.
2. Calculate all multiplications and divisions as they appear from left to right.
3. Change all subtraction operations into addition operations by changing the sign of numbers to be subtracted.
4. Add the remaining numbers.

If you are careful and do only one operation on each line, you will often have several steps. (But each step will be easy, and you will be surer of getting the right answer.)

Keep in mind that for every calculation involving a signed number, you must apply the rules for that operation to determine the sign and the magnitude for each result.

EXAMPLE 1   Simplify $(3 - 5 + 10) + 2(-3 - 4) - (6 - 1)$.

1.	$(3 - 5 + 10) + 2(-3 - 4) - (6 - 1) =$	Compute within parentheses.
2.	$(+8) + 2(-7) - (+5) =$	Multiply.
3.	$(8) - 14 - (+5) =$	Change subtraction.
4.	$(8) - 14 + (-5) =$	Add same signs.
5.	$(+8) + (-19) = -11$	

EXAMPLE 2   Simplify $(17 - 8) - 3(7 - 5)$.

    1. $(17 - 8) - 3(7 - 5) =$ Compute within parentheses.

    2. $(9) - 3(2) =$ Multiply.

    3. $9 - 6 = +3$ ❏

        Whenever there is a *minus sign and a number in front of a parenthesis*, consider the minus as that number's negative sign when you multiply it by the value of the calculation within parentheses (as we did in Example 2). The sign of that product is the sign of the number you add in the later steps.

EXAMPLE 3   Simplify $2(7 - 3) + 3(5 - 7) - 6(-8 - 1 + 5)$.

    1. $2(7 - 3) + 3(5 - 7) - 6(-8 - 1 + 5) =$ Compute within parentheses.

    2. $2(+4) + 3(-2) - 6(-4) =$ Multiply.

    3. $+8 - 6 + 24 =$ Add same signs.

    4. $+32 - 6 = +26$ ❏

EXAMPLE 4   Simplify $5 - 3(8 - 3 - 1) + 2(3 - 7)(-2 - 3)$.

    1. $5 - 3(8 - 3 - 1) + 2(3 - 7)(-2 - 3) =$ Compute within parentheses.

    2. $5 - 3(+4) + 2(-4)(-5) =$ Multiply.

    3. $5 - 12 + (-8)(-5) =$ Multiply.

    4. $5 - 12 + 40 =$ Add same signs.

    5. $45 - 12 = +33$ ❏

EXAMPLE 5   Simplify $-15 \div 3 + 7(8 - 3 - 1)$.

    1. $-15 \div 3 + 7(8 - 3 - 1) =$ Compute within parentheses.

    2. $\underline{-15 \div 3} + \underline{7(+4)} =$ Divide and multiply.

    3. $-5 + 28 = +23$ ❏

EXERCISES 2.5A

Follow the order of operations and simplify.

  1. $(-8 + 4 + 1)(7 - 9 - 2)(+6 - 3) =$ **2.** $3(-5 - 1) + 6(2 - 3 - 4) =$

  3. $6(8 - 7 - 5) + 4(3 - 7 - 1) - 6(8 - 3 + 1) =$

  4. $-7 + 6(-2 + 1 - 3) - 5(2 - 1) =$

  5. $-7(8 - 5 - 2) - 2(15 - 18 + 5)(4 - 3 - 2) =$

  6. $-15 \div 3 + 4(5 - 2 - 7) =$

  7. $(-5 - 3 + 4)(-2 - 3) + (-6 - 3 + 5)(-1 + 3) - (6 - 3 - 1) =$

  8. $-6 + 10 \div (-2) + 5(3 - 1 - 3)(4 - 2) =$

  9. $3(5 - 4 - 3) + 6(5 - 2) + 5(4 + 1 - 3)(-3 + 1) =$

  10. $-7 - 6(-5 - 2) - 4(-2 - 1 + 5)(3 + 1 - 1) =$

  11. $4 - 3(9 - 4 - 2) + 2(5 - 1 - 2)(-1 - 2 - 2) =$

  12. $18 \div (-6) + 4(-1 - 1) - (5 - 2 + 1) =$

Check your answers on page 720. ✓

Evaluating Formulas

As we did in Section 1.5, let's evaluate formulas; but now we can let signed numbers represent the values of the letters.

Recall that the formula $D = (R)(T)$ tells us that distance is the product of the rate multiplied by the time. During a physical exercise session, we might consider time spent resting as negative.

EXAMPLE 6 After Patti exercised, she calculated that since she began working on the bicycle machine at 10 A.M. and finished at 2 P.M., she had worked out for 4 hours. Her rate during the session was constantly 20 miles per hour. She was impressed with the notion that she had pedaled the equivalent of 80 miles. Then she remembered that she had rested several times. Her rests totaled 1 hour. Let's represent the time she rested as negative and correct her calculation.

$$D = 80 \text{ mi} + \text{corrected } [(R)(T)]$$
$$= 80 \text{ mi} + (20 \text{ mph})(-1 \text{ hr})$$
$$= 80 \text{ mi} - 20 \text{ mi}$$
$$= 60 \text{ mi} \qquad ❑$$

In algebra you'll run into more complicated formulas. The letters sometimes represent negative numbers.

EXAMPLE 7 Evaluate $5a + 4b - 3c$ if $a = -1$, $b = -2$, and $c = -5$.

$$5a + 4b - 3c =$$
$$5(-1) + 4(-2) - 3(-5) =$$
$$-5 - 8 + 15 =$$
$$-13 + 15 = +2 \qquad ❑$$

EXAMPLE 8 Evaluate $3(a + b) + a(c + a) + (a)(c)$ if $a = -2$, $b = +3$, and $c = -5$.

$$3(a + b) + a(c + a) + (a)(c) =$$
$$3[(-2) + (+3)] + (-2)[(-5) + (-2)] + (-2)(-5) =$$
$$3[+1] + (-2)[-7] + (-2)(-5) =$$
$$3 + 14 + 10 = 27 \qquad ❑$$

EXERCISES 2.5B

Evaluate each formula if $a = -2$, $b = +3$, and $c = -5$.

13. $a + 2b + 3c =$ **14.** $7a - 4b - 3c =$

15. $a + (b)(c) =$ **16.** $3 + a(b + c) =$

17. $5 - b(a + c) =$ **18.** $2(a + b) + 3(a - c) =$

19. $b + c(a + b) =$ **20.** $c - 3(b + a) =$

21. $2(a + c) - 3(5 + b)(a + 1) =$

22. $10 + 3(a + b + c) - (a + b + c) =$

23. $2(a + c) - 7(a + b)(4 + c + 1) =$

24. $7 + a(1 + b + c) + 2(a + b + c) =$

Check your answers on page 720. ✓

 **Using Signed Numbers to Compute Real-Life Problems**

When using signed numbers to compute real-life problems, you have to assign a plus or minus sign to the numbers representing opposite situations.

In problems related to banking, transactions involving deposits and receiving interest are *positive,* and transactions involving withdrawals, writing checks, and bank charges are *negative.*

EXAMPLE 1 Last month my bank reported that I had made the following transactions: deposited $125; wrote checks for $12, $25, and $78; received $3 interest; withdrew $80; and had a $10 service charge for writing a bad check.

What is the net effect on my balance from those transactions?

To find the answer, we must represent each transaction by a signed number and then add them all. They translate into:

$$+125 - 12 - 25 - 78 + 3 - 80 - 10 =$$
$$(125 + 3) + (-12 - 25 - 78 - 80 - 10) =$$
$$(+128) + (-205) = -77$$

So the net effect was a *decrease* of $77.   ❑

EXAMPLE 2 If my bank balance before the transactions reported in Example 1 was $308, what is my balance now?

To find the present balance, we add the effect of the transactions to the previous balance.

Balance now = previous balance + net effect of transactions
Balance now = ($308) + ($-77)
$$= \$231$$ ❑

In sports statistics, we consider *gains* as *positive* and *losses* and penalties as *negative.*

EXAMPLE 3 What is Stella's net gain if she carried the ball for three gains of 4 yards each, two gains of 7 yards each, two losses of 3 yards each, was penalized once for 10 yards, and had one great run for a gain of 32 yards?

To find the net gain, we represent each gain as a positive number and each loss as a negative number. Then we find their sum.

Those plays translate into this calculation:

$$3(+4 \text{ yd}) + 2(+7 \text{ yd}) + 2(-3 \text{ yd}) + (-10 \text{ yd}) + (+32 \text{ yd}) =$$
$$12 \text{ yd} + 14 \text{ yd} - 6 \text{ yd} - 10 \text{ yd} + 32 \text{ yd} =$$
$$(12 + 14 + 32) \text{ yd} + (-6 - 10) \text{ yd} =$$
$$+58 \text{ yd} - 16 \text{ yd} = +42 \text{ yd}$$

So Stella had a net gain of 42 yards.   ❑

EXERCISES 2.6

1. What is the net effect of these transactions: deposit of $45; checks written for $30, $10, and $35; withdrawal of $60; deposit of $150; and receive $2 interest?

2. What is the final balance if an account begins with $180 and includes the following transactions: checks written for $20, $15, $40, and $18; deposits of $60 and $103; withdrawal of $50; and bank charge of $15?

3. What is your final balance if your account opens with $100 and you write five checks for $15 each, deposit a $10 birthday gift and a $70 tax refund, and receive $2 in interest from the bank?

4. What is Rick's total yardage gained if these are his statistics: gain of 3 yd, gain of 12 yd, loss of 4 yd, loss of 1 yd, gain of 5 yd, and gain of 3 yd?

5. What is Nebraska's total yardage gained if these are its statistics: 8 yd completed pass, 5 yd run, two runs of 3 yd, loss of 4 yd, 12 yd pass, gain of 7 yd, and loss of 5 yd due to overaggressive blocking?

6. What is the net effect on a particular stock if it registered the following fluctuations in points on the stock market: up 2, up 1, down 5, down 1, up 3, down 1, down 4, and up 3?

7. What is the final height of a plant that started out 7 inches high after it had the following growth spurts and pruning cutbacks: grew 3 in., cut 5 in., grew 4 in., cut 2 in., grew 2 in., grew 6 in., and cut 3 in.?

8. What is the final temperature if at the first reading it was 60 degrees and then had the following fluctuations: up 3°, up 1°, down 4°, up 2°, up 5°, up 1°, and down 6°?

9. How much money do I end up with if I start off with $20 and then pay three debts of $5 each, receive a $10 gift and pay $7, receive $8 each from four friends and pay $8 to one friend?

10. What is the net effect of the following fluctuations in weight: lost 3 lb, lost 1 lb, gained 2 lb, gained 3 lb, lost 4 lb, lost 2 lb three times, lost 7 lb, and gained 2 lb.

11. One day several neighbors and I set up a garage sale. If I started out with $5 in my pocket, how much money did I have at the end of the day after making the following transactions: sold a lamp for $7, bought some shorts for $2, bought a set of cups for $4, sold a shirt for $1, sold two games for $3 each, sold a picture for $2, bought a car compass for $4, sold my bowling shoes for $5, bought three puzzles for $2 each, bought a dog bed for $7, and sold a yo-yo that shines in the dark for $1?

12. When I bought my fish tank, I stocked it with six neon tetras, a male and a female swordtail, and one catfish to keep the tank bottom clean. Four of the tetras died. I bought two angel fish and two more tetras. Somebody ate a swordtail, three more tetras died, and one survived out of the three baby swordtails that were born. How many fish do I have now?

13. A farmer started out with a herd of 43 cows. Over the year 8 calves were born, 10 cows were slaughtered, 2 cows were lost, and 5 new calves were purchased. How many cows does she have now?

14. A notorious slumlord bought one building with 8 units, two buildings with 4 units each, and three buildings with 6 units each. He subdivided 3 large units, making each one into 2 units, had mysterious fires

that destroyed 14 units, sold 6 units to the tenants at great profit as condominiums, and finally was forced to sell the remaining units to the nonprofit neighborhood land trust. How many units were sold to the land trust?

15. Phil's army squad started out with 30 men. They experienced the following changes: 5 men transferred out, 8 new recruits came, 10 men died, 16 men deserted, 4 were discharged, and 3 more transferred in. How many men are in the squad now?

Check your answers on page 720.

Glossary

absolute value   The distance a number is from zero on the number line. There is no reference to direction (positive or negative), so the absolute value is always given as a positive number. The notation to represent a number's absolute value is | |.

Example: The absolute value of -7 is

$$|-7| = 7.$$

magnitude   A number's absolute value; that is, the distance it is from zero.

Example: The magnitude of -13 is 13.

natural number   An integer larger than zero.

Example: 1, 2, 3, ... are natural numbers.

negative number   A number less than zero. Negative numbers are found to the left of zero on the number line. They are always written with a negative sign $(-)$ in front of them.

Example: -3, -15, and -84 are negative numbers.

opposites   Numbers that have the same magnitude but different signs. When added, their sum is equal to zero.

Example: 7 and -7 are opposites.

order of operations   A law in mathematics that, in problems involving more than one operation, specifies the order in which the operations must be performed:

1. Compute within parentheses.
2. Multiply and divide (from left to right).
3. Change subtraction into addition by changing the sign of the number to be subtracted.
4. Add the remaining numbers.

Example: $3 + 4(-5 + 2) =$
$$3 + 4(-3) =$$
$$3 - 12 = -9$$

positive number   A number greater than zero.

Example: 1, 3, 8, 73, ... are positive numbers.

sign   The indication of a number's direction from zero. Numbers greater than zero are positive and are written with a plus $(+)$ sign or no sign. Numbers less than zero are negative and are always written with a negative $(-)$ sign.

Example: The sign of -8 is negative.
The sign of 14 is positive.

signed number   Any number with a specified direction and magnitude.

Chapter 2 Review

Do each of the problems. After you are confident that you have done them as accurately as possible, compare your answers with those in the back of the book. If any are wrong, go back to the section where the problem came from (indicated to the left of the problem) and review the section.

Once you understand all the sections, you'll be sure to learn the skills solidly and remember how to do them if you practice them a little bit more. Turn to the Supplementary Exercises and do all the problems from any section where you had difficulty on these review exercises.

2.1
1. What is the sign of -18?

2. What is the magnitude of -18?

3. What is $|-32|$?

Combine these numbers.

2.2
4. $(-4) + (-17) =$

5. $(-3) + (-4) + (-6) + (-8) =$

6. $-7 - 2 - 2 - 5 - 3 =$

7. $(6) + (-19) =$

8. $(5) + (-8) + (1) + (-4) + (-2) + (6) =$

9. $-3 + 2 + 4 - 5 - 3 + 1 - 2 =$

2.3
10. $(-6) - (-11) =$

11. $(-13) - (10) =$

12. $(3) - (-2) + (-5) - (7) - (-8) + (4) =$

2.4
13. $(-3)(-9) =$

14. $(30) \div (-6) =$

15. $(-1)(2)(-3)(-1)(-2)(-1) =$

2.5
16. $3 + 2(-3 - 1 - 1) - 4(5 - 4 - 3) =$

17. What is the value of J if $J = a + 2b - 3(a + b + c)$, and $a = -1$, $b = -5$, and $c = 2$?

2.6
18. What is the final balance in a checking account if it opens with a balance of $150 after the following transactions: checks written for $15, $20, $45, and $31; deposits of $25, $10, and $73; receipt of $3 interest; and service charge of $10?

Answers

1. _____

2. _____

3. _____

4. _____

5. _____

6. _____

7. _____

8. _____

9. _____

10. _____

11. _____

12. _____

13. _____

14. _____

15. _____

16. _____

17. _____

18. _____

Check your answers on page 720.

Supplementary Exercises

Do all the problems in every section that involve a skill in which you now lack complete mastery. With a little more practice you will achieve that sense of really understanding the topics, and you'll remember how to do these problems.

2.1   **1.** What is the sign of -58?

2. What is the magnitude of -137?

3. What is $|-61|$?

4. What is $|78|$?

Combine these numbers.

2.2   **5.** $(-4) + (-17) =$ **6.** $(-23) + (-57) =$

7. $(-5) + (-1) + (-4) + (-7) =$ **8.** $-8 - 1 - 2 - 3 - 4 =$

9. $(14) + (-35) =$

10. $(3) + (-6) + (2) + (-5) + (-4) + (7) =$

11. $-2 + 7 + 1 - 6 - 4 + 8 - 5 =$ **12.** $-6 - 3 + 5 - 2 + 7 - 1 + 3 =$

2.3   **13.** $(-8) - (-15) =$ **14.** $(-17) - (12) =$

15. $(30) - (-18) =$

16. $(1) - (-6) + (-4) - (2) - (-7) + (3) =$

17. $(5) - (7) + (-2) - (4) - (-8) - (1) =$

2.4   **18.** $(-2)(-6) =$ **19.** $(-7)(4) =$

20. $(-24) \div (8) =$ **21.** $(-15) \div (-3) =$

22. $(-2)(5)(-1)(-1)(-4)(1) =$ **23.** $(-1)(-1)(3)(-1)(1)(-7)(-1) =$

2.5   **24.** $1 + 3(-2 - 1 - 2) - 2(4 - 2 - 3) =$

25. $5 - 2(3 - 4 - 1)(5 + 2 - 4) =$

26. What is the value of M if $M = a - 3b - 2(a + b + c)$, and $a = -2$, $b = -1$, and $c = 3$?

27. What is the value of Y if $Y = 2x - 2y + 3(x + w)$, and $x = -1$, $y = -3$, and $w = 5$?

2.6   **28.** What is the final balance in a checking account if it opens with a balance of $300 after the following transactions: checks written for $35, $57, $40, and $145; deposits of $70, $40, and $100; receipt of $7 interest; and service charge of $15?

Check your answers on page 721.

Answers

1. _____

2. _____

3. _____

4. _____

5. _____

6. _____

7. _____

8. _____

9. _____

10. _____

11. _____

12. _____

13. _____

14. _____

15. _____

16. _____

17. _____

18. _____

19. _____

20. _____

21. _____

22. _____

23. _____

24. _____

25. _____

26. _____

27. _____

28. _____

Chapter 2 Test

1. What is the sign of -29?

2. What is $|-813|$?

Combine these numbers.

3. $-16+5$

4. $-15-3$

5. $(-5) + (-11)$

6. $(7)+(-15)$

7. $(9) + (-5) + (-3) + (-5) + (7)$

8. $(-7) + (8) + (-15) - (3) - (-5) + (-1)$

9. $-4+3+1-5-9-3$

10. $(-6)(-2)$

11. $(-18) \div (-3)$

12. $(4)(-7)$

13. $(20) \div (-2)$

14. $(-5)(-1)(-1)(2)(3)(-1)$

15. $3 + 7(-4 - 2 + 1) - 15$

16. $12 \div (-3) + 4(8 - 5 - 3)(2 + 1 - 5) - 18$

17. What is the value of K if $a = -1$, $b = 2$, and $c = -2$?

$$K = 3a - 5b - 3c + 4(a + b + c)$$

Solve this problem.

18. What is the final balance in an account that has an opening balance of $191 after the following transactions: checks written for $50, $25, $7, $86, $15, and $26; deposits of $20 and $50; and service charges $3?

Answers

1. _____

2. _____

3. _____

4. _____

5. _____

6. _____

7. _____

8. _____

9. _____

10. _____

11. _____

12. _____

13. _____

14. _____

15. _____

16. _____

17. _____

18. _____

Check your answers on page 721.

Cumulative Review
CHAPTERS 1 AND 2

1. $506 + 23 + 58{,}170 =$

2. $50{,}218 - 3{,}605 =$

3. $14 - 15 - 18 + 7 - 11 + 16 =$

4. $(23) - (-45) =$

5. $(9) - (-7) + (-12) - (14) - (-6) + (10) =$

6. $(-9)(-6) =$

7. $(-24) \div (6) =$

8. $(-5)(-1)(1)(-3)(2)(1)(-1) =$

9. $|-228| =$

10. $|817| =$

11. $7 - 3(-1 - 4 - 2) + 7(-12 + 9 - 1) =$

12. $1 + 2(-7 + 4 - 1)(-3 - 1 - 1) - 3(13 - 15 - 2) =$

13. $15{,}645 \div 15 =$

14. $37 \times 43 =$

15. What is the perimeter of a triangle with sides 10 in., 6 in., and 8 in.?

16. What is the area of a rectangle with width 2 ft and length 8 ft?

17. What is the volume of a rectangular solid with height 5 in. and square base with 3 in. sides?

18. How tall is a plant that was 6 in. high when bought, then grew 3 in., was cut back 2 in., grew 4 in., was cut back 3 more in., grew 5 in., and was cut back 1 in.?

19. What was the team's net yardage if the statistics were as follows: 12 yd gain, 4 yd gain, 20 yd gain, 5 yd penalty, 2 yd loss, 34 yd gain, 18 yd gain, 3 yd gain, 6 yd gain, 15 yd penalty, and finally a 7 yd gain?

20. What is the balance in a bank account that opens with $125 after the following transactions: checks written for $30, $10, $7, and $55; deposits of $25, $30, and $25; and service charge of $5?

Answers

1. _____

2. _____

3. _____

4. _____

5. _____

6. _____

7. _____

8. _____

9. _____

10. _____

11. _____

12. _____

13. _____

14. _____

15. _____

16. _____

17. _____

18. _____

19. _____

20. _____

Check your answers on page 721.

3 Fractions

 Concept of a Fraction

Representation of Part of a Whole by a Fraction

A fraction is one way to represent a number that is only a *part of an integer*.

EXAMPLE 1 When a pizza is to be shared equally by four people, each person's share is less than one whole pizza and can be described by a fraction. If the pizza is cut into four equal parts, each slice is called one-fourth of a pizza. This fraction is written as $\frac{1}{4}$.

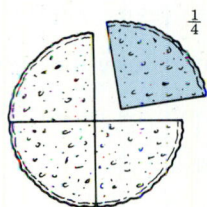

❑

EXAMPLE 2 If you eat three slices of a pizza that was cut into eight equal parts, you will have eaten three-eighths $\left(\frac{3}{8}\right)$ of a pizza.

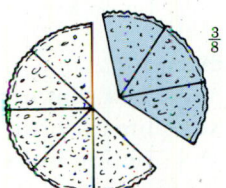

❑

A fraction is written with one number above another.

The number on top is called the **numerator**. It represents the quantity of pieces involved. In $\frac{3}{8}$, 3 is the numerator.

The number on the bottom is called the **denominator**. It represents the amount of equal parts that the whole was divided into. In $\frac{3}{8}$, 8 is the denominator.

The denominator determines the *size* of each piece. Cutting a pizza into four pieces yields *bigger* slices than cutting it into eight slices. The more people who share a pizza, the smaller each person's share.

Fractions, like people, have first and second names. A fraction's *first name* is its *numerator* and *second name* its *denominator*. That is why $\frac{3}{8}$ is called *three eighths*.

EXAMPLE 3 Two thirds $\left(\frac{2}{3}\right)$ represents the fraction that results from dividing an item into three equal parts and using two of the parts.

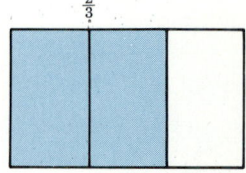

In Section 1.4 we mentioned that writing one number over another was a notation for dividing whole numbers. Thus every *fraction* can be thought of as a *division* problem.

EXAMPLE 4 Five sevenths $\left(\frac{5}{7}\right)$ represents each person's share when five pizzas are to be divided by seven people.

The result of that division is a number (of pizzas for each person) that is less than 1, since there are fewer pizzas than people.

A fair way to divide five pizzas equally among seven people is to cut each pizza into seven equal pieces. Each slice is one-seventh $\left(\frac{1}{7}\right)$ of a pizza. Each person could take one slice from each of the five pizzas. Each person's total share would be those five slices, which can be described as five sevenths $\left(\frac{5}{7}\right)$ of a pizza.

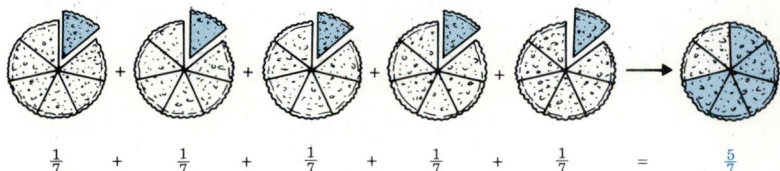

Signed Fractions

We know that every fraction can be thought of as a division problem. If either the numerator or the denominator is negative (and the other one is positive), the fraction represents the division of numbers with *different signs*. In Section 2.4 we learned that the resulting quotient is *negative*. So if either number in a fraction is negative, we consider the *sign* of that fraction to be *negative*.

If both the numerator and the denominator are negative, that fraction represents the division of two negatives, which is positive. And so the fraction is *positive*.

If the *answer* to a fractional computation has a minus sign in the numerator, in the denominator, or in both, determine the *sign of the fraction*. Leave the answer expressed with a negative sign in front of the fraction if it is negative. (If it turns out to be positive, remember that no sign is needed.)

EXAMPLE 5 $\dfrac{-2}{7} = -\dfrac{2}{7}; \ \dfrac{3}{-4} = -\dfrac{3}{4}; \ \dfrac{-5}{-8} = \dfrac{5}{8}$

⇨ *Note*: When working with a **negative fraction** *during a computation*, it is usually best to consider the *numerator* as negative and the denominator as positive.

Generally, number lines are written with markings to indicate whole numbers and are labeled with integers. Fractions can be illustrated on a number line too.

A fraction of less than 1 is called a **proper fraction**. The numerator is smaller than the denominator. That is, fewer parts are used than the amount into which the whole was divided. They appear on the number line between 0 and 1 if positive or between 0 and -1 if negative.

EXAMPLE 6   The proper fractions $\dfrac{7}{10}$ and $-\dfrac{3}{10}$ are illustrated below.   ❏

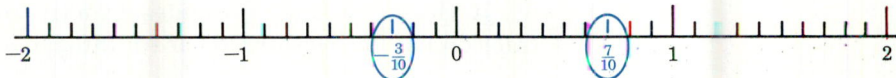

Representing 1 as a Fraction

You probably have already noticed that the easiest number to multiply by is 1. Multiplying any number by 1 never changes the number.

EXAMPLE 7 $35{,}684 \times 1 = 35{,}684$ ❏

In fractional notation, 1 can be expressed many different ways. Whenever you cut a whole into parts and use *all* the parts, that fraction is 1.

In other words, if the numerator and the denominator of a fraction are the same, that fraction is equal to 1.

EXAMPLE 8   Here are a few ways of expressing 1 as a fraction: $\dfrac{2}{2}; \dfrac{7}{7}; \dfrac{58}{58}; \dfrac{103}{103}; \dfrac{690}{690}; \dfrac{-3}{-3}; \dfrac{-15}{-15}$

❏

Improper Fractions

Whenever the numerator is larger than the denominator, the fraction has a magnitude larger than 1 (since there are more pieces than the amount into which the whole was divided). Such fractions are called **improper fractions**.

EXAMPLE 9 $\dfrac{3}{2}, \dfrac{5}{3}, \dfrac{7}{5}$ are improper fractions.   ❏

Since improper fractions are larger than 1, they can be expressed as the combination of an integer and a proper fraction. An improper fraction expressed as such a combination is called a **mixed number**. It is easier for us to picture the size of improper fractions when they are expressed as mixed numbers.

To convert improper fractions to mixed numbers, just consider the fraction as representing the division of its numerator by its denominator.

EXAMPLE 10   Express $\dfrac{5}{3}$ as a mixed number.

$\dfrac{5}{3}$ represents $5 \div 3$.

$$\begin{array}{r} 1 \\ 3\overline{)\,5} \\ -3 \\ \hline 2 \end{array}$$

So $\dfrac{5}{3}$ can be expressed as $1\dfrac{2}{3}$.

❏

Mixed numbers appear on number lines between integers.

EXAMPLE 11 $3\frac{2}{5}$ lies between 3 and 4.

$-2\frac{3}{5}$ lies between -2 and -3.

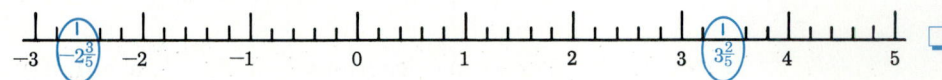

If the denominator divides exactly into the numerator, the improper fraction can be represented as an integer.

EXAMPLE 12 $\frac{12}{3}$ is an improper fraction representing $12 \div 3 = 4$.

Any integer can be expressed as a fraction by writing it over the denominator of 1.

EXAMPLE 13   7 can be expressed as the improper fraction $\frac{7}{1}$.

The denominator of a fraction can never be zero. What would it mean to take five pizzas and divide them by zero people?

EXERCISES 3.1

State the name of each fraction shaded in each figure.

1.    2.

3.    4.

5. 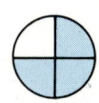   6.

7.    8.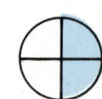

9.    10.

11. What is the numerator of $\frac{7}{13}$?

12. What is the denominator of $\frac{7}{13}$?

Determine the sign of each fraction.

13. $\dfrac{-6}{7}$ **14.** $\dfrac{4}{-9}$ **15.** $\dfrac{-2}{-15}$

16. $\dfrac{-8}{11}$ **17.** $\dfrac{3}{-8}$ **18.** $\dfrac{-1}{-5}$

Identify whether each fraction is proper or improper.

19. $\dfrac{3}{4}$ **20.** $\dfrac{4}{3}$

21. $\dfrac{6}{3}$ **22.** $\dfrac{7}{8}$

State what integer each fraction represents.

23. $\dfrac{21}{3}$ **24.** $\dfrac{17}{17}$

25. $\dfrac{87}{87}$ **26.** $\dfrac{40}{5}$

Check your answers on page 721. ✓

3.2 Computing with Fractions

Addition and Subtraction of Fractions with the Same Denominator

We can visualize what is happening when we add or subtract fractions if we keep in mind what the numerator and the denominator represent. As long as the denominators are the same, the problem represents combining pieces that are the *same size*. To find the total, just combine the quantities involved, which are *represented by the numerators*. The denominator represents the size of the pieces, which does not change by adding or subtracting; it should remain as the denominator of the answer.

EXAMPLE 1 $\dfrac{3}{7} + \dfrac{1}{7} =$

$\dfrac{(3+1)}{7} = \dfrac{4}{7}$

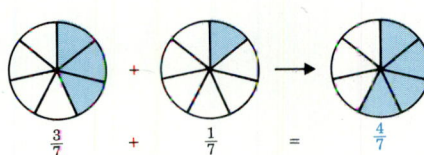

We can summarize these observations as follows:

To add (any quantity of) fractions that have the *same denominators*, just *add* their *numerators* and *keep* their *denominators*.

EXAMPLE 2 $\dfrac{1}{9} + \dfrac{5}{9} + \dfrac{2}{9} =$

$\dfrac{(1+5+2)}{9} = \dfrac{8}{9}$

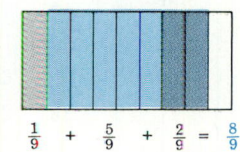

To subtract fractions with the *same denominators,* just *subtract* their *numerators.*

EXAMPLE 3 $\dfrac{5}{11} - \dfrac{2}{11} =$

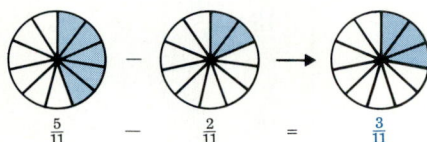

$\dfrac{(5-2)}{11} = \dfrac{3}{11}$

We get the same result for the problem in Example 3 whether we consider that problem the *subtraction of* $\dfrac{2}{11}$ *from* $\dfrac{5}{11}$ or the *addition of negative* $\dfrac{2}{11}$ *to* $\dfrac{5}{11}$. If any fractions are negative, consider the numerator negative. We can add their numerators by using the rules of signed numbers that we learned in Section 2.2.

EXAMPLE 4   Combine $\dfrac{1}{20} + \dfrac{3}{20} - \dfrac{5}{20} - \dfrac{6}{20} + \dfrac{7}{20} - \dfrac{1}{20}.$

Since this list of fractions has signs between the fractions, we are expected to add; consider the signs as the signs of the numerators.

$$\dfrac{(1 + 3 - 5 - 6 + 7 - 1)}{20} =$$

$$\dfrac{[(1 + 3 + 7) + (-5 - 6 - 1)]}{20} =$$

$$\dfrac{[(+11) + (-12)]}{20} = -\dfrac{1}{20}$$

EXAMPLE 5   Combine $\left(-\dfrac{5}{11}\right) - \left(-\dfrac{3}{11}\right).$

$$\dfrac{[(-5) - (-3)]}{11} =$$

$$\dfrac{[(-5) + (+3)]}{11} = -\dfrac{2}{11}$$

EXERCISES 3.2A

Combine the fractions.

1. $\dfrac{2}{5} + \dfrac{2}{5} =$

2. $\dfrac{5}{10} + \dfrac{4}{10} =$

3. $\dfrac{8}{11} + \dfrac{2}{11} =$

4. $\dfrac{15}{17} - \dfrac{3}{17} =$

5. $\dfrac{6}{13} - \dfrac{2}{13} =$

6. $\dfrac{3}{14} + \dfrac{1}{14} + \dfrac{5}{14} =$

7. $\dfrac{4}{19} + \dfrac{5}{19} + \dfrac{2}{19} =$

8. $\dfrac{1}{10} + \dfrac{3}{10} + \dfrac{1}{10} + \dfrac{2}{10} =$

9. $\dfrac{3}{17} + \dfrac{1}{17} - \dfrac{6}{17} =$

10. $-\dfrac{4}{19} - \dfrac{1}{19} - \dfrac{3}{19} - \dfrac{7}{19} =$

11. $\dfrac{7}{23} - \dfrac{9}{23} - \dfrac{2}{23} + \dfrac{5}{23} =$

12. $\dfrac{4}{31} - \dfrac{20}{31} - \dfrac{2}{31} + \dfrac{8}{31} =$

13. $\dfrac{7}{11} - \left(-\dfrac{2}{11}\right) =$

14. $\left(-\dfrac{4}{7}\right) - \left(\dfrac{2}{7}\right) =$

15. $\left(-\dfrac{9}{13}\right) - \left(-\dfrac{1}{13}\right) =$

16. $\left(\dfrac{12}{29}\right) - \left(\dfrac{17}{29}\right) =$

Check your answers on page 721. ✓

Multiplication

In English the word *multiply* suggests an increase. When we multiply any number by proper fraction, though, the magnitude of their product is always smaller than that number. The word **of** is often used instead of *times* to indicate multiplication by a fractional part of a number. For example, when asked to determine what $\dfrac{1}{2}$ of 3 is, we are expected to multiply them.

We can visualize what is happening in a multiplication problem if we express it as an addition problem.

EXAMPLE 6 What is $\dfrac{1}{2}$ of 3?

$\dfrac{1}{2}$ of 3 means $\dfrac{1}{2} \times 3$.

We can let the integer (3) represent the number of groups, and the fraction $\dfrac{1}{2}$ represent the amount in each group.

$$\dfrac{1}{2} \times 3 =$$

$$\dfrac{1}{2} + \dfrac{1}{2} + \dfrac{1}{2} =$$

$$\dfrac{(1+1+1)}{2} = \dfrac{3}{2}$$

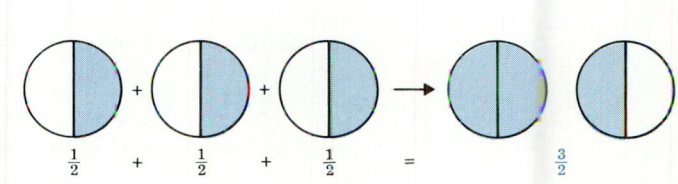

❑

EXAMPLE 7 What is $\dfrac{2}{3}$ of 5?

$$\dfrac{2}{3} \times 5 =$$

$$\dfrac{2}{3} + \dfrac{2}{3} + \dfrac{2}{3} + \dfrac{2}{3} + \dfrac{2}{3} =$$

$$\dfrac{(2+2+2+2+2)}{3} = \dfrac{10}{3}$$

$$\dfrac{10}{3}$$

❑

(Remember, a whole number can always be expressed as a fraction by writing it over 1.)

EXAMPLE 8 $5 = \dfrac{5}{1}; \; 4 = \dfrac{4}{1}; \; 13 = \dfrac{13}{1}$ ❏

We could have achieved the same results in Examples 6 and 7 if we directly *multiplied the numerators together* and *multiplied the denominators together*. Example 7 would have looked like this:

$$5 \times \frac{2}{3} =$$

$$\frac{5}{1} \times \frac{2}{3} =$$

$$\frac{(5 \times 2)}{(1 \times 3)} = \frac{10}{3}$$

An easy way to remember the rule for multiplying fractions is to recall that it is the most straightforward one you could think of.

> To multiply two fractions:
>
> Multiply the numerators to get the product's numerator.
> Multiply the denominators to get the product's denominator.

We can visualize this rule by using a number line. When we cut something into two equal parts, each part is one half the original amount. So $\dfrac{1}{2}$ of 3 represents one of the parts that result from cutting three things into two equal parts.

EXAMPLE 9 $\dfrac{1}{2}$ of 3 is shaded on this number line.

$\dfrac{1}{2}$ of $3 = 1\dfrac{1}{2}$

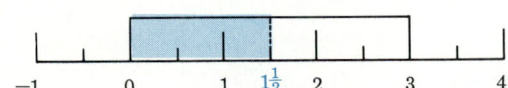

❏

Similarly, $\dfrac{1}{2}$ of $\dfrac{1}{4}$ is one of the equal parts that result from cutting one-fourth of a thing into two equal parts.

EXAMPLE 10 $\dfrac{1}{2}$ of $\dfrac{1}{4}$ is shaded on this number line.

$\dfrac{1}{2}$ of $\dfrac{1}{4} = \dfrac{1}{8}$

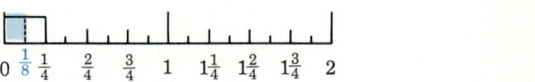

❏

EXAMPLE 11 If you were using a cookie recipe calling for $\dfrac{1}{4}$ cup of shortening, and you only wanted to make $\dfrac{1}{2}$ as much cookie batter as the recipe called for, you would need to calculate $\frac{1}{2}$ of $\frac{1}{4}$ to know how much shortening to use.

$$\frac{1}{2} \text{ of } \frac{1}{4} =$$

$$\left(\frac{1}{2}\right) \times \left(\frac{1}{4}\right) =$$

$$\frac{(1 \times 1)}{(2 \times 4)} = \frac{1}{8}$$

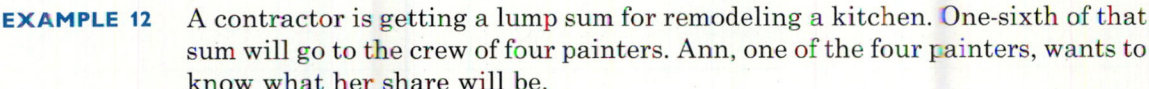

You would use $\frac{1}{8}$ cup of shortening.

EXAMPLE 12 A contractor is getting a lump sum for remodeling a kitchen. One-sixth of that sum will go to the crew of four painters. Ann, one of the four painters, wants to know what her share will be.

Each painter receives $\frac{1}{4}$ of the $\frac{1}{6}$.

$$\frac{1}{4} \text{ of } \frac{1}{6} =$$

$$\left(\frac{1}{4}\right) \times \left(\frac{1}{6}\right) =$$

$$\frac{(1 \times 1)}{(4 \times 6)} = \frac{1}{24}$$

So Ann will receive $\frac{1}{24}$ of the remodeling charge.

EXAMPLE 13 Five people equally shared the one-third of the pie that was left over after dinner. How much pie did each one eat?

Each one ate $\frac{1}{5}$ of $\frac{1}{3}$ of a pie.

$$\frac{1}{5} \text{ of } \frac{1}{3} =$$

$$\left(\frac{1}{5}\right)\left(\frac{1}{3}\right) =$$

$$\frac{(1 \times 1)}{(5 \times 3)} = \frac{1}{15}$$

So each ate $\frac{1}{15}$ of a pie.

EXAMPLE 14 What is $\frac{3}{4}$ of $\frac{5}{7}$?

$$\frac{3}{4} \times \frac{5}{7} =$$

$$\frac{(3 \times 5)}{(4 \times 7)} = \frac{15}{28}$$

Find the products.

17. $\dfrac{1}{4} \times \dfrac{1}{5} =$ **18.** $\dfrac{2}{3} \times \dfrac{2}{5} =$ **19.** $\dfrac{1}{2}$ of $\dfrac{3}{5} =$ **20.** $\dfrac{3}{8}$ of $\dfrac{1}{2} =$

21. $\dfrac{1}{3} \times \dfrac{5}{11} =$ **22.** $\dfrac{4}{7}$ of $\dfrac{1}{3} =$ **23.** $\dfrac{3}{4} \times \dfrac{3}{7}$ **24.** $\dfrac{2}{5} \times \dfrac{3}{7} =$

25. $\dfrac{3}{4}$ of $5 =$ **26.** $\dfrac{1}{3}$ of $7 =$

Check your answers on page 721. ✓

3.3 Factoring

Factors and Primes

Factoring is a tool that makes it easier to compute with fractions. A **factor** of a number is a number that divides into it exactly, with no remainder.

EXAMPLE 1   2 is a factor of 10, and 5 is a factor of 10, since $2 \times 5 = 10$. ❑

A **prime** is a number that has only *itself* and 1 as factors.

EXAMPLE 2   2 is a prime. 3 is a prime. 17 is a prime.

4 is not a prime ($4 = 2 \times 2$). 6 is not a prime ($6 = 2 \times 3$). ❑

EXAMPLE 3   6 is a factor of 12 but is *not* a prime factor.

2 *is* a prime factor of 12. ❑

The number 1 is a special number that by definition is not considered to be a prime.

Factoring a number completely means writing it as the product of its prime factors.

EXAMPLE 4   To factor 15 completely, express it as $15 = (3)(5)$. ❑

The easiest way to begin factoring a large number is to find *one* factor. Divide that factor into the number to find another factor. If those two factors are *not primes,* continue to factor them until all the factors are prime.

Since factoring large numbers often requires repeating the process of factoring several times, I find it helpful to use a factoring tree, such as the one shown in Example 5, to keep track of the work.

EXAMPLE 5   Factor 24 completely.

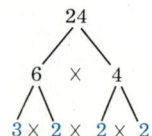

❑

You can *begin* factoring a number many different ways, but you will always end up with the same list of prime factors. For example, to factor 36, one person might begin with $36 = 9 \times 4$, while someone else might begin with $36 = 6 \times 6$ or

3 × 12 or 2 × 18. Below are several different factoring trees for 36. Notice that they all end up with the *same correct list of prime factors*.

EXAMPLE 6   Factor 36 completely.

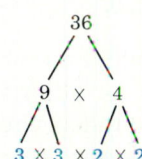 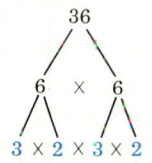 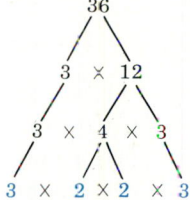 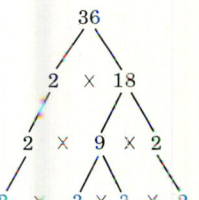

The *order* of the list of prime factors is not important. The prime factors of 36 are two 2s and two 3s. I usually list the prime factors of a number from smallest to largest.

Always check whether your list of prime factors is correct by multiplying them together. Their product should be the original number you were factoring.

EXAMPLE 7   Factor 50 completely.

So 50 = (2)(5)(5).

Check: (2)(5)(5) =

(10)(5) = 50

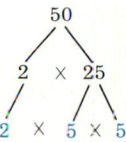

Hints for Factoring

You can quickly determine whether 2, 5, or 10 is a factor of a number just by looking at the number.

2 is a factor of every *even number*. Remember, an even number is an integer that has a 0, 2, 4, 6, or 8 as its units digit.

EXAMPLE 8   2 *is* a factor of 37,186. 2 is *not* a factor of 63,247.

5 is a factor of every number that has a 0 or a 5 as its units digit. Notice that when you list the numbers from the fives table (0, 5, 10, 15, 20, 25, 30, . . .), they all end in 5 or 0.

EXAMPLE 9   5 *is* a factor of 790. 5 *is* a factor of 1,385.
5 is *not* a factor of 861; 502; 753; 52,384; 56; 40,217; 38; or 4,389.

10 is a factor of every number that has a 0 as its units digit. Notice that when you list the numbers from the tens table (0, 10, 20, 30, 40, . . .), they all end in 0.

EXAMPLE 10   10 *is* a factor of 34, 720.
10 is *not* a factor of 1,061; 412; 73; 984; 25; 40,386; 47; 318; or 179.

There is an easy way to test whether 3 is a factor of a number without trying to divide it by 3.

3 is a factor of a number if the sum of its digits is a number that 3 divides into exactly. So to test for a factor of 3 in a number, add its digits and check whether 3 divides into that small number.

EXAMPLE 11   Is 3 a factor of 151,611?

The sum of the digits is $1 + 5 + 1 + 6 + 1 + 1 = 15$.

3 is a factor of 15.

So 3 must be a factor of 151,611.

EXAMPLE 12 Is 3 a factor of 23,801,210?

The sum of the digits is $2 + 3 + 8 + 0 + 1 + 2 + 1 + 0 = 17$.

3 is not a factor of 17.

So 3 cannot be a factor of 23,801,210. ❏

Recognizing one of the easy-to-find factors gets you started. If 2, 3, 5, or 10 is not a factor of the number, test for other factors by dividing them into the original number. If there is no remainder, you have found a factor; if there is a remainder, you have not found a factor (keep looking).

EXAMPLE 13 Factor 210.

10 is a factor since 210 ends in 0.

$$210 = 10 \times 21$$

The sum of the digits in 21 $(2 + 1) = 3$, so 3 is a factor of 21.

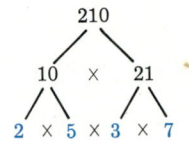

$$210 = (2)(5)(3)(7)$$

Check: $(2)(5)(3)(7) =$
$(10)(21) = 210$ ✓ ❏

If one number is a factor of a second number, we can multiply the first number by some integer to produce the second. We say the second number is a **multiple** of the first number.

In other words, we can say that 5 is a *factor* of 10 or that 10 is a *multiple* of 5. Both statements express the same relationship. It's like saying "She is my mother" or "I am her daughter." Sometimes we find it more convenient to talk about factors, sometimes about multiples.

EXAMPLE 14 Since 5 is a factor of 210, we can say that 210 is a multiple of 5. ❏

EXERCISES 3.3

Factor each number into a product of its prime factors. Check by multiplying them together.

1. 35	**2.** 14	**3.** 70
4. 66	**5.** 28	**6.** 40
7. 100	**8.** 300	**9.** 81
10. 48	**11.** 42	**12.** 180
13. 540	**14.** 1,000	**15.** 750
16. 480	**17.** 315	**18.** 105
19. 243	**20.** 680	

Check your answers on page 721. ✓

3.4 Simplifying Fractions

There are many different ways to express the same fractional part, but there is one name that is considered the **simplest** or **most reduced** form for each fraction. In Exercises 3.1 we saw that fractions with the names $\frac{1}{2}, \frac{4}{8}, \frac{2}{4}$, and $\frac{3}{6}$ all represented the same part of the circle; $\frac{1}{2}$ is the simplest name for that fraction.

The *simplest* form of a fraction is the one involving the smallest numbers in the numerator and the denominator.

When an answer to a computation is a fraction, you are always expected to *express it in its simplest form.*

You might have once been told to reduce fractions by dividing the numerator and the denominator by some number. That method works well to simplify fractions if you know a number that divides into both the numerator and the denominator. I have a method for simplifying fractions that will always work, even if you are not sure you know a number that divides into the numerator and the denominator.

> To simplify a fraction:
>
> Factor the numerator and the denominator into a product of prime factors. Cancel any factor that appears in both the numerator and the denominator.

EXAMPLE 1 Simplify $\frac{6}{10}$.

$$\frac{6}{10} =$$
$$\frac{2 \cdot 3}{2 \cdot 5} =$$
$$\frac{\cancel{2} \cdot 3}{\cancel{2} \cdot 5} = \frac{3}{5}$$

So $\frac{3}{5}$ is the simplied form of $\frac{6}{10}$.

When we canceled the 2s in Example 1, we were really dividing the numerator and the denominator by 2. Finding the factors of two numbers is easier than finding a factor common to both.

EXAMPLE 2 Simplify $\frac{30}{45}$.

$$\frac{30}{45} =$$
$$\frac{(2 \cdot 3 \cdot 5)}{(3 \cdot 3 \cdot 5)} =$$
$$\frac{(2 \cdot {}^1\cancel{3} \cdot {}^1\cancel{5})}{(3 \cdot {}^1\cancel{3} \cdot {}^1\cancel{5})} = \frac{2}{3}$$

So $\frac{2}{3}$ is the simplied form of $\frac{30}{45}$.

If *all* the factors in either the numerator or the denominator cancel out, that numerator or denominator becomes the number **1**. This is so because 1 is always another factor, though we don't usually write it out: $3 = 3 \times 1$.

EXAMPLE 3   Simplify $\dfrac{18}{90}$.

$$\frac{18}{90} =$$

$$\frac{(2 \cdot 3 \cdot 3)}{(2 \cdot 3 \cdot 3 \cdot 5)} =$$

$$\frac{(^1\cancel{2} \cdot {}^1\cancel{3} \cdot {}^1\cancel{3})}{(^1\cancel{2} \cdot {}^1\cancel{3} \cdot {}^1\cancel{3} \cdot 5)} = \frac{1}{5} \qquad \square$$

So $\dfrac{1}{5}$ is the simplied form of $\dfrac{18}{90}$

If the denominator becomes 1, that fraction simplifies to an integer and is expressed without the denominator.

EXAMPLE 4   Simplify $\dfrac{42}{14}$.

$$\frac{42}{14} =$$

$$\frac{(2 \cdot 3 \cdot 7)}{(2 \cdot 7)} =$$

$$\frac{(\cancel{2} \cdot 3 \cdot \cancel{7})}{(\cancel{2} \cdot \cancel{7})} =$$

$$\frac{3}{1} = 3 \qquad \square$$

So 3 is the simplified form of $\dfrac{42}{14}$

The only way to be sure that your fraction is as simplified as possible is to see that no common factor is left in the numerator and the denominator.

You can check for careless errors. If two fractions are really equal, when you multiply the numerator of one with the denominator of the other you should get the same product as when you multiply the other numerator and denominator together. Multiplying numerators and denominators across the equal signs is called **cross multiplication**.

EXAMPLE 5   Use cross multiplication to check Example 3.

Does $\dfrac{18}{90} = \dfrac{1}{5}$?

If $\dfrac{18}{90} \bowtie \dfrac{1}{5}$, then $(18)(5)$ should equal $(1)(90)$.

$(18)(5) = 90$ ✓ and $(1)(90) = 90$. ✓ $\qquad \square$

So we were right: $\dfrac{18}{90} = \dfrac{1}{5}$.

EXERCISES 3.4 _____

Simplify each fraction. Check your answers by cross multiplication.

1. $\dfrac{6}{15}$ 2. $\dfrac{10}{12}$ 3. $\dfrac{14}{20}$ 4. $\dfrac{12}{30}$

5. $\dfrac{25}{45}$ 6. $\dfrac{35}{49}$ 7. $\dfrac{20}{60}$ 8. $\dfrac{38}{19}$

9. $\dfrac{40}{24}$ 10. $\dfrac{28}{70}$ 11. $\dfrac{26}{78}$ 12. $\dfrac{84}{140}$

13. $\dfrac{26}{130}$ 14. $\dfrac{144}{200}$ 15. $\dfrac{252}{270}$ 16. $\dfrac{80}{240}$

17. $\dfrac{180}{588}$ 18. $\dfrac{175}{252}$

Check your answers on page 721.

3.5 Multiplication and Division of Fractions

Multiplication

As we saw in Section 3.2, in order to multiply fractions you simply multiply the numerators to find the numerator of the product, and multiply the denominators to find the denominator of the product.

Since it is easier to factor small numbers than it is to factor large numbers, and since the answer to a fractional computation must be in its most reduced form, I suggest that you factor and cancel before multiplying the fractions together.

You may cancel any factor from the numerator with that same factor from the denominator; they can come from the same or different fractions. The problem can be written as _one fraction_: The numerator is expressed as the product of the prime factors of each numerator, and the denominator is expressed as the product of the prime factors of each denominator.

EXAMPLE 1 Multiply $\dfrac{6}{15}$ by $\dfrac{14}{20}$.

$$\frac{6}{15} \times \frac{14}{20} =$$

$$\frac{2 \cdot 3}{3 \cdot 5} \times \frac{2 \cdot 7}{2 \cdot 2 \cdot 5} =$$

$$\frac{2 \cdot 3 \cdot 2 \cdot 7}{3 \cdot 5 \cdot 2 \cdot 2 \cdot 5} =$$

$$\frac{\cancel{2} \cdot \cancel{3} \cdot \cancel{2} \cdot 7}{\cancel{3} \cdot 5 \cdot \cancel{2} \cdot \cancel{2} \cdot 5} =$$

$$\frac{7}{5 \cdot 5} = \frac{7}{25}$$

So $\dfrac{6}{15} \times \dfrac{14}{20} = \dfrac{7}{25}$.

Note these three advantages of factoring and canceling before multiplying:

1. The numbers you factor are smaller before multiplying.
2. The numbers you end up needing to multiply are smaller.
3. The answer will always be in the most simplified form.

To multiply fractions:

1. Factor all numerators and denominators into a product of *prime* factors.
2. Cancel every pair of factors appearing in *both* the numerator and the denominator.
3. Multiply the remaining factors together in the numerator to find the numerator of the answer.
4. Multiply the remaining factors together in the denominator to find the denominator of the answer.

If any of the fractions to be multiplied are negative, be sure to use the rules of signed numbers to determine what the sign of the product will be. Remember, the product of every two negative numbers is positive. Determine the sign first, and if it is negative, place the negative sign before the fraction in each step.

EXAMPLE 2   Determine the product of $\left(-\dfrac{12}{18}\right) \times \left(\dfrac{27}{40}\right)$.

$$-\frac{2\cdot 2\cdot 3}{2\cdot 3\cdot 3} \times \frac{3\cdot 3\cdot 3}{2\cdot 2\cdot 2\cdot 5} =$$

$$-\frac{\cancel{2}\cdot\cancel{2}\cdot\cancel{3}\cdot\cancel{3}\cdot 3\cdot 3}{\cancel{2}\cdot\cancel{3}\cdot\cancel{3}\cdot\cancel{2}\cdot 2\cdot 2\cdot 5} =$$

$$-\frac{(3\cdot 3)}{(2\cdot 2\cdot 5)} = -\frac{9}{20}$$

So $-\dfrac{12}{18} \times \dfrac{27}{40} = -\dfrac{9}{20}$.

We can multiply many fractions together easily by using this method. If the sign of the product is positive, no sign is necessary.

EXAMPLE 3

$$-\frac{2}{9} \times \frac{6}{10} \times \left(-\frac{5}{7}\right) =$$

$$+\frac{2}{3\cdot 3} \times \frac{2\cdot 3}{2\cdot 5} \times \frac{5}{7} =$$

$$\frac{\cancel{2}\cdot 2\cdot\cancel{3}\cdot\cancel{5}}{\cancel{3}\cdot 3\cdot\cancel{2}\cdot\cancel{5}\cdot 7} =$$

$$\frac{2}{3\cdot 7} = \frac{2}{21}$$

So $-\dfrac{2}{9} \times \dfrac{6}{10} \times \left(-\dfrac{5}{7}\right) = \dfrac{2}{21}$.

I wrote the "+" in the first step to remind you that I had considered the sign and found it positive. After that, I did not bother to write the positive sign.

EXERCISES 3.5A

Multiply the fractions. Make sure that your answers are in the simplest form.

1. $\dfrac{10}{21} \times \dfrac{28}{18} =$

2. $\dfrac{25}{40} \times \dfrac{24}{35} =$

3. $\left(-\dfrac{22}{30}\right) \times \dfrac{54}{66} =$

4. $\left(-\dfrac{12}{30}\right) \times \dfrac{45}{42} =$

5. $\left(-\dfrac{20}{21}\right) \times \left(-\dfrac{49}{50}\right) =$

6. $\left(-\dfrac{27}{44}\right) \times \left(-\dfrac{55}{60}\right) =$

7. $\dfrac{14}{26} \times \left(-\dfrac{65}{70}\right) =$

8. $\dfrac{24}{80} \times \left(-\dfrac{56}{100}\right) =$

9. $\left(-\dfrac{1}{5}\right) \times \dfrac{2}{3} \times \dfrac{15}{18} =$

10. $\dfrac{4}{7} \times \left(-\dfrac{6}{8}\right) \times \left(-\dfrac{14}{20}\right) =$

11. $\left(-\dfrac{12}{15}\right) \times \left(-\dfrac{20}{9}\right) \times \left(-\dfrac{2}{6}\right) =$

12. $\left(-\dfrac{80}{14}\right) \times \left(-\dfrac{21}{30}\right) \times \dfrac{5}{12} =$

13. $\dfrac{24}{30} \times \dfrac{27}{40} \times \left(-\dfrac{15}{36}\right) =$

14. $\left(-\dfrac{10}{24}\right) \times \left(-\dfrac{6}{25}\right) \times \left(-\dfrac{4}{7}\right) =$

Check your answers on page 721. ✓

Division and Reciprocals

Before discussing how to divide fractions, we must define the term **reciprocal**. The **reciprocal** of a number is the number it must be multiplied by to produce the number 1. The easiest way to find the reciprocal of a fraction is to turn it upside down.

You can check whether you have found a number's reciprocal by multiplying the original number and the reciprocal together. Their product should be 1.

EXAMPLE 4   The reciprocal of $\dfrac{2}{3}$ should be $\dfrac{3}{2}$.

Check: Does $\dfrac{2}{3} \times \dfrac{3}{2} = 1$?

$$\dfrac{2}{3} \times \dfrac{3}{2} =$$

$$\dfrac{\overset{1}{\cancel{2}} \cdot \overset{1}{\cancel{3}}}{\underset{1}{\cancel{3}} \cdot \underset{1}{\cancel{2}}} = 1 \quad \checkmark$$

 ❑

If you keep in mind that whole numbers can always be expressed as fractions by writing them over 1, you can easily find reciprocals of integers too.

EXAMPLE 5   Find the reciprocal of 2.

2 can be written as $\dfrac{2}{1}$.

The reciprocal of $\dfrac{2}{1}$ should be $\dfrac{1}{2}$.

Check: Does $\frac{2}{1} \times \frac{1}{2} = 1$?

$$\frac{\cancel{2} \cdot 1}{1 \cdot \cancel{2}} = 1 \quad \checkmark$$

To *divide* by a fraction means to *multiply* by its *reciprocal*.

EXAMPLE 6 Divide $\frac{2}{3}$ by $\frac{5}{7}$.

$$\frac{2}{3} \div \frac{5}{7} =$$

$$\frac{2}{3} \times \frac{7}{5} =$$

$$\frac{2 \cdot 7}{3 \cdot 5} = \frac{14}{15}$$

In order to understand the division of fractions, let's review what the division of whole numbers means. The expression $6 \div 2 = 3$ means either to

separate 6 items into groups of 2 items in each (which makes 3 groups)

or to

separate 6 items into 2 groups (with 3 items in each group).

The divisor represents either the quantity in each group or the number of groups. When an integer is divided by a fraction, it is easier to visualize what is happening if we let the fraction represent the quantity in each group.

So $8 \div \frac{1}{2}$ represents separating 8 items into groups with $\frac{1}{2}$ item in each. The quotient represents the number of such groups.

EXAMPLE 7 $8 \div \frac{1}{2}$ might represent 8 dollars' worth of change divided into groups of a half-dollar in each.

The quotient would be the resulting number of groups.

$$8 \div \frac{1}{2} =$$

$$\frac{8}{1} \times \frac{2}{1} = 16$$

So, 8 dollars divides into 16 half-dollars.

EXAMPLE 8 $\frac{3}{4} \div \frac{1}{4}$ could represent our having $\frac{3}{4}$ of a cake left over and wanting to cut it into pieces that were each $\frac{1}{4}$ of a whole cake.

The quotient would represent how many pieces we would get.

$$\frac{3}{4} \div \frac{1}{4} =$$

$$\frac{3}{4} \times \frac{4}{1} =$$

$$\frac{3 \cdot \cancel{4}}{\cancel{4} \cdot 1} = 3$$

So we would get 3 pieces.

⇒ *Note*:   In Example 8 I noticed that there was a factor of 4 in both the numerator and the denominator after I had converted the problem into a multiplication problem. So I canceled the 4s without factoring completely to primes.

Since the point is to simplify the fraction by canceling all pairs of factors appearing in both the numerator and the denominator, you do not have to factor to primes *if you notice nonprime factors that will cancel.*

Remember, when *dividing* fractions just replace the divisor by its *reciprocal* and *multiply*. You can only *factor* and *cancel* in a *multiplication* problem. (No factoring and canceling while it looks like a division problem.)

EXAMPLE 9   Simplify $\dfrac{12}{20} \div \dfrac{15}{30}$.

$$\frac{12}{20} \div \frac{15}{30} =$$

$$\frac{12}{20} \times \frac{30}{15} =$$

$$\frac{2 \cdot 2 \cdot 3}{2 \cdot 2 \cdot 5} \times \frac{2 \cdot 3 \cdot 5}{3 \cdot 5} =$$

$$\frac{\cancel{2} \cdot \cancel{2} \cdot \cancel{3} \cdot 2 \cdot 3 \cdot \cancel{5}}{\cancel{2} \cdot \cancel{2} \cdot \cancel{5} \cdot \cancel{3} \cdot 5} =$$

$$\frac{2 \cdot 3}{5} = \frac{6}{5}$$

When working with whole numbers, we used multiplication to check division problems. We can do the same thing with fractions. If the quotient is correct, when multiplied by the divisor it will produce the dividend.

EXAMPLE 10   Let's use multiplication to check Example 9.

Does $\dfrac{12}{20} \div \dfrac{15}{30} = \dfrac{6}{5}$?

Check: The answer is correct if $\dfrac{6}{5} \times \dfrac{15}{30} = \dfrac{12}{20}$.

$$\frac{6}{5} \times \frac{15}{30} =$$

$$\frac{2 \cdot 3}{5} \times \frac{3 \cdot 5}{2 \cdot 3 \cdot 5} =$$

$$\frac{\cancel{2} \cdot \cancel{3} \cdot 3 \cdot \cancel{5}}{5 \cdot \cancel{2} \cdot \cancel{3} \cdot \cancel{5}} = \frac{3}{5}$$

We found $\dfrac{6}{5} \times \dfrac{15}{30}$ to be $\dfrac{3}{5}$. Did we make a mistake?

Not necessarily. $\dfrac{12}{20}$ is not in simplified form. Let's reduce it.

$$\frac{12}{20} =$$

$$\frac{\cancel{2} \cdot \cancel{2} \cdot 3}{\cancel{2} \cdot \cancel{2} \cdot 5} = \frac{3}{5} \quad \checkmark$$

So we were right !

Don't forget to determine the sign of your answer. If it is negative, write the sign in front of the fraction.

EXAMPLE 11 What is the quotient of $\left(-\dfrac{3}{10}\right) \div \dfrac{12}{25}$?

$$-\frac{3}{10} \div \frac{12}{25} =$$

$$-\frac{3}{10} \times \frac{25}{12} =$$

$$-\frac{3 \cdot 5 \cdot 5}{2 \cdot 5 \cdot 2 \cdot 2 \cdot 3} =$$

$$-\frac{\cancel{3} \cdot \cancel{5} \cdot 5}{2 \cdot \cancel{5} \cdot 2 \cdot 2 \cdot \cancel{3}} = -\frac{5}{8}$$

EXERCISES 3.5B

Change each division problem into a multiplication problem by replacing the divisor by its reciprocal. Then factor and cancel, leaving the answer in its most simplified form.

15. $\dfrac{2}{5} \div \dfrac{1}{4} =$

16. $\dfrac{1}{7} \div \dfrac{3}{4} =$

17. $\left(-\dfrac{2}{3}\right) \div \dfrac{1}{5} =$

18. $\left(-\dfrac{3}{5}\right) \div \dfrac{2}{7} =$

19. $\left(-\dfrac{4}{5}\right) \div \left(-\dfrac{6}{10}\right) =$

20. $\left(-\dfrac{12}{15}\right) \div \left(-\dfrac{20}{30}\right) =$

21. $\dfrac{9}{12} \div \left(-\dfrac{8}{15}\right) =$

22. $\left(-\dfrac{8}{14}\right) \div \left(-\dfrac{20}{49}\right) =$

23. $\left(-\dfrac{6}{24}\right) \div \left(-\dfrac{9}{10}\right) =$

24. $\dfrac{18}{25} \div \dfrac{15}{35} =$

25. $\left(-\dfrac{22}{32}\right) \div \left(-\dfrac{33}{20}\right) =$

26. $\left(-\dfrac{36}{50}\right) \div \dfrac{63}{75} =$

27. $\dfrac{40}{56} \div \dfrac{24}{50} =$

28. $\left(-\dfrac{16}{27}\right) \div \left(-\dfrac{42}{60}\right) =$

Check your answers on page 721.

Representing Division with Fractions over Fractions

The fraction $\dfrac{\frac{2}{3}}{\frac{4}{5}}$ may look strange to you. In Section 3.1 we pointed out that the *fractional notation* is often used to represent *division*. So the fraction just expressed represents the division problem $\dfrac{2}{3} \div \dfrac{4}{5}$. Remember, the *divisor* is always written as the *denominator*.

EXAMPLE 12   Simplify $\dfrac{\frac{2}{3}}{\frac{4}{5}}$.

$$\frac{\frac{2}{3}}{\frac{4}{5}} =$$

$$\frac{2}{3} \div \frac{4}{5} =$$

$$\frac{2}{3} \times \frac{5}{4} =$$

$$\frac{2 \times 5}{3 \times 2 \times 2} =$$

$$\frac{\cancel{2} \times 5}{3 \times \cancel{2} \times 2} = \frac{5}{6}$$

☐

EXERCISES 3.5C

Simplify each problem by expressing it as a division problem. Replace each divisor by its reciprocal so that the problem becomes a multiplication problem. Then factor and cancel.

29. $\dfrac{\frac{1}{2}}{\frac{3}{5}} =$

30. $\dfrac{\frac{4}{5}}{\frac{12}{10}} =$

31. $\dfrac{-\frac{4}{9}}{\frac{1}{12}} =$

32. $\dfrac{\frac{21}{30}}{-\frac{35}{40}} =$

33. $\dfrac{-\frac{10}{12}}{\frac{6}{14}} =$

34. $\dfrac{-\frac{33}{49}}{-\frac{77}{70}} =$

Check your answers on page 721. ✓

3.6 New Names

In order for us to add or subtract fractions, the fractions must have the same denominator. If their denominators are the same, they represent parts that are the same size, and so to add we just add the numerators. If two fractions with different denominators are to be added or subtracted, they must be rewritten with new names in which the denominators are the same. For example, $\dfrac{1}{2} + \dfrac{1}{4}$ cannot be added in that form; but $\dfrac{2}{4} + \dfrac{1}{4}$ can be added $\left(\dfrac{2}{4}\text{ is a new name for }\dfrac{1}{2}\right)$.

There is an easy way to find new names for fractions. We combine two important properties of numbers:

The number 1, multiplied by any number, produces that same number.

EXAMPLE 1 $1 \times 7 = 7$; $1 \times 15 = 15$; $1 \times \dfrac{2}{3} = \dfrac{2}{3}$ ❑

Every fraction with the same numerator and denominator is equal to 1.

EXAMPLE 2 $1 = \dfrac{2}{2}$ or $\dfrac{3}{3}$ or $\dfrac{7}{7}$ or $\dfrac{13}{13}$ or $\dfrac{28}{28}$ and so on.   ❑

> To find a new name for a fraction without changing its value:
>
> Multiply the fraction by a fraction equal to 1 (*with the same numerator and denominator*).

EXAMPLE 3   We can find a new name for $\dfrac{2}{3}$ by multiplying it by $\dfrac{5}{5}$.

$$\frac{2}{3} =$$

$$\frac{2}{3} \times 1 =$$

$$\frac{2}{3} \times \frac{5}{5} = \frac{10}{15}$$ ❑

So $\dfrac{10}{15}$ is another fractional way to express $\dfrac{2}{3}$. If we cut a cake into 3 large pieces and eat 2 of them, it is the same as having cut the cake into 15 small pieces and eating 10 of them.

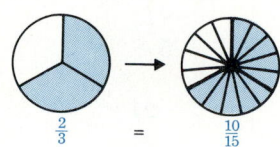

$$\frac{2}{3} \qquad = \qquad \frac{10}{15}$$

EXAMPLE 4   Find another name for $\dfrac{2}{3}$.

This time let's multiply by $\dfrac{7}{7}$.

$$\frac{2}{3} =$$

$$\frac{2}{3} \times \frac{7}{7} = \frac{14}{21}$$ ❑

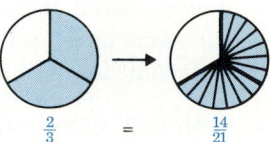

$$\frac{2}{3} \qquad = \qquad \frac{14}{21}$$

$\dfrac{14}{21}$ is another name for $\dfrac{2}{3}$. If we had divided the cake into 21 tiny pieces and eaten 14 of them, we would have done the same damage as eating 2 of the 3 large portions.

Finding new names for fractions is really the reverse (or the inverse) of simplifying fractions as we did in Section 3.4.

EXAMPLE 5   If we were to simplify $\frac{14}{21}$, we would obtain $\frac{2}{3}$.

$$\frac{14}{21} =$$

$$\frac{2 \cdot 7}{3 \cdot 7} =$$

$$\frac{2 \cdot \cancel{7}}{3 \cdot \cancel{7}} = \frac{2}{3}$$

Usually you find a new name because you want to express the fraction with a particular denominator. Since you can use any number to make 1 (as long as it is used in both the numerator and the denominator), use the number that will multiply your old denominator to give you the new denominator you want. To find that number, divide the denominator you want by the old denominator.

EXAMPLE 6   Change $\frac{2}{3}$ to a new name so that its denominator will be 24.

What number times 3 will produce 24?

24 divided by 3 is 8, so 8 is the number you want.

So use 8 to make $1 : 1 = \frac{8}{8}$.

$$\frac{2}{3} =$$

$$\frac{2}{3} \times \frac{8}{8} = \frac{16}{24}$$

EXAMPLE 7   Find a new name for $\frac{3}{5}$ that has a denominator of 20.

$20 \div 5 = 4$, so express 1 as $\frac{4}{4}$.

$$\frac{3}{5} =$$

$$\frac{3}{5} \times \frac{4}{4} = \frac{12}{20}$$

EXERCISES 3.6

Find new names for each fraction with the requested denominators.

1. $\frac{1}{2}$, with denominator 12

2. $\frac{1}{3}$, with denominator 15

3. $\frac{3}{4}$, with denominator 24

4. $\frac{2}{5}$, with denominator 30

5. $\frac{3}{7}$, with denominator 35

6. $\frac{5}{8}$, with denominator 48

7. $\frac{2}{9}$, with denominator 72

8. $\frac{5}{6}$, with denominator 42

9. $\frac{3}{5}$, with denominator 45

10. $\frac{4}{7}$, with denominator 63

Check your answers on page 721.

You are now capable of changing the name of any fraction.

3.7 Addition of Fractions with Different Denominators

Finding New Common Denominators

If the fractions to be added have different denominators, you must figure out what name to use for each fraction so that they all have the same denominator. This skill is called finding a **new common denominator**. (Swen Nater, a basketball player on the L.A. Lakers, says he will name his next daughter Carmen Denomin. Do you think that will make it easier for her to learn to add fractions?)

Since we find new names for fractions by *multiplying* the same number by the fraction's numerator and denominator, the common denominator must be a multiple of the original denominator.

So finding a common denominator for two fractions is really a question of finding a common multiple of the two original denominators—that is, a number that is a multiple of *both* denominators. We will look at three kinds of problems that you will run into when finding common multiples.

When One Denominator Is a Multiple of the Other

Case 1. The denominator of one fraction may already be a multiple of the denominator of the other fraction. In this case the *larger* denominator is the common multiple. You don't have to look any farther, and you can use this number for your common denominator. Only the fraction with the smaller denominator needs a name change.

EXAMPLE 1   Add $\frac{1}{5} + \frac{1}{10}$.

10 is a multiple of 5, so 10 is the common denominator.

We only have to change the name of $\frac{1}{5}$ to a fraction with the denominator 10.

$10 \div 5$ is 2, so use $\frac{2}{2}$ for 1.

$$\frac{1}{5} =$$

$$\frac{1}{5} \times \frac{2}{2} = \frac{2}{10}$$

Now we can add:

$$\frac{1}{5} + \frac{1}{10} =$$

$$\frac{2}{10} + \frac{1}{10} = \frac{3}{10}$$

If the computation with fractions involves signed numbers, the answer might be negative.

EXAMPLE 2 Subtract $\dfrac{1}{12} - \dfrac{2}{3}$.

12 is a multiple of 3, so 12 is the common denominator. Only $\dfrac{2}{3}$ needs to have its name changed.

$$\frac{2}{3} =$$

$$\frac{2}{3} \times \frac{4}{4} = \frac{8}{12}$$

Now we can combine them:

$$\frac{1}{12} - \frac{2}{3} =$$

$$\frac{1}{12} - \frac{8}{12} = -\frac{7}{12}$$

Answers to addition and subtraction problems must always be checked to see whether they can be simplified.

EXAMPLE 3 Add $\dfrac{1}{4} + \dfrac{3}{20}$.

20 is a multiple of 4, so 20 is the common denominator. $20 \div 4 = 5$, so use $\dfrac{5}{5}$ for 1.

$$\frac{1}{4} =$$

$$\frac{1}{4} \times \frac{5}{5} = \frac{5}{20}, \qquad \text{so} \quad \frac{1}{4} + \frac{3}{20} =$$

$$\frac{5}{20} + \frac{3}{20} = \frac{8}{20}, \quad \text{which reduces to:}$$

$$\frac{\cancel{2} \cdot \cancel{2} \cdot 2}{\cancel{2} \cdot \cancel{2} \cdot 5} = \frac{2}{5}$$

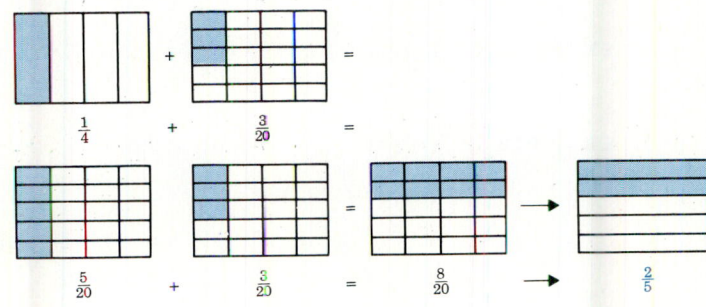

EXERCISES 37A

Find the common multiple of the denominators for the fractions. Change the name of any fraction that does not have the common multiple as its denominator. Then add the fractions as indicated and simplify the answers.

1. $\dfrac{4}{5} + \dfrac{3}{20} =$ **2.** $\dfrac{4}{7} + \dfrac{1}{21} =$ **3.** $\dfrac{3}{5} - \dfrac{2}{15} =$

4. $\dfrac{3}{10} - \dfrac{7}{30} =$ **5.** $-\dfrac{7}{10} - \dfrac{1}{2} =$ **6.** $\dfrac{14}{15} - \dfrac{1}{3} =$

7. $\dfrac{2}{3} + \dfrac{1}{9} =$ **8.** $\dfrac{4}{30} - \dfrac{4}{5} =$ **9.** $-\dfrac{3}{4} - \dfrac{5}{12} =$

10. $-\dfrac{6}{7} + \dfrac{1}{14} =$

Check your answers on page 721.

Lowest Common Multiple

In some problems, finding the common multiple is not quite so easy: Neither denominator is a multiple of the other. Their common multiple will have to be an entirely different number.

One can always find a common multiple of two numbers by multiplying them together. So a common multiple of 2 and 5 is 10: $(2)(5) = 10$. Ten is a multiple of 2 because $5 \times 2 = 10$, and 10 is a multiple of 5 because $2 \times 5 = 10$. Ten is not the *only* common multiple of 2 and 5, however.

10 is a common multiple of 2 and 5.

20 also is a common multiple of 2 and 5.

70 also is a common multiple of 2 and 5.

We can use any of these common multiples for our common denominator, but the computation is easiest if we use the *smallest* number—that is, the **lowest common multiple (LCM)**. When we use it as a new denominator for two fractions, it can be called their **lowest common denominator (LCD)** or the **lowest common multiple of their denominators**.

The two remaining kinds of problems deal with finding the LCM of their denominators.

Prime Denominators

Case 2. If the numbers are primes, their *product* will always be their lowest common multiple.

EXAMPLE 4   The LCM of the denominators in $\dfrac{2}{3}$ and $\dfrac{1}{5}$ is their product, 15. $3 \times 5 = 15$.

EXAMPLE 5   To add $\dfrac{2}{3} + \dfrac{1}{5}$, we must change each fraction's name so that its new denominator will be 15.

$$\dfrac{2}{3} \times \dfrac{5}{5} = \dfrac{10}{15} \qquad \dfrac{1}{5} \times \dfrac{3}{3} = \dfrac{3}{15}$$

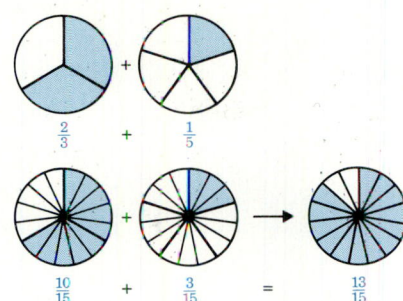

So $\dfrac{2}{3} + \dfrac{1}{5} =$

$\dfrac{10}{15} + \dfrac{3}{15} = \dfrac{13}{15}$, which cannot be reduced. ❑

EXERCISES 3.7B

Find the lowest common multiple of the denominators. Change each fraction to a new name with its LCM as the denominator. Add the fractions. Then reduce the answers.

11. $\dfrac{1}{3} + \dfrac{1}{2} =$

12. $\dfrac{2}{5} + \dfrac{1}{2} =$

13. $\dfrac{2}{7} - \dfrac{1}{2} =$

14. $\dfrac{1}{3} - \dfrac{1}{11} =$

15. $-\dfrac{5}{7} - \dfrac{1}{3} =$

16. $-\dfrac{3}{5} - \dfrac{1}{13} =$

17. $-\dfrac{4}{7} + \dfrac{1}{3} =$

18. $\dfrac{1}{7} + \dfrac{1}{5} =$

19. $\dfrac{7}{11} - \dfrac{1}{2} =$

20. $\dfrac{6}{7} - \dfrac{1}{5} =$

Check your answers on page 721. ✓

LCM of Nonprime Denominators

Case 3. If the denominators are not primes and one is not the multiple of the other, their product will not always be the *lowest* common multiple. Their product will be a common multiple, but not always the smallest one.

The best way to find the LCM of the denominators is to *factor* both denominators into primes. To be a common multiple, a number will have to be a product of these prime factors. But the *lowest* common multiple will not repeat any of these factors unnecessarily.

EXAMPLE 6 To add $\dfrac{1}{10}$ and $\dfrac{5}{6}$, you can multiply 10×6 to get a common multiple of their denominators: 60.

$$\dfrac{1}{10} + \dfrac{6}{6} = \dfrac{6}{60} \qquad \dfrac{5}{6} \times \dfrac{10}{10} = \dfrac{50}{60}$$

So
$$\frac{1}{10} + \frac{5}{6} =$$

$$\frac{6}{60} + \frac{50}{60} =$$

$$\frac{56}{60} =$$

$$\frac{2 \cdot 2 \cdot 2 \cdot 7}{2 \cdot 2 \cdot 3 \cdot 5} =$$

$$\frac{\cancel{2} \cdot \cancel{2} \cdot 2 \cdot 7}{\cancel{2} \cdot \cancel{2} \cdot 3 \cdot 5} = \frac{14}{15}$$

But you could have found a *smaller* common multiple of 10 and 6, 30.

30 is a multiple of 10 because $10 \times 3 = 30$.

30 is a multiple of 6 because $6 \times 5 = 30$.

$$\frac{1}{10} = \frac{3}{30} \quad \text{and} \quad \frac{5}{6} = \frac{25}{30}$$

So
$$\frac{1}{10} + \frac{5}{6} =$$

$$\frac{3}{30} + \frac{25}{30} =$$

$$\frac{28}{30} =$$

$$\frac{2 \cdot 2 \cdot 7}{2 \cdot 3 \cdot 5} =$$

$$\frac{\cancel{2} \cdot 2 \cdot 7}{\cancel{2} \cdot 3 \cdot 5} = \frac{14}{15} \qquad \square$$

The second calculation used smaller numbers and was easier to simplify than the first. That is why it is worth taking the time to find the *lowest* common multiple of the denominators.

To find the LCM of two numbers, list *all* the factors of *one* number. Include in that list any factor of the other number that is not already listed, and multiply them together.

EXAMPLE 7   Find the LCM of 10 and 6 by factoring.

$$10 = (2)(5) \quad \text{and} \quad 6 = (2)(3)$$

A multiple of 10 must contain its factors: 2 and 5.

A multiple of 6 must contain its factors: 2 and 3.

So a common multiple of 10 and 6 must contain factors 2, 5, and 3.

The LCM of 10 and 6 contains *only* those necessary factors.

The LCM of 10 and 6 is $(2)(3)(5) = 30$. $\qquad \square$

You should check to see that your LCM really is a multiple of both numbers. If they each divide exactly into the LCM, it is a multiple.

EXAMPLE 8   Check to see that 30 is a multiple of both 10 and 6.

$$30 \div 10 = 3 \quad \checkmark \quad \text{and} \quad 30 \div 6 = 5 \quad \checkmark \qquad \square$$

Many people find new denominators by trial and error. That's okay with small numbers. But the factoring method makes it easy to find the LCM of large numbers. And later, when you work with algebraic fractions (in Chapter 14), you must understand how to find the LCM of factors. You will have an easier time when you get to algebra if you master the technique now, when working with small numbers.

EXAMPLE 9   Find the LCM of 15 and 21.

$15 = (3)(5)$ and $21 = (3)(7)$

$$\text{Their LCM} = (3)(5)(7)$$
$$= 105$$

Check: $105 \div 15 = 7$ ✓   and $105 \div 21 = 5$ ✓

EXERCISES 3.7C

Find the LCM for each pair of numbers. Check to see that each of the numbers divides exactly into the LCM.

21. 14 and 6   **22.** 15 and 10

23. 6 and 21   **24.** 10 and 14

25. 35 and 15   **26.** 10 and 55

27. 21 and 10   **28.** 14 and 15

Check your answers on page 721. ✓

LCM of Numbers Involving Repeated Factors

Some numbers, when factored, have a prime factor that appears more than once.

EXAMPLE 10 $4 = (2)(2)$; $18 = (2)(3)(3)$; $20 = (2)(2)(5)$;
$8 = (2)(2)(2)$; $24 = (2)(2)(2)(3)$; $27 = (3)(3)(3)$

Each factor must appear in the LCM *as many times as* it appears in the number where it appears the most.

EXAMPLE 11   Find the LCM of 18 and 24.

$$18 = (2)(3)(3) \quad \text{and} \quad 24 = (2)(2)(2)(3)$$

Since 3 appears *twice* as a factor in 18, it must appear twice in the LCM.

Since 2 appears *three times* as a factor in 24, it must appear three times in the LCM.

$$\text{LCM} = (2)(2)(2)(3)(3) = 72$$

Check: $72 \div 18 = 4$ ✓   and $72 \div 24 = 3$ ✓

If a factor appears *fewer* times in the LCM than it does in one of the numbers, then that number will not divide into the LCM. So the LCM is *not* really a common multiple.

If a factor appears *more* times in the LCM than it does for any of the numbers, then the LCM is larger than necessary and is just a *common multiple* rather than the lowest common multiple.

EXERCISES 37D

Find the LCM for the numbers. Check to see that each number divides exactly into your answer.

29. 8 and 10

30. 6 and 9

31. 12 and 9

32. 8 and 12

33. 12 and 18

34. 20 and 24

35. 24 and 27

36. 18 and 27

37. 20 and 16

38. 45 and 36

39. 24 and 30

40. 40 and 24

Check your answers on page 721. ✓

Adding Fractions with Different Denominators

If the denominators of the fractions you are adding are different, find their LCM. Change each fraction's name so that its new denominator is that LCM. Then you can add the fractions, but be sure to leave the answer in its most reduced form.

EXAMPLE 12 Combine $\dfrac{1}{6} - \dfrac{1}{4}$.

$6 = 3 \cdot 2$ and $4 = 2 \cdot 2$, so LCM is $3 \cdot 2 \cdot 2 = 12$.

$$\frac{1}{6} = \qquad \text{and} \qquad \frac{1}{4} =$$

$$\frac{1}{6} \times \frac{2}{2} = \frac{2}{12} \qquad \qquad \frac{1}{4} \times \frac{3}{3} = \frac{3}{12}$$

So

$$\frac{1}{6} - \frac{1}{4} =$$

$$\frac{2}{12} - \frac{3}{12} = -\frac{1}{12}$$

EXAMPLE 13 Combine $\dfrac{7}{12} + \dfrac{3}{10}$.

$12 = 2 \cdot 2 \cdot 3$ and $10 = 2 \cdot 5$, so LCM $= 2 \cdot 2 \cdot 3 \cdot 5 = 60$.

$$\frac{7}{12} \times \frac{5}{5} = \frac{35}{60} \qquad \text{and} \qquad \frac{3}{10} \times \frac{6}{6} = \frac{18}{60}$$

So

$$\frac{7}{12} + \frac{3}{10} =$$

$$\frac{35}{60} + \frac{18}{60} = \frac{53}{60}$$

Whenever possible, it's a good idea to check your answer to a computation. You can always check an addition problem: If you subtract from your answer one of the numbers you have added, the result should be the other number.

EXAMPLE 14   To check that $\dfrac{7}{12} + \dfrac{3}{10} = \dfrac{53}{60}$, see whether $\dfrac{53}{60} - \dfrac{3}{10} = \dfrac{7}{12}$.

Since 60 is a multiple of 10, the LCM of 60 and 10 is 60.

$$\frac{3}{10} \times \frac{6}{6} = \frac{18}{60}$$

So

$$\frac{53}{60} - \frac{3}{10} =$$

$$\frac{53}{60} - \frac{18}{60} =$$

$$\frac{35}{60} =$$

$$\frac{\cancel{5} \cdot 7}{\cancel{5} \cdot 12} = \frac{7}{12} \quad ✓$$

Checking Whether the Answer Makes Sense

Before actually checking your answer, it's a good idea to think about whether your answer makes sense. You know that you can add only the numerators of fractions when the denominators are the same, but people often make the careless error of adding both the numerators and the denominators. If you think about the size of the piece that each fraction represents, you can usually catch that error.

EXAMPLE 15   If you made the mistake of adding both the numerators and the denominators in adding $\dfrac{2}{3} + \dfrac{3}{4}$, you would get the wrong answer of $\dfrac{5}{7}$. Here is what should happen when you think about that answer. Both $\dfrac{2}{3}$ and $\dfrac{3}{4}$ represent more than $\dfrac{1}{2}$; so when added, their sum should be more than 1. Yet $\dfrac{5}{7}$ is less than 1. Something must be wrong!

We all make careless errors. You will avoid most of them if you perform computations as we have done in the examples, one step at a time. And you will catch the few you do make if you ask whether the answers make sense and then check your work.

EXERCISES 37E

For each problem do the following: (1) Factor each denominator; (2) find its LCM; (3) change each fraction to a new name with the LCM as the denominator; (4) add the numerators when all the fractions have the same denominator; and (5) reduce the answers, if possible.

41. $\dfrac{1}{10} + \dfrac{5}{18} =$ **42.** $\dfrac{7}{12} + \dfrac{4}{30} =$ **43.** $\dfrac{3}{8} - \dfrac{5}{12} =$ **44.** $\dfrac{7}{9} - \dfrac{1}{12} =$

45. $-\dfrac{4}{15} - \dfrac{3}{10} =$ **46.** $\dfrac{6}{10} - \dfrac{3}{14} =$ **47.** $\dfrac{5}{9} - \dfrac{4}{15} =$ **48.** $-\dfrac{5}{14} - \dfrac{4}{21} =$

49. $\dfrac{5}{18} + \dfrac{10}{27} =$ **50.** $\dfrac{7}{20} + \dfrac{5}{24} =$

Check your answers on page 722.   ✓

Adding More Than Two Fractions

If you are adding more than two fractions, the new denominator will be the lowest common multiple of *all* the denominators.

If all the denominators are different prime numbers, the LCM will be their *product*.

EXAMPLE 16 Add $\dfrac{2}{3} + \dfrac{1}{5} - \dfrac{2}{7}$.

3, 5, and 7 are all primes, so the LCM is $3 \cdot 5 \cdot 7$, or 105.

$$\frac{2}{3} = \qquad\qquad \frac{1}{5} = \qquad\qquad \frac{2}{7} =$$

$$\frac{2}{3} \times \frac{35}{35} = \frac{70}{105} \qquad \frac{1}{5} \times \frac{21}{21} = \frac{21}{105} \qquad \frac{2}{7} \times \frac{15}{15} = \frac{30}{105}$$

So

$$\frac{2}{3} + \frac{1}{5} - \frac{2}{7} =$$

$$\frac{70}{105} + \frac{21}{105} - \frac{30}{105} =$$

$$\frac{70 + 21 - 30}{105} =$$

$$\frac{91 - 30}{105} = \frac{61}{105}$$

❑

If the denominators are not all primes, see whether one denominator is a multiple of *all* the others. Change the names of fractions until all have the same denominator and can be added.

EXAMPLE 17 Add $\dfrac{7}{12} + \dfrac{1}{3} - \dfrac{3}{4}$.

12 is a multiple of 3 and a multiple of 4, so 12 is the LCM.

$$\frac{1}{3} = \qquad\qquad \frac{3}{4} =$$

$$\frac{1}{3} \times \frac{4}{4} = \frac{4}{12} \qquad \frac{3}{4} \times \frac{3}{3} = \frac{9}{12}$$

So

$$\frac{7}{12} + \frac{1}{3} - \frac{3}{4} =$$

$$\frac{7}{12} + \frac{4}{12} - \frac{9}{12} =$$

$$\frac{7 + 4 - 9}{12} =$$

$$\frac{11 - 9}{12} =$$

$$\frac{2}{12} = \frac{1}{6}$$

❑

If the denominators are not all primes, and no denominator is a multiple of *all* the others, then you must find the LCM by factoring.

EXAMPLE 18 Add $\dfrac{3}{10} - \dfrac{2}{15} - \dfrac{1}{12}$.

$10 = 2 \cdot 5$, $15 = 3 \cdot 5$, and $12 = 2 \cdot 2 \cdot 3$, so the LCM is $2 \cdot 2 \cdot 3 \cdot 5 = 60$.

$$\frac{3}{10} = \qquad \frac{2}{15} = \qquad \frac{1}{12} =$$

$$\frac{3}{10} \times \frac{6}{6} = \frac{18}{60} \qquad \frac{2}{15} \times \frac{4}{4} = \frac{8}{60} \qquad \frac{1}{12} \times \frac{5}{5} = \frac{5}{60}$$

$$\text{So } \frac{3}{10} - \frac{2}{15} - \frac{1}{12} =$$

$$\frac{18}{60} - \frac{8}{60} - \frac{5}{60} =$$

$$\frac{18 - 8 - 5}{60} =$$

$$\frac{18 - 13}{60} =$$

$$\frac{5}{60} =$$

$$\frac{\cancel{5}}{\cancel{5} \cdot 12} = \frac{1}{12}$$

To add fractions:

1. Find the LCM of all the denominators and change the names of all fractions as necessary so that each has the LCM as its denominator.

2. Add fractions with the same denominator by adding numerators. If there are more than two fractions, combine numerators of the same sign first. Then combine the positive and negative numbers, subtracting the smaller magnitude and keeping the sign of number with the larger magnitude.

3. Reduce the answer by factoring the numerator and denominator and canceling any factor common to both.

EXERCISES 3.7F

Combine the fractions.

51. $\dfrac{4}{7} - \dfrac{1}{3} + \dfrac{1}{5} =$

52. $\dfrac{1}{2} - \dfrac{1}{3} - \dfrac{1}{5} =$

53. $-\dfrac{2}{3} - \dfrac{4}{6} - \dfrac{1}{2} =$

54. $\dfrac{3}{5} + \dfrac{1}{4} - \dfrac{1}{2} =$

55. $\dfrac{3}{8} - \dfrac{2}{5} - \dfrac{1}{10} =$

56. $\dfrac{1}{3} + \dfrac{1}{4} + \dfrac{1}{6} =$

57. $-\dfrac{3}{35} - \dfrac{2}{7} - \dfrac{1}{5} =$

58. $\dfrac{3}{20} - \dfrac{1}{2} - \dfrac{1}{5} =$

59. $\dfrac{3}{10} + \dfrac{1}{6} + \dfrac{4}{15} =$

60. $-\dfrac{2}{5} - \dfrac{1}{10} - \dfrac{1}{4} =$

61. $\dfrac{3}{10} + \dfrac{1}{6} + \dfrac{1}{4} =$

62. $\dfrac{1}{4} - \dfrac{4}{9} + \dfrac{7}{12} =$

63. $\dfrac{3}{4} - \dfrac{1}{8} - \dfrac{1}{10} =$

64. $\dfrac{5}{12} + \dfrac{3}{8} - \dfrac{1}{10} =$

Check your answers on page 722.

 ## Mixed Numbers

Mixed Numbers and Improper Fractions

If we wanted to split four small pizzas evenly among three people, we could cut each of the pizzas into thirds. Each person's share would be one-third of each of the four pizzas.

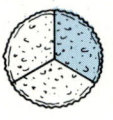 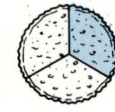 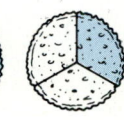 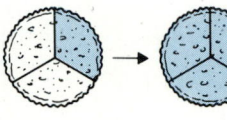 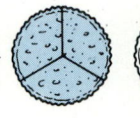

$$\text{So } \dfrac{1}{3} \text{ of } 4 =$$

$$\dfrac{1}{3} + \dfrac{1}{3} + \dfrac{1}{3} + \dfrac{1}{3} = \dfrac{4}{3}$$

Each person's share is *more than one* whole pizza, since there are more pizzas than there are people.

$$\left(\dfrac{1}{3} + \dfrac{1}{3} + \dfrac{1}{3} \right) + \dfrac{1}{3} =$$

$$\dfrac{3}{3} + \dfrac{1}{3} = 1\dfrac{1}{3}$$

As we mentioned in Section 3.1, a *mixed number* is the combination of an integer and a proper fraction. The value of a mixed number is always larger than 1.

EXAMPLE 1 $1\dfrac{1}{3}, 4\dfrac{2}{5}$, and $7\dfrac{4}{9}$ are mixed numbers.   ❏

A fraction like $\dfrac{4}{3}$, which is larger than 1, is called an *improper fraction*. The fact that $\dfrac{4}{3}$ is larger than 1 means that the numerator (the number of parts being discussed) is more than the denominator (the number of parts that one whole was divided into).

EXAMPLE 2 $\dfrac{4}{3}, \dfrac{22}{5}$, and $\dfrac{67}{9}$ are improper fractions.   ❏

Mixed numbers and improper fractions are two different ways to represent a number that is more than 1 but includes a fractional part. On a number line they lie between integers. Every mixed number can be expressed as an improper fraction.

Converting Mixed Numbers into Improper Fractions

A mixed number really represents the sum of the integer and the proper fractional parts. So mixed numbers can be converted into improper fractions by writing the integer as a fraction with the same denominator as the proper fraction and then adding them.

EXAMPLE 3   Express $1\frac{1}{3}$ as an improper fraction.

$$1\frac{1}{3} =$$

$$1 + \frac{1}{3} =$$

$$\frac{3}{3} + \frac{1}{3} = \frac{4}{3}$$

$1\frac{1}{3} = \frac{4}{3}$

There is a shortcut for this procedure:

Multiply the integer by the denominator of the fraction.

Add that product to the numerator.

The result is the *magnitude* of the numerator in the improper fraction.

The denominator will be the same as it was in the denominator of the proper fraction in the mixed number.

If the mixed number was negative, the improper fraction will be negative also.

EXAMPLE 4   Express $-7\frac{4}{9}$ as an improper fraction.

$$-7\frac{4}{9} =$$

$$-\frac{7(9) + 4}{9} =$$

$$-\frac{63 + 4}{9} = -\frac{67}{9}$$

EXERCISES 3.8A

Convert each mixed number into an improper fraction.

1. $1\frac{1}{5}$ **2.** $3\frac{2}{5}$ **3.** $5\frac{1}{2}$ **4.** $4\frac{3}{7}$

5. $2\frac{3}{4}$ **6.** $7\frac{2}{9}$ **7.** $6\frac{3}{8}$ **8.** $8\frac{1}{5}$

Check your answers on page 722.   ✓

Converting Improper Fractions into Mixed Numbers

Every fraction can be thought of as the division of its numerator by its denominator. To convert an improper fraction into a mixed number, perform that division. If the denominator divides exactly into the numerator, the improper fraction is an integer.

EXAMPLE 5   The improper fraction $\dfrac{12}{4}$ is the division problem $12 \div 4 = 3$.

So $\dfrac{12}{4} = 3$.   ❑

If the denominator does not divide exactly into the numerator, there will be a remainder. That remainder is still "waiting" to be divided by the denominator. We express this remaining division as a proper fraction, with the remainder as the numerator.

EXAMPLE 6   If we want to divide seven cookies equally among three people, we say that each person's share is seven-thirds $\left(\dfrac{7}{3}\right)$. To represent that amount as a mixed number, we say that $\dfrac{7}{3}$ means 7 divided by 3.

$$\begin{array}{r} 2 \\ 3\overline{)\ 7} \\ -6 \\ \hline 1 \end{array}$$

Each person gets two whole cookies, (that accounts for six of the cookies), and the remaining one is split into thirds. So each person gets two and one-third $\left(2\dfrac{1}{3}\right)$ cookies.

❑

EXAMPLE 7   Express $\dfrac{11}{4}$ as a mixed number.

$$\dfrac{11}{4} \rightarrow \begin{array}{r} 2 \\ 4\overline{)\ 11} \\ -8 \\ \hline 3 \end{array}$$

So $\dfrac{11}{4} = 2\dfrac{3}{4}$.   ❑

Answers are usually expressed as mixed numbers rather than improper fractions since it is easier to recognize where mixed numbers appear on the number line. The more comfortable you are with an answer, the more likely you will be to think about your answer and to find any careless errors.

EXERCISES 3.8B

Convert each improper fraction into a mixed number.

9. $\dfrac{7}{3} =$ **10.** $\dfrac{5}{2} =$ **11.** $\dfrac{13}{4} =$ **12.** $\dfrac{16}{3} =$

13. $\dfrac{15}{2} =$ **14.** $\dfrac{23}{5} =$ **15.** $\dfrac{35}{6} =$ **16.** $\dfrac{43}{7} =$

Check your answers on page 722. ✓

Multiplication and Division of Mixed Numbers

To *multiply* or *divide* mixed numbers, you must express them as *improper* fractions. Then multiply or divide as you would with any fraction. When you study algebra in Chapter 13, you will see why this procedure is necessary.

Remember, to divide fractions replace the divisor (the fraction following the division symbol) by its reciprocal and then multiply.

To multiply fractions with the least amount of calculation and simplifying, factor the numerator and the denominator and cancel any factors common to both. Then multiply all remaining factors in the numerator to obtain the numerator of your answer, and multiply all remaining factors in the denominator to obtain the denominator.

Note that you do not need a common denominator to multiply or divide fractions; you only need that to add or subtract them.

EXAMPLE 8   Multiply these mixed numbers. Reduce the answer if possible, and express it as a mixed number.

$$1\frac{1}{4} \times 3\frac{1}{5} =$$

$$\frac{5}{4} \times \frac{16}{5} =$$

$$\frac{\cancel{5} \cdot \cancel{4} \cdot 4}{\cancel{4} \cdot \cancel{5}} =$$

$$\frac{4}{1} = 4 \qquad ❏$$

Recall that you cannot factor and cancel until you represent the problem as a *multiplication* computation. Express the answers to computations with mixed numbers as mixed numbers, not as improper fractions.

If either or both of the numbers are negative, be careful to use the rules of signed numbers to determine the correct sign of the answer.

EXAMPLE 9   Divide these mixed numbers and express the answer as a reduced mixed number.

$$3\frac{1}{2} \div \left(-2\frac{1}{3}\right) =$$

$$\frac{7}{2} \div \left(-\frac{7}{3}\right) =$$

$$\frac{7}{2} \times \left(-\frac{3}{7}\right) =$$

$$-\frac{\cancel{7} \cdot 3}{2 \cdot \cancel{7}} =$$

$$-\frac{3}{2} = -1\frac{1}{2} \qquad ❏$$

_____

Perform the multiplication or division that is indicated and express the answers as reduced proper fractions, mixed numbers, or integers.

17. $1\dfrac{1}{5} \times 5\dfrac{1}{2} =$ **18.** $-1\dfrac{1}{2} \times \left(-4\dfrac{2}{3}\right) =$

19. $2\dfrac{2}{3} \times 3\dfrac{1}{4} =$ **20.** $\left(-2\dfrac{1}{3}\right) \times 3\dfrac{2}{5} =$

21. $1\dfrac{1}{7} \div \left(-2\dfrac{3}{4}\right) =$ **22.** $5\dfrac{3}{4} \div 2\dfrac{1}{2} =$

23. $\dfrac{3\frac{2}{3}}{5\frac{1}{2}} =$ **24.** $\dfrac{3\frac{1}{4}}{2\frac{1}{6}} =$

25. $\left(-4\dfrac{1}{2}\right) \times \left(-2\dfrac{1}{5}\right) =$ **26.** $3\dfrac{1}{2} \div 1\dfrac{5}{6} =$

27. $2\dfrac{3}{4} \div 1\dfrac{1}{6} =$ **28.** $\left(-4\dfrac{1}{3}\right) \times 2\dfrac{1}{2} =$

Check your answers on page 722.

Long Division of Whole Numbers with Fractional Remainders

In Section 1.4 we reviewed long division of whole numbers. Now that we have studied how to compute with fractions and mixed numbers, we can look at examples of long division in which the divisor does not divide exactly into the dividend. In such cases, we say that the dividend is *not* a multiple of the divisor.

When the dividend is not a multiple of the divisor, there will be a remainder. This remainder can be expressed either as a fraction or as a decimal. In this section we will represent the remainder as a fraction. (In Section 4.4 we will express it as a decimal.)

EXAMPLE 10   Find the quotient of $162,451 \div 8$.

$$
\begin{array}{r}
20,306 \\
8\overline{)162,451} \\
16 \\
\overline{02} \\
00 \\
\overline{2\,4} \\
2\,4 \\
\overline{05} \\
00 \\
\overline{51} \\
48 \\
\overline{3}
\end{array}
$$

When we divided 162,451 by 8, we found the answer to be 20,306 with 3 left over as the remainder.

The 3 that is left over is to be divided by 8, which we can express as the fraction $\frac{3}{8}$.

So we say the quotient is $20,306\frac{3}{8}$.

We can check by multiplying our quotient by the divisor. Their product should be the dividend.

Check: Does $(8)\left(20,306\frac{3}{8}\right) = 162,451$?

To multiply, we represent the mixed number $20,306\frac{3}{8}$ as an improper fraction.

$$20,306\frac{3}{8} = \frac{(20,306)(8) + 3}{8}$$

$$= \frac{162,448 + 3}{8}$$

$$= \frac{162,451}{8}$$

So

$$(8)\left(20,306\frac{3}{8}\right) =$$

$$\left(\frac{8}{1}\right)\left(\frac{162,451}{8}\right) =$$

$$\left(\frac{\cancel{8}}{1}\right)\left(\frac{162,451}{\cancel{8}}\right) = 162,451 \quad \checkmark$$

EXERCISES 3.8D

Perform each long division, Check your answers by representing mixed-number quotients as improper fractions and multiplying them by the divisors.

29. $346 \div 23 =$ **30.** $6667 \div 31 =$

31. $5078 \div 25 =$ **32.** $3022 \div 15 =$

33. $165,157 \div 16 =$ **34.** $298,273 \div 14 =$

35. $451,687 \div 15 =$ **36.** $261,355 \div 13 =$

Check your answers on page 722. $\checkmark$

Addition of Mixed Numbers

Mixed numbers can be added in the form either of mixed numbers or of improper fractions. In most cases the calculation is simplest when the numbers are expressed as mixed numbers.

That is certainly true when the sum involves integers and only *one mixed number*. Write the mixed number as the sum of its integer and its fractional part, and add only the integers.

EXAMPLE 11   Add $3 + 5\frac{1}{4}$.

$$3 + 5\frac{1}{4} =$$

$$3 + 5 + \frac{1}{4} = 8\frac{1}{4}$$ ❏

If the mixed number is negative, remember that *both* the integer and the fractional part are *negative*. If the resulting integer and fractional part have the *same sign,* they can be combined into a mixed number.

EXAMPLE 12   Add $-4\frac{1}{3} + 1$.

$$-4\frac{1}{3} + 1 =$$

$$-4 - \frac{1}{3} + 1 =$$

$$-3 - \frac{1}{3} = -3\frac{1}{3}$$ ❏

It is also easiest to keep the numbers expressed as mixed numbers when the fractional parts all have the *same denominator.* Combine the integers and combine the fractional parts. If the numbers have different signs, you might want to group together those with the same signs.

EXAMPLE 13   Add $7\frac{1}{4} + 4\frac{3}{4} - 5\frac{1}{4} - 3\frac{2}{4} + \frac{1}{4}$.

$$7\frac{1}{4} + 4\frac{3}{4} - 5\frac{1}{4} - 3\frac{2}{4} + \frac{1}{4} =$$

$$7 + \frac{1}{4} + 4 + \frac{3}{4} - 5 - \frac{1}{4} - 3 - \frac{2}{4} + \frac{1}{4} =$$

$$(7 + 4) + (-5 - 3) + \left(\frac{1}{4} + \frac{3}{4} + \frac{1}{4}\right) + \left(-\frac{1}{4} - \frac{2}{4}\right) =$$

$$+11 - 8 + \frac{5}{4} - \frac{3}{4} =$$

$$+3 + \frac{2}{4} =$$

$$+3\frac{2}{4} = 3\frac{1}{2}$$ ❏

When the fractional parts of the mixed numbers have different denominators, first group and add the integers together. Then find the LCM of the denominators, change the names of the fractions so that all have the same denominator, and combine them.

EXAMPLE 14   Add $1\frac{1}{2} + 2\frac{1}{3}$.

$$1\frac{1}{2} + 2\frac{1}{3} =$$

$$(1 + 2) + \left(\frac{1}{2} + \frac{1}{3}\right) =$$

$$3 + \frac{1}{2} + \frac{1}{3} =$$

The LCM of 2 and 3 is $(2)(3) = 6$.

$$\frac{1}{2} = \frac{3}{6} \quad \text{and} \quad \frac{1}{3} = \frac{2}{6}$$

$$3 + \left(\frac{3}{6} + \frac{2}{6} \right) = 3\frac{5}{6}$$

❏

You must be very careful when the fractional part ends up with a sign different from that of the integer. At that point you will be less likely to make a careless error if you convert the integer into an improper fraction to combine the two parts. Then convert your answer into a mixed number, which is the more familiar form.

EXAMPLE 15   Add $-7 + 4\frac{2}{5}$.

$$-7 + 4\frac{2}{5} =$$

$$-7 + 4 + \frac{2}{5} =$$

$$-3 + \frac{2}{5} = \qquad\qquad -3 = -3 \times \frac{5}{5} = -\frac{15}{5}$$

$$-\frac{15}{5} + \frac{2}{5} =$$

$$-\frac{13}{5} = -2\frac{3}{5}$$

❏

Your work will be simplified if you combine fractions with the same denominator. Then find the LCM and change *all* fractions to ones with the same denominator. To simplify further, I always combine numbers with the same sign first.

EXAMPLE 16   Add these numbers:

$$3\frac{9}{10} - 2\frac{4}{5} + 1 - 6\frac{2}{5} + \frac{3}{5} - \frac{7}{10} =$$

$$3 + \frac{9}{10} - 2 - \frac{4}{5} + 1 - 6 - \frac{2}{5} + \frac{3}{5} - \frac{7}{10} =$$

$$(3 - 2 + 1 - 6) + \left(\frac{9}{10} - \frac{7}{10} \right) + \left(-\frac{4}{5} - \frac{2}{5} + \frac{3}{5} \right) =$$

$$(4 - 8) + \left(\frac{9}{10} - \frac{7}{10} \right) + \left(-\frac{6}{5} + \frac{3}{5} \right) =$$

$$(-4) + \left(\frac{2}{10} \right) + \left(-\frac{3}{5} \right) =$$

$$(-4) + \frac{1}{5} - \frac{3}{5} =$$

$$-4 - \frac{2}{5} = -4\frac{2}{5}$$

❏

These problems can be quite complicated, so do them slowly and carefully. Perform only one calculation in each step.

EXAMPLE 17 Add these numbers:

$$3\frac{2}{3} - 1\frac{3}{4} + 2 + \frac{1}{4} + 4\frac{1}{5} =$$

$$3 + \frac{2}{3} - 1 - \frac{3}{4} + 2 + \frac{1}{4} + 4 + \frac{1}{5} =$$

$$(3 - 1 + 2 + 4) + \left(\frac{2}{3} - \frac{3}{4} + \frac{1}{4} + \frac{1}{5}\right) =$$

$$(9 - 1) + \left(\frac{2}{3} - \frac{2}{4} + \frac{1}{5}\right) =$$

$$(8) + \left(\frac{2}{3} - \frac{1}{2} + \frac{1}{5}\right) = \qquad \text{The LCM of 3, 2, and 5}$$
$$\text{is } 3 \cdot 2 \cdot 5, \ 30.$$

So
$$\frac{2}{3} = \qquad -\frac{1}{2} = \qquad \frac{1}{5} =$$

$$\frac{2}{3} \times \frac{10}{10} = \frac{20}{30}, \qquad -\frac{1}{2} \times \frac{15}{15} = -\frac{15}{30}, \qquad \frac{1}{5} \times \frac{6}{6} = \frac{6}{30}.$$

The problem becomes

$$(8) + \left(\frac{20}{30} - \frac{15}{30} + \frac{6}{30}\right) =$$

$$8 + \left(\frac{26}{30} - \frac{15}{30}\right) =$$

$$8 + \frac{11}{30} = 8\frac{11}{30}. \qquad \square$$

EXERCISES 3.8E

Add the mixed numbers and express your answers as mixed numbers.

37. $2\frac{1}{3} + 3\frac{1}{2} =$ **38.** $1\frac{1}{5} + 2\frac{3}{5} =$

39. $3\frac{1}{4} + 4\frac{1}{5} =$ **40.** $4\frac{7}{8} - 1\frac{1}{6} =$

41. $-5\frac{4}{7} - 2\frac{1}{3} =$ **42.** $4\frac{1}{5} - 1\frac{1}{2} =$

43. $5\frac{1}{3} - 2\frac{3}{4} =$ **44.** $-4\frac{2}{7} + 1\frac{3}{7} + 3\frac{1}{7} =$

45. $-6\frac{3}{8} - 1\frac{1}{2} + 2\frac{1}{4} =$ **46.** $-5\frac{1}{2} + 2\frac{1}{3} - 3\frac{5}{6} =$

47. $2\frac{4}{7} + 5\frac{2}{7} - 1\frac{3}{7} =$ **48.** $-5\frac{1}{2} - 1\frac{5}{6} - 2\frac{1}{3} =$

Check your answers on page 722.

3.9 Comparing and Rounding off Fractions

Rounding off Fractions

When calculating with lots of fractions and mixed numbers, it is helpful to know how to round off to the **nearest integer** (that is, the integer closest on the number line). It is much easier to compute when we have only integers to deal with. Whenever a rough idea of the answer will suffice, rounding off allows us to get the information we need quickly and painlessly.

EXAMPLE 1 Round off $\dfrac{9}{10}$ to the nearest integer.

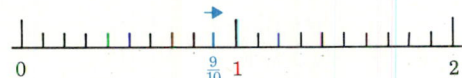

We would round $\dfrac{9}{10}$ up to 1. ❑

To round off a fraction, just compare it to $\dfrac{1}{2}$ and determine whether it is larger or smaller than $\dfrac{1}{2}$.

A fraction that is *larger* than $\dfrac{1}{2}$ rounds *up* to 1.

A fraction that is *smaller* than $\dfrac{1}{2}$ rounds *down* to 0.

A fraction that is equal to $\dfrac{1}{2}$ is just as close to 0 as to 1. (Usually $\dfrac{1}{2}$ is rounded up to 1.)

Inequality symbols simplify our notation when we compare numbers of different sizes:

> is a symbol that means "is greater than."

< is a symbol that means "is less than."

If you think of the inequality symbol as the point of an arrow, the point always faces the smaller number.

EXAMPLE 2 7 > 3 means "7 is greater than 3."

2 < 5 means "2 is less than 5." ❑

Remember the denominator represents the number of equal parts that the whole is divided into, and the numerator represents the number of those parts involved in this fraction. So:

If the numerator of a fraction = (equals) half the denominator, that fraction is *equal* to $\dfrac{1}{2}$.

If the numerator > (is more than) half the denominator, that fraction is *larger* than $\dfrac{1}{2}$.

If the numerator < (is less than) half the denominator, that fraction is *smaller* than $\frac{1}{2}$.

EXAMPLE 3   Round off $\frac{3}{8}$ to the nearest integer.

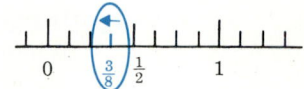

Half the denominator (8) is 4; the numerator 3 < 4, so $\frac{3}{8} < \frac{1}{2}$.

$\frac{3}{8}$ rounds down to 0.

EXAMPLE 4   Round off $\frac{7}{10}$ to the nearest integer.

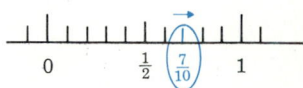

Half the denominator (10) is 5; the numerator 7 > 5, so $\frac{7}{10} > \frac{1}{2}$.

$\frac{7}{10}$ rounds up to 1.

EXAMPLE 5   Round off $\frac{2}{9}$ to the nearest integer.

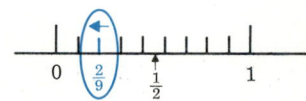

Half the denominator (9) is $4\frac{1}{2}$; the numerator $2 < 4\frac{1}{2}$, so $\frac{2}{9} < \frac{1}{2}$.

$\frac{2}{9}$ rounds down to 0.

Rounding off Mixed Numbers

Mixed numbers appear on the number line between the integer in the mixed number and the next larger integer.

EXAMPLE 6 $5\frac{3}{8}$ is located on the number line.

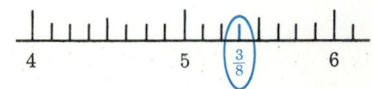

Remember the denominator represents the number of equal parts that the whole is divided into, and the numerator represents the number of those parts involved in this fraction. So:

> To round off a mixed number, just look at the fractional part:
>
> If the fraction is $< \frac{1}{2}$, the mixed number rounds *down* to the integer in that mixed number.
>
> If the fraction is $> \frac{1}{2}$, the mixed number rounds *up* to the next integer.

EXAMPLE 7   Round off $5\frac{3}{8}$ to the nearest integer.

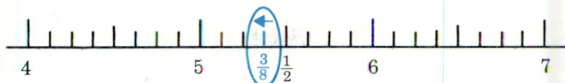

$\frac{3}{8} < \frac{1}{2}$, so $5\frac{3}{8}$ rounds down to 5.   ❑

EXAMPLE 8   Round off $2\frac{7}{10}$ to the nearest integer.

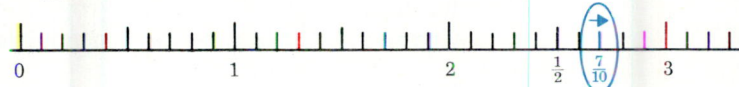

$\frac{7}{10} > \frac{1}{2}$, so $2\frac{7}{10}$ rounds up to 3.   ❑

EXERCISES 3.9A

Round off each number to the nearest integer.

1. $\frac{1}{6}$ 2. $\frac{5}{6}$ 3. $\frac{3}{4}$ 4. $\frac{5}{12}$

5. $\frac{5}{9}$ 6. $\frac{3}{7}$ 7. $\frac{2}{13}$ 8. $\frac{6}{15}$

9. $1\frac{5}{8}$ 10. $3\frac{3}{14}$ 11. $7\frac{4}{15}$ 12. $8\frac{7}{9}$

13. $\frac{7}{19}$ 14. $\frac{6}{17}$

Check your answers on page 722.   ✓

Using Common Sense to Round Off

When we round off fractions outside of classrooms, we should use the procedures outlined above, but we must also use common sense.

EXAMPLE 9   I figure that I will need $2\frac{1}{3}$ loaves of bread to make sandwiches for the picnic.

Even though $2\frac{1}{3}$ is closer to 2 than to 3, I'd better not round down to 2 and buy only two loaves of bread. I won't have enough bread. It would be better to round up to 3 and have $\frac{2}{3}$ loaf left over.   ❑

EXAMPLE 10   I have enough money to buy $3\frac{4}{5}$ record albums. Even though that rounds up to 4, I'd better pick out only three albums. I don't have enough money for four. The store owner won't be very impressed if I argue that I learned to round off to the nearest integer in this book.   ❑

Comparing Fractions

Another useful skill in math courses as well as in our everyday lives is to be able to compare fractions and easily recognize which one is larger. In math courses, that skill helps us to check quickly whether our answers make sense.

In life, it helps us compare unit prices or do projects that involve fractional measurements.

In the last section we saw how to compare fractions to $\frac{1}{2}$. It is easier to compare a fraction to $\frac{1}{2}$ than to any other fraction, but now we will see that it's not very difficult to compare two fractions even if neither is $\frac{1}{2}$.

If two fractions have the *same denominator*, then they represent pieces of the *same size*. The numerator represents the number of pieces, so the fraction with the larger numerator is the larger fraction.

EXAMPLE 11 $\frac{7}{10}$ is larger than $\frac{3}{10}$: $\frac{7}{10} > \frac{3}{10}$.

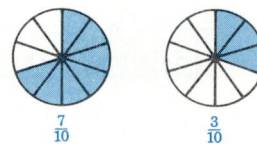

$\frac{7}{10}$ $\frac{3}{10}$

If two fractions have the same numerator, then they represent the *same quantity* of pieces. But the *denominator* determines the *size* of each piece. Remember, the smaller the denominator, the larger the piece (the fewer people we have to share a pizza with, the larger our share). So the fraction with the smaller denominator is the larger fraction when the numerators are the same.

EXAMPLE 12 $\frac{5}{7}$ is larger than $\frac{5}{11}$: $\frac{5}{7} > \frac{5}{11}$.

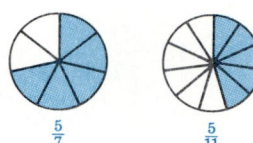

$\frac{5}{7}$ $\frac{5}{11}$

If neither the numerators nor the denominators in two fractions are the same, it may be harder to tell which is larger just by looking. Sometimes rounding off will help.

EXAMPLE 13   Which is larger, $\frac{1}{3}$ or $\frac{5}{7}$?

Since $\frac{1}{3}$ rounds down to 0 and $\frac{5}{7}$ rounds up to 1, $\frac{5}{7}$ is definitely larger.

When rounding off doesn't work, find the lowest common multiple of the denominators, as we did when we added fractions in Section 3.7. Once the two fractions are expressed with the same denominator, then the one with the larger numerator is the larger fraction.

EXAMPLE 14   Which fraction is larger, $\frac{2}{3}$ or $\frac{4}{5}$?

The denominators 3 and 5 are both prime, so their LCM is $(3)(5) = 15$.

$$\frac{2}{3} = \qquad\qquad \frac{4}{5} =$$

$$\frac{2}{3} \times \frac{5}{5} = \frac{10}{15} \qquad \frac{4}{5} \times \frac{3}{3} = \frac{12}{15}$$

Since $\dfrac{12}{15} > \dfrac{10}{15}$, $\rightarrow \dfrac{4}{5} > \dfrac{2}{3}$.

EXAMPLE 15 Which fraction is largest, $\dfrac{2}{3}, \dfrac{5}{8}$, or $\dfrac{7}{9}$?

$3 = 3, 8 = 2 \cdot 2 \cdot 2, 9 = 3 \cdot 3$, so LCD is $2 \cdot 2 \cdot 2 \cdot 3 \cdot 3 = 72$.

$$\frac{2}{3} = \qquad \frac{5}{8} = \qquad \frac{7}{9} =$$

$$\frac{2}{3} \times \frac{24}{24} = \frac{48}{72} \qquad \frac{5}{8} \times \frac{9}{9} = \frac{45}{72} \qquad \frac{7}{9} \times \frac{8}{8} = \frac{56}{72}$$

Since $\dfrac{56}{72}$ has the largest numerator, $\dfrac{7}{9}$ is the largest fraction.

EXERCISES 3.9B

Which is the larger fraction in each pair below?

15. $\dfrac{2}{5}, \quad \dfrac{3}{5}$ **16.** $\dfrac{3}{7}, \quad \dfrac{1}{7}$

17. $\dfrac{1}{7}, \quad \dfrac{1}{3}$ **18.** $\dfrac{4}{9}, \quad \dfrac{4}{13}$

19. $\dfrac{4}{5}, \quad \dfrac{2}{5}$ **20.** $\dfrac{3}{7}, \quad \dfrac{3}{19}$

21. $\dfrac{6}{11}, \quad \dfrac{6}{7}$ **22.** $\dfrac{2}{9}, \quad \dfrac{7}{9}$

23. $\dfrac{4}{13}, \quad \dfrac{11}{13}$ **24.** $\dfrac{8}{21}, \quad \dfrac{4}{21}$

25. $\dfrac{2}{15}, \quad \dfrac{2}{3}$ **26.** $\dfrac{1}{2}, \quad \dfrac{1}{5}$

27. $\dfrac{2}{5}, \quad \dfrac{3}{4}$ **28.** $\dfrac{3}{7}, \quad \dfrac{2}{5}$

29. $\dfrac{3}{8}, \quad \dfrac{4}{9}$ **30.** $\dfrac{4}{5}, \quad \dfrac{7}{9}$

31. $\dfrac{2}{7}, \quad \dfrac{4}{15}$ **32.** $\dfrac{5}{12}, \quad \dfrac{3}{8}$

Which fraction is largest?

33. $\dfrac{3}{4}, \quad \dfrac{5}{6}, \quad \dfrac{7}{12}$ **34.** $\dfrac{1}{3}, \quad \dfrac{2}{8}, \quad \dfrac{2}{7}$

35. $\dfrac{5}{6}, \quad \dfrac{11}{12}, \quad \dfrac{2}{3}$ **36.** $\dfrac{7}{10}, \quad \dfrac{3}{4}, \quad \dfrac{2}{3}$

Check your answers on page 722.

 **Using Fractions to Compute Real-Life Problems**

The most important reason to know how to compute with fractions is that in our real lives (outside of classrooms) we need to make lots of computations that involve fractional data.

In order to use the skills learned in this chapter to perform those computations, you must first recognize which operation—*addition, subtraction, multiplication,* or *division*—is involved. If you read carefully the problems that follow, I think you will recognize which operation is involved.

To begin, write each problem as a calculation. Read it over to make sure it represents the question in each problem.

EXAMPLE 1   I was having a hard time weighing my dog, Aurie, until I convinced a friend to help. Julie held the dog; together they weighed $201\frac{1}{4}$ pounds. Then I weighed Julie, who alone weighed $132\frac{1}{2}$ pounds.

How much does the dog weigh?

The dog's weight is the difference between their combined weight and Julie's weight, so I must subtract her weight from their combined weight to find his weight.

$$\text{Aurie's weight} = \text{combined weight} - \text{Julie's weight}$$

$$= 201\frac{1}{4}\,\text{lb} - 132\frac{1}{2}\,\text{lb}$$

$$= \left(201 + \frac{1}{4} - 132 - \frac{1}{2}\right)\text{lb}$$

$$= (201 - 132) + \left(\frac{1}{4}\quad\frac{1}{2}\right)\text{lb}$$

$$= (69) + \left(\frac{1}{4} - \frac{2}{4}\right)\text{lb}$$

$$= (69) + \left(-\frac{1}{4}\right)\text{lb}$$

$$= 68\frac{3}{4}\,\text{lb}$$

EXAMPLE 2   The carpenters are replacing all the broken clapboards on my house so that the house can be painted. Here are the measurements of the boards they removed:

$$3\frac{1}{2}\text{ feet, }5\frac{1}{4}\text{ feet, }6\frac{1}{2}\text{ feet, }5\text{ feet, }4\frac{1}{2}\text{ feet, }7\frac{3}{4}\text{ feet, and }2\frac{1}{2}\text{ feet.}$$

What is the total amount of lumber I need to buy to replace those boards?

The key word here is *total,* which directs me to add (total lumber = sum of all the pieces).

$$\text{Total} = \left(3\frac{1}{2} + 5\frac{1}{4} + 6\frac{1}{2} + 5 + 4\frac{1}{2} + 7\frac{3}{4} + 2\frac{1}{2}\right)\text{ft}$$

$$= (3 + 5 + 6 + 5 + 4 + 7 + 2) + \left(\frac{1}{2} + \frac{1}{4} + \frac{1}{2} + \frac{1}{2} + \frac{3}{4} + \frac{1}{2}\right)$$

$$= (32) + \left(\frac{4}{2} + \frac{4}{4}\right)$$

$$= (32 + 2 + 1)$$

$$= 35 \text{ ft}$$

EXAMPLE 3   A recipe calls for $\frac{3}{4}$ cup of butter for one batch of brownies. In an attempt to keep from eating too many, David decides to bake only $\frac{1}{3}$ of a batch. How much butter should he use?

In $\frac{1}{3}$ of a batch, he would use $\frac{1}{3}$ *of* $\frac{3}{4}$ cup of butter, which is calculated by multiplying by $\frac{1}{3}$.

$$\frac{1}{3} \text{ of } \frac{3}{4} = \frac{1}{3} \times \frac{3}{4}$$

$$= \frac{1 \times \cancel{3}}{\cancel{3} \times 4}$$

$$= \frac{1}{4}$$

So David needs $\frac{1}{4}$ cup of butter.

EXAMPLE 4   Adam bought a roll of 35 feet of crepe paper streamers to make banners for a party. If each banner is $2\frac{1}{2}$ feet long, how many banners can he make from this roll?

We need to know how many times $2\frac{1}{2}$ divides into 35.

$$\text{The number of banners} = 35 \div 2\frac{1}{2}$$

$$= \frac{35}{1} \div \frac{5}{2}$$

$$= \frac{35}{1} \times \frac{2}{5}$$

$$= \frac{7 \cdot \cancel{5} \cdot 2}{1 \cdot \cancel{5}}$$

$$= 14$$

He will be able to cut exactly 14 banners.

EXAMPLE 5 The weather report said the temperature should rise to around 20 degrees Celsius today. I need to convert that figure into the Fahrenheit scale to know whether to wear a sweater. The Fahrenheit-Celsius formula is

$$F° = \left(\frac{9}{5}\right)C° + 32.$$ I'll replace $C°$ by $20°$.

$$F = \left(\frac{9}{5}\right)(20°) + 32$$

$$= \frac{9 \cdot \cancel{5} \cdot 4}{\cancel{5}} + 32$$

$$= 36 + 32$$

$$= 68°$$

Since 20 degrees Celsius translates into 68 degrees Fahrenheit, I'd better bring a sweater. ❑

EXAMPLE 6 Denise, Jill, and Tracy are sharing a large bag of peatmoss for their different-sized gardens. Denise has used $\frac{1}{3}$ of the bag, and Jill has used $\frac{1}{4}$ of the bag. How much is left for Tracy's share?

We must add the amount that Denise and Jill have used, and subtract that sum from 1 (which represents the whole bag).

The portion of a bag that is left is

$$1 - \left(\frac{1}{3} + \frac{1}{4}\right) =$$ The LCM of 3 and 4 = 12.

$$1 - \left(\frac{4}{12} + \frac{3}{12}\right) =$$

$$\frac{12}{12} - \frac{7}{12} = \frac{5}{12}, \text{ or almost half a bagful.}$$ ❑

EXERCISES 3.10

Determine what kind of computation is involved and answer each question. Round off only when told to do so.

1. What is the total of all these weights:

$$4\frac{1}{2} \text{ lb, } 3 \text{ lb, } 5\frac{1}{4} \text{ lb, } 1 \text{ lb, } 2\frac{3}{4} \text{ and } 1\frac{1}{2} \text{ lb?}$$

2. How many feet of curtain material will I need for all these lengths:

$$6\frac{1}{5} \text{ ft, } 4\frac{7}{10} \text{ ft, } 3 \text{ ft, } 2\frac{1}{2} \text{ ft, } 1\frac{9}{10} \text{ ft, and } 3\frac{1}{10} \text{ ft?}$$

3. If I triple a recipe calling for $\frac{3}{4}$ cup of milk, how much milk do I need?

4. A casserole recipe looks very good, but it came from a catering cookbook and feeds two dozen people. I want to make $\frac{1}{4}$ of that amount. How many cups of noodles do I need to use if the original recipe calls for 7 cups of noodles?

5. I have a board that is 16 ft long. How much will be left if I cut off pieces of the following lengths:

$$4\frac{1}{3} \text{ ft}, 2\frac{1}{2} \text{ ft}, 5\frac{1}{4} \text{ ft}, \text{ and } 1\frac{1}{3} \text{ ft}?$$

How many whole feet would that figure round off to?

6. A clothing-store owner encourages profit sharing among her workers. She figures that workers should keep $\frac{1}{5}$ of the total of their sales. How much does a worker who sells $340 worth of clothes get to keep?

7. After working for a total of 24 hours at a temporary job shredding the papers of a politician under investigation for corruption, we were told that we had just completed $\frac{1}{4}$ of the task. (a) What fraction of the job is left to do? (b) How much longer should it take us to finsish destroying the evidence? (c) How long will the entire job take?

8. It takes 40 minutes of turning a crank to get the ice cream solution frozen into hard, cold ice cream. If 5 people divide the work equally, how long must each turn the crank on the freezer?

9. The measurements of lengths of wood I need for a project are

$$5\frac{1}{4} \text{ cm}, 3\frac{1}{2} \text{ cm}, 7\frac{3}{4} \text{ cm}, 2\frac{1}{2} \text{ cm}, \text{ and } 6\frac{1}{4} \text{ cm}.$$

(a) What is the total length of wood I need? (b) How much will be left over from a board that is 30 cm long?

10. A friend volunteered to pay $\frac{1}{3}$ of all the expenses of the party. The costs were $20, $4, $7, $15, and $5. (a) What is the total cost? (b) How much will it cost her?

11. Luke is $3\frac{1}{4}$ ft taller than Jesse. If Luke is $6\frac{1}{2}$ ft tall, how tall is Jesse?

12. The tailor has 50 ft of black leather. He figures he needs $4\frac{1}{3}$ ft for each pair of leather pants. How many pairs of pants can he cut from that material?

13. The co-op purchased $48\frac{1}{2}$ lb of rice. It is packaged in bags that each hold $1\frac{1}{3}$ lb. How many bags can be filled with rice?

14. A South American visitor was confused to read that the next day's temperature would be 77° Fahrenheit. Translate that into Celsius for him with this formula:

$$C = \frac{5}{9}(F - 32)$$

15. When I left the table to make a phone call, our pitcher of beer was still $\frac{2}{3}$ full. When I got back, it looked as if Rick had drunk half of that. How much of the pitcher did he drink while I was gone?

16. We pay $\frac{1}{5}$ of our earnings for taxes each year. (a) In a year when we make $18,000, how much goes to pay taxes? (b) How much is left for us to spend?

17. If we average 45 mph on a trip, how far have we traveled in $3\frac{2}{3}$ hours?

18. Sonya figures she can save $\frac{1}{6}$ of the money she makes each week. How much is saved when she earns $42?

19. Ari reports that he has read $\frac{5}{9}$ of the book. Mike says that he has read $\frac{7}{12}$ of the same book. Who has read more of the book?

20. The café cuts each pie into 6 pieces. If $8\frac{5}{6}$ pies are consumed in a day, how many pieces were sold?

21. The teacher graded 5 exams in $\frac{3}{4}$ of an hour. If she continues to grade at the same rate, how long will it take her to grade 30 exams?

Check your answers on page 722.

 **Unit Cancellation Conversion**

Conversion Fractions

Usually when numbers are involved in our lives outside of classrooms, they appear in terms of some unit of measurement. We use such measures as 5 inches, 7 feet, 32 degrees, $3\frac{1}{2}$ yards, 7 ounces, 5 pounds.

In fact, when we talk about measurements, the numbers are completely meaningless unless we know which unit is involved. I was once told that the price of a hand-carved box in a yard sale was 2. After some debate with myself about whether it was worth it, I reached into my pocket for $2. The vendor glared at me and said that the price was $200. Lack of clarity about units can be embarrassing!

You will find it handy to use your knowledge about fractions to convert from one unit of measurement to another.

EXAMPLE 1 It is not much help to know that a recipe calls for 3 pints of milk if you have only a 1 cup measure and no pint containers. You will have to convert the recipe into cups to use your utensil.

EXAMPLE 2 You may have calculated the dimensions of a room in terms of square yards, but if the store marks its carpet sizes according to square feet, you will have to convert your measurements into square feet too. ❑

Unit cancellation conversion is the process of setting up the units of measurement so that there is cancellation of the undesired unit when converting from one unit of measurement to another.

To convert units, you need a chart showing how one unit relates to another. In Table 3.1 I have listed several of the most commonly used conversion relationships from the **English measurement system** (the one most commonly used in the United States).

Table 3.1 English units of measure

Length	Time	Volume	Weight
12 inches = 1 foot	60 seconds = 1 minute	3 teaspoons = 1 tablespoon	16 ounces = 1 pound
3 feet = 1 yard	60 minutes = 1 hour	2 tablespoons = 1 ounce	2,000 lb = 1 ton
5,280 feet = 1 mile	24 hours = 1 day	8 ounces = 1 cup	
	12 months = 1 year	2 cups = 1 pint	
	10 years = 1 decade	2 pints = 1 quart	
		4 quarts = 1 gallon	

> To convert from one unit of measurement to another:
>
> Find the relationship in the table between the two units. Since those two representations are equal, you can make two fractions, *each having a value of 1,* by using one as the numerator and the other as the denominator.

EXAMPLE 3 3 feet = 1 yard,

$$\text{so} \quad \frac{3\text{ feet}}{1\text{ yard}} = 1 \quad \text{and} \quad \frac{1\text{ yard}}{3\text{ feet}} = 1.$$ ❑

These fractions are called **conversion fractions**. You can multiply them by the original data without changing the value of the data—because multiplying with these fractions is the same as multiplying with 1. This is really the same thing we did in Section 3.6 when we first learned to find new (but equal) names for fractions.

You can cancel *units* from the numerator and denominator just as you canceled factors in Section 3.4. For example, to convert 7 yards into feet, use a conversion fraction that will cancel yards.

EXAMPLE 4 To convert 7 yards into feet, I'll use $\dfrac{3\text{ feet}}{1\text{ yard}}$.

$$7\text{ yards} =$$

$$7\text{ yards}\left[\frac{3\text{ feet}}{1\text{ yard}}\right] =$$

$$7\ \cancel{\text{yards}}\left[\frac{3\text{ feet}}{1\ \cancel{\text{yard}}}\right] =$$

$$(7)(3)\text{ feet} = 21\text{ feet}$$ ❑

I like using conversion fractions, in contrast to the way I was taught to do conversions in school. That was to *memorize* the situations in which we *multiply* to convert a unit versus those in which we *divide* to convert a unit. I could never remember which one to do.

With conversion fractions you don't have to memorize which operation to perform. Just use the form of the fraction in which the unit you want to remove is on the bottom, and the new unit you want to convert into is on top. Canceling unit names the way you cancel numbers helps you make sure you are doing it right.

EXAMPLE 5 Convert 9 pints into quarts.

The relationship between pints and quarts is 2 pints = 1 quart.

The conversion fraction that will cancel pints and replace them with quarts is $\dfrac{1 \text{ quart}}{2 \text{ pints}}$.

$$9 \text{ pints} =$$

$$(9 \text{ pints})\left[\frac{1 \text{ quart}}{2 \text{ pints}}\right] =$$

$$9 \,\cancel{\text{pints}}\left[\frac{1 \text{ quart}}{2 \,\cancel{\text{pints}}}\right] =$$

$$\frac{9}{2} \text{ quarts} = 4\frac{1}{2} \text{ quarts}$$

Conversion fractions can be used any time you want to change the unit of measurement. If you travel to another country, you can use them to convert money values as long as you are told the relationship between their unit and the dollar.

Since the world's economy changes so frequently, I'll make up a fantasy example to keep this book from becoming outdated too quickly.

EXAMPLE 6 In the land of Chocolateberg people use a unit of money called a cocochip. The money exchanger tells us that 3 dollars are worth 8 cocochips.

How many cocochips will I get for 150 dollars?

The conversion fraction to cancel dollars and leave me in cocochips is $\dfrac{8 \text{ cocochips}}{3 \text{ dollars}}$.

$$\text{So } 150 \text{ dollars} =$$

$$150 \text{ dollars}\left[\frac{8 \text{ cocochips}}{3 \text{ dollars}}\right] =$$

$$150 \,\cancel{\text{dollars}}\left[\frac{8 \text{ cocochips}}{3 \,\cancel{\text{dollars}}}\right] =$$

$$\frac{(150)(8)}{3} \text{ cocochips} = 400 \text{ cocochips}$$

Conversion fractions are particularly helpful when our data involve fractions.

EXAMPLE 7   Convert $\dfrac{3}{4}$ cup into ounces.

The relationship between cups and ounces is 8 ounces = 1 cup.

The conversion fraction that will cancel cups and leave us in ounces is $\dfrac{8 \text{ ounces}}{1 \text{ cup}}$.

$$\frac{3}{4} \text{ cup} =$$

$$\left(\frac{3}{4}\text{ cup}\right)\left(\frac{8 \text{ ounces}}{1 \text{ cup}}\right) =$$

$$\left(\frac{3\,\cancel{\text{cup}}}{4}\right)\left(\frac{8 \text{ ounces}}{1\,\cancel{\text{cup}}}\right) =$$

$$\left(\frac{3}{4}\right)(8)\text{ounces} =$$

$$\frac{3 \cdot 2 \cdot 4}{4} \text{ ounces} =$$

$$(3 \cdot 2) \text{ ounces} = 6 \text{ ounces}$$

EXAMPLE 8   Convert $10\dfrac{1}{2}$ inches into feet.

Since 12 inches = 1 foot, use $\dfrac{1 \text{ foot}}{12 \text{ inches}}$ to cancel inches for feet.

Represent $10\dfrac{1}{2}$ as an improper fraction.

$$10\frac{1}{2} \text{ inches} =$$

$$\frac{21}{2} \text{ inches} =$$

$$\left(\frac{21}{2}\text{ inches}\right)\left(\frac{1 \text{ foot}}{12 \text{ inches}}\right) =$$

$$\left(\frac{21\,\cancel{\text{inches}}}{2}\right)\left(\frac{1 \text{ foot}}{12\,\cancel{\text{inches}}}\right) =$$

$$\frac{21 \text{ feet}}{(2)(12)} =$$

$$\left(\frac{\cancel{3} \cdot 7 \text{ feet}}{2 \cdot \cancel{3} \cdot 4}\right) =$$

$$\frac{7}{2 \cdot 4} \text{ feet} = \frac{7}{8} \text{ foot}$$

⟹ *Note*: We use the singular or plural form of the units as we need it. In Example 8 we used both *foot* and *feet*. When we realized that our answer was less than a foot, we switched back to singular.

EXERCISES 3.11A _____

Determine what the conversion fraction should be for the units involved. Use it to convert each expression into the new unit of measurement.

1. 6 tablespoons to teaspoons

2. 20 pints to quarts

3. 5 days to hours

4. 3 miles to feet

5. 40,000 pounds to tons

6. 9 feet to inches

7. 11 years to months

8. 15 cups to pints

9. 100 minutes to hours

10. 15 quarts to pints

11. 40 ounces to pounds

12. 7 pints to cups

Convert each expression into the requested unit.

13. $\frac{2}{3}$ foot into inches

14. $4\frac{1}{2}$ tablespoons into teaspoons

15. $3\frac{1}{4}$ miles into feet

16. $5\frac{3}{4}$ pounds into ounces

17. $5\frac{1}{2}$ quarts into pints

18. $3\frac{1}{3}$ gallons into quarts

19. $3\frac{1}{4}$ hours into minutes

20. $4\frac{2}{3}$ inches into feet

Check your answers on page 722.

Using More Than One Conversion Fraction

Conversion from any of the units in Table 3.1 to any of the other units in the same category (length, time, volume, or weight) can be done without expanding the table to express each unit in terms of every other unit. For example, you can convert from inches into yards—even though their relationship is not actually listed in the table—by using more than one conversion fraction. You use one conversion fraction to go from inches to feet, and then another conversion fraction to go from feet to yards.

In general, to convert by using a table like Table 3.1, use as many conversion fractions as necessary to go from the original unit through all the intermediate units listed until the new unit is involved. Set them up so that the original unit cancels, and then the unit it is converted into is canceled, and so on until the unit left is the one you want.

EXAMPLE 9 Convert 72 inches into yards.

Table 3.1 gives us relationships between inches and feet, and feet and yards. They are 12 inches = 1 foot; 3 feet = 1 yard.

So to convert inches into yards, use the conversion fractions $\frac{1 \text{ foot}}{12 \text{ inches}}$, which will cancel the original inches and replace them with feet, and $\frac{1 \text{ yard}}{3 \text{ feet}}$, which will cancel the feet and leave the answer in terms of yards.

$$72 \text{ inches} =$$

$$72 \text{ inches} \cdot \frac{1 \text{ foot}}{12 \text{ inches}} \cdot \frac{1 \text{ yard}}{3 \text{ feet}} =$$

$$72 \ \cancel{\text{inches}} \cdot \frac{1 \ \cancel{\text{foot}}}{12 \ \cancel{\text{inches}}} \cdot \frac{1 \text{ yard}}{3 \ \cancel{\text{feet}}} =$$

$$\frac{(2)(3)(12) \text{ yards}}{(12)(3)} =$$

$$\frac{(2)\cancel{(3)}\cancel{(12)} \text{ yards}}{\cancel{(12)}\cancel{(3)}} = 2 \text{ yards}$$

So 72 inches = 2 yards. ❏

EXAMPLE 10   Convert 3 quarts into ounces.

The necessary relationships are 1 quart = 2 pints; 1 pint = 2 cups; 1 cup = 8 ounces.

The necessary conversion fractions to cancel quarts, then pints, then cups, leaving the answer expressed in terms of ounces, are used below:

$$3 \text{ quarts} =$$

$$(3 \text{ quarts}) \cdot \frac{(2 \text{ pints})}{(1 \text{ quart})} \cdot \frac{(2 \text{ cups})}{(1 \text{ pint})} \cdot \frac{(8 \text{ ounces})}{(1 \text{ cup})} =$$

$$(3 \ \cancel{\text{quarts}}) \cdot \frac{(2 \ \cancel{\text{pints}})}{(1 \ \cancel{\text{quart}})} \cdot \frac{(2 \ \cancel{\text{cups}})}{(1 \ \cancel{\text{pint}})} \cdot \frac{(8 \text{ ounces})}{(1 \ \cancel{\text{cup}})} =$$

$$(3)(2)(2)(8) \text{ ounces} = 96 \text{ ounces}$$

So 3 quarts = 96 ounces. ❏

EXERCISES 3.11B

Convert each expression into the new unit requested.

21. 2 yards into inches

22. 5 hours into seconds

23. 2 days into minutes

24. 3 tons into ounces

25. 5 quarts into cups

26. 16 pints into gallons

27. 4 miles into yards

28. 5 days into seconds

29. $3\frac{1}{2}$ pints into ounces

30. $3\frac{1}{4}$ days into minutes

Check your answers on page 722. ✓

3.12 Perimeter and Area Applications of Dimensional Analysis and Fractions

Computations Involving Fractions

In Section 1.7 we discussed how to compute the perimeter and area of rectangles, triangles, and squares. Review that section before proceeding.

Now we can find the perimeter and area when the dimensions involve fractions.

EXAMPLE 1   Find the perimeter of a rectangle with width $4\frac{1}{2}$ inches and length $6\frac{1}{2}$ inches.

$$\text{Perimeter}_{\text{rectangle}} = 2(\text{width}) + 2(\text{length})$$

$$= \left[2\left(4\frac{1}{2}\right) + 2\left(6\frac{1}{2}\right) \right] \text{in.}$$

$$= \left[\cancel{2}\left(\frac{9}{\cancel{2}}\right) + \cancel{2}\left(\frac{13}{\cancel{2}}\right) \right] \text{in.}$$

$$= [9 + 13] \text{ in.}$$

$$= 22 \text{ in.}$$

$4\frac{1}{2}$ in.

$6\frac{1}{2}$ in.

EXAMPLE 2   Find the perimeter of a triangle with sides $3\frac{1}{4}$ ft, $4\frac{1}{2}$ ft, and $7\frac{1}{3}$ ft.

Convert all fractions to twelfths.

$$\text{Perimeter}_{\text{triangle}} = \text{side}_1 + \text{side}_2 + \text{side}_3$$

$$= 3\frac{1}{4} + 4\frac{1}{2} + 7\frac{1}{3} \text{ ft}$$

$$= 3\frac{3}{12} + 4\frac{6}{12} + 7\frac{4}{12} \text{ ft}$$

$$= (3 + 4 + 7) + \frac{3 + 6 + 4}{12} \text{ ft}$$

$$= 14 + \frac{13}{12} \text{ ft}$$

$$= 14 + 1\frac{1}{12} \text{ ft}$$

$$= 15\frac{1}{12} \text{ ft}$$

$3\frac{1}{4}$ ft $4\frac{1}{2}$ ft

$7\frac{1}{3}$ ft

EXAMPLE 3   Find the perimeter of a square with sides $2\frac{3}{4}$ inch.

$$\text{Perimeter}_{\text{square}} = 4(\text{side})$$

$$= 4\left(2\frac{3}{4}\right) \text{in.}$$

$$= \cancel{4}\left(\frac{11}{\cancel{4}}\right) \text{in.}$$

$$= 11 \text{ in.}$$

The careless error most frequently made when calculating area is to forget that area *must be expressed in* **square units**. To express the area of an object in feet rather than in square feet is a serious mistake. It suggests that you do not understand that area represents the number of *square feet* it would take to cover the object. The unit of measurement is a very important part of the answer.

Remember, a square foot is a 1 foot by 1 foot square.

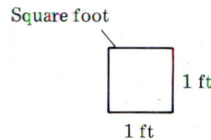

Square foot

1 ft

1 ft

EXAMPLE 4   If the area of a rectangle is 16 square feet (16 sq ft), it would take 16 tiles, each 1 foot by 1 foot, to cover the rectangle.

1 ft

1 ft —

1	2	3	4	5	6	7	8
9	10	11	12	13	14	15	16

EXAMPLE 5   Find the area of a rectangle with width $2\frac{1}{2}$ inches and length $5\frac{3}{4}$ inches.

$$\text{Area}_{\text{rectangle}} = (\text{width})(\text{length})$$

$$= \left(2\frac{1}{2}\text{ in.}\right)\left(5\frac{3}{4}\text{ in.}\right)$$

$$= \left(\frac{5}{2}\right)\left(\frac{23}{4}\right)(\text{in.})(\text{in.})$$

$$= \frac{5 \cdot 23}{2 \cdot 4}\text{ sq. in.}$$

$$= \frac{115}{8}\text{ sq. in.}$$

$$= 14\frac{3}{8}\text{ sq. in.}$$

$2\frac{1}{2}$ in.

$5\frac{3}{4}$ in.

EXAMPLE 6   Find the area of a square with sides $4\frac{2}{3}$ feet.

$$\text{Area}_{\text{square}} = (\text{side})(\text{side})$$

$$= \left(4\frac{2}{3}\text{ ft}\right)\left(4\frac{2}{3}\text{ ft}\right)$$

$$= \left(\frac{14}{3}\right)\left(\frac{14}{3}\right)\text{sq. ft}$$

$$= \left(\frac{14 \cdot 14}{3 \cdot 3}\right)\text{sq. ft}$$

$$= \frac{196}{9}\text{ sq. ft}$$

$$= 21\frac{7}{9}\text{ sq. ft}$$

$4\frac{2}{3}$ ft

$4\frac{2}{3}$ ft

EXAMPLE 7   Find the area of a triangle with base $4\frac{1}{2}$ inches and height 6 inches.

$$\text{Area}_{\text{triangle}} = \left(\frac{1}{2}\right)(\text{base})(\text{height})$$

$$= \left(\frac{1}{2}\right)\left(4\frac{1}{2}\text{ in.}\right)(6\text{ in.})$$

$$= \left(\frac{1}{2}\right)\left(\frac{9}{2}\right)\left(\frac{6}{1}\right)\text{ sq. in.}$$

$$= \frac{9 \cdot \cancel{2} \cdot 3}{\cancel{2} \cdot 2}\text{ sq. in.}$$

$$= \frac{27}{2}\text{ sq. in.}$$

$$= 13\frac{1}{2}\text{ sq. in.}$$

6 in.

$4\frac{1}{2}$ in.

EXERCISES 3.12A

Compute the perimeters and areas requested.

1. What is the perimeter of a rectangle with width $5\frac{1}{3}$ in. and length $8\frac{2}{3}$ in.?

2. What is the perimeter of a triangle with sides $2\frac{1}{5}$ ft, $4\frac{2}{5}$ ft, and 6 ft?

3. What is the perimeter of a square with sides $4\frac{1}{2}$ m?

4. What is the perimeter of a rectangle with width $3\frac{1}{2}$ ft and length $5\frac{2}{3}$ ft?

5. What is the perimeter of a triangle with sides $5\frac{2}{3}$ in., $3\frac{1}{2}$ in., and $4\frac{1}{6}$ in.?

6. What is the perimeter of a square with sides $7\frac{2}{3}$ ft?

7. What is the perimeter of a triangle with sides $8\frac{2}{5}$ cm, $5\frac{1}{4}$ cm, and $5\frac{2}{3}$ cm?

8. What is the area of a rectangle with width $2\frac{1}{3}$ ft and length $5\frac{2}{3}$ ft?

9. What is the area of a square with sides $12\frac{3}{4}$ in.?

10. What is the area of a triangle with base $5\frac{1}{2}$ m and height $4\frac{2}{3}$ m?

11. What is the area of a square with sides $8\frac{2}{5}$ ft?

12. What is the area of a triangle with base $\frac{3}{4}$ in. and height $\frac{5}{8}$ in.?

13. What is the perimeter of a triangle with sides $\frac{4}{12}$ in., $\frac{3}{4}$ in., and $\frac{5}{6}$ in.?

14. What is the perimeter of a rectangle with width $\frac{4}{5}$ cm and length $\frac{7}{10}$ cm?

Check your answers on page 722. ✓

Converting All Dimensions into the Same Unit of Measurement

To calculate the perimeter of an object, make sure that all the sides are expressed in the *same unit of measurement*. If they are not, use conversion fractions to convert them into the unit you choose for all the dimensions.

EXAMPLE 8 Find the perimeter of a triangle with sides 8 inches, $2\frac{1}{2}$ feet, and 3 feet.

Since most of the sides are expressed in feet, let's choose that unit as the one to use. We must convert 8 inches into feet.

$$(8 \text{ in.})\left(\frac{1 \text{ ft}}{12 \text{ in}}\right) =$$

$$\left(\frac{2 \cdot 4 \text{ in}}{1}\right)\left(\frac{\text{ft}}{3 \cdot 4 \text{ in}}\right) = \frac{2}{3} \text{ ft}$$

$$\text{Perimeter}_{\text{triangle}} = \text{side}_1 + \text{side}_2 + \text{side}_3$$

So perimeter $= \frac{2}{3} \text{ ft} + 2\frac{1}{2} \text{ ft} + 3 \text{ ft}$

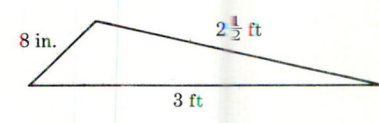

$$= \left(\frac{4}{6} + 2\frac{3}{6} + 3\right) \text{ft}$$

$$= 5\frac{7}{6} \text{ ft}$$

$$= 5 + 1\frac{1}{6} \text{ ft}$$

$$= 6\frac{1}{6} \text{ ft} \qquad \qquad \square$$

When computing the area of an object, also make sure that all dimensions are expressed in the same unit. If they are not, convert them into the same unit before computing.

EXAMPLE 9 Find the area of a rectangle with width $4\frac{1}{2}$ feet and length $\frac{3}{4}$ yard.

Let's change $\frac{3}{4}$ yard into feet.

$$\frac{3}{4} \text{ yd} =$$

$$\left(\frac{3}{4} \text{ yd}\right)\left(\frac{3 \text{ ft}}{1 \text{ yd}}\right) = \frac{9}{4} \text{ ft}$$

$$\text{Area}_{\text{rectangle}} = (\text{length})(\text{width})$$

$$= \left(\frac{9}{4}\,\text{ft}\right)\left(4\frac{1}{2}\,\text{ft}\right)$$

$$= \left(\frac{9}{4}\,\text{ft}\right)\left(\frac{9}{2}\,\text{ft}\right)$$

$$= \frac{81}{8}\,\text{sq ft}$$

$$= 10\frac{1}{8}\,\text{sq ft}$$

4 ½ ft

¾ yd

If all three sides of a triangle are expressed in different units, it is usually easiest to express all three in terms of the smallest unit.

EXAMPLE 10   Find the perimeter of a triangle with sides 5 inches, $\frac{3}{4}$ foot, and $\frac{1}{3}$ yard.

Let's convert each side into inches.

$$\frac{3}{4}\,\text{ft} = \qquad\qquad\qquad \frac{1}{3}\,\text{yd} =$$

$$\left(\frac{3}{4}\,\text{ft}\right)\left(\frac{12\,\text{in.}}{1\,\text{ft}}\right) = \qquad\qquad \left(\frac{1}{3}\,\text{yd}\right)\left(\frac{36\,\text{in.}}{1\,\text{yd}}\right) =$$

$$\left(\frac{3}{4}\,\cancel{\text{ft}}\right)\left(\frac{12\,\text{in.}}{1\,\cancel{\text{ft}}}\right) = \qquad\qquad \left(\frac{1}{3}\,\cancel{\text{yd}}\right)\left(\frac{36\,\text{in.}}{1\,\cancel{\text{yd}}}\right) =$$

$$\frac{3\cdot 3\cdot 4}{4}\,\text{in.} = 9\,\text{in.} \qquad\qquad \frac{\cancel{3}\cdot 12}{\cancel{3}}\,\text{in.} = 12\,\text{in.}$$

So
$$\text{perimeter} = 5\,\text{in.} + \frac{3}{4}\,\text{ft} + \frac{1}{3}\,\text{yd}$$

$$= 5\,\text{in.} + 9\,\text{in.} + 12\,\text{in.}$$

$$= 26\,\text{in.}$$

EXERCISES 3.12B

Before computing the perimeter or area requested, use dimensional analysis to express all dimensions in the same unit of measurement.

15. What is the perimeter of a rectangle with width $3\frac{1}{2}$ ft and length 4 in.?

16. What is the perimeter of a triangle with sides 10 in., $2\frac{1}{2}$ ft, and 3 ft?

17. What is the perimeter of a rectangle with width 2 ft and length $7\frac{2}{3}$ yd?

18. What is the area of a triangle with base 8 in. and height $1\frac{1}{3}$ ft?

19. What is the perimeter of a rectangle with width 6 in. and length 2 yd?

20. What is the area of a rectangle with width 3 in. and length $4\frac{2}{3}$ ft?

21. What is the perimeter of a triangle with sides 10 in., $2\frac{1}{3}$ ft, and 1 yd?

22. What is the perimeter of a triangle with sides 40 in., $1\frac{2}{3}$ ft, and $1\frac{1}{2}$ yd?

23. What is the perimeter of a triangle with sides 6 in., $5\frac{3}{4}$ ft, and 2 yd?

24. What is the area of a triangle with base $\frac{3}{4}$ ft and height $\frac{2}{3}$ yd?

Check your answers on page 722. ✓

Glossary

conversion fractions   Fractions whose numerators and denominators are two equal forms of the same measurement, expressed in different units. So conversion fractions always equal 1.

Example: $\frac{3\text{ feet}}{1\text{ yard}}$ and $\frac{1\text{ yard}}{3\text{ feet}}$ are conversion fractions.

cross multiplication   Process of multiplying the numerator of one fraction by the denominator of an *equal* fraction, and multiplying the denominator of the first fraction by the numerator of the other fraction. The products are always equal (if the fractions are equal). Cross multiplication can be used to check whether a fraction was correctly reduced.

Example: Since $\frac{5}{10}$ and $\frac{1}{2}$ are equal, they can be cross multiplied:

$$\frac{5}{10} \diagdown\!\!\!\diagup \frac{1}{2}$$

$$5(2) = 10(1)$$

$$10 = 10 \quad ✓$$

denominator   Bottom number in a fraction. It represents the amount of equal parts that the whole was divided into and therefore determines the *size* of each piece.

Example: In $\frac{3}{8}$, 8 is the denominator.

division   Multiplication by the divisor's reciprocal.

Example: $\frac{2}{3} \div \frac{3}{4}$ means $\frac{2}{3} \times \frac{4}{3} = \frac{8}{9}$

English measurement system   System of measurement most commonly used in the United States and United Kingdom. It expresses length in terms of inches, feet, yards, and miles; weight in terms of ounces, pounds, and tons; and volume in terms of teaspoons, tablespoons, ounces, cups, pints, quarts, gallons, and bushels.

even number   A number with 0, 2, 4, 6, or 8 in the units place value.

Example: 3,176 is an even number.

factor: A number that divides exactly into a certain number.

Example: 2 is a factor of 10.

factoring a number completely   Expressing a number as the product of its prime factors.

Example: When factoring 10 we express it as $10 = 2 \cdot 5$.

fraction   One way to represent a number that is a *part of an integer*.

improper fractions   Fractions whose numerators are larger than their denominators. Their magnitudes are larger than 1 since there are more pieces than the whole was divided into, and so they can be expressed as mixed numbers.

Example: $\frac{7}{5}$ is an improper fraction.

inequality symbols   Arrows ($>$ and $<$) indicating that one number is larger than another. They always point toward the smaller number.

Example: $7 > 3$ indicates that 7 is larger than 3, and $2 < 9$ indicates that 2 is smaller than 9.

lowest common denominator The smallest number that is a multiple of the denominators of two or more fractions. Also called the lowest common multiple of the denominators.

Example: 12 is the lowest common denominator of $\frac{1}{4}$ and $\frac{5}{6}$.

lowest common multiple The smallest number that is a multiple of two or more numbers.

Example: 12 is the lowest common multiple of 4 and 6.

mixed numbers Numbers made up of an integer and a proper fraction. They appear between integers on the number line and can be expressed as improper fractions.

Example: $5\frac{1}{3}$ is a mixed number.

multiple One number is a multiple of a second number if the second number can be multiplied by some integer to produce the first number. The second number would be a factor of the first.

Example: 10 is a multiple of 5.

numerator Top number in a fraction. It represents the *quantity* of pieces involved.

Example: In $\frac{3}{8}$, 3 is the numerator.

of A word that indicates multiplication.

Example: $\frac{1}{2}$ of 8 means $\left(\frac{1}{2}\right)(8) = 4$

prime A number that has no factors other than itself and 1.

Example: 2, 3, 5, 7, 11, 13, 19, and 23 are prime numbers.

proper fractions Fractions whose numerators are smaller than their denominators. Their magnitudes are less than 1 since there are fewer pieces than the amount into which the whole was divided.

Example: $\frac{5}{7}$ is a proper fraction.

reciprocal Upside-down version of a fraction. The product of a fraction and its reciprocal is always 1.

Example: The reciprocal of $\frac{3}{5}$ is $\frac{5}{3}$.

reduced fraction A fraction with no factors common to both the numerator and the denominator. Fractions should always be expressed in their reduced, or simplest, form.

Example: $\frac{3}{4}$ is a reduced fraction.

unit cancellation conversion Process of setting up the units of measurement so that there is cancellation of the undesired unit when converting from one unit of measurement to another.

Chapter 3 Review

Do each of the problems. After you are confident that you have done them as accurately as possible, compare your answers with those in the back of the book. If any are wrong, go back to the section where the problem came from (indicated to the left of the problem) and review the section.

Once you understand all the sections, you'll be sure to learn the skills solidly and remember how to do them if you practice them a little bit more. Turn to the Supplementary Exercises and do all the problems from any section where you had difficulty on these review exercises.

Answers

3.1

1. What is the name of this fraction?

2. What is the sign of $\dfrac{-3}{-5}$?

3. Is $\dfrac{23}{27}$ a proper or an improper fraction?

4. What integer does $\dfrac{80}{4}$ represent?

3.2

5. $\dfrac{8}{13} + \dfrac{1}{13} =$

6. $\dfrac{19}{23} - \dfrac{7}{23} =$

7. $\dfrac{3}{8} - \left(-\dfrac{2}{8}\right) =$

8. $\dfrac{3}{5}$ of $\dfrac{1}{7} =$

3.3

9. Factor 220 into a product of prime factors.

3.4

10. Reduce $\dfrac{36}{60}$.

3.5

11. $\dfrac{12}{42} \times \dfrac{56}{60} =$

12. $-\dfrac{8}{9} \times \dfrac{12}{14} \times \dfrac{7}{10} =$

13. $\left(\dfrac{-15}{21}\right) \div \left(\dfrac{-12}{14}\right) =$

14. $\dfrac{\frac{3}{5}}{\frac{7}{8}}$

3.6

15. What is the name of the fraction that is equal to $\dfrac{3}{4}$ and has a denominator of 20?

3.7

16. $\dfrac{2}{3} + \dfrac{7}{12} =$

17. $\dfrac{5}{7} - \dfrac{1}{3} =$

18. What is the LCM of 10 and 14?

19. $\dfrac{3}{10} + \dfrac{5}{14} =$

20. $-\dfrac{7}{8} + \dfrac{2}{14} =$

21. $-\dfrac{1}{4} - \dfrac{1}{3} - \dfrac{1}{6} =$

Answers

1. _____
2. _____
3. _____
4. _____
5. _____
6. _____
7. _____
8. _____
9. _____
10. _____
11. _____
12. _____
13. _____
14. _____
15. _____
16. _____
17. _____
18. _____
19. _____
20. _____
21. _____

Continued

Chapter 3 Review   *(Cont.)*

3.8   22. Convert $2\dfrac{1}{4}$ into an improper fraction.

23. Convert $\dfrac{18}{5}$ into a mixed number.

24. $3\dfrac{4}{5} \times 1\dfrac{1}{2} =$

25. $30{,}017 \div 23 =$

26. $7\dfrac{2}{3} - 4\dfrac{1}{2} =$

27. $-1\dfrac{2}{5} - 4\dfrac{1}{4} - 3\dfrac{3}{10} =$

3.9   Round off to the nearest integer.

28. $\dfrac{3}{14}$

29. $5\dfrac{8}{9}$

30. Which is larger, $\dfrac{7}{12}$ or $\dfrac{17}{30}$?

3.10   31. If a cake recipe calls for $2\dfrac{1}{2}$ cups of flour, how much flour should be used to bake a cake that is $\dfrac{1}{3}$ the size of the cake in the recipe?

3.11   Use Table 3.1 on page 107.

32. Convert $\dfrac{2}{3}$ foot into yards.

33. Convert $4\dfrac{1}{2}$ pints into ounces.

3.12   34. What is the perimeter of a rectangle with width $5\dfrac{1}{2}$ ft and length $3\dfrac{1}{4}$ ft?

35. What is the area of a triangle with base $\dfrac{2}{3}$ ft and height $4\dfrac{1}{4}$ in.?

Answers
22.
23.
24.
25.
26.
27.
28.
29.
30.
31.
32.
33.
34.
35.

Check your answers on page 722.

Supplementary Exercises

Do all the problems in every section that involves a skill in which you now lack complete mastery. With a little more practice you will achieve that sense of really understanding the topics, and you will remember how to do these problems.

3.1
1. What is the name of this fraction?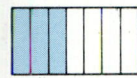

2. What is the sign of $\dfrac{-7}{9}$?

3. Is $\dfrac{34}{25}$ a proper or an improper fraction?

4. What integer does $\dfrac{65}{5}$ represent?

3.2
5. $\dfrac{4}{17} - \dfrac{5}{17} =$

6. $\dfrac{26}{31} + \dfrac{3}{31} =$

7. $\dfrac{5}{9} - \left(-\dfrac{3}{9}\right) =$

8. $\left(-\dfrac{5}{13}\right) - \left(-\dfrac{4}{13}\right) =$

9. $\dfrac{4}{7}$ of $\dfrac{5}{9} =$

3.3
10. Factor 180 into a product of prime factors.

3.4
11. Reduce $\dfrac{24}{34}$.

3.5
12. $\dfrac{10}{32} \times \dfrac{20}{35} =$

13. $-\dfrac{4}{5} \times \dfrac{10}{15} \times \dfrac{1}{6} =$

14. $\left(-\dfrac{12}{25}\right) \div \left(\dfrac{45}{60}\right) =$

15. $\dfrac{\dfrac{2}{7}}{\dfrac{3}{5}} =$

3.6
16. What is the name of the fraction that is equal to $\dfrac{2}{7}$ and has a denominator of 42?

3.7
17. $\dfrac{5}{6} + \dfrac{4}{15} =$

18. $-\dfrac{4}{5} - \dfrac{1}{7} =$

19. What is the LCM of 15 and 20?

20. $-\dfrac{7}{15} + \dfrac{3}{20} =$

21. $-\dfrac{7}{10} - \dfrac{42}{15} =$

22. $-\dfrac{1}{2} + \dfrac{2}{5} - \dfrac{3}{10} =$

Continued

Answers

1. _____
2. _____
3. _____
4. _____
5. _____
6. _____
7. _____
8. _____
9. _____
10. _____
11. _____
12. _____
13. _____
14. _____
15. _____
16. _____
17. _____
18. _____
19. _____
20. _____
21. _____
22. _____

Supplementary Exercises *(Cont.)*

3.8   23. Convert $3\frac{2}{7}$ into an improper fraction.

24. Convert $\frac{13}{4}$ into a mixed number.

25. $5\frac{2}{3} \times 2\frac{2}{5} =$

26. $28,217 \div 26 =$

27. $-7\frac{2}{3} - 4\frac{1}{2} =$

28. $7\frac{2}{3} - 4\frac{1}{2} - 3\frac{5}{6} =$

3.9   Round off to the nearest integer.

29. $\frac{7}{12}$ 30. $3\frac{5}{13}$

31. Which is larger, $\frac{9}{13}$ or $\frac{14}{25}$?

3.10   32. Emily is saving to buy a ski package on sale for $150. If she has already saved $\frac{2}{5}$ of the amount she needs, how much does she now have?

3.11   Use Table 3.1 on page 107.

33. Convert $\frac{3}{4}$ pint into quarts.

34. Convert $2\frac{3}{4}$ feet into inches.

35. Convert $5\frac{2}{3}$ yards into inches.

3.12   36. What is the perimeter of a rectangle with width $4\frac{2}{3}$ ft and length $3\frac{3}{4}$ ft

37. What is the area of a triangle with base $\frac{3}{4}$ ft and height $7\frac{1}{2}$ in.?

23. _____

24. _____

25. _____

26. _____

27. _____

28. _____

29. _____

30. _____

31. _____

32. _____

33. _____

34. _____

35. _____

36. _____

37. _____

Check your answers on page 722.

Chapter 3 Test

Express your answers in the most reduced form, except for Problem 3.

1. Write 360 as the product of its prime factors.

2. Express $\dfrac{36}{40}$ in its most reduced form.

3. Find a new name for $\dfrac{5}{6}$ with a denominator of 24.

Determine which fraction in each group is largest.

4. $\dfrac{5}{11}, \dfrac{5}{7}, \dfrac{5}{31}$

5. $\dfrac{3}{5}, \dfrac{7}{9}, \dfrac{8}{15}$

Combine the fractions in each group.

6. $-\dfrac{7}{20} + \dfrac{5}{20} - \dfrac{4}{20}$

7. $\dfrac{4}{15} + \dfrac{2}{3} - \dfrac{1}{5}$

8. $\dfrac{5}{12} + \dfrac{7}{15}$

9. $-\dfrac{20}{36} \times \dfrac{18}{25}$

10. $\dfrac{\frac{2}{9}}{\frac{42}{90}}$

11. $5\dfrac{1}{2} \div 3\dfrac{2}{3}$

12. $-4\dfrac{1}{4} - 1\dfrac{1}{2}$

13. $\left(-\dfrac{6}{7}\right) \times \left(-\dfrac{4}{9}\right) \times \left(-\dfrac{1}{2}\right) =$

14. $-\dfrac{7}{12} - \left(-\dfrac{1}{3}\right) =$

Round off to the nearest integer.

15. $4\dfrac{1}{4}$

16. $\dfrac{4}{5}$

17. $3\dfrac{2}{9}$

18. What is the total length of ribbon needed to make ribbons of the following lengths:

$4\dfrac{1}{3}$ in., $5\dfrac{1}{2}$ in., $1\dfrac{3}{4}$ in., $7\dfrac{1}{2}$ in., $4\dfrac{1}{4}$ in., $5\dfrac{1}{3}$ in., and $\dfrac{2}{3}$ in.?

19. Convert $4\dfrac{2}{3}$ yards into inches.

20. What is the perimeter of a triangle with sides $4\dfrac{2}{3}$ in., $6\dfrac{1}{2}$ in., and $8\dfrac{1}{6}$ in.?

21. What is the area of a rectangle with width $\dfrac{3}{4}$ ft and length 7 in.?

Check your answers on page 723.

Answers	
1.	
2.	
3.	
4.	
5.	
6.	
7.	
8.	
9.	
10.	
11.	
12.	
13.	
14.	
15.	
16.	
17.	
18.	
19.	
20.	
21.	

Cumulative Review
CHAPTERS 1–3

Combine these numbers.

1. $-4 + 7 + 1 - 3 - 6 + 2 =$

1. _____

2. $(-16) - (-7) + (-5) - (14) + (12) - (-11) =$

2. _____

3. $-3 - 7 - 3 - 4 - 1 - 8 - 4 - 2 =$

3. _____

4. $(-4)(-1)(-1)(2)(-1)(1)(-3) =$

4. _____

5. $7 + 2(8 - 5 - 1)(1 - 2 + 3) =$

5. _____

6. $-13 - 4(-7 + 3 + 2)(-1 - 1 - 1) - (14 - 19 + 5) =$

6. _____

7. $(-24) \div (-4) =$

8. $\dfrac{4}{15} - \dfrac{2}{15} - \dfrac{7}{15} + \dfrac{1}{15} =$

7. _____

8. _____

9. $\left(-\dfrac{8}{15}\right)\left(-\dfrac{5}{12}\right) =$

10. $\dfrac{3}{10} - \dfrac{4}{5} + \dfrac{1}{6} =$

9. _____

11. $3\dfrac{2}{5} - 5\dfrac{1}{2} - 1\dfrac{2}{3} =$

12. $\left(-4\dfrac{2}{3}\right)\left(\dfrac{3}{4}\right) =$

10. _____

11. _____

13. $\dfrac{\left(-2\dfrac{3}{5}\right)}{\left(-1\dfrac{1}{3}\right)} =$

14. $\dfrac{3}{4} + \dfrac{1}{2}\left(\dfrac{1}{6} + \dfrac{5}{6} - \dfrac{3}{6}\right) =$

12. _____

13. _____

Round off to the nearest integer.

14. _____

15. $6\dfrac{7}{8}$

16. $\dfrac{11}{40}$

15. _____

Evaluate these formulas if $P = -4$, $M = -1$, and $K = -5$.

16. _____

17. $3P - 5M + 2K =$

17. _____

18. $(P + M) + 3K =$

18. _____

19. What is the perimeter of a triangle with sides $4\dfrac{1}{2}$ ft, $7\dfrac{1}{2}$ ft, and $5\dfrac{1}{2}$ ft?

19. _____

20. _____

20. What is the area of a triangle with base 10 in. and height $\dfrac{3}{4}$ ft?

Check your answers on page 723. ✓

4 Decimals and Percents

 Concept of the Decimal System

In the last chapter we learned how to represent parts of whole numbers, or integers, with fractions. Another way to represent parts of whole numbers is to use **decimals**. Any fractional part of a whole number can be expressed in either form.

Decimal notation is commonly used to represent money. Also the statistical data we read about in newspapers or magazines are expressed in percents, which are a type of decimal.

It is important to master computation in both fractional and decimal forms. In Section 4.7 we will learn to convert from one form to the other.

Place Value

In Section 1.1 we learned that we could represent all possible numbers by using only the ten digits: 0, 1, 2, 3, 4, 5, 6, 7, 8, and 9. The value of the number depends on the place where each digit appears; that is, the number is the product of the digit multiplied by the value of its place.

In whole numbers the **place values** (from right to left) are **ones**, **tens**, **hundreds**, **thousands**, and so on. Hence the number 3,548 represents 3 thousands, 5 hundreds, 4 tens, and 8 ones.

$$3,548$$

Thousands
Hundreds
Tens
Ones

If we look at the place values from left to right, we might observe that if we divide each place value by 10, the result will be the value of the place to its right.

1,000 divided by 10 is 100.
100 divided by 10 is 10.
10 divided by 10 is 1.

By continuing this pattern we can represent digits in place values that are smaller than one. To determine the value of the next place to the right, just divide the former place value by 10.

When we studied fractions, we found that we could represent the division of 1 by 10 by the fraction $\frac{1}{10}$. In Section 3.5 we learned that to divide by 10 is the same as to multiply by $\dfrac{1}{10}$, and in Section 3.2 we learned that the word "of"

represents multiplication. So if we calculate $\dfrac{1}{10}$ of each place value, we will

know the value of the next place to the right.

The place to the right of the *ones place* has the value

$$\frac{1}{10} \text{ of } 1 =$$

$$\frac{1}{10}(1) = \frac{1}{10}, \text{ or one tenth}.$$

The next place to the right of the *tenths place* has the value

$$\frac{1}{10} \text{ of } \frac{1}{10} =$$

$$\left(\frac{1}{10}\right)\left(\frac{1}{10}\right) = \frac{1}{100}, \text{ or one hundredth}.$$

The next place to the right of the *hundredths place* has the value

$$\frac{1}{10} \text{ of } \frac{1}{100} =$$

$$\frac{1}{10}\left(\frac{1}{100}\right) = \frac{1}{1,000}, \text{ or one thousandth}.$$

The next place to the right of the *thousandths place* has the value

$$\frac{1}{10} \text{ of } \frac{1}{1,000} =$$

$$\left(\frac{1}{10}\right)\left(\frac{1}{1,000}\right) = \frac{1}{10,000}, \text{ or one ten-thousandth—and so on}.$$

We use a dot called a **decimal point** to separate the ones place value from the tenths place value. The decimal point separates the integer of a mixed number from the decimal part of the number. So the number 0.296 represents

$$0.296.$$

We can use fractions to determine the value of the decimal 0.296.

The 2 is in the tenths place, so it has the value $2\left(\dfrac{1}{10}\right)$.

The 9 is in the hundredths place, so it has the value $9\left(\dfrac{1}{100}\right)$.

The 6 is in the thousandths place, so it has the value $6\left(\dfrac{1}{1,000}\right)$.

$$0.296 = 2\left(\frac{1}{10}\right) + 9\left(\frac{1}{100}\right) + 6\left(\frac{1}{1,000}\right)$$

$$0.296 = \frac{2}{10} + \frac{9}{100} + \frac{6}{1,000}$$

The LCM for the denominators of these fractions is 1,000. We must find new names for $\frac{2}{10}$ and $\frac{9}{100}$ with the denominator 1,000.

$$\frac{2}{10} = \qquad\qquad \frac{9}{100} =$$

$$\frac{2}{10} \times \frac{100}{100} = \frac{200}{1,000} \qquad \frac{9}{100} \times \frac{10}{10} = \frac{90}{1,000}$$

Hence

$$0.296 = \frac{200}{1,000} + \frac{90}{1,000} + \frac{6}{1,000},$$

$$0.296 = \frac{296}{1,000}.$$

So the decimal 0.296 is the same as the fraction $\frac{296}{1,000}$ (two hundred ninety six, thousandths).

Naming Decimals

Think of a decimal as having two names, just as a fraction does. The first name is the *number* you see to the *right of the decimal point*. Thus for 0.296 the first name is *two hundred ninety six*.

The second name indicates the *place value of the last digit on the right*. Thus for 0.296 the second name is *thousandths* (the place value of 6).

EXAMPLE 1   0.13 is read as thirteen *hundredths*.

0.013 is read as thirteen *thousandths*.

0.0013 is read as thirteen *ten-thousandths*.   ❑

Each decimal can be expressed by many different decimal representations, just as each fraction can be represented by many different fractional numbers. Adding zeros to the right of a decimal number gives it a new name but *does not change its value*.* The *most simplified* decimal form is one with no zeros to the right. (The exception is that, when writing a decimal to express money, we usually represent both the tenths and the hundredths digits, even if the hundredths digit is zero: for example, we write $3.50 instead of $3.5.)

EXAMPLE 2   0.4 is read as 4 *tenths*.

0.40 is read as 40 *hundredths*.

0.400 is read as 400 *thousandths*.   ❑

By representing each decimal as a fraction and reducing it, we will see that all three decimals represent the same number. The decimal's first name is the fraction's numerator; the decimal's second name is the fraction's denominator.

0.4 (4 tenths) is $\frac{4}{10} = \frac{2}{5}$.

0.40 (40 hundredths) is $\frac{40}{100} = \frac{2}{5}$.

0.400 (400 thousandths) is $\frac{400}{1,000} = \frac{2}{5}$.   ❑

---

* The zeros do affect the number's precision when it is used in measurement. Thus 0.50 inch indicates measurement with a more precise instrument than 0.5 inch represents.

We can also visualize this relationship in terms of money.

EXAMPLE 3   Ten pennies are ten hundredths of a dollar ($0.10).

One dime is one tenth of a dollar ($0.1).

But both represent the same amount, 10 cents. So 0.1 and 0.10 are decimals with different names but equal values.   ❏

As with fractions, we prefer to represent decimals by the simplest names. It is easier to simplify a decimal than to simplify a fraction. Just write the decimal without any zeros to the right.

EXAMPLE 4   Which is the simplest form:

0.015; 0.0150; 0.01500; 0.015000; 0.015000; or 0.0150000?

The simplest form of all those equal decimals is 0.015.   ❏

Mixed Numbers

A number with digits on both sides of the decimal point is a mixed number, just like those we studied in Section 3.8. If you read the decimal point as "**and**," and you do not use the word "and" in reading the rest of the number, a listener can easily tell when you have finished with the whole number and have started reading the decimal part.

EXAMPLE 5   3,518.296 is read as "three thousand five hundred eighteen *and* two hundred ninety six thousandths."   ❏

Thousands Hundreds Tens Ones Tenths Hundredths Thousands

EXAMPLE 6   43.0009 is read as "forty-three and nine ten-thousandths."   ❏

You will not have to memorize any rules for computing with decimals if you understand how to read the name of the decimal and know that the value of a digit depends on its place value.

If *no decimal point* is written in a number, it is understood to be to the right of the last digit.

EXAMPLE 7   13 = 13.0   ❏

When a decimal has no integer, we usually put a zero in the ones place. The zero helps us notice the decimal point; it is so small that without the zero we might overlook it. The dot really changes the value of a number. Wouldn't you rather get 15 dollars than 0.15 dollars? The zero helps us distinguish 15 cents ($0.15) from 15 dollars ($15).

EXERCISES 4.1

Put the first name of each decimal in the first blank, and the second name in the second blank.

EXAMPLE 8   0.314   three hundred fourteen _____   thousandths ___   ❏

1. 0.573   _____   _____

2. 0.03   _____   _____

3. 0.008   _____   _____

  4. 0.0037 _____ _____

  5. 0.0004 _____ _____

  6. 0.068 _____ _____

State the name of each number.

  7. 56.19 _____

  8. 23.014 _____

  9. 6.34 _____

  10. 15.7 _____

  11. 15.007 _____

  12. 5.52 _____

  13. 11.0052 _____

  14. 60.052 _____

Check your answers on page 723.

4.2 Rounding Off

Often numbers are stated more accurately than necessary. For example, we might want to round off figures to an estimate, or to what is often called a ball-park figure. Instead of saying that there are 32,981 seats in a stadium, we might prefer to state that there are approximately 33,000 seats. We know the exact number, but the approximate number is easier to say and to write and will do for most purposes. The general rule for rounding off whole numbers follows.

To round off whole numbers:

Decide to what place value you want to round off.

Put an arrow after the digit in that place value.

Look at the digit to the right of the arrow (it is the first digit you will drop, because it is next to the place value that you are rounding off to).

Round *down* to the number in front of the arrow if the first digit to drop is *less than* 5.

Round *up* to the number that is one larger than the number in front of the arrow if the first digit to drop is 5 *or more*.

Replace all dropped digits with zeros.

EXAMPLE 1   A stadium has a capacity of 32,981, and we want to round off that figure to the nearest *thousand*.
32,981 is between 32,000 and 33,000.
The digit in the thousands place is 2.
So we write **32↓981**.
We look at the digit after the thousands place, which is 9.
32,500 is halfway between 32,000 and 33,000.
9 > 5, so 32,981 > 32,500, or it is past the halfway point and is closer to 33,000.
We round up to 33,000.   ❏

EXAMPLE 2   We want to round off 802,734 to the nearest hundred.
802,734 is between 802,700 and 802,800.
So we write **802,7↓34**.
We look at the 3.
3 < 5, so 802,734 < 802,750 and is closer to 802,700.
We round down to 802,700.   ❏

EXAMPLE 3   Round off 4,687 to the nearest thousand.
The digit in the thousands place is 4.
The question is, do we round up to 5,000 or down to 4,000?
We write 4↓687.
The digit after the 4 is 6, and 6 > 5.
So we round up to 5,000.   ❏

We usually represent money with decimals. Often an answer to a problem involving money is more exact than is necessary.

EXAMPLE 4   If a store sells six oranges for 89 cents and you want to know how much it would cost to buy one orange, you would divide .89 by 6.

$$\begin{array}{r} .1483\ldots \\ 6\,)\overline{.8900} \end{array}$$

By the time you reach the fourth decimal place, you realize that the smallest unit in our money system is a penny, which is worth 0.01 of a dollar. So you only need to calculate the price of one orange to the closest cent (or hundredth of a dollar).

$0.14↓83 rounds up to $0.15.

You would expect to pay 15 cents for one orange.   ❏

In school, when we are expected to round off numbers we are given instructions about how precise the answer should be. Outside of school, when we round off numbers the rule is: Keep as much of the answer as is useful.

On Federal Income Tax forms, the government lets us round off to the nearest dollar.

A carpenter building a bookcase may find only measurements up to one-eighth of an inch relevant, because she cannot cut any more accurately than that with the saw.

> To round off a decimal:
> Drop the digits to the right of the place value that you rounded off to.

Do not replace them with zeros as we do when rounding off integers.

EXAMPLE 5   Round off 56.0837 to the nearest hundredths place.
Write **56.08**↓37.
3 < 5, so round 56.0837 down to 56.08.   ❏

EXAMPLE 6   Round off 56.0837 to the nearest thousandths place.
Write **56.083**↓7.
7 > 5, so round 56.0837 up to 56.084.   ❏

In everyday experience, when we round off we cannot always follow the procedure shown above. Markets make it a practice to charge any fraction of a cent, no matter how small, as a whole cent. For example, they would round off $3.680000003 to $3.69.

As was indicated in Section 3.9, you must always use common sense when rounding off. If your gas tank holds 14.8 gallons of gas, you cannot round that figure up to 15 gallons. If you did, the excess would spill over.

If the government has budgeted funds to supply food stamps to 15.75 million families, that number cannot be rounded up to 16 million; there would not be enough money to pay for the extra quarter-million.

If we calculate that we need 5.2 buses to transport everyone to the picnic, we had better not round that figure down to 5 buses or we will not have enough vehicles to get everyone there.

So use the rule sensibly. Be sure that your answer makes sense for the problem you are working on.

EXAMPLE 7   We are planning a leaflet campaign for our neighborhood and have been told by the census office that 2,387 families are living here.

The printer insists on orders in terms of thousands.

How many leaflets should we order?

Round 2,387 to the nearest thousand.

2↓387 is closer to 2,000 than to 3,000.

But if we want every family to get a leaflet, we had better order 3,000.   ❏

EXERCISES 4.2

Round off each number.
Round to the nearest tenth.

1. 5.17	2. 14.835	3. 20.00743	4. 0.8605
_____	_____	_____	_____

Round to the nearest hundredth.

5. 5.3187	6. 638.927	7. 0.00397	8. 70.07644
_____	_____	_____	_____

Round to the nearest thousandth.

9. 5.3187	10. 638.9271	11. 0.00397	12. 53,280.06581
_____	_____	_____	_____

Round to the nearest integer.

13. 15.81	14. 27.028	15. 150.073	16. 67.09
_____	_____	_____	_____

Round to the nearest hundred.

17. 34,680   **18.** 5,807   **19.** 26,383   **20.** 45,832,081

_____   _____   _____   _____

21. Round off each number to the nearest dollar: $5.67, $23.08, $7.80, $13.05, $25.37, and $1.80.

22. For each of these numbers, 3.003, 23.764, 408.527, 34.186, 7.093, and 186.919,
a) round off to the nearest hundredth;
b) round off to the nearest tenth.

23. There are 8,368 students enrolled in the school.
a) We want to print an announcement encouraging the students to participate in a rally protesting tuition increases. If we must order by the thousand, how many announcements should we order? Be sure there are enough for each student. (Round up to the nearest thousand.)
b) If we estimate that 1 out of every 100 students will require a tetanus shot this year, how many doses of vaccine should be ordered? (Round up to the nearest hundred.)
c) If 1 out of every 10 students will need a tutoring session this week, how many sessions will be needed? (Round up to the nearest ten.)

24. The capacity of the sports arena is 63,182.
a) Tickets are printed by the thousand. How many tickets must be ordered?
b) The number of security guards hired depends on how many hundreds of fans the management expects to attend. When an event is sold out, what attendance number is used? (Round up to the nearest hundred.)
c) The concession manager estimates how much of each item she should order based on how many customers she anticipates. She uses numbers rounded off to the nearest ten. When the arena is sold out, what number of customers should she use? (Round off to the nearest ten.)

Check your answers on page 723.

 **Addition and Subtraction of Decimals**

When we add whole numbers, we line them up so as to add digits with the same place value—we add ones to ones, tens to tens, and so on. The same is true for decimals. They must be lined up such that

tenths are added to tenths,

hundredths are added to hundredths,

thousandths are added to thousandths, and so on.

With whole numbers, we write the numbers such that the digits on the right line up according to place value. That arrangement works for whole numbers because the digit on the right is in the ones place in all whole numbers. But the place value farthest to the right in two decimal numbers *may not* be the same place value in each number.

EXAMPLE 1 In 6.83 the rightmost place value is *hundredths*.
In 27.5 the rightmost place value is *tenths*.
In 3.856 the rightmost place value is *thousandths*. ❏

We cannot add these numbers by lining up the digits on the right (the 5 below the 3, or the 6 below the 3 or 5); we would be adding numbers representing different place values.

But there is an easy way to be sure of adding numbers from the same place value:

Line the numbers up such that the decimal point in one number is above the decimal point in the other number.

To avoid careless errors, you can add zeros at the end of the decimals so that they all have the same number of digits after the decimal point. Most errors are caused by not writing the numbers above each other and then carelessly adding digits from the wrong place values because the numbers were written horizontally in the problem.

EXAMPLE 2 Add 6.83 + 27.5 + 3.856.

Rewrite the numbers as

$$
\begin{array}{c c}
6.83 & 6.830 \\
27.5 & \rightarrow\ 27.500 \\
\underline{3.856} & \underline{3.856} \\
& 38.186
\end{array}
$$
❏

We learned in Section 2.2 that when we add numbers with different signs, we subtract the smaller magnitude from the larger one and keep the sign of the number with the larger magnitude.

To simplify the computation with decimals, write the larger magnitude first. Then write the smaller magnitude under it so that the decimal points line up. It is very easy to make a careless error when subtracting, especially if you must borrow and no digit is written in the place value. To minimize mistakes, add zeros in the top figure so that both figures have the same number of digits after the decimal point.

EXAMPLE 3 Add −6 + 4.032.
First write the + 4.032 beneath the −6, and remember that −6 is " − 6."—with a decimal point.

To subtract we must borrow.

$$
\begin{array}{c c}
-6. & \rightarrow\ -6.000 \\
\underline{+4.032} & \underline{+4.032}
\end{array}
$$

The problem becomes

$$
\begin{array}{r}
9\ 9 \\
-{}^{5}6.{}^{1}0{}^{1}0{}^{1}0 \\
\underline{+\ 4.\ 0\ 3\ 2} \\
-\ 1.\ 9\ 6\ 8
\end{array}
$$
❏

We can visualize what borrowing means in a subtraction problem if the numbers represent money.

EXAMPLE 4   Suppose I have $52.70 in my wallet; that is, 5 ten-dollar bills, 2 one-dollar bills, and 7 dimes. If I buy some snacks worth $4.32 from a vendor who has no change, what can I do?

I need more pennies and more one-dollar bills than I now have. I'd go get change somewhere; I'd exchange 1 ten-dollar bill for 10 one-dollar bills, and 1 dime for 10 pennies. Then I could pay for the snacks exactly. I would still have $52.70, but now that would be

Subtract

$$\$52.70 \rightarrow \quad 4 \text{ tens,} \quad 12 \text{ ones,} \quad 6 \text{ dimes, and} \quad 10 \text{ pennies.}$$
$$\underline{4.32 \rightarrow \qquad\qquad -4 \text{ ones,} -3 \text{ dimes, and} \; -2 \text{ pennies.}}$$
$$4 \text{ tens,} \quad 8 \text{ ones,} \quad 3 \text{ dimes, and} \quad 8 \text{ pennies} = \$48.38.$$

Exchanging one bill or coin for another is what we do with place values whenever we borrow:

$$
\begin{array}{c}
5\,2\,.\,7\,0 \rightarrow \\
-\quad 4\,.\,3\,2 \\
\hline
\end{array}
\qquad
\begin{array}{c}
^{4}\!5^{12}2.^{6}\!7^{1}0 \\
-\quad 4\,.\,3\,2 \\
\hline
4\,8\,.\,3\,8 \quad \checkmark
\end{array}
$$

If there are more than two numbers to add, and they have different signs, combine the positives first and then the negatives. Then subtract the results, as we did in Section 2.2 with integers.

EXAMPLE 5   Add $3.2 - 5.04 - 7.6 + 8.211 - 5.003$.

Positives:	$+3.2$	Negatives:	-5.04
	$+8.211$		-7.6
	$\downarrow$		-5.003
			$\downarrow$

$$
\begin{array}{r}
+3.200 \\
+8.211 \\
\hline
+11.411
\end{array}
\qquad
\begin{array}{r}
-5.040 \\
-7.600 \\
-5.003 \\
\hline
-17.643
\end{array}
$$

$$
\begin{array}{r}
+\,11.411 - 17.643 = -17.643 \\
+11.411 \\
\hline
-\quad 6.232
\end{array}
$$

So $3.2 - 5.04 - 7.6 + 8.211 - 5.003 = -6.232$.

Be careful when adding decimals.

EXERCISES 4.3

Rewrite each problem so that one number is above the other; line up the decimal points; put in zeros to match the place values; then find the answers.

1. $6.25 + 205.9 =$ **2.** $365.8 + 1.202 =$

3. $14.87 - 7.3 =$ **4.** $36.08 - 59.7 =$

5. $-13.4 + 8.19 =$ **6.** $26.3 - 9.53 =$

7. $16.08 - 4.89 =$ 8. $67.3 - 46.72 =$

9. $-3.4 - 15 - 6.03 =$ 10. $-8.62 - 15.4 - 0.034 =$

11. $5.8 + 3.002 - 9.1 + 12 =$ 12. $7 + 12.3 + 5.008 + 16.78 =$

13. $8.3 - 5.02 - 6.3 - 0.02 + 2.78 + 13.09 =$

14. $4 - 6.04 + 7.41 + 4.5 - 2.95 - 16 + 6.34 =$

15. $-6.78 - 13.4 - 8.09 + 3.46 - 1.8 + 10.7 =$

16. $-7.01 - 18.4 - 9.2 - 14 - 0.399 - 4.37 =$

17. $45.03 + 15.9 + 6.008 + 12.3 + 32 + 18.068 =$

18. $25.6 - 83.07 - 94.005 - 0.03 + 41.04 + 7.002 - 0.03 + 44.2 =$

19. $8.006 - 3.45 + 15.8 - 1.027 - 34.5 + 18.96 + 0.071 =$

20. $0.0043 + 58.365 - 12.03 - 6.0911 + 4.67 - 37.009 + 0.041 =$

21. Find the sum of these figures, and round it off to the nearest dollar: $5.67, $23.08, $7.80, $13.05, $25, $1.80.

22. a) Find the sum of these numbers: 3.003, 23.7, 408.52, 34.18, 7.09, and 186.
    b) Round off the sum to the nearest hundredth.
    c) Round off the sum to the nearest tenth.

Check your answers on page 723.

Multiplication of Decimals

Determining Where to Put the Decimal Point in the Product

To multiply decimals, it is *not* necessary to line them up with the decimal points above each other, as in addition. The numbers can first be multiplied without concern for the decimal point. Then, after you have obtained the digits for the product, use the following guide.

> To determine where the decimal point belongs in the product:
>
> Count the total number of digits to the *right* of the decimal points in the numbers you are multiplying.
>
> Be sure that the answer has that total number of digits after its decimal point.

EXAMPLE 1 Find the product of 5.73 × 1.2.

$$\begin{array}{r} 5.73 \\ \times\ 1.2 \\ \hline 1146 \\ 573 \\ \hline 6876 \end{array}$$ (before putting in the decimal point)

5.73 has two digits to the right of the decimal point.

1.2 has one digit to the right of the decimal point.

So there should be a total of three (2 + 1) digits to the right of the decimal point in their product. Put the decimal point between the 6 and the 8, which will leave three digits to the right of the decimal point in their product.

$$5.73 \times 1.2 = 6.876$$ ❑

That mechanical rule works fine, but you should always check that your answer makes sense. It is easy to make a careless error. As a check, round off the mixed decimals to the nearest integer. Be sure that your answer is *close to* that product.

EXAMPLE 2 We found the product of 5.73 × 1.2 to be 6.876.

5.73 is almost 6.   1.2 is a little more than 1.

Their product should be close to 6 × 1 = 6.

The digits in the product of 5.73 and 1.2 are 6876 (before putting in the decimal point).

Of some possible places to put the decimal point, such as

0.6876, 6.876, 68.76, 687.6, or 6876,

6.876 is the one closest to 6. ❑

If any of the numbers you are multiplying are negative, be sure that you use the rules of signed numbers from Section 2.4 to determine the correct sign.

Since we know that the product of any number with zero is always zero, we can use a shortcut when multiplying decimals with zeros in the *leftmost places*. We do not bother writing down those lists of zeros as we multiply. But when we place the decimal point in our answer, we put in as many zeros as we need to the *left* of the digits.

EXAMPLE 3 Find the product of (−0.17) × (0.003).

The sign of the product will be *negative* because these two numbers have different signs.

$$\begin{array}{r} -0.17 \\ \times\ 0.003 \\ \hline -51 \end{array}$$ (before putting in the decimal point)

The answer should have five digits to the right of the decimal point (−0.17 has two, and 0.003 has three). So we put in three zeros and get the answer − .00051. If we put a zero *in front* of the decimal point for emphasis, we would write the answer as −0.00051. ❑

EXERCISES 4.4A

Multiply the numbers. Determine the sign of their product, and be sure to put the decimal point in the correct position.

1. 5.41 × 6.3 =   **2.** 1.5 × 3.4 =

3. $-8.3 \times 0.61 =$

4. $-9.02 \times -0.81 =$

5. $-0.14 \times -0.005 =$

6. $10.007 \times 2.5 =$

7. $62.08 \times -0.14 =$

8. $-5.4 \times -0.003 =$

9. $13.005 \times -0.002 =$

10. $-0.0004 \times 0.023 =$

11. $-6.001 \times -0.0032 =$

12. $-14.002 \times 0.037 =$

13. $103.05 \times 0.0021 =$

14. $-200.31 \times -0.0014 =$

Check your answers on page 723.

Advantages of Multiplying by 10 in the Decimal System

Did you ever notice what happens when you multiply any number by 10?

EXAMPLE 4   Find the product of 58.73×10.

To determine the digits in the answer, write the digits without any decimal points first.

$$\begin{array}{r} 58.73 \\ \times \quad 10 \\ \hline 0000 \\ 5873 \quad\;\; \\ \hline 58730 \end{array}$$ (before putting in decimal point)

There should be two digits to the right of the decimal point in the product of 58.73×10.

So $58.73 \times 10 = 587.30$. ❏

If you multiply a *decimal* or *decimal mixed number* by 10, it has the effect of moving the decimal point one place to the right in the number. So

$$58.73 \times 10 =$$
$$58.73 \to 587.3.$$

If you multiply an *integer* by 10, it has the effect of adding a zero to the right of the number.

EXAMPLE 5   Find the product of 231×10.

$$\begin{array}{r} 231 \\ \times \quad 10 \\ \hline 000 \\ 231 \quad\;\; \\ \hline 2310 \end{array}$$ So $231 \times 10 = 2,310$. ❏

Since every integer has an understood decimal point to the right of its last digit, we could write 231 as 231.0. When we multiplied 231 by 10 we got 2310; so we could say that (as with decimals and mixed decimals) the decimal point moved one place to the right.

EXAMPLE 6 231×10 is the same as $231.0 \times 10 = 2,310$. ❏

With that observation, we can simply say:

Whenever a number is multiplied by 10, the decimal point is moved one place to the right.

A notation called **exponents** is used when we multiply by the same number more than once. The exponent represents the *number of times* we wish to multiply. The exponent (which is expressed as a small number) is written above and to the right of the multiplying number. For instance, we could express that we want to multiply a number by 10 twice as: 10^2.

EXAMPLE 7   To multiply a number by 10 three times, we write 10^3. The effect is the decimal point in that number moves 3 places to the right.

To multiply a number by 10 five times, we write 10^5. The effect is the decimal point in that number moves 5 places to the right.

To multiply a number by 10 nine times, we write 10^9. The effect is the decimal point in that number moves 9 places to the right.   ❑

EXAMPLE 8   Multiply 18,392.076 by 10 twice.

$$18{,}392.076 \times 10^2 =$$
$$18.392.076 = 1{,}839{,}207.6$$ ❑

EXAMPLE 9   Multiply 436.004378 by 10 four times.

$$436.004378 \times 10^4 =$$
$$436.004378 = 4{,}360{,}043.78$$ ❑

EXERCISES 4.4B

Find the products.

15. Multiply 376 by 10. | **16.** Multiply 67.03 by 10.

17. Multiply 5.0043 by 10. | **18.** Multiply 0.0084 by 10.

19. Multiply 47 by 10 twice. | **20.** Multiply 164 by 10 twice.

21. Multiply 3.076 by 10 twice.

22. Multiply 0.0718 by 10 three times.

23. Multiply 28,062 by 10 three times.

24. Multiply 462.09 by 10 three times.

25. Multiply 3,807.25 by 10 four times.

26. $340.076 \times 10^2 =$ | **27.** $5.0083 \times 10^2 =$

28. $23.5478 \times 10^3 =$ | **29.** $0.00647 \times 10^3 =$

30. $0.000384 \times 10^4 =$ | **31.** $0.000007012 \times 10^4 =$

32. $0.0067015 \times 10^4 =$ | **33.** $54.00238 \times 10^5 =$

Check your answers on page 723.

 **Division of Decimals**

Determining Where to Put the Decimal Point in the Quotient

If we use long division to divide a decimal or a decimal mixed number by an *integer*, the decimal point goes directly above the place where it appears in the dividend.

EXAMPLE 1 Find the quotient of $32.89 \div 23$.

$$
\begin{array}{r}
1.43 \\
23 \overline{)\ 32.89} \\
-23 \\
\hline
9\,8 \\
-\ 9\,2 \\
\hline
69 \\
-\ 69 \\
\hline
0
\end{array}
$$

So $32.89 \div 23 = 1.43$.

To make life easy for ourselves, we should always adjust any division-of-decimals problem so that the divisor is an integer. Then we will always know where to put the decimal point in the answer.

Suppose the divisor is a decimal. We can make any decimal into a whole number by moving the decimal point until it is to the right of the number's last digit. As we saw in Section 4.4, each time we move the decimal point one place to the right, we are multiplying by 10.

We cannot move the decimal point only in the divisor, for that would change our problem. But we can move the decimal point in the divisor if we also move it the same number of places in the dividend.

The reason we can do so is that we could represent the division problem as a fraction. The dividend would become the numerator, and the divisor would become the denominator. Since every move of the decimal point is the same as multiplication by 10, when we move the decimal point in both numbers it is as though we were multiplying the numerator and denominator of the fraction by 10 each time. What we are doing, in fact, is multiplying the fraction by 10/10, or 100/100, and so on, which is the same as multiplying by 1. Multiplying by 1 does not change the value of a number. We learned that this is a valid manipulation of fractions in Section 3.6.

EXAMPLE 2 Without changing the value of the problem, convert $34.68 \div 3.4$ so that the divisor (3.4) becomes an integer.

Let's represent that division problem as a fraction: $34.68 \div 3.4 = \dfrac{34.68}{3.4}$.

$$
\frac{34.68}{3.4} \times \frac{10}{10} = \frac{346.8}{34}
$$

Now you can see that the following rule is a legitimate procedure. To set up a division problem so that the divisor is an integer:

Move the decimal point until it is to the right of all digits in the divisor.

Then move the decimal point the *same number of places in the dividend*.

EXAMPLE 3 Find the quotient of $34.68 \div 3.4$.

$$3.4\overline{)34.68} \quad \text{becomes} \quad 34.\overline{)346.8}$$

$$
\begin{array}{r}
10.2 \\
34 \overline{)\,346.8} \\
-34 \\
\hline
06 \\
-\,00 \\
\hline
6\,8 \\
-\ \ 6\,8 \\
\hline
0
\end{array}
$$

So $34.68 \div 3.4 = 10.2$. ❑

You can always check division problems for careless errors. Multiplying the quotient by the divisor should produce the dividend. Use the problem's *original divisor* and expect to produce the problem's *original dividend*. That way, if you made a mistake anywhere, even in moving the decimal point, you will find it.

EXAMPLE 4 Check that $34.68 \div 3.4 = 10.2$.

$$
\begin{array}{r}
1\,0.2 \\
\times\ \ \ 3.4 \\
\hline
4\,0\,8 \\
3\,0\,6 \\
\hline
3\,4.6\,8 \quad \checkmark
\end{array}
$$

❑

In Section 3.8 we saw that when a divisor does not divide evenly into the dividend, we can express the remainder as a fraction. We can also express it as a decimal.

If the dividend is an integer, insert a decimal point after the ones place. Represent each decimal place with a zero, and keep dividing until there is no remainder.

When you divide with enough place values, eventually there will be no remainder or else a repeating pattern of digits will appear in the quotient. In Section 4.7 we will look at examples in which the pattern repeats when we convert from fractions to decimals.

If the dividend is already a decimal or a mixed decimal, just continue to put zeros into each decimal place and divide until there is no remainder. Remember, you can always add zeros to the end of a decimal without changing its value: $1 = 1.0 = 1.00000000$. So you do *not* need to add any zeros to the divisor to match the ones you are adding to the dividend.

In Section 3.8, Example 10, we calculated $162,451 \div 8$ to be $20,306\frac{3}{8}$. Let's calculate that quotient again, but this time we will represent the remainder as a decimal.

EXAMPLE 5   Find the quotient of $162,451 \div 8$.

$$
\begin{array}{r}
20\,306.375 \\
8\overline{)162,451.000} \\
16 \\
\hline
02 \\
00 \\
\hline
2\,4 \\
2\,4 \\
\hline
05 \\
00 \\
\hline
51 \\
48 \\
\hline
3\,0 \\
2\,4 \\
\hline
60 \\
56 \\
\hline
40 \\
40 \\
\hline
0
\end{array}
$$

So, when computed as a decimal, $162,451 \div 8 = 20,306.375$.

Don't forget that when a division problem involves at least one negative number, you must use the rules for signed numbers that we learned in Section 3.5 to determine the sign of the quotient.

Once the decimal point appears in the quotient, there must be a digit in each place value of the quotient. If the divisor does not divide into the first digit(s) of the dividend, you must put a zero in that place(s).

EXAMPLE 6   Find the quotient of $-0.00312 \div 0.15$.

I'll write the divisor, 0.15, without the zero (.15), since I will be making it into an integer first anyway. The quotient will be negative, because the signs are different, so I will begin by writing the minus sign into the quotient, and then I will not have to think about the sign during the computation.

$$
.15\overline{)-0.00\,312} \quad \rightarrow \quad 15\overline{)-0.312}
$$

I will add zeros after the 2 if there is a remainder.

$$
\begin{array}{r}
-.0208 \\
15\overline{)-0.3120} \\
30 \\
\hline
12 \\
00 \\
\hline
120 \\
120 \\
\hline
0
\end{array}
$$

So $-0.00312 \div 0.15 = -0.0208$.

Check: Does $(-0.0208) \times (0.15) = -0.00312$?

$$
\begin{array}{r}
-0.0208 \\
\times\ 0.15 \\
\hline
1040 \\
208 \\
\hline
-3120
\end{array}
$$
  (before putting in the decimal point)

There should be six digits after the decimal point.
So $(-0.0208) \times (0.15) = -0.003120$ or -0.00312. ✓

EXERCISES 4.5A

Divide the numbers to find their quotient. Check your answer by multiplying the quotient by the *original* divisor. Be careful about the signs. Represent any remainders as *decimals*, as we did in Examples 5 and 6.

1. $3.71 \div 0.7 =$

2. $0.804 \div 0.6 =$

3. $292.1 \div 2.3 =$

4. $24.448 \div 0.08 =$

5. $-24{,}144 \div 1.2 =$

6. $0.0002376 \div (-0.18) =$

7. $-0.0000049 \div (-0.007) =$

8. $-0.000615 \div (-0.03) =$

9. $0.0332 \div 0.08 =$

10. $-0.0052 \div 0.16 =$

Check your answers on page 723. ✓

Advantages of Dividing by 10 in the Decimal System

It is just as easy to *divide* by 10 as to multiply by 10 in the decimal system. Let's try it and see what happens.

EXAMPLE 7   Divide 24.078 by 10. I will let you work out the long-division steps yourself.

$$
\begin{array}{r}
2.4078 \\
10\overline{)24.0780}
\end{array}
$$

So $24.078 \div 10 = 2.4078$. ❑

Try as many examples as you wish. You will see that every time you divide a number by 10, the decimal point moves one place to the left.

You might have expected that result, because when we multiply by 10, the decimal point moves one place to the right, and we know that multiplication and division are inverse operations.

There is no need to memorize which way to move the decimal point. Just think about it. If you divide by 10, the number should get smaller (which is the result of moving the decimal point to the left); if you multiply by 10, the number should get larger (which is the result of moving the decimal point to the right).

EXAMPLE 8   34.8 divided by 10 is $3.4.8 = 3.48$.

126.85 divided by 10 twice is $1.26.85 = 1.2685$.

4.63 divided by 10 three times is $.004.63 = 0.00463$.

73.05 multiplied by 10 three times is $73.050. = 73{,}050$. ❑

Exponents can also be used to represent repeated division by the *same* number. **Negative exponents** represent division. If we write a number *times* **10 raised to a negative exponent**, we mean **divide by ten** that many times.

EXAMPLE 9 10^{-2} means divide by 10 twice.

$$4{,}538.06 \times 10^{-2} =$$
$$4{,}538.06 = 45.3806$$ ❑

EXAMPLE 10   We can represent 507,183 divided by 10 four times as

$$507{,}183 \times 10^{-4} =$$
$$507{,}183. = 50.7183.$$ ❑

Use the observations made in this section to easily multiply and divide by 10.

11. $24.58 \times 10 =$

12. 304.92 divided by $10 =$

13. $7{,}890.65 \times 10^{2} =$

14. $5{,}340.17 \times 10^{-2} =$

15. $5.000092 \times 10^{4} =$

16. $0.0073 \times 10^{6} =$

17. $43{,}070 \times 10^{-3} =$

18. $65.008 \times 10^{-2} =$

19. $5{,}083.75 \times 10^{-3} =$

20. $0.0052 \times 10^{-4} =$

Check your answers on page 723.   ✓

4.6 Order of Operations and Evaluation of Formulas with Decimals

Order of Operations

Now that you know how to add, subtract, multiply, and divide with decimals, you can compute problems involving more than one of those operations.

Remember, when there is more than one operation the correct order for computing is:

1. Simplify all computations within parentheses.
2. Perform all multiplication and division operations as they appear from left to right in the problem.
3. Add together the remaining numbers.

EXAMPLE 1   Simplify $5(3.2 + 1.6 + 0.03) + 3.2(1.7)$.

$$5(3.2 + 1.6 + 0.03) + 3.2(1.7) =$$
$$5(\ \ ) + 3.2(1.7) =$$
$$5(4.83) + 3.2(1.7) =$$
$$24.15 + 5.44 = 29.59$$ ❑

When you are adding more than two numbers that do not have the *same sign*, I recommend that you combine the positive numbers together and the negative numbers together. Then combine the two results.

EXAMPLE 2 Simplify $4.3 + 2(-3.5 - 2.8 - 1.07) - 3(4.5 - 7.8)$.

$$4.3 + 2(-3.5 - 2.8 - 1.07) - 3(4.5 - 7.8) =$$
$$4.3 + 2(\quad) - 3(\quad) =$$
$$4.3 + 2(-7.37) - 3(-3.3) =$$
$$4.3 - 14.74 + 9.9 =$$
$$+ 14.2 - 14.74 = -0.54$$

EXAMPLE 3 Simplify $-6.0 \div 1.2 + 4.3 - 0.56$.

$$-6.0 \div 1.2 + 4.3 - 0.56 =$$
$$-5 + 4.3 - 0.56 =$$
$$-5.56 + 4.3 = -1.26$$

EXERCISES 4.6A

Simplify each problem according to the order of operations.

1. $4.3(1.4 + 0.06 + 0.2) + 7.1(0.06 + 0.3) =$

2. $6 - 2.3(1.5) + 8.03 =$

3. $25.2 \div 0.04 - 40.2(-3.2 - 0.56) =$

4. $4.3(0.281 - 0.06 - 3.4) - 2.6 - 0.03(-5.4) =$

5. $6.28 - 0.47 + 3.1(-0.06 - 0.3 - 4.21) =$

6. $(-6.8) \div (-0.02) + 3.571 - 0.68 - 5.2 =$

7. $3.02(5.1 - 0.07) - 0.3(-9.03 - 3.4) =$

8. $7.68 - 0.5(4.05 - 7.2) - 9.1 - 0.006 =$

9. $12 - 4(3.02 - 4.3 - 0.09) - 2.1(-0.03) =$

10. $6.2 + 3(-0.4) - 0.5(-3.04 + 4.8) =$

Check your answers on page 723. ✓

Evaluating Formulas with Decimal Data

You can evaluate formulas even when the data are expressed in decimals. Just replace each letter, every time it appears, with its number value written within parentheses. Follow the order of operations in your computation.

EXAMPLE 4 Evaluate $3a - 5a + 7$ if $a = 2.4$.

$$3a - 5a + 7 =$$
$$3(2.4) - 5(2.4) + 7 =$$
$$7.2 - 12 + 7 =$$
$$+ 14.2 - 12 = 2.2$$

EXAMPLE 5   Evaluate $3[a + b + c] - 2[3b]$ if $a = 1.4$, $b = 0.5$, and $c = -2.6$.

$$3[a + b + c] - 2[3b] =$$
$$3[(1.4) + (0.5) + (-2.6)] - 2[3(0.5)] =$$
$$3[1.9 - 2.6] - 2[1.5] =$$
$$3[-0.7] - 2[1.5] =$$
$$-2.1 - 3.0 = -5.1$$ ❑

EXERCISES 4.6B

Evaluate each formula if $a = 3.5$, $b = 0.02$, and $c = -0.2$.

11. $a + b + c =$ **12.** $2a + 3b + 5c =$

13. $2a + 2b + 2c =$ **14.** $-2a - 2b - 2c =$

15. $4a - 3b + 7c =$ **16.** $3a - 5b + 3c =$

17. $2(a + b) + c =$ **18.** $4a + 7(b + c) =$

19. $2(a + b + c) =$ **20.** $2(a + b + c) - 3a - 2c =$

Check your answers on page 723. ✓

4.7 Percents and Conversion to Decimals and Fractions

Percents

It seems that when numbers come up in conversation, they are usually mentioned in terms of percent. A politician tells us his programs caused the economy to grow by 6 percent; a doctor says that she believes 40 percent of cancers are caused at work; the weather forecaster says that there is a 60 percent chance of rain; I complain that 90 percent of the time our papergirl misses our front porch with the morning newspaper; sportcasters say that 74 percent of a tennis player's first serves are good.

I saw a cartoon strip in which one man, complaining about his wages, said, "Thirty-five percent goes for the house, 50 percent for food, 10 percent for clothes, and 45 percent for this and that." The other responded, "But that's 140 percent!"

What does **percent** mean? The name gives us a good idea. A penny is called a cent, and it represents *one-hundredth* of a dollar.

"Percent" means per hundred.

When I say that 80 percent of the picnics I schedule are ruined by rain, I mean that if I scheduled 100 picnics, I would expect rain during 80 of them. I may not really know that this statement is accurate, but that is what the words mean. When the man in the cartoon strip calculated that he spent 140 percent of his wages, he meant that he spent $140 for every $100 he earned. You could not do that without getting into debt, but much of the economy does run on debt.

Percents are represented by the symbol (%).

EXAMPLE 1   15 percent is written as 15% and means 15 per hundred.   ❑

EXAMPLE 2   If the bank pays us 4% interest on our savings account, that means they will pay us 4 cents for every dollar that we keep in our account. ❑

EXAMPLE 3   An advertisement claiming that a drink contains 38% pure fruit juice means that 38 drops of pure fruit juice went into making every 100 drops of the drink. ❑

EXAMPLE 4   When economists tell us that the inflation rate has been 5% over the past year, they mean that an item that cost $100 last year now costs $5 more. ❑

EXAMPLE 5   When the union reports that our wages have lost 10% of their buying power over the last two years, it means that $90 today will purchase what $100 did two years ago. ❑

Fortunately, to understand percents and to compute with them, you do not have to learn anything new. Every percent can be represented by a fraction or a decimal, and you already know how to compute with them!

So all you have to do is learn to convert percents into fractions and decimals. Then I will show you how to convert between fractions and decimals, and you can use whichever form you prefer for any calculation involving *parts of integers*.

Converting Percents to Decimals

The word "per" represents *division*. If a bakery sells cookies for $1.50 *per dozen*, we must *divide* $1.50 by 12 to find the price of each cookie. As "percent" means *per hundred*, the percent symbol represents *division by a hundred*.

Since $100 = (10)(10)$, to divide by 100 is the same as to divide by 10 twice. As we saw in Section 4.5, the decimal point moves *two places to the left* when we divide by 10 twice.

EXAMPLE 6   15% represents 15 *divided by* 100.

$$15 \div 100 =$$
$$15. \div 100 = 0.15$$ ❑

To represent any percent as a decimal:

Remove the percent symbol (%). Move the decimal point two places to the left.

Remember, if no decimal point is written, it is assumed to occur after the last digit.

EXAMPLE 7   Convert 7% into a decimal.
7% represents 7 *divided by* 100.

$$7 \div 100 =$$
$$07. \div 100 = 0.07$$ ❑

EXAMPLE 8   Convert 86% into a decimal.

$$86\% =$$
$$86. \div 100 = 0.86$$ ❑

EXAMPLE 9   Convert 47.8% into a decimal.

$$47.8\% =$$
$$47.8 \div 100 = 0.478$$ ❑

Banks often list their interest rates in percents like 6.3%. What does that number represent?

EXAMPLE 10   Convert 6.3% into a decimal.

$$6.3\% =$$
$$06.3 \div 100 = 0.063$$

So 6.3% is the same as 0.063. That means the bank will pay us 6.3 cents for every $1.00 that we keep in the account. We do not have coins representing tenths of a cent, so a better way to express that is to say the bank will pay us $6.30 for every $100 that we keep in the account.

EXAMPLE 11   100% = 100.% → 1.00.   So 100% is the same as 1.

Percents greater than 100 are larger than 1 and become mixed decimals.

EXAMPLE 12   Translate 240% into a decimal.

$$240\% =$$
$$240. \div 100 = 2.40$$

EXAMPLE 13   Translate 0.3% into a decimal.

$$0.3\% =$$
$$00.3 \div 100 = 0.003$$

EXAMPLE 14   Note the difference between 7% and 70%.

$$70\% = 70.\% \to 0.70$$
$$7\% = 07.\% \to 0.07$$

EXERCISES 4.7A

Translate each percent into a decimal.

1. 4% =	**2.** 17% =
3. 41% =	**4.** 15% =
5. 30% =	**6.** 45% =
7. 3.5% =	**8.** 12.6% =
9. 0.2% =	**10.** 0.05% =
11. 125% =	**12.** 340% =
13. 0.0037% =	**14.** 0.00014% =
15. 0.008% =	**16.** 6% =
17. 60% =	**18.** 600% =

Check your answers on page 723.

Converting Percents into Fractions

We can convert percents into fractions by representing division by 100 as a *fraction with a denominator of* 100.

EXAMPLE 15 Represent 7% as a fraction.

$$7\% = \frac{7}{100}$$

As with any fraction, always reduce your answer to its simplest form.

EXAMPLE 16 Represent 15% as a fraction.

$$15\% = \frac{15}{100}$$

$\frac{15}{100}$ can be reduced:

$$\frac{15}{100} =$$

$$\frac{\cancel{5} \cdot 3}{\cancel{5} \cdot 20} = \frac{3}{20}$$

So $15\% = \frac{3}{20}$.

We often see percents appearing with decimal points in the number. But a fraction with decimals is too complicated. Such a fraction must be simplifed so that we can easily recognize the amount it represents.

As we saw in Section 4.4, moving the decimal point one place to the right is the same as multiplying by 10. As long as we multiply both the numerator and the denominator of a fraction by 10 the same number of times, the value of the fraction remains the same.

We will simplify fractions that contain decimal points by using those observations. Move the decimal point to the right of all digits, and move it the *same number of places in both the numerator and the denominator.*

Write the decimal point to the right of 100 (in the denominator). Then add enough zeros to the right of the decimal point so that *both the numerator and the denominator have the same number of digits after their decimal point before you move it.*

EXAMPLE 17 Represent 6.3% as a fraction.

$$6.3\% = \frac{6.3}{100}$$

Since 6.3 has one decimal place, we want to move the decimal point *one* place to the right. *Add one zero after the understood decimal point in* 100.

$$\frac{6.3}{100} =$$

$$\frac{6.3}{100.0} = \frac{63}{1,000}$$

EXAMPLE 18 Represent 0.03% as a fraction.

$$0.03\% =$$

$$\frac{0.03}{100} =$$

$$\frac{0.03}{100.00} = \frac{3}{10,000}$$

In Example 12 we saw that 240% represented a mixed number when converted into a decimal. Now let's represent it as a fraction and see if it becomes a mixed number again.

EXAMPLE 19 $240\% = \dfrac{240}{100}$. That is an improper fraction that can be reduced.

$$\frac{240}{100} =$$

$$\frac{\cancel{2} \cdot 12 \cdot \cancel{10}}{\cancel{2} \cdot 5 \cdot \cancel{10}} =$$

$$\frac{12}{5} = 2\frac{2}{5}$$

So 240% is the same as $2\dfrac{2}{5}$ or 2.4.   ❏

EXERCISES 4.7B

Convert each percent into a fraction, and reduce it if possible.

19. 4% =   **20.** 17% =

21. 41% =   **22.** 15% =

23. 30% =   **24.** 45% =

25. 3.5% =   **26.** 12.6% =

27. 0.2% =   **28.** 0.05% =

29. 125% =   **30.** 340% =

Check your answers on page 724.   ✓

While we are talking about converting from one form to another, we might as well see how to convert between fraction and decimal notation.

Converting Decimals into Fractions

When we introduced decimals in Section 4.1, we discussed reading the name of a decimal. Once you have read the first and second names of a decimal, you have also read the names of its fractional equivalent. Remember, in Section 3.1 we discussed the names of fractions.

The *first name* is the fraction's *numerator*.
The *second name* is the fraction's *denominator*.

EXAMPLE 20   Convert 0.417 into a fraction.
We read 0.417 as <u>four hundred seventeen</u>, *thousandths*.

So $0.417 = \dfrac{417}{1000}$.   ❏

Always reduce fractions to their simplest forms.

EXAMPLE 21   Convert 0.75 to a fraction.
We read 0.75 as <u>seventy five</u>, *hundredths*.

So $0.75 = \dfrac{75}{100}$, which can be reduced.

$$\dfrac{75}{100} =$$

$$\dfrac{\cancel{25} \cdot 3}{\cancel{25} \cdot 4} = \dfrac{3}{4}$$

So $0.75 = \dfrac{3}{4}$.   ❏

EXERCISES 4.7C

Convert each decimal into a fraction and reduce it.

31. $0.3 =$ **32.** $0.6 =$

33. $0.32 =$ **34.** $0.01 =$

35. $0.003 =$ **36.** $0.0043 =$

37. $0.045 =$ **38.** $0.125 =$

Check your answers on page 724. ✓

Converting Fractions into Decimals

If the *denominator* of a fraction is the name of a decimal *place value*, then you can convert it into a decimal by reading the name of the fraction, as we did to convert from decimals to fractions.

The fraction's *first name* becomes the number you see after the decimal point.

The fraction's *second name* becomes the name of the place value in which the rightmost digit is written.

EXAMPLE 22 Convert $\dfrac{69}{1000}$ into a decimal.

We read $\dfrac{69}{1000}$ as <u>sixty nine</u>, *thousandths*.

So the digit 9 of 69 is written in the thousandths place value:

$$\dfrac{69}{1000} = 0.069$$ ❏

Remember, the names of the decimal place values are tenth $\left(\dfrac{1}{10}\right)$, hundredth $\left(\dfrac{1}{100}\right)$, thousandth $\left(\dfrac{1}{1000}\right)$, ten-thousandth $\left(\dfrac{1}{10,000}\right)$, and so on. Those denominators are called **powers of ten** and can be produced by repeatedly multiplying 1 by 10.

If the denominator of a fraction is not a power of ten, we cannot convert it into a decimal simply by stating the name of the fraction. We convert it into a decimal by using the following rule.

> To convert any fraction into a decimal:
>
> Remember that every fraction represents the *division of its numerator by its denominator*.
>
> To divide, put the *understood* decimal point to the right of the numerator, add zeros, and divide by the denominator. (For some fractions there will be no remainder after a few steps.)

EXAMPLE 23 Convert $\frac{3}{8}$ into a decimal. Remember that $\frac{3}{8}$ represents $3 \div 8$.

$$
\begin{array}{r}
.375 \\
8\,)\,\overline{3.000} \\
-2\,4 \\ \hline
60 \\
-56 \\ \hline
40 \\
-40 \\ \hline
0
\end{array}
$$

So $\frac{3}{8} = 0.375$. ❑

In Section 3.8, Example 10, we found that the quotient of 162,451 divided by 8 was $20,306\frac{3}{8}$. We did the same division again in Section 4.5, Example 5, and found the quotient to be 20,306.375. We now see that these truly are two different forms of the *same* number. So the answer is the same whether we compute it as a fraction or as a decimal.

When we divide by an integer, there will either be no remainder after a few steps or there will always be a remainder—but its digits will repeat in a continual pattern. In problems where there is no remainder, we could say that there is an infinite pattern of zeros.

When you are dividing and keep getting a remainder, you should continue until you clearly see the pattern of repeating digits.

EXAMPLE 24 Convert $\frac{1}{3}$ to a decimal. $\frac{1}{3}$ is $1 \div 3$.

$$
\begin{array}{r}
.333,\text{ etc.} \\
3\,)\,\overline{1.000000} \\
-9 \\ \hline
10 \\
-9 \\ \hline
10
\end{array}
$$
❑

Several notations are used to represent a repeating pattern of digits. The one most universally used is a bar placed over the repeating pattern.

EXAMPLE 25 We represent $\frac{1}{3}$ as the decimal $0.\overline{3}$. ❑

Continue adding zeros and dividing until you see the pattern repeat at least twice.

EXAMPLE 26 Convert $\frac{3}{11}$ to a decimal. $\frac{3}{11} = 3 \div 11$.

$$
\begin{array}{r}
.2727, \text{ etc.} \\
11\overline{)3.0000000} \\
-2\,2 \\
\hline
80 \\
-77 \\
\hline
30 \\
-22 \\
\hline
80 \\
-77 \\
\hline
30, \text{ etc.}
\end{array}
$$

So $\frac{3}{11} = 0.\overline{27}$. ❑

A decimal written without a bar over the digits implies that there are zeros in the next place values. So if those place values have digits and you do not represent that fact with a bar, you will be representing a number different from the one intended. If a decimal has a pattern of repeating digits, always represent the pattern with a bar over the repeating digits.

EXAMPLE 27 4.316 represents the decimal 4.3160000000, etc.
4.3$\overline{16}$ represents the decimal 4.3161616161616, etc. ❑

EXERCISES 47D

Convert each fraction into a decimal. If there is a remainder of repeating digits, represent that pattern with a bar over the repeating digits.

39. $\frac{1}{5} =$ **40.** $\frac{17}{20} =$

41. $\frac{7}{25} =$ **42.** $\frac{5}{8} =$

43. $\frac{13}{40} =$ **44.** $\frac{15}{25} =$

45. $\frac{51}{125} =$ **46.** $\frac{47}{125} =$

47. $\frac{1}{11} =$ **48.** $\frac{3}{9} =$

49. $\frac{7}{12} =$ **50.** $\frac{11}{15} =$

51. $\frac{17}{22} =$ **52.** $\frac{7}{33} =$

Check your answers on page 724. ✓

Converting Decimals and Fractions into Percents

We often wish to express numbers as percents. I'll show you how to convert decimals into percents, since this conversion is the most commonly needed one and is also the easiest. To convert fractions into percents, simply change the fractions into decimals and then into percents.

Converting decimals into percents is exactly the reverse of converting percents into decimals.

Remember, to convert *percents into decimals*:

Move the decimal point *two* places to the *left*.
Remove the percent symbol (%).

So to convert *decimals into percents*:

Add the percent symbol (%).
Move the decimal point *two* places to the *right*.

Be sure to insert zeros if you do not have two digits to the right of the decimal point.

EXAMPLE 28 Convert each decimal into a percent.

$$0.73 \rightarrow 73\%$$
$$0.8 = 0.80 \rightarrow 80\%$$
$$0.08 \rightarrow 8\%$$
$$0.008 \rightarrow 0.8\%$$
$$3.47 \rightarrow 347\%$$
$$0.138 \rightarrow 13.8\%$$

EXERCISES 47E

Convert each decimal into a percent.

53. $0.41 =$ **54.** $0.58 =$

55. $0.03 =$ **56.** $0.157 =$

57. $0.072 =$ **58.** $0.0024 =$

59. $0.04 =$ **60.** $0.004 =$

61. $0.0037 =$ **62.** $0.037 =$

Check your answers on page 724.

4.8 Computing Percents

To Calculate the Percent of a Number

Computations involving percents often involve calculating the **percent of a number**. For example, we might be asked to compute 7% of 50. As mentioned before, the word "of" instructs us to multiply.

To calculate the percent of a number:

Convert the percent into a decimal or fraction and multiply.

EXAMPLE 1 What is 7% of 50?

(I will compute this problem as a fraction and then as a decimal. You decide which way you prefer; it is not necessary to do the computations both ways unless you want to use the second method as a check on your accuracy.)

<table>
<tr><td>As a fraction:</td><td>As a decimal:</td></tr>
<tr><td>7% of 50 =</td><td>7% of 50 =</td></tr>
<tr><td>$\left(\dfrac{7}{100}\right)\left(\dfrac{50}{1}\right) =$</td><td>$(0.07)(50) = 3.5$</td></tr>
<tr><td>$\dfrac{7 \cdot \cancel{50}}{\cancel{50} \cdot 2} =$</td><td></td></tr>
<tr><td>$\dfrac{7}{2} = 3\dfrac{1}{2}$</td><td></td></tr>
</table>

So 7% of 50 is $3\dfrac{1}{2}$ or 3.5. ❏

From now on I will compute the problems only as decimals, since I think that way is easier.

EXAMPLE 2 What is 15% of 300?

$$(0.15)(300) = 45$$ ❏

EXAMPLE 3 What is 6.3% of 80?

$$(0.063)(80) = 5.04$$ ❏

EXAMPLE 4 What is 210% of 43?

$$(2.10)(43) = 90.3$$ ❏

EXERCISES 4.8A

Compute the percents. (The answers were computed as decimals.)

1. 4% of 38 =	**2.** 18% of 97 =
3. 13% of 75 =	**4.** 34% of 250 =
5. 75% of 800 =	**6.** 84% of 2,340 =
7. 9.3% of 215 =	**8.** 12.5% of 500 =
9. 6.8% of 450 =	**10.** 15.4% of 860 =
11. 180% of 76 =	**12.** 240% of 900 =
13. 150% of 375 =	**14.** 400% of 235 =
15. 0.9% of 43 =	**16.** 0.04% of 180 =
17. 0.5% of 850 =	**18.** 0.08% of 500 =

Check your answers on page 724. ✓

Applications of Percents

Percents are used to figure taxes, discounts, interest, and concentration of chemical solutions, as well as for many other applications. For example, **taxes** are set up according to a rate described as a percentage. If a state has a restaurant meal tax, as Massachusetts does, it is defined as a certain amount the buyer must pay (which goes to the state treasury for tax) on every dollar of restaurant food purchased. The tax rate for meals in Massachusetts is 5%. That means the buyer must pay 5 cents on every 100 cents (per dollar) spent when eating out.

EXAMPLE 5 How much is charged for tax if the rate is 5% and the cost of lunch for two people is $7.80?

$$5\% \text{ of } 7.80 =$$

$$(0.05)(7.80) = 0.39 = \text{The tax is 39 cents.} \qquad \square$$

Percents are used to indicate how much you will save when stores are having sales. Stores usually set a rate of **discount** that is described as a percent.

EXAMPLE 6 Cheepo's has a year-end sale in which all items are discounted 20%. What is the discount on a barbecue that normally sells for $68.95?

$$20\% \text{ of } \$68.95 =$$

$$(0.20)(\$68.95) = \$13.79 \qquad \square$$

Interest is another rate described as a percent. The bank may pay us interest at 6% for investing our savings with them or they may charge us 16% if we need to borrow money. They calculate the interest as **compound interest**, which means they calculate it every day, week, or month. Such interest is charged on the interest and is complicated to calculate. But in this book, when we discuss interest we mean **simple interest**, which is calculated by year. That means we just multiply the interest rate by the money invested or borrowed and then multiply the product by the number of years involved.

EXAMPLE 7 How much simple interest will Joy earn in 1, 2, and 5 years for investing $860.40 at 6%?

The interest each year will be 6% of $860.40. Calculate the interest to three decimal places and round it off to the nearest cent.

$$6\% \text{ of } \$860.40 =$$

$$(0.06)(\$860.40) = \$51.624 \qquad \square$$

When numbers are rounded off, the answers are approximations rather than precise results. The symbol $\approx$ is used to represent an approximate or rounded off answer.

In 1 year the interest will be ($51.624)(1) or $\approx$ $51.62.
In 2 years the interest will be ($51.624)(2) = $103.248 or $\approx$ $103.25.
In 5 years the interest will be ($51.624)(5) = $258.120 = $258.12. $\qquad \square$

In chemistry the strength, or **concentration**, of a solution is usually described by a percentage, too. For example, you may work with a 34% salt solution. The solution is a mixture of salt and water in which the concentration of salt is 34 parts out of every 100 parts. So for 100 grams of that solution, 34 grams would be salt and the rest (66 grams) would be water.

EXAMPLE 8 How much salt is in 80 grams of a 34% salt solution?

$$34\% \text{ of } 80 \text{ g} =$$
$$(0.34)(80 \text{ g}) = 27.2 \text{ g}$$

EXERCISES 4.8B

Find each of the percents requested.

19. What is the 5% meal tax on a meal costing $6.80?

20. What is the discount on an $80 coat during a 15% sale?

21. What is the 8% simple interest on $280 for 1 year?

22. How much salt is in 200 g of a 43% salt solution?

23. What is the 5% meal tax on $40.80?

24. What is a 15% tip on a $68.00 bill?

25. What is the discount when a $60 sweater is on sale for 20% off?

26. What is the 18% simple interest on $580
(a) for 1 year? (b) for 3 years? (c) for 5 years?

27. How much salt is in 840 g of a 70% salt solution?

28. How much salt is in 6,800 g of a 46% salt solution?

29. What is the discount when the price of $2,500 for a car is reduced by 10%?

30. What is the 20% tip for a $48.60 meal?

Check your answers on page 724.

Real-life Applications of Decimals and Percents

There are many real-life uses of decimals and percents, ranging from the markings on a standard medical thermometer (98.6 degrees is the average normal body temperature) to laws about the alcoholic content of beer ("3.2% beer" is a household term in some western states). Problems involving dollars and cents, gasoline purchases, and sports averages usually involve decimals.

In each problem the key issue is recognizing how to set up the problem for calculating. You must decide whether to add, subtract, multiply, divide, or perform more than one of those operations. I hope that by thinking through the information given and understanding what you are being asked to find, you will easily recognize how to calculate the answer.

EXAMPLE 1 What is the total cost of the following purchases:

$2.83 for a chicken, $1.12 for milk, $0.83 for potatoes, $0.89 for lettuce, $1.04 for eggs, $2.46 for cheese, $0.68 for rolls, and $1.30 for brussels sprouts?

To find the total, we *add* all the items.

$$
\begin{aligned}
\text{Total cost} = \ &\$2.83 \\
&1.12 \\
&0.83 \\
&0.89 \\
&1.04 \\
&2.46 \\
&0.68 \\
&\underline{1.30} \\
&\$11.15
\end{aligned}
$$

The total cost of the purchases is $11.15. ❏

EXAMPLE 2 How much change should we get from a $20.00 bill to pay for the purchases described in Example 1?

Change is the difference between the amount we owe and the amount we pay, so we *subtract* the $11.15 from $20.00.

$$
\begin{aligned}
\text{Change} = \ &\$20.00 \\
&\underline{-\ 11.15} \\
&\$\ 8.85
\end{aligned}
$$

We should receive $8.85 in change. ❏

EXAMPLE 3 How much does it cost to buy 3.2 pounds of rice if each pound costs $0.83?

To find the cost of the rice, we must *multiply* the cost of each pound by the number of pounds purchased.

$$
\begin{aligned}
\text{Cost of rice} &= (3.2)(\$0.83) \\
&= \$2.656
\end{aligned}
$$

We must round off this number to the nearest cent (hundredth).

$$\$2.65{\downarrow}6 \text{ rounds up to } \$2.66.$$

So the rice costs $2.66. ❏

EXAMPLE 4 When the team was down by one point and there were only 13 seconds left to play, the coach wanted her best shooter to take the last shot. She looked toward Rhoda, the tournament high scorer with 53 baskets. The team statistician pointed out that even though Judy had made only 32 baskets, she was more likely than Rhoda to sink the crucial basket. Rhoda had attempted 86 shots to Judy's 47 attempts, so her field goal percentage was higher.

The field goal percentage is calculated by *dividing* the number of successful shots by the number of attempted shots. Field goal percentages (FGP) are usually calculated to four decimal places and then rounded off to three decimal places (thousandths). They indicate that at this rate the player will make that number of baskets out of 1,000 attempts.

What was each woman's field goal percentage?

$$\text{Rhoda's FGP} = \frac{53}{86} \quad \text{or} \quad 86\overline{)53.0000,} \quad \overset{\downarrow}{.6162} \text{ which rounds down to } .616.$$

$$\text{Judy's FGP} = \frac{32}{47} \quad \text{or} \quad 47\overline{)32.0000,} \quad \overset{\downarrow}{.6808} \text{ which rounds up to } .681.$$

So the coach picked Judy on the basis of her .681 FGP over Rhoda with her .616 FGP. Judy missed! Statistics cannot foretell the future. ❏

EXAMPLE 5 I want to calculate how much mileage I got per gallon from my last tankful of gas. The odometer read 12,341 miles when I last filled up the tank, and it reads 12,570 miles now. It took 8.3 gallons to refill the tank. How many miles per gallon (mpg) did I get?

First *subtract* the mileage on the odometer at the first fill-up from the mileage on the odometer now. That tells how many miles I drove on this tankful of gas.

$$\begin{array}{r} \text{Miles traveled} = \quad 12{,}570 \text{ mi} \\ -\,12{,}341 \text{ mi} \\ \hline 229 \text{ mi} \end{array}$$

Then *divide* that number of miles by the number of gallons used for driving. (Since the tank was full at the first reading, the number of gallons needed to fill it again now is the number of gallons used.) Round off the answer to the nearest tenth.

$$229.0 \text{ mi} \div 8.3 \text{ gal} = 2290 \text{ mi} \div 83 \text{ gal}$$

$$\begin{array}{r} 27.59 \\ 83\overline{)2290.00} \text{ mpg} \end{array}$$

27.5↓9 mpg rounds up to 27.6 mpg.
So my car got 27.6 miles per gallon on that tankful. ❑

EXAMPLE 6 If restaurant meals are taxed at the rate of 5%:

a) What would the tax be for a meal costing $14.35? (A 5% tax means that 5 cents for each dollar spent is charged additionally on the bill for tax.)

To find the amount of tax, *multiply* the tax rate (5%) by the cost of the meal ($14.35), after representing the tax by its decimal equivalent.

$$\text{Tax} = 5\%(\$14.35) \qquad \begin{array}{r} \$14.35 \\ \times \ \ 0.05 \\ \hline \$0.7175 \end{array} \qquad (5\% = 0.05)$$

Money must always be rounded off to the nearest cent (hundredth). So 0.71↓75 rounds up to $0.72. The tax is 72 cents.

b) What would the total be for the meal and tax?
The total bill is the sum of the food and tax costs, so *add* them.

$$\begin{array}{r} \text{Total meal and tax cost} = \$14.35 \\ +0.72 \\ \hline \$15.07 \end{array}$$

c) What would the grand total cost be if you left a 15% tip on the meal and tax cost?

$$15\% \text{ of } \$15.07 =$$
$$(0.15)(15.07) = \$2.2605$$

$2.26↓05 rounds down to $2.26.

$$\begin{aligned} \text{Total cost} &= (\text{meal} + \text{tax}) + \text{tip} \\ &= \$15.07 + \$2.26 \\ &= \$17.33 \end{aligned}$$
❑

EXAMPLE 7 A store is having a sale where every item is discounted by 15%. (That means 15 cents on every dollar of each price is subtracted as a discount.)

a) What is the amount of discount on a $9.75 shovel?

$$\text{Discount} = 15\% \text{ of } \$9.75$$
$$= 0.15 \times 9.75$$

$$\begin{array}{r} \$9.75 \\ \times \quad .15 \\ \hline 4875 \\ 975 \quad \\ \hline 14625 \rightarrow \$1.4625 \end{array}$$

$1.46^{\downarrow}25$ is rounded down to $1.46 for the discount.

b) What is the sale price of the shovel?

The sale price is the original price minus the discount. So *subtract* the discount from the original price.

$$\begin{array}{r} \text{Sale price} = \quad \$9.75 \text{ original price} \\ -1.46 \text{ discount} \\ \hline \$8.29 \text{ sale price} \end{array}$$

The sale price of the shovel is $8.29.   ❏

EXAMPLE 8   An economist estimates that the cost of living will increase by 8% over the next three years. That means the cost of items will be 8% more than they are now. How much should a package of gum cost in three years if it costs 35 cents now?

The increase in cost will be 8% of $0.35.

$$\text{Increase} = (0.08)(\$0.35)$$
$$= \$0.028$$

That number rounds up to 3 cents ($0.02^{\downarrow}8 \rightarrow \$0.03$).

$$\text{Price in three years} = \text{price now } + \text{increase}$$
$$= \$0.35 + \$0.03$$
$$= \$0.38$$

We should expect a package of gum to cost 38 cents in three years.   ❏

EXERCISES 4.9

Set up the appropriate calculation to answer each question.

1. Fran needs to find the total mileage traveled for her weekly travel expense voucher. She recorded the following daily mileages: Monday, 48.7 mi; Tuesday, 36.2 mi; Wednesday, 51.4 mi; Thursday, 70.8 mi; and Friday, 10.3 mi. What was her total mileage for the week?

2. Bob's purchases cost $7.80, $3.14, $6, $1.19, $4.20, 38 cents, $6.92, and $7.50.
   a) What is the cost of those purchases?
   b) How much change should Bob get back from $50?

3. How much change should you expect from a $20 bill if you made the following purchases: $3.98 for a shirt, $12.50 for a pair of pants, and $2.34 for some socks?

4. I made the following selections at the grocery store: tuna fish for $1.12, bread for 85¢, milk for 53¢, fruit for $1.16, and chips for 89¢. Find (a) the total cost and (b) the amount of change I should expect from a $10 bill.

5. If oil was for sale at the price of three cans for $4.00, how much should it cost me to buy just one can?

6. If soda sells for $1.50 for two bottles, how much should it cost me to buy seven bottles?

7. How much should it cost to buy 3 lb of peanuts and 4 lb of raisins if each pound of peanuts costs 75¢ and each pound of raisins costs $1.20?

8. My odometer read 14,504 mi at the last fill-up and now reads 14,788.8.
   a) How many miles did I travel between fill-ups?
   b) If it takes 8.9 gal to fill the tank, how many miles per gallon did the car go?

9. After a trip of 868 mi I wanted to calculate my expenses. If my car gets 31 mpg and gas cost me $1.25 per gallon, how much did it cost for the gas?

10. On my last trip I traveled 250 mi. My car gets 32 mpg, gas cost me $1.20 per gallon, and I had to pay the following tolls: 25¢, 35¢, $1.30, and 25¢? (Round off number of gallons used to tenths before calculating gas cost.)
    a) How much did gas cost?
    b) What was the total of those costs?

11. My car gets 18 mpg. An additive to the gas is advertised as increasing gas mileage. What mileage should I expect if I figure that with the additive I get 1.5 times as many miles per gallon as I get without it?

12. How much change do I get from $20 after making the following purchases: three packages of hot dogs, each costing $1.85; two packages of buns, each costing 80¢, and four bottles of juice, each costing $1.85, with a tax of 75¢?

13. If Johanna got 45 hits out of the 125 times she came up to bat this season, what is her batting average?

14. Ron had only 143 base hits even though he came up to bat 500 times. What is his batting average?

15. How much profit did Betty's business make this week if her expenses were $284.96 for inventory, $107.50 for rent, $31.14 for utilities, and $203.41 for payroll, and if her income was $123.52 on Monday, $186.91 on Tuesday, $114.36 on Wednesday, $186.09 on Thursday, $136.18 on Friday, and $206.52 on Saturday?

16. What is the balance in Phil's account if he opened with $200 and made the following transactions: he wrote checks for $25, $13.86, $9.80, $43.17, $120, $20, $7.35, $18.64, $9.07, and $78.94; deposited $50, $246.18, and $35.08; received $2.19 interest; and paid $15 in service charges.

17. What is the total bill if we go to a gas station, fill up the tank with 12.8 gal of gas that costs $1.19 a gallon, and buy a bottle of windshield cleaner for $1.98 and 2 qt of oil for 79¢ each?

18. Frank's car gets 32.5 mpg.
    a) How many miles can he travel on 9.5 gal of gas?
    b) If his odometer now reads 43,288 mi, what will the reading be when he has used 9.5 gal? (Round off the answer to the nearest whole number.) If his gas gauge breaks, he can use these calculations to know when to fill up his 10 gal tank.

19. Pat bought a television set advertised for $250. She put $25 down and agreed to pay $13.54 a month for 20 payments.
    a) What was her total cost?
    b) How much more did it cost her to buy the television set on this installment schedule compared to the advertised price?

20. Ann filled the tank of her car by adding 12.2 gal of gas. The odometer now reads 61,058 mi and was at 60,732 mi the last time she filled the tank. How many miles did she get per gallon on that last tankful? (Round off the answer to nearest tenth.)

21. Peter bought a chain saw that normally sold for $380 when the store was having a sale in which everything was discounted at the rate of 18%.
    a) What was the discount on the chain saw?
    b) What was the sale price?

22. The bill for our barbecue dinner celebrating the Celtics' victory in the NBA Championship comes to $43.86.
    a) How much tip should we leave if we agree that the waiter deserves to be tipped at the rate of 20% of our meal cost?
    b) If the restaurant adds a 5% tax to our bill of $43.86, what will be the tax?
    c) What will be the total cost of our celebration?

23. When Carol bought a house for $32,850, she had to come up with a 15% down payment.
    a) How much was the down payment?
    b) What was the balance she owed on her mortgage, not counting interest?

24. a) How much would the tax be on a restaurant bill of $7.60 if the meal tax rate was 5%?
    b) What would the total bill be if it was a self-service restaurant where no tip was involved?

25. a) How much would the tax be on a restaurant bill of $34.20 if the meal tax rate was 5%?
    b) What would the total bill be if it was a self-service restaurant where no tip was involved?

26. What is the sale price of a coat originally tagged at $85 if it is on sale with a 12% discount?

27. I bought items for the following amounts: $4.25, $3.50, $2.80, $6.35, and $1.30. None of them was a taxable item, but the store was having a clearance sale and cutting 25% off all purchases. What was my final bill?

28. Liz takes home $1,086 per month and figures that she can afford to spend 25% of her monthly income for rent. How much rent can she afford to pay each month?

29. Bill spent $2,058 last year on home improvements. The Internal Revenue Service recognized 43% of those costs as energy-saving investments. How much can Bill list on his tax return as having been spent on energy-saving repairs?

30. How much will Henry have to pay at the end of the year if he borrows $800 at 14% simple interest?

31. Roberta's instructor said that she needed to earn at least 64% of the possible 400 points from exams in her course in order to pass. How many points must Roberta earn to ensure a passing grade?

32. A poll reported that 72% of baseball fans think the National League is superior to the American League. How many people indicated this viewpoint if 42,800 people were polled?

33. Out of 500 people questioned about their favorite professional sports, 42% thought basketball was the best, 14% preferred hockey, 21% preferred baseball, 17% preferred football, 2% preferred soccer, and the rest preferred tennis.
    a) What percent preferred tennis? (*Hint*: The total percent should add up to 100%.)
    b) How many people preferred each sport?

    _____basketball   _____hockey   _____baseball
    _____football   _____soccer   _____tennis

34. Harold's dinner in the local deli comes to $14.80.
    a) How much is the tax if there is a 5% meal tax?
    b) How much tip should he leave if he wants to tip the waiter 15% of the meal cost? (The meal cost does not include tax.)
    c) What is the total cost of Harold's dinner?

35. My friend Lou, who is an economist, estimates that the cost of living will increase 2% next year. If he is right, how much will a quart of ice cream cost next year if it costs $1.50 now?

36. How much salt is in 150 g of a 24% salt solution?

37. If Ernie Banks got 679 hits after 2,356 times at bat, what was his batting average (to three decimal places)?

38. Three friends shared an apartment. Their joint monthly bills were $325 for rent, $178.34 for food, $87.62 for utilities, and $43.18 for the phone.
    a) What is each person's exact share?
    b) Round each person's share to the nearest dollar.

39. The bakery advertises chocolate chip cookies on sale at 85 cents a dozen (12 cookies).
    a) How much should they sell one for?
    b) How much will they sell one for?

Check your answers on page 724.

GLOSSARY

concentration of a solution   Amount of an ingredient in a solution, usually defined as a percent.

   Example: A 34% salt solution is a mixture of salt and water in which the concentration of salt is 34 parts out of every 100 parts. So for every 100 drops in a 34% salt solution, 34 drops are salt and the rest (66) are water.

decimal point   Dot used to separate the ones place value from the tenths place value in a decimal number. It separates the integer of a mixed number from the decimal part of the number.

   Example: The number 43.25 is the combination of the integer 43 and the decimal 25-hundredths.

discount   Amount taken off the price of an item when it is on sale. It is usually described as a percent.

   Example: If a store advertises that it is slashing 25% off all prices during a sale, an item normally selling for $80.00 would have 25% of $80.00, or $20.00, as its discount.

interest   Charge for using money, usually described as a percent.

   Example: A bank may pay us 6% interest for use of the money that we have deposited in a savings account.

compound interest   Form of interest used by most banks and calculated every day, week, or month.

simple interest   Form of interest calculated every year (and used in problems in this book). To find simple interest, multiply the rate of interest by the amount of money being used, and multiply that product by the number of years the money is used.

name of a decimal   The number appearing after the decimal point is the first name of a decimal; The place value of the last digit on the right is the second name.

   Example: For the decimal 0.34, thirty four is the first name and hundredths is the second name.

percent   Notation used to represent numbers that involve parts of whole numbers. "Percent" means per hundred. Percents can be expressed as decimals or as fractions.

   Example: 17% means 0.17 or $\frac{17}{100}$.

percent of a number   Product of the percent multiplied by the number.

   Example: 17% of $50 means (0.17)($50) = $8.50.

powers of ten   Numbers 10, 100, 1,000, 10,000, and so on that can be produced by repeatedly multiplying by 10.

tax   Amount usually defined as a percent rate that we pay to the government.

   Example: In Massachusetts there is a 5% meal tax, so a charge of 5 cents for each dollar of a restaurant bill is added to the cost of a meal. The tax on a $3.80 meal is 5% of $3.80, or $0.19.

Chapter 4 Review

Do each of the problems. After you are confident that you have done them as accurately as possible, compare your answers with those in the back of the book. If any are wrong, go back to the section where the problem came from (indicated to the left of the problem) and review the section.

Once you understand all the sections, you will be sure to learn the skills solidly and remember how to do them if you practice them a little bit more. Turn to the Supplementary Exercises and do all the problems from any section where you had difficulty on these review exercises.

4.1   1. What is the name of 0.0568?

4.2   2. Round off 769.093 to the nearest tenth.

   3. Round off 5,300.8612 to the nearest hundredth.

Combine and simplify.

4.3   4. $17.03 + 8.4 + 5.008 + 23.78 =$

   5. $607.3 - 298.07 =$

   6. $84.5 - 12.08 - 6.1 + 25.36 - 0.067 =$

4.4   7. $5.02 \times 40.6 =$

   8. $(-18.03) \times (-3.2) =$

   9. $(-0.0005) \times (-0.007) =$

   10. Multiply 34.0097 by 10 twice.

4.5   11. $8.00032 \div 0.02 =$

   12. $-65.442 \div 1.3 =$

   13. Divide 60.005 by 10 three times.

4.6   14. $5(1.2) + 3(-0.7) - 2.01(-0.04) =$

   15. $3.8 + 2(0.3 - 2.01 - 0.85 + 3.06) =$

   16. Evaluate $5a - 2b + c$ if $a = 0.2$, $b = -3.5$, and $c = -0.1$.

4.7   17. Convert to a decimal: 3.8%.

   18. Convert to a fraction: 38%.

   19. Convert to a fraction: 0.068.

   20. Convert to a decimal: $\dfrac{9}{30}$.

   21. Convert to a percent: 0.072.

   22. Convert to a percent: $\dfrac{5}{8}$.

Answers

1. _____

2. _____

3. _____

4. _____

5. _____

6. _____

7. _____

8. _____

9. _____

10. _____

11. _____

12. _____

13. _____

14. _____

15. _____

16. _____

17. _____

18. _____

19. _____

20. _____

21. _____

22. _____

4.8   **23.** What is 18 % of 495?

24. What is the 5 % meal tax on $24.80?

4.9   **25.** How much change did Ralph receive from $50 if he bought a shirt for $13.95, three pairs of socks for $1.25 a pair, and a sweater for $28.50?

26. Julie wanted to buy a $48,500 house and had to make a down payment of 8 % of the cost to get a mortgage from the bank.
a) How much was the down payment?
b) What was the balance she owed on her mortgage?

Answers

23. _____

24. _____

25. _____

26a. _____

b. _____

Check your answers on page 724.

Supplementary Exercises

Do all the problems in every section that involves a skill in which you now lack complete mastery. With a little more practice you will achieve that sense of really understanding the topics, and you will remember how to do these problems.

4.1   1. What is the name of 0.341?

4.2   2. Round off 28.52 to the nearest tenth.

3. Round off 618.0174 to the nearest hundredth.

Combine and simplify.

4.3   4. $54.72 + 16.9 + 84.03 + 6.082 =$

5. $4,210.96 - 3,088.065 =$

6. $183.4 - 53.07 - 4.31 + 78.02 - 0.04 =$

4.4   7. $10.06 \times 50.2 =$

8. $(20.07) \times (-7.4) =$

9. $(-0.008) \times (-0.0003) =$

10. Multiply 88.003 by 10 twice.

4.5   11. $5.0004 \div 0.03 =$

12. $-34.561 \div 1.7 =$

13. Divide 128.013 by 10 four times.

4.6   14. $3(1.5) - 2(-0.8) - 1.05(-0.03) =$

15. $2.7 + 4(0.5 - 1.07 - 0.43 + 6.02) =$

16. Evaluate $4a + b - 2c + c$ if $a = -0.1$, $b = -5.7$, and $c = -2.6$.

4.7   17. Convert to a decimal: 7.1%.

18. Convert to a fraction: 42%.

19. Convert to a fraction: 0.028.

20. Convert to a decimal: $\dfrac{3}{8}$.

21. Convert to a percent: 0.307.

22. Convert to a percent: $\dfrac{7}{20}$.

4.8   23. What is 34% of 206?

24. What is the 5% meal tax on $19.30?

Answers

1. _____
2. _____
3. _____
4. _____
5. _____
6. _____
7. _____
8. _____
9. _____
10. _____
11. _____
12. _____
13. _____
14. _____
15. _____
16. _____
17. _____
18. _____
19. _____
20. _____
21. _____
22. _____
23. _____
24. _____

4.9   **25.** What is Hank's batting average if he got 168 hits after 300 times at bat?

Answers

26. A newspaper article reported that 34% of those polled felt that their town should cut the proposed school budget rather than have a tax increase. If 1,650 people were polled, how many had this view?

25. _____

26. _____

Check your answers on page 724.

Chapter 4 Test

Write the name of each number.

1. 0.0268

2. 17.092

Convert these numbers into an equivalent

3. fraction—(a) 0.062 (b) 18% (c) 84.5%

4. decimal—(a) $\dfrac{36}{100}$ (b) $\dfrac{38}{54}$ (c) 7.3%

5. percent—(a) $\dfrac{5}{100}$ (b) 0.375 (c) 0.0091

Round off to the nearest

6. tenth—6.839

7. hundredth—421.03761

8. thousandth—0.08333

Combine these numbers.

9. $26.07 + 5.21 + 0.893 + 10.09 =$

10. $73.008 - 96.4 =$

11. $3.41 \times 0.006 =$

12. $(-1.0563) \div (0.21) =$

13. $3(5.3) + 7.2(0.6 - 0.3 - 0.45) =$

14. 9% of 670 =

Solve these problems.

15. How much change from $5.00 do you get if you purchase one roasted chicken when they are advertised at three for $3.81?

16. Susan has borrowed $500 at 8% simple interest per year.
   a) How much interest will she owe in one year?
   b) How much would she have to pay if she paid off this debt after one year?
   c) How much interest will she owe in five years?
   d) How much would she have to pay if she paid off this debt after five years?

17. Evaluate $a + 2b - 3c + 5(a + b)$ if $a = 2.3$, $b = -0.4$, and $c = -0.1$.

Check your answers on page 724. ✓

Answers

1. _____

2. _____

3a. _____

b. _____

c. _____

4a. _____

b. _____

c. _____

5a. _____

b. _____

c. _____

6. _____

7. _____

8. _____

9. _____

10. _____

11. _____

12. _____

13. _____

14. _____

15. _____

16a. _____

b. _____

c. _____

d. _____

17. _____

Cumulative Review

CHAPTERS 1–4

Combine these numbers.

Answers

1. $28 - 35 - 70 + 6 - 17 - 4 + 20 =$

1. _____

2. $(28) - (14) - (-30) + (-38) - (26) =$

2. _____

3. $(-1)(1)(-3)(-1)(1)(-4)(-2) =$

3. _____

4. $\dfrac{17}{20} + \dfrac{3}{20} - \dfrac{7}{20} - \dfrac{1}{20} + \dfrac{4}{20} =$

4. _____

5. _____

5. $\left(\dfrac{8}{25}\right) \times \left(\dfrac{35}{20}\right) =$

6. _____

6. $\left(-\dfrac{5}{7}\right) + \left(-\dfrac{3}{5}\right) + \left(-\dfrac{2}{9}\right) =$

7. _____

8. _____

7. $\dfrac{2}{3} + \dfrac{1}{5} + \dfrac{1}{2} =$

9. _____

8. $\dfrac{3}{10} + \dfrac{5}{12} - \dfrac{4}{15} =$

10. _____

11. _____

9. $2\dfrac{1}{5} + 3\dfrac{2}{5} + 1\dfrac{4}{5} + 4\dfrac{1}{5} =$

12. _____

10. $\left(2\dfrac{3}{4}\right) \times \left(1\dfrac{2}{3}\right) =$

13. _____

14. _____

11. $\left(-\dfrac{15}{21}\right) \div \left(\dfrac{14}{35}\right) =$

15. _____

12. $\dfrac{1}{2} + 3\left(\dfrac{5}{6} - \dfrac{1}{6} + \dfrac{1}{6}\right) =$

16a. _____

13. $6.17 - 14.09 - 9.2 - 0.025 + 0.05 =$

b. _____

14. $(0.006) \times (-3.05) =$

17a. _____

15. $8.3\% \text{ of } 600 =$

b. _____

Round off to the nearest

18. _____

16. integer—(a) $6\dfrac{7}{8}$ (b) $\dfrac{2}{7}$

19. _____

17. hundredth—(a) 3.0683 (b) 0.04088

Evaluate these formulas if $x = -1$, $y = -2$, and $w = 2$.

18. $3x - 4y - 2w =$

19. $5 + 3(x + y + w) =$

Continued

Cumulative Review (*Cont.*)

20. $2.4x - 0.3y - 1.1w =$

Answers

21. What is the perimeter of a triangle with sides 4.3 cm, 7.0 cm, and 5.9 cm?

22. What is the area of a rectangle with width $4\frac{2}{3}$ ft and length $7\frac{1}{2}$ ft?

23. What is the sale price of a $40 sweater during a 15% discount sale?

20. _____

21. _____

22. _____

23. _____

Check your answers on page 724.

5 Polynomials

5.1 Algebraic Terms

Concept of a Term

Algebra is a branch of mathematics in which letters take the place of numbers whose values we do not know.

The earlier chapters showed that if you understand what the notation of numbers and arithmetic operations represents, then the rules for computing with numbers make sense. The rest of the book introduces notation for expressions involving letters. If you understand what the notation of algebraic expressions and operations represent, then rules for computing with letters will also make sense.

In language we put letters together into words. In algebra the basic building blocks we make out of letters are called **terms**. We have seen that to represent distance traveled we use the term RT. To find the area of a triangle we use the term $\left(\dfrac{1}{2}\right)bh$. These terms can also be put together into more complicated expressions. Algebraic expressions are built out of terms, just as sentences are built out of words.

Terms are products of numbers. In Section 1.3 we learned several notations to indicate products of numbers.

EXAMPLE 1 To represent the product of 3 multiplied by 5, we could write

$$3 \times 5, \ 3 \cdot 5, \ 3(5), \ (3)5, \ \text{or} \ (3)(5).$$ ❏

Algebraic terms involve letters that represent numbers. Therefore you should *never* use $\times$ to represent multiplication in algebra. The letter x is often used to represent an unknown number in an algebraic expression, so using $\times$ to represent multiplication in algebra would be very confusing.

The notation to express the product of numbers that are represented by letters is very simple. Writing a number next to a letter or writing a letter next to another letter, with *no operation symbol* between them, always represents multiplication

EXAMPLE 2 The product of 3 multiplied by a is written as $3a$.

The product of 7 multiplied by x and by w is written as $7xw$. ❏

Recall that in Section 3.3 we defined **factors** of a given number as the numbers that are multiplied together to produce that given number. The factors of a term are the numbers and/or letters that are multiplied together to produce that term. The factors of a term can be only numbers, only letters, or a combination of letters and numbers.

EXAMPLE 3 7, 3x, 4ab, w, and mnx are examples of algebraic terms. ❏

A term made up only of a number is called a **constant**. The term 7 in Example 3 is a constant; that is, its value is constantly the number 7.

When a term has one or more letters for factors, the letters determine the *kind of term* it is. For example, we consider 3x to be an x term; w, a w term; and 4ab, an ab term. The number factors in such terms are called numerical **coefficients**.

EXAMPLE 4 In the term 3x, 3 is the numerical coefficient of x. 3x represents the fact that we have *three x*'s.

$$3x = x + x + x$$

In the term 4ab, 4 is the numerical coefficient of ab. 4ab represents the fact that we have *four ab*'s.

$$4ab = ab + ab + ab + ab$$ ❏

When no number is expressed in a term, the numerical coefficient is understood to be 1. The coefficient 1 is usually not written out in a term.

EXAMPLE 5 mnx represents *one mnx*; it is *not* written as 1 mnx. ❏

EXERCISES 5.1A

State the coefficient and kind of each term.

		Coefficient	Kind of term
Example:	5aw	5	aw
1.	7xy		
2.	3ac		
3.	−2x		
4.	yw		
5.	16abc		
6.	−4pwm		

Check your answers on page 724. ✓

Exponents (Exponential Notation)

If a factor is repeated in a term, recognizing how many times it appears may be difficult. A special notation was invented to make it easier to read such terms, as well as to work with them. This notation is called **exponential notation**.

An **exponent** represents the *number of times* that a factor is repeated in a term. The factor being repeated is called the **base**. An exponent is written above and to the right of its base. An exponent is also called a **power**.

EXAMPLE 6   The term y^4 represents the *base* y and the *exponent, or power*, 4.

$$y^4 \text{ represents } yyyy.$$ ❑

EXAMPLE 7 5^3 represents $(5)(5)(5)$. ❑

EXAMPLE 8 $3x^2y^3$ represents $3xxyyy$. ❑

EXAMPLE 9 $2a^2b^4cd^3$ represents $2aabbbbcddd$. ❑

If a factor is expressed without an exponent, it is understood that the factor appears only once. (The exponent 1, is usually not written.) Here is how to read expressions with exponents:

x^2 is read as "x squared" or "x to the power 2."

x^3 is read as "x cubed" or "x to the power 3."

x^4 is read as "x to the power 4."

x^5 is read as "x to the power 5."

In general, x^n is read "x to the power n."

EXAMPLE 10   In exponential notation $2aaa$ is written as $2a^3$. ❑

EXAMPLE 11   In exponential notation

$$(6)(6)mnnnnppppppp \text{ is written as } 6^2mn^4p^7.$$ ❑

EXERCISES 5.1B

Use exponential notation to express each term

7. $3aab$ **8.** $5xxxyy$

9. $-2abbbcc$ **10.** $8kkkkkmm$

11. $(3)(3)(3)(3)aaaaaa$ **12.** $(5)(5)wwwxyyyyy$

13. $(7)(7)(7)xxyyyywwwww$ **14.** $(2)(2)(2)(2)(2)ppmmmw$

Check your answers on page 724. ✓

Expressing Terms without Using Exponents

If an exponent appears in a term with more than one factor, it applies *only* to the factor to its left. If we wanted to apply it to any other factors we would have to enclose them in parentheses and write the exponent outside the parentheses.

EXAMPLE 12   In the term $3a^2bc^4$,

the exponent 2 applies only to a,

the exponent 4 applies only to c.

So the term $3a^2bc^4$ can be written without exponents as: $3aabcccc$. ❑

EXAMPLE 13   In the term $5(ab)^2$ the exponent 2 applies to all the factors grouped in the parentheses. So the term $5(ab)^2$ represents: $5(ab)(ab)$. ❑

We will discuss exponents again in Chapter 11, where we learn to simplify expressions involving exponents.

EXERCISES 5.1C

Express each term without using exponents.

15. $2x^3$

16. $5xy^3$

17. $7a^2b^3$

18. $4aw^5y^2$

19. $6x^2y^3w$

20. $3a^4bc^2$

21. $8p^3m^2n^5$

22. $9ab^3c^2$

Check your answers on page 725.

Evaluating Exponents

To evaluate a term involving exponents, we must be told the numerical value of each letter. Since an exponent indicates the number of times that the base factor appears in the product, we can write the term without exponents and substitute the proper number for each base factor that is a letter. Be sure to put parentheses around the number when substituting it for the letter.

EXAMPLE 14   Evaluate x^2 if $x = 5$.

$$x^2 =$$
$$xx =$$
$$(5)(5) = 25$$

EXAMPLE 15   Evaluate y^3 if $y = -2$.

$$y^3 =$$
$$yyy =$$
$$(-2)(-2)(-2) =$$
$$(4)(-2) = -8$$

When evaluating expressions, we must always apply exponents to their bases *before those bases can be multiplied by other factors.*

EXAMPLE 16   Evaluate $3w^2$ if $w = -7$.

$$3w^2 =$$
$$3(-7)^2 = \qquad \text{Apply the exponent first: } (-7)^2 = 49.$$
$$3(49) = 147 \quad \text{Then multiply.}$$

Remember, the exponent applies only to the factor to its left, not to any other factors or signs unless they are enclosed within parentheses.

EXAMPLE 17   Evaluate $-k^4$ if $k = -1$.

$$-k^4 =$$
$$-[kkkk] =$$
$$-[(-1)(-1)(-1)(-1)] =$$
$$-[+1] = -1$$

You can either evaluate each factor raised to its exponent, as we did in Example 16, or apply the exponent to its base by rewriting the base as a factor the number of times indicated by its exponent, as I will do in Example 18. The second procedure will leave you with a list of many factors; multiply them out in pairs. You might want to rearrange the order to simplify your computation.

EXAMPLE 18   Evaluate $5x^2y^3$ if $x = -4$ and $y = -2$.

$$5x^2y^3 = \quad \text{Two factors of } -4; \text{ three factors of } -2.$$
$$5(-4)^2(-2)^3 =$$
$$5(-4)(-4)(-2)(-2)(-2) = \quad \text{I'll rearrange these.}$$
$$\underline{(-4)(-2)}\,\underline{(-4)(-2)}\,\underline{(-2)(5)} = \quad \text{The product must be negative}$$
$$-(8)(8)(10) = \quad \text{because there are five negatives.}$$
$$-(64)(10) = -640 \qquad \square$$

Now let's do Example 18 again, applying the exponents to their bases first.

EXAMPLE 18   Evaluate $5x^2y^3$ if $x = -4$ and $y = -2$.

$$5x^2y^3 =$$
$$5\underline{(-4)^2}\,\underline{(-2)^3} =$$
$$\underline{5(16)}(-8) = \quad \text{The product is negative.}$$
$$-(80)(8) = -640 \qquad \square$$

Which method did you prefer? Use the one you find easiest.

When we want to group factors together we usually use a pair of parentheses ( ), but we could also use a pair of brackets [ ], or a pair of braces { }. I use brackets or braces in expressions I am evaluating because when I substitute the number value for each letter, I always write parentheses around the number. It is easier to understand what is happening in each step if we use different notations.

In Example 19, the parentheses around -7 indicate that it is to be multiplied by 3, and the brackets around $3w$ indicate that both of those factors are raised to the power 2.

EXAMPLE 19   Evaluate $[3w]^2$ if $w = -7$.

$$[3w]^2 =$$
$$[3\underline{(-7)}]^2 =$$
$$[-21]^2 =$$
$$[-21][-21] = 441 \qquad \square$$

Parentheses and brackets are an important punctuation. When they are used in an expression, they can change the problem. Note the difference between $3w^2$ in Example 16 and $[3w]^2$ in Example 19.

EXERCISES 5.1D _____

Evaluate each term.

23. $4m^2$ if $m = 3$ **24.** $2a^3$ if $a = 3$

25. $5x^2$ if $x = -2$ **26.** $4p^3$ if $p = -2$

27. $7h^4$ if $h = -1$ **28.** $8j^7$ if $j = -1$

29. $-b^2$ if $b = 7$ **30.** $(-b)^2$ if $b = 7$

31. $-x^3$ if $x = -2$ **32.** $(-x)^3$ if $x = -2$

33. $2a^2b^2$ if $a = 3$ and $b = 2$

34. $3ab^3$ if $a = -5$ and $b = -2$

35. $2x^3y^2$ if $x = -1$ and $y = 3$

36. $4a^2b^3$ if $a = -3$ and $b = 2$

37. $7a^2b^5$ if $a = 3$ and $b = -1$

38. $(2x)^2$ if $x = 3$

39. $(3y)^2$ if $y = -2$

40. $(2ab)^2$ if $a = 1$ and $b = -3$

41. $(3xy)^3$ if $x = 2$ and $y = -2$

42. $5m^3$ if $m = -2$

43. $(5m)^3$ if $m = -2$

44. $2x^2$ if $x = -3$

45. $(2x)^2$ if $x = -3$

Check your answers on page 725.

Degree of a Term

The **degree** of a term is the quantity of letter factors in the term. Recognizing the degree of a term will simplify our discussion about manipulating expressions later in this book.

EXAMPLE 20

The degree of $3x$ is one.

The degree of $7ab$ is two.

The degree of $5x^2$ ($5xx$) is two.

The degree of $3ab^2c^5$ ($3abbccccc$) is eight.

EXERCISES 5.1E

What is the degree of each term?

46. $4y$ **47.** $6abc$

48. $7x^2$ **49.** $3w^3$

50. $14m^8$ **51.** $10w^{15}$

52. $2x^3y$ **53.** $5ab^2c^4$

54. $8xw^3z^2$ **55.** $12m^3p^2n$

Check your answers on page 725.

5.2 Polynomials and Combining Like Terms

Concept of a Polynomial

If we took an inventory of my attic and listed things as we happened to find them, part of the list might look like this:

3 books, 4 lamps, 1 chair, 2 tables, 5 books, 1 lamp, 2 chairs

To get a clear idea of what is in my attic, it would make sense to collect all the books in one pile, all the lamps in another pile, chairs in another, and tables in yet another. In our list we could group similar things by using parentheses:

(3 books and 5 books) + (4 lamps and 1 lamp)

+ (1 chair and 2 chairs) + (2 tables)

We could simplify our list by totaling each group:

(8 books) + (5 lamps) + (3 chairs) + (2 tables)

In algebra we always want to simplify lists of different terms, just as we did with our inventory list, by combining similar terms.

A single term is called a **monomial**. (The prefix "mono" means one; the suffix "nominal" means term.)

A sum of two terms is called a **binomial**. (The prefix "bi" means two.)

A sum of three terms is called a **trinomial**. (The prefix "tri" means three.)

A sum of two or more terms can be be called a **polynomial**. (The prefix "poly" means many.)

EXAMPLE 1 $5x$ is a monomial.

$5x + 3y$ is a binomial.

$5x + 3y - 7z$ is a trinomial.

$5x + 3y - 7z + w$ is a polynomial (and so are the last two examples). ☐

Degree of a Polynomial

The degree of a polynomial is the degree of the term with the largest quantity of letter factors.

EXAMPLE 2   The degree of the polynomial

$3x - 5y + 2w - 5xw + 7wy + 4xwy$ is three;

$3ab + 5bcd - 6abcd$ is four;

$x^3 - 2x^2 + 5x - 3$ is three, because $x^3 = xxx$. ☐

Combining Like Terms

To simplify a polynomial, we combine like terms. Terms are alike if their letter factors are exactly the same. To combine them, we add their coefficients.

We could represent our original inventory list of items in the attic as an algebraic polynomial by using the first letters of the items as the letter factors of each term.

EXAMPLE 3   Simplify that polynomial.

$3B + 4L + C + 2T + 5B + L + 2C =$

$(3B + 5B) + (4L + L) + (C + 2C) + 2T = 8B + 5L + 3C + 2T$ ☐

Combining like terms is not hard; It is also not hard to make a careless error if you are not well organized and concentrating on what you are doing. I suggest that you group similar terms in parentheses, as I did in Example 3, or that you mark similar terms in some way. For instance, you can underline, double underline, check, or circle similar terms.

EXAMPLE 4   Simplify the polynomial $2a + 7b + 6a + 4a + 3b$.

$$2\underline{a} + 7\underline{b} + 6\underline{a} + 4\underline{a} + 3\underline{b} =$$
$$\underline{(2a + 6a + 4a)} + \underline{(7b + 3b)} = 12a + 10b \qquad \square$$

Remember, if no coefficient is expressed, it is understood to be 1. Also note that constant terms are like terms and should be combined into a single constant term.

EXAMPLE 5   Simplify $7m + 4 + 3n + n + 9 + m$.

$$7\underline{m} + 4 + 3\underline{\underline{n}} + \underline{\underline{n}} + 9 + \underline{m} =$$
$$\underline{(7m + m)} + \underline{\underline{(3n + n)}} + (4 + 9) =$$
$$(7m + 1m) + (3n + 1n) + (4 + 9) = 8m + 4n + 13 \qquad \square$$

When you list the terms in a polynomial, the order of the terms does not matter. When checking your answers, make sure that you have the same number of terms as I do and that each of your terms is in my list.

EXERCISES 5.2A

Simplify each polynomial by combining like terms.

1. $4a + 7b + a + 4a + 3b =$

2. $9x + 7w + 2w + 8x + 3w =$

3. $7c + 5m + 3c + 7k + 8c + 2m + 6k =$

4. $9n + 3u + 4u + 8f + 3n + 11f + 7u =$

5. $6k + 3p + 5k + 8h + 4k + 5h + 11p =$

6. $6a + a + 4b + 3a + 7b + b =$

7. $4c + c + 8 + 2c + 1 + 3 =$

8. $9w + 4x + 3 + 7x + x + 8w + 1 =$

9. $4h + 7k + h + 12 + 3k + 10h + k =$

10. $12b + 13 + 5k + 2 + 3b + k + 7 =$

11. $7u + 5 + u + 48 + 36w + 17u + 3 =$

12. $19 + 19x + w + 19w + 19 + 17x + 5 + x =$

13. $3m + 14 + 11n + 10 + n + 18m + 7 + m + 1 =$

14. $7x + 3w + 5 + y + 3w + 5 + x + 5 + 4y + 3x =$

Check your answers on page 725.   ✓

Combining Positive and Negative Like Terms

In the previous section all the terms in the polynomials were positive. Terms can also be negative. The sign of the term is always written to the left of the term. The easiest way to combine negative and positive terms is to consider the sign of the term to belong to the coefficient. Remember the rules for adding signed numbers that we learned in Section 2.2.

To simplify polynomials by combining like terms:

1. Group together the like terms.

2. In each group of like terms, add the coefficients with the same signs first (add the magnitudes and keep the signs).

3. Then add the resulting coefficients of like terms with different signs (subtract the coefficient with the smaller magnitude from the coefficient with the larger magnitude, and keep the sign of the coefficient with the larger magnitude.)

EXAMPLE 6 Simplify $5w - x + 15 - 3x - 8 - 7w + 5x + w - 2w$.

$$5w - x + 15 - 3x - 8 - 7w + 5x + w - 2w =$$
$$(5w - 7w + w - 2w) + (-x - 3x + 5x) + (15 - 8) =$$
$$(5w - 7w + w - 2w) + (-x - 3x + 5x) + (15 - 8) =$$
$$(6w - 9w) + (-4x + 5x) + (15 - 8) = -3w + x + 7 \quad \square$$

EXERCISES 5.2B

Simplify each polynomial by combining like terms.

15. $4a - 7b + b - 7a + a =$

16. $m - 5k + 11 - 4 - 6m + 2k =$

17. $27 - x - 14 - 7x - 3x =$

18. $p - 7w + 3 - 5w - 8p - 15 =$

19. $-14 - 3w + w + 7 - 3 =$

20. $5x - 3x + 7 - 11 - x + 4 + 2x =$

21. $3n - m - 5m - 7n + 3m - n =$

22. $3k - 4p - k + 7p + 5p - 6k + k =$

23. $4w - 11 - 5w - 2w - 4 - 3 =$

24. $y - 7w + 8y - w - 4w - 9y + y =$

25. $2z - w - 3z + 5z + 6w + w =$

26. $8a - 3b - 2c + c - 5b - 13a - a =$

27. $5k + k - 6g - 2g + g - 10k =$

28. $15u + 13w - w + 5w - 4u + u =$

29. $18 - 17r + 23 - 15r + 7r - 12 + 4w - 16 + r - 13w =$

30. $24h + 16j - 10 + 3h - 19j - j + 4h - 13 - 7h =$

Check your answers on page 725. ✓

Terms with More Than One Letter Factor

Some terms have *more than one letter factor*. Such terms should be *combined if they have the same letter factors* and are therefore alike.

EXAMPLE 7 Simplify by combining all xy terms and all xw terms.

$$3xy - \underline{xw} + 4\underline{xw} + 4\underline{xy} + 6\underline{xw} - 5\underline{xy} =$$
$$(3xy + 4xy - 5xy) + (-xw + 4xw + 6xw) =$$
$$(7xy - 5xy) + (-xw + 10xw) = 2xy + 9xw \qquad \square$$

Since the letter factors represent numbers that are being multiplied, the order of the letters does not matter. An xy term is the **same** kind of term as a yx term.

EXAMPLE 8 Assume that x represents 3 and y represents 5.

$$xy = \qquad yx =$$
$$(3)(5) = 15 \qquad (5)(3) = 15$$

So $xy = yx$. $\square$

The terms xy and yx should be combined when you simplify a polynomial. It does not matter if you express the resulting term as xy or yx. I usually try to leave the factors in each term in alphabetical order.

EXAMPLE 9 Simplify. (Note that ab and ba terms are the same; bc and cb terms are the same; and ac and ca terms are the same.)

$$3ab + 5bc + 7ac + 4bc + ba + 2ca + 7bc + 6ab =$$
$$3ab + 5bc + \mathbf{7ac} + 4bc + ba + \mathbf{2ca} + 7bc + 6ab =$$
$$(3ab + ba + 6ab) + (5bc + 4bc + 7bc) + \mathbf{(7ac + 2ca)} = 10ab + 16bc + 9ac \quad \square$$

EXERCISES 5.2C

Simplify each polynomial.

31. $2ab + 2ac + 7ac + 5ab + ab =$

32. $4mn + 3pr + pr + 5mn - 2pr =$

33. $5xw + 7 + 9ax + xw + 5 + 4ax + ax =$

34. $xw + xy + yw + 3xy + 7yw + xw =$

35. $4mn + 3mk + 5 + 2nk + 3mk + 1 + 7mn + nk =$

36. $4xy + 2xw - 5xw - 3xy + 7xy - xw + 3xw =$

37. $5ab + 4ac + 2ab + 6ca + ba =$

38. $xy + ab + 5ba + yx + 3xy =$

39. $12xy + 7xw + 3yx + 5xw + xy - 7wy + 2wx =$

40. $ab - 3ac + 2ba + ba - 4ca + 5ac - 4ba =$

41. $4xw + 7wy + 4wx + 6yw - 2wy - wx =$

42. $4mn + mp - 3nm + 2nm - 6pm - nm + 6mp =$

43. $12wx + 3 - 5 + 3wy + 4yw + xw + wx + 3 =$

44. $17abc + 5bcd - 7bac - 2cab + bdc - 4cdb + abc =$

45. $11wxy + 4zwx + 5xwz + 7ywx - xyw + wzx =$

46. $14jkm + 5jmn - kmj + 3mnj - 7jkm + jmk =$

Check your answers on page 725. ✓

Terms with Repeated Letter Factors

It is possible to have terms in which a letter factor is repeated. Terms are alike only if they have exactly the same combination of letter factors. If a factor is repeated, it must appear in terms the same number of times for the terms to be alike and be combined.

EXAMPLE 10   The terms x, xx, and xxx are all different. ❑

EXAMPLE 11   Simplify $3x + 5xx - 7xxx + xx - 4x + 3xxx$.

$$3\underline{x} + 5\underline{xx} - 7\boldsymbol{xxx} + \underline{xx} - 4\underline{x} + 3\boldsymbol{xxx} =$$
$$(3x - 4x) + (5xx + xx) + (-7\boldsymbol{xxx} + 3\boldsymbol{xxx}) = -x + 6xx - 4xxx \quad ❑$$

We usually use exponential notation to represent letter factors that are repeated in a term. Remember, terms can be combined only if their bases and exponents are exactly alike.

EXAMPLE 12   Simplify $2x + 3x^2 + 5x^3 + 7x^2 + 4x + 12x^3 + x$.

$$2\underline{x} + 3\underline{x^2} + 5\boldsymbol{x^3} + 7\underline{x^2} + 4\underline{x} + 12\boldsymbol{x^3} + \underline{x} =$$
$$(2x + 4x + x) + (3x^2 + 7x^2) + (5\boldsymbol{x^3} + 12\boldsymbol{x^3}) = 7x + 10x^2 + 17x^3 \quad ❑$$

EXERCISES 5.2D

Simplify each polynomial by combining like terms.

47. $8xx + 2xxx + 7xxx + 9xx + 13xx - 2xxx =$

48. $5x + 3xx + 4xxx + 2xx + 7x + 8xxx - xx =$

49. $14yy + 5yyyy + 11yyyy - 4yy - 3yyyy + 2yy =$

50. $6x + 3yy - 7xx + 5y + 8x - 2yy + 5xx + xx =$

51. $7ww + 3w - 5w + 12x + 7xx - 2ww + 8xx + w =$

52. $12y + 15xx + 7yy + 3x + 12y + 4xx - 8y + 3yy =$

53. $5w^2 + 3w + 5w + w^2 + 3w^2 + 2w =$

54. $6m^3 + 3m + 5m^3 + 4m^2 + 7m + m^3 =$

55. $5x + 3y + 2x^2 + 7y^2 + 8y - x + 4y^2 - x^2 =$

56. $4m^3 + 5w - 4m + 7m^3 + w + 5m + 2w^3 + w^3 =$

57. $3x^2 + 5y + 2y^2 - y + 4x^2 + 7x + 4y + x - 6y^2 =$

58. $5x^2 + 4 - 3x + 7 + x + 5x^2 + 8 - x^2 - 1 =$

59. $3p + 4g - 2 + 8g^2 - 4 + p^2 - 5p + 7g - g^2 =$

60. $1 - 4k + 3k^2 + m - 5k + 2m - 3m^2 + 5 - 2k^2 + 3 =$

Check your answers on page 725. ✓

Combining the Most Complicated Terms

The most complicated looking polynomials have terms with more than one letter and with a variety of exponents. In order to determine whether terms are alike and can be combined, you must look at each term very carefully. If the terms are alike, the base and exponent for each letter factor must be exactly the same.

EXAMPLE 13 x^2y^3 and x^3y^2 are not alike and cannot be combined.

x^2y^3 and y^3x^2 are alike and can be combined. ❑

Remember, the order of factors does not matter, but the base and exponents must be exactly the same in each term to be alike and be combined.

EXAMPLE 14 Simplify $3x^2y^3 + 5y^2x^3 + 2x^3y^2 + y^3x^2 + 2x^2y^3$.

$$\underline{3x^2y^3} + \underline{5y^2x^3} + \underline{\underline{2x^3y^2}} + \underline{y^3x^2} + \underline{2x^2y^3} =$$
$$(\underline{3x^2y^3} + \underline{y^3x^2} + \underline{2x^2y^3}) + (\underline{5y^2x^3} + \underline{2x^3y^2}) = 6x^2y^3 + 7x^3y^2$$ ❑

Be very careful when you are combining like terms. Combine only those terms that are alike. Make sure that you do combine terms that are alike, even though they may look different because of the order of the terms; all like terms must be combined to leave polynomials simplified. One of the polynomials in the exercises below cannot be simplified at all.

Check your answers. If you made a mistake, seeing the correct answer will usually allow you to tell what you did wrong.

EXERCISES 5.2E

Simplify each polynomial by combining like terms. Be careful!

61. $5x^2y^3 + x^3y^2 + 3x^3y^2 + y^3 + x^2 + 7x^2y^3 =$

62. $9xy^3 + x^3y + 4xy^3 + x^3y + 7yx^3 + 4xy^3 =$

63. $5xy^2 + 4x^2y + 7x^2y^2 + 2yx^2 + 5y^2x + 3x^2y^2 + 5xy^2 =$

64. $4x + 5x^2 + 4xy^2 + 7x^2 + 8x^2y + 4x + 9x^2 + 7yx^2 + 4y^2x =$

65. $2x^2 + 7x + 4 + 9x^3y + 5 + 3x + 4yx^3 + 7x^2 + 4xy^3 - x + 7 =$

66. $5w + 8 + 4wx^2 + 3 + 4w + x^2 + wx^2 + 9 + 4w + 5x^2w =$

67. $4x + 7y + 5x^2 + 4 + 2 + 9x + 3y + 7x^2 + 5y^2 - x + 3y + 1 =$

68. $4x^2w + 5w + 5w^2x + 7x^2 + 4w^2 + 9 + x =$

69. $3xw^2 + 5wx^2 + 7x^2w + w^2x - 3w^2x + 5wx =$

70. $x^3y^3 - 3xy^3 + 5x^3y^3 - y^3x - y^3x^3 =$

Check your answers on page 725. ✓

5.3 Adding and Subtracting Polynomials

Addition

Now that you know how to combine like terms to simplify polynomials, you know all that is necessary in order to add polynomials.

> ***To add polynomials:***   Combine the like terms from each polynomial.

EXAMPLE 1   Add $(5x^2 + 3x + 4) + (7x^2 + x - 6)$.

$$(5\underline{x^2} + 3\underline{x} + 4) + (7\underline{x^2} + \underline{x} - 6) =$$
$$(5x^2 + 7x^2) + (3x + x) + (4 - 6) = 12x^2 + 4x - 2 \qquad \square$$

EXAMPLE 2   Add $(2a + 3b + 5c) + (3a + 7b - c)$

$$(2\underline{a} + 3\underline{b} + 5c) + (3\underline{a} + 7\underline{b} - c) =$$
$$(2a + 3a) + (3b + 7b) + (5c - c) = 5a + 10b + 4c \qquad \square$$

EXERCISES 5.3A

Add the polynomials.

1. $(5a + 3b + c) + (2a + 2b + 2c) =$

2. $(3a + b + 2c) + (7a + 6b + c) =$

3. $(3x + 7y + 5) + (2x + y + 8) =$

4. $(x^2 + 3x + 2) + (x^2 + 5x + 7) =$

5. $(4x^2 + 7x + 3) + (x^2 - 2x - 3) =$

6. $(y^2 + 7y + 5) + (y^2 - 2y - 3) =$

7. $(x^5 + 7x^4 + x^3 + 5x^2 + 9) + (x^4 + 4x^3 + 7x^2 + x + 2) =$

8. $(6x^3 + 7x^2 - 4x + 8) + (3x^3 + 5x - 4) =$

9. $(15x^4 + 3x^2 + 5x + 4) + (7x^3 - x^2 - 8x - 12) =$

10. $(7y^3 - 5y^2 + y + 9) + (y^3 - 2y^2 - 7y + 2) =$

Check your answers on page 725.

Subtraction

To subtract polynomials, we apply the same definition of subtraction that we used in Section 2.3 to subtract signed numbers. To *subtract* a signed number is the same as to *add its opposite*. So a problem asking you to subtract a polynomial can be made into a problem of adding polynomials. Remember that, just as the *opposite of a number* is found by changing the sign of the number, the opposite of a term is obtained by changing the sign of the term.

> ***To subtract polynomials:***
>
> **1.** Change the sign of every term in the polynomial that is being subtracted.
>
> **2.** Add the resulting polynomial to the first one by combining like terms.

EXAMPLE 3 Subtract $(3a + 5b) - (2a + b)$.

$$(3\underline{a} + 5\underline{\underline{b}}) + (-2\underline{a} - \underline{\underline{b}}) =$$
$$(3a - 2a) + (5b - b) = a + 4b$$ ❑

EXAMPLE 4 Subtract $(7x + 9y + 8w) - (2x - 3y + 5w)$.

$$(7x + 9y + 8w) - (2x - 3y + 5w) =$$
$$(7\underline{x} + 9\underline{\underline{y}} + 8\boldsymbol{w}) + (-2\underline{x} + 3\underline{\underline{y}} - 5\boldsymbol{w}) =$$
$$(7\underline{x} - 2\underline{x}) + (9\underline{\underline{y}} + 3\underline{\underline{y}}) + (8\boldsymbol{w} - 5\boldsymbol{w}) = 5x + 12y + 3w$$ ❑

EXERCISES 5.3B

Subtract the polynomials by changing each to an addition problem. Then combine and simplify.

11. $(7a + 6b) - (a + 3b) =$

12. $(7x + 4y) - (3x + y) =$

13. $(15m + 7n) - (11m + 4n) =$

14. $(6k + p) - (2k + p) =$

15. $(7x - 3w) - (2x - w) =$

16. $(9m - 5n) - (6m - 3n) =$

17. $(11a - 6b) - (-2a + 4b) =$

18. $(12r - 3p) - (-5r + p) =$

19. $(6x + 3y - z) - (x - 4y + 7z) =$

20. $(4a - 3b - c) - (5a + 2b - 3c) =$

21. $(x^5 + 4x^2 - 4) - (2x^5 - 3x^2 + 1) =$

22. $(x^3 + 2x^2 - 5x) - (x^3 + 5x^2 - x) =$

23. $(3x^4 - 5x^3 + x^2 - 6x) - (2x^3 + 4x^2 + 7x - 1) =$

24. $(2y^3 - 7y^2 + y - 1) - (5y^2 + 6y + 4) =$

Check your answers on page 725. ✓

5.4 Dividing Polynomials by Monomials

In this section I show you how to divide polynomials by monomials. Later I will show you how some polynomials are divided by other polynomials (Chapter 14).

To divide a polynomial by a monomial, *divide every term in the polynomial by the monomial. In Chapter 3 you learned how to simplify numerical fractions and understood that the fraction notation represents division. Since each term is a product of factors, to divide a term by another term is the same as simplifying any fraction.*

To simplify a fraction, factor the numerator and the denominator. Any factors common to both the numerator and the denominator are canceled.

EXAMPLE 1 Simplify the numerical fraction $\dfrac{12}{30}$.

$$\frac{12}{30} =$$

$$\frac{\cancel{2} \cdot 2 \cdot \cancel{3}}{\cancel{2} \cdot \cancel{3} \cdot 5} = \frac{2}{5} \qquad \square$$

Letters in algebraic terms represent numbers and can be canceled just as numbers are canceled in numerical fractions.

EXAMPLE 2 Simplify the fraction $\dfrac{14abc}{6b}$.

$$\frac{14abc}{6b} =$$

$$\frac{\cancel{2} \cdot 7a\cancel{b}c}{\cancel{2} \cdot 3\cancel{b}} = \frac{7ac}{3} \qquad \square$$

If factors are written with exponents, rewrite them without exponents before canceling.

EXAMPLE 3 Simplify $\dfrac{10x^2y^3}{15xy^4}$.

$$\frac{2\cdot 5xxyyy}{3\cdot 5xyyyy}=$$

$$\frac{2\cdot\cancel{5}\cancel{x}x\cancel{y}\cancel{y}\cancel{y}}{3\cdot\cancel{5}\cancel{x}\cancel{y}\cancel{y}\cancel{y}y}=\frac{2x}{3y}\qquad\square$$

> **To divide a polynomial by a monomial:** Divide each term of the polynomial by the monomial.

EXAMPLE 4 Divide $(3ab - 6bc + 9ac)$ by $3a$.

$$\frac{(3ab - 6bc + 9ac)}{3a}=$$

$$\frac{3ab}{3a} - \frac{6bc}{3a} + \frac{9ac}{3a}=$$

$$\frac{\cancel{3}a b}{\cancel{3}a} - \frac{2\cdot\cancel{3}bc}{\cancel{3}a} + \frac{\cancel{3}\cdot 3ac}{\cancel{3}a} = b - \frac{2bc}{a} + 3c \qquad\square$$

When you divide a polynomial by a term, the result should have the same number of terms that the original polynomial had. So when the trinomial in Example 4 is divided by the monomial, the result is 3 terms.* Check that your answer has the same number of terms that the given polynomial has.

EXAMPLE 5 Divide $2x^2 + 10x - 4$ by $2x$.

$$\frac{(2x^2)}{2x} + \frac{(10x)}{2x} + \frac{(-4)}{2x}=$$

$$\frac{2xx}{2x} + \frac{2\cdot 5x}{2x} - \frac{2\cdot 2}{2x}=$$

$$\frac{\cancel{2}\cancel{x}x}{\cancel{2}\cancel{x}} + \frac{\cancel{2}\cdot 5\cancel{x}}{\cancel{2}\cancel{x}} - \frac{\cancel{2}\cdot 2}{\cancel{2}x} = x + 5 - \frac{2}{x} \qquad\square$$

EXAMPLE 6 Divide $3xy^3 - x^2y^2 + 15x^3y + 1$ by $5xy$.

$$\frac{3xyyy}{5xy} - \frac{xxyy}{5xy} + \frac{3\cdot 5xxxy}{5xy} + \frac{1}{5xy}=$$

$$\frac{3\cancel{x}yyy}{5\cancel{x}y} - \frac{\cancel{x}x\cancel{y}y}{5\cancel{x}\cancel{y}} + \frac{3\cdot\cancel{5}\cancel{x}xx\cancel{y}}{\cancel{5}\cancel{x}\cancel{y}} + \frac{1}{5xy}=$$

$$\frac{3yy}{5} - \frac{xy}{5} + 3xx + \frac{1}{5xy} = \frac{3y^2}{5} - \frac{xy}{5} + 3x^2 + \frac{1}{5xy} \qquad\square$$

---

* If a letter factor(s) exists in the denominator of a term or group of terms, the result is *not* considered to be a *polynomial*. This distinction is important in calculus and other advanced mathematics courses.

EXERCISES 5.4

Simplify each term.

1. $\dfrac{4y}{4} =$

2. $\dfrac{15xw}{3} =$

3. $\dfrac{21a}{7} =$

4. $\dfrac{2b}{b} =$

5. $\dfrac{12cd}{c} =$

6. $\dfrac{20x^2y}{2x} =$

7. $\dfrac{24ax^3}{6x} =$

8. $\dfrac{40abc^2}{8a^2c} =$

9. $\dfrac{4x^2y}{xy^2} =$

10. $\dfrac{10ab^2}{ab^3} =$

11. Divide $5w - 10$ by 5.

12. Divide $4a + 12$ by 2.

13. Divide $6a^2 - 8a + 2$ by 2.

14. Divide $6x^2 + 9x - 15$ by 3.

15. Divide $2y^2 + 6$ by y.

16. Divide $5x^2 - 15x$ by $3x$.

17. Divide $4x^2 - 6x + 12$ by $3x$.

18. Divide $12w^2 - 15w + 4$ by $2w$.

19. Divide $3x^2 - 4x + 1$ by x^3.

20. Divide $5a^2 + a - 5$ by a^4.

21. Divide $4abc + 14a^2bc^2 - 10a^3b^2c$ by $2ab$.

22. Divide $5xw^2 - 15x^3w^2 + 20x^4w$ by $10xw$.

Check your answers on page 725.

5.5 Distributive Law: Multiplying and Factoring

Multiplying Terms

Terms are simply products of numbers, some of which might be represented by letters. To multiply terms, multiply what you can. You can switch the order of the factors around, because when you multiply numbers, their order does not affect the answer.

> *To multiply terms:*
>
> 1. Regroup the factors, writing all the coefficients next to each other and all similar letter factors next to each other.
> 2. Multiply the coefficients together to find the coefficient of the product of the terms.
> 3. You cannot multiply letters until you know what numbers they represent. All you can do is write the letter factors next to each other.
> 4. If a letter factor is repeated, you can use exponential notation to make the term appear simpler.

EXAMPLE 1 Multiply $(2x)(3xy)$.

$$(2x)(3xy) =$$
$$(2)(3)\underline{xxy} = 6x^2y$$ ❑

If a term is negative, consider the negative sign to be the sign of the coefficient.

EXAMPLE 2 Multiply $(-4xy)(3yz)$.

$$(-4xy)(3yz) =$$
$$(-4)(3)x\underline{yy}z = -12xy^2z$$ ❑

If any of the factors already have exponents, you might want to rewrite them without exponents to figure out how many of each factor there will be in the term. Then put your answer back into exponential notation.

EXAMPLE 3 Multiply $(-5wx^2y^3)(-2w^2xy^2)$.

$$(-5wx^2y^3)(-2w^2xy^2) =$$
$$(-5)wxxyyy(-2)wwxyy =$$
$$(-5)(-2)\underline{(w)(ww)}\boldsymbol{(xx)(x)}(yyy)(yy) = +10w^3x^3y^5$$ ❑

EXERCISES 5.5A

Multiply the terms.

1. $(2a)(3bc) =$

2. $(5x)(4wy) =$

3. $(5kl)(7mn) =$

4. $(-7ab)(4cd) =$

5. $(4m)(-3pr) =$

6. $(-3x)(-5y) =$

7. $(-2ab)(dk) =$

8. $(-ab)(4c) =$

9. $(2ab)(-7cb) =$

10. $(-4xy)(-5yw) =$

11. $(3xy)(-5yz) =$

12. $(6ab)(4ac)(6bc) =$

13. $(-2x^2wz)(4y^2zw^2) =$

14. $(-2x^2yw)(6xw^3) =$

15. $(6klm^3)(5k^2pm^2) =$

16. $(9x^2y)(4x^3y^2w) =$

17. $(4r^3st)(-3rp^2s) =$

18. $(-10xy^3w^2)(3x^2w^2) =$

Check your answers on page 725. ✓

Multiplying Polynomials by Monomials (Distributive Law)

To represent the fact that we wish to multiply a polynomial by 1 term, we write the polynomial within parentheses and we write the term next to the parentheses.

EXAMPLE 4   To represent the fact that we wish to multiply 2 by $a + b + c$, we write

$$2(a + b + c).$$ ❏

We could interpret $2(a + b + c)$ as representing the fact that we have two groups with an "$a + b + c$" in each group. As we have seen before, multiplication problems can be transformed into addition problems.

So
$$2(a + b + c) =$$
$$(\underline{a} + \underline{\underline{b}} + \boldsymbol{c}) + (\underline{a} + \underline{\underline{b}} + \boldsymbol{c}) =$$
$$(\underline{a + a}) + (\underline{\underline{b + b}}) + \boldsymbol{(c + c)} = 2a + 2b + 2c$$

If the multiplying term were a large number or a letter, it would not be possible to determine the product by changing it into an addition problem.

We get the same result if we multiply the multiplying term by *each* term of the polynomial. Because that term gets *distributed* among the terms within parentheses, we call this process the **distributive law**.

The product of a polynomial multiplied by 1 term will always be a polynomial with the same number of terms that the original polynomial had.

We will use the distributive law to multiply in the examples below.

EXAMPLE 5
$$2(a + b + c) =$$
$$\underline{2(a)} + \underline{2(b)} + \underline{2(c)} = 2a + 2b + 2c$$ ❏

EXAMPLE 6
$$37(x + w + z) = 37x + 37w + 37z$$ ❏

EXAMPLE 7
$$5(y + 4) =$$
$$\underline{5(y)} + \underline{5(4)} = 5y + 20$$ ❏

EXAMPLE 8
$$14(x + 1) =$$
$$\underline{14(x)} + \underline{14(1)} = 14x + 14$$ ❏

EXAMPLE 9
$$a(x + y + w) = ax + ay + aw$$ ❏

EXERCISES 5.5B

Use the distributive law to multiply the terms.

19. $6(a + b) =$ **20.** $3(a + b) =$

21. $9(x + 7) =$ **22.** $2(r + w) =$

23. $7(m + n) =$ **24.** $5(k + 3) =$

25. $4(k + m) =$ **26.** $8(w + 1) =$

27. $6(m + 1) =$ **28.** $4(x + 3) =$

29. $2(a + b + c) =$ **30.** $5(x + y + z) =$

31. $8(x + y + z + w) =$ **32.** $3(m + n + p + r) =$

33. $a(b + c + d) =$ **34.** $a(x + y + w) =$

35. $x(m + n + k) =$ **36.** $x(a + b + c) =$

37. $n(a + b + g + h) =$ **38.** $r(m + n + p) =$

39. $w(a + d + g + k) =$ **40.** $y(a + x + b + w) =$

Check your answers on page 725. ✓

Multiplying More-Complicated Polynomials (Distributive Law)

Now let's look at multiplying polynomials by the distributive law when the terms are more complex.

I suggest that you write the monomial next to each term of the polynomial within parentheses before you begin multiplying the terms. I put a "+" between indicated products, because the polynomial is a *sum* of terms. In the next step I multiply out the coefficients, keeping the rules of signed numbers in mind. I write the correct sign to the left of each term after I find each product. These procedures help eliminate careless errors that are frequently made when the terms involve negative signs and/or coefficients.

EXAMPLE 10
$$5(x - y - z) =$$
$$\underline{5(x)} + \underline{5(-y)} + \underline{5(-z)} = 5x - 5y - 5z \qquad ❑$$

EXAMPLE 11
$$3(2a - 5b - 7c) =$$
$$\underline{3(2a)} + \underline{3(-5b)} + \underline{3(-7c)} = 6a - 15b - 21c \qquad ❑$$

EXAMPLE 12
$$-5(2x - 3y + z) =$$
$$\underline{(-5)(2x)} + \underline{(-5)(-3y)} + \underline{(-5)(z)} = -10x + 15y - 5z \qquad ❑$$

EXAMPLE 13
$$3a(2ab - 5abc - bc) =$$
$$\underline{(3a)(2ab)} + \underline{(3a)(-5abc)} + \underline{(3a)(-bc)} =$$
$$\underline{(3)(2)(aab)} + \underline{(3)(-5)(aabc)} + \underline{(3)(-1)(abc)} =$$
$$\underline{(6aab)} + \underline{(-15aabc)} + \underline{(-3abc)} = 6a^2b - 15a^2bc - 3abc \qquad ❑$$

EXAMPLE 14
$$-3x(5x^3 - 2x^2 - 1) =$$
$$\underline{(-3x)(5xxx)} + \underline{(-3x)(-2xx)} + \underline{(-3x)(-1)} =$$
$$\underline{(-3)(5)xxxx} + \underline{(-3)(-2)xxx} + \underline{(-3)(-1)x} = -15x^4 + 6x^3 + 3x \qquad ❑$$

EXERCISES 5.5C

Use the distributive law to multiply.

41. $5(a - b) =$ **42.** $3(x - y) =$

43. $4(x + y - w) =$ **44.** $5(a - b - c) =$

45. $x(a + b - c) =$ **46.** $a(x - y + z) =$

47. $6(x - 6) =$ **48.** $8(x - 2) =$

49. $7(w - 5) =$ **50.** $6(b - 3) =$

51. $5(x + 9y - 4z) =$ **52.** $4(a + 3b - c) =$

53. $8(3a - 4b + c) =$ **54.** $7(2x - 4y + 3w) =$

55. $3(11a - 4b - 7c) =$ **56.** $6(3m - 4n - 7p) =$

57. $-2(3x + 5y - 4z) =$ **58.** $-3(2a - 5b - c) =$

59. $-5(-3d + 4g - 5h + k) =$ **60.** $-4(-3x + y - 4z + 5w) =$

61. $a(a^2 + 5a + 7) =$ **62.** $x(7x^2 - 4x - 3) =$

63. $2y(y^2 - 3y - 2) =$ **64.** $5m(m^2 - m + 2) =$

65. $3w(2w^2 - 5w + 1) =$ **66.** $2x(-4x^2 - 3x + 5) =$

67. $-2b(5ab - 3c - 3) =$ **68.** $-4k(3mk - 5n + 3) =$

69. $-3x(x^2 - 5x - 2) =$ **70.** $-6p(3p^3 - 4mp^2 + 2mp) =$

Check your answers on page 726.

Understanding Multiplication of Large Numbers

We can use the distributive law to simplify the computation in multiplication problems with large numbers.

EXAMPLE 15   Use the distributive law to multiply 13 by 213.

$$13(213) =$$

$$13(200 + 10 + 3) =$$

$$13(200) + 13(10) + 13(3) =$$

$$2600 + 130 + 39 = 2769$$

This demonstration helps to clarify why we were taught that, when multiplying large numbers, we should multiply one number by each digit of the other and then add the results. Remember, we were told that we should move one place to the left when writing the products for each digit.

EXAMPLE 16   Multiply 13 by 213 the old way.

$$
\begin{array}{r}
13 \\
\times 213 \\
\hline
39 \\
13 \\
26 \\
\hline
2769
\end{array}
$$

You may not have understood why you did that. Now that you know about place values, you can see that the 1 in 213 is really 10, which is why the product of 13 multiplied by 1 is moved to the left to begin in the tens place. The 2 in 213 is really 200, which is why the product of 13 multiplied by 2 is moved to the left to begin in the hundreds place.

Factoring by the Distributive Law

In Section 3.3, you learned how to factor numbers. Factoring a number means writing it as a product of the prime numbers that divided exactly into the number. We call those prime numbers the **prime factors** of the number.

EXAMPLE 17 Factor 30: $30 = 2 \cdot 3 \cdot 5$.

We say that 2, 3, and 5 are the prime factors of 30. ❑

In algebra, we factor polynomials. Now that you know how to use the distributive law to multiply through polynomials, you can use the law *backward* to factor some polynomials.

Factoring a polynomial by the distributive law means reversing the multiplication process we have just been doing. It means rewriting polynomials as the product of a **term** times a **simpler polynomial**. Each of these is a *factor* of the original polynomial.

We call the term, the polynomial's **monomial factor**; and we call the simpler polynomial the polynomial's **polynomial factor**.

EXAMPLE 18 You know that $2a + 2b + 2c = 2(a + b + c)$. So the factors of $2a + 2b + 2c$ are 2 and $(a + b + c)$.

2 is the *monomial* factor

$(a + b + c)$ is the *polynomial* factor ❑

EXAMPLE 19 $3x(x^2 - 2x + 5) = 3x^3 - 6x^2 + 15x$. So the *factors* of $3x^3 - 6x^2 + 15x$ are

$3x$, the *monomial* factor, and

$(x^2 - 2x + 5)$, the *polynomial* factor. ❑

To factor a polynomial, start by identifying the factors of the first term. Then see whether any of these factors is also a factor of *every* other term in that polynomial. If so, then that common factor (or factors) is the monomial factor of the polynomial.

EXAMPLE 20 7 is the common (monomial) factor of $7a + 7b + 7c$. ❑

To find the polynomial factor, we have to figure out what terms we must multiply the monomial factor by to produce each term in the original polynomial. Don't forget that there must be the same number of terms in the polynomial factor as there are in the original polynomial.

EXAMPLE 21 To factor $7a + 7b + 7c$, we must determine what three terms will make up the polynomial factor. Since 7 is the common factor, we must decide:

first term: what term, multiplied by 7, will produce $7a$;

second term: what term, multiplied by 7, will produce $7b$;

third term: and what term, multiplied by 7, will produce $7c$.

The first term is a, the second term is b, and the third term is c. The polynomial factor is $(a + b + c)$.

So $7a + 7b + 7c$ factored is $7(a + b + c)$. ❑

You can always **check** to see whether you have factored a polynomial correctly: When you multiply the correct factors, their product will be the original polynomial.

EXAMPLE 22 Factor $ab + ac + ad + ae$.

The common factor of $ab + ac + ad + ae$ is a. This polynomial has four terms, so the polynomial factor will also have four terms.

The first term is b, since $a(b) = ab$.

The second term is c, since $a(c) = ac$.

The third term is d, since $a(d) = ad$.

The fourth term is e, since $a(e) = ae$.

The polynomial factor is $(b + c + d + e)$.

So, the factors of '$ab + ac + ad + ae$' are a and $(b + c + d + e)$.

Check: $a(b + c + d + e) =$
$$a(b) + a(c) + a(d) + a(e) = ab + ac + ad + ae \quad \checkmark$$

If any term is negative, be careful about the signs. The sign of the term in the polynomial factor that will produce that term must also be *negative*, if the monomial factor is positive. Otherwise their product can not be negative.

EXAMPLE 23   Factor $4x + 4y - 4w$.

4 is the common factor.

$(x + y - w)$ is the polynomial factor.

$$4x + 4y - 4w = 4(x + y - w)$$

Check: $4(x + y - w) = 4x + 4y - 4w \quad \checkmark$

If *all* terms in a polynomial are *negative*, then the common factor is *negative*. When it is factored out, *all* terms in the polynomial factor will be *positive*. Otherwise their product could not be negative.

EXAMPLE 24   In $-3x - 3y - 3w$, -3 is the common factor.

$(x + y + w)$ is the polynomial factor.

$$-3x - 3y - 3w = -3(x + y + w)$$

Check: $-3(x + y + w) = -3x - 3y - 3w \quad \checkmark$

EXERCISES 5.5D

Factor each polynomial into a common factor and a polynomial factor. Check your answers by multiplying them by the distributive law.

71. $7a - 7b =$ | **72.** $3x - 3y - 3w =$

73. $-2a - 2b - 2c =$ | **74.** $5p - 5r + 5w =$

75. $6y - 6d - 6c =$ | **76.** $-5a - 5b - 5c =$

77. $-3a - 3b - 3c =$ | **78.** $8v - 8w + 8y - 8x =$

79. $ay + by + cy =$ | **80.** $pk + pr - pm =$

81. $3ry + 5rm - 7rc =$ | **82.** $6pn + 5nr - 2wn =$

83. $4ac - 3cd + bc =$ | **84.** $10xh - 3hj + kh - 2h =$

85. $6ab - bc - 5db =$ **86.** $3aw - 5bw + 7w + xw =$

87. $2d - 3ad + cd =$ **88.** $-4ax - 3xb - cx - 9wx =$

Check your answers on page 726. ✓

Factoring to Prime Polynomial Factors

Now that you understand the basic idea about factoring polynomials, let's look at some polynomials that are more complicated than those we factored in the previous examples.

EXAMPLE 25   Factor $7ab + 7ac + 7ad$.

7 is a common factor of all the terms. The terms needed for the polynomial factor are ab, ac, and ad. So,

$$7ab + 7ac + 7ad = 7(ab + ac + ad).$$ ❏

The polynomial factor $(ab + ac + ad)$ in this example is not prime. All the terms still have a *common* factor of a. If we had looked carefully at the polynomial $7ab + 7ac + 7ad$ we would have found two common factors: 7 and a. Each of these factors appears in the first term and in all the others. When more than one common factor appears in a polynomial, use the product of *all the common factors* as the *monomial factor*. If you do not put *all* the common factors into the monomial factor, you will not get a *prime* polynomial factor.

EXAMPLE 26   Factor $7ab + 7ac + 7ad$ so that the polynomial factor is prime.

Because 7 and a are both common factors, the monomial factor is $7a$.

The polynomial factor is $(b + c + d)$. So,

$$7ab + 7ac + 7ad = 7a(b + c + d).$$

Check: $7a(b + c + d) = 7ab + 7ac + 7ad$ ✓ ❏

EXAMPLE 27   Factor $3xy + 3xz$.

3 and x are both common factors, so the monomial factor is $3x$.

The polynomial factor is $(y + z)$. So,

$$3xy + 3xz = 3x(y + z).$$ ❏

As the polynomials get more complicated, it is not always obvious, just by looking at the polynomial, what the terms must be in the prime polynomial factor. There is a way to always determine what each term in the prime polynomial factor must be:

Divide each term in the original polynomial by the monomial common factor.

This works because those quotients are the terms that the **monomial factor** must be multiplied by to produce the original polynomial. (For example, in a polynomial with the common factor of $2m$, to find the term that will produce $2mn$, we divide $2mn$ by $2m$. The quotient is n, the term which multiplied by $2m$ produces $2mn$.)

EXAMPLE 28   Factor $2mn - 2mp - 2mk$.

The monomial common factor is $2m$.

We divide each term by $2m$ to find the polynomial factor's terms.

$$\frac{2mn}{2m} = \qquad -\frac{2mp}{2m} = \qquad -\frac{2mk}{2m} =$$

$$\frac{2mn}{2m} = n \qquad -\frac{2mp}{2m} = -p \qquad -\frac{2mk}{2m} = -k$$

So the polynomial factor is $(n - p - k)$:

$$2mn - 2mp - 2mk = 2m(n - p - k).$$ ❑

When *all* the factors in a term are in the monomial factor and you divide that term by the monomial factor, their quotient is 1. People often make a careless error in such problems by forgetting to list the term 1 in the prime polynomial factor.

EXAMPLE 29   Factor $5axy - 5awy + 5ay$.

5, a, and y are all common factors, so the monomial factor is $5ay$.

$$\frac{5axy}{5ay} = x \qquad -\frac{5awy}{5ay} = -w \qquad +\frac{5ay}{5ay} = 1$$

So, $5axy - 5awy + 5ay = 5ay(x - w + 1).$ ❑

EXERCISES 5.5E

Factor each polynomial into a product of prime factors. Check your answers by multiplying them by the distributive law.

89. $2ab + 2ac + 2ad =$

90. $5ar + 5rw + 5kr =$

91. $7wd - 7dh - 7xd =$

92. $15aq - 15bq + 15cq =$

93. $8cb - 8cd + 8cf - 8cw + 8cg =$

94. $3abc + 5abd =$

95. $9as - 9cs + ds - 9ps =$

96. $-7abc - 7abd - 7abf =$

97. $6pmn - 6prn - 6pkn + 5pwn =$

98. $4ja + 4jb - 4jc + 3jd =$

99. $3xyw + 3ywz + 3yw =$

100. $9abc - 9acd + 9ac =$

Check your answers on page 726. ✓

Factoring Polynomials with Non-prime Coefficients

If the coefficients are not prime, the numerical common factor may not be obvious when you first look at the polynomial. Factoring the coefficients to primes should make all common factors obvious.

EXAMPLE 30   Factor $15x + 9y - 6w$.

$$15x + 9y - 6w = 3 \cdot 5x + 3 \cdot 3y - 2 \cdot 3w$$

Now we can recognize the common factor of 3:

$$3 \cdot 5x + 3 \cdot 3y - 2 \cdot 3w = 3(5x + 3y - 2w).$$

When we factored numbers in Chapter 3, we left them in factored form. But the final answer to a factored polynomial should be expressed with the coefficients written as one number; the numbers should not be left in factored form.

EXAMPLE 31   Factor $30a + 50b - 10c$.

$$30a + 50b - 10c =$$
$$2 \cdot 3 \cdot 5a + 2 \cdot 5 \cdot 5b - 2 \cdot 5c =$$
$$(2 \cdot 5)(3a + 5b - c) = 10(3a + 5b - c).$$

Check: $$10(3a + 5b - c) =$$
$$10(3a) + 10(5b) + 10(-c) = 30a + 50b - 10c \quad \checkmark$$

EXAMPLE 32   Factor $12ab + 30ac - 24ad + 6af$.

$$2 \cdot 2 \cdot 3ab + 2 \cdot 3 \cdot 5ac - 2 \cdot 2 \cdot 2 \cdot 3ad + 2 \cdot 3af =$$
$$2 \cdot 3a(2b + 5c - 2 \cdot 2d + f) = 6a(2b + 5c - 4d + f)$$

Check: $$6a(2b + 5c - 4d + f) =$$
$$6a(2b) + 6a(5c) + 6a(-4d) + 6a(f) = 12ab + 30ac - 24ad + 6af \quad \checkmark$$

EXERCISES 5.5F

Factor each polynomial. Check your answers by multiplying them by the distributive law.

101. $10a + 60b =$

102. $30x + 15y - 60w =$

103. $16f + 20g + 28k =$

104. $28p + 21k - 42m =$

105. $24r - 16p + 16pr =$

106. $24hk + 16km - 32kr =$

107. $45ab - 30db + 15pb =$

108. $12pqr + 30pkr - 42pmr =$

Check your answers on page 726. $\checkmark$

Factoring Polynomials Involving Exponents

If a polynomial has any letter factors repeated, they are probably written with exponents. Until you get comfortable dealing with exponents, the easiest way to find common factors without making any mistakes is to rewrite the terms without exponents. Take out common factors as many times as they appear in *all the terms*.

EXAMPLE 33   Factor $3x^2 + 5x$.

$$3x^2 + 5x =$$
$$3xx + 5x = x(3x + 5)$$

Check: $$x(3x + 5) =$$
$$\underline{x(3x)} + \underline{x(5)} = 3x^2 + 5x \quad \checkmark$$ ❑

EXAMPLE 34   Factor $2x^2y^3 + 3x^4y - 5x^3y^3$.

$$2xxyyy + 3xxxxy - 5xxxyyy =$$
$$xxy(2yy + 3xx - 5xyy) = x^2y(2y^2 + 3x^2 - 5xy^2)$$

Check: $$x^2y(2y^2 + 3x^2 - 5xy^2) =$$
$$x^2y(2y^2) + x^2y(3x^2) + x^2y(-5xy^2) = 2x^2y^3 + 3x^4y - 5x^3y^3 \quad \checkmark$$ ❑

EXAMPLE 35   Factor $6a^3b^5 - 10a^2b^4 + 4a^5b^3$.

$$2 \cdot 3aaabbbbb - 2 \cdot 5aabbbb + 2 \cdot 2aaaaabbb =$$
$$2aabbb(3abb - 5b + 2aaa) = 2a^2b^3(3ab^2 - 5b + 2a^3)$$

Check: $$2a^2b^3(3ab^2 - 5b + 2a^3) =$$
$$2a^2b^3(3ab^2) + 2a^2b^3(-5b) + 2a^2b^3(2a^3) = 6a^3b^5 - 10a^2b^4 + 4a^5b^3 \quad \checkmark$$ ❑

EXERCISES 5.5G

Factor a monomial from each polynomial. Check your answers by multiplying them by the distributive law.

109. $7y^2 + 8y =$ **110.** $9w + 5w^2 =$

111. $7a^2 - 3a =$ **112.** $5m^2 - 4m =$

113. $3x^2 + 12x =$

114. $4y^2 - 20y =$

115. $6w^2 - 15w =$

116. $10m^2 + 35m =$

117. $x^3 + 4x^2 + 5x =$

118. $2y^3 - 7y^2 + 4y =$

119. $x^3y + 5x^2y - 3xy =$

120. $w^3x^2 - 2w^2x^3 + 4wx =$

121. $2m^3n + 6m^2n + 10mn =$

122. $12a^4b^3 - 6a^3b^2 + 9ab =$

123. $21w^5y^3 - 15w^3y^2 + 12w^4y^5 =$

124. $10xy^4 - 40x^3y^3 - 15x^2y^2 =$

Check your answers on page 726.   ✓

5.6   Order of Operations with Polynomials

When we computed numerical problems that involved more than one operation, we learned the following law for the order of operations:

1. Simplify within the parentheses.
2. Perform all multiplication and division.
3. Perform all addition and subtraction.

Now that you know how to add and subtract polynomials and to multiply polynomials by monomials, you can simplify expressions that involve more than one of these operations. You must always follow the order of operations when working with polynomials as well as with numerical problems, because the letters in polynomials represent numbers.

EXAMPLE 1   Simplify $2(x^2 + 3x + 5) + 7(x^2 - 2x - 1)$.

$$2(x^2 + 3x + 5) + 7(x^2 - 2x - 1) =$$
$$2(x^2) + 2(3x) + 2(5) + 7(x^2) + 7(-2x) + 7(-1) =$$
$$2x^2 + 6x + 10 + 7x^2 - 14x - 7 = 9x^2 - 8x + 3 \qquad \square$$

In these problems we usually do not have to perform any computation within the original parentheses. You should always check whether simplification is possible; but if a polynomial within parentheses has no similar terms, there is nothing to simplify. As in Example 1, if the polynomial has nothing to simplify, the first step is to multiply the terms by the polynomial.

If there is a minus sign to the left of a number written outside a set of parentheses, as in $3(x^2 - 4x - 1) - 6(2x^2 - 5x + 4)$, remember that you can interpret it as the *sign* of 6. You can remove the second set of parentheses most easily by multiplying each term within it by -6.

EXAMPLE 2   Simplify $3(x^2 - 4x - 1) - 6(2x^2 - 5x + 4)$.

$$3(x^2 - 4x - 1) - 6(2x^2 - 5x + 4) =$$
$$3(x^2) + 3(-4x) + 3(-1) + (-6)(2x^2) + (-6)(-5x) + (-6)(4) =$$
$$3x^2 - 12x - 3 - 12x^2 + 30x - 24 = -9x^2 + 18x - 27$$
$$\square$$

If a problem includes a polynomial within a set of parentheses that has only a minus sign to the left (no number of letter), you have two choices for interpreting the problem: The minus sign to the left of a set of parentheses represents (1) *multiplying the terms within the parentheses by* -1, or *subtracting* the polynomial. In either case, the result would be that the sign of every term within the parentheses would be changed. All you need to do to remove the parentheses from that polynomial is to *change the sign of every one of its terms*.

EXAMPLE 3   Simplify $4(x^2 - 2x + 1) - (3x^2 - x + 5)$.

$$4(x^2 - 2x + 1) - (3x^2 - x + 5) =$$
$$4(x^2) + 4(-2x) + 4(1) - 3x^2 + x - 5 =$$
$$4x^2 - 8x + 4 - 3x^2 + x - 5 = x^2 - 7x - 1 \qquad \square$$

EXAMPLE 4   Simplify $a(3a - 5b) - b(a - 2ab)$.

$$a(3a - 5b) - b(a - 2ab) =$$
$$a(3a) + a(-5b) + (-b)(a) + (-b)(-2ab) =$$
$$3aa - 5ab - ab + 2abb =$$
$$3aa - 6ab + 2abb = 3a^2 - 6ab + 2ab^2 \qquad \square$$

EXAMPLE 5   Simplify $2y(y^2 - 5y - 3) - (y^2 - 1) + y(3y^2 - y)$.

$$2y(y^2 - 5y - 3) - (y^2 - 1) + y(3y^2 - y) =$$
$$2y(yy) + 2y(-5y) + 2y(-3) - y^2 + 1 + y(3yy) + y(-y) =$$
$$2y^3 - 10y^2 - 6y - y^2 + 1 + 3y^3 - y^2 = 5y^3 - 12y^2 - 6y + 1 \qquad \square$$

EXERCISES 5.6

Combine these polynomials. First multiply them by the terms written outside the parentheses. Then combine like terms.

1. $4(x^2 + 7x + 3) + 5(x^2 + x - 1) =$

2. $3(2x^2 + x - 4) + 7(3x^2 + 4x - 3) =$

3. $6(3x^2 + 5x - 4) - 2(x^2 + x + 5) =$

4. $2(5x^2 - x + 1) - 3(x^2 + 2x - 8) =$

5. $5(x^2 + 6x + 3) + 2(x^2 - x + 7) =$

6. $3(x^2 - 6x - 4) + 5(x^2 + 2x - 4) =$

7. $2(x^2 + 3x + 1) + 3(x^2 - 2x + 5) + 4(x^2 + x - 1) =$

8. $5(x^3 + 2x + 3) + 4(x^2 - x + 2) + 5(x^3 + 4x^2 - 1) =$

9. $2(x^2 + 4x + 1) - 3(x^2 + x - 2) + 4(2x^2 + x + 3) =$

10. $7(x^3 + x^2 - x + 2) - 3(x^3 - 2x^2 + 5x + 1) - 2(x^3 - x^2 + x - 4) =$

11. $-3(2x^2 - 5x + 4) - 2(5x^2 - x + 1) =$

12. $-7(3x^2 - x - 4) - (x^2 + x - 5) =$

13. $a(2a + 4b) + a(7a - 2b) =$

14. $b(7a + 3c) + a(2b - 5c) =$

15. $x(2x + 2y - 3) + y(x - 7y + 4) =$

16. $2a(a - 5b + 3c) + 5b(2a - 4b + c) =$

17. $3w(2x + 4y - 3z) - 2y(4x + 3w - z) =$

18. $5r(r^2 - 2r + 4) - (3r^3 + 5r^2 - 7r + 4) =$

19. $2k(k^2 - 3k + 1) - 3k(2k^2 - 2k + 4) =$

20. $4m(m^2 - 3m + 5) - 2(m^3 - m^2 + 6m + 2) =$

Check your answers on page 726. ✓

5.7 Evaluating Expressions

Evaluating Polynomials

You can evaluate polynomials if you are told the number value that each letter in the polynomial represents. If a letter appears in a polynomial more than once, it *always* represents the same number every time it occurs.

To evaluate a polynomial, replace a letter with its numerical value at every occurrence of the letter. In order not to lose any of the instructions given through the problem's notation, you must replace the letters very carefully. The best way to avoid errors is to replace each letter by a *number written within parentheses*. Then follow the order of operations to compute the answer.

EXAMPLE 1 Evaluate $3w + 5$ if $w = 4$.

$$3w + 5 =$$
$$\underline{3(4)} + 5 =$$
$$12 + 5 = 17 \qquad ❏$$

Without the parentheses, the problem would erroneously look like $34 + 5$ (which is 39) rather than the *product of 3 multiplied by* 4 added to 5 (which is 17). Be sure to *place parentheses around each number* before substituting it into a polynomial to replace a letter.

EXAMPLE 2 Evaluate $y + 7x - 5w$ if $y = 3$, $x = -2$, and $w = -1$.

$$y + 7x - 5w =$$
$$(3) + \underline{7(-2)} - \underline{5(-1)} =$$
$$\underline{3} - 14 + \underline{5} =$$
$$8 - 14 = -6 \qquad ❏$$

After you substitute numerical values into polynomials for the letters, the polynomials become order-of-operations problems. You should *apply any exponent* that appears in a term to its base *before* performing any other *multiplication*. That way you will know what number is to be multiplied by the coefficient.

So the *order of operations* becomes:

1. Simplify within the parentheses.
2. Apply exponents to their bases.
3. Perform all multiplication and division.
4. Change subtraction to addition.
5. Perform all addition.

(Some people remember this order by making up a sentence with words that begin with those letters: pemdsa—*please excuse my dear silly aunt!*)

EXAMPLE 3   Evaluate $3x^3 - 5x^2 + 7x - 1$ if $x = 2$.

$$3x^3 - 5x^2 + 7x - 1 =$$

2. $[(2)^3 = 2 \cdot 2 \cdot 2 = 8]$ $3(2)^3 - 5(2)^2 + 7(2) - 1 =$
   $[(2)^2 = 2 \cdot 2 = 4]$

3. $3(8) - 5(4) + 7(2) - 1 =$

5. $24 - 20 + 14 - 1 =$

5. $+38 - 21 = +17$ ❏

Be particularly careful when evaluating expressions that involve negative values.

EXAMPLE 4   Evaluate $2x^3 - 5x^2 - x - 4$ if $x = -3$.

$$2x^3 - 5x^2 - x - 4 =$$

2. $[(-3)^3 = (-3)(-3)(-3) = -27]$ $2(-3)^3 - 5(-3)^2 - (-3) - 4 =$
   $[(-3)^2 = (-3)(-3) = +9]$

3. $2(-27) - 5(9) - (-3) - 4 =$

4. $-54 - 45 - (-3) - 4 =$

5. $-54 - 45 + 3 - 4 =$

5. $-103 + 3 = -100$ ❏

Evaluating Polynomials as a Check

Now that you know how to evaluate polynomials, you can use this skill to check the accuracy of your work.

When you simplify an expression, to *check* your answer you can substitute any numerical value that you choose for the letters. You should make the work as easy for yourself as possible. Choose small, positive whole numbers, such as 2 or 3, for the letters. Do not use 0 or 1, because they have too many special properties.

If the simplified expressions *really are equal* to the original ones, then each will produce the same value when you evaluate them with the same choice of numbers for their letters.

EXAMPLE 5   In Section 5.6, Example 1, we saw that

$$2(x^2 + 3x + 5) + 7(x^2 - 2x - 1) = 9x^2 - 8x + 3.$$

Let's check whether those two expressions really are equal.

Let $x = 2$.
$$2[x^2 + 3x + 5] + 7[x^2 - 2x - 1] =$$
$$2[(2)^2 + 3(2) + 5] + 7[(2)^2 - 2(2) - 1] =$$
$$2[4 + 3(2) + 5] + 7[4 - 2(2) - 1] =$$
$$2[4 + 6 + 5] + 7[4 - 4 - 1] =$$
$$2[15] + 7[-1] =$$
$$30 - 7 = 23 \quad \checkmark$$

Again, let $x = 2$.
$$9x^2 - 8x + 3 =$$
$$9(2)^2 - 8(2) + 3 =$$
$$9(4) - 8(2) + 3 =$$
$$36 - 16 + 3 = 23 \quad \checkmark$$

Those expressions must be equal because both have the value 23 when x is equal to 2.   ❏

Evaluation of polynomials is a powerful tool. You can use it to check your work during exams to determine whether you have made any errors. Go back to the earlier sections in this chapter and check your work by substituting small numerical values of your choice.

EXERCISES 57A

Evaluate each polynomial. Follow the order of operations.

1. $3x - 7$ if $x = 4$

2. $2y + 6$ if $y = -5$

3. $4x - 3y + 7$ if $x = -3$ and $y = 4$

4. $3x - 5y - 11$ if $x = 6$ and $y = -2$

5. $5x - 4y - w$ if $x = 2$, $y = -1$ and $w = 3$

6. $6a - 3b + 4c + 7$ if $a = -4$, $b = -2$, and $c = -1$

7. $x^2 - 5x + 2$ if $x = 3$

8. $x^2 - 6x - 5$ if $x = 4$

9. $y^2 - 6y + 3$ if $y = -2$

10. $y^2 - 5y - 4$ if $y = -4$

11. $2w^2 - 5w + 7$ if $w = 3$

12. $3w^2 - 3w - 5$ if $w = 4$

13. $5a^2 - 6a - 4$ if $a = -2$

14. $4a^2 + 7a - 1$ if $a = -3$

15. $x^3 - 3x^2 + 5x - 4$ if $x = 2$

16. $x^3 + 5x^2 - x + 6$ if $x = 3$

17. $4x^5 - 3x^4 + 7x^3 + x^2 - 5x + 2$ if $x = -1$

18. $y^7 - 6y^4 + 3y^3 - 11y^2 + 2y - 5$ if $y = -1$

Check your answers on page 726. ✓

Real-Life Applications of Polynomials

Polynomials can be used to represent real-life situations. By using a polynomial to represent the inventory of items in my attic, we could apply our skill in evaluating polynomials. We could have a yard sale with those items and could replace each letter with the price assigned to each item. By evaluating the polynomial, we could project our potential income from the sale.

EXAMPLE 6 How much money would we make if we sold all the items represented by the polynomial in Section 5.2, Example 3?

That polynomial is $8B + 5L + 3C + 2T$.

The prices are $2 per book ($B$), $4 per lamp ($L$), $5 per chair ($C$), and $7 per table ($T$).

$$8B + 5L + 3C + 2T =$$
$$8(\$2) + 5(\$4) + 3(\$5) + 2(\$7) =$$
$$\$16 + \$20 + \$15 + \$14 = \$65 \qquad \square$$

EXAMPLE 7   What is the financial outcome of the following situation? Several friends and I combined resources in a yard sale. I sold some of my rummage, but I bought some of theirs.

I sold 3 shirts (S), sold 5 records (R), bought a jacket (J), bought 2 chairs (C), sold 4 pictures (P), and bought 7 games (G).

I can represent those transactions with this polynomial (selling is positive, because I receive the money; buying is negative, because I pay the money):

$$3S + 5R - J - 2C + 4P - 7G$$

The prices were $2 per shirt, $1.50 per record, $10 per jacket, $6 per chair, $0.50 per picture, and $1.25 per game.

My transactions represent the following:

$$3(\$2) + 5(\$1.50) - (\$10) - 2(\$6) + 4(\$0.50) - 7(\$1.25) =$$
$$\$6 + \$7.50 - \$10 - \$12 + \$2.00 - \$8.75 = -\$15.25$$

So it cost me $15.25 to have the yard sale. It's a good thing that I don't try to support myself through yard sales! $\qquad \square$

EXERCISES 5.7B

Represent each situation with a polynomial. Evaluate the polynomials to answer the questions.

19. What was the total amount of sales if 5 chairs, 3 lamps, and 7 scarves were sold, and each chair cost $5, each lamp $2, and each scarf $0.40?

20. How much money did we make in the bake sale if we sold 7 cakes, 4 pies, 8 loaves of bread, and 12 cups of juice? Each cake sold for $3, each pie for $4, each loaf for $2, and each cup of juice for $0.50.

21. How much money was raised at the carnival if the merry-go-round sold 80 tickets, the whip sold 100 tickets, and the pony ride sold 25 tickets? Merry-go-round tickets cost $0.25, whip tickets cost $0.50, and pony ride tickets cost $0.60.

22. What was my grocery bill if I bought 9 TV dinners for $2.50 each, 2 cakes for $1.80 each, 10 bottles of soda for $0.65 each and 3 bags of chips for $0.89 each?

23. What was my financial outcome from the yard sale if I sold 3 records for $0.50 each, sold 4 sweaters for $2 each, bought a toy for $5, sold 5 puzzles for $0.60 each, bought 2 pairs of pants for $3 each, bought a jacket for $10, and sold 3 books for $0.10 each?

24. What was my financial outcome from the yard sale if I sold 8 records for $0.50, sold 2 sweaters for $2 each, bought a toy for $2, sold 9 puzzles for $0.85 each, bought 3 pairs of pants for $4 each, bought a jacket for $14, and sold 7 books for $0.10 each?

25. How much profit did I make if I sold 12 cakes for $6 each, 8 pies for $7 each, and 4 loaves of bread for $3 each, if flour cost me $4, nuts $6, and fruit $5?

26. What was the profit from selling 9 books for $0.50 each, 2 lamps for $3.50 each, 7 shirts for $2.00 each, 5 games for $0.40 each, and 15 photos for $2 each, if it cost me $10 to rent the booth?

27. What is my financial outcome if I bought 3 sweaters for $18 each, 2 pairs of pants for $15 each, 2 belts for $2 each, and a jacket for $35 and sold 3 albums for $2 each, a coat for $8, and 4 hats for $3 each?

Check your answers on page 726.

GLOSSARY

addition of polynomials   Operation involving the combination of like terms.

Example: Add $(5x^2 + 3x + 4) + (7x^2 + x - 6)$.

$$(5\underline{x^2} + 3\underline{x} + 4) + (7\underline{x^2} + \underline{x} - \mathbf{6}) =$$
$$\underline{(5x^2 + 7x^2)} + \underline{(3x + x)} + \mathbf{(4 - 6)} = 12x^2 + 4x - 2$$

base   Repeated factor in a term, when expressed in exponential notation. It is written below and to the left of the exponent.

Example: y^4 is a term with base y.

binomial   Sum of two terms. (Prefix "bi" means two.)

Example: $5x + 3y$ is a binomial.

coefficient   Number factor in an algebraic term.

Example: In the term $3x$, 3 is the coefficient.

constant   Term made up of only a number.

Example: The term 7 is a constant.

degree of a term   Quantity of letter factors in a term.

Example: The degree of $7ab$ is two.

degree of a polynomial   Degree of the term with the most letter factors.

Example: The degree of the polynomial

$$3x - 5y + 2w - 5xw + 7wy + 4xwy \text{ is three}.$$

distributive law   Rule that provides a shortcut for multiplying one term by a polynomial. It states that the product is obtained by multiplying that term by each term of the polynomial.

Example: Multiply $2(a + b + c)$ by the distributive law.

$$2(a + b + c) =$$
$$\underline{2(a)} + \underline{2(b)} + \underline{2(c)} = 2a + 2b + 2c$$

evaluate   Find the number that an algebraic expression represents. To evaluate an expression, we must be told the numerical value(s) of the letter(s).

Example: Evaluate x^2 if $x = 5$.

$$x^2 =$$
$$(5)^2 =$$
$$\underline{(5)(5)} = 25$$

exponent   Number written above a factor that is repeated in the term. Represents the number of times that the factor is repeated.

Example: y^4 is a term with the exponent 4.
y^4 represents $yyyy$.

exponential notation   Method of writing a term that contains one or more repeated letter factors. Frequency of the factor's occurrence is expressed by an exponent rather than by writing that factor each time it occurs.

Example: The term $5aabbb$ written in exponential notation is $5a^2b^3$.

factors   Numbers that are multiplied to produce a given number. Factors of a term are the numbers and letters that are multiplied to produce the term.

Example: The factors of $3a^2b$ are 3, a, a, and b.

like terms   Terms that have the same letter factors. Like terms can be combined by adding their coefficients.

Example: $3xy + 7xy$ can be combined into $10xy$.

monomial   Single term. (Prefix "mono" means one; suffix "nomial" means term.)

Example: $5x$ is a monomial.

multiplying terms To multiply terms:

Multiply the coefficients together to find the coefficient of their product.

Write the letter factors next to each other.

If a letter factor is repeated, use exponential notation to indicate the number of times that the factor appears in the term.

Example: Multiply $(2x)(3xy)$.

$$(2x)(3xy) =$$
$$(2)(3)xxy = 6x^2y$$

Order of Operations with Polynomials

1. Simplify within the parentheses.
2. Apply exponents to their bases.
3. Perform all multiplication and division.
4. Change subtraction to addition.
5. Perform all addition.

Example: Simplify $2(x^2 + 3x + 5) - (x^2 - 2x - 1)$.

$$2(x^2 + 3x + 5) - (x^2 - 2x - 1) =$$
$$2(x^2) + 2(3x) + 2(5) - (x^2 - 2x - 1) =$$
$$2x^2 + 6x + 10 - x^2 + 2x + 1 =$$
$$2x^2 + 6x + 10 - x^2 + 2x + 1 = x^2 + 8x + 11$$

polynomial Sum of two or more terms that have no letter factors in the denominator. (Prefix "poly" means many.)

Example: $5x + 3y - 7z + w$ is a polynomial.

power Another name for exponent.

Example: y^4 is a term with the exponent, or power, 4.

subtraction of polynomials Operation in which we add the opposite of each term in the subtracted polynomial to any like terms in the other polynomial.

Example: Subtract $(3a + 5b) - (2a + b)$.

$$(3a + 5b) + (-2a - b) =$$
$$(3a - 2a) + (5b - b) = a + 4b$$

term Product of numbers. In an algebraic term some of the numbers are represented by letters.

Example: 7, w, mnx, $3x$, and $4ab$ are algebraic terms.

trinomial Sum of three terms. (Prefix "tri" means three.)

Example: $5x + 3y - 7z$ is a trinomial.

Chapter 5 Review

Do each of the problems. After you are confident that you have done them as accurately as possible, compare your answers with those in the back of the book. If any are wrong, go back to the section where the problem came from (indicated to the left of the problem) and review the section.

Once you understand all the sections, you will be sure to learn the skills solidly and remember how to do them if you practice them a little bit more. Turn to the Supplementary Exercises and do all the problems from any section where you had difficulty on these review exercises.

5.1

1. State the coefficient of the term $-8pr$.

2. Write in exponential notation: $6xyyyzz$.

3. Write without using exponents: $3m^3np^4$.

4. Evaluate $5xy^3$ if $x = 4$ and $y = -2$.

5. What is the degree of $8ab^3c^2$?

5.2

Simplify these polynomials.

6. $m + 3k + 4 + 8 + k + 5m + 1 =$

7. $p - 4k + 3m - k + 3m + p - 2m + 5p + 2k =$

8. $5ac + 3bc - ab + 6cb + 2ba - 7ca + 4ac =$

9. $a^3 - 4b + b^3 + 7a - 5b - 9a^3 + 6b^3 - 2a =$

10. $4xy - 3xy^2 + y^3x - 7yx + 4xy^3 + 2y^2x - 3xy^2 + 5xy =$

5.3

Combine these polynomials.

11. $(y^3 + 4y^2 - y - 1) + (4y^3 - 9y + 13) =$

12. $(3x - 5y + 7xy - 5) - (6x - y - 3yx + 8) =$

5.4

13. Simplify $\dfrac{(12ab^3)}{(8ab)}$.

14. Divide $(15x^2y - 6xy + 12y^2)$ by $3y$ and simplify.

5.5

Determine the products and simplify.

15. $7(x^2 - 5x + 1) =$

16. $(-xy)(5yw) =$

17. $-2w(w^2 - 4w + 7) =$

Factor.

18. $6ab - 6ac + 6ad =$

19. $10x^2y - 25xy^3 + 30x^4y^2 =$

Answers

1. _____

2. _____

3. _____

4. _____

5. _____

6. _____

7. _____

8. _____

9. _____

10. _____

11. _____

12. _____

13. _____

14. _____

15. _____

16. _____

17. _____

18. _____

19. _____

5.6   Simplify.   Answers

   20. $4(x^2 - 2x + 3) - 5(2x^2 - x - 3) =$

   21. $7(4a - 3b) - (6a - b) - 2a(-5 + 3b) =$

5.7   Evaluate.

   22. $a - 2b - 5c$ if $a = 4$, $b = -3$, and $c = -1$

   23. $y^3 + 3y^2 - 5y + 3$ if $y = -2$

   24. What is the net effect of the following transactions on my finances?

   I sell 2 chairs, sell 1 hat, buy 3 cups, buy a record, sell 5 shirts, and buy 2
   scarves. The prices are $5 per chair, $2 per hat, 50 cents per cup, $2 per record,
   $1 per shirt, and 30 cents per scarf. (Set up a polynomial and evaluate it to
   answer this question.)

20. _____

21. _____

22. _____

23. _____

24. _____

Check your answers on page 726.

Supplementary Exercises

Do all the problems in every section that involves a skill in which you now lack complete mastery. With a little more practice you will achieve that sense of really understanding the topics, and you will remember how to do these problems.

5.1

1. State the coefficient of the term $-13mh$.

2. Write in exponential notation: $5aabccccc$.

3. Write without using exponents: $7a^2b^4c^3$.

4. Evaluate $-3x^2y^3$ if $x = 2$ and $y = -1$.

5. What is the degree of $4x^5yw^3$?

5.2 Simplify these polynomials.

6. $x + 5y + 3 + y + 7 + 4x + 2y =$

7. $3a + 4b - c + 7b - 8a - 4c + b + 6a =$

8. $3nm - nm^2 - m^3n + 4mn + 4nm^3 + 6m^2n - 2mn^2 + 7nm =$

9. $a^3 - 4b + b^3 + 7a - 5b - 9a^3 + 6b^3 - 2a =$

10. $2wz - 5wz^2 - 3w^3z + 2zw + zw^3 + 7z^2w - 4wz^2 - 6zw =$

5.3 Combine these polynomials.

11. $(n^3 + 3n^2 - 4n + 11) + (5n^3 - 6n^2 + n - 15)$

12. $(2a - 6b + 3ab - 5) - (6a - b + 4ba + 8)$

5.4

13. Simplify. $\dfrac{(15mk^4)}{(5mk)}$.

14. Divide $(14a^2b - 6ab + 12b^2)$ by $2b$ and simplify.

5.5 Determine the products and simplify.

15. $4(y^2 - 3y - 1) =$

16. $(-ab)(-4ac) =$

17. $-3x(x^2 - 8x - 5) =$

Factor.

18. $5xy - 5xw - 5zx =$

19. $12a^2b - 15ab^3 + 30a^3b^4 =$

5.6 Simplify.

20. $2(x^2 + 3x + 5) - 4(3x^2 - 7x + 6) =$

21. $3(x - 2xy) - (5x - y) - 2x(4 + 3y) =$

5.7 Evaluate.

22. $m - 3n - 2k$ if $m = 5$, $n = -2$, and $k = -1$

Answers

1. _____

2. _____

3. _____

4. _____

5. _____

6. _____

7. _____

8. _____

9. _____

10. _____

11. _____

12. _____

13. _____

14. _____

15. _____

16. _____

17. _____

18. _____

19. _____

20. _____

21. _____

22. _____

23. $a^3 + 4a^2 - 2a - 1$ if $a = -3$

24. What is the outcome of my transactions at a yard sale if I bought a hat for $3, sold a bowl for $0.75, sold a camera for $5, sold a purse for $4, bought a puzzle for $1, bought a sandwich for $1.50, and sold a lamp for $6? (Set up a polynomial and evaluate it to answer this question.)

Answers

23. _____

24. _____

Check your answers on page 726.

Chapter 5 Test

1. Express without using exponents: $7ab^2c^3$.

Simplify.

2. $5a - 3 + 7b + 5 - 4b + a + 2 =$

3. $6xw - 3 + 5wy + 2wx - 3wy + 2 - yw =$

4. $y^3 - 5y + 2y^2 - y + 6y^3 + 5y =$

5. $2a^3b^2 - 5ba^3 + b^2a^2 - 7b^2a^3 + ab^3 - 2a^2b^2 =$

6. $(7a^3 + 5a^2 - a + 6) + (2a^3 - 3a^2 - 6a - 12) =$

7. $(9w^3 - 5w^2 + 6w + 3) - (2w^3 + 7w^2 - 4w - 1) =$

8. $(-4xy)(-3x^2y^5) =$

9. $4(3x^3 - 5x + 6) =$

10. $2a(5a^3 + 7a^2b - 5a^2b^2 + ab^3 - 9b^2) =$

11. $7(2c - 5a + 6f + 8) + 4(3c + a - 2f - 13) =$

12. $2(w^3 - 4w^2 + 6w + 7) - 5(w^3 + 6w^2 - 5w + 3) =$

13. $x(4 + 2x - 3y) - 5y(2 + x - 4y) =$

14. $\dfrac{(24ab^3c^7)}{(6b^3c^4)} =$

15. Divide $12x^3 - 15x^2y + 9xy^2$ by $3x$ and simplify.

Factor.

16. $8mw - 4mp - mk =$

17. $14xy^3 - 7x^2y^2 + 21x^4y^3 =$

Evaluate.

18. $6a - 3b + 5$ if $a = 2$ and $b = -5$

19. $3x^2 - 5x + 2$ if $x = 4$

20. $2w^2 + 4w + 1$ if $w = -2$

21. What is the result of my transactions if I sold 5 cakes for \$2 each, 4 pies for \$3 each, and 7 loaves of bread for \$1 each and bought flour for \$4, nuts for \$6, and fruit for \$3? (Set up a polynomial and evaluate it to answer this question.)

Answers

1. _____

2. _____

3. _____

4. _____

5. _____

6. _____

7. _____

8. _____

9. _____

10. _____

11. _____

12. _____

13. _____

14. _____

15. _____

16. _____

17. _____

18. _____

19. _____

20. _____

21. _____

Check your answers on page 727.

Cumulative Review
Chapters 1–5

Combine these numbers.

1. $43 - 82 + 56 + 19 - 37 - 40 =$

2. $(-16) - (38) - (-50) + (-28) - (-23)$

3. $(-10)(-37) =$

4. $(-1)(-1)(1)(1)(-2)(-3)(-1)(-2)(-3)(-1) =$

5. $\dfrac{13}{15} - \dfrac{8}{15} - \dfrac{1}{15} =$

6. $\dfrac{1}{2} + \dfrac{3}{5} =$

7. $\dfrac{7}{9} - \dfrac{5}{12} =$

8. $-3\dfrac{1}{4} - 1\dfrac{2}{3} =$

9. $\left(\dfrac{18}{35}\right)\left(-\dfrac{42}{63}\right) =$

10. $3.2 + 8.37 - 15.9 + 20.06 - 8.006 =$

11. $36.048 \div 1.2 =$

12. 12.4% of $6800 =$

Round off to the nearest integer.

13. $\dfrac{7}{20}$ 14. $\dfrac{13}{18}$

15. $4\dfrac{5}{6}$ 16. $7\dfrac{2}{5}$

17. Round off to the nearest hundredth: 560.094.

18. Round off to the nearest tenth: 340.271.

Evaluate if $m = -2$, $k = -1$, and $p = -7$.

19. $3m - 5k + p =$ 20. $m^2 =$ 21. $-m^2 =$

22. $m^3 =$ 23. $-m^3 =$ 24. $k^7 =$

Continued

Answers

1. _____
2. _____
3. _____
4. _____
5. _____
6. _____
7. _____
8. _____
9. _____
10. _____
11. _____
12. _____
13. _____
14. _____
15. _____
16. _____
17. _____
18. _____
19. _____
20. _____
21. _____
22. _____
23. _____
24. _____

Cumulative Review (*Cont.*)

25. $k^{12} =$ **26.** $m^2 + m^3 + k^7 =$ Answers

27. $3m^2 - m^2 + 4mp + k^3 =$

Simplify.

28. $3(p^3 - 5p^2 + 7p - 1) - (4p^3 - 2p^2 + 5p - 1) =$

29. $3y^4 - 15y^3 + 6y^2 - 12y$ divided by $3y =$

30. How much must we pay at the end of one year if we borrow $700 at 9% simple interest per year?

25. _____

26. _____

27. _____

28. _____

29. _____

30. _____

Check your answers on page 727. ✓

6 Solving Linear Equations and Inequalities

 **Equations and Solutions**

Until now we have been manipulating algebraic expressions without knowing what number the letter stands for unless someone has given us that information. Now we are going to learn how to determine the value of the letters by *solving equations*.

Equations are statements asserting that two expressions are equal. Algebraic equations involve algebraic expressions.

EXAMPLE 1 $2x + 1 = 7$ is an algebraic equation.   ❏

The **solution** to an equation is the numerical value of the letter(s) that makes both expressions (the left and right sides) equal.

In Sections 6.2 through 6.4 you will learn methods for solving equations. However you solve them, you must check your solutions by substituting your number values for the letters and making sure that both expressions in the equation are really equal.

To check whether a proposed number is a solution to an equation, evaluate the expressions on both sides of the equation. If you get the same value for both, then your number is a solution.

EXAMPLE 2   Is $x = 1$ the solution to $2x + 1 = 7$?

$$\text{If } x = 1, \text{ then } 2x + 1 =$$
$$2(1) + 1 =$$
$$2 + 1 = 3$$

When $x = 1$, $2x + 1 \neq 7$. (The symbol $\neq$ means not equal.) So 1 is *not* the solution.   ❏

EXAMPLE 3   Is $x = 3$ the solution to $2x + 1 = 7$?

$$\text{If } x = 3, \text{ then } 2x + 1 =$$
$$2(3) + 1 =$$
$$6 + 1 = 7 \quad \checkmark$$

When $x = 3$, $2x + 1 = 7$. So 3 *is* the solution.   ❏

EXAMPLE 4 Is $x = -5$ the solution to $3x - 12 = 3$?

$$\text{If } x = -5, \text{ then } 3x - 12 =$$
$$\underline{3(-5) - 12 =}$$
$$-15 - 12 = -27$$

When $x = -5$, $3x - 12 \neq 3$. So -5 is *not* the solution.

EXAMPLE 5 Is $x = 4$ a solution to $x^2 - 3x = 4$?

$$\text{If } x = 4, \text{ then } x^2 - 3x =$$
$$(4)^2 - 3(4) =$$
$$16 - 12 = 4 \quad \checkmark$$

So $x = 4$ *is* a solution.

In these examples, the expression on the right side was just a number. So we only had to evaluate the expression on the left side and check whether it was equal to that number.

Some equations have algebraic expressions on *both* sides. To check a solution to such equations, evaluate each side separately. If each expression results in the same value, then the proposed number is a solution.

EXAMPLE 6 Is $x = -7$ the solution to $2x + 4 = 5x + 25$?

$$\text{If } x = -7, \text{ then } 2x + 4 = \qquad\qquad 5x + 25 =$$
$$2(-7) + 4 = \qquad\qquad 5(-7) + 25 =$$
$$-14 + 4 = -10 \;\checkmark \qquad -35 + 25 = -10 \;\checkmark$$

So $x = -7$ *is* the solution to $2x + 4 = 5x + 25$.

EXERCISES 6.1

Check whether these numbers are solutions to the equations.

1. $x = 3$; $x - 6 = -3$ **2.** $a = 5$; $7a - 20 = 15$

3. $b = -3$; $2b + 1 = 7$ **4.** $w = -7$; $3w - 5 = 26$

5. $k = 2$; $-4k + 8 = 16$ **6.** $k = -2$; $-4k + 8 = 16$

7. $m = 2$; $3m - 5 = 7m - 15$ **8.** $y = 3$; $4y + 1 = 3y + 4$

9. $u = -5$; $8u - 3 = 5u - 18$ **10.** $s = -1$; $7s + 5 = s - 3$

11. $p = 2$; $p^2 + 3 = 7$ **12.** $r = -1$; $r^2 + 3r = 5r - 2$

13. $n = -3$; $n^2 - 2n = 5n - 7$ **14.** $z = 4$; $2z^2 - 17 = 3z + 3$

Check your answers on page 727. ✓

There are equations that have *no* solutions. In life we sometimes have problems without solutions; why should algebra be any different?

EXAMPLE 7 There can be no solution to $x + 5 = x + 3$.

You can never get the same result by adding 5 to some number that you get by adding 3 to the same number. ❏

All the equations in this book that you are asked to solve *will* have solutions. In this chapter we will solve only one type of equation: *equations that involve polynomials of degree one, with only one letter*. These equations can have only one solution. So here you will learn to find *the* solution to each equation. Equations that involve expressions of degree one are called **linear** equations. You will see why they are called linear in Chapter 8, where you will learn to represent the solutions to linear equations with two letters by the graphs of these equations, which are lines.

Example 5 above is a nonlinear equation: It has two solutions, 4 and -4. You will learn to solve it and similar equations in Chapter 15.

6.2 Developing a Strategy for Solving Linear Equations

There are many simple equations that you can "solve in your head." That is, if you understand what the expressions represent and what it means for a number to be a solution, you can determine the solution just by *thinking about it*.

EXAMPLE 1 The equation $x + 7 = 10$ represents the statement that some number (x), when added to 7, has the sum of 10.

The number that does so is 3.

Check: Does $x + 7 = 10$ when $x = 3$?

$$x + 7 =$$
$$(3) + 7 = 10 \checkmark$$
❏

Anytime you think you can solve an equation without writing out your work, do so. *But be sure to check whether your solution is right.*

Some equations are hard to solve in your head. They may involve large numbers, several negative signs, fractions, decimals, or too many terms to keep track of easily.

There is a **strategy** for solving all linear equations. Recall that solving an equation means to find the number value of the letter. So do whatever legitimate manipulations you can to the equation until you get the letter alone on one side of the equation. If the letter is x, for example, you must manipulate terms until the equation reads $x = $ [*some number*].

What can you legitimately do to an equation? You can add, subtract, multiply, or divide it by any number* as long as you do the same thing to *both sides of the equation*.

How do you know what you *want to do to a side*? Just ask yourself why the letter is not alone. Then add, subtract, multiply, or divide the side by the term that will help to get the letter alone. Let's look at different cases separately.

---

* The only exception to this rule is that you can never divide by *zero*, which is an undefined operation.

Solving Equations with a Constant Added to the Letter Term

Case I

If a letter term is not alone because some number is *added* to it, just add the *opposite of that number* to both sides of the equation. This strategy will remove the number from the letter side of the equation.

EXAMPLE 2 Solve $y - 15 = 71$.

y is not alone because -15 is added to it.

Add $+15$ (the opposite of -15) to both sides.

$$
\begin{array}{rcr}
y - 15 = & & 71 \\
+\,15 & & +15 \\
\hline
y \quad\;\; = & & \mathbf{86}
\end{array}
$$

Check: Does $y - 15 = 71$ when $y = 86$?

$$
\begin{aligned}
y - 15 &= \\
(86) - 15 &= 71 \quad \checkmark
\end{aligned}
$$

EXAMPLE 3 Solve $w + 378 = 293$.

w is not alone because 378 is added to it.

Add -378 (the opposite of $+378$) to both sides.

$$
\begin{array}{rcr}
w + 378 = & & 293 \\
-\,378 & & -378 \\
\hline
w \quad\;\; = & & \mathbf{-85}
\end{array}
$$

Check: Does $w + 378 = 293$ when $w = -85$?

$$
\begin{aligned}
w + 378 &= \\
(-85) + 378 &= 293 \quad \checkmark
\end{aligned}
$$

EXAMPLE 4 Solve $x - 18 = -35$.

Add $+18$ to both sides.

$$
\begin{array}{rcr}
x - 18 = & & -35 \\
+\,18 & & +18 \\
\hline
x \quad\;\; = & & \mathbf{-17}
\end{array}
$$

Check: Does $x - 18 = -35$ when $x = -17$?

$$
\begin{aligned}
x - 18 &= \\
(-17) - 18 &= -35 \quad \checkmark
\end{aligned}
$$

EXAMPLE 5 Solve $y + 3.7 = 18.2$.

Add -3.7 to both sides.

$$
\begin{array}{rcr}
y + 3.7 = & & 18.2 \\
-\,3.7 & & -3.7 \\
\hline
y \quad\;\; = & & \mathbf{+14.5}
\end{array}
$$

Check: Does $y + 3.7 = 18.2$ when $y = 14.5$?

$$
\begin{aligned}
y + 3.7 &= \\
(14.5) + 3.7 &= 18.2 \quad \checkmark
\end{aligned}
$$

EXAMPLE 6 Solve $z - \dfrac{2}{3} = \dfrac{1}{12}$.

Add $+\dfrac{2}{3}$ to both sides.

$$z - \frac{2}{3} = \frac{1}{12}$$

$$+\frac{2}{3} \quad +\frac{2}{3}$$

$$z \quad = \frac{1}{12} + \frac{2}{3} \qquad \text{Represent } \frac{2}{3} \text{ as } \frac{8}{12}.$$

$$z = \frac{1}{12} + \frac{8}{12}$$

$$z = \frac{9}{12} \qquad \frac{9}{12} \text{ reduces to } \frac{3}{4}.$$

$$z = \frac{3}{4}$$

Check: Does $z - \dfrac{2}{3} = \dfrac{1}{12}$ when $z = \dfrac{3}{4}$?

$$z - \frac{2}{3}$$

$$\left(\frac{3}{4}\right) - \frac{2}{3} = \qquad \text{12 is the 1 cm for 3 and 4.}$$

$$\frac{9}{12} - \frac{8}{12} = \frac{1}{12} \quad \checkmark \qquad \frac{3}{4} = \frac{9}{12}; \ \frac{2}{3} = \frac{8}{12}. \qquad \square$$

EXERCISES 6.2A

In Exercises 1 through 10, state what you will add to each side of the equation. Then do so to solve the equation, and check each answer.

1. $x - 48 = 70$ Add: _____

2. $x - 22 = 418$ Add: _____

3. $y - 31 = 509$ Add: _____

4. $w + 73 = 896$ Add: _____

5. $k - 108 = 978$ Add: _____

6. $m + 135 = 859$ Add: _____

7. $r - 213 = 1{,}084$ Add: _____

8. $p - 347 = 906$ Add: _____

9. $f + 815 = 3{,}467$ Add: _____

10. $a - 391 = 5{,}082$ Add: _____

Solve each equation and check to see whether you have really found the solution.

11. $m - 17 = -41$

12. $k - 35 = -803$

13. $58 + h = -90$

14. $-23 = x + 10$

15. $-34 = y + 20$

16. $11 = j - 27$

17. $k + 135 = -405$

18. $-293 = p - 501$

19. $10.3 = x + 6.02$

20. $y - 5.34 = 18.61$

21. $w - 4.03 = 7.6$

22. $a + 32.008 = 9.2$

23. $\dfrac{7}{5} = b + \dfrac{3}{5}$

24. $k - \dfrac{2}{9} = \dfrac{4}{9}$

25. $m - \dfrac{1}{3} = \dfrac{2}{5}$

26. $w + \dfrac{3}{8} = \dfrac{7}{12}$

27. $r + \dfrac{3}{10} = \dfrac{7}{15}$

28. $\dfrac{8}{15} = p + \dfrac{3}{20}$

29. $x + 2\dfrac{1}{2} = 7\dfrac{1}{2}$

30. $y - 3\dfrac{1}{2} = 1\dfrac{2}{3}$

Check your answers on page 727.

Solving Equations Where the Letter Term Has a Coefficient

Case II

If a letter term is not alone because some integer (or decimal) is *multiplied* by it, just *divide* both sides of the equation by that coefficient.

Recall that multiplication and division are *inverse* operations. If you multiply a number by 3 and then divide by 3, for example, the result is the original number. *You can undo multiplication by dividing by the same number.* The number multiplied by the letter is called its coefficient. Dividing by the coefficient will remove it from the expression.

EXAMPLE 7 You can probably solve the equation $3x = 21$ in your head. Let's use the *strategy* to solve it. x is multiplied by 3; so divide both sides by 3.

$$3x = 21$$

$$\frac{3x}{3} = \frac{21}{3}$$

$$\frac{\cancel{3}x}{\cancel{3}} = \frac{7 \cdot \cancel{3}}{\cancel{3}}$$

$$x = \mathbf{7}$$

Check: Does $3x = 21$ when $x = 7$?

$$3x =$$

$$3(7) = 21 \quad$$

The strategy is particularly helpful when the solution is a fraction or a mixed number. Such equations are hard to do without writing out the steps. If the solution is an improper fraction, express it as a mixed number for the answer, but convert it back to the improper-fraction form for the check. *Always express fractional answers in their most reduced form.*

EXAMPLE 8 Solve $15y = 3$.

(Students doing this problem in their heads often make the careless error of dividing by 3 and thinking the solution is 5. $15y \neq 3$ when y is 5!)

Divide both sides by 15.

$$\frac{\cancel{15}y}{\cancel{15}} = \frac{3}{15}$$

$$y = \frac{3}{15}, \text{ which can be reduced.}$$

$$y = \frac{1}{5}$$

Check:

$$15y =$$

$$15\left(\frac{1}{5}\right) = 3 \quad \checkmark$$

If the coefficient is negative, be sure to divide by that *negative* number.

EXAMPLE 9 Solve $-4w = 26$.

Divide both sides by -4.

$$\frac{-4w}{-4} = \frac{26}{-4}$$

$$\frac{-\cancel{4}w}{-\cancel{4}} = \frac{\cancel{2} \cdot 13}{-2 \cdot \cancel{2}}$$

$$w = -\frac{13}{2} \text{ or } -6\frac{1}{2}$$

Check:

$$-4w =$$

$$-4\left(-\frac{13}{2}\right) =$$

$$\left(\frac{-2 \cdot \cancel{2}}{1}\right)\left(-\frac{13}{\cancel{2}}\right) =$$

$$\frac{(-2)(-13)}{1} = +26 \quad \checkmark$$

EXAMPLE 10 Solve $-7a = -7$.

Divide both sides by -7.

$$\frac{-7a}{-7} = \frac{-7}{-7}$$

$$a = +1$$

Check: $-7(+1) = -7 \quad \checkmark$

EXAMPLE 11 Solve $-8p = 8$.

Divide both sides by -8.

$$\frac{-8p}{-8} = \frac{8}{-8}$$

$$p = -1$$

Check:

$$-8p =$$

$$-8(-1) = 8 \quad \checkmark$$

EXAMPLE 12 Solve $3.5x = 2.485$.

Divide both sides by 3.5.

$$\frac{3.5x}{3.5} = \frac{2.485}{3.5} \qquad 3.5)\overline{2.485} \to 35)\overline{24.85}$$

$$x = 0.71$$

$$\begin{array}{r} .71 \\ 35)\overline{24.85} \\ \underline{24\ 5} \\ 35 \\ \underline{35} \end{array}$$

Check:

$$3.5x =$$

$$(3.5)(0.71) = 2.485 \quad \checkmark$$

When the coefficient of the letter is a fraction, there is a shortcut you can use to make the problem easier. Recall from Section 3.5 that *dividing by a fraction* is the same as *multiplying by its reciprocal*.

> **To remove a fractional coefficient from an equation:** Multiply both sides by the fraction's reciprocal.

Turn a fraction upside down, and that is the reciprocal!

EXAMPLE 13 Solve $\left(\dfrac{2}{3}\right)y = \dfrac{1}{5}$.

We want to remove the coefficient $\dfrac{2}{3}$.

The reciprocal of $\dfrac{2}{3}$ is $\dfrac{3}{2}$, so multiply both sides by $\dfrac{3}{2}$.

$$\left(\frac{3}{2}\right)\left(\frac{2}{3}\right)y = \left(\frac{3}{2}\right)\left(\frac{1}{5}\right)$$

$$\left(\frac{3}{2}\right)\left(\frac{2}{3}\right)y = \frac{3 \cdot 1}{2 \cdot 5}$$

$$y = \frac{3}{10}$$

Check:

$$\left(\frac{2}{3}\right)y =$$

$$\left(\frac{2}{3}\right)\left(\frac{3}{10}\right) =$$

$$\left(\frac{2}{3}\right)\left(\frac{3}{2 \cdot 5}\right) = \frac{1}{5} \quad \checkmark$$

If the equation contains mixed numbers, convert them into improper fractions to compute.

EXAMPLE 14   Solve $\left(2\dfrac{1}{2}\right)w = 3\dfrac{1}{4}$.

The improper fraction that represents $2\dfrac{1}{2}$ is $\dfrac{5}{2}$.

The reciprocal of $\dfrac{5}{2}$ is $\dfrac{2}{5}$, so multiply both sides by $\dfrac{2}{5}$. Also convert $3\dfrac{1}{4}$ into $\dfrac{13}{4}$.

$$\left(2\frac{1}{2}\right)w = 3\frac{1}{4}$$

$$\frac{5}{2}w = \frac{13}{4}$$

$$\left(\frac{2}{5}\right)\left(\frac{5}{2}\right)w = \left(\frac{2}{5}\right)\left(\frac{13}{4}\right)$$

$$\left(\frac{\cancel{2}}{\cancel{5}}\right)\left(\frac{\cancel{5}}{\cancel{2}}\right)w = \left(\frac{\cancel{2}}{5}\right)\left(\frac{13}{\cancel{2}\cdot 2}\right)$$

$$w = \frac{13}{10} \text{ or } 1\frac{3}{10}$$

Check:

$$\left(2\frac{1}{2}\right)w =$$

$$\left(2\frac{1}{2}\right)\left(\frac{13}{10}\right) =$$

$$\left(\frac{\cancel{5}}{2}\right)\left(\frac{13}{2\cdot\cancel{5}}\right) =$$

$$\frac{13}{2\cdot 2} =$$

$$\frac{13}{4} = 3\frac{1}{4} \quad \checkmark$$

EXERCISES 6.2B

Solve each equation and check your solutions.

31. $2x = 16$ **32.** $3y = 852$

33. $5x = 75$ **34.** $-7w = 49$

35. $8w = 128$ **36.** $7y = 196$

37. $-9y = 162$ **38.** $11w = 385$

39. $10x = 537$ **40.** $-15z = 105$

41. $5y = 3$ **42.** $-3x = 2$

43. $9m = -40$ **44.** $4k = 11$

45. $5x = 33$ **46.** $-8m = 50$

47. $-6w = 38$ **48.** $7r = -23$

49. $1.4x = 0.028$ **50.** $0.06y = 900$

51. $0.21w = 1.05$ **52.** $0.003a = 451.02$

53. $\left(\dfrac{1}{3}\right)m = 5$ **54.** $\left(\dfrac{4}{5}\right)b = \dfrac{1}{2}$

55. $\left(\dfrac{2}{7}\right)c = \left(\dfrac{3}{14}\right)$ **56.** $\left(\dfrac{2}{3}\right)k = \dfrac{11}{12}$

57. $\left(2\dfrac{1}{2}\right)y = 15$ **58.** $\left(3\dfrac{1}{4}\right)w = \dfrac{7}{8}$

59. $\left(1\dfrac{1}{3}\right)a = 3\dfrac{1}{2}$ **60.** $\left(3\dfrac{1}{4}\right)w = \left(1\dfrac{2}{3}\right)$

61. $-8.3r = 0.0581$ **62.** $0.04w = -48{,}176$

Check your answers on page 727. ✓

Solving Equations Where the Letter Term is Divided by a Number

Case III

If a letter term is not alone because it is being divided by a number, just multiply both sides of the equation by that number. This strategy will remove the denominator from the letter side of the equation.

EXAMPLE 15   Solve $\dfrac{x}{5} = 17$.

x is divided by 5, so multiply both sides of the equation by 5.

$$5\left(\frac{x}{5}\right) = 5(17)$$

$$\cancel{5}\left(\frac{x}{\cancel{5}}\right) = 5(17)$$

$$x = \mathbf{85}$$

Check:

$$\frac{x}{5} =$$

$$\frac{85}{5} = 17 \;\; ✓$$

EXAMPLE 16   Solve $\dfrac{y}{0.3} = 20$.

Multiply both sides of the equation by 0.3.

$$(0.3)\left(\frac{y}{0.3}\right) = (0.3)(20)$$

$$(\cancel{0.3})\left(\frac{y}{\cancel{0.3}}\right) = (0.3)(20)$$

$$y = \textbf{6.0}$$

Check:

$$\frac{y}{0.3} =$$

$$\frac{6.0}{0.3} = 20 \quad \checkmark \qquad 0.3\overline{)6.0} \to 3\overline{)60}^{\,20}$$

By the way, what is $\left(\frac{1}{5}\right)x$?

$$\left(\frac{1}{5}\right)x =$$

$$\left(\frac{1}{5}\right)\left(\frac{x}{1}\right) = \frac{x}{5}$$

So the equation $\frac{x}{5} = 17$ is the same as $\left(\frac{1}{5}\right)x = 17$.

If we expressed Example 15 as $\left(\frac{1}{5}\right)x = 17$, the equation would be like Examples 13 and 14 in Case II. We would solve it by multiplying both sides by the reciprocal of $\frac{1}{5}$, which is $\frac{5}{1}$ or 5. Note that either way, we solve the equation by multiplying both sides by 5.

EXERCISES 6.2C

Solve each equation and check your solutions.

63. $\frac{x}{4} = 7$ **64.** $\frac{y}{5} = 11$

65. $\frac{w}{7} = 15$ **66.** $\frac{m}{-3} = 21$

67. $\frac{p}{-2} = -19$ **68.** $\frac{k}{-4} = -17$

69. $\frac{r}{7} = -41$ **70.** $\frac{q}{-5} = -12$

71. $\frac{t}{8} = -70$ **72.** $\frac{f}{-3} = -54$

73. $\frac{x}{0.4} = 7$ **74.** $\frac{y}{0.03} = 8$

75. $\frac{m}{1.2} = 0.05$ **76.** $\frac{k}{3.01} = 0.2$

Check your answers on page 727. ✓

 **Solving Linear Equations That Require More Than One Step**

Solving Linear Equations with Constants and Coefficients

In the equation $4x + 2 = 14$, there are two reasons why x is not alone:

1. x is multiplied by a coefficient, 4.

2. The x term $(4x)$ has a constant term, (2), added to it.

You could deal with either situation first, but the computation will be much easier if you start by eliminating the constant term. Once you are left with only an x term on one side of the equation, you can solve for x by dividing out the coefficient.

EXAMPLE 1 Solve $4x + 2 = 14$.

To remove the constant term 2, you must add its *opposite*, -2, to both sides.

$$\begin{array}{rcl} 4x + 2 &=& 14 \\ -2 && -2 \\ \hline 4x &=& 12 \end{array}$$

To remove the 4, divide both sides by 4.

$$\frac{4x}{4} = \frac{12}{4}$$

$$\frac{4x}{4} = \frac{4 \cdot 3}{4}$$

12 could be factored into $4 \cdot 3$ rather than into primes if you remember that 4 is a factor of 12.

$$\frac{\cancel{4}x}{\cancel{4}} = \frac{\cancel{4} \cdot 3}{\cancel{4}}$$

$$x = \mathbf{3}$$

Check: Does $4x + 2 = 14$ when $x = 3$?

$$\begin{array}{r} 4x + 2 = \\ 4(3) + 2 = \\ \hline 12 + 2 = 14 \quad \checkmark \end{array}$$

> *Strategy for solving equations:*
>
> 1. Remove the constant term by adding its opposite to each side.
> 2. Remove the coefficient of the letter term by dividing each side by that coefficient.
> 3. Check the solution by substituting it back into the equation.

EXAMPLE 2 Solve $5x - 3 = 32$.

Add $+3$ to both sides.

$$\begin{array}{rcl} 5x - 3 &=& 32 \\ +3 && +3 \\ \hline 5x &=& 35 \end{array}$$

Divide both sides by 5.

$$\frac{5x}{5} = \frac{35}{5}$$

$$\frac{\cancel{5}x}{\cancel{5}} = \frac{\cancel{5} \cdot 7}{\cancel{5}}$$

$$x = \mathbf{7}$$

Check: Does $5x - 3 = 32$ when $x = 7$?

$$5x - 3 =$$
$$\underline{5(7) - 3 =}$$
$$35 - 3 = 32 \quad \checkmark$$

EXAMPLE 3   Solve $4 = 3a + 10$.

$$4 = 3a + 10$$
$$\underline{-10 \qquad -10}$$
$$-6 = 3a$$
$$\frac{-6}{3} = \frac{\cancel{3}a}{\cancel{3}}$$
$$-2 = \boldsymbol{a}$$

Check: Does $3a + 10 = 4$ when $a = -2$?

$$3a + 10 =$$
$$\underline{3(-2) + 10 =}$$
$$-6 + 10 = 4 \quad \checkmark$$

Solutions are not always integers. Be especially careful when checking solutions that are decimals, fractions, or mixed numbers.

EXAMPLE 4   Solve $3y + 13 = 15$.

Add -13 to both sides.

$$3y + 13 = \quad 15$$
$$\underline{-13 \qquad -13}$$
$$3y \quad = \quad 2$$

Divide both sides by 3.

$$\frac{3y}{3} = \frac{2}{3}$$
$$\frac{\cancel{3}y}{\cancel{3}} = \frac{2}{3}$$
$$y = \boldsymbol{\frac{2}{3}}$$

Check: Does $3y + 13 = 15$ when $y = \frac{2}{3}$?

$$3y + 13 =$$
$$3\left(\frac{2}{3}\right) + 13 =$$
$$\cancel{3}\left(\frac{2}{\cancel{3}}\right) + 13 =$$
$$2 + 13 = 15 \quad \checkmark$$

Don't get intimidated by linear equations that involve decimals. Use the *strategy* so that the problem reduces to computing a few simple steps.

EXAMPLE 5   Solve $3x - 4.2 = 6.6$.

First add $+4.2$ to both sides.

$$3x - 4.2 = \quad 6.6$$
$$\underline{+4.2 \qquad +4.2}$$
$$3x \quad = \quad 10.8$$

Then divide both sides by 3.

$$\frac{\cancel{3}x}{\cancel{3}} = \frac{10.8}{3} \qquad 3\overline{)10.8}^{\,3.6}$$

$$x = \mathbf{3.6}$$

Check: Does $3x - 4.2 = 6.6$ when $x = 3.6$?

$$3x - 4.2 =$$
$$3(3.6) - 4.2 =$$
$$10.8 - 4.2 = 6.6 \quad \checkmark$$

EXERCISES 6.3A

Solve each equation and check your solutions.

1. $3x + 1 = 10$ **2.** $2x - 9 = 25$

3. $30 = 3x - 15$ **4.** $5x + 4 = 19$

5. $5y + 18 = 3$ **6.** $30 = 4x + 2$

7. $8x + 4 = -20$ **8.** $3w - 4 = -10$

9. $25 = 2x + 9$ **10.** $6x - 5 = 55$

11. $6y + 2 = -4$ **12.** $8x + 2 = 42$

13. $-15 = 4y - 3$ **14.** $46 = 7x + 4$

15. $2x - 4 = 48$ **16.** $23x - 7 = 108$

17. $14x - 12 = 142$ **18.** $7w + 5 = -16$

19. $3x + 7 = 20$ **20.** $5x - 2 = 19$

21. $4x - 8 = 21$ **22.** $7x + 13 = 50$

23. $2x - 17 = 40$ **24.** $3x - 14 = 60$

25. $3x + 1 = 17$ **26.** $6x + 5 = 36$

27. $9x - 4 = 44$ **28.** $2x + 7 = 58$

29. $4x - 1.7 = 0.7$ **30.** $3y + 0.43 = 3.40$

31. $0.05w + 6.3 = 7.95$ **32.** $0.02a - 18.4 = 6.16$

Check your answers on page 727. $\checkmark$

Solving Linear Equations That Involve Polynomials

Here is where the previous chapter's work on simplifying polynomials begins to pay off. When solving linear equations, always simplify the expression on each side of the equation as much as possible before removing the constant term or coefficient.

If any like terms appear on the *same side* of an equation, combine them first.

EXAMPLE 6 Solve $3x + 5x + 7 + 3 + x = 100$.

$$3x + 5x + 7 + 3 + x = 100$$
$$(3x + 5x + x) + (7 + 3) = 100$$
$$9x + 10 = 100 \qquad \text{Add } -10 \text{ to both sides.}$$
$$\underline{ -10 \quad -10}$$
$$9x = 90 \qquad \text{Divide both sides by 9.}$$
$$\frac{9x}{9} = \frac{90}{9}$$
$$\frac{\not{9}x}{\not{9}} = \frac{(\not{9} \cdot 10)}{\not{9}}$$
$$x = \mathbf{10}$$

Check: Does $3x + 5x + 7 + 3 + x = 100$ when $x = 10$?

$$3(10) + 5(10) + 7 + 3 + (10) =$$
$$30 + 50 + 7 + 3 + 10 = 100 \quad \checkmark$$

⇒ ***Note*:** Always check by substituting into the *original* equation. You can never tell in which step you might have made an error. If your solution solves the original equation, you can be sure it is correct.

If parentheses appear on either side of the equation, you must use the distributive law to remove them before you can combine like terms.

EXAMPLE 7 Solve $3(2y - 1) = 21$.

$$3(2y - 1) = 21$$
$$3(2y) + 3(-1) = 21$$
$$6y - 3 = 21 \qquad \text{Add } +3 \text{ to both sides.}$$
$$\underline{ + 3 = +3}$$
$$6y = 24 \qquad \text{Divide both sides by 6.}$$
$$\frac{6y}{6} = \frac{6 \cdot 4}{6}$$
$$\frac{\not{6}y}{\not{6}} = \frac{\not{6} \cdot 4}{\not{6}}$$
$$y = \mathbf{4}$$

Check: Does $3[2y - 1] = 21$ when $y = 4$?

$$3[2y - 1] =$$
$$3[2(4) - 1] =$$
$$3[8 - 1] =$$
$$3[7] = 21 \quad \checkmark$$

EXAMPLE 8 Solve $2(x + 3) + 5(x - 7) = 41$.

$$2(x + 3) + 5(x - 7) = 41$$
$$2x + 2(3) + 5x + 5(-7) = 41$$
$$2x + 6 + 5x - 35 = 41$$
$$7x - 29 = 41$$
$$\frac{+ 29}{7x} = \frac{+ 29}{70}$$
$$\frac{7x}{7} = \frac{70}{7}$$
$$\frac{\cancel{7}x}{\cancel{7}} = \frac{\cancel{7} \cdot 10}{\cancel{7}}$$
$$x = 10$$

Check: Does $2[x + 3] + 5[x - 7] = 41$ when $x = 10$?

$$2[x + 3] + 5[x - 7] =$$
$$2[(10) + 3] + 5[(10) - 7] =$$
$$2[13] + 5[3] =$$
$$26 + 15 = 41 \quad \checkmark$$

EXAMPLE 9 Solve $3x + 5(x - 3) - (4x - 1) + x + 21 = -23$.

$$3x + 5(x - 3) - (4x - 1) + x + 21 = -23$$
$$3x + 5x + 5(-3) - 4x + 1 + x + 21 = -23$$
$$3x + 5x - 15 - 4x + 1 + x + 21 = -23$$
$$(3x + 5x - 4x + x) + (-15 + 1 + 21) = -23$$
$$5x + 7 = -23$$
$$\frac{- 7}{5x} = \frac{- 7}{-30}$$
$$\frac{5x}{5} = \frac{-5 \cdot 6}{5}$$
$$\frac{\cancel{5}x}{\cancel{5}} = \frac{-\cancel{5} \cdot 6}{\cancel{5}}$$
$$x = -6$$

Check: Does $3x + 5[x - 3] - [4x - 1] + x + 21 = -23$ when $x = -6$?

$$3(-6) + 5[(-6) - 3] - [4(-6) - 1] + (-6) + 21 =$$
$$3(-6) + 5[(-6) - 3] - [-24 - 1] - 6 + 21 =$$
$$3(-6) + 5[-9] - [-25] - 6 + 21 =$$
$$-18 - 45 + 25 - 6 + 21 =$$
$$(-18 - 45 - 6) + (+25 + 21) =$$
$$-69 + 46 = -23 \quad \checkmark$$

EXERCISES 6.3B

Simplify each equation; then solve and check.

33. $4x + 7x + 8 + 2 = 76$

34. $6w + 5 + 4w + 3 - 2w = 64$

35. $9y + y - 4 - 5 - 3y = 26$

36. $13x + 7 - 15 - 4x + 1 = 20$

37. $3x - 2 + 9x + 7 = 29$

38. $z + z + 8 + 3 + 5z = 25$

39. $8m + 5m + 4 - 11 - 2m = 48$

40. $w - 5w + 6 - 14 + 8w = 20$

41. $x - 5 + 3x - x + 1 + 2x = 31$

42. $7k - 3 - 5k + 1 + k - 2k = 27$

43. $4m + m + 3 + 7m - 5 - 4m = 70$

44. $w - 5 + 3 - 4w - w + 7 = 49$

45. $0.5x + 3.2 + 2.3x = 11.6$

46. $0.03y + 0.06 - 0.31 + 4.2y = 20.9$

47. $1.3a - 4.16 + 0.7a - 2.43 = 3.41$

48. $8.3b + 0.04 + 0.23 - 2.5b = 11.87$

49. $2(x + 4) + 3(x + 2) = 29$

50. $5(x + 1) + 2(x - 1) = 38$

51. $7(x - 3) + 4(x - 1) = 19$

52. $4(x + 4) + 3(x + 1) + 2(x - 2) = 51$

53. $2(3x + 2) + 3(5x + 1) = 49$

54. $7(2x - 3) + 5(3x + 5) = 33$

55. $3(4x + 1) - 4x + 7 + x = 28$

56. $6(5x - 12) - 8x - 13 + x + 18 = 48$

57. $8x + 16 - 7x + 3(3x - 1) = -7$

58. $6 + 3(5x - 1) + 2x = 37$

59. $9(2x - 4) - 3(2x + 5) = -15$

60. $7(3x - 7) - 4(5x - 1) = -42$

61. $8(2x + 1) - 3(4x - 9) = 55$

62. $7x - (x + 5) - 3x + 7 = 5$

63. $8 - 2(7x - 1) - 4x + 3 - x = -82$

64. $3(x - 5) - 2(x + 7) + 9x = 11$

65. $5(x - 1) - 3(2x - 4) + 2x - 4 = 3$

66. $6(3x - 1) - 2(5x + 3) = 28$

67. $2(3x + 4) - (5x + 1) - 4(x - 7) = -1$

68. $2(x - 1) + 3(x - 4) - (x - 7) = 25$

Check your answers on page 727. ✓

Solving Linear Equations That Involve Fractions

If a linear equation involves fractions, you can begin your work by getting rid of the fractions. This step is particularly useful if the problem involves more than one fraction.

Let's first look at an example that involves just one fraction. Remember, fractions can be interpreted as representing division (of the numerators by the denominators). As we saw in Section 6.2, the denominators are canceled when we multiply *both sides* of the equation by the denominator.

Multiplying both sides of an equation means multiplying *every term* on each side. So if more than one term appears on a side, use the **distributive law** to multiply the denominator by every term.

EXAMPLE 10 Solve $\dfrac{x}{3} + 4 = 8$

Multiply both sides by 3.

$$3\left(\frac{x}{3} + 4\right) = 3(8)$$

$$\not{3}\left(\frac{x}{\not{3}}\right) + \underline{3(4)} = \underline{3(8)}$$

$$x + 12 = 24$$

$$\frac{-12}{x} = \frac{-12}{12}$$

Check: Does $\dfrac{x}{3} + 4 = 8$ when $x = 12$?

$$\frac{x}{3} + 4 =$$

$$\frac{12}{3} + 4 =$$

$$\frac{2 \cdot 2 \cdot \not{3}}{\not{3}} + 4 =$$

$$4 + 4 = 8 \quad ✓$$

❑

Suppose there are two or more fractions with different denominators, as in $\frac{x}{6} - \frac{1}{2} = \frac{1}{3}$. You could clear the equation of fractions by multiplying three times, once by each denominator, but it is better to use a shortcut.

The shortcut is to find the *lowest common multiple* (LCM) of the denominators, just as we did in Chapter 3 when adding and subtracting fractions. The lowest common multiple of denominators is the smallest number that all the denominators will divide into without producing a remainder. (Review Section 3.7 if you are not familiar with LCMs.)

Shortcut for solving equations involving fractions:

1. Find the LCM of all denominators in the equation.
2. Multiply both sides of the equation by the LCM to clear the equation of fractions.
3. Add the opposite of the constant term to both sides.
4. Divide both sides by the coefficient of the letter term.

EXAMPLE 11   Solve $\frac{x}{6} - \frac{1}{2} = \frac{1}{3}$.

The LCM of 6, 2, and 3 is 6. Multiply *both sides* of the equation by 6.

$$6\left(\frac{x}{6} - \frac{1}{2}\right) = 6\left(\frac{1}{3}\right)$$

$$6\left(\frac{x}{6}\right) + 6\left(-\frac{1}{2}\right) = 6\left(\frac{1}{3}\right)$$

$$\frac{6x}{6} + 2\cdot 3\left(-\frac{1}{2}\right) = \frac{2\cdot 3}{3}$$

$$
\begin{array}{rcr}
x - 3 = & & 2 \\
+ 3 & & +3 \\
\hline
x & = & 5
\end{array}
$$

Check: Does $\frac{x}{6} - \frac{1}{2} = \frac{1}{3}$ when $x = 5$?

$$\frac{x}{6} - \frac{1}{2} =$$

$$\frac{(5)}{6} - \frac{1}{2} =$$

To combine, use $\frac{3}{6}$ for $\frac{1}{2}$.

$$\frac{5}{6} - \frac{3}{6} =$$

$$\frac{2}{6} = \frac{1}{3} \quad \checkmark$$

EXAMPLE 12   Solve $\frac{2}{5}y + \frac{1}{5} = 11$.

The only denominator is 5, so multiply both sides of the equation by 5.

$$5\left[\frac{2}{5}y + \frac{1}{5}\right] = 5[1]$$

$$5\left[\frac{2}{5}y\right] + 5\left[\frac{1}{5}\right] = 5[1]$$

$$\not{5}\left[\frac{2}{\not{5}}y\right] + \not{5}\left[\frac{1}{\not{5}}\right] = 5[1]$$

$$2y + 1 = 5$$
$$\underline{-1 \quad\quad -1}$$
$$2y \quad\;\; = \;\; 4$$

$$\frac{2y}{2} = \frac{4}{2}$$

$$y = \mathbf{2}$$

Check: Does $\dfrac{2}{5}y + \dfrac{1}{5} = 1$ when $y = 2$?

$$\frac{2}{5}y + \frac{1}{5} =$$

$$\frac{2}{5}(2) + \frac{1}{5} =$$

$$\frac{4}{5} + \frac{1}{5} =$$

$$\frac{5}{5} = 1 \quad \checkmark$$

EXAMPLE 13 Solve $\dfrac{2}{3}w - \dfrac{1}{2} = 5$.

The LCM of 3 and 2 is their product (6), so multiply each term on both sides by 6. All denominators will cancel because they are factors of 6.

$$6\left[\frac{2}{3}w - \frac{1}{2}\right] = 6[5]$$

$$6\left[\frac{2}{3}w\right] + 6\left[-\frac{1}{2}\right] = 6[5]$$

$$(2 \cdot \not{3})\left[\frac{2}{\not{3}}w\right] + (\not{2} \cdot 3)\left[-\frac{1}{\not{2}}\right] = 6[5]$$

$$4w - 3 = 30$$
$$\underline{+3 \quad\quad +3}$$
$$4w \quad\;\; = \;\; 33$$

$$\frac{4w}{4} = \frac{33}{4}$$

$$w = \frac{\mathbf{33}}{\mathbf{4}} \text{ or } 8\frac{1}{4}$$

Check: Does $\dfrac{2}{3}w - \dfrac{1}{2} = 5$ when $w = 8\dfrac{1}{4}$?

Use the improper form, $\dfrac{33}{4}$, to check.

$$\left(\frac{2}{3}\right)\left(\frac{33}{4}\right) - \frac{1}{2} =$$

$$\frac{\cancel{2} \cdot \cancel{3} \cdot 11}{\cancel{3} \cdot \cancel{2} \cdot 2} - \frac{1}{2} =$$

$$\frac{11}{2} - \frac{1}{2} =$$

$$\frac{10}{2} = 5 \;\checkmark \qquad \qquad \square$$

Note that we could have used this method to solve the equation $z - \dfrac{2}{3} = \dfrac{1}{12}$ in Example 6 of Section 6.2. We will show you that in Example 14, below.

EXAMPLE 14   Solve $z - \dfrac{2}{3} = \dfrac{1}{12}$.

The LCM of 3 and 12 is 12, so multiply both sides by 12.

$$12\left[z - \frac{2}{3}\right] = 12\left[\frac{1}{12}\right]$$

$$12z + 12\left(-\frac{2}{3}\right) = 12\left[\frac{1}{12}\right]$$

$$12z - \frac{4 \cdot \cancel{3} \cdot 2}{\cancel{3}} = \cancel{12}\left[\frac{1}{\cancel{12}}\right]$$

$$12z - 8 = 1$$
$$\underline{ + 8 \quad +8}$$
$$12z \;= \;9$$

$$\frac{12z}{12} = \frac{9}{12}$$

$$z = \frac{3}{4} \;\checkmark \qquad\qquad \square$$

Which method do you prefer? Use the one you find easiest when solving an equation that contains fractions.

EXERCISES 6.3C

Find the LCM of all the denominators in each equation. Multiply all terms on each side by that LCM to remove the fractions. Then proceed to solve each equation and check your solutions.

69. $\dfrac{w}{3} - \dfrac{1}{2} = \dfrac{1}{6}$ **70.** $\dfrac{y}{5} + \dfrac{1}{3} = \dfrac{14}{15}$

71. $\dfrac{3}{4} - \dfrac{x}{3} = \dfrac{1}{12}$ **72.** $\dfrac{x}{5} - \dfrac{1}{2} = \dfrac{1}{10}$

73. $\dfrac{2}{7}m - \dfrac{1}{7} = \dfrac{4}{7}$ **74.** $\dfrac{4}{9}b + \dfrac{2}{9} = \dfrac{8}{9}$

75. $\dfrac{x}{3} + \dfrac{x}{6} = \dfrac{1}{2}$

76. $\dfrac{y}{5} - \dfrac{2y}{15} = \dfrac{2}{15}$

77. $\dfrac{p}{3} + \dfrac{p}{6} = 6$

78. $\dfrac{k}{2} + \dfrac{3k}{8} = 7$

79. $\dfrac{y}{2} + \dfrac{y}{4} = \dfrac{1}{2}$

80. $\dfrac{5w}{3} + \dfrac{w}{2} = \dfrac{13}{15}$

Check your answers on page 727.

Review

To solve linear equations that require more than one step:

1. Clear the equation of any fractions by multiplying both sides of the equation by the lowest common multiple of all the denominators. (If the equation simply consists of an integer on one side and a letter term with a fractional coefficient on the other side, just *multiply* both sides by the *reciprocal* of that coefficient to find the solution.)

2. Simplify the polynomial on each side of the equation. Use the distributive law to remove parentheses and then combine like terms.

3. If an equation involves a constant term and a coefficient on the *same side*, remove the constant term first by adding its opposite to both sides of the equation.

4. To remove an integer (or decimal) coefficient, divide both sides of the equation by that coefficient.

5. Always check your solution by substituting into the *original* equation.

EXERCISES 6.3D

Solve each equation. Then check your solutions by substituting them into the original equations.

81. $\left(\dfrac{3}{2}\right)x = 9$

82. $-\dfrac{3}{5}y = 24$

83. $w - 437 = 809$

84. $3m - 58 = -78$

85. $5a - 17 = 12a + 18$

86. $84 - 8r = 23r - 71$

87. $13v + 14 = 9v + 17$

88. $6p - 28 = 11p - 25$

89. $k + 17 - 5k + 1 + 2k = 68$

90. $4j + 3 - 9j + j - 19 = 12$

91. $3(s - 5) + 12 - 6s - (8s - 7) = 37$

92. $2(3z - 1) + 4z - 8 - 4(2z - 1) = -16$

93. $\dfrac{5}{8} - \dfrac{x}{3} = -\dfrac{1}{24}$

94. $\left(\dfrac{y}{2}\right) + \left(\dfrac{3y}{4}\right) - \left(\dfrac{5y}{8}\right) = 5$

Check your answers on page 727.

 ## Solving Linear Equations with Letter Terms on Both Sides

Solving Equations with a Binomial on Each Side

Some linear equations contain letter terms on both sides. The first step in solving such equations is to decide on which side of the equation you want the letter terms to end up. Then the other side of the equation is where the number terms should appear.

It does not really matter which side you choose for the letter terms or the number terms. But the final computation will be a little easier if you allow the letter terms to appear on the side that contained the most letters to begin with. By the most letters, I mean the largest positive coefficient.

For example, in the equation $3x + 5 = 2x - 9$, I would choose the left side for letter terms because the coefficient 3 on the left side is larger than the coefficient 2 on the right side.

Once you decide which side is which, you will want to remove letter terms from the number side. Add the *opposites* of these *letter terms* to both sides. In $3x + 5 = 2x - 9$, that would mean removing the letter term $2x$ from the right side by adding $-2x$ to both sides.

(Once you remove the letter term from the number side, the equation will be just like those you solved at the beginning of Section 6.3.)

Proceed similarly to remove the constant term from the letter side. To remove $+5$ from the left side, add -5 to both sides.

EXAMPLE 1   Solve $3x + 5 = 2x - 9$.

Choose the left side for letter terms, because $3x$ is more than $2x$. To remove $2x$ from the right side, add $-2x$ to both sides.

$$
\begin{array}{rcr}
3x + 5 = & & 2x - 9 \\
-2x & & -2x \\
\hline
x + 5 = & & -9 \\
-5 & & -5 \\
\hline
x \quad = & & -14
\end{array}
$$

Check: Does $3x + 5 = 2x - 9$ when $x = -14$?

$$
\begin{array}{cc}
3x + 5 = & 2x - 9 = \\
(3)(-14) + 5 = & 2(-14) - 9 = \\
-42 + 5 = -37 \ \checkmark & -28 - 9 = -37 \ \checkmark
\end{array}
$$

EXAMPLE 2   Solve $7x - 5 = 11x + 15$.

Choose the right side for letter terms, because $11x$ is more than $7x$. To remove $7x$ from the left side, add $-7x$ to both sides.

$$
\begin{array}{rcl}
7x - 5 = & & 11x + 15 \\
-7x & & -7x \\
\hline
-5 = & & 4x + 15 \\
-15 & & -15 \\
\hline
-20 = & & 4x \\
\dfrac{-20}{4} = & & \dfrac{4x}{4} \\
-5 = & & x
\end{array}
$$

Add -15 to both sides to remove $+15$ from the right side.

Divide by 4.

Check: Does $7x - 5 = 11x + 15$ when $x = -5$?

$$
\begin{array}{ll}
7x - 5 = & 11x + 15 = \\
\underline{7(-5) - 5 =} & \underline{11(-5) + 15 =} \\
-35 - 5 = -40 \;\checkmark & -55 + 15 = -40 \;\checkmark
\end{array}
$$

If any coefficient is negative, be careful when choosing the letter side. Remember, all negative numbers are less than zero and therefore smaller than any positive number. If the signs of the coefficients are different, always choose the side where the coefficient is positive for letter terms.

EXAMPLE 3   Solve $2x + 5 = -6x + 21$.

Choose the left side for letter terms. To remove $-6x$ from the right side, add $+6x$ to both sides.

$$
\begin{array}{rcr}
2x + 5 = & -6x + 21 \\
\underline{+6x} & \underline{+6x} \\
8x + 5 = & 21 \\
\underline{-5} & \underline{-5} \\
8x \;\;\; = & 16
\end{array}
$$

$$\frac{8x}{8} = \frac{16}{8}$$

$$\frac{\cancel{8}x}{\cancel{8}} = \frac{\cancel{8}\cdot 2}{\cancel{8}}$$

$$x = \mathbf{2}$$

Check: Does $2x + 5 = -6x + 21$ when $x = 2$?

$$
\begin{array}{ll}
2x + 5 = & -6x + 21 = \\
\underline{2(2) + 5 =} & \underline{-6(2) + 21 =} \\
4 + 5 = 9 \;\checkmark & -12 + 21 = 9 \;\checkmark
\end{array}
$$

If both coefficients are negative, remember that the term with the smaller magnitude is the larger number (-3 is larger than -4).

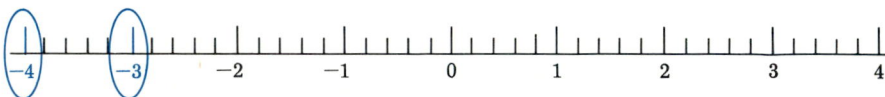

EXAMPLE 4   Solve $-4y + 7 = -3y + 5$.

Since -3 is larger than -4, choose the right side for the letter terms. To remove $-4y$ from the left side, add $+4y$ to both sides.

$$
\begin{array}{rcr}
-4y + 7 = & -3y + 5 \\
\underline{+4y} & \underline{+4y} \\
7 = & y + 5 \\
\underline{-5} & \underline{-5} \\
\mathbf{2} = & y
\end{array}
$$

Check: Does $-4y + 7 = -3y + 5$ when $y = 2$?

$$-4y + 7 = \qquad\qquad -3y + 5 =$$
$$\underline{-4(2) + 7 =} \qquad\qquad \underline{-3(2) + 5 =}$$
$$-8 + 7 = -1 \quad \checkmark \qquad -6 + 5 = -1 \quad \checkmark$$

EXERCISES 6.4A

Solve each equation and check your solutions.

1. $5x + 1 = 2x + 28$
2. $7x + 4 = 3x + 36$

3. $7x - 6 = x + 48$
4. $3x - 1 = 8x + 34$

5. $x - 4 = 7x + 8$
6. $8x + 3 = 5x + 15$

7. $9x - 11 = 14x + 9$
8. $6x + 4 = 2x + 36$

9. $5x - 6 = 7x + 8$
10. $x + 4 = 7x - 20$

11. $4x + 7 = -x + 37$
12. $3x + 1 = -5x + 25$

13. $-4x - 5 = 2x + 7$
14. $6x - 15 = -2x + 17$

15. $-x - 3 = 7x + 5$
16. $4x + 1 = -7x + 34$

17. $6 - 2x = 12 - 5x$
18. $4 - 3x = -6x + 34$

19. $7 - 4x = 13 - x$
20. $13 - 8x = -22 - 3x$

Check your answers on page 728. $\checkmark$

Solving Complicated Equations with Letter Terms on Both Sides

Remember always to simplify the polynomial on each side of the equation before removing any terms from a side. Your computation will be easier, and you will make fewer errors, if you simplify each side first.

EXAMPLE 5 Solve $2w + 3 + 7w = 5w + 30 + 5$.

$$2w + 3 + 7w = 5w + 30 + 5 \qquad \text{Combine like terms.}$$
$$9w + 3 = 5w + 35 \qquad \text{Choose the left side for letter}$$
$$\underline{-5w \qquad\qquad -5w} \qquad \text{terms. Remove } 5w \text{ from the right}$$
$$4w + 3 = +35 \qquad \text{by adding } -5w \text{ to both sides.}$$

$$4w + 3 = +35$$
$$\underline{-3 \qquad\qquad -3} \qquad \text{Remove } +3 \text{ from the left by}$$
$$4w = +32 \qquad \text{adding } -3 \text{ to both sides.}$$

$$\frac{4w}{4} = \frac{32}{4}$$

$$\boldsymbol{w = 8}$$

Check: Does $2w + 3 + 7w = 5w + 30 + 5$ when $w = 8$?

$$2w + 3 + 7w = \qquad\qquad 5w + 30 + 5 =$$
$$\underline{2(8) + 3 + 7(8) =} \qquad\qquad \underline{5(8) + 30 + 5 =}$$
$$16 + 3 + 56 = 75 \quad \checkmark \qquad 40 + 30 + 5 = 75 \quad \checkmark$$

EXAMPLE 6 $5(x + 3) + 3 = 2(3x - 1) + 5x + 8.$

$$5(x + 3) + 3 = 2(3x - 1) + 5x + 8$$
$$5x + 5(3) + 3 = 2(3x) + 2(-1) + 5x + 8$$
$$5x + 15 + 3 = 6x - 2 + 5x + 8$$
$$5x + 18 = 11x + 6$$
$$\underline{-5x \qquad\qquad -5x}$$
$$18 = 6x + 6$$
$$\underline{-6 \qquad\quad -6}$$
$$12 = 6x$$
$$\frac{12}{6} = \frac{6x}{6}$$
$$\frac{\cancel{6} \cdot 2}{\cancel{6}} = \frac{\cancel{6}x}{\cancel{6}}$$
$$\mathbf{2 = x}$$

Check: Does $5[x + 3] + 3 = 2[3x - 1] + 5x + 8$ when $x = 2$?

$$5[x + 3] + 3 = \qquad\qquad 2[3x - 1] + 5x + 8 =$$
$$5[(2) + 3] + 3 = \qquad\qquad 2[3(2) - 1] + 5(2) + 8 =$$
$$5[(2) + 3] + 3 = \qquad\qquad 2[6 - 1] + 5(2) + 8 =$$
$$5[5] + 3 = \qquad\qquad\quad 2[5] + 5(2) + 8 =$$
$$25 + 3 = 28 \quad\checkmark \qquad\qquad 10 + 10 + 8 = 28 \quad\checkmark \qquad \square$$

Remember always to check by substituting your solution into the *original equation*.

EXERCISES 6.4B _____

Solve each equation and check your solutions.

21. $7x + 10 + 3x = 5x + 13 + 17$

22. $32 + 8 + 9x - 3x = 4x + 61 - x$

23. $14 + x + 6x + 1 = x + 80 + x - 15$

24. $3x + 1 + x - 8 = 2x + 3 - 8x$

25. $4(3x - 1) - 2(x + 3) + x = 7(x + 3) - 11$

26. $7(2x - 1) + 10x - 11 = 4x + x + 21 + 6 + 4x$

27. $2x - 3(2x + 1) = 5(x - 4) + x - 13$

28. $5(3x - 2) - 7x + 1 = 6 - (3x - 5) + 2$

29. $8(x - 1) - 5x + 13 = 3(4x - 9) + 5$

30. $x + 3 + 3(7x - 5) = 6(x - 5) + 34$

Check your answers on page 728. $\checkmark$

More Solving of Equations Involving Fractions

If the equation contains a fraction, you can clear it out by multiplying *both sides* of the equation by the denominator, just as you did with simpler equations in Section 6.3.

EXAMPLE 7   Solve $\dfrac{x}{2} + 5 = 3x - 5$.

Multiply both sides by 2.

$$2\left(\frac{x}{2} + 5\right) = 2(3x - 5)$$

$$\cancel{2}\left(\frac{x}{\cancel{2}}\right) + \underline{2(5)} = \underline{2(3x)} + \underline{2(-5)}$$

$$
\begin{array}{rcl}
x + 10 & = & 6x - 10 \\
-x & & -x \\
\hline
10 & = & 5x - 10 \\
+10 & & +10 \\
\hline
20 & = & 5x \\
\dfrac{20}{5} & = & \dfrac{5x}{5} \\
4 & = & x
\end{array}
$$

Check: Does $\dfrac{x}{2} + 5 = 3x - 5$ when $x = 4$?

$$
\begin{array}{ll}
\dfrac{x}{2} + 5 = & 3x - 5 = \\[2mm]
\dfrac{(4)}{2} + 5 = & 3(4) - 5 = \\[2mm]
2 + 5 = 7 \; \checkmark & 12 - 5 = 7 \; \checkmark
\end{array}
$$

If the equation has more than one denominator, multiply *both sides* by the LCM of all the denominators.

EXAMPLE 8   Solve $\dfrac{x}{2} + 4 = \dfrac{x}{3} + 3$.

The LCM of 2 and 3 is 6, so multiply both sides by 6.

$$6\left(\frac{x}{2} + 4\right) = 6\left(\frac{x}{3} + 3\right)$$

$$6\left(\frac{x}{2}\right) + \underline{6(4)} = 6\left(\frac{x}{3}\right) + \underline{6(3)}$$

$$
\begin{array}{rcl}
3x + 24 & = & 2x + 18 \\
-2x & & -2x \\
\hline
x + 24 & = & +18 \\
-24 & & -24 \\
\hline
x & = & -6
\end{array}
$$

Check: Does $\dfrac{x}{2} + 4 = \dfrac{x}{3} + 3$ when $x = -6$?

$$\dfrac{x}{2} + 4 = \qquad\qquad \dfrac{x}{3} + 3 =$$

$$\dfrac{(-6)}{2} + 4 = \qquad\qquad \dfrac{(-6)}{3} + 3 =$$

$$-3 + 4 = +1 \;\checkmark \qquad\qquad -2 + 3 = +1 \;\checkmark$$

EXERCISES 6.4C

Solve each equation and check your solutions.

31. $\dfrac{x}{2} + 5 = 8$ **32.** $\dfrac{x}{6} + 7 = 19$

33. $\dfrac{x}{7} + 11 = 13$ **34.** $\dfrac{x}{4} + 9 = 7$

35. $\dfrac{x}{3} + 8 = 19$ **36.** $\dfrac{x}{9} - 7 = -2$

37. $\dfrac{x}{4} - 3 = 2$ **38.** $\dfrac{x}{5} - 11 = 14$

39. $\dfrac{x}{2} - 3 = \dfrac{x}{5}$ **40.** $\dfrac{y}{3} + 5 = \dfrac{y}{2} + 3$

41. $\dfrac{2x}{4} - 1 = \dfrac{x}{3}$ **42.** $\dfrac{3y}{6} + 4 = \dfrac{y}{7} - 1$

43. $\dfrac{-y}{6} + 5 = \dfrac{5y}{10} - 3$ **44.** $\dfrac{w}{3} + 4 = \dfrac{w}{2} + 3$

Check your answers on page 728.

Review

To solve linear equations with letter terms on both sides:

1. Clear the equation of any fractions by multiplying both sides of the equation by the lowest common multiple of all the denominators.

2. Simplify the polynomial on each side of the equation. Use the distributive law to remove parentheses and then combine like terms.

3. Choose one side for letter terms and the other side for number terms. To get all the letter terms on one side and number terms on the other side, remove terms by adding their opposites to both sides.

4. Solve the remaining equation by dividing both sides by the coefficient of the letter term.

EXERCISES 6.4D

Solve each equation and check your solutions by substituting them into the original equations.

45. $\dfrac{3}{8}y = 9$ **46.** $-\dfrac{2}{7}k = -6$

47. $p + 362 = 9p + 178$

48. $\dfrac{x}{2} + 5 = \dfrac{x}{6} + 9$

49. $\dfrac{y}{3} - 1 = \dfrac{y}{4}$

50. $37 - 2n = 59 - 4n$

51. $y - 5 + 3y - 7 + 2y = 9y - 1 + 5 + 3y - 4$

52. $3(5c - 2) + 4 - 2(c - 1) = 5(2c - 3)$

53. $4b + 3 - 2b + 7 - 6b - 5 = 8b - (2b - 6) - 21$

54. $\dfrac{T}{7} + \dfrac{3}{14} = \dfrac{T}{14} + \dfrac{5}{14}$

55. $\dfrac{k}{2} + 7 = \dfrac{k}{5} + 4$

56. $4(3h - 2) + 5h - 10 = 9 - 2(7h + 1) + 6h$

Check your answers on page 728.

6.5 Inequalities

Reading Inequalities

So far, both in arithmetic and in algebra we have been dealing with statements of equality, such as

$$2 + 2 = 4, \quad \frac{3}{6} = \frac{1}{2}, \quad \frac{4}{5} = 0.8, \quad 3x + 5 = 11 \text{ [if } x = 2\text{], and so on.}$$

Now suppose we want to describe the relation between expressions that are not equal, such as

$$(2 + 2) \text{ and } (5), \frac{4}{5} \text{ and } 0.9, (3x + 5) \text{ and } (12) \text{ [if } x = 2\text{], and so on?}$$

For this purpose we use an expression called an **inequality**. A strict inequality is a statement asserting that two expressions are *not* equal. Strict inequalities are represented by the symbols

$$> \quad \text{and} \quad <.$$

If two expressions are not equal, one must be the larger and the other must be the smaller expression. Think of the symbol as an arrow: The arrow points toward the smaller expression.

EXAMPLE 1 The relation between 3 and 5 can be written as

$$3 < 5 \quad \text{or} \quad 5 > 3.$$

These statements mean the same thing and so can be read as either

"3 is less than 5" or "5 is greater than 3."

If one side of an inequality is a letter, you will find it easier to work with the inequality if you read the letter side first.

EXAMPLE 2 The statement $a > 5$ is read as "*a is greater than 5.*"

Read $x < 8$ as "*x is less than 8.*"

Read $3 > y$ as "*y is less than 3.*"

Read $7 < w$ as "*w is greater than 7.*"   ❑

The symbols $<$ and $>$ are called **strict** inequality symbols because they mean that one expression is definitely larger or smaller than the other.

The symbols $\leq$ and $\geq$ are called **nonstrict** inequality symbols. The line beneath the arrows means that the expressions could be equal.

EXAMPLE 3 The statement $x \leq 8$ is read as

"*x is less than or equal to 8.*"   ❑

EXAMPLE 4 The statement $x \geq 2$ is read as

"*x is greater than or equal to 2.*"   ❑

EXERCISES 6.5A

Translate the following inequality statements into sentences.

1. $m < 10$ 2. $5 > y$

3. $12 > n$ 4. $p > 14$

5. $8 \leq z$ 6. $x \geq 3$

7. $4 > h$ 8. $23 \geq x$

Check your answers on page 728. ✓

Illustrating Inequalities on a Number Line

The inequality notation is useful for describing groups of numbers. Groups of numbers are called **sets**. Instead of talking about solutions to inequalities, we talk about their **solution sets**. These sets include not only integers but also all the mixed number fractions and irrational numbers (such as $\sqrt{2}$; see Section 12.1) between those integers. So there is an infinite (unending) list of numbers in each set.

A simplified algebraic inequality contains a letter on one side that represents all the numbers in the solution set. It contains a number on the other side that is the boundary for the set of numbers that make up the solution set.

In a *strict inequality* the *boundary* is *not part* of the *solution set*.

EXAMPLE 5 $y < 7$ (*y* is less than 7) represents all numbers less than 7, but it does not include the boundary number 7.   ❑

EXAMPLE 6 $4 \geq w$ (*w* is less than or equal to 4) represents all numbers less than 4, and it includes the boundary number 4 too.   ❑

You can illustrate the infinite solution set of an inequality on a number line. We say we are **graphing the solution** to an inequality when we illustrate it on a number line. Place a circle on the boundary number and draw an arrow in the direction of the numbers included in the solution set. The arrow notation indicates that the set continues along an infinite number line in that direction.

Point toward the numbers larger than the boundary if the inequality represents numbers larger than the boundary number.

EXAMPLE 7   Graph the solution set of $x > 2$ on a number line.

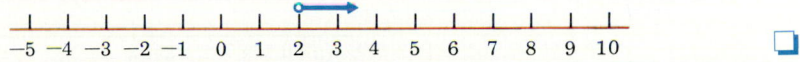

Point toward the numbers smaller than the boundary if the inequality represents numbers smaller than the boundary number.

EXAMPLE 8   Illustrate the solution set of $x < 5$ on a number line.

If the boundary number is not part of the solution set, the circle is not filled in. In Examples 7 and 8 the boundary numbers are not part of the solution set, so the circles are not filled in.

If a boundary number is part of the solution set, the circle on the number line is filled in. Boundaries are included only in nonstrict inequalities.

EXAMPLE 9   Graph $x \leq 4$.

EXAMPLE 10   Graph $x \geq -5$.

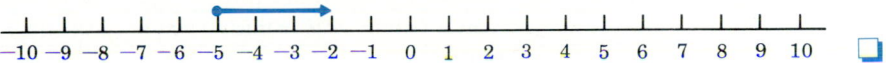

EXERCISES 6.5B

Graph the solution set of each inequality on a number line. Fill in the circle if the boundary is included; do not fill it in if the boundary is not included.

9. $a \geq 2$ 10. $b < 4$

11. $c \leq -1$ 12. $d > -3$

13. $e < -5$ 14. $f \geq 0$

15. $g \leq 7$ 16. $h > 1$

Check your answers on page 728.   ✓

Solving and Graphing Single Inequalities

To solve inequalities means to find all the numerical values for the letters that make the inequality statement true.

You solve inequalities by doing whatever legitimate manipulations you can to isolate the letter on one side of the inequality—just as you do when you solve equations.

When you add any number, negative or positive, to both sides of an inequality, you are changing both sides equally. The side that was larger remains larger after you add any number to both sides.

EXAMPLE 11   Solve $x + 7 < 15$.

Add -7 to both sides to remove $+7$ from the left side.

$$\begin{array}{r} x + 7 < 15 \\ -7 \quad -7 \\ \hline x \qquad < 8 \end{array}$$

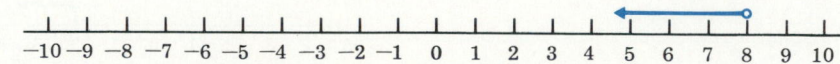

The solution set includes all numbers less than 8.   ❑

If you multiply or divide both sides by the same *positive* number, then the side that was larger is still larger. Isn't twice a large number more than twice a small number? Isn't a third of a large number more than a third of a small number?

EXAMPLE 12   Solve $6 > 2x$.

Divide both sides by 2 to remove the coefficient 2 from the right side.

$$\frac{6}{2} > \frac{2x}{2}$$
$$3 > x$$

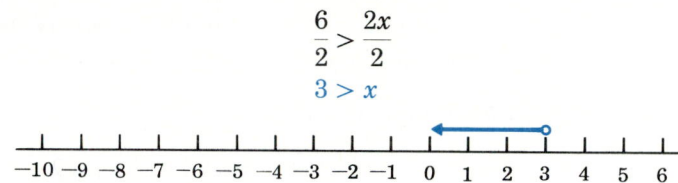

The solution set includes all numbers less than 3.   ❑

To solve an inequality where the letter side contains both a constant term and a coefficient, remove the constant term first, just as you do when you solve equations.

EXAMPLE 13   Solve $3x - 8 > 16$.

$$\begin{array}{r} 3x - 8 > 16 \\ +8 \quad +8 \\ \hline 3x \qquad > 24 \end{array}$$
$$\frac{\cancel{3}x}{\cancel{3}} > \frac{8 \cdot \cancel{3}}{\cancel{3}}$$
$$x > 8$$

All numbers greater than 8 are included in the solution set.   ❑

EXAMPLE 14   Solve $4x - 1 \geq 19$ and graph the solution set.

$$\begin{array}{r} 4x - 1 \geq 19 \\ +1 \quad +1 \\ \hline 4x \qquad \geq 20 \end{array}$$
$$\frac{\cancel{4}x}{\cancel{4}} \geq \frac{20}{4}$$
$$x \geq 5$$

❑

EXERCISES 6.5C _____

Solve and graph each inequality.

17. $2x - 5 < 15$ **18.** $3x + 4 > 22$

19. $5x + 1 \geq 36$ **20.** $2x - 7 \leq 21$

21. $6x + 5 > 35$ **22.** $7x - 2 \geq 19$

23. $3x - 8 < 22$ **24.** $4x + 3 \leq 15$

25. $5x - 6 > 44$ **26.** $2x + 17 < 31$

Check your answers on page 728.

Multiplying or Dividing an Inequality by a Negative Number

Let's see what happens to an inequality when we multiply both sides by a negative number. We will use a numerical inequality so that we can see the result clearly.

EXAMPLE 15 $5 > 3$ Let's multiply both sides by -2.

$$(-2)(5) \qquad (-2)(3)$$
$$-10 \quad < \quad -6$$

The left side of the inequality was larger when we began; but after we multiplied both sides by -2, the right side became the larger side.

So $5 > 3$, but $(-2)(5) < (-2)(3)$.   ❑

EXAMPLE 16 $6 < 15$ Let's divide both sides by -3.

$$\frac{6}{-3} \qquad \frac{15}{-3}$$
$$-2 \quad > \quad -5$$

The right side was larger when we began, but after we divided both sides by -3, the left side became the larger side.

So $6 < 15$, but $\dfrac{6}{-3} > \dfrac{15}{-3}$.   ❑

Multiplying or dividing by a negative number changes the sign of a number and so reverses its direction from zero. That is why the side that was larger becomes smaller after both sides are multiplied or divided by a negative number.

Whenever both sides of an inequality are multiplied or divided by a negative number, the inequality symbol is reversed.

If you need to multiply or divide both sides by a negative number to remove a negative coefficient from the letter side, remember to reverse the inequality symbol.

If the original inequality was a _strict inequality_, it remains a strict inequality no matter what is done to both sides.

If the original inequality was a _nonstrict inequality_, it remains a nonstrict inequality no matter what is done to both sides.

EXAMPLE 17   Solve $-2x + 7 < 21$.

$$
\begin{array}{rcl}
-2x + 7 & < & 21 \\
-7 & & -7 \\
\hline
-2x & < & 14
\end{array}
$$

Adding a negative number does *not* affect the inequality symbol.

$$
\frac{-2x}{-2} > \frac{14}{-2}
$$

Multiplying by -2 reverses the inequality symbol.

$$
x > -7
$$

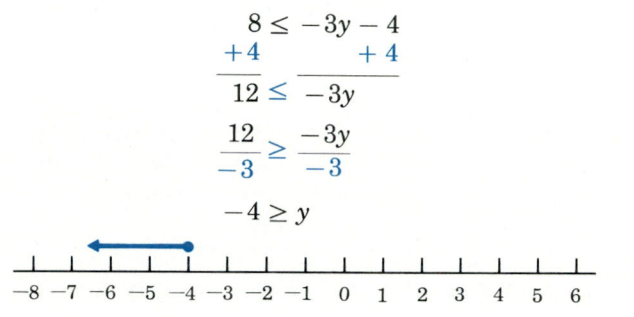

EXAMPLE 18   Solve $8 \le -3y - 4$.

$$
\begin{array}{rcl}
8 & \le & -3y - 4 \\
+4 & & +4 \\
\hline
12 & \le & -3y
\end{array}
$$

$$
\frac{12}{-3} \ge \frac{-3y}{-3}
$$

$$
-4 \ge y
$$

EXERCISES 6.5D

Solve and graph each inequality.

27. $-3x < 15$ **28.** $-5x > 20$

29. $-4x \le 24$ **30.** $14 \ge -2x$

31. $-3x - 1 > 5$ **32.** $-4x + 1 \ge -23$

33. $-5x + 2 \le 37$ **34.** $-6x + 4 < 28$

35. $-7x - 14 > 0$ **36.** $-43 < -8x - 3$

Check your answers on page 728.

Solving and Graphing Double Inequalities

A statement that includes two inequality symbols is called a **double inequality**. An example of a double inequality is $3 < x < 7$.

The easiest way to understand what a double inequality represents is to read the letter part first.

EXAMPLE 19 $3 < x < 7$ is read as "x is greater than 3 and less than 7." So x is all the numbers between 3 and 7. Both boundaries, 3 and 7, are excluded.

Double inequalities represent a segment on the number line with two boundaries. This segment contains an infinite amount of numbers. You cannot simply list the few integers—you must also remember all the mixed numbers, like 3.7, $4\frac{1}{7}$, and 5.0013, and the irrational numbers, like $\sqrt{10}$ that lie between the integers.

As with single inequalities, fill in the circle at a boundary if it is included; do not fill it in if the boundary is excluded. Shade the segment on the number line between the boundaries.

EXAMPLE 20 $-5 \le x < 7$ represents all numbers *between* -5 and 7, including -5 and excluding 7. The solution set is graphed below.

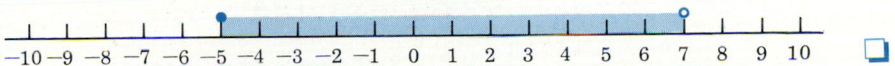

EXAMPLE 21 $-4 \le x \le 3$ represents all numbers between -4 and 3. Both boundaries, -4 and 3, are included. The solution set is graphed below.

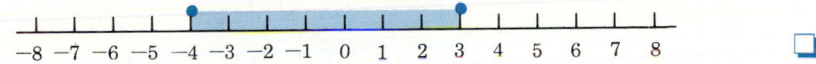

To solve a double inequality, use the same strategy that you use to solve equations and single inequalities. Decide what you must do to isolate the letter term first. Whatever computation will accomplish this step must be done to all *three* segments of the inequality.

EXAMPLE 22   Solve $4 < x + 3 \le 10$.

$$\begin{array}{ll} 4 < x + 3 \le 10 & \text{Add } -3 \text{ to all three segments.} \\ \underline{-3 \qquad -3 \quad -3} & \\ 1 < x \qquad \le 7 & \end{array}$$

The solution set is all the numbers between 1 and 7, including 7 but excluding 1. It is graphed below.

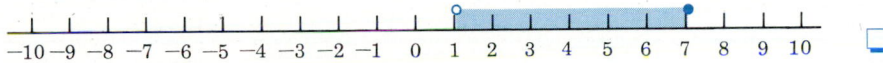

If the letter has a coefficient, remove it after isolating the letter term.

EXAMPLE 23   Solve $-4 \le 3x - 1 < 11$.

$$\begin{array}{ll} -4 \le 3x - 1 < 11 & \text{First add } +1 \text{ to all three segments.} \\ \underline{+1 \qquad +1 \quad +1} & \\ -3 \le 3x \qquad < 12 & \text{Now divide all three segments by } +3. \\ \dfrac{-3}{3} \le \dfrac{3x}{3} \qquad < \dfrac{12}{3} & \\ -1 \le x \qquad < 4 & \end{array}$$

The solution set is all numbers between -1 and 4, including -1 but excluding 4.

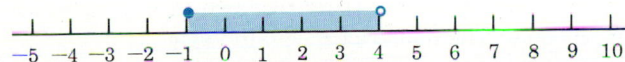

If you multiply or divide by a negative number, reverse both inequality symbols.

EXAMPLE 24   Solve $-11 \le -2x + 3 \le 15$.

$$\begin{array}{ll} -11 \le -2x + 3 \le 15 & \text{Add } -3 \text{ to all three segments.} \\ \underline{-3 \qquad\quad -3 \quad -3} & \\ -14 \le -2x \qquad \le 12 & \text{Divide all three segments by } -2, \\ \dfrac{-14}{-2} \ge \dfrac{-2x}{-2} \ge \dfrac{12}{-2} & \text{which will reverse both inequality} \\ & \text{symbols.} \end{array}$$

$7 \ge x \ge -6$, which can be expressed as $-6 \le x \le 7$.

The solution set is all numbers between -6 and 7, including both -6 and 7.

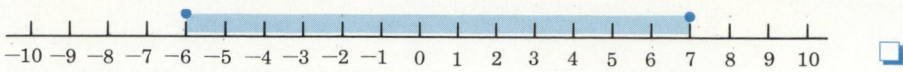

EXERCISES 6.5E

Solve and graph each inequality.

37. $1 < x - 5 < 8$ **38.** $13 > x + 4 > 3$

39. $3 < x + 4 \leq 10$ **40.** $10 > x - 3 \geq -4$

41. $6 < 2x \leq 14$ **42.** $3 \leq 3x < 12$

43. $5 < 2x - 3 < 15$ **44.** $-2 \leq 3x + 4 \leq 19$

45. $16 > 5x + 1 > -9$ **46.** $0 \geq 4x + 8 > -12$

47. $-7 \leq 4x + 5 \leq 21$ **48.** $-11 < 3x - 5 \leq 4$

49. $0 < -2x - 6 < 8$ **50.** $-5 < -3x + 1 \leq 13$

Check your answers on page 728.

Glossary

equation   Statement asserting that two expressions are equal.

   Example: $2x + 1 = 7$ is an algebraic equation.

inequality (strict)   Statement asserting that two expressions are not equal. The inequality symbols, $>$ and $<$, always point toward the smaller expression. (These symbols are used for expressions that are never equal, called **strict** inequalities.)

   Example: $3 < 7$ is read as "3 is less than 7" or "7 is greater than 3."

nonstrict inequality   Statement asserting that two expressions can be equal. The nonstrict inequality symbols are $\geq$ and $\leq$.

   Example: $x \leq 5$ is read as "x is less than or equal to 5."

double inequality   Statement that involves two inequality symbols. Read the expression containing the letter first.

   Example: $1 < x < 9$ is read as "x is greater than 1 and smaller than 9." It describes the set of numbers between 1 and 9 (not including either 9 or 1).

sets   Groups of numbers. The solution to an inequality is a group of numbers, so we call it a solution set.

   Example: The set of all numbers larger than 3 is the solution set of the inequality $3 < x$.

solution to an equation   Value of the letter that makes the expressions on the left and right sides of the equation equal.

   Example: $x = 3$ is the solution to $2x + 1 = 7$.

$$\text{If } x = 3, \text{ then } 2x + 1 =$$
$$2(3) + 1 =$$
$$6 + 1 = 7$$

strategy   Method for solving all linear equations. Solving means finding the number value of the letter, so do whatever legitimate manipulations you can to the equation until you get the letter alone on one side of the equation. If the letter is x, for example, manipulate terms until the equation reads $x = $ [*some number*].

1. Clear the equation of any fractions by multiplying both sides of the equation by the lowest common multiple of all the denominators. If the equation consists only of an integer on one side and a letter

term (with a fractional coefficient) on the other side, just multiply both sides by the reciprocal of that coefficient to obtain the solution.

2. Simplify each side. Remove any parentheses by using the distributive law and then combine like terms.

3. Choose one side for the letter terms and the other side for the constants. Remove terms from one side by adding their opposites to both sides.

4. Remove the coefficient of the resulting letter term by dividing both sides of the equation by that coefficient.

5. Always check your solution by substituting it into the *original* equation.

Example:
$$5(x + 3) + 3 = 2(3x - 1) + 5x + 8$$
$$\underline{5(x + 3)} + 3 = \underline{2(3x - 1)} + 5x + 8$$
$$5x + \underline{5(3)} + 3 = \underline{2(3x)} + 2(-1) + 5x + 8$$
$$5x + \underline{15} + \underline{3} = \mathbf{6x} - \underline{2} + \mathbf{5x} + \underline{8}$$
$$5x + 18 = 11x + 6$$
$$\underline{-5x} \qquad \underline{-5x}$$
$$18 = 6x + 6$$
$$\underline{-6} \qquad \underline{-6}$$
$$12 = 6x$$
$$\frac{12}{6} = \frac{6x}{6}$$
$$\mathbf{2 = x}$$

Check: Does $5[x + 3] + 3 = 2[3x - 1] + 5x + 8$ when $x = 2$?

$$5[x + 3] + 3 = \qquad\qquad 2[3x - 1] + 5x + 8 =$$
$$5[(2) + 3] + 3 = \qquad\qquad 2[3(2) - 1] + 5(2) + 8 =$$
$$5[5] + 3 = \qquad\qquad 2[6 - 1] + 5(2) + 8 =$$
$$25 + 3 = 28 \;\checkmark \qquad\qquad 2[5] + 5(2) + 8 =$$
$$10 + 10 + 8 = 28 \;\checkmark$$

Chapter 6 Review

Do each of the problems. After you are confident that you have done them as accurately as possible, compare your answers with those in the back of the book. If any are wrong, go back to the section where the problem came from (indicated to the left of the problem) and review the section.

Once you understand all the sections, you will be sure to learn the skills solidly and remember how to do them if you practice them a little bit more. Turn to the Supplementary Exercises and do all the problems from any section where you had difficulty on these review exercises.

6.1
1. Is $x = 7$ the solution to $3x + 5 = 25$?

2. Is $y = 4$ the solution to $5 - 2y = 3y - 15$?

Solve each equation and check your solutions by substituting them into the original equations.

6.2
3. $m + 12 = -15$

4. $a - 3.7 = 18.04$

5. $7m = -35$

6. $24k = 6$

7. $4.3h = 0.129$

8. $\dfrac{x}{3} = 12$

9. $\dfrac{y}{0.05} = -3.1$

6.3
10. $3x - 2 = 13$

11. $8y + 6 = -42$

12. $0.2m + 5.3 = 16.7$

13. $2x + 5 - 7 + x = 19$

14. $-4(z + 5) = 12$

15. $1 + 2(w - 1) - (w + 4) = 7$

16. $\dfrac{x}{2} + \dfrac{x}{5} = 14$

6.4
17. $7p - 4 = 3p + 16$

18. $8 - 5r = 2r + 1$

19. $3f - 5 + 2 - 7f - 1 = f - 3 + 5 + f$

20. $2x - (7x - 3) + 4 = 4x - 3 + 2(x - 5) - x$

Answers

1. _____
2. _____
3. _____
4. _____
5. _____
6. _____
7. _____
8. _____
9. _____
10. _____
11. _____
12. _____
13. _____
14. _____
15. _____
16. _____
17. _____
18. _____
19. _____
20. _____

6.5   Solve and graph each inequality.

21. $5x < 30$

−10 −9 −8 −7 −6 −5 −4 −3 −2 −1 0 1 2 3 4 5 6 7 8 9 10

22. $y - 4 > 3$

−10 −9 −8 −7 −6 −5 −4 −3 −2 −1 0 1 2 3 4 5 6 7 8 9 10

23. $2k + 5 > 13$

−10 −9 −8 −7 −6 −5 −4 −3 −2 −1 0 1 2 3 4 5 6 7 8 9 10

24. $6 - 3w < 12$

−10 −9 −8 −7 −6 −5 −4 −3 −2 −1 0 1 2 3 4 5 6 7 8 9 10

25. $0 < z + 7 \leq 10$

−10 −9 −8 −7 −6 −5 −4 −3 −2 −1 0 1 2 3 4 5 6 7 8 9 10

26. $-5 \leq 2x - 3 \leq 9$

−10 −9 −8 −7 −6 −5 −4 −3 −2 −1 0 1 2 3 4 5 6 7 8 9 10

21. _____
22. _____
23. _____
24. _____
25. _____
26. _____

Check your answers on page 729.

Supplementary Exercises

Do all the problems in every section that involves a skill in which you now lack complete mastery. With a little more practice you will achieve that sense of really understanding the topics, and you will remember how to do these problems.

6.1

1. Is $x = -4$ the solution to $7x - 5 = 23$?

2. Is $y = 3$ the solution to $6 - 4y = 3y - 15$?

Solve each equation and check your solutions by substituting them into the original equations.

6.2

3. $m - 31 = -15$

4. $a + 4.6 = 23.07$

5. $-6m = -30$

6. $40k = 4$

7. $-2.9h = -0.232$

8. $\dfrac{x}{7} = 21$

9. $\dfrac{y}{0.06} = -5.4$

6.3

10. $5x - 8 = 22$

11. $4y + 9 = -27$

12. $0.5m + 4.6 = 20.1$

13. $3x - 4 - 1 + x = 11$

14. $-6(z - 1) = 12$

15. $8 - 3(w - 1) - (w + 5) = 30$

16. $\dfrac{x}{3} + \dfrac{x}{5} = 8$

6.4

17. $8p - 1 = 6p + 23$

18. $6 - 7r = 4r - 27$

19. $4f - 3 + 2 - 5f - 1 = f - 6 + 1 + f$

20. $2x - 2(4x - 3) + 15 = 5x - 2 - (x - 5) - x$

Answers

1. _____

2. _____

3. _____

4. _____

5. _____

6. _____

7. _____

8. _____

9. _____

10. _____

11. _____

12. _____

13. _____

14. _____

15. _____

16. _____

17. _____

18. _____

19. _____

20. _____

6.5   Solve and graph each inequality.

21. $6x < 30$

$$\begin{array}{cccccccccccccccccccccc} \text{|} & \text{|} & \text{|} & \text{|} & \text{|} & \text{|} & \text{|} & \text{|} & \text{|} & \text{|} & \text{|} & \text{|} & \text{|} & \text{|} & \text{|} & \text{|} & \text{|} & \text{|} & \text{|} & \text{|} & \text{|} \\ -10 & -9 & -8 & -7 & -6 & -5 & -4 & -3 & -2 & -1 & 0 & 1 & 2 & 3 & 4 & 5 & 6 & 7 & 8 & 9 & 10 \end{array}$$

21. _____

22. _____

22. $y - 7 > 1$

$$\begin{array}{cccccccccccccccccccccc} \text{|} & \text{|} & \text{|} & \text{|} & \text{|} & \text{|} & \text{|} & \text{|} & \text{|} & \text{|} & \text{|} & \text{|} & \text{|} & \text{|} & \text{|} & \text{|} & \text{|} & \text{|} & \text{|} & \text{|} & \text{|} \\ -10 & -9 & -8 & -7 & -6 & -5 & -4 & -3 & -2 & -1 & 0 & 1 & 2 & 3 & 4 & 5 & 6 & 7 & 8 & 9 & 10 \end{array}$$

23. _____

24. _____

23. $2k + 5 \le 13$

$$\begin{array}{cccccccccccccccccccccc} \text{|} & \text{|} & \text{|} & \text{|} & \text{|} & \text{|} & \text{|} & \text{|} & \text{|} & \text{|} & \text{|} & \text{|} & \text{|} & \text{|} & \text{|} & \text{|} & \text{|} & \text{|} & \text{|} & \text{|} & \text{|} \\ -10 & -9 & -8 & -7 & -6 & -5 & -4 & -3 & -2 & -1 & 0 & 1 & 2 & 3 & 4 & 5 & 6 & 7 & 8 & 9 & 10 \end{array}$$

25. _____

26. _____

24. $7 - 4w \ge 23$

$$\begin{array}{cccccccccccccccccccccc} \text{|} & \text{|} & \text{|} & \text{|} & \text{|} & \text{|} & \text{|} & \text{|} & \text{|} & \text{|} & \text{|} & \text{|} & \text{|} & \text{|} & \text{|} & \text{|} & \text{|} & \text{|} & \text{|} & \text{|} & \text{|} \\ -10 & -9 & -8 & -7 & -6 & -5 & -4 & -3 & -2 & -1 & 0 & 1 & 2 & 3 & 4 & 5 & 6 & 7 & 8 & 9 & 10 \end{array}$$

25. $-2 \le z + 5 < 11$

$$\begin{array}{cccccccccccccccccccccc} \text{|} & \text{|} & \text{|} & \text{|} & \text{|} & \text{|} & \text{|} & \text{|} & \text{|} & \text{|} & \text{|} & \text{|} & \text{|} & \text{|} & \text{|} & \text{|} & \text{|} & \text{|} & \text{|} & \text{|} & \text{|} \\ -10 & -9 & -8 & -7 & -6 & -5 & -4 & -3 & -2 & -1 & 0 & 1 & 2 & 3 & 4 & 5 & 6 & 7 & 8 & 9 & 10 \end{array}$$

26. $0 \le 3x - 6 \le 15$

$$\begin{array}{cccccccccccccccccccccc} \text{|} & \text{|} & \text{|} & \text{|} & \text{|} & \text{|} & \text{|} & \text{|} & \text{|} & \text{|} & \text{|} & \text{|} & \text{|} & \text{|} & \text{|} & \text{|} & \text{|} & \text{|} & \text{|} & \text{|} & \text{|} \\ -10 & -9 & -8 & -7 & -6 & -5 & -4 & -3 & -2 & -1 & 0 & 1 & 2 & 3 & 4 & 5 & 6 & 7 & 8 & 9 & 10 \end{array}$$

Check your answers on page 729.

Chapter 6 Test

1. Is $x = -1$ the solution to $3x - 5 = -2$?

2. Is $y = 3$ the solution to $5y - 4 = 2 + 3y$?

3. Is $w = 5$ in the solution set of $2w - 1 > 5$?

Solve and check each equation.

4. $r - 76 = 24$ **5.** $-12p = 60$

6. $m + 16.9 = 8.03$ **7.** $2.4d = 5.04$

8. $\dfrac{y}{8} = 5$ **9.** $\dfrac{w}{0.04} = -5.2$

10. $4j - 5 = 27$ **11.** $0.3a + 4.5 = 8.1$

12. $4x - 5 + x - 3 - 7x + 1 = 7$ **13.** $-3(2y - 5) = -3$

14. $4w - 3(7w - 2) + 8w = 24$ **15.** $\dfrac{z}{3} - \dfrac{z}{4} = 5$

16. $8b + 6 = 12b - 6$

17. $c - 7 + 6c - 1 - 5c + 4 = 2c - 8 + 5c + 14$

18. $6k - (3k - 2) + 8k - 4 = 7(2k - 4) + 5$

Solve and graph each inequality.

19. $6n > 30$

$$\overset{\text{|}}{-10}\,\overset{\text{|}}{-9}\,\overset{\text{|}}{-8}\,\overset{\text{|}}{-7}\,\overset{\text{|}}{-6}\,\overset{\text{|}}{-5}\,\overset{\text{|}}{-4}\,\overset{\text{|}}{-3}\,\overset{\text{|}}{-2}\,\overset{\text{|}}{-1}\,\overset{\text{|}}{0}\,\overset{\text{|}}{1}\,\overset{\text{|}}{2}\,\overset{\text{|}}{3}\,\overset{\text{|}}{4}\,\overset{\text{|}}{5}\,\overset{\text{|}}{6}\,\overset{\text{|}}{7}\,\overset{\text{|}}{8}\,\overset{\text{|}}{9}\,\overset{\text{|}}{10}$$

20. $p - 7 < -4$

$$\overset{\text{|}}{-10}\,\overset{\text{|}}{-9}\,\overset{\text{|}}{-8}\,\overset{\text{|}}{-7}\,\overset{\text{|}}{-6}\,\overset{\text{|}}{-5}\,\overset{\text{|}}{-4}\,\overset{\text{|}}{-3}\,\overset{\text{|}}{-2}\,\overset{\text{|}}{-1}\,\overset{\text{|}}{0}\,\overset{\text{|}}{1}\,\overset{\text{|}}{2}\,\overset{\text{|}}{3}\,\overset{\text{|}}{4}\,\overset{\text{|}}{5}\,\overset{\text{|}}{6}\,\overset{\text{|}}{7}\,\overset{\text{|}}{8}\,\overset{\text{|}}{9}\,\overset{\text{|}}{10}$$

21. $4a + 3 \le 15$

$$\overset{\text{|}}{-10}\,\overset{\text{|}}{-9}\,\overset{\text{|}}{-8}\,\overset{\text{|}}{-7}\,\overset{\text{|}}{-6}\,\overset{\text{|}}{-5}\,\overset{\text{|}}{-4}\,\overset{\text{|}}{-3}\,\overset{\text{|}}{-2}\,\overset{\text{|}}{-1}\,\overset{\text{|}}{0}\,\overset{\text{|}}{1}\,\overset{\text{|}}{2}\,\overset{\text{|}}{3}\,\overset{\text{|}}{4}\,\overset{\text{|}}{5}\,\overset{\text{|}}{6}\,\overset{\text{|}}{7}\,\overset{\text{|}}{8}\,\overset{\text{|}}{9}\,\overset{\text{|}}{10}$$

22. $-8m < 40$

$$\overset{\text{|}}{-10}\,\overset{\text{|}}{-9}\,\overset{\text{|}}{-8}\,\overset{\text{|}}{-7}\,\overset{\text{|}}{-6}\,\overset{\text{|}}{-5}\,\overset{\text{|}}{-4}\,\overset{\text{|}}{-3}\,\overset{\text{|}}{-2}\,\overset{\text{|}}{-1}\,\overset{\text{|}}{0}\,\overset{\text{|}}{1}\,\overset{\text{|}}{2}\,\overset{\text{|}}{3}\,\overset{\text{|}}{4}\,\overset{\text{|}}{5}\,\overset{\text{|}}{6}\,\overset{\text{|}}{7}\,\overset{\text{|}}{8}\,\overset{\text{|}}{9}\,\overset{\text{|}}{10}$$

23. $0 < k + 8 < 15$

$$\overset{\text{|}}{-10}\,\overset{\text{|}}{-9}\,\overset{\text{|}}{-8}\,\overset{\text{|}}{-7}\,\overset{\text{|}}{-6}\,\overset{\text{|}}{-5}\,\overset{\text{|}}{-4}\,\overset{\text{|}}{-3}\,\overset{\text{|}}{-2}\,\overset{\text{|}}{-1}\,\overset{\text{|}}{0}\,\overset{\text{|}}{1}\,\overset{\text{|}}{2}\,\overset{\text{|}}{3}\,\overset{\text{|}}{4}\,\overset{\text{|}}{5}\,\overset{\text{|}}{6}\,\overset{\text{|}}{7}\,\overset{\text{|}}{8}\,\overset{\text{|}}{9}\,\overset{\text{|}}{10}$$

24. $-4 \le 2n - 8 \le 4$

$$\overset{\text{|}}{-10}\,\overset{\text{|}}{-9}\,\overset{\text{|}}{-8}\,\overset{\text{|}}{-7}\,\overset{\text{|}}{-6}\,\overset{\text{|}}{-5}\,\overset{\text{|}}{-4}\,\overset{\text{|}}{-3}\,\overset{\text{|}}{-2}\,\overset{\text{|}}{-1}\,\overset{\text{|}}{0}\,\overset{\text{|}}{1}\,\overset{\text{|}}{2}\,\overset{\text{|}}{3}\,\overset{\text{|}}{4}\,\overset{\text{|}}{5}\,\overset{\text{|}}{6}\,\overset{\text{|}}{7}\,\overset{\text{|}}{8}\,\overset{\text{|}}{9}\,\overset{\text{|}}{10}$$

Check your answers on page 729.

Answers

1. _____
2. _____
3. _____
4. _____
5. _____
6. _____
7. _____
8. _____
9. _____
10. _____
11. _____
12. _____
13. _____
14. _____
15. _____
16. _____
17. _____
18. _____
19. _____
20. _____
21. _____
22. _____
23. _____
24. _____

Cumulative Review
Chapters 1–6

Simplify.

1. $(-29) - (58) + (-73) - (-62) + (24) =$

2. $(-1)(-1)(1)(-2)(-3)(-1)(-1)(-5) =$

3. $8 - 2(-3 - 5 + 4) - 5(-7 + 3)(-6 + 8 - 3) =$

4. $\dfrac{5}{12} - \dfrac{1}{12} + \dfrac{7}{12} - \dfrac{1}{12} =$

5. $\dfrac{5}{6} + \dfrac{1}{4} =$

6. $\left(\dfrac{2}{9}\right)\left(\dfrac{5}{7}\right) =$

7. $\left(\dfrac{10}{21}\right) \div \left(\dfrac{35}{45}\right) =$

8. $3.008 + 15.9 + 28 + 0.03 + 7.002 =$

9. $(23.4)(10.05) =$

10. What is 8.4% of 607.50?

11. $3x + 5y - 7 + y - 9 + 7x + 2 - 6y =$

12. $4.3x^2 + 2.1x + 3.5 - 1.8x - 4.6x^2 + 6.9 =$

13. $(8x^2 - 5x - 4) - (x^2 - x + 9) =$

14. $3(5ab - a + 4b) - 6(2ab - 4a - 3b) =$

15. $2x(5x - 4y - 1) - 3y(2x - y - 4) =$

16. $(3x^2 - 15x + 9)$ divided by $3x =$

Evaluate each expression if $a = -2$, $b = 3$, and $c = -1$.

17. a^2 **18.** a^3 **19.** $5ab^2$ **20.** abc

21. $5ab - 3ac + 6b - 5a + 2bc$

Factor each polynomial.

22. $15abc - 10ac + 35abd - 20acd$

23. $6xy^2 - 12x^2y - 15xy + 9x^3y^2$

Answers

1. _____
2. _____
3. _____
4. _____
5. _____
6. _____
7. _____
8. _____
9. _____
10. _____
11. _____
12. _____
13. _____
14. _____
15. _____
16. _____
17. _____
18. _____
19. _____
20. _____
21. _____
22. _____
23. _____

Continued

Cumulative Review (*Cont.*)   Answers

Solve and check.

24. $7a - 5 = 4a - 6$ 24. _____

25. $3n - 5(2n + 1) - 6n + 7 = n - (3n - 5) + 4n + 27$ 25. _____

26. Solve and graph $-6 \leq 2x - 6 < 4$.   26. _____

```
 |   |   |   |   |   |   |   |   |   |   |   |   |   |   |   |   |   |   |   |   |
-10 -9 -8 -7 -6 -5 -4 -3 -2 -1  0  1  2  3  4  5  6  7  8  9  10
```

Check your answers on page 729.

7 Applications of Linear Equations

7.1 Solving Percent Equations

There are lots of situations where, in trying to figure out the answer to some problem, we end up doing difficult arithmetic gymnastics in our heads. Or else we guess, hoping to find the answer by trial and error. In many cases we could get the answer more quickly by setting up the problem as a simple linear equation and solving the equation. This chapter deals with some *practical applications* of linear equations. In the first section we discuss problems involving percents that can be solved as equations.

Suppose you are in an appliance store where a $275 air conditioner is on sale at $20 off. You know that another store is offering the same air conditioner at 15 percent off. Which is the better deal? To find out, you need to know what percent $20 is of $275.

Suppose you are being graded on an exam that was an essay question. Your instructor says that you covered only 86 percent of the important points, and she circles 43 major points that you covered. How many points did she think there were? In other words, 43 is 86% of what number?

A basketball player was pleased to learn that he had hit 85 percent of his outside shots last night. If he made 17 baskets from the outside, how many times did he attempt an outside shot? That is, 17 is 85% of what number?

By using equations it is possible to set up these problems easily without becoming confused about which number is divided by which number. We know that if one number (call it N_1) is some percentage (call it $P\%$) of another number (call it N_2), that relationship is expressed by the equation $N_1 = P\%$ of N_2. (Writing a small number slightly below a letter is a notation used to distinguish the representation of one number from another; we call that number the letter's *subscript*. Read N_1 as N sub-one!)

When we are told that *one number* is *some percent* of *another number*, and we know the value of two out of those three things, we can always find the value of the third. We would do so by substituting the known values into $N_1 = P\%$ of N_2 and solving for the unknown letter. If we are told the value of $P\%$, we must convert it into a decimal or a fraction. Remember that $P\%$ of N_2 means that $P\%$ is *multiplied* by N_2.

EXAMPLE 1   Jack made 43 points on the essay test and learned he had made 86 percent of the possible points. How many points could have been made on the test?

We restate that problem as 43 is 86% of some number. We see that N_1 is 43 and $P\%$ is 86%, so we solve the equation for N_2 (the number of points that could have been made on the test), which we do not know.

$$43 = 86\% \text{ of } N_2$$
$$43 = 0.86N_2 \qquad \text{Divide both sides by 0.86.}$$
$$\frac{43}{0.86} = \frac{0.86N_2}{0.86}$$
$$50 = N_2$$

So it was possible to make 50 points on the test. ❑

EXAMPLE 2 Meg has put together all the cash she can spare for a down payment on a house. She has $2,940. The bank requires anyone getting a mortgage to make a down payment of 7 percent of the house cost. How expensive a house can Meg afford to buy?

We know that 7 percent of the house cost has to be $2,940. So we can express this problem as the equation 7% of $H = \$2,940$ and solve it for H (the house cost).

$$0.07H = \$2,940$$
$$\frac{0.07H}{0.07} = \frac{\$2,940}{0.07} \qquad \text{Divide both sides by 0.07.}$$
$$H = \$42,000$$

So Meg can afford to pay as much as $42,000 for a house. ❑

EXAMPLE 3 The Lowd Sownd store advertised that it was slashing $50 off the price of a $250 stereo. Tom wanted to compare this sale with the one at Cheep Noises, where all items were being discounted 23 percent in celebration of the store's twenty-third anniversary. What is the percent of discount offered at Lowd's?

We know that $50 is *some percent* of $250. So we can represent this problem as the equation $\$50 = P\% \text{ of } \$250$.

$$\$50 = P\%(\$250) \qquad \text{Solve for } P\% \text{ by dividing both sides by \$250.}$$
$$\frac{\$50}{\$250} = \left[P\% \frac{\$250}{\$250} \right]$$
$$0.20 = P\% \qquad \text{Express 0.20 as a percent: } 0.20 \to 20\%.$$

So Lowd's is offering only a 20% discount. Tom would be better off shopping at Cheep Noise. ❑

EXAMPLE 4 In Clever, Iowa, a group of concerned citizens planned to introduce a workplace ban on smoking in local restaurants at their next town meeting. They figured that with 4,200 registered voters, they needed at least 525 folks to attend the town meeting for there to be a quorum. What percent of registered voters is required for a quorum in that town?

$$525 = P\% \text{ of } 4,200$$
$$525 = P\%(4,200) \qquad \text{Divide both sides by 4,200.}$$
$$\frac{525}{4,200} = \frac{P\%(4,200)}{4,200}$$
$$0.125 = P\% \qquad \text{Express 0.125 as a percent: } 0.125 = 12.5\%$$

So the town requires 12.5% of registered voters for a quorum. ❑

EXAMPLE 5   The sales rep. calculates her percent of increase* in sales from one year to another as

$$P = \frac{(S_2 - S_1)}{S_1}, \text{ where}$$

S_1 represents sales from last year,

S_2 represents sales from this year.

a. What is her percent of increase in sales if at this date last year her sales were $227,059 and they are $318,794 now?

$$P = \frac{318,794 - 227,059}{227,059}$$

$$= \frac{91,735}{227,059}$$

$$= 0.4040139$$

$$\approx 40\%\dagger$$

Her percent of increase in sales is approximately 40%, so if she continues at this rate, she will have 40% more sales this year than last year.

b. Companies use percent of increase to estimate the total sales for the year, called **projected sales**.

Projected sales $= S + P\%$ of S, where

S represents last year's sales,

P represents percent of increase in last year's sales.

If the rep.'s sales last year were $285,190, how much are her sales projected to be this year?

Projected sales $= (\$285,190) + 40\%(\$285,190)$

$$= \$285,190 + \$114,076$$

$$= \$399,266$$

So if her rate of sales continues at this pace, she should expect her total sales for the year to be $399,266.   ❏

EXERCISES 7.1

Set up an equation for each problem and solve for the unknown.

1. How many times did Joe practice his tennis serve if he says that 40% of his tries were good and he got 26 serves in?

2. When we played squash last night, Patty calculated that I used my backhand 62% of the time. She counted 186 shots hit backhand, so how may shots did I hit altogether?

3. If a bank requires anyone getting a mortgage to make a down payment of 12% of the house cost, what is the price of the house Ben can buy if he has $7,200 available for his down payment?

---

* If her sales are less this year than they were last year, she will have a percent of **decrease** in sales.

† The symbol ≈ means "approximately equal."

4. The bank informed Hal that he would need a down payment of at least $4,680 to buy a $26,000 home. What percent of the cost does the bank require as a down payment?

5. The quality control inspector found 25,020 items unsatisfactory on the conveyor belt. If she calculated that number to be just 3% of the total items inspected, how many items had she inspected?

6. What percent of the problems were correct if Roy got 18 out of 25 correct?

7. The Computernut Store promised a 19.3% discount to students. Abbey was delighted to find out she had saved $617.60 by purchasing her computer from them. How much did the computer cost non-students?

8. Rose figured that because her monthly rent of $265 was 22% of her monthly income, she could afford to stay in her apartment. How much is her monthly income (to the nearest whole dollar)?

9. Greg's Bakery spent $428.80 on ingredients one month and figured that this amount was 13.4% of the monthly income. How much does the bakery make each month?

10. Dick's team won 37 of 50 games this season. What percent of all games did they win?

11. The author received a royalty check for $6,720. The publisher said her book's sales had been $56,000. What percent of the sales did the author earn in royalties?

12. Frieda completed 30% of the task when she had called 144 of the names on her list.
    a. How many names are on her list?
    b. How many more people must she call to get the job done?

13. By the time Nick saved $179.20, he had 64% of the money he needed for the repairs on his car. How much did it cost him to fix the car?

14. Margarita weighed 140 lb after 2 weeks of dieting. This is 87.5% of her weight before she began the diet. How much did she weigh then?

15. The instructors adjusted the final exam curve so that 12.5% of the students received an A. If 155 A's were recorded, how many students took the exam?

16. The car dealer tells me that 2.5% of the cars he gets have sunroofs. I want one with a sunroof. If there are 42 buyers ahead of me waiting for cars with sunroofs, how many cars have to come into the dealership before I get the one I want? (I get the 43rd car with a sunroof.)

17. Doctors estimated that 6.4% of the population of Itchville came down with a rash. If 512 people reported that they got the rash, what is the population of Itchville?

18. Only 2% of the people at the 1985 Sports Trivia Convention were alive when the Chicago Cubs last won a pennant, in 1945. If 72 of the participants were alive in 1945, how many people attended the convention?

19. A newspaper poll discovered that 34,040 Bostonians claimed to have attended the seventh championship game where the Celtics beat the Lakers for the 1984 NBA Championship. But the reporter calculated that this was 230% of the capacity at Boston Garden, so a lot of the people were lying. What is the capacity?

20. After Watergate, only 4.2% of those asked admitted they had voted for Nixon. If 1,260 people made this admission, how many people were polled?

21. a. What is the percent of increase in sales if at this date last year the rep.'s sales were $308,071 and her sales now are $327,296? Round off to the nearest percent.
    b. What is the projected increase in sales if her total sales last year were $409,463?
    c. What are the projected sales for this year if last year's sales were $409,463? Round off to the nearest dollar.

22. a. What is the percent of decrease in sales if at this date last year the rep.'s sales were $327,135 and her sales now are $234,850? Round off to the nearest percent.
    b. What is the projected decrease in sales this year if last year's sales were $390,250?
    c. What are the projected sales for this year if last year's sales were $390,250? Round off to the nearest dollar.

Check your answers on page 729.

7.2 Formulas

Evaluating Formulas

Some equations represent relationships among concepts we frequently use. We let certain letters represent these concepts, and we call the equations formulas.

Distance, rate, and time
One example of such a relationship is the one among distance, rate, and time. We use it so automatically that sometimes we are not even aware that it is a formula.

If you were driving at the rate of 50 miles per hour for 3 hours, without thinking about it you would know you had gone 150 miles. What you did subconsciously was to use the formula

$$\text{Distance} = \text{Rate} \times \text{Time} \qquad [D = RT].$$

EXAMPLE 1 How far did we go if we drove at the rate of 50 miles per hour for 3 hours?

$$D = \text{Rate} \times \text{Time}$$

$$= \left(\frac{50 \text{ miles}}{\text{hour}}\right)(3 \text{ hours})$$

$$= \left(\frac{50 \text{ miles}}{\cancel{\text{hour}}}\right)(3 \cancel{\text{hours}})$$

$$= (50)(3) \text{ miles}$$

$$= 150 \text{ miles} \qquad \square$$

Temperature

Another relationship expressed as a formula concerns temperature. Two methods of measuring temperature are Fahrenheit and Celsius. Some people refer to Celsius as centigrade. Most countries use Celsius—we are beginning to hear more about this method in the United States.

In the United States, temperature is usually expressed in terms of Fahrenheit, and so that is the method most familiar to us. When the radio announcer says it is 95 degrees, we assume that the reference is to Fahrenheit, and we recognize that it is a very warm day.

On the Fahrenheit scale, the number used to express the temperature for freezing water is 32, and the number for boiling water is 212. In Celsius the number used to express the temperature for freezing water is 0, and the number for boiling water is 100. Celsius, or centigrade, is based on a system of measuring temperature on a scale from 0–100. (Notice the metric prefix "centi.")

Fahrenheit	Celsius
212°	100°
104°	40°
98.6°	37°
95°	35°
32°	0°

The freezing temperature is 32 degrees more in Fahrenheit than in Celsius, but the boiling temperature is 112 degrees more in Fahrenheit than in Celsius. So it isn't possible to convert from one to the other just by adding some constant number.

The formula to convert from Fahrenheit to Celsius is

$$C = \frac{5}{9}(F - 32°).$$

EXAMPLE 2   What would the temperature be in Celsius when it is 95 degrees Fahrenheit?

$$C = \left(\frac{5}{9}\right)(95° - 32°)$$

$$= \frac{5(63)°}{9}$$

$$= \frac{5(7)(\cancel{9})°}{\cancel{9}}$$

$$= 35°$$

So 95°F is the same as 35°C.

   If a radio announcer said the temperature was 35 degrees, it would sound like a cold day to someone familiar only with the Fahrenheit scale; but 35 degrees Celsius is the same hot day as 95 degrees Fahrenheit. It is very important to mention the unit Celsius or Fahrenheit, and not just state the number of the temperature. Celsius is being used in the United States more and more.
   The formula for converting Celsius to Fahrenheit is

$$F = \frac{9}{5}C + 32°.$$

We would use this formula if traveling in Europe so we could translate the stated temperature to the Fahrenheit scale that we are more familiar with.

EXAMPLE 3   What is the temperature on the Fahrenheit scale when it is 40 degrees on the Celsius scale?

$$F = \left(\frac{9}{5}\right)C + 32$$

$$= \left(\frac{9}{5}\right)(40°) + 32$$

$$= \frac{9}{5}\frac{(8)(5)}{1} + 32$$

$$= \frac{9}{\cancel{5}}\frac{(8)(\cancel{5})}{1} + 32$$

$$= 72 + 32$$

$$= 104°$$

So 40°C = 104°F.

Falling objects
Many algebraic formulas are used in the sciences. To give you practice in using such formulas, I will briefly introduce two common formulas from physics. Both of them have to do with falling objects.

Acceleration.   If an object is dropped (not pushed or thrown), two things affect the rate at which it falls: the force of gravity and the time that has gone by since it was dropped. **Gravity** is the attraction of objects to the center of the earth. Gravity is a force that makes objects fall toward earth with increasing speed, or **acceleration**.
   The acceleration caused by gravity has been calculated to be 32 feet per second, per second. That is, in the first second after an object is dropped, its speed increases from 0 feet per second to 32 feet per second. In the next second,

its speed increases by another 32 ft/sec to 64 ft/sec. In the third second, its speed increases by another 32 ft/sec to 96 ft/sec, and so on.

Algebraically, we can call the acceleration due to gravity g, and we can express gravity in the English system as 32 feet per second, per second as

$$g = \frac{32 \text{ ft/sec}}{\text{sec}}, \quad \text{which equals} \quad \frac{32 \text{ ft}}{\text{sec}} \cdot \frac{1}{\text{sec}} = \frac{32 \text{ ft}}{(\text{sec})(\text{sec})} \rightarrow \frac{32 \text{ ft}}{\text{sec}^2}$$

$$g = \frac{32 \text{ ft}}{\text{sec}^2}$$

or, in the metric system,

$$g = \frac{9.8 \text{ m}}{\text{sec}^2}.$$

Velocity. Now that we understand what g represents, we can look at the formula for the speed of falling objects. If we know how long it has been since an object was dropped, we can determine how fast it must be falling at this time. The technical term for speed, in physics, is **velocity**. The formula for the velocity of a falling object is

$V = gt$, where V is velocity, g is gravity, and t is time.

This means that the velocity at any time is the product of g by the amount of time since the object was dropped.

EXAMPLE 4 Find the velocity with which a free-falling object is traveling 8 seconds after it was dropped.

$$V = gt$$

$$= \frac{32 \text{ ft}}{\text{sec}^2} (8 \text{ sec})$$

$$= \frac{32 \text{ ft}}{(\text{sec})(\text{sec})} (8 \text{ sec})$$

$$= \frac{256 \text{ ft}}{\text{sec}}$$

or

$$\frac{9.8 \text{ m}}{(\text{sec})(\text{sec})} (8 \text{ sec}) =$$

$$\frac{(9.8)(8) \text{ m}}{\text{sec}} = \frac{78.4 \text{ m}}{\text{sec}}.$$

So the velocity is 256 ft/sec or 78.4 m/sec. ❑

Since 256 feet per second is equal to 175 miles per hour (or 281 kilometers per hour), it is clear why it is not a good idea to stand under objects that have been falling for an appreciable length of time.

If an object is pushed off with some force, however, its velocity will be even faster than if it is simply dropped. That is because it had some downward velocity even before gravity took over, called **initial velocity** and is abbreviated V_0. The formula for a pushed object's velocity after a given amount of time becomes

$V = V_0 + gt$ Use the English or metric expression for gravity to match the unit in which V_0 is expressed.

EXAMPLE 5   Find the velocity of a falling object after 2 seconds if it had an initial velocity of 3 feet per second.

$$V = \frac{3 \text{ ft}}{\text{sec}} + \frac{(32 \text{ ft})}{(\text{sec}^2)}(2 \text{ sec})$$

$$= \frac{3 \text{ ft}}{\text{sec}} + \frac{64 \text{ ft}}{\text{sec}}$$

$$= \frac{67 \text{ ft}}{\text{sec}} \qquad \square$$

EXAMPLE 6   Find the velocity with which an object is falling after 2 seconds if it had an initial velocity of 9 meters per second.

$$V = \frac{9 \text{ m}}{\text{sec}} + \frac{(9.8 \text{ m})}{(\text{sec}^2)}(2 \text{ sec})$$

$$= \frac{9 \text{ m}}{\text{sec}} + \frac{19.6 \text{ m}}{\text{sec}}$$

$$= \frac{28.6 \text{ m}}{\text{sec}} \qquad \square$$

Distance.   Those formulas tell us the velocity of an object at any instant. However, suppose we want to know the *distance* that an object will fall in a certain amount of time—say, for instance, 3 seconds. We know that $d = rt$, and r (rate) means the same thing as velocity (V). However, the velocity we would need is not the velocity at the *end* of 3 seconds but the *average* velocity during the 3 seconds. So we need to use a different formula.

I won't go through the complexities of how average velocity is determined. I will just present the resulting formula, commonly used in physics, for computing the distance that a free-falling object will travel in a given amount of time. That formula is

$$d = \frac{1}{2}gt^2, \text{ where } d \text{ is distance, } g \text{ is gravity, and } t \text{ is time.}$$

EXAMPLE 7   Find the distance in feet that a falling object will travel after 3 seconds.

$$d = \frac{1}{2}gt^2$$

$$= \frac{1}{2}\frac{(32 \text{ ft})}{(\text{sec}^2)}(3 \text{ sec})^2$$

$$= \frac{(32 \text{ ft})(9 \text{ sec}^2)}{2 \text{ sec}^2}$$

$$= \frac{32 \text{ ft }(9 \text{ sec}^2)}{2 \text{ sec}^2}$$

$$= \frac{(32)(9)}{2} \text{ ft}$$

$$= (16)(9) \text{ ft}$$

$$= 144 \text{ ft}$$

So a falling object travels 144 ft in 3 sec. $\qquad \square$

EXAMPLE 8 Find the distance in meters that a falling object travels in 2 seconds.

$$d = \frac{1}{2} gt^2$$

$$= \frac{1}{2} \frac{(9.8 \text{ m})}{(\text{sec}^2)} (2 \text{ sec})^2$$

$$= \frac{1}{2} \frac{(9.8) \text{ m } (2 \text{ sec})(2 \text{ sec})}{(\text{sec})(\text{sec})}$$

$$= \frac{(9.8)(\cancel{2})(2) \text{ m } (\cancel{\text{sec}})(\cancel{\text{sec}})}{(\cancel{2})(\cancel{\text{sec}})(\cancel{\text{sec}})}$$

$$= (9.8)(2) \text{ m}$$

$$= 19.6 \text{ m}$$

So a falling object travels 19.6 m in 2 sec. ❑

EXERCISES 7.2A

Use the appropriate formula to answer each question. Be sure that your units are properly set up so that your answer has the correct units.

1. How far did we travel after driving at an average rate of 45 mph for 6 hr?

2. How far did we ride on our bikes if we traveled at an average speed of 8 mph for $4\frac{1}{2}$ hr?

3. What is the temperature in Celsius when it is 113°F?

4. What is the temperature in Fahrenheit when it is 25°C?

5. How many feet has a falling object dropped after 4 sec?

6. At what velocity is a falling object traveling after 3 sec if it had an initial velocity of 2 m/sec?

7. At what velocity is a falling object traveling after 5 sec if it had an initial velocity of 4 ft/sec?

8. How far in meters will an object fall after 4 sec?

9. What would the temperature be in Celsius if it was 50° F?

10. What would the temperature be in Fahrenheit (to the nearest whole degree) if it was 14° C?

11. Approximately how many whole feet will an object fall after 3.2 sec?

12. What will the velocity of a falling object be after 4 sec if it has an initial velocity of 7 ft/sec?

Check your answers on page 730. ✓

Solving Formulas for Different Letters

There are times when a formula you need to use is not given in the most useful form. You can use the skills you learned in Chapter 6 to transform the formula into a more useful one.

Besides solving an equation in one letter for the numerical value of that letter, we can also solve an equation in more than one letter for an expression for one of those letters.

For example, $F = \frac{9}{5}C + 32°$ is an equation in two letters (F and C). In that form, we say it is solved for the letter F, or it is a *formula* for F.

As we know, temperature is usually reported in terms of the Fahrenheit scale in the United States. If we were visiting a country where the Celsius (or centigrade) scale was used to report temperature, we would need to convert Celsius numbers into Fahrenheit. We would use the formula $F = \frac{9}{5}C + 32$, where C stands for the temperature in Celsius.

If we wanted to explain what a temperature reported in Fahrenheit meant to someone more familiar with the Celsius scale, we could take the Fahrenheit formula and convert it into one that would give the Celsius temperature. That is, we would solve the formula $F = \frac{9}{5}C + 32°$ for C. To do so, we must isolate C on the right side of the equation. We would then find the formula for Celsius (in terms of Fahrenheit).

EXAMPLE 9 Solve $F = \frac{9}{5}C + 32°$ for C.

First remove the constant term from the right side by adding $-32°$ to both sides.

$$F = \frac{9}{5}C + 32°$$
$$\underline{-32° \qquad -32°}$$
$$F - 32° = \frac{9}{5}C$$

Now multiply both sides by $\frac{5}{9}$ to remove the coefficient $\frac{9}{5}$ from the right side.

$$\frac{5}{9}(F - 32°) = \frac{5}{9}\cdot\frac{9}{5}C$$
$$\frac{5}{9}(F - 32°) = C$$

So the formula for C is: $C = \frac{5}{9}(F - 32°)$

EXAMPLE 10 The distance (D) traveled is the product of the rate of speed (R) and the time (T): $D = RT$.

What is the formula for finding the rate of speed if we know the distance and time? Solve $D = RT$ for R.

$$D = RT$$
$$\frac{D}{T} = \frac{RT}{T}$$
$$\frac{D}{T} = R$$

So the rate is found by dividing the distance by the time.

What is the average rate of speed if we travel 280 miles over 7 hours?

$$R = \frac{280 \text{ miles}}{7 \text{ hours}}$$
$$= 40 \text{ miles per hour}$$

So the rate is 40 mph.

EXAMPLE 11   The area (A) of a triangle is $A = \dfrac{bh}{2}$, where b is the base and h is the height of the triangle. What is the formula for finding the base when we know the area and height of a triangle?

$$A = \frac{bh}{2}$$

$$2(A) = 2\left[\frac{bh}{2}\right]$$

$$2A = bh$$

$$\frac{2A}{h} = \frac{b\cancel{h}}{\cancel{h}}$$

$$\frac{2A}{h} = b$$

Use this formula to find the base of a triangle with an area of 24 square inches and a height of 8 inches.

$$b = \frac{2(24 \text{ in.}^2)}{8 \text{ in.}}$$

$$= \frac{2 \cdot \cancel{8} \cdot 3 \cdot \cancel{\text{in.}} \cdot \text{in.}}{\cancel{8} \cancel{\text{in.}}}$$

$$= 2 \cdot 3 \text{ in.}$$

$$= 6 \text{ in.}$$

So the base is 6 in.   ❑

EXAMPLE 12   Since the volume (V) of a rectangular cylindrical solid is the product of its width (W), length (L), and height (H), to find a formula for the height of a rectangular cylindrical solid we solve $V = WLH$ for H.

$$V = WLH$$

$$\frac{V}{WL} = \frac{WLH}{WL}$$

$$\frac{V}{WL} = H$$

We could use this formula to find height when we know the values of the solid's volume, width, and length.

What is the height of a rectangular cylindrical solid with a volume (V) of 360 cubic centimeters, a width (W) of 4 centimeters, and a length (L) of 5 centimeters?

$$H = \frac{V}{WL}$$

$$= \frac{(360 \text{ cm}^3)}{(4 \text{ cm})(5 \text{ cm})}$$

$$= -\frac{4 \cdot 5 \cdot 18 \text{ cm} \cdot \text{cm} \cdot \text{cm}}{4 \cdot 5 \text{ cm} \cdot \text{cm}}$$

$$= -\frac{\cancel{4} \cdot \cancel{5} \cdot 18 \cancel{\text{ cm}} \cdot \cancel{\text{cm}} \cdot \text{cm}}{\cancel{4} \cdot \cancel{5} \cancel{\text{ cm}} \cdot \cancel{\text{cm}}}$$

$$= 18 \text{ cm}$$

So the height is 18 cm.   ❑

EXAMPLE 13 $E = mc^2$ is a formula from Einstein's theory of relativity. It represents the equivalence between mass and energy. E represents energy, m represents mass, and c represents the speed of light. $E = mc^2$ allows you to compute the energy if you know an object's mass.

Let's find the formula for finding an object's mass if we know its energy. Solve the formula for m. Divide both sides by c^2.

$$E = mc^2$$

$$\frac{E}{c^2} = \frac{mc^2}{c^2}$$

$$\frac{E}{c^2} = m$$

EXAMPLE 14 $Q = I - P$ is a formula that economists use. Quantity of demand for goods or services (Q) is a function of the income of the general population (I) minus the price of goods or services (P). To derive a formula for income in terms of quantity of demand and price, add P to both sides.

$$\begin{array}{r} Q = I - P \\ + P \quad\ + P \\ \hline Q + P = I \end{array}$$

EXAMPLE 15   Another formula economists often use is $C = L + N + K$, where C represents total cost, L represents cost of labor, N represents cost of land, and K represents cost of capital (cost of borrowing money). To derive a formula for cost of labor in terms of the others, subtract N and K from both sides.

$$\begin{array}{r} C = L + N + K \\ - N - K \quad\ - N - K \\ \hline C - N - K = L \end{array}$$

In some formulas the same letter is used to represent different numbers. This might happen when data are recorded at different times, such as at the beginning of an experiment and again later. Subscripts must be used on the letter to clarify that it represents different numbers.

For example, the formula $N_f = N_o + 3k$ might represent the final number (N_f), and N_o could represent the initial number one started with. Letters with different subscripts should be treated as different kinds of terms; they cannot be combined together because they are **not like terms**.

EXAMPLE 16   Solve $N_f = N_o + 3k$ for N_o.

$$\begin{array}{r} N_f = N_o + 3k \\ - 3k \quad\quad\ - 3k \\ \hline N_f - 3k = N_o \end{array}$$

EXAMPLE 17   Solve $N_f = N_o + 3k$ for k.

$$\begin{array}{r} N_f = \quad N_o + 3k \\ - N_o \quad - N_o \\ \hline N_f - N_o = \quad 3k \end{array}$$

$$\frac{N_f - N_o}{3} = \frac{3k}{3}$$

$$\frac{N_f - N_o}{3} = k$$

EXERCISES 7.2B

Solve each formula for the requested letter.

13. $p = a + b + c$ for b

14. $c = 2\pi r$ for r

15. $a = lw$ for w

16. $a = \dfrac{bh}{2}$ for h

17. $k = 3mp$ for m

18. $R = a + b - c$ for b

19. $i = prt$ for r

20. $V = V_0 + at$ for V_0

21. $V = V_0 + at$ for a

22. $V = V_0 + at$ for t

23. $d = \dfrac{1}{2} gt^2$ for g

24. $m = 3ab + c$ for c

25. $k = 5a - b$ for a

26. $h = 3k - 5m$ for k

27. $x = y + w + z$ for w

28. $a = 2b - 3c + 5d$ for b

29. $p = 7m - 4k$ for m

30. $J = 2L + 8P$ for P

Check your answers on page 730.   ✓

7.3 Ratios and Proportions

Ratios

We frequently find ourselves comparing two groups of things. The mathematical expression used to compare two things is called **ratio**. The notation used in a ratio is :, which is read as "to."

EXAMPLE 1   If we noticed that there were 5 men and 7 women in a room, we might say the ratio of men to women is 5:7.   ❑

If we compared a to b, we would consider a the first term and b the second term of the ratio and would write it $a:b$.

The order of terms in a ratio is very important because ratios can be represented by fractions. The first term of a ratio becomes the numerator; the second term becomes the denominator.

EXAMPLE 2   5:7 can be represented as $\dfrac{5}{7}$.   ❑

Expressing ratios as fractions makes it easy to express the ratio in its simplest, or most reduced, form. Just as the fraction $\dfrac{2}{6}$ reduces to $\dfrac{1}{3}$, the ratio 2:6 reduces to 1:3.

For example, we can say that the ratio of people with blue eyes to people with brown eyes is 1:3, as long as there is 1 blue-eyed person for every 3 brown-eyed persons. We might actually have based this conclusion on finding 2 people with blue eyes and 6 with brown eyes; or there could have been 40 people with blue eyes and 120 people with brown eyes.

Ratios should always be expressed in their most reduced form.

To reduce a ratio:
1. Write the ratio as a fraction.
2. Factor the numerator and the denominator to primes.
3. Cancel all pairs of factors common to both the numerator and the denominator.
4. Multiply the remaining factors in the numerator, and multiply the remaining factors in the denominator.
5. Convert the fraction back to ratio form.

EXAMPLE 3 Reduce the ratio $12:18$.

$$12:18 =$$
$$\frac{12}{18} =$$
$$\frac{2 \cdot 2 \cdot 3}{2 \cdot 3 \cdot 3} =$$
$$\frac{\cancel{2} \cdot 2 \cdot \cancel{3}}{\cancel{2} \cdot \cancel{3} \cdot 3} =$$
$$\frac{2}{3} = 2:3$$

So $12:18$ reduces to $2:3$.

EXAMPLE 4 Reduce the ratio $40:700$.

$$40:700 =$$
$$\frac{40}{700} =$$
$$\frac{2 \cdot 2 \cdot 2 \cdot 5}{2 \cdot 2 \cdot 5 \cdot 5 \cdot 7} =$$
$$\frac{\cancel{2} \cdot \cancel{2} \cdot 2 \cdot \cancel{5}}{\cancel{2} \cdot \cancel{2} \cdot \cancel{5} \cdot 5 \cdot 7} =$$
$$\frac{2}{5 \cdot 7} =$$
$$\frac{2}{35} = 2:35$$

So $40:700$ reduces to $2:35$.

EXERCISES 7.3A

Reduce each ratio.

1. $3:12 =$　　2. $4:20 =$　　3. $5:35 =$

4. $10:14 =$　　5. $12:20 =$　　6. $24:80 =$

7. $33:165 =$　　8. $15:70 =$　　9. $30:210 =$

10. $250:450 =$

Check your answers on page 730.

Proportions

A **proportion** is a statement asserting that two ratios are equal. In Example 3 we saw that the ratios 12:18 and 2:3 were equal. So the equation 12:18 = 2:3 is a proportion.

There are four terms to a proportion. The first and fourth are called the **outside terms**; the second and third are called the **inside terms**.

All proportions have an interesting property.

> ***Proportion Property:***
> The produce of the *inside* terms of a proportion is always equal to the product of the *outside* terms.

EXAMPLE 5

$$\text{inside}$$
$$12:18 = 2:3$$
$$\text{outside}$$

The inside terms are 18 and 2; their product is 36. ✓

The outside terms are 12 and 3; their product is 36. ✓ ❑

When proportions are used in math and quantitative fields, the inside terms are called the **means**, and the outside terms are called the **extremes**.

If we represent ratios as fractions, we can say that we used the **proportion property** in Section 3.4 to check whether we correctly reduced a fraction. We called it **cross multiplication** in Chapter 3.

Cross multiplication

The product of the numerator of one fraction multiplied by the denominator of an equal fraction is equal to the product of the other numerator and denominator.

$$\text{If } \frac{12}{18} = \frac{2}{3}, \text{ then } (12)(3) = (2)(18).$$

$$36 = 36 \quad ✓$$

In a proportion, it does not matter which ratio is written on which side.

EXAMPLE 6 We could write the proportion from Example 5 as

$$\text{inside}$$
$$2:3 = 12:18$$
$$\text{outside}$$

The inside terms are 3 and 12; their product is 36. ✓

The outside terms are 2 and 18; their product is 36. ✓ ❑

In a proportion we can also switch the order of the terms in a ratio if we switch the order in *both* ratios. Just be sure that the first term in each ratio refers to the same category.

EXAMPLE 7 It is the same to say:

The ratio of 12 cats to 18 dogs is 2 cats:3 dogs

as to say:

The ratio of 18 dogs to 12 cats is 3 dogs:2 cats. ❑

That property will allow us an easy way to check whether we correctly reduced a ratio.

EXAMPLE 8 A politician found that 3,800 of the citizens polled preferred his position on U.S. policy in Central America, while 9,500 opposed it. Was he right to say that the ratio of support for him was 2:3?

If he was right, then 3,800:9,500 = 2:3.

> The inside terms are 9,500 and 2; their product is 19,000.
> The outside terms are 3,800 and 3; their product is 11,400.

Since the products of the inside and outside terms are not equal, the politician was wrong. (He probably took advantage of the fact that lots of people do not know how to work with ratios. Politicians often mislead us with their calculations. Now that you understand ratios and proportions, you will be able to check these things out for yourself.) ❏

EXAMPLE 9 Is the ratio of people in favor of cutting back the budget for social programs to those who oppose those cuts really 2:3 if 3,360 voted for the cuts and 5,040 voted against the cuts? If it is, then 3,360:5,040 = 2:3.

> The inside terms are 5,040 and 2; their product is 10,080.

> The outside terms are 3,360 and 3; their product is 10,080.

Yes, the ratio is 2:3. ❏

EXERCISES 7.3B

Use the proportion property to see whether the ratios are really equal.

11. Is 40:88 the same as 5:11? **12.** Is 74:296 the same as 2:7?

13. Is 74:296 the same as 1:4? **14.** Is 69:115 the same as 5:3?

15. Is 69:115 the same as 3:5?

16. Is 330:618 the same as 51:103?

17. Is 16,458:21,311 the same as 78:101?

18. Is 360,153:600,255 the same as 3:5?

19. Is 80,366:200,920 the same as 2:5?

20. Is 1,530,114:4,080,304 the same as 3:8?

Check your answers on page 730.

7.4 Solving Proportions

Strategy for Solving Proportions

We can use the proportion property—the fact that the product of the *inside* terms of a proportion is always equal to the product of the *outside* terms—to make the proportion into an equation and to solve it for an unknown term when we know the value of only three terms in a proportion.

EXAMPLE 1   Solve for n if $7{:}21 = 2{:}n$.

Since the product of the *inside* terms is equal to the product of the *outside* terms, the proportion

$$7{:}21 = 2{:}n$$

can be made into the equation

$$7n = 2(21)$$

$$7n = 42 \qquad \text{To solve for } n, \text{ divide both sides by 7.}$$

$$\frac{7n}{7} = \frac{42}{7}$$

$$n = 6$$

So the solution is $n = 6$.   ❑

There are two ways to check.

Check 1: If $n = 6$, see whether the product of the *inside* terms is equal to the product of the *outside* terms.

$$7{:}21 = 2{:}n$$

product of the *inside* terms is $2(21) = 42$ ✓

product of the *outside* terms is $7n =$

$$7(6) = 42 \quad ✓$$

Check 2: If $n = 6$, do both ratios reduce to the same fraction?

$$7{:}21 = 2{:}n$$

$$7{:}21 = \qquad\qquad 2{:}n =$$

$$\frac{7}{21} = \frac{1}{3} \quad ✓ \qquad 2{:}(6) =$$

$$\frac{2}{6} = \frac{1}{3} \quad ✓$$

To check, use whichever method you prefer.

EXAMPLE 2   Solve for k if $k{:}5 = 4{:}20$.

$$k{:}5 = 4{:}20 \text{ becomes}$$

$$20k = 4(5)$$

$$20k = 20$$

$$\frac{20k}{20} = \frac{20}{20}$$

$$k = 1$$

So the solution is $k = 1$.   ❑

Check: If $k = 1$: Inside product is $(5)(4) = 20$ ✓

Outside product is $20k =$

$$20(1) = 20 \quad ✓$$

Sometimes the ratios of the proportion we want to solve are expressed as fractions. In such problems, we state that those fractions are equal, and solve by directly using **cross multiplication**.

EXAMPLE 3   Solve $\dfrac{3}{x} = \dfrac{5}{40}$ for x.

By cross multiplication, we get

$$3(40) = 5x$$

$$\frac{3 \cdot 5 \cdot 8}{5} = \frac{5x}{5}$$

$$\frac{3 \cdot \cancel{5} \cdot 8}{\cancel{5}} = \frac{\cancel{5}x}{\cancel{5}}$$

$$3 \cdot 8 = x$$

$$24 = x$$

So the solution is $x = 24$.

Check: Does $\dfrac{5}{40} = \dfrac{3}{x}$ when $x = 24$?

$$\frac{5}{40} = \qquad\qquad \frac{3}{x} =$$

$$\frac{5}{5 \cdot 8} = \qquad\qquad \frac{3}{24} =$$

$$\frac{\cancel{5}}{\cancel{5} \cdot 8} = \frac{1}{8} \quad\checkmark \qquad\qquad \frac{3}{3 \cdot 8} =$$

$$\frac{\cancel{3}}{\cancel{3} \cdot 8} = \frac{1}{8} \quad\checkmark$$

Sometimes we are told what the ratio of two letters reduces to. If we know the value for one letter, we can substitute it in and solve the proportion for the other letter.

EXAMPLE 4   If $x:y = 4:20$ and $x = 7$, find y.

Let $x = 7$ and solve for y.

$$(7):y = 4:20$$

$$4y = 7(20)$$

$$\frac{4y}{4} = \frac{7 \cdot 4 \cdot 5}{4}$$

$$y = 7 \cdot 5$$

$$y = 35$$

So the solution is $y = 35$.

Check: If $y = 35$ and $x = 7$, does $x:y = 4:20$?

$$\text{Inside product is } 4y =$$

$$4(35) = 140 \quad\checkmark$$

$$\text{Outside product is } 20x =$$

$$20(7) = 140 \quad\checkmark$$

EXERCISES 7.4A

Solve each proportion and check your solutions.

1. $3:4 = k:20$ **2.** $2:9 = y:36$

3. $3:4 = 18:w$ **4.** $p:3 = 42:6$

5. $5:x = 35:49$ **6.** $2:h = 5:40$

7. $\dfrac{m}{84} = \dfrac{3}{63}$ **8.** $\dfrac{7}{70} = \dfrac{r}{170}$

9. $\dfrac{4}{28} = \dfrac{5}{n}$ **10.** $\dfrac{4}{26} = \dfrac{t}{130}$

11. If $a:b = 4:5$ and $a = 20$, find b.

12. If $m:p = 3:7$ and $p = 35$, find m.

13. If $2:9 = k:p$ and $k = 26$, find p.

14. If $5:2 = r:w$ and $w = 34$, find r.

15. If $7:6 = a:z$ and $a = 77$, find z.

16. If $h:b = 4:3$ and $b = 51$, find h.

17. If $j:5 = k:40$ and $k = 48$, find j.

18. If $p:6 = q:30$ and $p = 7$, find q.

Check your answers on 730.

Solving Proportions Involving Fractions and Decimals

Don't let problems involving fractions or decimals intimidate you. You can convert all proportions into linear equations, which you learned to solve in Chapter 6, if you know the value of three of the four terms.

EXAMPLE 5   Solve for y if $2.1:y = 0.3:10$.

$$2.1:y = 0.3:10$$

$$10(2.1) = 0.3y$$

$$21 = 0.3y \qquad \text{Divide both sides by 0.3.}$$

$$\frac{21}{0.3} = \frac{0.3y}{0.3}$$

$$\frac{21.0}{0.3} = \frac{0.3y}{0.3}$$

$$\frac{210}{3} = y$$

$$70 = y$$

Check: Does $2.1:y = 0.3:10$ when $y = 70$?

Inside product is $(0.3)y =$

$$(0.3)(70) = 21$$

Outside product is $(2.1)(10) = 21$

EXAMPLE 6   Solve for x if $\dfrac{3}{5}:x = 7:2$.

$$\dfrac{3}{5}:x = 7:2$$

$$2\left(\dfrac{3}{5}\right) = 7x$$

$$\left(\dfrac{2}{1}\right)\left(\dfrac{3}{5}\right) = 7x$$

$$\dfrac{6}{5} = 7x$$

Normally we would divide both sides by 7, but because this problem involves fractions, it is easier to multiply both sides by $\dfrac{1}{7}$.

$$\left(\dfrac{1}{7}\right)\left(\dfrac{6}{5}\right) = \left(\dfrac{1}{7}\right)(7x)$$

$$\dfrac{6}{35} = x$$

So the solution is $x = \dfrac{6}{35}$.

Check: If $x = \dfrac{6}{35}$, is $\dfrac{3}{5}:\left(\dfrac{6}{35}\right) = 7:2$ a proportion?

$$\text{Inside product is } \left(\dfrac{6}{35}\right)(7) =$$

$$\left(\dfrac{6}{7\cdot5}\right)\left(\dfrac{7}{1}\right) = \dfrac{6}{5} \quad \checkmark$$

$$\text{Outside product is } \left(\dfrac{3}{5}\right)(2) =$$

$$\left(\dfrac{3}{5}\right)\left(\dfrac{2}{1}\right) = \dfrac{6}{5} \quad \checkmark$$

If proportions involve mixed numbers, translate them into improper fractions before computing.

EXAMPLE 7   Solve for n if $3\dfrac{1}{2}:n = 35:40$.

$$3\dfrac{1}{2}:n = 35:40$$

$$\dfrac{7}{2}:n = 35:40$$

$$(40)\left(\dfrac{7}{2}\right) = 35n$$

$$\dfrac{2\cdot2\cdot2\cdot5\cdot7}{2} = 35n$$

$$(2\cdot2\cdot5\cdot7) = 35n \qquad \text{Divide both sides by 35, } (7\cdot5).$$

$$\dfrac{2\cdot2\cdot5\cdot7}{7\cdot5} = \dfrac{35n}{35}$$

$$2\cdot2 = n$$

$$4 = n$$

So the solution is $n = 4$.

Check: If $n = 4$, is $3\frac{1}{2}:n = 35:40$ a proportion?

$$\text{Inside product is } 35n =$$
$$(35)(4) = 140 \checkmark$$

$$\text{Outside product is } \left(3\frac{1}{2}\right)(40) =$$
$$\left(\frac{7}{2}\right)\left(\frac{40}{1}\right) =$$
$$\frac{7 \cdot \cancel{2} \cdot 20}{\cancel{2}} =$$
$$7 \cdot 20 = 140 \checkmark$$

EXERCISES 7.4B

Solve for the letter in each proportion.

19. $\frac{2}{3}:x = 7:4$ **20.** $\frac{1}{5}:y = 4:12$

21. $m:\frac{2}{3} = 2:5$ **22.** $3.4:h = 28:14$

23. $0.63:k = 51:17$ **24.** $r:2.5 = 14:70$

25. $1\frac{1}{2}:w = 2\frac{1}{3}:11\frac{2}{3}$ **26.** $2\frac{1}{5}:6\frac{3}{5} = x:2\frac{1}{4}$

27. $\frac{2}{3}:2\frac{2}{3} = 1\frac{1}{3}:p$ **28.** $\frac{4}{5}:h = 3\frac{1}{2}:10\frac{1}{2}$

Check your answers on page 730. ✓

Using Proportions to Solve Real-Life Problems

Many real-life problems can be set up as proportions. If you are given the ratio of one category to another and are told the value of one of those categories, you can set up a proportion to find the value of the other one.

Begin by writing a proportion with two identical ratios that contain only those two categories. Replace the categories with the information you are given and choose a letter to represent the category you want to find.

EXAMPLE 8 The ratio of blacks to whites in my community is 3:8. I want the racial representation on my staff to reflect the ratio in the community. There are 56 white workers on the staff now. I will expand the staff. How many black workers do I need to recruit?

The categories of the ratios are blacks and whites. Using those ratios, begin with the proportion

$$\text{Blacks:whites} = \text{blacks:whites.}$$

In one ratio substitute the values from the given ratio.

$$(3 \text{ blacks}):(8 \text{ whites}) = \text{blacks:whites}$$

Now substitute in the value that you are given for one of those categories. In this case it would be 56 whites.

Let B represent the number of blacks needed on the staff, and solve the proportion for B.

$$(3 \text{ blacks}):(8 \text{ whites}) = B:(56 \text{ whites})$$

$$\text{or} \quad 3:\underline{8} = \underline{B}:\underline{56}$$

$$(56)(3) = \underline{8B}$$

$$\frac{[(\not{8})(7)(3)]}{\not{8}} = \frac{(\not{8}B)}{\not{8}}$$

$$(7)(3) = B$$

$$21 = B$$

So I need to recruit 21 black workers. ❏

Check: If I recruit 21 blacks, to expand the staff of 56 whites, will the ratio of blacks (B) to whites (W) be 3:8?

$$B:W =$$

$$21:56 =$$

$$\frac{21}{56} =$$

$$\frac{(\not{7})(3)}{(\not{7})(8)} =$$

$$\frac{3}{8} = 3:8 \quad \checkmark$$

EXAMPLE 9 I am using a map with a scale of 2 inches representing 25 miles. How many miles long is a trip that measures 5 inches on the map?

The categories of the ratios are inches and miles. Using those ratios, begin with the proportion

$$\text{Inches:miles} = \text{inches:miles}.$$

In one ratio substitute the values from the given ratio.

$$(2 \text{ inches}):(25 \text{ miles}) = \text{inches:miles}$$

Now substitute in the value we were given, 5 inches. We want to know how many miles are represented by 5 inches, so call the unknown number of miles m, and solve the proportion.

$$(2 \text{ inches}):(25 \text{ miles}) = (5 \text{ inches}):(m)$$

$$(2 \text{ inches}):\underline{(25 \text{ miles})} = \underline{(5 \text{ inches})}:(m)$$

$$(2 \text{ inches})(m) = (5 \text{ inches})(25 \text{ miles})$$

You can divide inches away from both sides.

$$2m \, \frac{\cancel{\text{inches}}}{\cancel{\text{inches}}} = 125 \, \frac{\cancel{\text{inches}} \text{ miles}}{\cancel{\text{inches}}}$$

$$2m = 125 \text{ miles}$$

$$\frac{2m}{2} = \frac{125 \text{ miles}}{2} \qquad \text{Divide both sides by 2.}$$

$$m = 62\frac{1}{2} \text{ miles}$$

So that trip is actually $62\frac{1}{2}$ miles long. ❏

EXAMPLE 10   If boxes of crackers are on sale at 4 for $3.20, how much would 11 boxes cost?

Use the proportion Boxes:dollars = boxes:dollars.

We are given the ratio 4 boxes:3.20 dollars.

$$(4 \text{ boxes}):(3.20 \text{ dollars}) = \text{boxes:dollars}$$

We want to find out how many dollars 11 boxes will cost. Let d be the dollars.

$$(4 \text{ boxes}):(3.20 \text{ dollars}) = (11 \text{ boxes}):d$$

$$(4 \text{ boxes})d = (3.20 \text{ dollars})(11 \text{ boxes})$$

$$\frac{(4 \text{ boxes})d}{(4 \text{ boxes})} = \frac{(3.20 \text{ dollars})(11 \text{ boxes})}{(4 \text{ boxes})}$$

$$d = \frac{(3.20)(11) \text{ dollars}}{4}$$

$$d = \frac{(35.20) \text{ dollars}}{4}$$

$$d = 8.80 \text{ dollars}$$

So 11 boxes will cost $8.80   ❏

EXAMPLE 11   An inch is defined to be exactly 2.54 centimeters. Ron was told he needed 21.59 centimeters of material, but the store measured it in inches. How many inches did he need?

Use the proportion Inches:centimeters = inches:centimeters.

$$1 \text{ in.}:2.54 \text{ cm} = x \text{ in.}:21.59 \text{ cm}$$

$$(1 \text{ in.})(21.59 \text{ cm}) = (2.54 \text{ cm})(x \text{ in.})$$

$$\frac{(1 \text{ in.})(21.59 \text{ cm})}{2.54 \text{ cm}} = x \text{ in.}$$

$$\frac{21.59}{2.54} \text{ in.} = x \text{ in.}$$

$$8.5 \text{ in.} = x \text{ in.}$$

So Ron needed 8.5 in. of the material.   ❏

EXAMPLE 12   A sales rep. was trained to calculate how many days to spend with each account based on the ratio of the number of days to the amount of sales made. She spent 165 days in the field last year and sold $421,200 worth of merchandise. How many days should she plan to spend with the account where her sales were $47,800 last year?

Use the proportion Days:sales = days:sales.

$$165:\$421,200 = d:\$47,800$$

$$(165)(\$47,800) = \$421,200d$$

$$\frac{(165)(\$47,800)}{\$421,200} = d$$

$$\frac{7,887,000}{421,200} = d$$

$$18.72507 = d$$

Round off to the nearest whole day.

↓

$$18.72507 \rightarrow 19$$

So she should plan to spend 19 days with that account.   ❏

EXERCISES 7.4C

Solve each problem by setting up a proportion and solving for the unknown quantity.

29. If the ratio of men to women in a room is 2:3, and there are 86 men, how many women are there?

30. If a person is paid at the rate of $9 for 2 hr of work, how many hours would that person need to work to earn $72?

31. If a person earns $15 for 4 hr of work, how much will that person earn in 500 hr?

32. If a map is drawn to a scale of 2 in. representing 25 mi, how far apart are two cities that are separated by 14 in. on the map?

33. If a map is drawn so that 2 in. represents 15 mi, how long would an actual distance of 105 mi be on that map?

34. If a casserole for 6 people calls for 2 cups of noodles, how many cups of noodles should be used for a casserole to feed 10 people?

35. Carl and Peggy agree to divide the profits in a joint enterprise on a ratio of 3:4 (3 for Carl, 4 for Peggy). How much does Peggy make when Carl makes $5,700?

36. If the ratio of children to adults at a picnic is 7:5, how many adults are there if there are 42 children?

37. If the ratio of people preferring root beer to lemonade is 4:7, how many people prefer lemonade when 28 prefer root beer?

38. If the ratio of in-state residents to out-of-state residents at a school is 9:2, how many in-state residents are there when there are 7,460 out-of-state residents?

39. If the tax rate in a town is $245.80 for every $1,000 of property, what would the taxes be on a property valued at $27,000?

40. If 5 acres yield 300 bu of wheat, how many bushels should we expect to get from 20 acres?

41. If Luke can run in 3 min the same distance it takes Sonya to run in 7 min, then in a race that takes Luke 15 min, how long would Sonya take?

42. If soda is selling at three bottles for $2.10, how much would you expect five bottles to cost?

43. The ratio of foreign-made cars to American-made cars on my block is 3 to 10. How many cars are foreign made if 40 are American made?

44. If 1 lb is equivalent to 454 g, how many grams are there in 3.5 lb?

45. If 1 m is the same as 39.37 in., how many meters is 157.48 in.?

46. Jeff spent 140 days in the field as a sales rep. and sold $58,800 worth of merchandise. How many days should he plan to spend with an account that purchased $3,780 worth of merchandise?

47. Laura's sales last year were $288,000 for 180 days in the field. She calculated that to match those sales again, she should spend 7 days with the Dick's Disks account. How much were the sales for that account?

48. Henry calculated that he sold $5,340 worth of tools for every 20 days in the field. How many days must he work to sell $21,360 worth of tools?

Check your answers on page 730.

7.5 Solving Proportions That Require Setting Up New Ratios

The proportions we solved in Section 7.4 consisted of the two categories that were being compared in the given ratio. We were told the value of one category and asked to solve for the number in the other category.

Sometimes we are given the ratio of two categories, but instead of the value of either category, we are told the total in both categories and expected to find the value of each category. In such situations we must set up a new ratio to use in the proportion—the ratio of *one of the categories to the total*. Use whichever category you want to solve for first.

For example, if we know the ratio of Republicans to Democrats in some town and are told the total number of voters, we will be able to solve for the actual number of Republicans and of Democrats. If we choose to solve first for the number of Democrats, the ratio to use in the proportion is

Democrats:voters.

EXAMPLE 1   If the ratio of Republicans to Democrats is 2:3, what proportion should we set up to figure out how many of 735 voters are Democrats?

The proportion we set up to solve for Democrats is

Democrats:voters = Democrats:voters.

To solve a proportion you must know the value of three of the four terms. We can use the fact that the ratio of Republicans to Democrats is 2:3 to figure out the ratio of Democrats to voters. This ratio means that for every two Republicans there are three Democrats. So when there are three Democrats, there are five voters [3 Democrats + 2 Republicans = 5 voters].

So the ratio of Democrats to voters is 3:5. Now we can figure out how many of the 735 voters are Democrats.

EXAMPLE 2   If the ratio of Republicans to Democrats is 2:3, how many of the 735 voters are Democrats?

The proportion we set up to solve for Democrats is

Democrats:voters = Democrats:voters.

The ratio of Democrats to voters is 3:5. The number of voters is 735. Let D represent the number of Democrats. Replace the terms of the proportions with those values.

$$3:5 = D:735 \quad \text{Set the inside and the outside products equal.}$$
$$3(735) = 5D \quad \text{Divide both sides by 5.}$$
$$\frac{3(\cancel{5})147}{\cancel{5}} = \frac{\cancel{5}D}{\cancel{5}}$$
$$3(147) = D$$
$$441 = D$$

So there are 441 Democrats. ❏

Once you know the value of one category, the easiest way to find the value of the other category is to subtract the first one from the total. The total is the *total of those two categories*.

EXAMPLE 3   If the ratio of Republicans to Democrats is 2:3, and we know that 441 of the 735 voters are Democrats, how many are Republicans?

$$\text{Republicans} = \text{total voters} - \text{Democrats}$$
$$= 735 - 441$$
$$= \mathbf{294}$$ ❏

Check: 294 Republicans + 441 Democrats = 735 voters ✓

Is the ratio of Republicans to Democrats 2:3?

Does 2:3 = 294:441?

$$\text{Inside product is } 3(294) \quad = 882 \quad ✓$$
$$\text{Outside product is } 2(441) = 882 \quad ✓$$

When we know the ratio of two categories and the total quantity in both, we must set up the ratio of one category to the total.

1. Set up the proportion by using the ratio of one category to the total.
2. Replace three of the terms in the proportion with numbers from the data you are given. (The total in one ratio is the sum of amounts in each category.)
3. Choose a letter to represent the value of the category you want to find.
4. Solve for that letter, and subtract it from the total to find the value of the other category.

EXAMPLE 4   I'm cooking for a neighborhood festival. In my community the ratio of vegetarians to meat eaters is 3:8. I expect 77 neighbors to come to the festival. How many vegetarians should I expect?

The categories of the ratios are vegetarians and neighbors. Begin with this proportion;

$$\text{Vegetarian:neighbors} = \text{vegetarians:neighbors}$$

The fact that the ratio of vegetarians to meat eaters is 3:8 means that for every three vegetarians there are eight meat eaters. So for every three vegetarians there are 11 neighbors [3 vegetarians + 8 meat eaters = 11 neighbors]. In one ratio substitute 3:11. We were told that the number of neighbors is 77. Represent the number of vegetarians by V. We can now solve the proportion for V.

$$\underline{3}:\underline{11} = \underline{V}:\underline{77}$$
$$\underline{(3)(77)} = \underline{11\,V}$$
$$\frac{3\cdot 7\cdot \cancel{11}}{\cancel{11}} = \frac{\cancel{11}\,V}{\cancel{11}} \qquad \text{\color{blue}Divide both sides by 11.}$$
$$3\cdot 7 = V$$
$$21 = V$$

So I should expect 21 vegetarians to come to the festival. ❏

Since 21 of the 77 neighbors are vegetarians, the rest should be meat eaters.

$$77 - 21 = 56$$

So 56 meat eaters should come.

Check: If there are 21 vegetarians and 56 meat eaters, is the ratio of vegetarians (V) to meat eaters (M) really 3:8?

$$V:M =$$
$$21:56 =$$
$$\frac{\cancel{7}\cdot 3}{\cancel{7}\cdot 8} =$$
$$\frac{3}{8} = 3:8 \quad \checkmark$$

EXAMPLE 5   If the ratio of women to men at a party is 6:7, find out how many of the 234 guests are of each sex.

We could solve for either, but let's find out how many are women first. The proportion to use is Women:total = women:total.

Since the ratio of women to men at the party is 6:7, when there are six women, there is a total of 13, $(7 + 6)$ guests.

Let W represent the number of women. The proportion becomes $6:13 = W:234$.

Set the product of the inside terms equal to the product of the outside terms, and solve for W.

$$\underline{6}:\underline{13} = \underline{W}:\underline{234}$$
$$\underline{(6)(234)} = \underline{(13)\,W}$$
$$\frac{6\cdot 18\cdot \cancel{13}}{\cancel{13}} = \frac{\cancel{13}\,W}{\cancel{13}}$$
$$6\cdot 18 = W$$
$$108 = W$$

There are 108 women.

Now find out how many are men.

$$\text{Number of men} = \text{total} - \text{number of women}$$
$$= 234 - 108$$
$$= 126$$

So out of the 234 guests, 108 are women and 126 are men. ❏

Check: Is the ratio of women to men 6:7?

Does 6:7 = 108:126?

$$\text{Inside product is } 7(108) = 756 \quad \checkmark$$
$$\text{Outside product is } 6(126) = 756 \quad \checkmark$$

EXAMPLE 6 If 3 people are unemployed for every 28 people who are employed, in a community of 4,650 people how many are unemployed? Let u represent the number of unemployed.

$$\text{Unemployed:total residents} = \text{unemployed:total residents}$$
$$3:(3 + 28) = u:4650$$
$$3:\underline{\underline{31}} = u:\underline{4650}$$
$$\underline{3(4650)} = \underline{(31)}u$$
$$\frac{3 \cdot 4650}{31} = \frac{31u}{31}$$
$$\frac{3 \cdot \cancel{31} \cdot 150}{\cancel{31}} = \frac{\cancel{31}u}{\cancel{31}}$$
$$3 \cdot 150 = u$$
$$450 = u$$

So 450 are unemployed, and 4,200 (4,650 − 450) are employed. ❑

Check: If 450 people are unemployed and 4,200 are employed, is the ratio of unemployed to employed 3:28?

Does 3:28 = 450:4200?

$$\text{Inside product is } 28(450) = 12,600 \quad \checkmark$$
$$\text{Outside product is } 3(4200) = 12,600 \quad \checkmark$$

EXERCISES 7.5

Solve each problem by setting up a proportion involving the unknown quantity and the number of total items involved.

1. If the ratio of women to men is 4:3, (a) how many women are present in a group of 42 people? (b) How many men are there?

2. If the ratio of National League fans to American League fans is 7:2, (a) how many National League fans would be found in a group of 135 people? (b) How many American League fans are in that same group?

3. If a recipe calls for 2 cups of milk and 3 cups of water, (a) how many cups of liquid used were milk when an enlargement of the recipe was made calling for 15 cups of liquid? (b) How many cups were water?

4. To mix a pretty shade of purple, I need 3 drops of red dye for every 2 drops of blue. (a) How many drops of blue should I use when I'm using 40 drops of dye? (b) How many drops of red should I use?

5. If the ratio of friends preferring tennis to friends preferring golf is 9:2, out of 33 friends (a) how many prefer golf? (b) How many prefer tennis?

6. If the ratio of people preferring cold weather to hot weather is 3:5, out of 32 people (a) how many prefer hot weather? (b) How many prefer cold weather?

7. If I hit 3 good golf shots to every 2 poor golf shots, (a) how many good ones did I have when I took a total of 125 shots? (b) How many poor ones did I have?

8. If the ratio of friends owning cats to friends owning dogs is 6:5, (a) how many dog owners are there out of 44 friends who own pets? (b) How many cat owners are there? (Assume none own both a dog and a cat.)

9. In Toilville 5 adults are unemployed for every 120 who have jobs. Out of 2,125 adults (a) how many have jobs? (b) How many do not have jobs?

10. The ratio of graduates who found jobs to those who did not was 27 to 4. In a class of 1,860 students (a) how many got jobs? (b) How many did not get jobs?

11. On the last test only 3 students did problem no. 5 wrong for every 8 who did it right. If 44 students took the test, (a) how many did problem no. 5 right? (b) How many did problem no. 5 wrong?

12. The winner got 13 votes for every 9 that his opponent got. Out of 11,000 voters (a) how many voted for the winner? (b) How many voted for his opponent?

13. The ratio of guests who joined the sing-a-long to those who did not was 17 to 8. Out of 350 guests (a) how many joined the singing? (b) How many did not join the singing?

14. Last season our volleyball team scored 7 points for every 9 points scored against us. Out of 3,136 points scored (a) how many did we get? (b) How many did our opponents get?

15. The papergirl gets our paper on the steps 9 times for every 4 times that she misses. Out of 260 days how many times do I have to walk down the stairs to retrieve the paper?

16. I score 21 points in table tennis for every 18 that are scored against me. Out of 1,560 points (a) how many have I scored? (b) How many are scored against me?

17. The ratio of adults to children on my block is 8 to 13. If 294 people live on my block, (a) how many are adults? (6) How many are children?

18. If the ratio of letters to bills in my mail is 2 to 5, out of 280 pieces of mail (a) how many are letters? (b) How many are bills?

19. If the phone rings 7 times when I am in the bathtub for every 4 times when I am not, out of 275 calls how many times should I expect to be disturbed from a bath by the phone?

20. If the ratio of dinners that are freshly prepared to those that are leftovers is 8 to 5, out of 312 dinners how many are leftovers?

Check your answers on page 730.

Glossary

acceleration Rate at which a moving body's velocity (speed) increases per unit of time. It is always expressed in terms of a unit of length divided by a unit of time squared.

Example: 3 cm/sec² is an acceleration.

Celsius Temperature scale used in most countries other than the United States and Great Britain. On the Celsius scale 0° is the value for freezing water, and 100° is the value for boiling water. The formula for Celsius is $C = \left(\dfrac{5}{9}\right)(F - 32°)$.

Fahrenheit Temperature scale generally used in the United States. On the Fahrenheit scale 32° is the value for freezing water, and 212° is the value for boiling water. The formula for Fahrenheit is $F = \left(\dfrac{9}{5}\right)C + 32°$.

formula Equation in which letters represent certain values.

Example: The formula $P = 2W + 2L$ represents the fact that the perimeter (P) around a rectangle is equal to twice the width (W) plus twice the length (L).

gravity Force that makes objects fall with increasing speed toward earth. It has been calculated to be 32 ft/sec², or 9.8 m/sec².

projected sales Estimate of future sales based on past sales.

Example: We project the sales for each month next year according to the amount of sales each month this year.

proportion Statement asserting that two ratios are equal.

Example: The equation 12:18 = 2:3 is a proportion.

  inside terms Terms of a proportion immediately to the left and right of the equal sign. They are sometimes called the *means*.

outside terms First and last terms of a proportion. They are sometimes called the *extremes*.

Example: In the proportion 12:18 = 2:3, the inside terms are 18 and 2, and the outside terms are 12 and 3.

proportion property Attribute of a proportion wherein the product of the *inside* terms is always equal to the product of the *outside* terms.

Example: In the proportion 12:18 = 2:3, the product of the inside terms is 18(2) = 36, and the product of the outside terms is 12(3) = 36.

ratios Expressions used to compare two things. The notation, :, read as "to," is used in a ratio. Ratios should always be expressed in their most reduced form.

Example: If we noticed that there were five men and seven women in a room, we might say the ratio of men to women is 5:7.

Example: When 12 children and 42 adults are at a party, we say the ratio of children to adults is 2:7.

subscript Notation in which a small number is written slightly below a letter to distinguish it from other values of that letter. Letters with different subscripts represent different numbers and are not like terms.

Example: In N_1, the 1 is a subscript of N; N_1 is read as "N sub-one." N_1 and N_2 represent different numbers.

velocity Another word for speed. It is expressed in terms of a unit of distance divided by a unit of time.

Example: $30\ \dfrac{\text{miles}}{\text{hour}}$ is a velocity.

$\approx$ Symbol that means "approximately equal."

Chapter 7 Review

Do each of the problems. After you are confident that you have done them as accurately as possible, compare your answers with those in the back of the book. If any are wrong, go back to the section where the problem came from (indicated to the left of the problem) and review the section.

  Once you understand all the sections, you will be sure to learn the skills solidly and remember how to do them if you practice them a little bit more. Turn to the Supplementary Exercises and do all the problems from any section where you had difficulty on these review exercises.

7.1 **1.** Liz has saved $4,300 for the down payment on a house. The bank requires 8 % of the house cost as a down payment. How expensive a house can Liz afford to buy?

7.2 **2.** What is 45°C on the Fahrenheit scale?

    3. Solve $K = P + 4M$ for M.

7.3 **4.** Reduce the ratio 14:35.

    5. Is 317:2219 the same as 180:1260?

    6. Is 81:1,053 the same as 28:362?

7.4 **7.** Solve for n if $6:17 = n:68$.

    8. Solve for p if $\dfrac{2}{3}:p = 1\dfrac{1}{4}:8\dfrac{3}{4}$.

    9. Solve for a if $0.07:0.42 = a:6.06$.

   10. How many miles long is a route that measures 7 in. on a map with a scale of 2 in. representing 15 mi?

7.5 **11.** In a community of 1,440 adults, 5 people are unemployed for every 31 people who are employed. How many are unemployed?

Answers

1. _____

2. _____

3. _____

4. _____

5. _____

6. _____

7. _____

8. _____

9. _____

10. _____

11. _____

Check your work on page 730.

Supplementary Exercises

Do all the problems in every section that involves a skill in which you now lack complete mastery. With a little more practice you will achieve that sense of really understanding the topics, and you will remember how to do these problems.

7.1

1. Bill saved $40.80 when he bought his stereo on sale during the store's 12% discount sale. How much did that stereo cost before the sale?

7.2

2. What is 50° F on the Celsius scale?

3. Solve $R = 2A + 5J$ for J.

4. Solve $K = 2N + 5W$ for W.

7.3

5. Reduce the ratio $24:162$.

6. Is $293:1172$ the same as $519:2086$?

7. Is $43:258$ the same as $117:702$?

7.4

8. Solve for p if $7:43 = p:172$.

9. Solve for m if $19:m = 7:49$.

10. Solve for n if $\frac{3}{4}:n = 2\frac{1}{3}:11\frac{2}{3}$.

11. Solve for b if $0.15:1.2 = b:1.84$.

12. The camera store advertised that it would cost $5.10 to make three copies of a photo. How much should it cost to order five copies?

7.5

13. There are 3 women for every 58 men in the union. Out of 1,403 members, how many are women?

14. If there are 7 dogs for every 4 cats in my neighborhood, out of 132 pets (a) how many are dogs? (b) How many are cats?

Answers

1. _____

2. _____

3. _____

4. _____

5. _____

6. _____

7. _____

8. _____

9. _____

10. _____

11. _____

12. _____

13. _____

14a. _____

b. _____

Check your work on page 730. ✓

Chapter 7 Test

1. If a bank requires a person getting a mortgage to make a down payment of 9% of the house cost, what is the price of a house that Janine can buy if she has saved $6,000 for her down payment? (Round the answer to the nearest dollar.)

2. What is the value of P when $P = \dfrac{2M - 3(N - 2K)}{3L}$, if $M = -3$, $N = 7$, $K = -4$, and $L = -1$?

3. Solve $R = 8 - 3S + 2P$ for P.

4. Simplify $360:252$.

5. Is $80:35$ the same as $24:10.5$?

6. Solve $5:A = 17:51$ for A.

7. Solve $0.07:0.021 = N:0.069$ for N.

8. Solve $M:\dfrac{8}{9} = \dfrac{3}{5}:1\dfrac{1}{5}$ for M.

9. If a store sells five packages of popcorn for $2.30, how much should three packages cost?

10. If the ratio of new music to old is $2:9$ in my record collection, (a) how many of my 374 albums are old? (b) How many are new?

Answers
1. _____
2. _____
3. _____
4. _____
5. _____
6. _____
7. _____
8. _____
9. _____
10a. _____
b. _____

Check your work on page 730. ✓

Cumulative Review
CHAPTERS 1–7

Simplify.

1. $5 - 2(9 - 4 - 3 + 1) - 3(-2 - 3 - 4)(-13 + 21 - 4) =$

3. $\dfrac{2}{3} + \dfrac{3}{4} - \dfrac{1}{12} - \dfrac{1}{3} - \dfrac{1}{4} + \dfrac{7}{12} =$

3. $\left(\dfrac{6}{25}\right)\left(\dfrac{15}{28}\right)\left(\dfrac{21}{30}\right) =$

4. $\left(2\dfrac{3}{4}\right) \div \left(3\dfrac{5}{8}\right) =$

5. $5.6 + 8.03 - 14.9 + 0.006 - 0.034 =$

6. $x^2 - 5x - 3x^3 + x + 8x^3 - 7x^2 =$

7. $4.2a - 3b - 5.2c + a + 7.3b - 0.06c =$

8. $(7x^2 - 5x + 5) - (9x^2 - 4x + 3) =$

9. $2a(a^2 - 3a + 5) - 5a(a^2 - 2a - 1) =$

Evaluate.

10. 8.2% of $\$5,000 =$

11. $a^2 - 3ab - 5b^2$ if $a = 2$ and $b = -3$

Solve for the letter.

12. $3x - 5 + x - 8 = 7$

13. $7y - 11 = 12y + 4$

14. $3w - 2(4w - 1) = 8w - 15 - 3(2w + 1) + 6$

15. $2m - 6 \geq 10$

16. $0 < 5p + 3 \leq 28$

17. $12 : a = 114 : 19$

18. 14% of N is 51.80

19. The ratio of checks to deposits in my monthly bank statement was 17 to 3. If I made six deposits, how many checks did I write?

20. If the ratio of men to women at the meeting was 5 to 6, out of 352 people attending (a) how many men attended? (b) How many women attended?

Answers

1. _____

2. _____

3. _____

4. _____

5. _____

6. _____

7. _____

8. _____

9. _____

10. _____

11. _____

12. _____

13. _____

14. _____

15. _____

16. _____

17. _____

18. _____

19. _____

20a. _____

b. _____

Check your work on page 730.

 # Graphing Linear Equations and Inequalities

 ### Graphing on a Cartesian Coordinate System

Plotting Points on a Cartesian Coordinate System

Graphs are visual forms (pictures) of mathematical information. In this chapter we begin dealing with graphs that are pictures of algebraic equations.

An **equation** and its **graph** say the same thing in two different ways. The information contained in one is contained in the other—no more, no less, just like a sentence that is expressed in English and in Spanish. Because the two languages are different, some words or ideas may stand out in one language more than in the other, but the information is the same. Essentially, what we will be doing in this chapter is learning to translate—not from English to Spanish and from Spanish to English, but from equations to graphs and from graphs to equations.

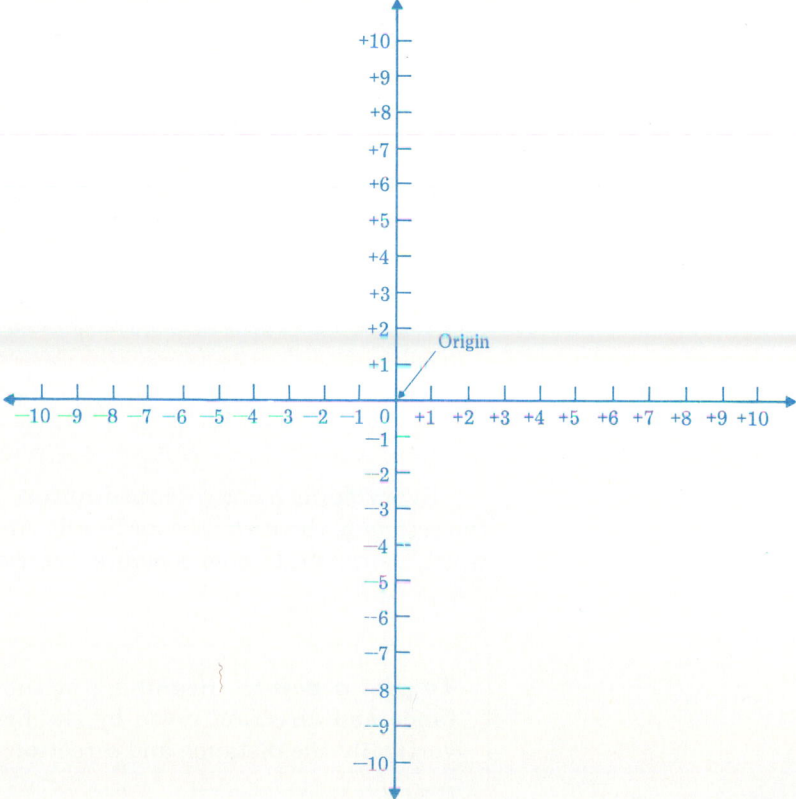

Before we can discuss the relationship between equations and graphs (Section 8.2), we must introduce the concepts of graphing. We experienced elements of graphing when we illustrated the solutions to inequalities on a number line (Section 6.5). The first step in graphing is plotting, or putting dots (called **points**), on a Cartesian coordinate system.

A **Cartesian coordinate system** is a grid made up of two perpendicular (right-angle) number lines. One is called the **horizontal axis**; the other is called the **vertical axis**. The two number lines intersect each other at their zeros. The place where the number lines meet is called the **origin**.

On the *horizontal axis*, numbers to the *right* of the *origin* are *positive*; numbers to the *left* of the origin are *negative*.

Horizontal axis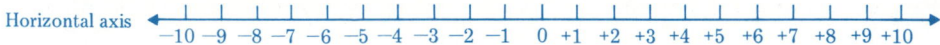

On the *vertical axis*, numbers *above* the *origin* are *positive*; numbers *below* the *origin* are *negative*.

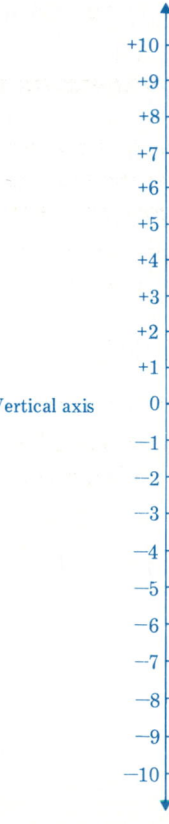

Every point has two **coordinates**: The *first* is the *horizontal* coordinate; the *second* is the *vertical* coordinate. An easy way to remember this is that the word "horizontal" comes before "vertical" alphabetically.

> ***To plot a point:*** Beginning at the origin, move horizontally the distance and direction given by the first coordinate. From there, proceed vertically the distance and direction given by the second coordinate.

(Examples 1–4 are plotted on the Cartesian coordinate system shown in Exercises 8.1A as points A, B, C, and D.)

EXAMPLE 1   To plot $A(2, 6)$: First move horizontally two spaces to the right (positive 2); then move up six spaces (positive 6).   ❑

EXAMPLE 2   To plot $B(-2, 3)$: First move horizontally two spaces to the left (negative 2); then move up three spaces (positive 3).   ❑

EXAMPLE 3   To plot $C(1, -3)$: First move one space to the right (positive 1); then move down three spaces (negative 3).   ❑

EXAMPLE 4   To plot $D(3, 0)$: First move three spaces to the right (positive); then move zero spaces (none) vertically (0).   ❑

Using the coordinates to find a point on a Cartesian coordinate system is similar to playing the game bingo or locating a particular place on a map.

When the bingo caller announces $G3$, we look at the top row of our card for the G; then we look in that column to see if we have a 3 on the card. That is like looking for the point with coordinates $(G, 3)$. But on a Cartesian coordinate system the numbers will appear in order, unlike the random way they appear on bingo cards.

To find the Smithsonian museum on our map of Washington, D.C. we look up the museum in the points-of-interest index. It says: Smithsonian K-17. So we look across the top of the map for the section marked K; then we look along the side of the map for the section marked 17. The Smithsonian museum can be found in that rectangle on the map. This procedure is like locating the point with coordinates $(K, 17)$.

EXERCISES 8.1A

1. Graph each of the points listed below as I did with points $A(2, 6)$, $B(-2, 3)$, $C(1, -3)$ and $D(3, 0)$. Use a dot to represent the point and a letter name to label it.

   $E(4, 1)$; $F(1, 4)$; $G(-1, 5)$; $H(2, -4)$; $I(-3, -2)$; $J(5, 0)$; $K(0, -4)$

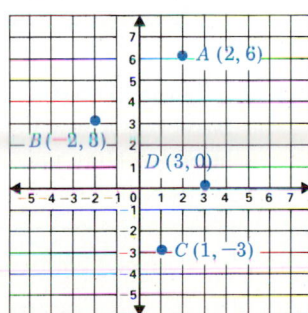

2. Plot each point listed here with a dot and a letter name. I have plotted point $A(1, 7)$ for you. Connect all the dots in alphabetical order and then connect K to A. Use the grid on the next page.

   $B(0, 5)$; $C(-3, 8)$; $D(-6, 4)$; $E(-7, 0)$; $F(-4, -6)$;

   $G(0, -10)$; $H(4, -6)$; $I(7, 0)$; $J(6, 4)$; $K(3, 8)$

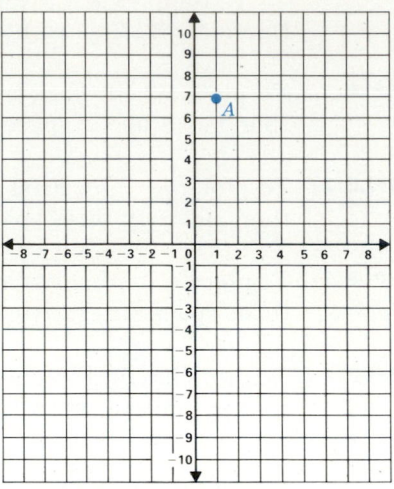

3. On the graph below, plot each of the following points. Label each dot by its letter name, connect the dots in alphabetical order, and connect P to A.

$$A(3, 0); B(4, 3); C(5, 5); D(5, 8); E(3, 7); F(2, 5);$$
$$G(0, 5); H(-2, 5); I(-3, 7); J(-4, 8); K(-5, 5); L(-4, 3);$$
$$M(-3, 0); N(-2, -3); O(0, -5); P(2, -3)$$

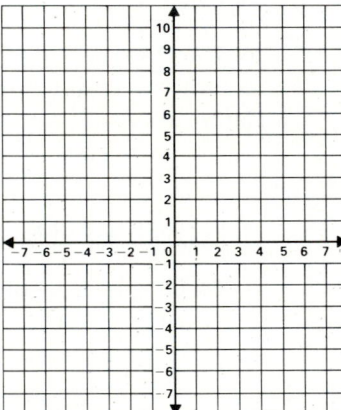

Check your answers on page 731.

Reading the Coordinates of a Point from a Graph

Now you know how to translate a pair of numbers into a point on a graph. But suppose that you are given the graph rather than the numbers. How do you translate a point into a pair of coordinates?

> *To read the coordinates of a point from a graph:*
>
> **1.** Find the **first coordinate** by counting the number of spaces the point is (horizontally) from the vertical axis. If the point is to the *right* of the vertical axis, its first coordinate is *positive*. If the point is to the *left* of the vertical axis, its first coordinate is *negative*.
>
> **2.** Find the **second coordinate** by counting the number of spaces the point is (vertically) from the horizontal axis. If the point is *above* the horizontal axis, its second coordinate is *positive*. If the point is *below* the horizontal axis, its second coordinate is *negative*.

EXAMPLE 5   Read the coordinates of point A.

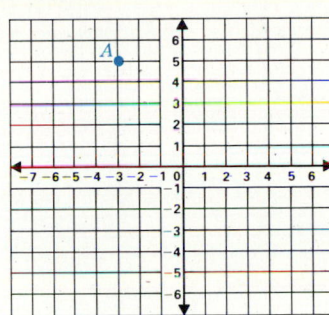

A is three spaces *left* of the vertical axis, so its first coordinate is -3.
A is five spaces *above* the horizontal axis, so its second coordinate is $+5$.
So A has coordinates $(-3, 5)$.   ❑

⟹ ***Note***:   A positive coordinate is written with no sign; the plus is understood. A negative coordinate must be written with a negative sign.

EXERCISES 8.1B

4. Find the coordinates for each point on the graph below.

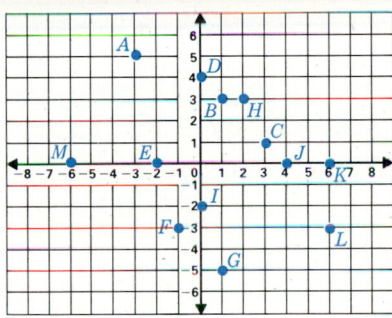

$A(\quad, \underline{\quad})$; $B(\underline{\quad}, \underline{\quad})$; $C(\underline{\quad}, \underline{\quad})$; $D(\underline{\quad}, \underline{\quad})$; $E(\underline{\quad}, \underline{\quad})$;
$F(\underline{\quad}, \underline{\quad})$; $G(\underline{\quad}, \underline{\quad})$; $H(\underline{\quad}, \underline{\quad})$; $I(\underline{\quad}, \underline{\quad})$; $J(\underline{\quad}, \underline{\quad})$;
$K(\underline{\quad}, \underline{\quad})$; $L(\underline{\quad}, \underline{\quad})$; $M(\underline{\quad}, \underline{\quad})$;

Check your answers on page 731.   ✓

Determining Whether Points Are on a Line

Illustration 8.1 is a Cartesian coordinate system with the graph of a line. In each example that follows we are given the coordinates of a point and asked to determine whether that point is on the line.

To determine whether a point is on a line:
Locate the point on the coordinate system.

If the line passes through the point, then it is *on the line*.

If the line does not pass through the point, then it is *not on the line*.

Illustration 8.1

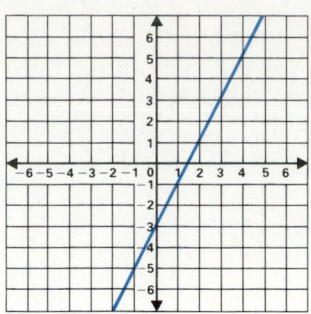

EXAMPLE 6 Is $(0, -3)$ on the line graphed in Illustration 8.1?
Yes it is.   ❏

EXAMPLE 7 Is $(1, 1)$ on the line graphed in Illustration 8.1?
No it is not.   ❏

EXAMPLE 8 Is $(-2, 1)$ on the line graphed in Illustration 8.1?
No it is not.   ❏

EXAMPLE 9 Is $(2, 1)$ on the line graphed in Illustration 8.1?
Yes it is.   ❏

On the graph below, I have plotted and circled the points from Examples 6–9 so that you can see whether they are on that line.

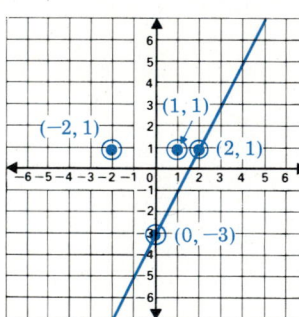

EXERCISES 8.1C

Determine which points with coordinates from Exercises 5–16 are on this line.

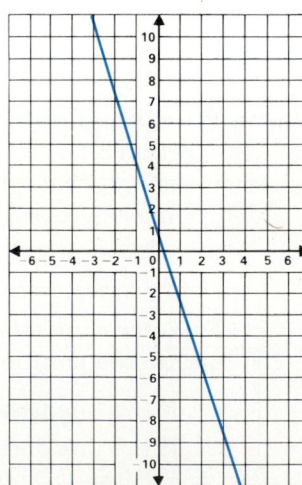

5. $(0, -1)$ **6.** $(0, 1)$

7. $(1, -2)$ **8.** $(1, 4)$

9. $(-2, 7)$ **10.** $(-2, -7)$

11. $(3, 4)$ **12.** $(3, -8)$

13. $(-1, -2)$ **14.** $(-1, 4)$

15. $(2, -5)$ **16.** $(2, 7)$

Check your answers on page 731.

More about the Cartesian Coordinate System

If either coordinate of a point is *zero*, the point lies on an *axis*. If the *first* coordinate is *zero*, the point will be on the *vertical* axis. If the *second* coordinate is *zero*, the point will be on the *horizontal* axis.

EXAMPLE 10 $(0, 5)$ is on the vertical axis.

$(3, 0)$ is on the horizontal axis.

$(0, 0)$ is on the origin.

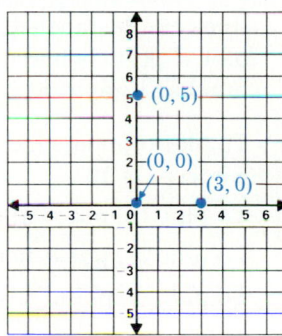

The two axes divide the Cartesian coordinate system into four sections called **quadrants**.

Quadrant 1 is on the *upper right*. To end up in quadrant 1, you must go to the *right* and then *up*. All points in this quadrant have positive first and second coordinates.

Quadrant 2 is on the *upper left*. To end up in quadrant 2, you must go to the *left* and then *up*. All points in this quadrant have negative first coordinates and positive second coordinates.

Quadrant 3 is on the *lower left*. To end up in quadrant 3, you must go to the *left* and then *down*. All points in this quadrant have negative first and second coordinates.

Quadrant 4 is on the *lower right*. To end up in quadrant 4, you must go to the *right* and then *down*. All points in this quadrant have positive first coordinates and negative second coordinates.

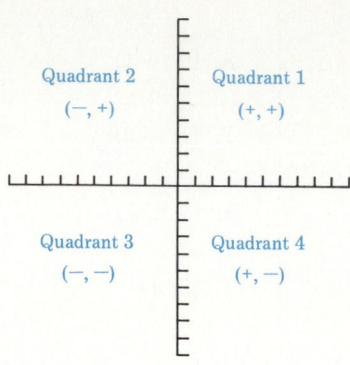

Even though axes are usually labeled with whole numbers, the coordinates of points can be fractions. They cannot be plotted or read as precisely as whole numbers can. We can only estimate fractional coordinates.

EXAMPLE 11   The coordinate $2\frac{1}{2}$ is plotted approximately halfway between 2 and 3.

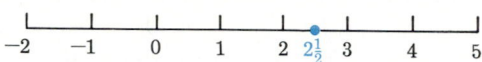

The coordinate $4\frac{1}{3}$ is plotted approximately one-third of the way between 4 and 5 (closer to 4 than to 5).

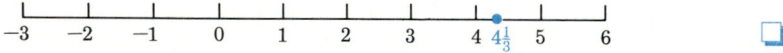

When the data we want to graph involve very large numbers, the number lines can be labeled so that each step represents 10 or 50 or 100 instead of 1. To plot a point where the coordinate is not one of the numbers labeled on the number line, treat it as you do fractions.

EXAMPLE 12   To plot numbers like 15, 23, and 39, we could use a number line where each step represents 10.

To plot 15, put it approximately *halfway* between **10** and **20**.

To plot 23, put it between **20** and **30**, but *closer to 20.*

To plot 39, put it between **30** and **40**, but *much closer to 40.*

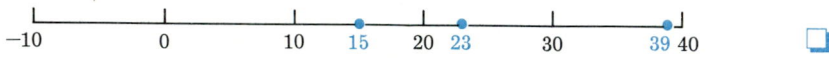

Determine the scales for the horizontal and vertical axes so that you can display your data as clearly and easily as possible. If you use a different scale for the horizontal axis than for the vertical axis, the graph will be distorted, but the reader will understand what is represented if the scales are labeled.

EXERCISES 8.1D

17. Read the coordinates from the following Cartesian coordinate system. Beneath the coordinates, write which quadrant the point is in or which axis it is on. I have done A and B for you.

    $A(\ 10\ , -20)$; $B(\ 0\ , -40)$; $C(\underline{\quad}, \underline{\quad})$;

       Quad 4   Vertical axis

$D(\underline{\quad},\underline{\quad})$; $E(\underline{\quad},\underline{\quad})$; $F(\underline{\quad},\underline{\quad})$;

$$\underline{\hspace{4cm}}\qquad\underline{\hspace{4cm}}\qquad\underline{\hspace{4cm}}$$

$G(\underline{\quad},\underline{\quad})$; $H(\underline{\quad},\underline{\quad})$;

$$\underline{\hspace{4cm}}\qquad\underline{\hspace{4cm}}$$

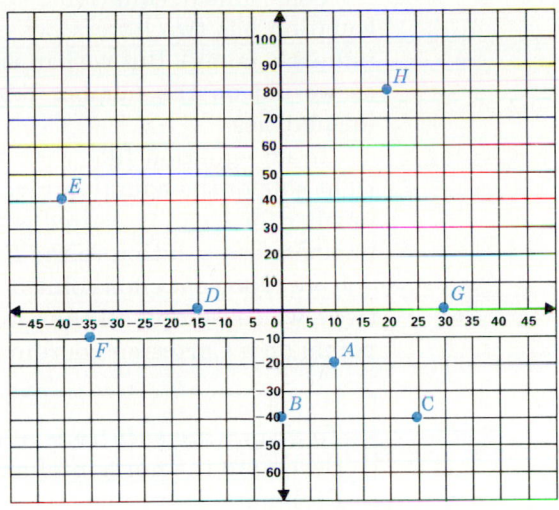

18. Plot these points approximately on the following coordinate system.

$$A\left(\frac{1}{2},\frac{1}{4}\right);\ B\left(-\frac{1}{4},\frac{1}{2}\right);\ C\left(0,3\frac{1}{2}\right);\ D\left(-\frac{1}{4},0\right);\ E\left(-3\frac{1}{4},1\frac{1}{2}\right);\ F\left(2\frac{1}{2},-2\frac{1}{4}\right)$$

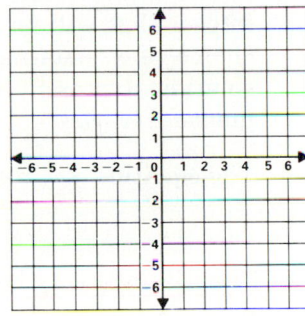

19. Plot these points approximately on the following coordinate system.

$$A(5,3),\ B(17,-25);\ C(53,-2);\ D(-36,-18);\ E(-49,23);\ F(-28,16)$$

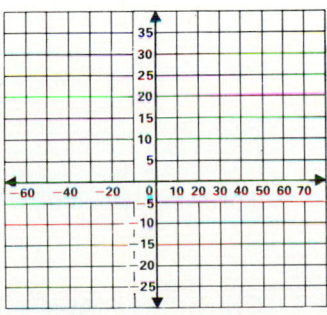

Check your answers on page 731.

8.2 Graphing Linear Equations

Finding Points That Represent Solutions for Equations

In Chapter 6 we learned how to solve linear equations that contained one letter. We found that each equation had exactly one solution.

Some linear equations, such as $y = x + 3$, contain two letters. When an equation has two letters, we cannot possibly list all the solutions (there are infinitely many). But we can use graphing to illustrate all the solutions, much as we did on the number line in Section 6.5 to illustrate the solutions to inequalities.

In an equation like $y = x + 3$, every time x is replaced by a number, the right side of the equation gives us the numerical value of y. Each choice of a number for x can be paired with the value it gives us for y. Each pair of such numbers is a solution to the linear equation.

If we represent each solution pair (x and y) as a point whose coordinates are (x, y), we can graph all the solutions to each linear equation by plotting these points on a Cartesian coordinate system.

> **To graph the solutions of a linear equation with two letters:**
> Designate the horizontal axis to represent one letter.
> Designate the vertical axis to represent the other letter.

If the letters are x and y, usually the x-axis is horizontal and the y-axis is vertical. Each axis should be labeled so that anyone looking at the graph will know what it represents.

To find a solution for an equation with x and y, choose any small, positive integer that will be easy to work with and let it be the value of x. For that choice of x, determine the value of y. The resulting pair of numbers is a solution. Represent that solution as the point whose coordinates are the pair of numbers (x, y).

EXAMPLE 1   The equation $y = x + 3$ has many solutions. Find three of them.

$$\text{Let } x = 1, \text{ then } y = (1) + 3$$
$$= 4$$

So $x = 1$ and $y = 4$ is a solution, and the coordinates of that point are $(1, 4)$.

$$\text{Let } x = 0, \text{ then } y = (0) + 3$$
$$= 3$$

So $x = 0$ and $y = 3$ is a solution, and the coordinates of that point are $(0, 3)$.

$$\text{Let } x = 2, \text{ then } y = (2) + 3$$
$$= 5$$

So $x = 2$ and $y = 5$ is a solution, and the coordinates of that point are $(2, 5)$.
Three solutions of $y = x + 3$ are

$$(1, 4), (0, 3), \text{ and } (2, 5).$$ ❑

EXAMPLE 2   Graph those three solutions to $y = x + 3$.

We have plotted them on the Cartesian coordinate system below.

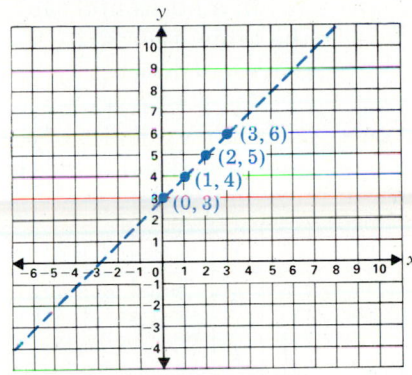

In Example 2, note that if you were to connect the points, you would get a straight line. That line is the *picture of all the infinite solutions* to the equation $y = x + 3$. Remember, the axes are number lines that extend infinitely in both directions.

If you read the coordinates of any other points on that line, you will find that they are also solutions to the equation $y = x + 3$.

EXAMPLE 3   The line passes through point $(3, 6)$. Check to see whether this is a solution to $y = x + 3$.

$(3, 6)$ represents the fact that when $x = 3$, $y = 6$.

Check whether it is a solution by substituting those values into the equation. See if the values make both sides equal.

$$\text{If } x = 3 \text{ and } y = 6, y = x + 3 \text{ becomes}$$
$$(6) = (3) + 3$$
$$6 = 6 \quad \checkmark$$

So $(3, 6)$ is a solution.

The graph of *every* linear equation is a **line**; that is, all the points that represent solutions to the equation form a straight line when they are plotted on a Cartesian coordinate system. This is why the equation is called *linear*. Once you have plotted three points, you can be pretty sure where the line is. (Two points are sufficient to draw the line, but using three is a way to check that you have not made any errors. If you did make a mistake, the points will not all lie on a straight line.)

Take a straightedge, connect the three dots, and you have the graph of the line that illustrates all the solutions to the equation. We call this procedure *sketching* the graph of the equation.

To find three points that represent solutions for equations:

Choose *any* three numbers for the values of x (I usually choose 0, 1, and 2 for x).

Substitute each of these values into the equation in place of x.

Solve the equation for y.

It is a good idea to record the coordinates of these solution points in a **table**. In a vertical table, list your choices of x values beneath the x in the table; then write the values you get for y next to each choice for the corresponding x.

In a horizontal table, list your choices of x values to the right of the x in the table; then write the values you get for y beneath each choice for the corresponding x.

EXAMPLE 4 Here are examples of a vertical table and a horizontal table. I have substituted in the values we found for three solutions to $y = x + 3$.

x	y
0	3
1	4
2	5

x	0	1	2
y	3	4	5

To graph a linear equation:

Find three pairs of solutions for x and y and plot each as a point.

After plotting the points, *remember to extend the line as far as possible in both directions.*

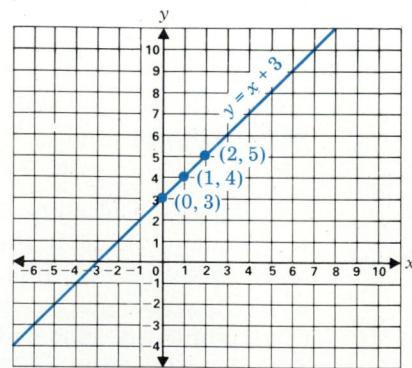

EXAMPLE 5 Use a vertical table to record the coordinates of points for the equation $y = 2x - 5$. Then sketch the graph.

$$\text{Let } x = 0, y = \underline{2(0)} - 5$$
$$= -5$$

$$\text{Let } x = 1, y = \underline{2(1)} - 5$$
$$= 2 - 5$$
$$= -3$$

$$\text{Let } x = 2, y = \underline{2(2)} - 5$$
$$= 4 - 5$$
$$= -1$$

x	y
0	-5
1	-3
2	-1

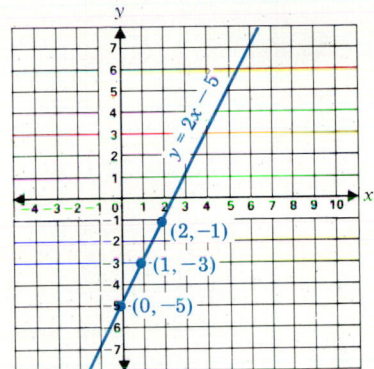

The graph of the solutions to every linear equation with two letters is always a line. If the letters are not x and y, decide which letter is represented on each axis. If a letter is isolated on one side of the equation, it is usually represented by the vertical axis. The **vertical** *axis always is represented by the* **second** *coordinate of the point.*

EXAMPLE 6   Sketch the graph of $B = 3A + 1$.

Let's choose A for the *horizontal* axis, so the values of A will be the *first* coordinates.
B then is the *vertical* axis, so the values of B will be the *second* coordinates.

$$\text{Let } A = 0, B = 3(0) + 1$$
$$= 1 \quad \text{So } (0, 1) \text{ is a solution.}$$

$$\text{Let } A = 1, B = 3(1) + 1$$
$$= 3 + 1$$
$$= 4 \quad \text{So } (1, 4) \text{ is a solution.}$$

$$\text{Let } A = 2, B = 3(2) + 1$$
$$= 6 + 1$$
$$= 7 \quad \text{So } (2, 7) \text{ is a solution.}$$

Those coordinates are recorded in the table below.

A	0	1	2
B	1	4	7

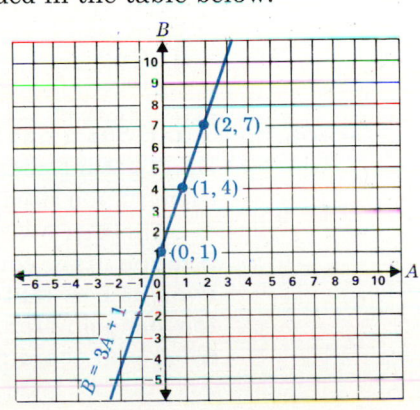

EXERCISES 8.2A

For each linear equation, find three solutions. Write them in a table as coordinates of points. Plot those points and sketch each graph on a piece of graph paper.

1. $y = x + 5$

x	y

2. $y = x - 3$

x	y

3. $y = 2x$

x	y

4. $y = 2x + 3$

x	y

5. $y = 7x - 10$

x	y

6. $y = x$

x	y

7. $y = -2x + 1$

x	y

8. $y = 3x$

x	y

9. $y = -5x$

x	y

10. $y = 3x - 4$

x	y

11. $b = -3a - 5$

x	y

12. $m = 6n - 5$

x	y

13. $y = 3x + 1$

x	y

14. $y = 4x - 3$

x	y

Check your answers on page 731.

Graphing Equations That Involve Fractions

Don't get intimidated if an equation involves fractions. You can compute with them. If you need a review, turn back to Chapter 3.

There is a way to make the computation as easy as possible. If the only fraction in the equation is the coefficient of x, choose the values of x to be 0 and two numbers that are multiples of the denominator. Those numbers will cancel the denominator and get rid of the fractions so you won't have to graph fractional coordinates.

EXAMPLE 7 Sketch the graph of $y = \left(\dfrac{2}{3}\right)x + 1$.

$$\text{Let } x = 0, \quad y = \left(\frac{2}{3}\right)(0) + 1$$

$$= 1 \quad \text{So } (0, 1) \text{ is a solution.}$$

$$\text{Let } x = 3, \quad y = \left(\frac{2}{3}\right)(3) + 1$$

$$= \left(\frac{2}{\cancel{3}}\right)\left(\frac{\cancel{3}}{1}\right) + 1$$

$$= 2 + 1$$

$$= 3 \quad \text{So } (3, 3) \text{ is a solution.}$$

$$\text{Let } x = 6, \quad y = \left(\frac{2}{3}\right)(6) + 1$$

$$= \left(\frac{2}{\cancel{3}}\right)\left(\frac{\cancel{3} \cdot 2}{1}\right) + 1$$

$$= (2 \cdot 2) + 1$$

$$= 4 + 1$$

$$= 5 \quad \text{So } (6, 5) \text{ is a solution.}$$

Those solutions are recorded in the table below.

x	y
0	1
3	3
6	5

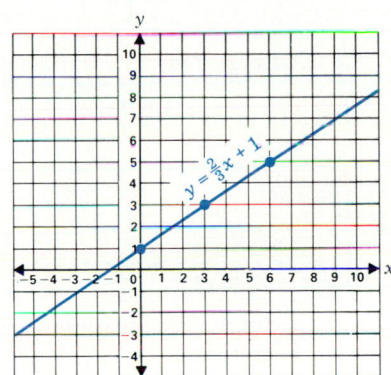

Sometimes it is more trouble to find values for x that will produce whole-number solutions than it is worth. If the y-coordinate is a fraction, you will have to estimate where the point is to be plotted. Be prepared to approximate the plotting of some points.

EXAMPLE 8 Sketch the graph of $y = \dfrac{1}{2}x - \dfrac{2}{3}$.

$$\text{Let } x = 0, \quad y = \left(\frac{1}{2}\right)(0) - \frac{2}{3}$$

$$= -\frac{2}{3} \quad \text{So } \left(0, -\frac{2}{3}\right) \text{ is a solution.}$$

$$\text{Let } x = 1, y = \frac{1}{2}(1) - \frac{2}{3}$$

$$= \frac{1}{2} - \frac{2}{3} \qquad \text{To combine, find LCD is 6}$$
$$\qquad\qquad \text{and replace with equal}$$
$$= \frac{3}{6} - \frac{4}{6} \qquad \text{fractions with denominator 6.}$$

$$= -\frac{1}{6} \qquad \text{So } \left(1, -\frac{1}{6}\right) \text{ is a solution.}$$

$$\text{Let } x = 2, y = \left(\frac{1}{2}\right)(2) - \frac{2}{3}$$

$$= \left(\frac{1}{\cancel{2}}\right)\left(\frac{\cancel{2}}{1}\right) - \frac{2}{3}$$

$$= 1 - \frac{2}{3}$$

$$= \frac{1}{3} \qquad \text{So } \left(2, \frac{1}{3}\right) \text{ is a solution.}$$

Those solutions are recorded in the table below.

x	0	1	2
y	$-\dfrac{2}{3}$	$-\dfrac{1}{6}$	$\dfrac{1}{3}$

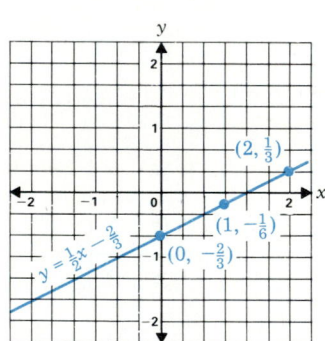

EXERCISES 8.2B

Find the coordinates for three points and sketch the graph of each equation.

15. $y = \left(\dfrac{3}{4}\right)x$

16. $y = \left(\dfrac{2}{3}\right)x$

17. $y = \left(\dfrac{1}{2}\right)x + 3$

18. $y = \left(\dfrac{1}{4}\right)x - 2$

19. $y = \left(\dfrac{2}{3}\right)x - 1$

20. $y = \left(\dfrac{2}{5}\right)x + 3$

21. $y = \left(\dfrac{3}{5}\right)x + 2$

22. $y = \left(\dfrac{3}{4}\right)x - 2$

23. $y = \left(\dfrac{1}{3}\right)x + \dfrac{2}{3}$

24. $y = \left(\dfrac{3}{5}\right)x - \dfrac{1}{5}$

25. $y = \left(\dfrac{2}{3}\right)x + \dfrac{1}{2}$

26. $y = \left(\dfrac{1}{4}\right)x + \dfrac{1}{3}$

Check your answers on page 732.

8.3 Graphing Horizontal and Vertical Lines

We learned in the last section that the graph of every linear equation with two letters is a line. We can extend that statement to say that the graph of every linear equation is a line

The graph of a linear equation with only one letter is also a line, but it is much easier to sketch than the graph of an equation with two letters.

Horizontal Lines

A point is on a line if the coordinates of that point are numbers that are solutions to the equation. If the equation is $y = 3$, then the value of y does not depend on the value of x. Every point with the y coordinate of 3 is a solution, no matter what the x-coordinate is. To graph the line, just record any three points in the table that have 3 as the second coordinate.

EXAMPLE 1 Sketch the graph of the equation $y = 3$.

$(0, 3)$, $(1, 3)$, $(2, 3)$, $(8, 3)$, $(-5, 3)$, and $(-9, 3)$ are all solutions to $y = 3$.

x	0	1	2	8	-5	-9
y	3	3	3	3	3	3

So the graph of $y = 3$ is:

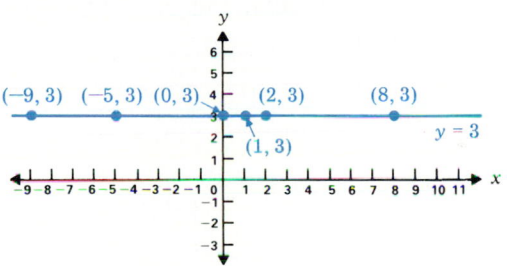

The graphs of linear equations with y as the only letter term will always be horizontal lines. That is because all the points on the lines will have the same value of y for their second (or vertical) coordinate. Try graphing $y = -4$!

Vertical Lines

In the linear equation $x = 5$, there is only one letter term, x. Every point with the x-coordinate of 5 is a solution, no matter what the y-coordinate is.

EXAMPLE 2 Sketch the graph of the equation $x = 5$.

The x-coordinate of all solutions must be 5.

Every point with a first coordinate of 5 is a solution.

$(5, 0)$, $(5, 1)$, $(5, 3)$, $(5, -2)$, and $(5, -7)$ are all points on the line that is the graph of $x = 5$.

x	5	5	5	5	5
y	0	1	3	-2	-7

So the graph of $x = 5$ is:

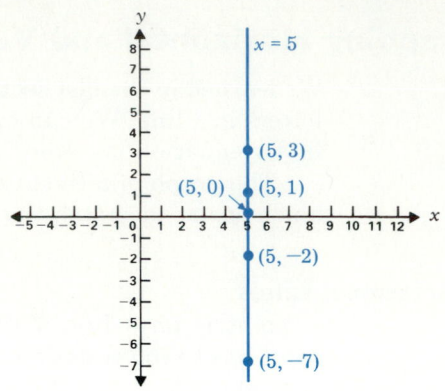

The graphs of linear equations with x as the only letter term will always be vertical lines. That is because all the points on the lines will have the same value of x for their first (or horizontal) coordinate. Try graphing $x = -2$!

Equations of the form $y = k$ (for some given number k) will all have horizontal lines for their graphs.

Equations of the form $x = c$ (for some given number c) will all have vertical lines for their graphs.

EXERCISES 8.3

Fill in a table with coordinates of three points for each equation, plot the points, and sketch the graph.

1. $y = 4$	**2.** $x = 2$
3. $y = -2$	**4.** $x = -3$
5. $y = 0$	**6.** $x = 0$
7. $x = -1$	**8.** $y = -1$
9. $x = -6$	**10.** $y = 2$

Check your answers on page 733.

8.4 Setting up Equations for Graphing and Function Notation

Setting up Equations for Graphing

The easiest way to fill in a table for graphing is to simplify the equation so it is solved for y. All the equations we graphed in Section 8.2 were solved for y. You learned how to solve equations for a particular letter in Section 7.2. Remember, solving for y means isolating y on one side of the equation.

EXAMPLE 1 $y = 3x - 5$; $y = 2x + 9$; $y = -5x + 7$; and $y = x - 3$ are all equations that are solved for y.

$y - 3x = 7$; $2y = 6x - 10$; $3y - 12 = 9x$; and $2y + 5x = 7$ are equations that are not solved for y.

EXERCISES 8.4A

Which of the following equations are solved for y?

1. $y = 3x + 5$ 2. $y - x = 7$

3. $y + 5 = 3x$ 4. $5x - 1 = y$

5. $x - y = 3$ 6. $y = 14 + 2x$

Check your answers on page 733.

Solving for y

When an equation is solved for y, it is easy to find solutions. If y is isolated on one side of the equation, we find the value of the y-coordinate of a solution simply by evaluating the expression on the other side of the equation for each choice of a value for x.

If an equation is not solved for y, add, subtract, multiply, and/or divide what is necessary on both sides of the equation to get the y by itself.

EXAMPLE 2   Solve $y - 3x = 7$ for y.

Add $3x$ to both sides of the equation to remove $-3x$ from the left side of the equation.

$$
\begin{array}{rcl}
y - 3x &=& 7 \\
+3x && +3x \\
\hline
y &=& 3x + 7 \text{ is that equation solved for } y.
\end{array}
$$

EXAMPLE 3   Solve $2y = 6x - 10$ for y.

Divide both sides by 2 to remove the 2 from the left side. (To divide the right side by 2, divide each term by 2.)

$$2y = 6x - 10$$

$$\frac{(\cancel{2}y)}{\cancel{2}} = \frac{(6x)}{2} - \frac{10}{2}$$

$$y = 3x - 5 \text{ is that equation solved for } y.$$

EXAMPLE 4   Solve $3y - 12 = 9x$ for y.

The 3 and the -12 must each be removed from the left side to solve the equation for y.

First get the y term alone by removing the constant term, -12. Add its opposite, $+12$, to both sides.

$$
\begin{array}{rcl}
3y - 12 &=& 9x \\
+12 && +12 \\
\hline
3y &=& 9x + 12
\end{array}
$$

Now remove the 3 (coefficient of y) by dividing both sides of the equation by 3.

$$\frac{(\cancel{3}y)}{\cancel{3}} = \frac{(9x)}{3} + \frac{12}{3}$$

$$y = 3x + 4 \text{ is that equation solved for } y.$$

EXAMPLE 5 Solve $2y + 5x = 7$ for y.

To solve for y, first remove the x term from the left side of the equation by adding its opposite, $-5x$, to both sides.

$$
\begin{array}{r}
2y + 5x = 7 \\
\underline{-5x -5x} \\
2y = -5x + 7
\end{array}
$$

Now remove the coefficient 2 by dividing both sides by 2.

$$\frac{2y}{2} = \frac{-5x}{2} + \frac{7}{2}$$

$$y = -\frac{5}{2}x + \frac{7}{2} \text{ is that equation solved for } y.$$ ❏

Function Notation

When a linear equation is solved for y, it is easy to see that the value of y is obtained by applying certain operations to the number we choose for x.

EXAMPLE 6 In the equation $y = 2x + 3$, y is defined as the result of *multiplying* x by 2 and then *adding* that result to 3. ❏

For each choice of x, those are the operations we apply to find the y-coordinate of that solution.

If one letter (y) is defined in terms of the other (x), so that for each choice of x there is exactly *one* value of y, we say that y is a **function** of x. Instead of calling the second letter y, we can represent it as $f(x)$. That notation is read as "function of x."

The equation $y = 2x + 3$ could be written as $f(x) = 2x + 3$. To graph it, we would label the vertical axis $f(x)$. Replace x with three numbers and evaluate the function for each one.

EXAMPLE 7 Sketch the graph of $f(x) = 2x + 3$.

$$
\begin{array}{lll}
f(0) = 2(0) + 3 & f(1) = 2(1) + 3 & f(2) = 2(2) + 3 \\
f(0) = 3 & f(1) = 2 + 3 & f(2) = 4 + 3 \\
 & f(1) = 5 & f(2) = 7
\end{array}
$$

Three solutions to $f(x) = 2x + 3$ are $(0, 3)$, $(1, 5)$, and $(2, 7)$.

x	0	1	2
$f(x)$	3	5	7

The graph is below.

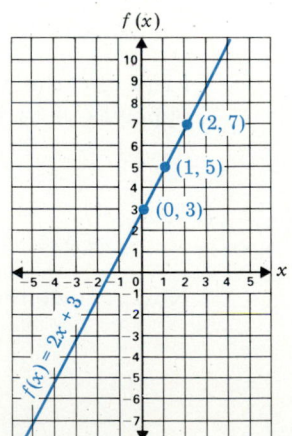

❏

Solve each equation for y; express each as a function of x; find the coordinates of three solutions; plot those points; and sketch the line on graph paper.

	Solved for y	Function of x	Points		

7. $y + 2x = 9$

x			
$f(x)$			

8. $y - 5x = -4$

x			
$f(x)$			

9. $y + 3 = x$

x			
$f(x)$			

10. $y - 4 = 3x$

x			
$f(x)$			

11. $2y = 6x - 4$

x			
$f(x)$			

12. $3y = 12x - 9$

x			
$f(x)$			

13. $2y + 6 = 4x$

x			
$f(x)$			

14. $5y - 15 = 10x$

x			
$f(x)$			

	Solved for y	Function of x	Points		
15. $4y + 12x = -8$			x		
			$f(x)$		
16. $3y - 6x = 12$			x		
			$f(x)$		
17. $2y - 3 = 5x$			x		
			$f(x)$		
18. $3y + 4 = 12x$			x		
			$f(x)$		
19. $5y - 1 = 3x$			x		
			$f(x)$		
20. $7y + 4 = 2x$			x		
			$f(x)$		

Check your answers on page 734.

8.5 Graphic Linear Inequalities

Strict Inequalities

We were introduced to inequalities in Section 6.5. Recall that the notation for inequalities consists of the symbols $<$ and $>$. The arrow always points toward the smaller side.

To graph a linear inequality, you first must graph the line that represents the solutions to the statement written as an equation. That line will be the boundary to the solution set of the inequality. The line divides the Cartesian coordinate system into two sets of points: those on one side of the line and those on the other side of it.

So begin by setting the two sides equal and then graph the line.

EXAMPLE 1 To graph $y > x + 3$, first graph the line $y = x + 3$. The solution set to $y > x + 3$ includes all the points (x, y) that are found on one side of the line.

All the points on the other side of the line are solutions to $y < x + 3$.

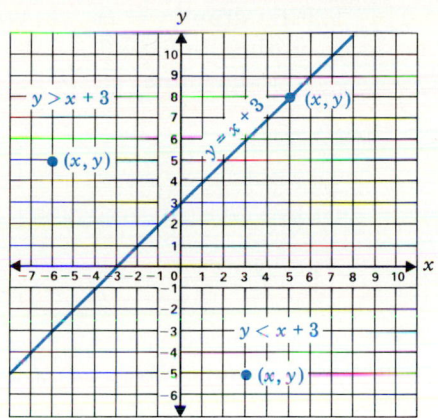

To determine which set of points contains the solutions to an inequality, choose any test point (that is not on the boundary line) you want and check whether it is a solution to the inequality.

Avoid careless errors by choosing a point that is far enough away from the line so you can be sure which portion you are testing.

EXAMPLE 2   Find and graph the solutions to $y > x + 3$ from Example 1.

Let's use (0, 10) as the test point; it is above the line. (0, 10) means that $x = 0$ and $y = 10$.

Is (0, 10) a solution to $y > x + 3$?

Is (10) > (0) + 3 true?

$10 > 3$ ✓

10 is greater than 3, so the points above the line are in the solution set.

To graph an inequality, shade the portion of the Cartesian coordinate system (above or below the line) representing the set of points that are solutions.

Since the points on the line make the two sides of the inequality equal, they cannot satisfy the definition of a strict inequality, in which one side is larger than the other). So the line is *not* part of the solution. This fact is indicated by dotting the line.

EXAMPLE 3   Sketch the graph of solutions to $y > x + 3$.

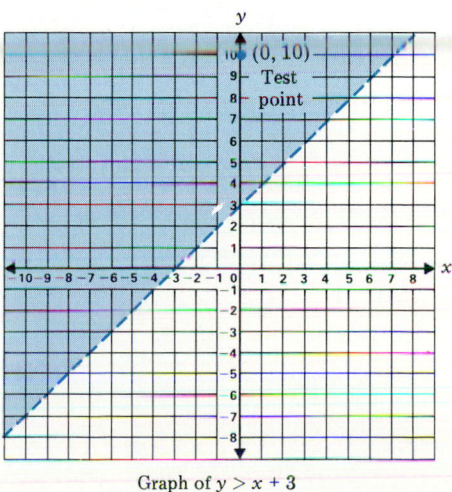

Graph of $y > x + 3$

EXAMPLE 4   Graph the solutions to $y < 2x - 5$.

First graph $y = 2x - 5$. We found in Example 5 of Section 8.2 that three points on this line are $(0, -5)$, $(1, -3)$, and $(2, -1)$. Plot these points and draw the boundary line.

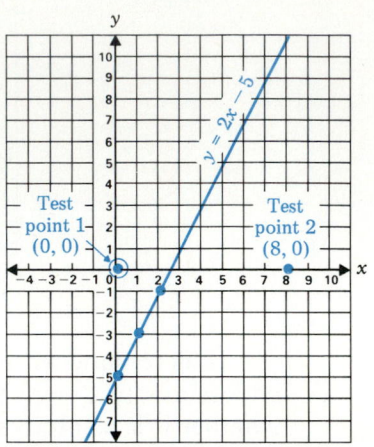

Choose $(0, 0)$ as a test point. It definitely is above the line. $(0, 0)$ means that $x = 0$ and $y = 0$.

Is $y < 2x - 5$ when $x = 0$ and $y = 0$?

Is $(0) < 2(0) - 5$?

$0 \not< -5$ ($\not<$ means NOT LESS than.)

But 0 is not less than -5, so $(0, 0)$ is not a solution. $(0, 0)$ is above the line, so the points *above* the line are *not solutions*. The solution set must be the points *below* the line.

To be sure, let's test a point from below the line. $(8, 0)$ is definitely below the line, so let $x = 8$ and $y = 0$.

Is $(0) < \underline{2(8)} - 5$?

$0 < 16 - 5$

$0 < 11$ ✓   Yes, 0 is less than 11.

Shade that portion and dot the line.

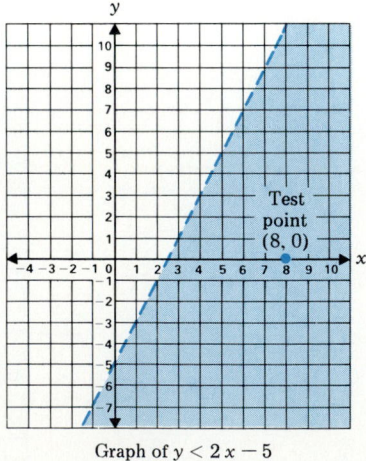

Graph of $y < 2x - 5$

EXERCISES 8.5A

Sketch the solution to each strict inequality.

1. $y > x - 4$ 2. $y < x + 5$

3. $y > 2x$ 4. $y < -5x$

5. $y > 3x - 1$ 6. $y < 2x + 1$

7. $y > -4x + 2$ 8. $y < -x + 4$

9. $y < -3x - 4$ 10. $y < -x - 3$

Check your answers on page 735.

Nonstrict Inequalities

The symbols $\leq$ and $\geq$ represent nonstrict inequalities. The solution set for nonstrict inequalities consists of all the solutions to the equation as well as all the solutions to the inequality. Nonstrict inequalities are graphed just like strict inequalities, except that the boundary line is part of the solution set. The points on the line represent solutions when the two sides of the inequality are equal.

To indicate that the *boundary line* is included as part of the solution, leave it solid after shading the set of points that are solutions to the inequality.

EXAMPLE 5 Graph the solutions to $B \leq 3A + 1$.

From Example 6 in Section 8.2, we know that $(0, 1)$, $(1, 4)$, and $(2, 7)$ are points (A, B) on the line $B = 3A + 1$.

Plot these points and draw the boundary line.

Let's use $(4, 0)$ as a test point; it is below the line. $(4, 0)$ means $A = 4$ and $B = 0$. Is that a solution to $B \leq 3A + 1$?

Is $(0) \leq 3(4) + 1$?

$0 \leq 12 + 1$

$0 \leq 13$ Yes, so the lower set of points is the solution.

The boundary line is part of the solution, so it is left solid.

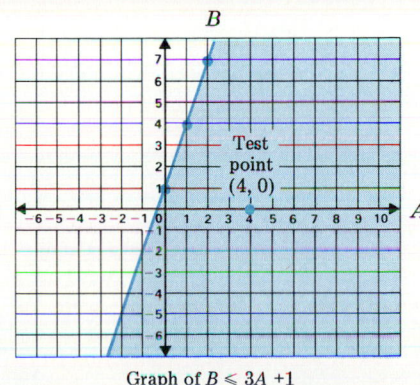

Graph of $B \leq 3A + 1$

> **To graph an inequality:**
> Graph the boundary line.
>
> Test a point to determine which side is the set of points that are solutions to the inequality.
>
> Shade the set of solution points.
> Dot the boundary line in a strict inequality. Leave the boundary line solid in a nonstrict inequality.

EXAMPLE 6   Graph the solutions to $x \geq 2$.

The boundary line ($x = 2$) is the vertical line of points, all with x-coordinate 2.

Test a point to the right of the boundary line. Use (5, 3), which means $x = 5$ and $y = 3$. Is $x \geq 2$ when $x = 5$ and $y = 3$?

Is (5) ≥ 2? Yes, so the solution is the set of points on the line and to its right. This is a nonstrict inequality, so the line is solid.

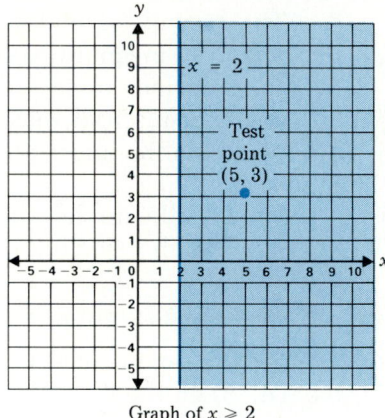

Graph of $x \geqslant 2$

EXAMPLE 7   Graph the solutions to $y < -4$.

The boundary line ($y = -4$) is the horizontal line of points with y-coordinate -4.

Test a point below the line. Use (1, -6): $x = 1$, $y = -6$.

Is $y < -4$ when $y = -6$?   There is no x to substitute 1 for.

Is (-6) < -4? Yes, -6 is less than -4, so the solution is the set of points below the line.

This is a strict inequality, so the line is dotted.

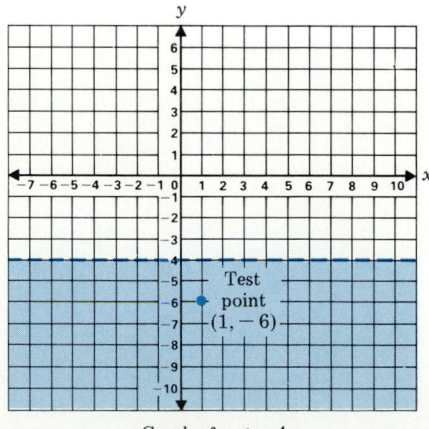

Graph of $y < -4$

EXERCISES 8.5B

Graph the solution set for each inequality on graph paper. Graph the line; test a point; shade in the correct side of points; and decide whether the boundary line is dotted or solid.

11. $y \geq x - 2$ 12. $y \leq x + 3$

13. $y \geq 3x$ 14. $y \leq -2x$

15. $B \leq 5A - 2$ 16. $M \geq 3N + 1$

17. $y \leq -2x + 3$ 18. $y \geq 2x - 2$

19. $k > 4n - 1$ 20. $r < 4p + 4$

21. $y \leq 3x - 3$ 22. $y > -x + 2$

23. $y < 6x - 8$ 24. $y \geq -4x - 5$

25. $y < 1$ 26. $y \geq -2$

27. $x \geq -6$ 28. $x < 7$

Check your answers on page 735. ✓

Intercepts of a Line

y-Intercept

The **y-intercept** of a line is the point where the line crosses the y-axis.

EXAMPLE 1

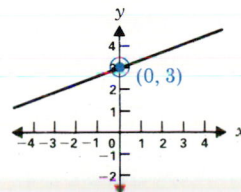

The y-intercept is $(0, 3)$

EXAMPLE 2

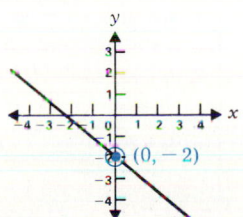

The y-intercept is $(0, -2)$

EXAMPLE 3

The *y*-intercept is (0, 0).

Since every *y*-intercept is a point on the *y*-axis, the *x-coordinate* is always *zero*. So the notation for a *y*-intercept can be simply to state the value of the *y*-coordinate. If the *y*-intercept for a line is the coordinates (0, 3), you can say that the *y*-intercept is 3.

EXERCISES 8.6A

State the *y*-intercept for each line

1.

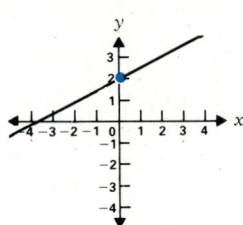

y-intercept: _____

2.

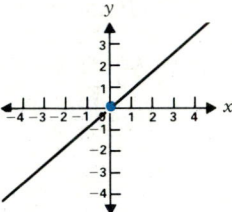

y-intercept: _____

3.

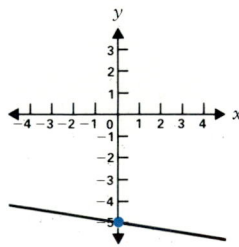

y-intercept: _____

4.

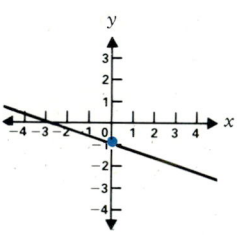

y-intercept: _____

Each line on the following grid is labeled with a letter. Consider that letter the name of each line. State what the *y*-intercept is for each line.

5. *A*: _____

6. *B*: _____

7. *C*: _____

8. *D*: _____

9. *E*: _____

10. *F*: _____

11. *G*: _____

12. *H*: _____

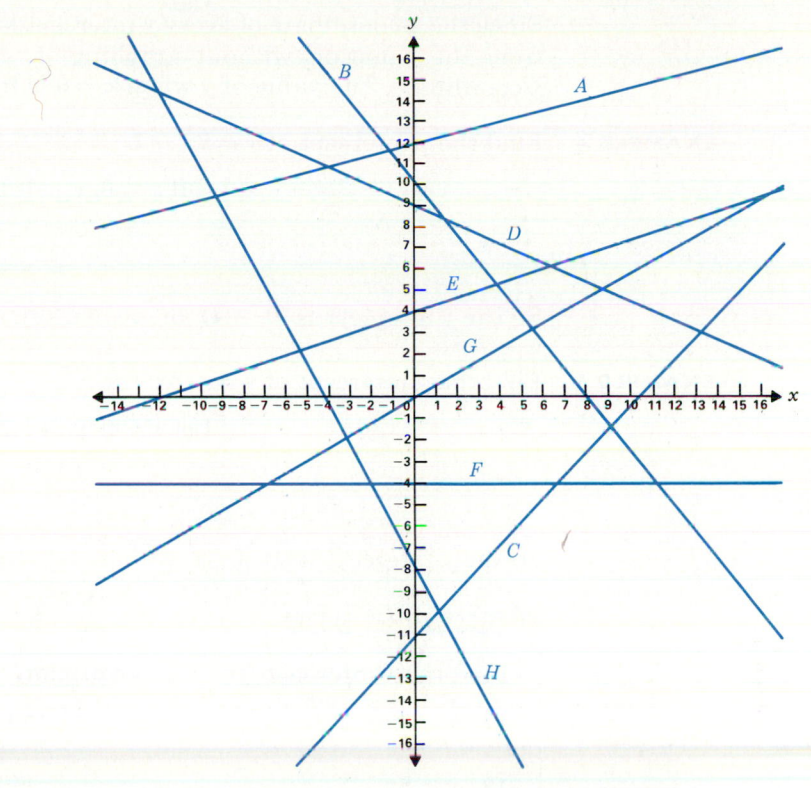

For each equation, find the coordinates of three solutions, plot the points, and sketch the graphs. Then state what the y-intercept is for each line.

13. $y = 3x + 5$

y-intercept: _____

x			
y			

14. $y = -2x + 1$

y-intercept: _____

x			
y			

15. $y = 7x - 4$

y-intercept: _____

x			
y			

16. $y = x - 5$

y-intercept: _____

x			
y			

Check your answers on page 736.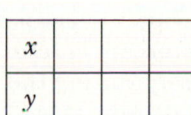

Finding the y-Intercept from an Equation

Since the x-coordinate of every y-intercept is zero, you can find the y-intercept from the equation without graphing it. Just let $x = 0$ and solve for the y-coordinate. The value of y when $x = 0$ is the y-intercept.

EXAMPLE 4 Find the y-intercept of $y = 7x - 4$.

$$\text{If } x = 0, y = \underline{7(0)} - 4$$
$$= 0 - 4$$
$$= -4$$

So the y-intercept is $(0, -4)$, or we can say it is -4. ❏

EXAMPLE 5 Find the y-intercept of $y = -3x + 2$.

$$\text{Let } x = 0, y = \underline{-3(0)} + 2$$
$$= 0 + 2$$
$$= 2$$

So the y-intercept is $(0, 2)$, or we can say it is 2. ❏

EXERCISES 8.6B

Find the y-intercept from each equation.

17. $y = 2x - 1$ **18.** $y = -5x + 3$

19. $y = 8x - 7$ **20.** $y = -3x$

21. $y = 7x$ **22.** $y = x$

23. $y = -4x$ **24.** $y = -x$

25. $y = -x + 9$ **26.** $y = 6x - 5$

Check your answers on page 737. ✓

x-Intercept

The **x-intercept** of a line is the point where the line crosses the x-axis.

EXAMPLE 6

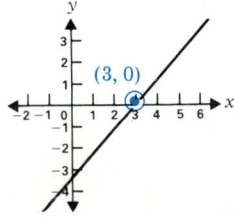

The x-intercept is $(3, 0)$. ❏

Since every point on the x-axis has y-coordinate 0, the *y-coordinate* of every x-intercept is *zero*. So, just as with y-intercepts, you can state the x-intercept by merely stating the number that is the x-coordinate. If the x-intercept of a line is the coordinates $(3, 0)$, you can say the x-intercept is 3.

EXAMPLE 7 If the x-intercept for a line is the coordinates $(7, 0)$, you can say the x-intercept is 7. ❏

EXERCISES 8.6C

Find the *x*-intercept for each line.

27.

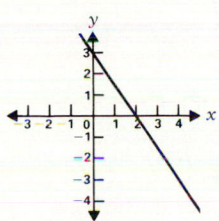

x-intercept: _____

28.

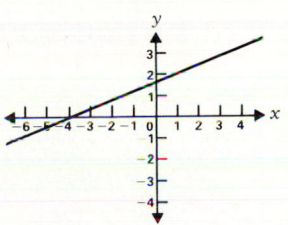

x-intercept: _____

29.

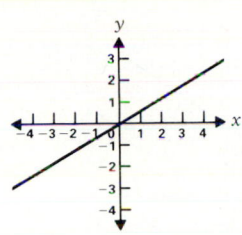

x-intercept: _____

30.

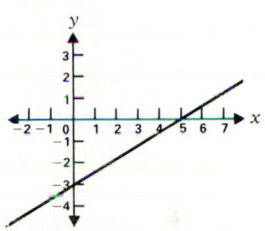

x-intercept: _____

For each line, state the coordinates of the *y*-intercept and the *x*-intercept.

EXAMPLE 8

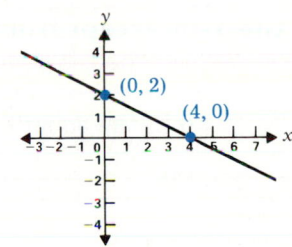

y-intercept is 2; (0, 2).

x-intercept is 4; (4, 0).

31.

y-intercept: _____

x-intercept: _____

32.

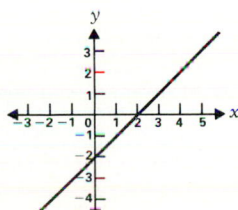

y-intercept: _____

x-intercept: _____

33.

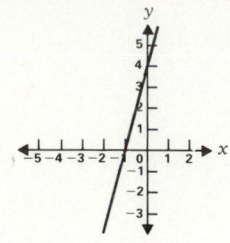

y-intercept: _____

x-intercept: _____

34.

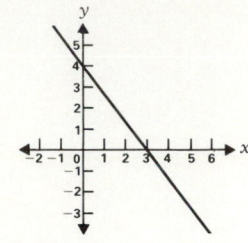

y-intercept: _____

x-intercept: _____

35.

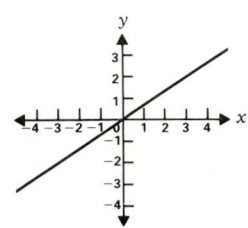

y-intercept: _____

x-intercept: _____

36.

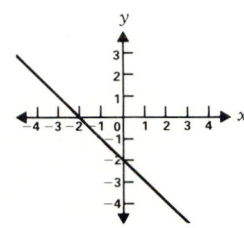

y-intercept: _____

x-intercept: _____

Check your answers on page 737.

Finding the x-intercept from an Equation

We can find the x-intercept of a linear equation without graphing the line. Since the y-coordinate of every x-intercept is zero, to find the x-intercept, let $y = 0$ in the equation and solve for x.

EXAMPLE 9   Find the x-intercept of $y = 2x - 8$.

Let $y = 0$.

$$(0) = 2x - 8$$
$$\underline{+8 \qquad\quad +8}$$
$$8 = 2x$$
$$\frac{8}{2} = \frac{\cancel{2}x}{\cancel{2}}$$
$$4 = x$$

So the x-intercept is 4, $(4, 0)$.   ❑

EXERCISES 8.6D

Find the x-intercept from each equation without graphing.

37. $y = 7x$ **38.** $y = 3x$

39. $y = x$ **40.** $y = -5x$

41. $y = x + 5$ **42.** $y = x - 3$

43. $y = x - 7$

44. $y = 3x - 6$

45. $y = -2x + 10$

46. $y = -5x - 15$

47. $y = 4x - 12$

48. $y = 3x - 2$

49. $y = -2x - 5$

Check your answers on page 737.

 **Slope of a Line**

Concept of Slope

Slope is a number that indicates the steepness of a line. The higher the number, the steeper the line. For example, a line with slope 7 is steeper than a line with slope 1.

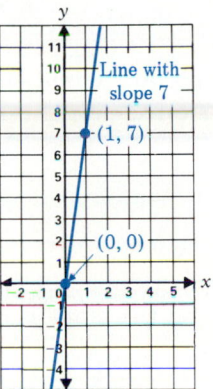

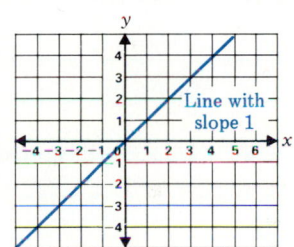

The slope tells us useful information about the line.

If the slope is *positive*, the line *goes up* as you look at it from left to right.

If the slope is *negative*, the line *goes down* as you look at it from left to right.

If the slope is *zero*, the line is *horizontal*.

If the slope is *undefined* (we say it has *no* slope), the line is *vertical*.

Where does this number (slope) come from?

To indicate the steepness of a line drive hit by a baseball player, we might say that the ball was caught by the leaping shortstop 7 feet off the ground at a point 110 feet from home plate.

To indicate the steepness of the Grand Canyon, we might say that the bottom is nearly 1 mile below the South Rim and 2 miles north of it.

In other words, we would describe how far something rises or falls vertically compared with how far it travels horizontally.

We do the same thing with the slope of the graph of a line. Suppose you move your finger from left to right, from one point to another along the line. The slope represents the ratio of the number of spaces you move it vertically for every space you move it horizontally.

Represent that ratio as a fraction to find the number that is the slope.

EXAMPLE 1   The line in the graph below *rises* three spaces for every one space moved to the right. The ratio of these moves is

$$3:1 =$$

$$\frac{3}{1} = 3.$$

So we say this line has slope 3. Notice that the line goes *up* as we read it from left to right, and its slope is *positive*.

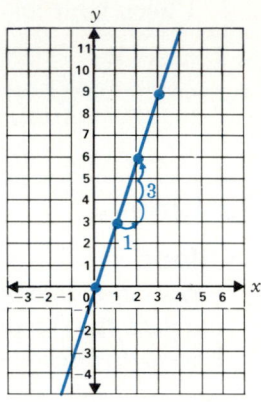

EXAMPLE 2   The line in the graph falls eight spaces while it moves one space to the right. The ratio of the distances is

$$-8:1 =$$

$$-\frac{8}{1} = -8.$$

So we say this line has slope -8. Notice that the line goes *down* as we read it from left to right, and the slope is *negative*.

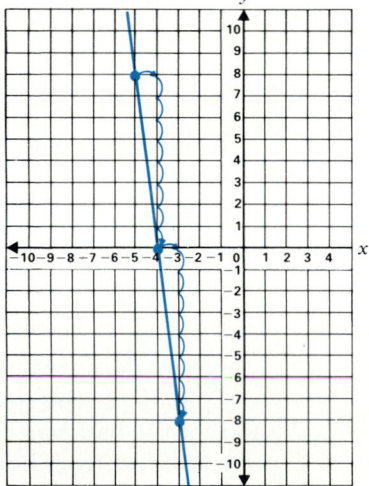

Computing the Slope from Two Points

Expressed in mathematical terms, the slope is the ratio of the change in the vertical distance between any two points to the change in the horizontal distance between them.

The change in the vertical distance between two points is often represented by the symbol (Δy) read as "delta y."

The change in the horizontal distance between two points is often represented by the symbol (Δx) read as "delta x."*

In general we can say that, for any two points on a line, (x_1, y_1) and (x_2, y_2), the slope of that line is the ratio

$$\Delta y : \Delta x, \text{ which is } \quad [(y_2) - (y_1)] = [(x_2) - (x_1)].$$

To find the vertical distance between two points: Subtract the y-coordinate of one from the y-coordinate of the other.

To find the horizontal distance between two points: Subtract the x-coordinate of one from the x-coordinate of the other (in the same order as you subtracted the y-coordinates).

Be careful to choose one point to be the one whose coordinates are subtracted from the other point.

It does not matter in what order you subtract the coordinates, as long as you do it in the same order for the x-coordinates as you do for the y-coordinates.

EXAMPLE 3   Find the slope of the line that passes through points $(1, 3)$ and $(4, 12)$.

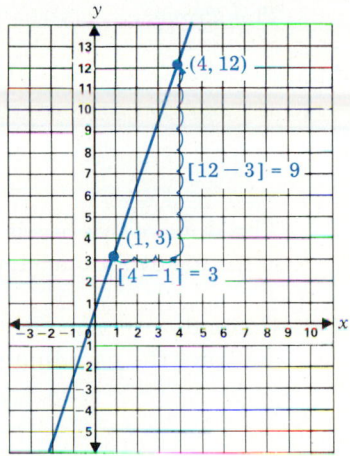

The change in the vertical direction is $+9$ as we move vertically from the y-coordinate 3 to the y-coordinate $12 : [(12) - (3) = 9]$.

The change in the horizontal direction is $+3$ as we move horizontally from the x-coordinate 1 to the x-coordinate $4 : [(4) - (1) = 3]$.

So the slope of the line containing $(1, 3)$ and $(4, 12)$ is

$$[(12) - (3) = 9] : [(4) - (1) = 3] =$$
$$[+9] \quad : \quad [+3] \quad =$$
$$\frac{9}{3} = 3,$$

The slope of this line is 3. That means we would expect its graph to be a line *going up three* steps for every movement of one step to the right.   ❏

In order to cut down on errors (especially if the coordinates are negative numbers) when computing the ratio, express each coordinate within parentheses.

---

* Δy and Δx are notations you will work with if you study calculus.

EXAMPLE 4 Determine the slope of the line through $(-2, -3)$ and $(1, 3)$.

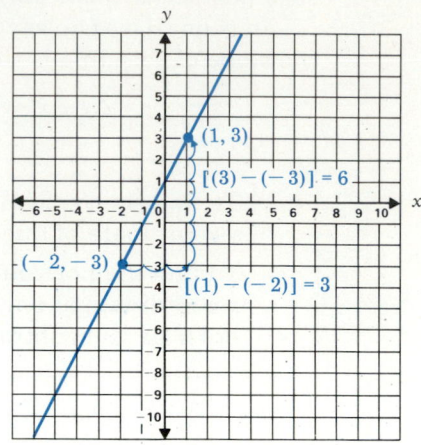

The change in the vertical direction from -3 to 3 is 6, $[(3) - (-3) = 3 + 3 = 6]$.
The change in the horizontal direction from -2 to 1 is 3, $[(1) - (-2) = 1 + 2 = 3]$.

The slope is

$$[(3) - (-3) = 6] : [(1) - (-2) = 3] =$$
$$[6] \qquad : \qquad [3] \qquad =$$
$$\frac{6}{3} = 2$$

The slope of this line is 2. So we would expect its graph to be a line going up two steps for every one step to the right. ❏

EXAMPLE 5 Compute the slope of the line passing through points $A(5, 2)$ and $B(3, 0)$. We subtract the coordinates of B *from* A.

$$[(2) - (0)] : [(5) - (3)] =$$
$$[2] \quad : \quad [2] \quad =$$
$$\frac{2}{2} = 1$$

The slope of this line is 1.

Let's compute the slope of the line again, this time subtracting the points in the other order. Subtract the coordinates of A *from* B: $\{A(5, 2) : B(3, 0)\}$.

$$[(0) - (2)] : [(3) - (5)] =$$
$$[0 - 2] : [3 - 5] \quad =$$
$$[-2] : [-2] \quad =$$
$$\frac{(-2)}{(-2)} = 1 \quad ✓$$

Computed either way, the slope is 1. ❏

There is a shortcut to computing the slope from two points: Express the ratio as a fraction right from the start.

We can compute the slope of the line passing through any two points on it (x_1, y_1) and (x_2, y_2) by simplifying the fraction that results from subtracting their y-coordinates in the numerator and subtracting their x-coordinates in the denominator.

The slope of a line through (x_1, y_1) and (x_2, y_2) is

$$\frac{\Delta y}{\Delta x} = \frac{\text{difference of the } y\text{-coordinates (rise)}}{\text{difference of the } x\text{-coordinates (run)}} = \frac{(y_2) - (y_1)}{(x_2) - (x_1)}.$$

Remember from our work with ratios that

$$[(y_2) - (y_1)] : [(x_2) - (x_1)] = \frac{(y_2) - (y_1)}{(x_2) - (x_1)}.$$

EXAMPLE 6   Find the slope of the line that passes through $(-2, 1)$ and $(-5, 10)$.

$$\text{Slope} =$$
$$\frac{[(10) - (1)]}{[(-5) - (-2)]} =$$
$$\frac{[10 - 1]}{[-5 + 2]} =$$
$$\frac{9}{-3} = -3$$

So the slope of this line is -3. We would expect its graph to be a line going *down* with the steepness of a line with slope 3. ❏

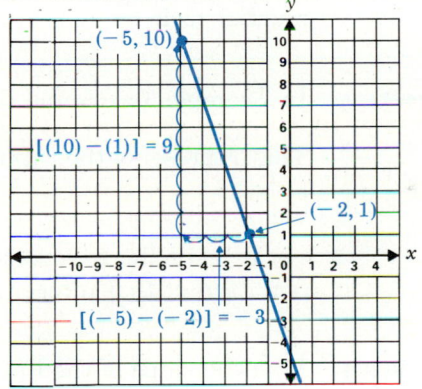

The slope is not always a whole number. When the slope is a fraction, always reduce it to its simplest form.

EXAMPLE 7   Find the slope of the line that passes through $(1, 3)$ and $(7, 5)$.

$$\text{Slope} =$$
$$\frac{[(5) - (3)]}{[(7) - (1)]} =$$
$$\frac{[5 - 3]}{[7 - 1]} =$$
$$\frac{2}{6} = \frac{1}{3}$$

The slope of this line is $\frac{1}{3}$. We would expect its graph to go up less steeply than a line with slope 1.

There are two ways to describe a line with slope $\frac{1}{3}$. We could say that to go from one point on it to another point we

move *one-third* of a space up for every *one* space moved to the right.

or

move *one* space up for every *three* spaces moved to the right.

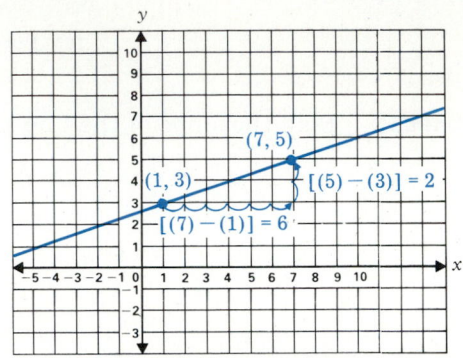

If the slope is an improper fraction, always express it as a fraction rather than a mixed number, since that notation represents the ratio of the change in the rise (indicated by its numerator) to the change in the run (indicated by its denominator). A whole-number slope really represents the ratio of that whole number to 1.

EXAMPLE 8 In Example 3 we found the slope to be 3.

That represents the ratio of 3 to 1, or $\frac{3}{1}$.

It is simpler to represent whole numbers without the denominator of 1, but don't forget that integer slopes also represent a ratio.

EXAMPLE 9 Find the slope of the line that passes through points $(3, 6)$ and $(5, 1)$.

$$\text{Slope} =$$

$$\frac{[(6) - (1)]}{[(3) - (5)]} =$$

$$\frac{[5]}{[-2]} = -\frac{5}{2}$$

The slope of this line is $-\frac{5}{2}$. Since $-\frac{5}{2}$ is the same as $-2\frac{1}{2}$, we would expect it to be a line that goes down a little steeper than a line with slope -2.

There are two ways to describe a line with slope $-\frac{5}{2}$. We could say that to go from one point on it to another point we

move *five-halves* of a space down for every *one* space moved to the right,

or

move *five* spaces down for every *two* spaces moved to the right.

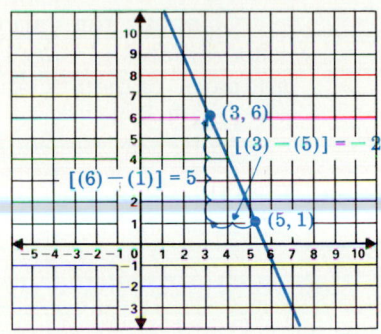

EXERCISES 8.7A

Find the slope of the line that passes through each pair of points.

1. $(3, 5), (7, 9)$ **2.** $(1, 5), (6, 10)$

3. $(0, 7), (1, 3)$ **4.** $(1, 6), (-2, 7)$

5. $(2, 3), (3, 7)$ **6.** $(4, 0), (9, 5)$

7. $(-1, 3), (1, 7)$ **8.** $(2, -1), (-4, 2)$

9. $(0, 4), (-2, 5)$ **10.** $(7, 0), (-1, 3)$

Check your answers on page 737. ✓

Slope as a Ratio of Any Two Points

It does not matter which two points on the line are used to compute the slope. This ratio will be the same for *any* two points on that line.

To illustrate that this ratio always reduces to the same number, I have sketched the line with equation $y = 2x + 1$ below. I have computed the ratio for two points in the first quadrant $[(0, 1)$ and $(2, 5)]$, and for two points in the third quadrant $[(-4, -7)$ and $(-1, -1)]$. Notice the slope is 2 in both computations. These points are on the line with equation

$$y = 2x + 1$$

x	0	2	-1	-4
y	1	5	-1	-7

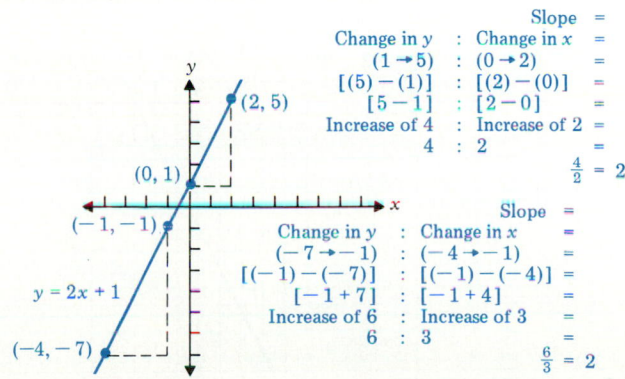

Compute the ratio of change in the y-coordinates to the change in x-coordinates for *any* two points on this line. You will see that for any and all pairs of points, that ratio is 2. The slope of the line with equation $y = 2x + 1$ is 2.

EXAMPLE 10   Find the slope of the line graphed below by determining the coordinates of two points on the line.

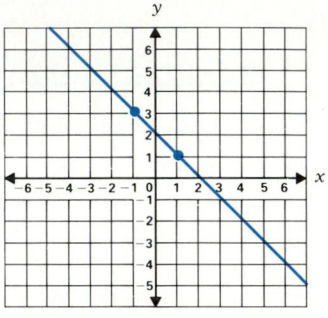

It looks like the coordinates of the two points shown on that line are $(-1, 3)$ and $(1, 1)$.

$$\text{Slope} = \frac{[(3) - (1)]}{[(-1) - (1)]}$$

$$= \frac{[3 - 1]}{[-1 - 1]}$$

$$= \frac{2}{(-2)}$$

$$= -1$$

So the slope is -1. ❑

EXERCISES 8.7B

Calculate the slopes of the following lines, choosing two points whose coordinates you can identify exactly.

11.

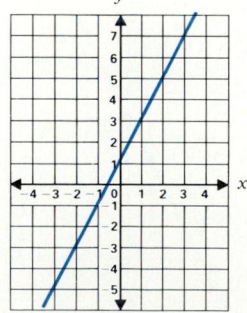

12.

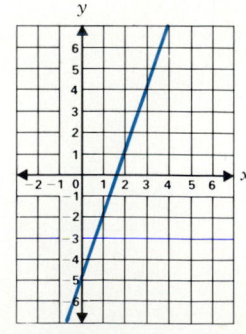

13.

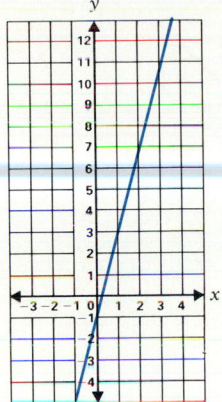

14.

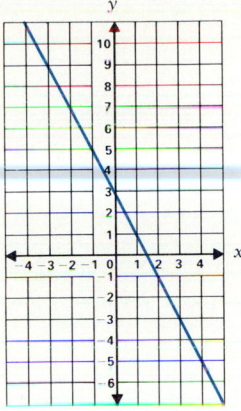

15.

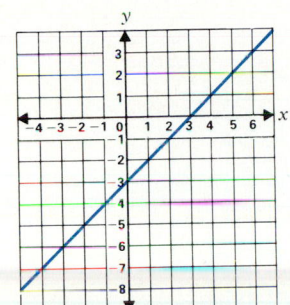

16.

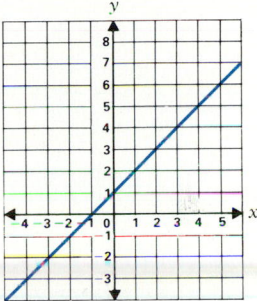

17.

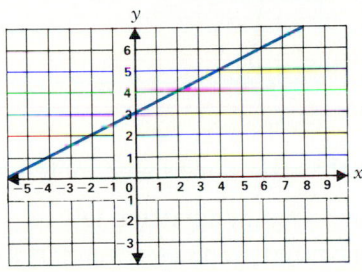

18.

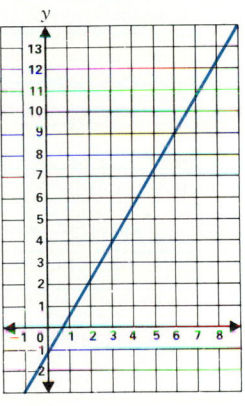

19.

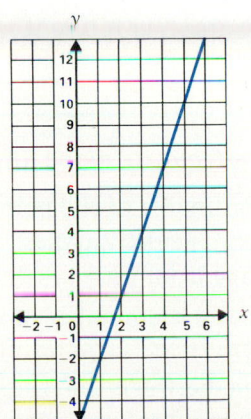

20.

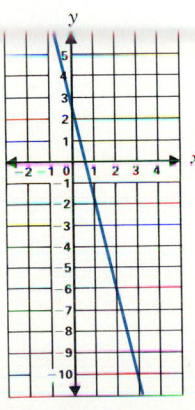

21.

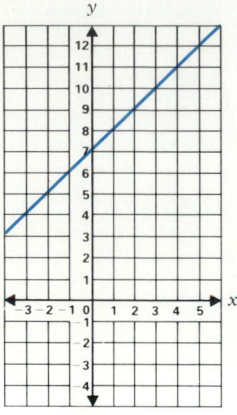

22.

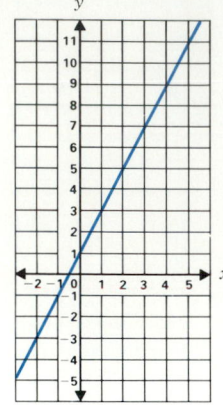

23.

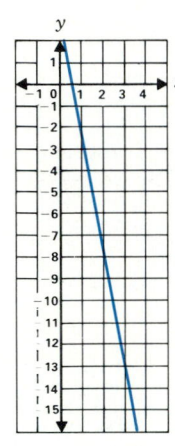

24.

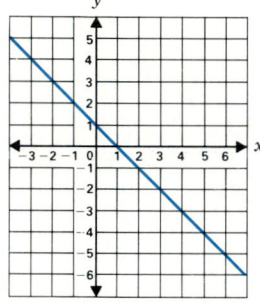

Check that the slope's sign matches the direction of the line.

Lines that go up (from left to right) have a positive slope.

Lines that go down (from left to right) have a negative slope.

Check your answers on page 737.

Graphing a Line from Its Slope

We can sketch the graph of a line if we know its *slope* and the coordinates of *one point on the line.*

> *To graph a line from its slope and one point:*
>
> Plot the coordinates of the point given.
>
> Move to the right the number of units in the denominator of the slope. (If slope is an integer, that denominator is 1.)
>
> Move vertically the number of units in the numerator of the slope. (If the slope is *positive*, go *up*; if it is *negative*, go *down*.)
>
> Draw the line between the original point and the point arrived at from those horizontal and vertical moves, and extend the line.

You can continue to find points on the line by proceeding from each point with the horizontal and vertical directions given by the slope.

EXAMPLE 11   Graph the line with slope $\frac{2}{3}$ that goes through point (1, 4).

Begin at (1, 4) and move *three* spaces to the *right* and *two* spaces up. That brings you to point (**4, 6**).

Draw the line between (1, 4) and (4, 6).

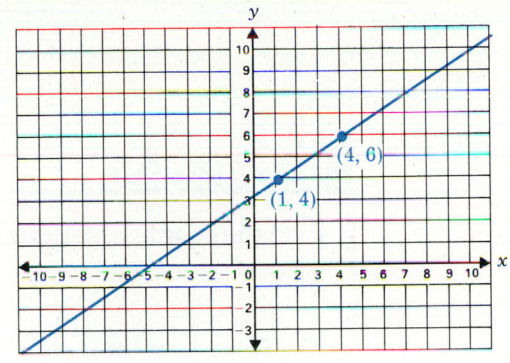

EXAMPLE 12   Graph the line with slope -5, that goes through point (2, 1). $\left(\text{Remember, slope } -5 \text{ is really } \frac{-5}{1}.\right)$

Begin at (2, 1), and move *one* space to the *right* and *five down*. That brings you to point (**3, −4**).

To find another point on the line, from (3, −4) move *one* step to the *right* and *five* spaces *down*. That brings you to point (**4, −9**).

Draw the line through (2, 1), (3, −4), and (4, −9), and extend it.

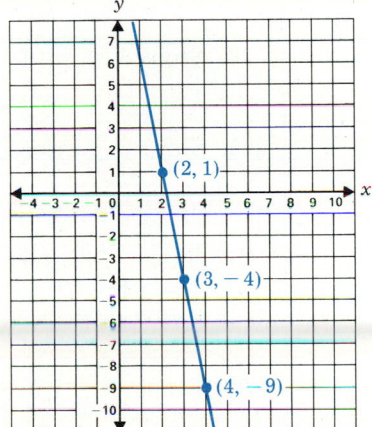

EXAMPLE 13   Graph the line with slope $-\frac{3}{4}$ that has y-intercept 2.

y-intercept 2 is point (0, 2). Begin there and move *four* spaces to the *right* and *three* spaces *down*. That brings you to point (**4, −1**).

A third point can be found. From (4, −1), move *four* spaces to the *right* and *three* spaces *down*. That brings you to point (**8, −4**).

Draw the line through these three points.

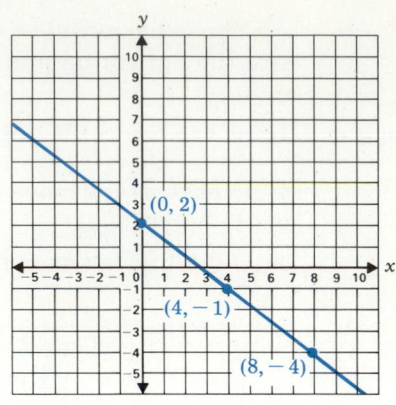

EXERCISES 8.7C

Find the coordinates of two more points for each line. Sketch the graph of each line on graph paper.

25. Slope $\frac{1}{3}$, point $(2, 1)$

26. Slope $\frac{2}{5}$, point $(1, -6)$

27. Slope $-\frac{2}{3}$, point $(4, 5)$

28. Slope $-\frac{3}{5}$, point $(4, 9)$

29. Slope 2, point $(-3, -4)$

30. Slope 3, point $(1, 1)$

31. Slope -4, point $(5, 8)$

32. Slope -1, point $(3, 4)$

33. Slope 1, point $(-3, 2)$

34. Slope -3, point $(5, 5)$

35. Slope 3, y-intercept 1

36. Slope -2, y-intercept 5

37. Slope $\frac{1}{3}$, y-intercept -2

38. Slope $\frac{3}{4}$, y-intercept 1

39. Slope $-\frac{2}{3}$, y-intercept 0

40. Slope $-\frac{2}{5}$, y-intercept 6

Check your answers on page 737. ✓

Slope of Horizontal and Vertical Lines

Let's look at the equation of a horizontal line. We know that if it is horizontal, all the points on the line must have the same y-(second) coordinate. As we learned in Section 8.3, the equation of a horizontal line must be "y equals some constant number."

EXAMPLE 14 $y = 3$ is the equation of a horizontal line.

Two of the points on this line are $(0, 3)$ and $(1, 3)$. The slope of the line through $(0, 3)$ and $(1, 3)$ is

$$\frac{3 - 3}{1 - 0} =$$

$$\frac{0}{1} = 0$$

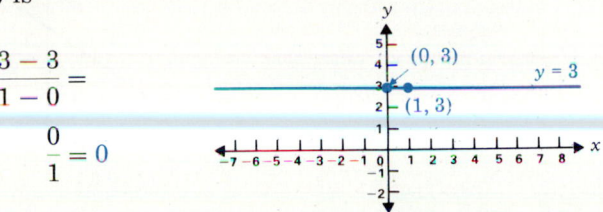

We see that the slope of *this horizontal line is zero*. In fact, the slope of any horizontal line will *always be zero. On any horizontal line, all points have the same y- (second) coordinate. So the difference between any two y-coordinates (the numerator of the slope) will be zero.*

It makes sense that the slope of every horizontal line is zero. A horizontal line is flat. It has no steepness, or, we say mathematically, it has a steepness of 0.

Now let's look at the slope of a vertical line. Remember that a vertical line is made up of points with the same *x*- (first) coordinate. As we learned in Section 8.3, the equation of a vertical line must be "*x* equals some constant number."

EXAMPLE 15 $x = 2$ is the equation of a vertical line.

Two of the points on this line are $(2, 0)$ and $(2, 1)$. The slope of the line containing $(2, 0)$ and $(2, 1)$ is $\dfrac{1 - 0}{2 - 2} = \dfrac{1}{0}$, which is not defined.

You cannot divide by zero. There is no slope.

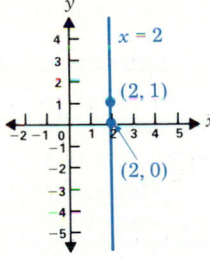

The slope of this vertical line does not exist because we cannot divide by zero. No slope of any vertical line will exist because in every vertical line the *x*-coordinates are all the same, so the difference in the *x*-coordinates (the denominator of the slope) will always be zero. Thus the denominator of the slope will be zero (which is not defined) for every vertical line.

Another way to say this is that a vertical line is the steepest line possible. There is no largest number, so there can be no number to indicate its steepness or slope.

To summarize:

A line with positive slope goes up as you read it from left to right.

A line with negative slope goes down as you read it from left to right.

A line with zero slope is a horizontal line.

A line with no slope is a vertical line.

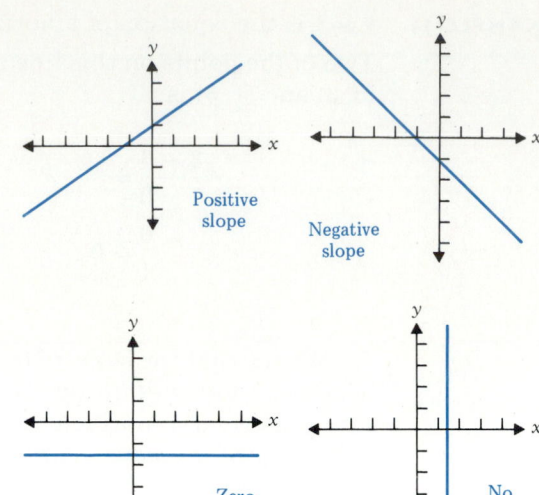

Positive slope

Negative slope

Zero slope

No slope

Calculating the Slope of a Line from Its Equation

If you recognize the equation of a horizontal or vertical line, you can determine the slope without any computation at all.

Horizontal lines are of the form $y = k$ (where k is the second coordinate of every point).

Vertical lines are of the form $x = c$ (where c is the first coordinate of every point).

EXAMPLE 16   Find the slope of the line $y = 4$.

Since the graph of $y = 4$ is a horizontal line, the slope must be 0.   ❑

However, it is also possible to determine the slope of any linear equation, whether or not you can easily visualize the graph. In fact, you do not even have to sketch the graph. Simply find the coordinates of any two points that are solutions to the equation and calculate the slope.

Remember that, in general, for points (x_1, y_1) and (x_2, y_2) the slope is $\dfrac{y_2 - y_1}{x_2 - x_1}$.

EXAMPLE 17   Find the slope of the line with equation $y = 5x - 2$.

$$\text{Let } x = 1, y = 5(1) - 2$$
$$= 5 - 2$$
$$= 3 \quad \text{So } (1, 3) \text{ is a solution.}$$

$$\text{Let } x = 2, y = 5(2) - 2$$
$$= 10 - 2$$
$$= 8 \quad \text{So } (2, 8) \text{ is a solution.}$$

Now calculate the slope of the line through $(1, 3)$ and $(2, 8)$.

$$\text{Slope} =$$
$$\frac{(8) - (3)}{(2) - (1)} =$$
$$\frac{5}{1} = 5$$

So the slope is 5.   ❑

EXAMPLE 18   Find the slope of the line $y = -3x + 7$.
Two solutions are $(0, 7)$ and $(1, 4)$.

$$\text{Slope} =$$
$$\frac{7 - 4}{0 - 1} =$$
$$\frac{3}{-1} = -3$$

So the slope is -3.

EXERCISES 8.7D

For each equation, find the coordinates of two points and determine the slope of the line.

41. $y = 2x - 3$

42. $y = -5x + 3$

43. $y = 6$

44. $x = -3$

45. $y = -4x$

46. $y = 3x - 1$

47. $y = -x + 6$

48. $y = 3x$

49. $y = x - 4$

50. $y = 2x - 5$

Check your answers on page 738.

8.8   Slope-Intercept Form of Linear Equations

Slope-Intercept Form

In Section 8.4 we learned how to solve a linear equation with two letters for one of the letters.

We say that the equation is in **slope-intercept form** when a linear equation with two letters is solved for one of the letters.

A general way to express a linear equation in slope-intercept form is $y = mx + b$, where m and b represent numbers.

EXAMPLE 1   The following equations are all in slope-intercept form:

$$y = 3x + 5, \text{ where } m = 3 \text{ and } b = 5;$$
$$y = -7 - 5x, \text{ where } m = -5 \text{ and } b = -7;$$
$$x - 3 = y, \text{ where } m = 1 \text{ and } b = -3.$$

EXERCISES 8.8A

Which of the following equations are in slope-intercept form?

1. $y = 3x + 5$

2. $y - x = 7$

3. $y + 5 = 3x$

4. $5x - 1 = y$

5. $x - y = 3$

6. $y = 12 + 2x$

Check your answers on page 738.

Using Slope-Intercept Form to Find the y-Intercept

Remember, the y-intercept of a line is the point on that line with x-coordinate equal to zero. When a linear equation is in slope-intercept form and you let $x = 0$, then the x term (mx) will equal zero. So when you solve for the y-coordinate of the y-intercept, you will find

$$\text{if } x = 0, \qquad y = \underline{m(0)} + b;$$
$$y = b.$$

> When the equation is solved for y (in slope-intercept form, $y = mx + b$), the y-intercept will always be the constant term b.

EXAMPLE 2 The y-intercept of $y = 7x - 4$ is -4. ❏

EXAMPLE 3 The y-intercept of $y = -3x + 1$ is 1. ❏

EXAMPLE 4 The y-intercept of $y = 5 + 6x$ is 5. ❏

Note that the reversed order of terms in Example 4 makes no difference; the y-intercept is always the constant term.

You can use this shortcut for finding the y-intercept of a line only if the linear equation is in slope-intercept form.

EXAMPLE 5 To find the y-intercept of $-3y - 15x = 9$ without graphing, solve the equation for y so it will be in slope-intercept form.

$$
\begin{array}{r}
-3y - 15x = 9 \\
\underline{\quad + 15x \qquad + 15x} \\
-3y \quad = 9 + 15x
\end{array}
$$

$$\frac{-3y}{-3} = \frac{9}{-3} + \frac{15x}{-3}$$

$$y = -3 - 5x$$

So the y-intercept is -3, and the line crosses the y-axis at the coordinates $(0, -3)$. ❏

There is no shortcut for finding the x-intercept of a linear equation as there was for finding the y-intercept. You have to let $y = 0$ and solve the equation for x as we did in Section 8.6.

EXERCISES 8.8B

Express each equation in slope-intercept form and find the y-intercept for each line.

7. $y - 5x = 9$ **8.** $2y = 10x - 14$

9. $3y + 12x = 15$ **10.** $2y - 6 = 8x$

11. $4y + 12x = 16$ **12.** $-5y = 30x - 15$

13. $x + y = -11$ **14.** $-4x + 5y = 3$

15. $2y - 3x = -7$ **16.** $-3y - 4x = 1$

Check your answers on page 738. ✓

Shortcut for Finding Slope from an Equation

There is a shortcut for finding the slope of a line from its linear equation that can save you a lot of time and work. In Section 8.7, Examples 17 and 18, we found that

the *slope* of $y = 5x - 2$ is 5;

the *slope* of $y = -3x + 7$ is -3.

> If a linear equation is solved for y (in slope-intercept form), the slope is always the coefficient of the x term.

Why does this shortcut work?

Slope is "the number of spaces you move vertically for every space you move horizontally from one point to another on a line." When the equation is in slope-intercept form, the coefficient of the x term is the amount y will increase for every increase of 1 for x.

EXAMPLE 6 In the equation $y = 2x + 1$, if we increase x by 1, we increase y by 2.

Note: Let $x = 0$, $y = 1$;   Notice that as x increases by 1,
let $x = 1$, $y = 3$;   from 0 to 1, y increases by 2, from 1 to 3.

let $x = 2$, $y = 5$.   Notice that as x increases by 1, from 1 to 2, y increases by 2, from 3 to 5.

So the slope of the line $y = 2x + 1$ is 2.   ❏

To use this shortcut, be sure the equation is in slope-intercept form.

EXAMPLE 7 The slope of $y = -8x + 4$ is -8.   ❏

EXAMPLE 8 The slope of $y = 9x - 13$ is 9.   ❏

EXAMPLE 9 Find the slope of $2y - 6x = 10$.

First, solve the equation for y to get it in slope-intercept form.

$$
\begin{array}{rcl}
2y - 6x &=& 10 \\
\underline{+\ 6x \qquad\quad} & & \underline{+\ 6x} \\
2y &=& 10 + 6x
\end{array}
$$

$$\frac{2y}{2} = \frac{10}{2} + \frac{6x}{2}$$

$$y = 5 + 3x$$

The slope is 3. Note that the slope is *not* -6, the coefficient of x in the original equation ($2y - 6x = 10$), which is *not* in slope-intercept form.   ❏

You cannot *read* the slope or y-intercept from the equation until it is put into the slope-intercept form.

EXERCISES 8.6C

Express each equation in slope-intercept form and find the slope.

17. $y = 2x - 5$ **18.** $y = -x + 7$

19. $y + 3 = x$ **20.** $y - 2x = 11$

21. $3y = 9x - 6$ **22.** $2y + 10 = -4x$

23. $5y + 20x = 10$ **24.** $-4y = 12x - 20$

25. $2y - 14x = 40$ **26.** $-2y + 10x = -28$

Check your answers on page 738. ✓

Advantages of Slope-Intercept Form

In Example 9 we saw that $2y - 6x = 10$ could be expressed in slope-intercept form as $y = 3x + 5$.

We learned that

the y-intercept of $y = 3x + 5$ is $+5$;

the slope of $y = 3x + 5$ is 3.

The name "slope-intercept form" was chosen because it is so easy to recognize the slope and the y-intercept when the equation is in that form.

You can now see that when an equation is in slope-intercept form, $y = mx + b$, the **slope is m** (the coefficient of x) and the **y-intercept** is b (the constant term).

EXAMPLE 10   Find the slope and y-intercept of $5y - 8x = 7$.

First solve for y to get in slope-intercept form.

$$5y - 8x = 7$$
$$\underline{+ 8x \qquad + 8x}$$
$$5y = 7 + 8x$$
$$\frac{5y}{5} = \frac{7}{5} + \frac{8x}{5}$$
$$y = \frac{7}{5} + \frac{8}{5}x$$

The slope is $\frac{8}{5}$.

The y-intercept is $\frac{7}{5}$ or $1\frac{2}{5}$: $\left(0, 1\frac{2}{5}\right)$. ❏

⟹ **Note:** Since the y-intercept is the y-coordinate of a point, if it is an improper fraction it is better to express it as a mixed number. In that form, we more easily will recognize where it is on the number line. That is why we expressed the y-intercept of $\frac{7}{5}$ as $1\frac{2}{5}$.

Remember, slope is never to be expressed as a mixed number.

EXERCISES 8.8D

Express each equation in slope-intercept form and find the slope and y-intercept.

27. $y = 3x - 5$

28. $y + 4 = 3x$

29. $y - 2x = 7$

30. $y + x = 9$

31. $y + 7x = 11$

32. $3y = 6x - 15$

33. $4y + 12 = -8x$

34. $-2y = 6x - 14$

35. $-y = 3x + 4$

36. $-2y + 8x = 30$

37. $3y - 4x = 1$

38. $2y = 3x + 7$

39. $5y - 6 = 2x$

40. $2y - x = 3$

Check your answers on page 738.

Sketching a Graph from Slope-Intercept Form

Another advantage of expressing linear equations in slope-intercept form is that you can sketch the line directly without substituting in values of x and solving for y to find the coordinates of three points.

From $y = mx + b$, you can read the coordinates of the y-intercept as $(0, b)$.

If m is a whole number, just move one space to the right from $(0, b)$ over to $(1, b)$ and then move vertically m spaces. That brings you to another point on the line. Continue to find other points on the line by moving one space to the right and m spaces vertically.

EXAMPLE 11 Sketch the graph of $y = 2x - 5$.

Begin at the y-intercept $(0, -5)$.

The slope m is $+2$, so move *one* space to the *right* and *up two* spaces to $(\mathbf{1, -3})$.

Find a third point from $(1, -3)$ by going *one* space to the *right* and *up two* spaces again. That brings you to $(\mathbf{2, -1})$.

Plot those three points and sketch the line.

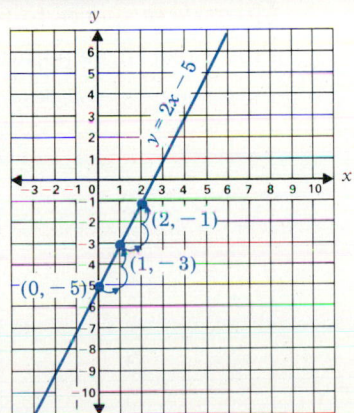

EXAMPLE 12 Sketch the graph of $y = -x + 4$.

Begin at the y-intercept $(0, 4)$.

The slope m is -1, so the *vertical* movement is *down*.

From $(0, 4)$ move *one* space to the *right* and *down one* space to **(1, 3)**.

From $(1, 3)$ move *one* space to the *right* and *down one* space again to **(2, 2)**.

Plot those three points and sketch the line.

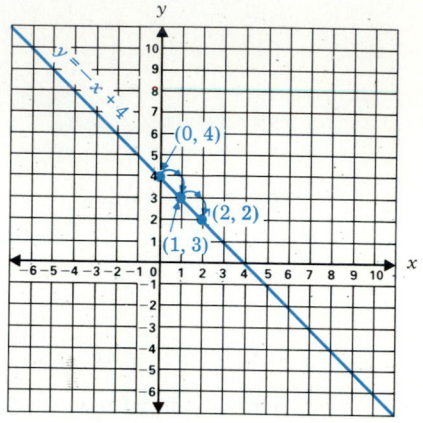

If m is a fraction, move to the *right* the number of units in the *denominator* and then move *vertically* the number of units in the *numerator*.

If the fraction is **negative**, always interpret the denominator as positive and the numerator as negative

> *To locate points from the directions given by the slope, remember:*
>
> Horizontal movement is *always to the right*.
>
> If the slope is *positive*, the *vertical* movement is *up*.
>
> If the slope is *negative*, the *vertical* movement is *down*.

EXAMPLE 13 Sketch the graph of $y = -\dfrac{2}{3}x + 1$.

The y-intercept is $(0, 1)$; the slope $\dfrac{-2}{3}$.

Move *three* spaces to the *right* and *down two* spaces to the next point **(3, −1)**.

Move *three* spaces to the *right* and *down two* spaces to next point **(6, −3)**.

Plot those three points and sketch the line.

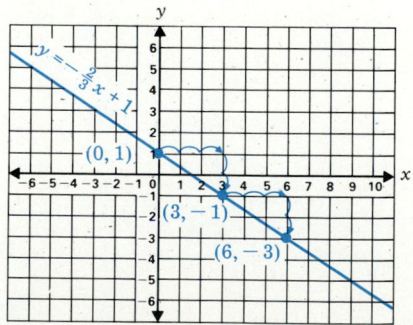

EXAMPLE 14   Sketch the graph of $y = \frac{1}{2}x$.

First, think of $y = \frac{1}{2}x$ as $y = \frac{1}{2}x + 0$.

Begin at the y-intercept $(0, 0)$. The slope is $\frac{1}{2}$.

From $(0, 0)$ move *two* spaces to the *right* and *up one* space to **(2, 1)**.
From $(2, 1)$ move *two spaces* to the *right* and *up one* space to **(4, 2)**.
Plot those three points and sketch the line.

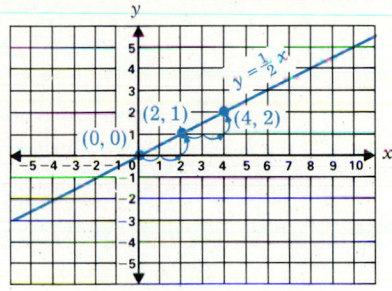

EXERCISES 8.8E

Sketch each line directly from the equation by plotting the y-intercept and moving to the next point, using the directions given by the slope. Plot three points for each line.

41. $y = 3x - 3$ **42.** $y = 2x + 1$

43. $y = -2x + 4$ **44.** $y = -4x + 7$

45. $y = x - 3$ **46.** $y = x + 1$

47. $y = -x$ **48.** $y = 3x$

49. $y = \frac{1}{3}x - 2$ **50.** $y = \frac{2}{3}x + 1$

51. $y = -\frac{3}{4}x$ **52.** $y = -\frac{2}{3}x + 5$

53. $y = \frac{2}{5}x$ **54.** $y = \frac{1}{2}x - 3$

Check your answers on page 738. ✓

 **Parallel Lines**

Concept of Parallel Lines

Lines that lie alongside of each other, never crossing, are called **parallel**. The two rails of a railroad track are parallel. So are the streets in many cities that are laid out like graphs. In Manhattan, for instance, the avenues all run north and south and do not cross each other; the numbered streets all run east and west and do not cross each other.

On one Cartesian coordinate system, it is possible to graph more than one line. If you plot two lines and one line is steeper than the other, the lines will eventually cross. If they have exactly the same steepness, they will not cross; they will be parallel. Since slope is a number that indicates the steepness of a line, two lines that have different slopes must eventually cross. (We will learn to find the place where they cross in Chapter 10.)

If lines have the *same slope*, they are **parallel**.

Let's sketch three lines that all have slope 3 on one Cartesian coordinate system.

$$y = 3x + 1, \ y = 3x + 7, \text{ and } y = 3x - 4$$

are all equations of lines in slope-intercept form, and so we can easily see that each of them has slope 3.

EXAMPLE 1

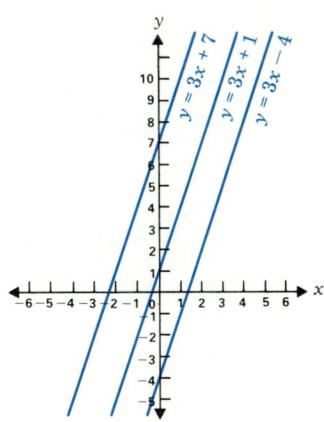

Since the constant term is different in each equation, the equations are not the same; they each have a different y-intercept. Even though our graph illustrates only segments of each line (theoretically speaking, they continue infinitely), there is enough in the sketch to see that these lines will always run alongside one another. They are parallel.

EXERCISES 8.9A

Using graph paper, sketch each pair of parallel lines on a separate Cartesian coordinate system.

1. $y = x - 5$
   $y = x + 4$

2. $y = -3x + 1$
   $y = -3x + 4$

3. $y = 4x - 5$
   $y = 4x - 1$

4. $y = 2x - 3$
   $y = 2x + 3$

Check your answers on page 739.

Using Slope-Intercept Form to Recognize Equations of Parallel Lines

To decide whether two linear equations represent parallel lines, put them both into slope-intercept form. If they have the same slope (coefficient of x) but a different y-intercept (constant term), they are parallel.*

EXAMPLE 2   Are the lines $y - 4x = 7$ and $2y = 6x + 1$ parallel?

$$\begin{array}{c} y - 4x = 7 \\ \underline{+\,4x \qquad +\,4x} \\ y = 7 + 4x \end{array} \qquad\qquad \begin{array}{c} 2y = 6x + 1 \\ \dfrac{2y}{2} = \dfrac{6x}{2} + \dfrac{1}{2} \\ y = 3x + \dfrac{1}{2} \end{array}$$

   The slope is 4.   The slope is 3.

No, they are not parallel. They have different slopes. ❑

EXAMPLE 3   Are the lines $y + 7x = 8$ and $3y = -21x + 15$ parallel?

$$\begin{array}{c} y + 7x = 8 \\ \underline{-\,7x \qquad -\,7x} \\ y = 8 - 7x \end{array} \qquad\qquad \begin{array}{c} 3y = -21x + 15 \\ \dfrac{3y}{3} = \dfrac{-21x}{3} + \dfrac{15}{3} \\ y = -7x + 5 \end{array}$$

   The slope is -7.   The slope is -7.

Yes, they are parallel. They have the same slope. ❑

EXERCISES 8.9B

Determine whether each pair of equations represents lines that are parallel.

5. $y = 3x + 5$ and $y = 5x + 3$ **6.** $y - 4 = 2x$ and $3y = 6x + 12$

7. $y - 5x = 7$ and $5y = -25x + 30$ **8.** $2y = 6x$ and $-3y = -9x + 7$

9. $4y = 8x - 20$ and $5y = 20x + 10$ **10.** $3y = 7$ and $x = 8$

Check your answers on page 739. ✓

Finding the Equations of Lines

Representing Lines with $y = mx + b$

When a linear equation is expressed in slope-intercept form, $y = mx + b$, you can determine many things about its line;

    The slope,

    The y-intercept,

    Coordinates of points on the line,

    The graph of the line.

---

* If they have the same slope and the same y-intercept, they are the same line.

You can also work this process in reverse. If you are given enough information about a line, you can determine what linear equation it represents.

Remember, a line is simply the set of all points (x, y) that are solutions to the equation $y = \underline{mx + b}$, where m represents the <u>slope</u> and b represents the <u>y-intercept</u>.

To find the equation of a line, all you need to know is the value of the line's slope and y-intercept. Replace m with the value of the slope and replace b with the value of the y-intercept in $y = mx + b$.

Let's look separately at each case in which you are given enough information about a line to find its equation.

When You Know the Slope and y-Intercept

If someone is nice enough to tell you the line's slope and y-intercept, you can substitute those values into $y = mx + b$ and immediately state the equation of the line without any computation.

EXAMPLE 1   What is the equation of the line with slope 3 and y-intercept 5?

Let $m = 3$, the slope, and $b = 5$, the y-intercept.

The equation of that line is $y = \underline{3}x + \underline{5}$.   ❏

EXAMPLE 2   What is the equation of the line with slope -7 that passes through point $(0, 2)$?

Point $(0, 2)$ is the line's y-intercept, since its x-coordinate is zero; so $b = 2$. The slope is -7, so $m = -7$.

The equation is

$$y = \underline{-7}x + \underline{2}.$$ ❏

EXAMPLE 3   What is the equation of the line with slope $\dfrac{3}{4}$ and y-intercept $\dfrac{2}{5}$?

The equation is $y = \dfrac{3}{4}x + \dfrac{2}{5}$.   ❏

EXAMPLE 4   What is the equation of the line with slope 0.3 and y-intercept -0.27?

The equation is $y = \underline{0.3x - 0.27}$.   ❏

EXERCISES 8.10A

Find the equations of the lines that have the following properties.

1. Slope 6, y-intercept $(0, 2)$.   Equation:_____

2. Slope -4, y-intercept $(0, 0)$.   Equation:_____

3. Slope -1, y-intercept $(0, 3)$.   Equation:_____

4. Slope 5, y-intercept $(0, -1)$.   Equation:_____

5. Slope -3, going through point $(0, 3)$.   Equation:_____

6. Slope 7, going through point $(0, -5)$.   Equation:_____

7. Slope 9, going through point $(0, 11)$.   Equation:_____

8. Slope $\frac{2}{3}$, y-intercept $\frac{1}{3}$. Equation:_____

9. Slope 0.7, y-intercept 0.04. Equation:_____

10. Slope $\frac{3}{5}$, y-intercept $-\frac{1}{4}$. Equation:_____

Check your answers on page 739. ✓

When You Know the Slope and Any Point

If you know a line's slope and the coordinates of one point on the line, you have enough information to find the equation. To find the equation that a line represents, all you need do is find the value of m (the slope) and b (the y-intercept) and substitute those values into the general form of a linear equation,

$$y = mx + b.$$

Substitute the value of the slope in the equation for m.

The coordinates of every point (x, y) on a line are solutions to its equation. So if you replace the first coordinate of the point for x and the second coordinate for y, this will leave b as the only letter in the equation. Solve the equation for b, and then you will know the equation of the line.

EXAMPLE 5 Find the equation of the line with slope 3, going through point $(1, 5)$.

The slope is 3, so $m = 3$. Therefore the equation must be $y = 3x + b$.

Since you know that $(1, 5)$ is a point on the line, that means $x = 1$ and $y = 5$ is one solution to the equation.

Replace x by 1 and y by 5 in $y = 3x + b$, and solve the equation for b.

$$y = 3x + b$$
$$(5) = 3(1) + b$$
$$5 = 3 + b$$
$$\underline{-3 -3 }$$
$$2 = b$$

Now that you know the value of m and b, you can write that equation as $y = 3x + 2$. ❑

EXERCISES 8.10B

Find the equations of the lines that have the following properties.

11. Slope 2, going through point $(2, 3)$. Equation:_____

12. Slope -5, going through point $(1, 7)$. Equation:_____

13. Slope 3, going through point $(-2, 4)$. Equation:_____

14. Slope -1, going through point $(-2, -4)$. Equation:_____

15. Slope 7, going through point $(-3, -3)$. Equation:_____

16. Slope 1, going through point $(8, 0)$.   Equation:_____

17. Slope -4, going through point $(-5, 1)$.   Equation:_____

18. Slope -3, going through point $(2, 0)$.   Equation:_____

Check your answers on page 739.   ✓

When You Have the Line's Graph

If a line is graphed on a Cartesian coordinate system where you can read the coordinates of at least two points with integer coordinates, you have enough information to find the equation.

Read the y-coordinate of the point where the line crosses the y-axis to find the value of b, if the line crosses where y is an integer.

Read the coordinates of any point on the line where the x- and y-coordinates are both integers.

Replace x by the x-coordinate and y by the y-coordinate of that point.

The equation $y = mx + b$ can now be solved for m. Once you know m, replace m and b by their values, and you have the equation of the line.

EXAMPLE 6

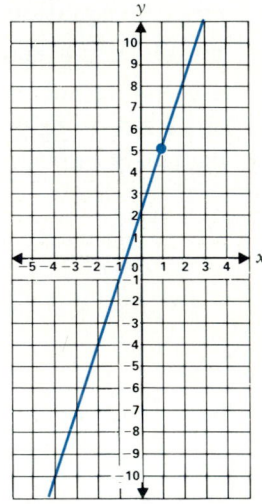

The y-intercept is $(0, 2)$ so $b = 2$. The line goes through the point $(1, 5)$ so let $x = 1$ and $y = 5$.

$y = mx + b$ becomes $(5) = m(1) + (2)$, which we can solve for m.

$$(5) = m(1) + (2)$$
$$5 = m + 2$$
$$\underline{-2 \qquad -2}$$
$$3 = m \qquad \text{and } b = 2, \text{ so } y = mx + b \text{ is}$$

$$y = \underline{3}x + \underline{2}, \text{ the equation of that line.} \qquad \square$$

EXERCISES 8.10C

Read the coordinates of the *y*-intercept and one other point from each line, and find the equation for each graph.

19.

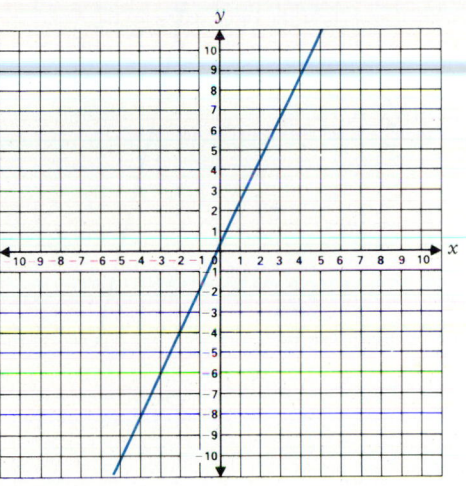

20.

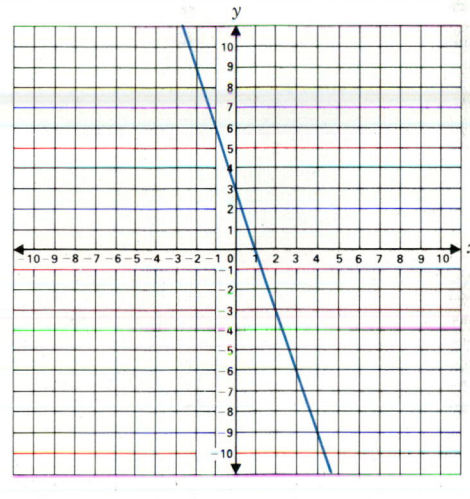

21.

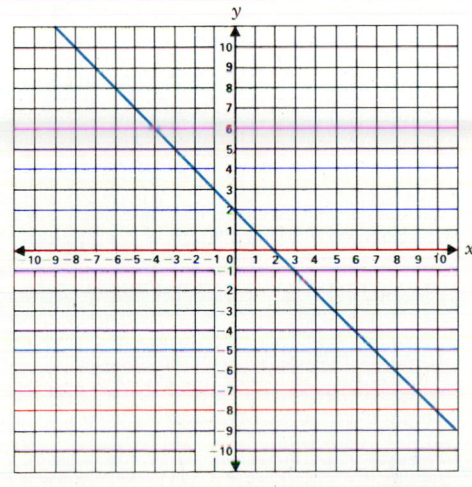

22.

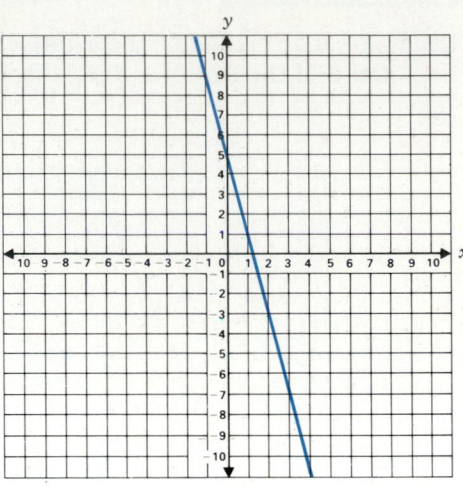

23.

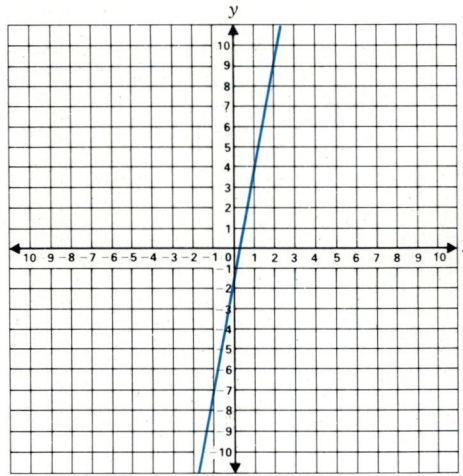

24.

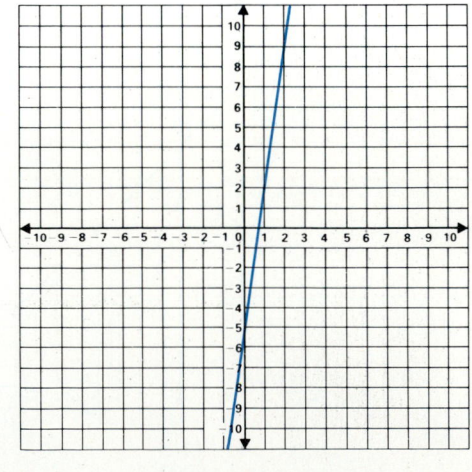

25.

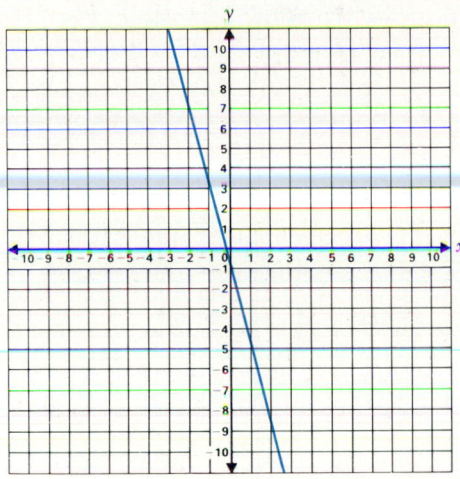

26.

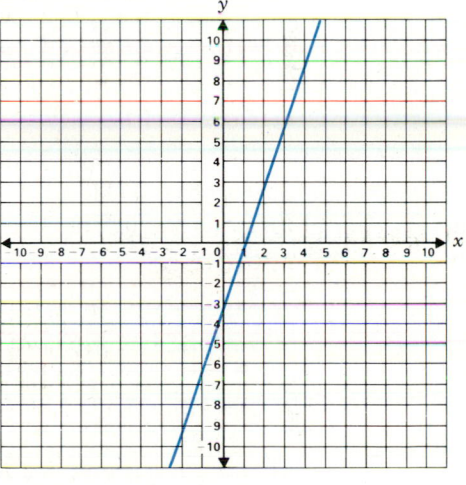

27.

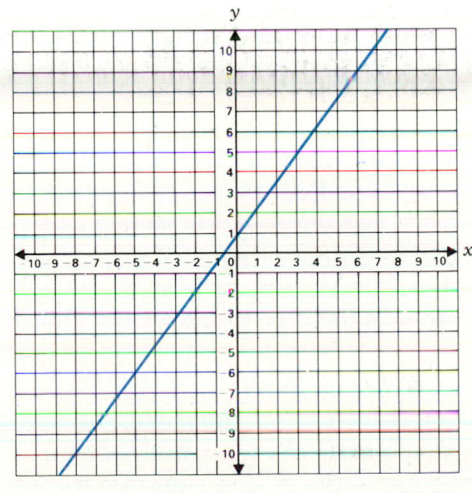

28.

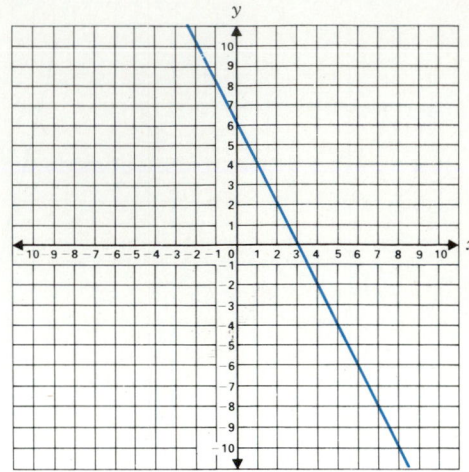

29.

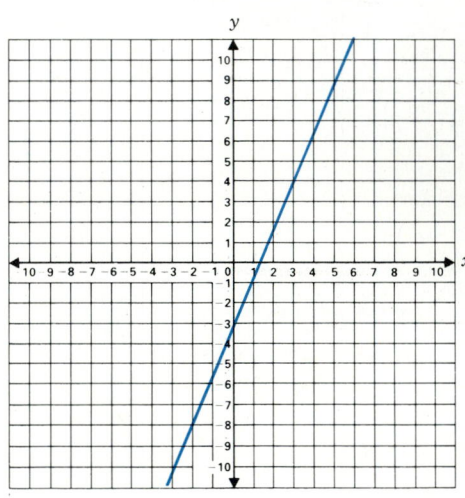

Check your answers on page 739.

When You Know Two Points

You can find the equation if you know the coordinates of two points on the line. Knowing the coordinates of two points on the line, you can calculate the slope (as we learned in Section 8.7).

Remember, the slope of the line passing through (x_1, y_1) and (x_2, y_2) is

$$\frac{y_2 - y_1}{x_2 - x_1}.$$

Once you know the slope, use the coordinates of one of the points given (x, y) to substitute for x and y in $y = mx + b$, and substitute the value of the slope for m. Solve that equation for b.

Again, once you know the values of m and b, you can state the equation.

EXAMPLE 7 Find the equation of the line that goes through points $(3, 5)$ and $(6, 11)$.

$$\text{Slope} = \frac{11 - 5}{6 - 3} = \frac{6}{3} = 2$$

So $m = 2$, and the equation is $y = 2x + b$.

Now this problem is just like Example 5. Use the coordinates of one of the points to solve for b. Let's use $(3, 5)$, so $x = 3$ and $y = 5$ in $y = 2x + b$.

$$y = 2x + b$$
$$(5) = \underline{2(3)} + b$$
$$5 = \quad 6 + b$$
$$\underline{-6 \quad \; -6}$$
$$-1 = b$$

The equation is $y = 2x - 1$.

$\Rightarrow$ ***Note:*** You can use this method if the y-intercept of a line is a fraction, since you will not be able to read the y-coordinate from the graph. In such a case, you still can find the equation if you can read the coordinates of any two points on the line. Then proceed as in Example 7.

EXERCISES 8.10D

Find the equation of the line that goes through each pair of points. (First calculate the slope.)

30. $(4, 7)$ and $(6, 13)$ **31.** $(5, 1)$ and $(2, 10)$

32. $(8, 3)$ and $(10, 5)$ **33.** $(1, 7)$ and $(5, 19)$

34. $(-3, 7)$ and $(1, 11)$ **35.** $(-2, -5)$ and $(-4, 3)$

36. $(-7, 1)$ and $(-2, 16)$ **37.** $(4, 0)$ and $(7, 3)$

Check your answers on page 740. ✓

When You Know a Point and the Equation of Parallel Lines

If you are told that two lines are parallel, then you know that they have the same slope. You can find the equation of a line if you know the equation of a line it is parallel to, and if you know the coordinates of one point on the line.

EXAMPLE 8 Find the equation of the line that is parallel to $y = 2x - 1$, which passes through point $(5, 4)$.

The slope of $y = 2x - 1$ is 2, so the equation we are trying to find has slope 2 also.

Let $m = 2$, and (as in Example 5) since $(5, 4)$ is a point on the line, let $x = 5$ and $y = 4$ and solve for b.

$$(4) = 2(5) + b$$
$$4 = \quad 10 + b$$
$$\underline{-10 \quad \; -10}$$
$$-6 = b$$

Since $m = 2$ and $b = -6$, the equation is $y = 2x - 6$.

EXERCISES 8.10E _____

Find the equations of the lines with the following properties.

38. Parallel to $y = 3x + 2$; passes through $(4, 1)$.   Equation:_____

39. Parallel to $y = 5x + 4$; passes through $(1, 4)$.   Equation:_____

40. Parallel to $y = -2x + 7$; passes through $(5, 4)$.   Equation:_____

41. Parallel to $y = -4x + 1$; passes through $(-1, 2)$.   Equation:_____

42. Parallel to $y = -x - 5$; passes through $(3, -7)$.   Equation:_____

43. Parallel to $y = -3x + 4$; passes through $(1, 5)$.   Equation:_____

Check your answers on page 740. ✓

Review

It is possible to find the equation of a line if we know at least one of the following:

1. The slope and y-intercept of the line.
2. The slope and coordinates of any point on the line.
3. The graph and we can read the coordinates of the y-intercept and one other point on the line, or the coordinates of two points.
4. The coordinates of any two points on the line.
5. The equation of a parallel line and the coordinates of one point on the line.

EXERCISES 8.10F _____

Find the equations of the lines with the following properties.

44. Slope 1, y-intercept -4.   Equation:_____

45. Slope $-\dfrac{2}{3}$, y-intercept $\dfrac{4}{5}$.   Equation:_____

46. Slope $-\dfrac{1}{4}$, passes through $(0, -1)$.   Equation:_____

47. Slope -5, passes through $(3, -1)$.   Equation:_____

48. Passes through $(1, -1)$ and $(0, -5)$.   Equation:_____

49. Passes through $(-2, 7)$ and $(-1, -4)$.   Equation:_____

50. Parallel to $y = -3x + 5$ and passes through $(-1, 3)$.
Equation:_____

51. Parallel to $y = x - 4$ and passes through $(-3, 3)$.   Equation:_____

Check your answers on page 740. ✓

EXAMPLE 7 Find the equation of the line that goes through points $(3, 5)$ and $(6, 11)$.

$$\text{Slope} =$$
$$\frac{11 - 5}{6 - 3} =$$
$$\frac{6}{3} = 2$$

So $m = 2$, and the equation is $y = 2x + b$.

Now this problem is just like Example 5. Use the coordinates of one of the points to solve for b. Let's use $(3, 5)$, so $x = 3$ and $y = 5$ in $y = 2x + b$.

$$y = 2x + b$$
$$(5) = 2(3) + b$$
$$\begin{array}{r} 5 = 6 + b \\ -6 \quad -6 \\ \hline -1 = b \end{array}$$

The equation is $y = 2x - 1$. ❏

⟹ *Note*: You can use this method if the y-intercept of a line is a fraction, since you will not be able to read the y-coordinate from the graph. In such a case, you still can find the equation if you can read the coordinates of any two points on the line. Then proceed as in Example 7.

EXERCISES 8.10D

Find the equation of the line that goes through each pair of points. (First calculate the slope.)

30. $(4, 7)$ and $(6, 13)$ **31.** $(5, 1)$ and $(2, 10)$

32. $(8, 3)$ and $(10, 5)$ **33.** $(1, 7)$ and $(5, 19)$

34. $(-3, 7)$ and $(1, 11)$ **35.** $(-2, -5)$ and $(-4, 3)$

36. $(-7, 1)$ and $(-2, 16)$ **37.** $(4, 0)$ and $(7, 3)$

Check your answers on page 740. ✓

When You Know a Point and the Equation of Parallel Lines

If you are told that two lines are parallel, then you know that they have the same slope. You can find the equation of a line if you know the equation of a line it is parallel to, and if you know the coordinates of one point on the line.

EXAMPLE 8 Find the equation of the line that is parallel to $y = 2x - 1$, which passes through point $(5, 4)$.

The slope of $y = 2x - 1$ is 2, so the equation we are trying to find has slope 2 also.

Let $m = 2$, and (as in Example 5) since $(5, 4)$ is a point on the line, let $x = 5$ and $y = 4$ and solve for b.

$$(4) = 2(5) + b$$
$$\begin{array}{r} 4 = 10 + b \\ -10 \quad -10 \\ \hline -6 = b \end{array}$$

Since $m = 2$ and $b = -6$, the equation is $y = 2x - 6$. ❏

EXERCISES 8.10E _____

Find the equations of the lines with the following properties.

38. Parallel to $y = 3x + 2$; passes through $(4, 1)$.   Equation:_____

39. Parallel to $y = 5x + 4$; passes through $(1, 4)$.   Equation:_____

40. Parallel to $y = -2x + 7$; passes through $(5, 4)$.   Equation:_____

41. Parallel to $y = -4x + 1$; passes through $(-1, 2)$.   Equation:_____

42. Parallel to $y = -x - 5$; passes through $(3, -7)$.   Equation:_____

43. Parallel to $y = -3x + 4$; passes through $(1, 5)$.   Equation:_____

Check your answers on page 740.   ✓

Review

> *It is possible to find the equation of a line if we know at least one of the following:*
>
> 1. The slope and y-intercept of the line.
> 2. The slope and coordinates of any point on the line.
> 3. The graph and we can read the coordinates of the y-intercept and one other point on the line, or the coordinates of two points.
> 4. The coordinates of any two points on the line.
> 5. The equation of a parallel line and the coordinates of one point on the line.

EXERCISES 8.10F _____

Find the equations of the lines with the following properties.

44. Slope 1, y-intercept -4.   Equation:_____

45. Slope $-\dfrac{2}{3}$, y-intercept $\dfrac{4}{5}$.   Equation:_____

46. Slope $-\dfrac{1}{4}$, passes through $(0, -1)$.   Equation:_____

47. Slope -5, passes through $(3, -1)$.   Equation:_____

48. Passes through $(1, -1)$ and $(0, -5)$.   Equation:_____

49. Passes through $(-2, 7)$ and $(-1, -4)$.   Equation:_____

50. Parallel to $y = -3x + 5$ and passes through $(-1, 3)$.
Equation:_____

51. Parallel to $y = x - 4$ and passes through $(-3, 3)$.   Equation:_____

Check your answers on page 740.   ✓

Glossary

axes

horizontal axis   Horizontal number line of a Cartesian coordinate system.

vertical axis   Vertical number line of a Cartesian coordinate system.

Example: (See **Cartesian coordinate system**, below.)

boundary   Line of the graph of an inequality representing the equation that results from setting the sides of the inequality equal.

Example: The boundary of the graph of $y < x + 3$ is the line represented by $y = x + 3$.

Cartesian coordinate system   Grid determined by two perpendicular number lines. One is called the *horizontal axis*; the other is called the *vertical axis*. The two number lines intersect each other at their zeros in the point called the *origin*.

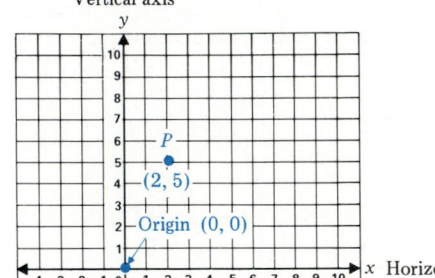

coordinates   Numbers that direct us where to plot a point on a Cartesian coordinate system. The first coordinate represents the horizontal distance from the origin; the second coordinate represents the vertical distance.

Example: The point P (2, 5) is located by moving two spaces to the right and then five spaces up from the origin. It is plotted in the example above.

function (notation)   When a linear equation is solved for one letter in terms of another, the first letter is said to be a function of the other. The notation is $f(x)$ if the letter in the expression is x. That notation is read as "*function of x.*"

Example: The equation $y = 2x + 3$ could be written in functional notation as $f(x) = 2x + 3$. Choose values of x to evaluate $f(x)$.

$$f(1) = 2(\underline{1}) + 3$$
$$f(1) = 2 + 3$$
$$f(1) = 5$$

graph of a linear equation   Linear equations in two letters (x and y, for instance) have infinitely many solutions, each of which can be represented as a point with coordinates (x, y). All the points that represent solutions to the equation form a straight line when they are plotted on a Cartesian coordinate system.

Example: Sketch the graph of $y = 2x - 5$.
(0, −5), (1, −3), and (2, −1) are three solutions.

x	y
0	−5
1	−3
2	−1

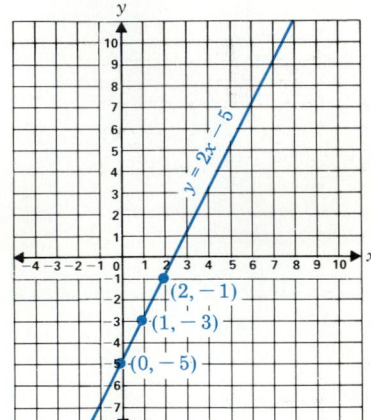

horizontal lines   All linear equations with y as the only letter term have horizontal lines for their graphs because all the points have the same value of y for their second coordinate. The slope of any horizontal line is *zero*.

Example: Below is the sketch of the graph of the equation $y = 3$. Every point with a second coordinate of 3 is a solution: (0, 3), (1, 3), (2, 3), (−5, 3), and (7, 3) all are points on the line that is the graph of $y = 3$.

x	0	1	2	−5	7
y	3	3	3	3	3

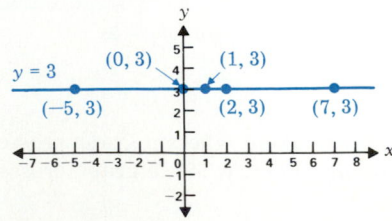

inequality

strict inequality   Statement that two expressions are not equal. The inequality symbol ($>$ or $<$) points toward the smaller expression. The solution is all points on one side of the **dotted boundary line** representing the equation (if both sides were set equal).

Example: $y < 2x - 5$ is a strict inequality.

nonstrict inequality   Statement that two expressions can be equal or not equal. The boundary line of the inequality is part of the solution, so it is graphed **solid**.

Example: $y \leq 2x - 5$ is a nonstrict inequality. Strict and nonstrict inequalities are graphed below.

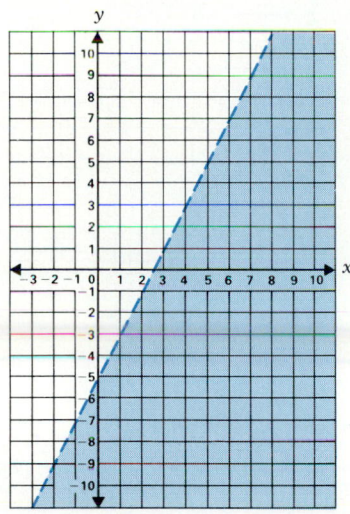

Graph of $y < 2x - 5$

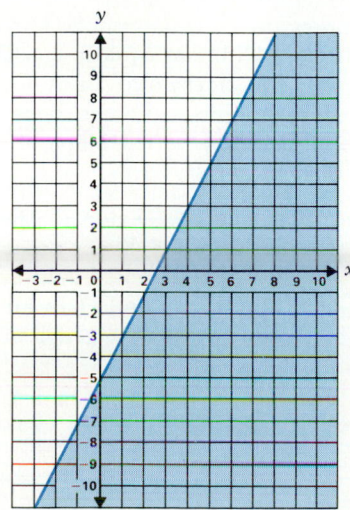

Graph of $y \leq 2x - 5$

intercepts

x-intercept   Point where a graph crosses the x-axis.

y-intercept   Point where a graph crosses the y-axis.

Example:

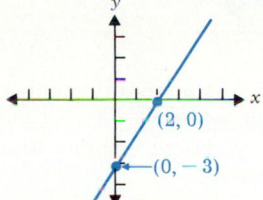

x-intercept is $(2, 0)$.
y-intercept is $(0, -3)$.

line   Graph of all the solutions to a linear equation.

origin   Point with coordinates $(0, 0)$ on a Cartesian coordinate system.

Example: (See **Cartesian coordinate system**).

parallel lines   lines that have the same slope and so when graphed on the same Cartesian coordinate system never cross.

Example: $y = 2x - 5$ and $y = 2x + 3$ are parallel lines (they have the same slope, 2).

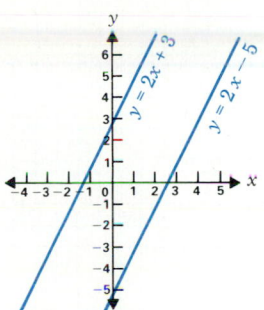

point   Dot located on the Cartesian coordinate system by two numbers, which are its coordinates.

To plot a point: Begin at the origin and move horizontally the distance and direction given by the first coordinate; then proceed vertically the distance and direction given by the second coordinate.

Example: To plot $P(2, 5)$: First proceed horizontally two spaces to the right (positive 2); then go up five spaces (positive 5). (See **Cartesian coordinate system**.)

quadrants   Four sections formed by the two axes of a Cartesian coordinate system.

Quadrant 1 is on the *upper right*; all points in this quadrant have *positive first* and *second coordinates* $(+, +)$

Quadrant 2 is on the *upper left*; all points in this quadrant have *negative first coordinates* and *positive second coordinates* $(-, +)$.

Quadrant 3 is on the *lower left*; all points in this quadrant have *negative first and second coordinates* $(-, -)$.

Quadrant 4 is on the *lower right*; all points in this quadrant have *positive first coordinates* and *negative second coordinates* $(+, -)$.

Example:

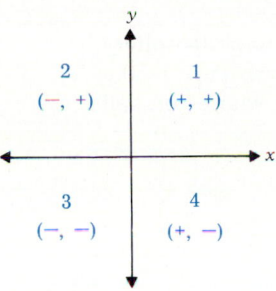

slope   Number that indicates the steepness of a line; the higher the number the steeper the line.

A line with *positive* slope goes *up* as you read from left to right.

A line with *negative* slope goes *down* as you read from left to right.

A line with *zero* slope is a *horizontal* line.

A line with *no* slope is a *vertical* line.

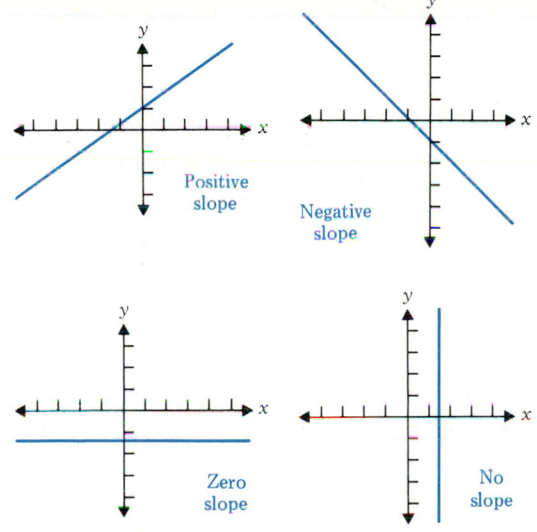

slope-intercept form   Form of a linear equation that is solved for one letter in terms of the other letter; represented as $y = mx + b$, where m and b are numbers.

vertical lines   All linear equations with x as the only letter term have vertical lines for their graphs. Because all the points have the same value of x for their first coordinate. A vertical line is the steepest line possible; there is no largest number, so there can be no number to indicate its steepness or slope.

Example: Below is the sketch of the graph of the equation $x = 5$. Every point with a first coordinate of 5 is a solution: $(5, 0)$, $(5, 1)$, $(5, 3)$, $(5, -2)$, and $(5, -7)$ are all points on the line that is the graph of $x = 5$.

x	5	5	5	5	5
y	0	1	3	-2	-7

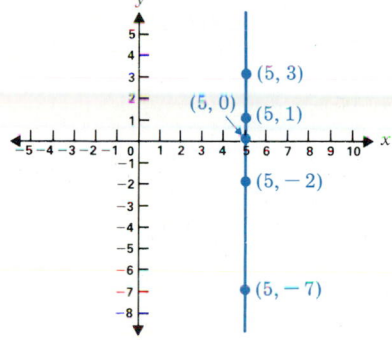

Chapter 8 Review

Do each of the problems. After you are confident that you have done them as accurately as possible, compare your answers with those in the back of the book. If any are wrong, go back to the section where the problem came from (indicated to the left of the problem) and review the section.

Once you understand all the sections, you will be sure to learn the skills solidly and remember how to do them if you practice them a little bit more. Turn to the Supplementary Exercises and do all the problems from any section where you had difficulty on these review exercises.

Answers

8.1

1. On the grid for problem 2, plot the following points:

$A(0, -3)$; $B(-1, 4)$; and $F\left(-2\dfrac{1}{2}, -\dfrac{3}{4}\right)$.

2. What are the coordinates of points C and D plotted below?

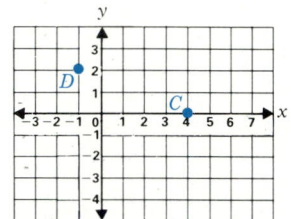

2. $C(\ \ ,\ \ )$ _____

$D(\ \ ,\ \ )$ _____

3. Is $P(2, 3)$ a point on the line graphed below?

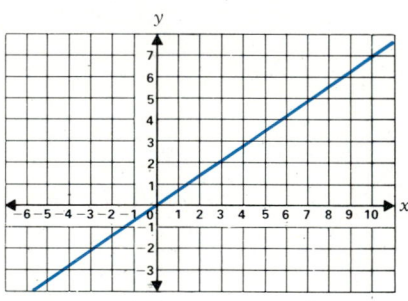

3. _____

8.2   Find the coordinates of three points that are solutions to the equations. Plot the points and sketch the graph of each equation.

4. $y = -3x + 1$

x			
y			

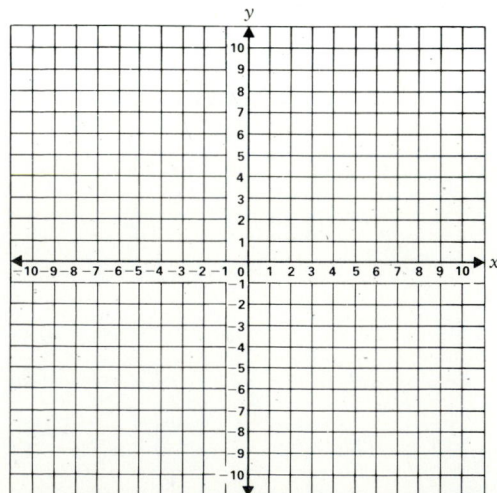

5. $y = \left(\dfrac{1}{2}\right)x - 5$

x			
y			

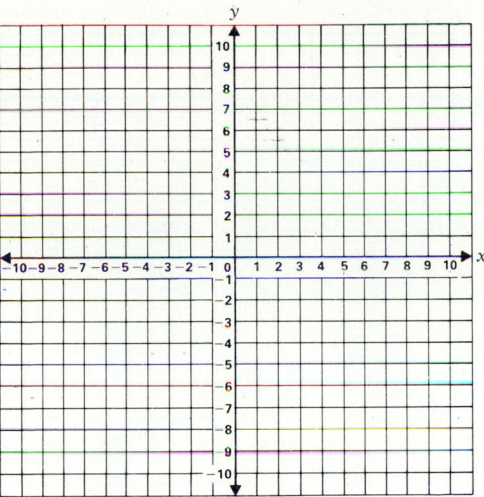

6. $y = \left(\dfrac{2}{3}\right)x - \left(\dfrac{1}{4}\right)$

x			
y			

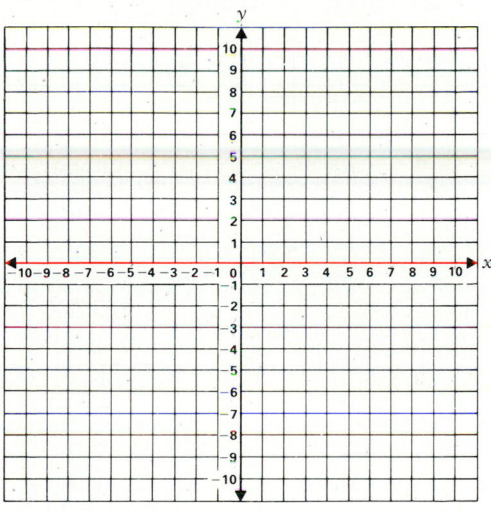

8.3   Sketch the graph of each equation.

7. $y = -4$

x			
y			

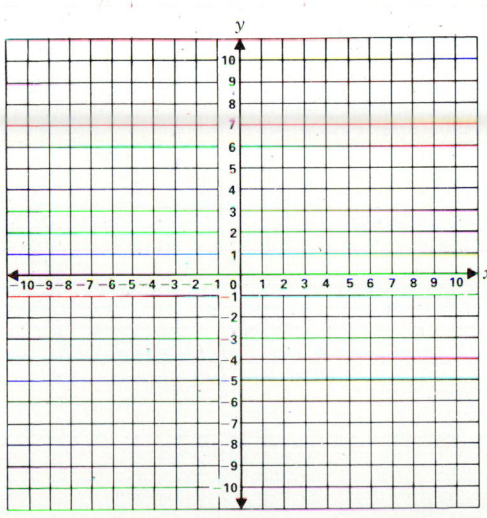

Continued

Chapter 8 Review *(Cont.)*

Answers

8. $x = 3$

x		
y		

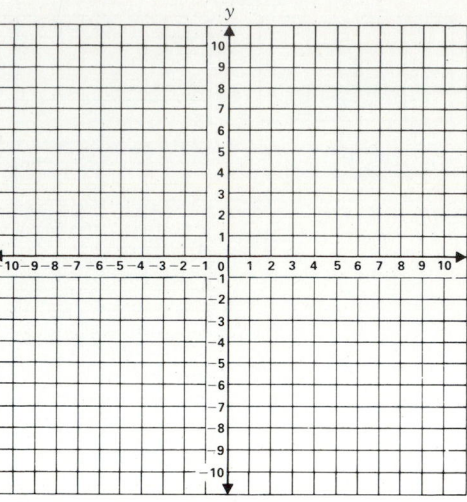

8.4   **9.** Solve for y; find the coordinates of three points; plot and graph: $5y - 10x = 20$ **9.** _____

x		
y		

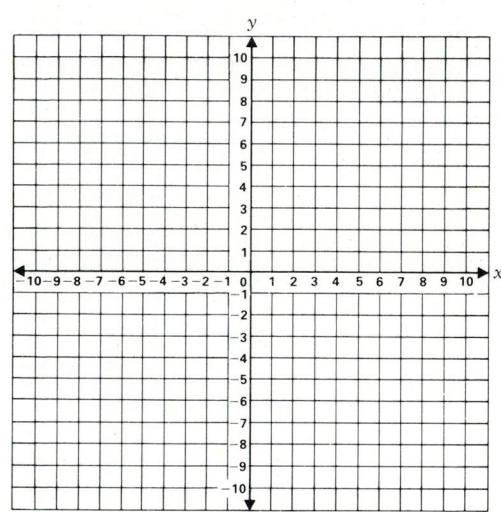

10. Find the coordinates of three points; plot and graph: $f(x) = 3x - 5$

x		
y		

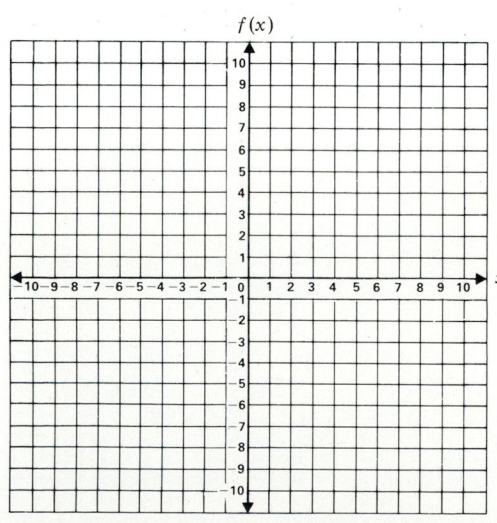

8.5 Graph the boundary line; test a point to determine which side of the line contains the points that are the solution set and then shade that side.

11. $y < x - 4$

x		
y		

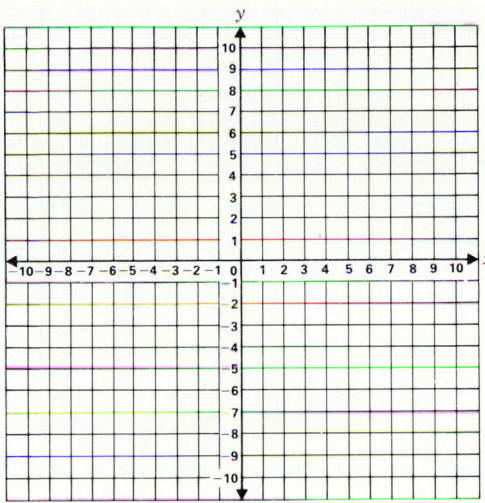

12. $y \geq 2x + 3$

x		
y		

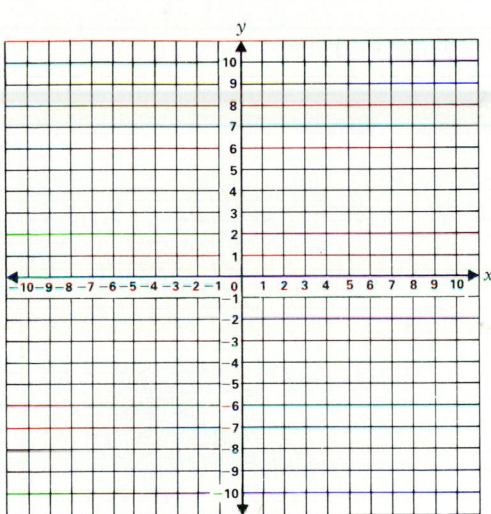

13. $y \geq 3$

x		
y		

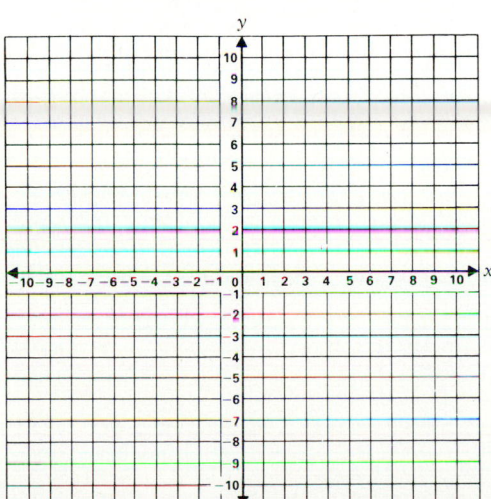

Continued

Chapter 8 Review (*Cont.*)

Answers

14. $x < 1$

x		
y		

8.6   **15.** Read the coordinates of the x- and y-intercepts.

15. x-int. _____

y-int. _____

16. Find the coordinates of the x- and y-intercepts of $y = 3x - 12$.

16. x-int. _____

y-int. _____

8.7   Find the slope of each line.

17. Line passing through points $(2, 5)$ and $(1, 1)$.

17. _____

18. Line passing through points $(-3, 4)$ and $(-1, 10)$.

18. _____

19. $y = -7x - 8$

19. _____

20. $y = 8$

20. _____

21. $x = -6$

21. _____

22. Graph the line with slope $\dfrac{3}{4}$ that passes through (0, 1).

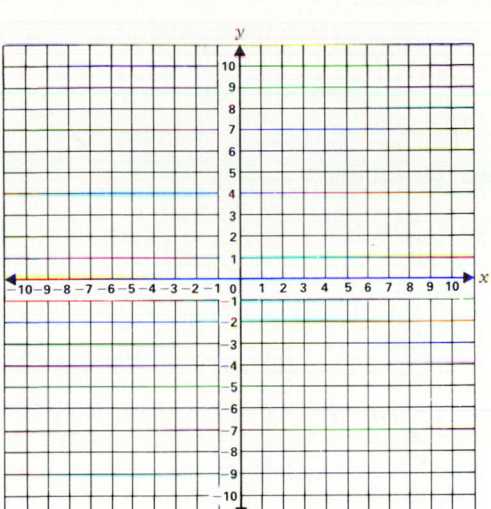

8.8   Find the slope and y-intercept of each equation.

23. $y = 4x - 3$

24. $3y - 12 = 15x$

25. Graph $y = 4x - 3$, using its slope and y-intercept.

23. _____

24. _____

25. _____

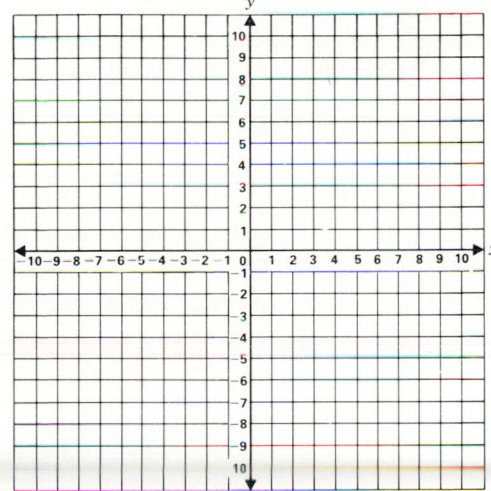

8.9   Determine whether each pair of equations represents parallel lines.

26. $y - 5x = 7$ and $y = -5x + 14$

27. $4y - x = 5$ and $y - 6 = \left(\dfrac{1}{4}\right)x$

8.10   In problems 28–33, find the equations of the lines with the following properties.

28. Has slope -7 and y-intercept 1.

26. _____

27. _____

28. _____

Continued

Chapter 8 Review (*Cont.*)

29. Has slope -1 and goes through $(0, -3)$.

29. _____

30. Has slope 4 and goes through $(1, -3)$.

30. _____

31. Goes through $(2, 5)$ and $(4, 13)$.

31. _____

32. Is parallel to $y = -3x + 5$ and goes through $(1, -5)$.

32. _____

33. Has the line sketched below:

33. _____

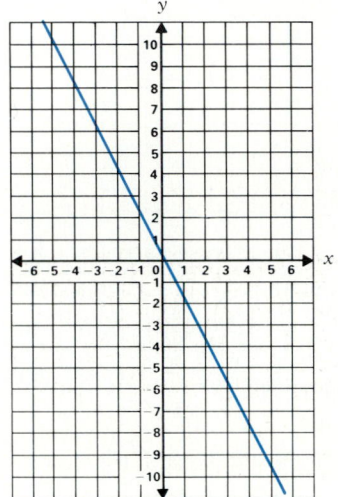

Check your answers on page 740.

Supplementary Exercises

Do all the problems in every section that involves a skill in which you now lack complete mastery. With a little more practice you will achieve that sense of really understanding the topics, and you will remember how to do these problems.

Answers

8.1

1. On graph paper plot the following points:
   $A(-2, 5)$, $B(6, 0)$, and $F\left(-\dfrac{1}{3}, 3\dfrac{3}{4}\right)$.

2. What are the coordinates of points C and D plotted below?

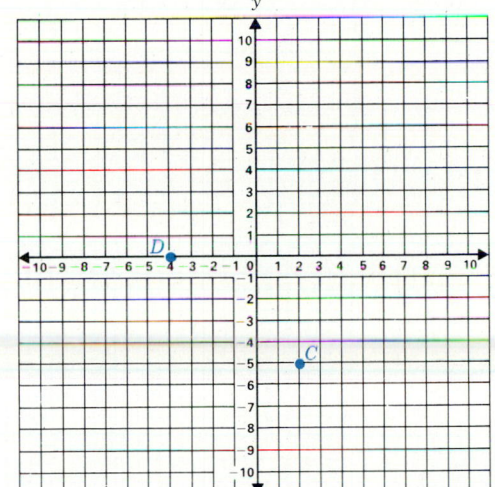

2. $C(\ \ , \ \ )$

$D(\ \ , \ \ )$

3. Is $P(-1, 2)$ a point on the line graphed below?

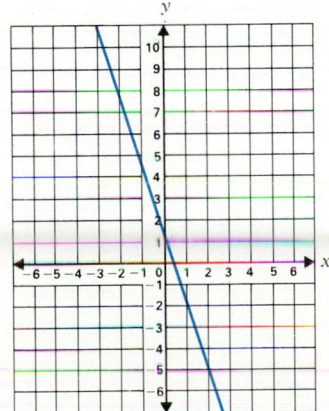

3. _____

Continued

Supplementary Exercises (*Cont.*)

8.2   Find the coordinates of three points that are solutions to the equations. Plot the points and sketch the graph of each equation.

4. $y = -2x - 1$

x			
y			

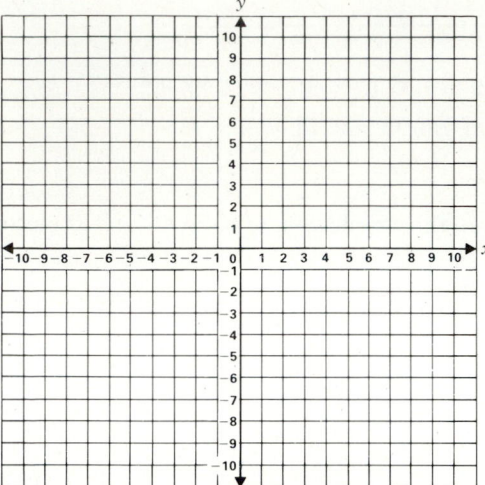

5. $y = \left(\dfrac{1}{3}\right)x - 2$

x			
y			

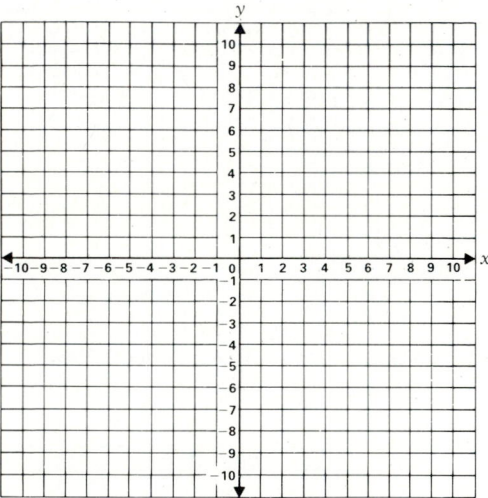

6. $y = \left(\dfrac{1}{4}\right)x + \left(\dfrac{2}{3}\right)$

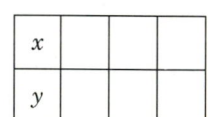

x			
y			

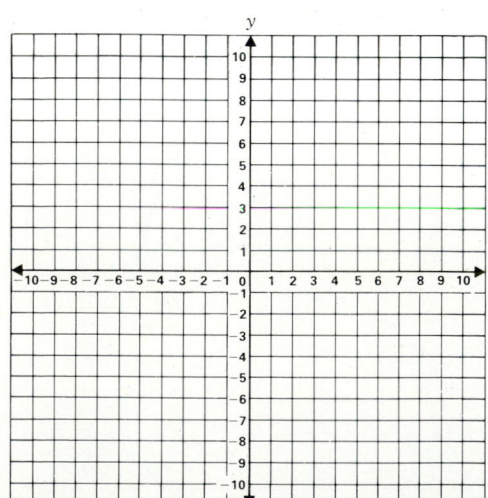

8.3   Sketch the graph of each equation.

7. $y = 1$

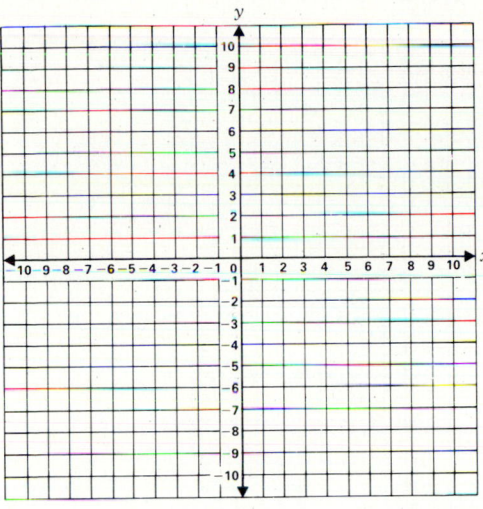

8. $x = -7$

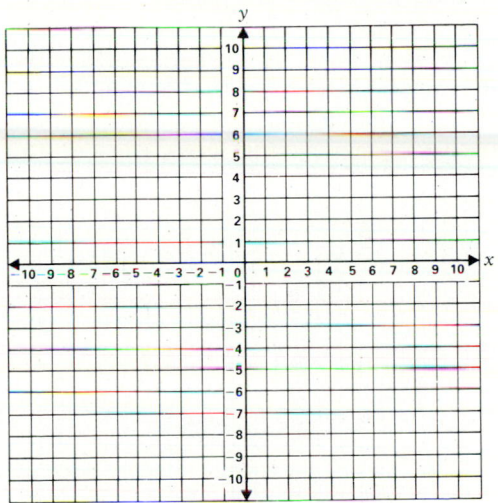

8.4   9. Solve for y; find the coordinates of three points; plot and graph: $6y - 18x = 30$ 9. _____

x		
y		

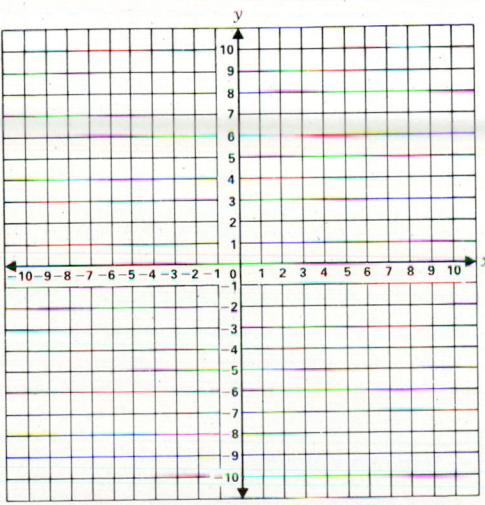

Continued

Supplementary Exercises (*Cont.*)

10. Find the coordinates of three points; plot and graph: $f(x) = -x + 4$

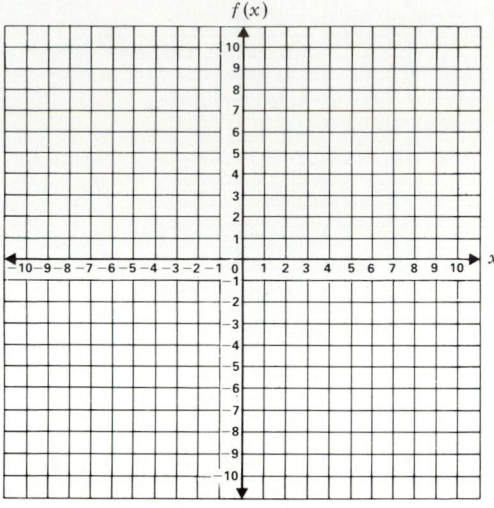

8.5 Graph the boundary line; test a point to determine which side of the line contains the points that are in the solution set and shade that side.

11. $y > x - 2$

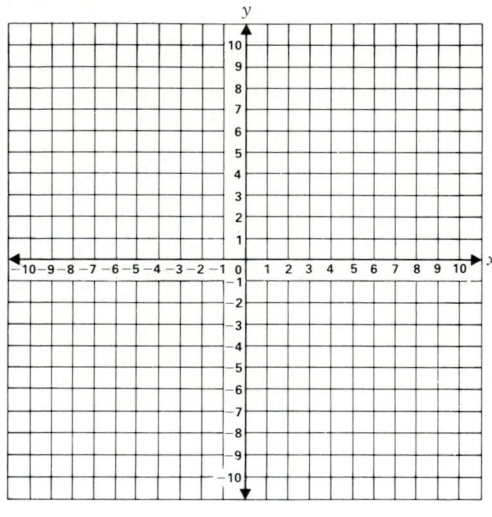

12. $y \le 3x + 5$

x			
y			

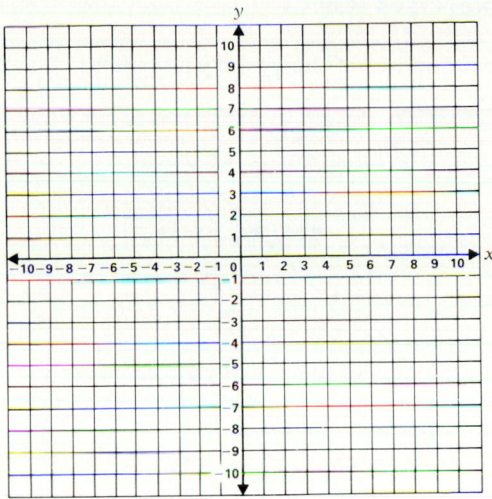

13. $x \ge 6$

x			
y			

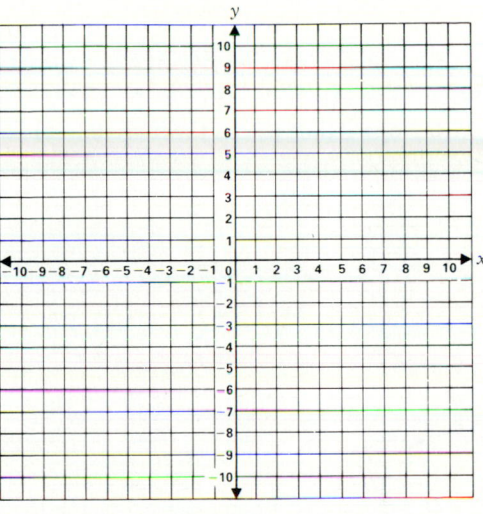

14. $x < -4$

x			
y			

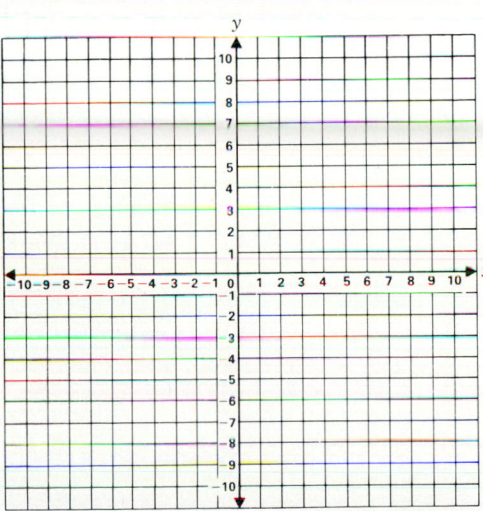

Continued

Supplementary Exercises *(Cont.)*

8.6   **15.** Read the coordinates of the *x*- and *y*-intercepts.

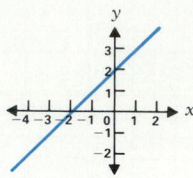

16. Find the coordinates of the *x*- and *y*-intercepts of $y = 5x - 35$.

8.7   Find the slope of each line.

17. Line passing through points $(1, 6)$ and $(3, 2)$.

18. Line passing through points $(-5, 3)$ and $(-2, 9)$.

19. $y = -11x + 5$

20. $y = -1$ **21.** $x = 7$

22. Graph the line with slope $\dfrac{2}{3}$ that passes through $(0, -2)$.

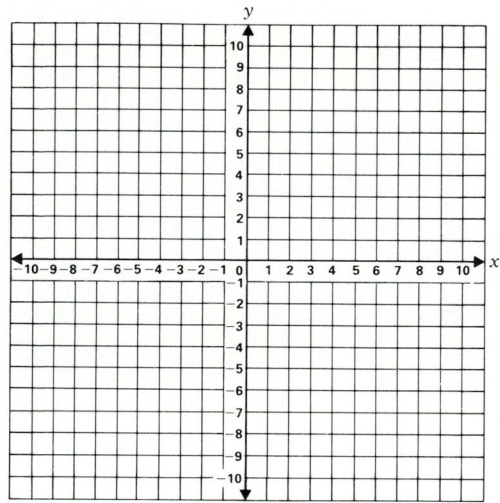

8.8   Find the slope and *y*-intercept of each equation.

23. $y = -7x + 2$ **24.** $2y - 10 = 8x$

15. *x*-int. _____

    y-int. _____

16. *x*-int. _____

    y-int. _____

17. _____

18. _____

19. _____

20. _____

21. _____

23. Slope _____

    y-int. _____

24. Slope _____

    y-int. _____

25. Graph $y = 3x - 5$, using its slope and y-intercept.

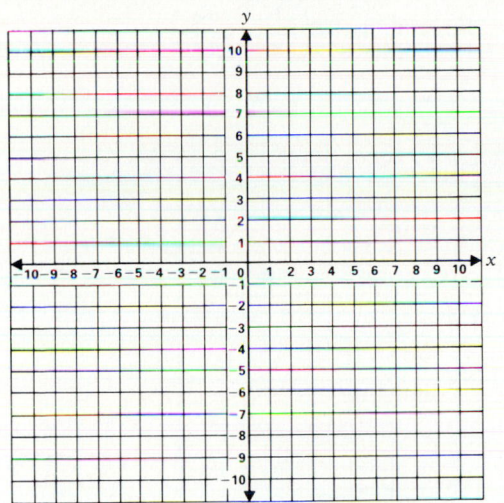

Answers

25. Slope _____

     y-int. _____

8.9   Determine whether each pair of equations represents parallel lines.

26. $y - 2x = 3$ and $y = -2x + 3$

26. _____

27. $3y - 2x = 1$ and $y + 5 = \left(\dfrac{2}{3}\right)x$

27. _____

28. _____

8.10   In problems 28–33, find the equations of the lines with the following properties.

28. Has slope -4 and y-intercept 5.

29. _____

29. Has slope 1 and goes through $(0, -4)$.

30. _____

30. Has slope -1 and goes through $(2, 6)$.

31. _____

31. Goes through $(4, 5)$ and $(1, 20)$.

32. _____

32. Is parallel to $y = -6x + 1$ and goes through $(-1, 4)$.

33. _____

33. Has the line sketched below:

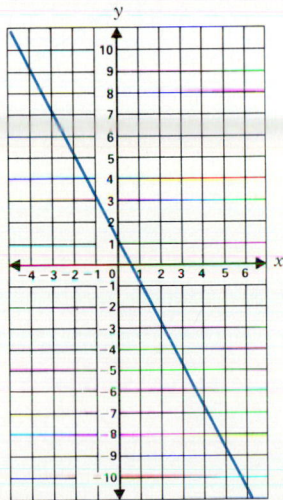

Check your answers on page 741. ✓

Chapter 8 Test

1. Plot these points on the Cartesian coordinate system below:

   $A(3, 0)$; $B(-2, 7)$; $C(4, -2)$; $D(0, -6)$; and $E\left(3\frac{1}{2}, -5\frac{3}{4}\right)$.

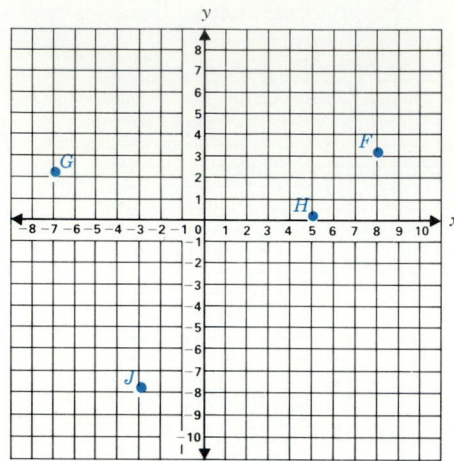

2. What are the coordinates of points F, G, H, and J plotted above?

Answer yes or no.

3. Is $P(-3, 4)$ on the line graphed below?

4. Is $Q(5, -3)$ on the line graphed below?

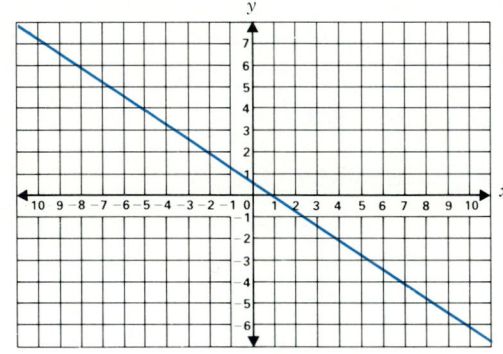

Answers

1. _____

2. F (,)

   G (,)

   H (,)

   J (,)

3. _____

4. _____

Find the coordinates of three points that are solutions to the equations. Plot the points and sketch the graph of each equation.

5. $y = 4x - 5$

x			
y			

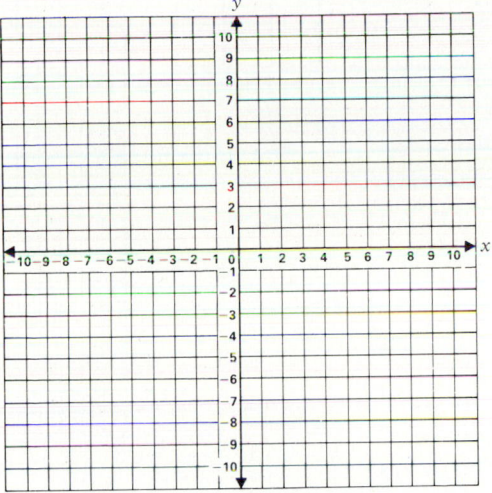

6. $y = -x + 4$

x			
y			

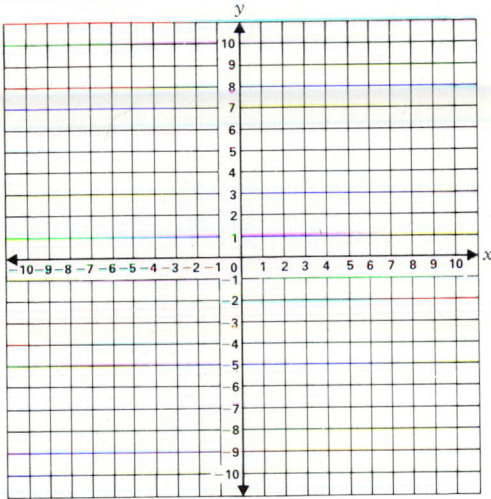

7. $y = \left(\dfrac{3}{4}\right)x - 1$

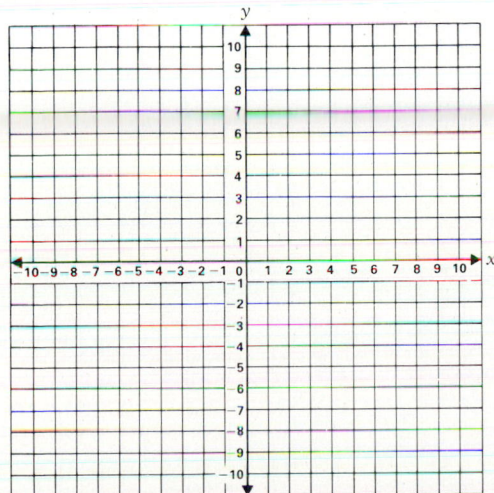

Continued

Chapter 8 Test (*Cont.*)

8. $y = \left(-\dfrac{1}{3}\right)x + \dfrac{2}{5}$

x			
y			

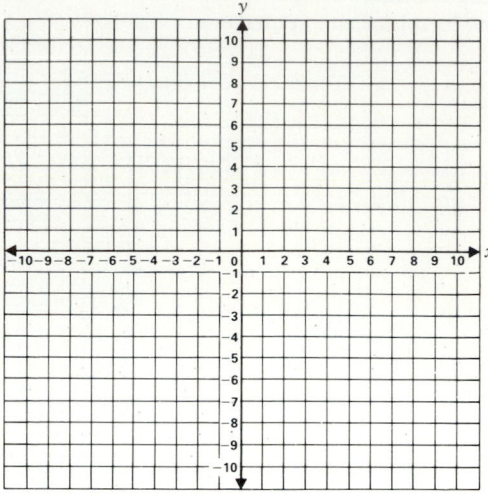

9. $y = -7$

x			
y			

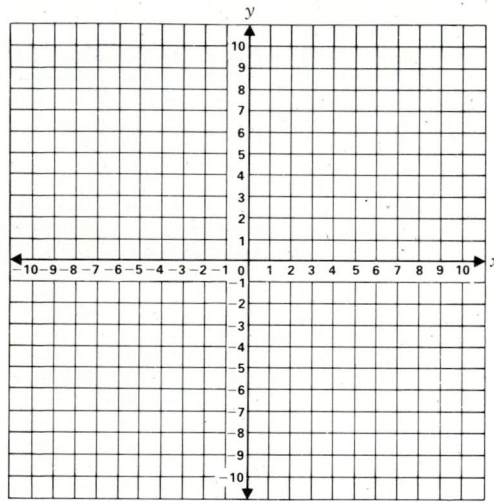

10. $x = 6$

x			
y			

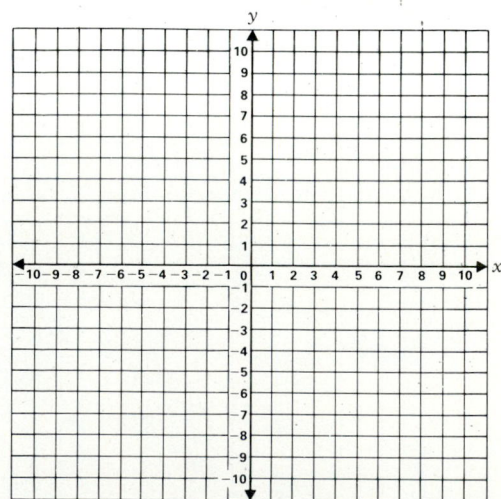

11. $B = -4A + 3$

A			
B			

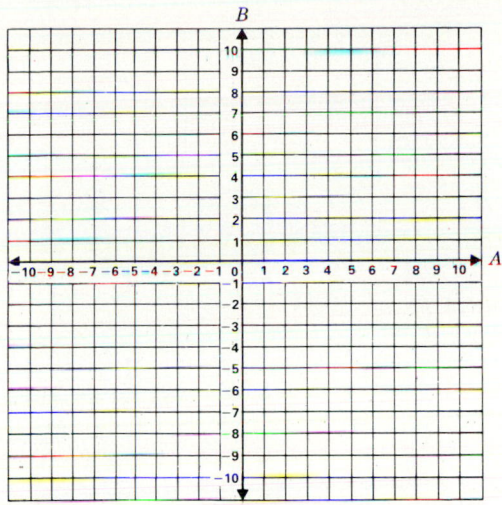

12. $f(x) = \left(-\dfrac{3}{5}\right)x$

x			
f(x)			

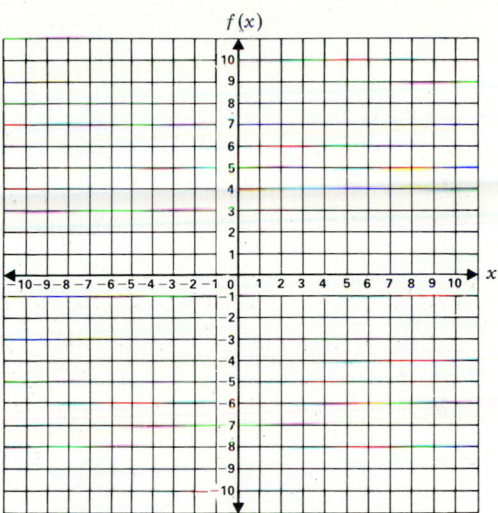

Solve for y.

13. $y - 8x = 7$

14. $7y + 10 = 6x$

Sketch the graph of each equation.

15. $y > 3x - 5$

x			
y			

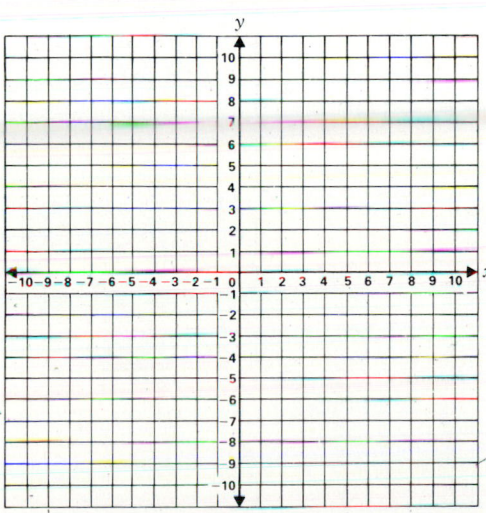

Continued

Chapter 8 Test (*Cont.*)

16. $y \le -2x + 3$

x			
y			

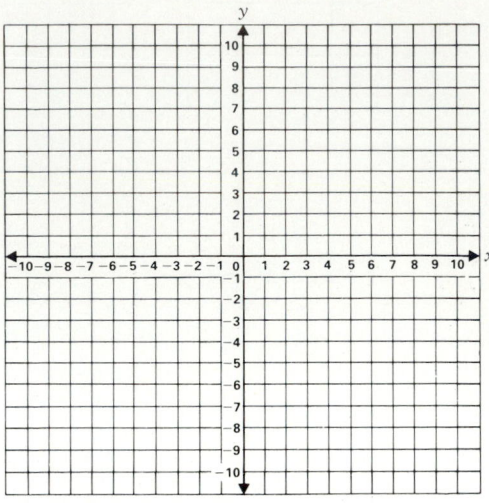

17. $x \ge -5$

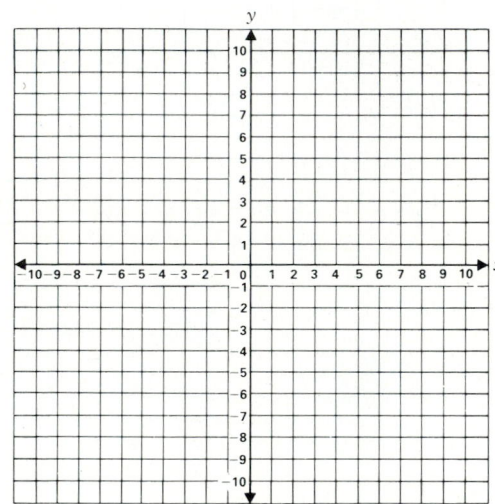

18. $y < 8$

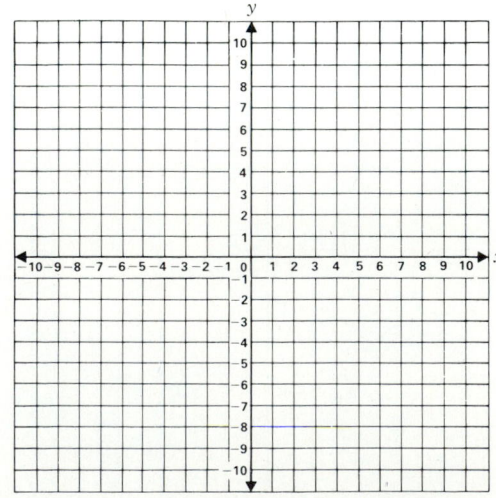

Cumulative Review (*Cont.*)

Sketch the graph of each equation.

23. $y = -2x - 2$

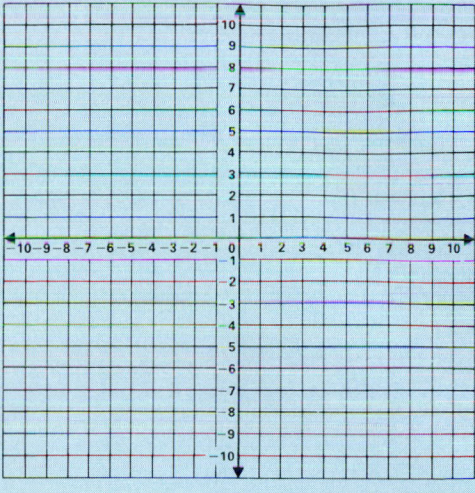

24. $3y - 3 = 12x$

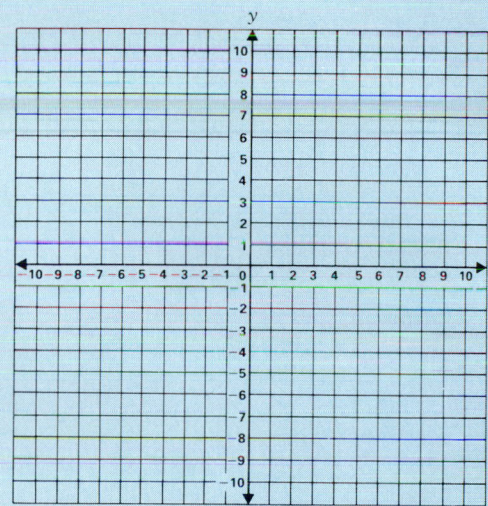

25. $f(x) \le 2x + 1$

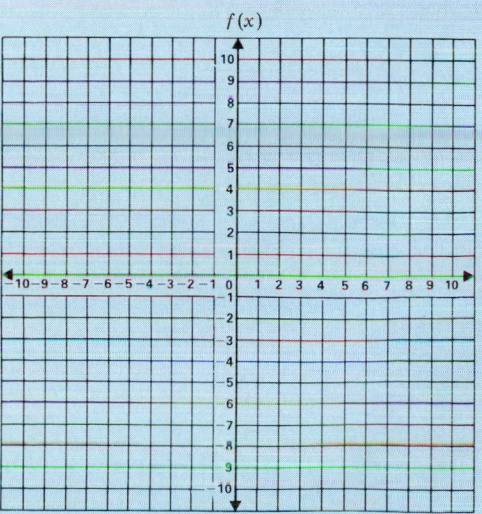

Check your answers on page 743.

---

19. Read the coordinates of the x- and y-intercepts.

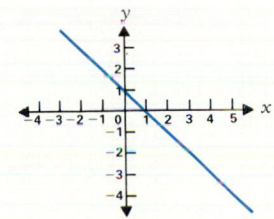

Find the coordinates of the slope and x- and y-intercepts of each equation.

20. $y = 4x - 12$ **21.** $3y - 6 = 15x$

22. $x = -3$ **23.** $y = 2$

In Problems 24–27, find the slope of the line passing through these points.

24. $(2, 7)$ and $(5, 13)$ **25.** $(-6, 3)$ and $(0, -3)$

26. $(3, 8)$ and $(7, 11)$ **27.** $(-6, 1)$ and $(-2, -5)$

28. Graph the line with slope $\dfrac{1}{2}$ that passes through $(-3, -5)$.

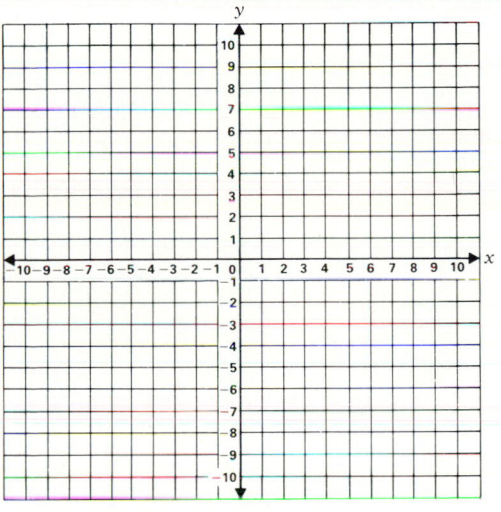

29. Are the lines $y = 3x - 7$ and $4y + 12x = -7$ parallel?

In problems 30–35, find the equations of the lines with the following properties.

30. Has slope 3 and y-intercept -6.

31. Has slope -5 and goes through $(0, 0)$.

32. Has slope 2 and goes through $(-1, -4)$.

33. Goes through $(-2, -1)$ and $(2, 7)$.

34. Is parallel to $y = 4x - 5$ and goes through $(1, 3)$.

Continued

Answers

19. x-int.

y-int.

20. Slope

y-int.

x-int.

21. Slope

y-int.

x-int.

22. Slope

y-int.

x-int.

23. Slope

y-int.

x-int.

24.

25.

26.

27.

29.

30.

31.

32.

33.

34.

Chapter 8 Test (*Cont.*)

Answers

35. Has the line graphed below:

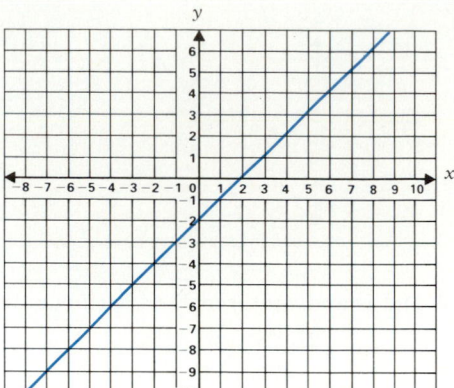

35. _____

Check your answers on page 742. ✓

Cumulative Review
Chapters 1–8

Answers

Simplify.

1. $8 - 2(-4 + 5 - 1)(8 - 2 - 3) + 5(-1 - 2 - 3) =$

2. $\dfrac{1}{2} - \dfrac{1}{3} + \dfrac{5}{6} - \dfrac{2}{3} + \dfrac{1}{2} - \dfrac{1}{6} =$

3. $\dfrac{3}{10} \div \dfrac{18}{15} =$

4. $0.6 - 0.34 + 4.8 - 17.3 - 0.009 + 23.04 =$

5. $y^5 - 4y^3 - 2y + 8y^5 - 6y^3 + 7y + y^3 =$

6. $4.1a - 3.5a^2 + 0.3a^3 - 9.2a^2 - 8.3a + 5.8a^3 =$

7. $4cb - 5ac + 7cb + ca - 6bc + 2ab - 3ca - 5ba =$

8. $(9x - 3y + w) - (5y - x - 8w) =$

9. $x(x^2 - 5x + 3) - 3x(-x^2 + 7x - 4) =$

Evaluate.

10. 14.7% of 3,400.

11. $x^2 - 5xy + 3yw$ if $x = -3$, $y = -1$, and $w = 4$.

12. The area of a triangle with base 4 ft and height 7 ft.

Solve for the letter and check by substitution.

13. $4y - 5 + 1 - y + 2y = 6$

14. $12w - 7 = 15w + 8$

15. $6a - 3(a - 1) + 2a - 5 = 3(3a - 2) + a - 1$

16. $a : 7 = 12 : 28$

17. 8.3% of N is \$464.80.

18. $3a - 5bc = d$; solve for a.

Solve and graph on a number line.

19. $4k + 3 < 19$

20. $0 \le 2w - 6 < 10$

21. The ratio of bills to letters in my daily mail is usually 5 to 2. By the time I have received eight letters, how many bills do I have?

22. Factor: $3xy - 15yw + 21yz$.

1. _____

2. _____

3. _____

4. _____

5. _____

6. _____

7. _____

8. _____

9. _____

10. _____

11. _____

12. _____

13. _____

14. _____

15. _____

16. _____

17. _____

18. _____

19. _____

20. _____

21. _____

22. _____

Continued

Answers

19. Read the coordinates of the x- and y-intercepts.

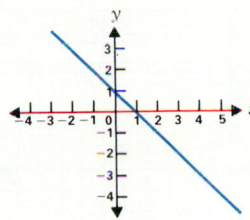

19. x-int. _____

_____ y-int. _____

20. Slope _____

_____ y-int. _____

_____ x-int. _____

Find the coordinates of the slope and x- and y-intercepts of each equation.

20. $y = 4x - 12$

21. $3y - 6 = 15x$

21. Slope _____

_____ y-int. _____

22. $x = -3$

23. $y = 2$

_____ x-int. _____

In Problems 24–27, find the slope of the line passing through these points.

24. $(2, 7)$ and $(5, 13)$

25. $(-6, 3)$ and $(0, -3)$

22. Slope _____

26. $(3, 8)$ and $(7, 11)$

27. $(-6, 1)$ and $(-2, -5)$

_____ y-int. _____

28. Graph the line with slope $\dfrac{1}{2}$ that passes through $(-3, -5)$.

_____ x-int. _____

23. Slope _____

_____ y-int. _____

_____ x-int. _____

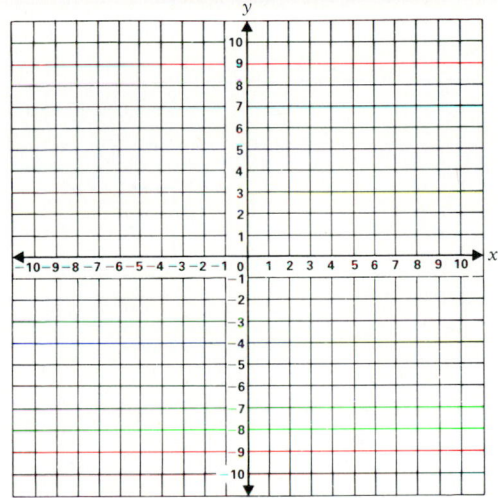

24. _____

25. _____

26. _____

27. _____

29. Are the lines $y = 3x - 7$ and $4y + 12x = -7$ parallel?

29. _____

In problems 30–35, find the equations of the lines with the following properties.

30. Has slope 3 and y-intercept -6.

30. _____

31. Has slope -5 and goes through $(0, 0)$.

31. _____

32. Has slope 2 and goes through $(-1, -4)$.

32. _____

33. Goes through $(-2, -1)$ and $(2, 7)$.

33. _____

34. Is parallel to $y = 4x - 5$ and goes through $(1, 3)$.

34. _____

Continued

Chapter 8 Test (*Cont.*)   Answers

35. Has the line graphed below:

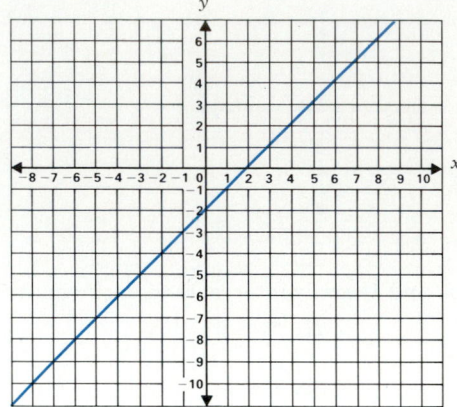

35. _____

Check your answers on page 742.

Cumulative Review
Chapters 1–8

Answers

Simplify.

1. $8 - 2(-4 + 5 - 1)(8 - 2 - 3) + 5(-1 - 2 - 3) =$

2. $\dfrac{1}{2} - \dfrac{1}{3} + \dfrac{5}{6} - \dfrac{2}{3} + \dfrac{1}{2} - \dfrac{1}{6} =$

3. $\dfrac{3}{10} \div \dfrac{18}{15} =$

4. $0.6 - 0.34 + 4.8 - 17.3 - 0.009 + 23.04 =$

5. $y^5 - 4y^3 - 2y + 8y^5 - 6y^3 + 7y + y^3 =$

6. $4.1a - 3.5a^2 + 0.3a^3 - 9.2a^2 - 8.3a + 5.8a^3 =$

7. $4cb - 5ac + 7cb + ca - 6bc + 2ab - 3ca - 5ba =$

8. $(9x - 3y + w) - (5y - x - 8w) =$

9. $x(x^2 - 5x + 3) - 3x(-x^2 + 7x - 4) =$

Evaluate.

10. 14.7% of $3,400$.

11. $x^2 - 5xy + 3yw$ if $x = -3$, $y = -1$, and $w = 4$.

12. The area of a triangle with base 4 ft and height 7 ft.

Solve for the letter and check by substitution.

13. $4y - 5 + 1 - y + 2y = 6$

14. $12w - 7 = 15w + 8$

15. $6a - 3(a - 1) + 2a - 5 = 3(3a - 2) + a - 1$

16. $a{:}7 = 12{:}28$

17. 8.3% of N is $\$464.80$.

18. $3a - 5bc = d$; solve for a.

Solve and graph on a number line.

19. $4k + 3 < 19$

20. $0 \le 2w - 6 < 10$

21. The ratio of bills to letters in my daily mail is usually 5 to 2. By the time I have received eight letters, how many bills do I have?

22. Factor: $3xy - 15yw + 21yz$.

1. _____

2. _____

3. _____

4. _____

5. _____

6. _____

7. _____

8. _____

9. _____

10. _____

11. _____

12. _____

13. _____

14. _____

15. _____

16. _____

17. _____

18. _____

19. _____

20. _____

21. _____

22. _____

Continued

Cumulative Review *(Cont.)*

Sketch the graph of each equation.

23. $y = -2x - 2$

x			
y			

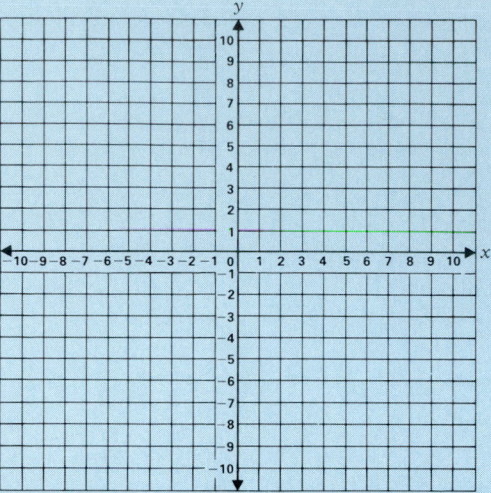

24. $3y - 3 = 12x$

x			
y			

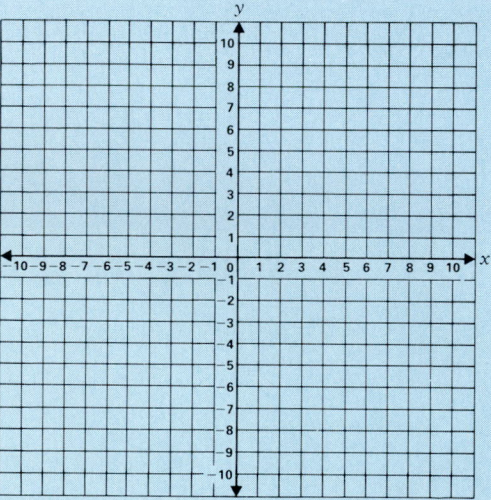

25. $f(x) \leq 2x + 1$

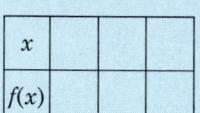

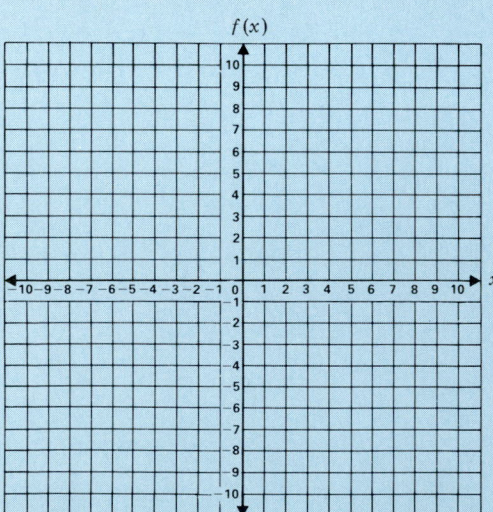

Check your answers on page 743.

9 Introduction to Word Problems

9.1 Using Algebraic Expressions and Tables to Set Up Word Problems

Translating Data into Algebraic Expressions

If algebra is like a language, as claimed in Chapter 5, we should be able to use it to discuss something. In fact, algebraic expressions are used all the time to talk about real-life situations. In business, in science, in construction, in engineering, and in social science research, for example, algebraic equations are used to state what we know and what we want to find out. By solving the equations with the tools we learned in Chapter 6, we can find the answers we need.

To use algebra to solve real-life problems, we must be able to represent the elements of these problems by algebraic expressions. That means learning to translate back and forth between problems stated in words and problems stated in algebra. To keep the expressions and equations *simple*, I will not include many of the most useful but complex applications, such as equations that can tell us when a space probe will intercept the orbit of Jupiter or how many subscriptions a new magazine needs to break even. Some of the first word problems in this chapter are more like riddles than like real day-to-day problems. But they provide the tools for many real-life applications that you might encounter. By the end of the chapter you will be solving problems that you can apply to your own life.

This chapter will teach you easy ways to represent real-life situations by algebraic expressions and equations. Without algebra, many problems would have to be solved through trial and error. That is okay if the first trial works and there is no error. But usually there are numerous trials, and so it takes longer to get the answers than it would if we used algebra and represented unknown quantities by algebraic expressions instead.

The key concept is that some unknown quantity in every problem should be represented by a single letter—often that letter is x. I will use x to represent unknown numbers in this section. We have already had some experience with this representation in Chapter 7. Recall the problems where there was one unknown quantity, which we represented by a letter in an equation that was a direct translation of the situation. So relax; you are already good at this skill.

Here we will deal with situations that involve more than one unknown quantity. It will be necessary to find some way to represent each of them, but we only know how to solve equations that involve *one* letter. So call one unknown quantity x (or any other letter that you prefer), and represent additional unknown quantities by algebraic expressions that include that same letter. It is usually easiest to let x represent the smallest quantity or the one that the others are compared to.

The first step to setting up word problems is to become comfortable in using algebraic expressions to represent unknown quantities.

Call one of the unknown quantities x (or any letter you prefer), and represent each of the other unknown quantities by algebraic expressions that include the letter x. It is usually easiest to let x represent the smallest quantity or the one that the others are compared to.

EXAMPLE 1   If I know there are two unknown quantities, and one is 3 larger than the other, then I can represent the smaller term by x and the larger one by $x + 3$.   ❏

EXAMPLE 2   If one unknown quantity is twice as large as the other, then I can represent one term by x and the other by $2x$.   ❏

EXAMPLE 3   If one unknown quantity is 1 less than twice the other, then I can represent one by x and the other by $2x - 1$.   ❏

EXAMPLE 4   One-half of 5 less 3 times x is $\frac{1}{2}(3x - 5)$.   ❏

EXERCISES 9.1A

Write a simple algebraic expression that involves x to represent the number that is:

EXAMPLE   21 less than x.   **Ex.** $x - 21$

1. 5 more than x.   1. _____

2. 4 less than x.   2. _____

3. 7 more than x.   3. _____

4. 11 less than x.   4. _____

5. 123 less than x.   5. _____

6. 86 more than x.   6. _____

7. 5 times as great as x.   7. _____

8. 16 times as great as x.   8. _____

9. Half as big as x.   9. _____

10. One-third as big as x.   10. _____

11. One-fifth as big as x.   11. _____

12. One-fifteenth as big as x.   12. _____

13. 3 greater than twice x.   13. _____

14. 5 less than twice x.   14. _____

15. 18 greater than 5 times as big as x.   15. _____

16. 1 less than 4 times as big as x.   16. _____

17. One-half of 1 more than x. 17. _____

18. One-third of 7 less than x. 18. _____

19. One-fifth of 2 more than 3 times x. 19. _____

20. One-eighth of 5 less than twice x. 20. _____

Check your answers on page 744. ✓

Using Tables to Organize Data

Algebra students have often felt that "doing word problems" was the most difficult part of learning algebra. Yet no topic is more important if you want to use your algebra skills in further study or work. Fortunately, word problems do not have to be so hard to do. By the end of this chapter, you will easily be solving some pretty complicated ones.

The part that seems hard is determining *what equation* must be solved. In all other aspects of algebra you are given the equation to solve, but with word problems *you* have to set up the equation. This section will introduce the procedure of recording data in tables that will make it easy to see what the equations should be.

One kind of situation that is readily translated into algebraic phrases involves problems about people who are of different ages. These problems are like the questions children often ask: "Linda, when will you be exactly twice as old as I am?" The pieces of information we have are my age and the age of my young friend Gregory. I am 39 and Gregory is 7. Now—take a minute to try and solve the problem without algebra.

Maybe you solved it by trial and error. Here is how to do it by using algebra to represent the facts: if we represent Gregory's age by x, we will have to find an expression (using x) for my age.

EXAMPLE 5 I am 32 years older than Gregory $(39 - 7 = 32)$, so my age is 32 more than his. Let x represent Gregory's age when I am twice as old as he is. My age will be $x + 32$.

No matter what else may happen to us, this relationship will never change. Whatever Gregory's age is (x), my age will always be 32 years more $(x + 32)$.

Now he wants to know about the time when

[Linda's age] = 2 times [Gregory's age].

To find that, we substitute the algebraic expressions that represent our ages and get

$[x + 32] = 2[x]$, which we can solve for x.

$$
\begin{array}{r}
x + 32 = 2x \\
-x \qquad -x \\
\hline
32 = x
\end{array}
$$

x represents Gregory's age, so he will be 32.

$x + 32$ represents my age, so I will be $(32) + 32 = 64$.

Sure enough, when he is 32, I will be 64. I will be twice as old as he is. ❑

To give you practice representing unknown quantities by algebraic expressions, I will give you data about the ages of Fred, Marge, and myself. You do not need to set up any equations yet; just become good at translating facts into algebraic terms.

If we represent *my* age by x, find expressions (using x) for the ages of Fred and Marge.

EXAMPLE 6   Fred is 7 years older than I am; so his age is 7 more than mine.

Let's represent Fred's age as $x + 7$.

Marge is 4 years younger than I am; so her age is 4 less than mine.

Let's represent Marge's age as $x - 4$.   ❏

Some problems involve dealing with people's ages at another time. Once you know how to express someone's age now, how do you represent what that person's age was 15 years ago? You subtract 15 from his or her present age.

EXAMPLE 7   If my age now is x, 15 years ago my age was $x - 15$.

If Fred's age now is $x + 7$, 15 years ago his age was $(x + 7) - 15$, which simplifies to $x - 8$.

If Marge's age now is $x - 4$, 15 years ago her age was $(x - 4) - 15$, which simplifies to $x - 19$.   ❏

Once you know how to express someone's age now, what must you do to express what that person's age will be in 10 years? You add 10 to his or her present age.

EXAMPLE 8   If my age now is x, in 10 years my age will be $x + 10$.

If Fred's age now is $x + 7$, in 10 years his age will be $(x + 7) + 10$, which simplifies to $x + 17$.

If Marge's age now is $x - 4$, in 10 years her age will be $(x - 4) + 10$, which simplifies to $x + 6$.   ❏

When a problem involves more than one unknown quantity, it is helpful to use a table to keep track of what algebraic expressions you are using to represent each of them.

Table I in the following exercises is used to record the expressions for Fred's, Marge's, and my present ages. The first line is filled in with the expressions for each of our ages that were used in Example 6.

Table II in the exercises is used to record the expressions for our ages 15 years ago. The first line is filled in with the expressions for each of our ages that were used in Example 7.

Table III is used to record the expressions for our ages in 10 years. The first line is filled in with the expressions for each of our ages that were used in Example 8.

EXERCISES 9.1B

Fill in each line of the tables to correspond with the information about our ages provided in statements 21–25. Simplify each expression as much as possible, as in the examples.

Let x represent my age now in each exercise, and find the expressions for the ages of Fred and Marge. For each line, assume that the names refer to a different group of people.

21. Fred is 9 years older than I am, and Marge is 1 year younger than I.

22. Fred is 2 years younger than I am, and Marge is 8 years older than I.

23. Fred is 1 year older than I am, and Marge is 5 years older than Fred.

24. Fred is 3 years younger than I am, and Marge is 5 years older than Fred.

25. Fred is twice as old as I am, and Marge is a year younger than Fred.

Table I   Age now

	Me	Fred	Marge
Ex. 6	x	$x + 7$	$x - 4$
21.			
22.			
23.			
24.			
25.			

Table II   Age 15 years ago

	Me	Fred	Marge
Ex. 7	$x - 15$	$x + 7 - 15 = x - 8$	$x - 4 - 15 = x - 19$
21.			
22.			
23.			
24.			
25.			

Table III   Age in 10 years

	Me	Fred	Marge
Ex. 8	$x + 10$	$x + 7 + 10 = x + 17$	$x - 4 + 10 = x + 6$
21.			
22.			
23.			
24.			
25.			

Check your answers on page 744.

Setting up Equations to Solve Word Problems

Now you know a lot about how much older or younger we are than one another. But you do not know our ages. Suppose we do not want to tell you that outright. But we agree to tell you the *sum of our ages*. The sum of our ages right now is 108. With an equation, you can use this new piece of information to find out how old each of us is.

The sum of our ages is the result of adding our ages together. So the statement that the sum of our ages is 108 can be translated into the equation:

(My age now) + (Fred's age now) + (Marge's age now) = 108.

To make an algebraic equation that you can solve, replace each of our ages by the algebraic expression that represents our age (as given in Example 6, page 386).

[My age now] + [Fred's age now] + [Marge's age now] = 108

becomes

$$[x] + [x + 7] + [x - 4] = 108,$$

an equation you can solve for x. Group together like terms.

$$(x + x + x) + (7 - 4) = 108$$
$$3x + 3 = 108$$
$$\underline{ - 3 \quad -3}$$
$$3x = 105$$
$$\frac{(\cancel{3}x)}{\cancel{3}} = \frac{105}{3}$$
$$x = 35$$

Since x represents my age, that means I am 35 . To find the ages of Fred and Marge, replace x by 35 in the expressions for their ages.

$$\text{Fred's age:} \quad x + 7 =$$
$$(35) + 7 = 42 \quad \text{So Fred is 42.}$$
$$\text{Marge's age:} \quad x - 4 =$$
$$(35) - 4 = 31 \quad \text{So Marge is 31.}$$

To check these solutions, go back to the original information and see whether it is true that our ages add up to 108.

Does $(35) + (42) + (31) = 108$?   Yes, so those are our ages.

EXERCISES 9.1C

For each number, use the expressions from Table I to write an equation that says the sum of our ages now is 108 . For each line, assume that the names refer to a different group of people.

EXAMPLE $[x] + [x + 7] + [x - 4] = 108$

I.   21. _____

22. _____

23. _____

24. _____

25. _____

Use the data in Table II to write equations saying that the sum of our ages 15 years ago was 63.

EXAMPLE $[x - 15] + [x - 8] + [x - 19] = 63$

II. 21. _____

22. _____

23. _____

24. _____

25. _____

Use the data in Table III to write equations saying that the sum of our ages in 10 years will be 138.

EXAMPLE $[x + 10] + [x + 17] + [x + 6] = 138$

III. 21. _____

22. _____

23. _____

24. _____

25. _____

Check your answers on page 744. ✓

9.2 Translating Real-Life Situations into Linear Equations

I remember thinking that word problems were really difficult the first time I studied algebra. I thought I had to be very creative in setting up a new equation for each problem. I finally grew confident that I could do them when I realized that there are several categories of word problems, and all problems of a category are set up with similar equations. Once I recognized what kind of problem I faced, I could set it up.

In the next sections we will discuss some categories of problems that can easily be solved by setting up a linear equation. In later chapters I will introduce you to more categories.

I know that many of the examples in this section are simple enough that you might be able to answer the questions without writing algebraic expressions and equations. But the goal of this chapter is to teach you the skill of writing expressions and equations for harder problems that you cannot solve in your head. So please do not defeat the purpose by doing these problems without equations.

This section presents problems that include a sentence that can easily be translated directly into a linear equation. Remember, since the only equations we have learned to solve at this point are linear with *one letter*, we can use only one letter to represent unknown quantities.

Solving Age Problems

EXAMPLE 1   David is 4 years older than Rachel is. The sum of their ages is 56. How old is each of them?

We can represent Rachel's age by x. David's age is then represented by $x + 4$. Here is a simple table to keep track of those expressions.

Person	Age
Rachel	x
David	$x + 4$

The equation that represents the statement: "The sum of their ages is 56" is

$$[\text{Rachel's age}] + [\text{David's age}] = 56,$$

which becomes the following algebraic equation when we replace those phrases by the expressions we chose to represent them.

$$[x] + [x + 4] = 56, \quad \text{which we can solve for } x.$$

$$2x + 4 = 56$$
$$\underline{-4 \qquad -4}$$
$$2x = 52$$
$$\frac{2x}{2} = \frac{52}{2}$$
$$x = 26$$

x represents Rachel's age, so Rachel is 26.

David's age is represented by

$$x + 4 =$$
$$(26) + 4 = 30, \quad \text{so David is 30.} \qquad \square$$

Check: Does Rachel's age plus David's age equal 56?

$$26 + 30 = 56 \quad \checkmark$$

Note that to check we did not substitute the values into the equation. Doing so would simply determine whether we had correctly solved that equation. We want to check whether we actually found the correct answer to the question, and part of that procedure is making sure we set up the right equation.

This way to check word problems is easier and more effective than substituting into the equation. Just see whether the answer makes the situation discussed in the problem true.

EXAMPLE 2   Ruth is 5 years older than Brad. In 10 years the sum of their ages will be 55. How old is each of them now?

Let Brad's age now be x. Ruth's age now is $x + 5$.

The table should include a column for expressions representing their ages now and a column for expressions representing their ages in 10 years.

Person	Age now	Age in 10 years
Brad	x	$x + 10$
Ruth	$x + 5$	$x + 15$

The equation representing the fact that in 10 years the sum of their ages will be 55 is

[Brad's age in 10 years] + [Ruth's age in 10 years] = 55,

which becomes

$$[x + 10] + [x + 15] = 55.$$

We can solve that equation.

$$(x + 10) + (x + 15) = 55$$
$$2x + 25 = 55$$
$$\underline{-25 \qquad -25}$$
$$2x \qquad\qquad 30$$

$$\frac{2x}{2} = \frac{30}{2}$$
$$x = 15$$

x represents Brad's age now, so he is 15 years old. Ruth's age now is $x + 5$, which would be $(15) + 5 = 20$. So Brad is 15 and Ruth is 20.   ❏

Check: Will the sum of their ages in 10 years be 55? Find how old each of them will be in 10 years.

Brad will be $(15) + 10 = 25$ Ruth will be $(20) + 10 = 30$

The sum of their ages in 10 years will be $25 + 30 = 55$.   ✓

You are now ready to tackle a problem that at first may sound quite complicated.

EXAMPLE 3   Ellie and Kate noticed that Kate's age in 9 years will be twice what Ellie's was 13 years ago. Kate is 3 years older than Ellie. How old will they be a year from now? Don't panic; this problem would be hard only if you had to do it in your head.

Set up a table with columns representing

age now, age in 9 years, and age 13 years ago

to find the expressions we need to set up an equation.

Person	Age now	Age in 9 years	Age 13 years ago
Ellie	x		$x - 13$
Kate	$x + 3$	$(x + 3) + 9 = x + 12$	

Note that we only need to represent *Ellie's* age 13 years ago and *Kate's* age in 9 years.

The equation we can set up is

[Kate's age in 9 years] = 2[Ellie's age 13 years ago].
$$[x + 12] = 2[x - 13]$$
$$x + 12 = 2x - 26$$
$$\underline{-x \qquad\qquad -x}$$
$$12 = x - 26$$
$$\underline{+26 \qquad +26}$$
$$38 = x$$

Looking at the table, we see that x represents Ellie's age now, so Ellie is 38. Kate's age now is

$$x + 3 =$$
$$(38) + 3 = 41.$$

We want to know how old they will each be a year from now. To find those figures add 1 to their ages now.

Ellie will be $38 + 1 = 39$ and Kate will be $41 + 1 = 42$.   ❑

Check: Will Kate's age in 9 years be twice Ellie's age 13 years ago?

Kate's age in 9 years =	Ellie's age 13 years ago =
her age *now* +9 =	her age *now* −13 =
(41) + 9 = 50	(38) − 13 = 25

50 is twice 25: [50 = 2(25)]. ✓

To set up age problems easily:

1. Let x represent someone's age now.

2. Use the information given to find an expression with x to represent the age now of everyone else in the problem.

3. If the problem discusses anyone's age at some time other than now, find an expression for the age then based on the expression for the age now.

4. Set up a table with a column to record the expression for age at each of those times.

5. Translate the sentence that describes the relationships of their ages into an equation. Replace each phrase with the algebraic expression representing it. That equation will only involve the letter x, so you can solve it for x.

6. Once you know the value of x, replace x by that number in the expression for each person's age *now* to learn the value of his or her age now.

7. Read the problem carefully to see what age was asked for and use the values of the ages now to find it.

If someone tells you the relationship of several people's ages and gives you one other fact about their ages (like the sum or that one is 3 more than twice the other), you can always figure out everyone's age.

EXAMPLE 4   Sam is **11 years older** than Frank. Dick is **19 years older** than Sam. They were taken aback when they noticed that the sum of Frank's and Dick's present ages and Sam's age a year ago was equal to 100. How old is each now?

Let x represent Frank's age, since he is the youngest.

Person	Age now	Age a year ago
Frank	x	
Sam	$x + 11$	$(x + 11) - 1 = x + 10$
Dick	$(x + 11) + 19 = x + 30$	

The equation is

[Frank's age now] + [Dick's age now] + [Sam's age a year ago] = 100.

$$[x] + [x + 30] + [x + 10] = 100$$
$$3x + 40 = 100$$
$$\underline{\quad -40 \qquad -40}$$
$$3x = 60$$
$$\frac{3x}{3} = \frac{60}{3}$$
$$x = 20$$

Looking back at the table, we see that x represents Frank's age now.

Frank's age now is $x = 20$.

Sam's age now is $x + 11 = 31$.

Dick's age now is $x + 30 = 50$.

Check: Is the sum of Frank's and Dick's present ages and Sam's age a year ago equal to 100? (Sam's age a year ago is $31 - 1 = 30$.)

$$(20) + (50) + (30) = 100 \quad \checkmark$$

EXERCISES 9.2A

Fill in the table and set up and solve an equation to find the solution to each age problem.

1. Holly is 4 years older than Chris. If the sum of their ages is 48, how old are they?

Person	Age
Chris	
Holly	

2. Carlos is 8 years older than Maria. If the sum of their ages is 90, how old are they?

Person	Age
Maria	
Carlos	

3. Harriet is 3 years younger than Rosa. In 5 years the sum of their ages will be 49. How old are they now?

Person	Age now	Age in 5 years
Harriet		
Rosa		

4. Bob is 4 years older than David. The sum of their ages 10 years ago was 64. How old are they now?

Person	Age now	Age 10 years ago
David		
Bob		

5. Sal is 5 years older than Margie. In 4 years the sum of their ages will be 69. Hold old are they now?

Person	Age now	Age in 5 years
Margie		
Sal		

6. Phil is twice as old as Ann. In 2 years the sum of their ages will be 100. How old are they now?

Person	Age now	Age in 2 years
Ann		
Phil		

7. Barbara is 2 years older than her sister. Last year the sum of their ages was 54. How old are they now?

Person	Age now	Age last year
Sister		
Barbara		

8. Jane is 3 years older than Judy. The sum of their ages in 12 years will be 55. How old were they 3 years ago?

Person	Age now	Age in 12 years
Judy		
Jane		

9. Archie is 7 years older than Mabel. His age in 20 years will be twice what her age was 3 years ago. How old are they now?

Person	Age now	Age in 20 years	Age 3 years ago
Mabel			
Archie			

10. Nick is 5 years older than Tim. In 7 years Nick will be twice as old as Tim was 3 years ago. How old are they now? How old will they be in 5 years?

Person	Age now	Age in 7 years	Age 3 years ago
Tim			
Nick			

11. Jill is twice as old as Adam. Jesse is 4 years older than Jill. In 2 years the sum of their ages will be 100. How old is each now? How old were they 5 years ago?

Person	Age now	Age in 2 years
Adam		
Jill		
Jesse		

12. Luke is 6 years older than Sonya. Rick is twice as old as Sonya was 5 years ago. In 3 years the sum of their ages will be 105. How old is each now? How old will they be in 10 years?

Person	Age now	Age 5 years ago	Age in 3 years
Sonya			
Luke			
Rick			

Check your answers on page 744.

Money and Other Problems

This section will not teach you how to solve the money problems you are probably most interested in, such as how to have enough of it. But it will teach you how to solve riddles that involve various amounts and kinds of coins. And it will provide helpful experience in setting up word problems.

If you are told the total number of coins, you can set up money problems in much the same way as age problems. Represent the quantity of one kind of coin by a letter. Choose the kind of coin that the quantity of another is compared with. Find expressions using that letter for all the other kinds of coins and record these expressions in a table like the one used in the next example.

EXAMPLE 5 I have some money in my pocket consisting of nickels, dimes, and quarters. I have 5 more dimes than nickels and 1 fewer quarter than nickels Altogether I have 13 coins in my pocket. How many of each kind do I have?

Let's represent the number of nickels by n since the numbers of dimes and quarters are compared to the number of nickels.

Kind of coin	Quantity
Nickels	n
Dimes	$n + 5$
Quarters	$n - 1$

[Quantity of nickels] + [quantity of dimes] + [quantity of quarters] = 13

The equation is $[n] + [n + 5] + [n - 1] = 13$.

Solve for n.

$$(n) + (n + 5) + (n - 1) = 13$$
$$3n + 4 = 13$$
$$\underline{-4 \qquad -4}$$
$$3n = 9$$
$$n = 3$$

n represents the quantity of nickels, so there are 3 nickels

$n + 5$ represents the quantity of dimes, so there are $(3) + 5 =$ 8 dimes

$n - 1$ represents the quantity of quarters, so there are $(3) - 1 =$ 2 quarters

Check: Is there a total of 13 coins? $3 + 8 + 2 = 13$ ✓

EXAMPLE 6 When I was preparing to make a tape from some of my favorite records, I noticed that the Frankie Lymon song was 1 minute longer than the Roy Orbison song, and the Holly Near selection was 1 minute less than twice as long as Roy's song.

If the total time for those three songs is 12 minutes, how long is each? Let R be the number of minutes for Roy Orbison's song, since the others are compared to his.

Singer	Number of minutes long
Roy Orbison	R
Frankie Lymon	$R + 1$
Holly Near	$2R - 1$

So the equation is

[Time of Roy's] + [time of Frankie's] + [time of Holly's] = 12.

$$[R] + [R + 1] + [2R - 1] = 12$$
$$4R = 12$$
$$R = 3$$

Roy Orbison's song is R, or 3 minutes long.

Frankie Lymon's song is $R + 1$, or 4 minutes long.

Holly Near's song is $2R - 1$, or 5 minutes long.

Check: Do those times add up to 12 minutes? $3 + 4 + 5 = 12$ ✓

EXERCISES 9.2B

Make a table; then set up an equation and solve it to answer each question. Check your solutions.

13. I have 1 more dime than nickels, 4 fewer quarters than nickels, and a total of 33 coins. How many of each coin do I have?

14. I have 3 times as many dimes as nickels, 1 more quarter than nickels, and 6 coins altogether. How many of each coin do I have?

15. I have 4 more nickels than pennies, 1 more dime than nickels, 1 fewer quarter than pennies, and 40 coins altogether. How many of each coin do I have?

16. I have 3 times as many nickels as pennies, as many dimes as pennies, 1 more quarter than nickels, and 41 coins altogether. How many of each coin do I have?

17. I have twice as many dimes as nickels, 2 more quarters than dimes, and a total of 37 coins. How many of each coin do I have?

18. Eleanor cut three pieces from an 8 ft piece of lumber. The second was twice as long as the first, and the third was 1 ft more than twice the second. How long was each?

19. Zale's favorite song is 2 min longer than Rachel's. But Jonah's song is twice as long as Zale's. If their three songs take a total of 18 min, how long is each?

20. Emily noticed that her drive to work and back took 12 min longer on Monday than on Tuesday, and on Wednesday the drive was 3 times as long as on Tuesday (because of a bad storm). The total time spent driving those three days was 3 hr and 32 min (or 212 min). How long did she drive each day?

21. The receptionist noticed that she answered twice as many calls on Monday as on Tuesday, and 5 fewer than 3 times as many on Wednesday as on Tuesday. If she answered a total of 133 calls on those three days, how many came each day?

22. Julie, a therapist, scheduled 7 more sessions in December than in November, and 10 fewer than twice as many in January as in November. If she scheduled a total of 237 sessions during that time, how many were scheduled each month?

23. Maury took twice as many pictures on his trip to Africa than he did on his trip to Italy. He took 15 fewer pictures in California than he did in Africa. If he took a total of 385 pictures on all those trips, how many did he take on each?

24. Harold caught 3 more fish on Saturday than he did on Friday, and twice as many on Sunday as on Saturday. How many did he catch each day if he caught a total of 37 fish on those three days?

Check your answers on page 744.

More Real-Life Problems That Translate into Linear Equations

In most of the problems we have just solved, I purposely withheld some information to make a riddle. For example, I could have simply told you how many quarters and nickels I had instead of giving you all that complex information. But you will often run into situations where simple information has been suppressed, either purposely or by accident. This is especially true with financial and statistical information. In the following simple problems, you might be able to find the missing information with some kind of arithmetic rather than by setting up an equation. But the problems show what equations can do. Learn how to use equations to solve problems easily.

For each problem there is only one unknown quantity, which you should represent by some letter. Then translate the information that describes the situation into an equation.

EXAMPLE 7   A detective is searching for a car with stolen jewels hidden in it. She thinks it passed through a certain used-car lot via a phony sale. The lot owner will not say how many cars he moved that week, but he agrees to show the detective some payroll information.

The records show that the dealer's only salesperson gets $120 a week plus $50 commission for each car sold. On the week in question, she got paid $470. How many cars were sold?

Let C represent the number of cars sold.

[Total earnings] equal [$120] + [$50 per car] translates to

$$[\$470] = [\$120] + [\$50C]$$
$$\underline{-120 \qquad -120}$$
$$\$350 = \$50C$$
$$\frac{\$350}{\$50} = \frac{\$50C}{\$50}$$
$$7 = C$$

She sold seven cars that week.   ◻

Check: Would her earnings come to $470 if she sold seven cars?

Commission for selling seven cars is $50(7) = $350.

Commission plus salary is $350 + $120 = $470.

EXAMPLE 8   You are doing research on a company, and you would like to know how many shares of stock it has issued over the years. You cannot find that piece of information, but you do find a magazine article about its profit performance this year. The article says that the company made $44,000,000; reinvested $12,000,000 in new plants and machinery; distributed $8,000,000 through an employee profit-sharing plan; and distributed the rest to its stockholders. Stockholders received $80 for each share they own. From this information, figure out the number of shares of stock.

Let S represent the number of shares of stock. The information,

[Profits] = [reinvestment] + [profit sharing] + [$80 per share of stock], translates into the following equation.

$$[\$44,000,000] = [\$12,000,000] + [\$8,000,000] + [\$80S]$$

$$
\begin{aligned}
\$44,000,000 &= \$20,000,000 + \$80S \\
-\$20,000,000 & \quad -\$20,000,000 \\
\hline
\$24,000,000 &= \$80S
\end{aligned}
$$

$$\frac{\$24,000,000}{\$80} = \frac{\$80S}{\$80}$$

$$300,000 = S$$

So there are 300,000 shares of stock. ❑

Check: Do the profits total $44,000,000 if there are 300,000 shares of stock?

$$
\begin{aligned}
\$80(300,000) =&\ \$24,000,000 \\
&\ \$12,000,000 \\
+&\ \$\ 8,000,000 \\
\hline
&\ \$44,000,000 \quad \checkmark
\end{aligned}
$$

EXERCISES 9.2C

Set up an equation and solve it to answer each question. Check your answers.

25. Lill works as a saleswoman at Comfy Coats and earns $100 a week salary plus $12 commission on each coat she sells. How many coats did she sell the week her earnings were $280?

26. Bill earns $50 a week in salary plus $15 on each sale. How many sales did he make the week his earnings came to $470?

27. Susan earns $80 a week plus $30 on each sale. How many sales did she make the week her earnings came to $470?

28. The local utility company earned $12,000,000; reinvested $4,000,000; distributed $1,000,000 back to the consumers; and returned the rest to the stockholders. If each share earned $50, how many shares of stock were there?

29. The Weesharedaprofits Company paid profit sharing back to each of the 30 employees. The dividend for each share of stock was $25 more than the profit sharing for each employee. If there were 100 shares of stock and the company reinvested $20,000 from the $55,000 profits earned before paying dividends and profit sharing, how much did each employee receive?

30. The Co-op earned $211,000 last year and reinvested $43,000. Twice as much profit sharing was paid to each employee as was paid in dividends for each stock to the shareholders. How much did each employee and each stock owner earn if there were 180 workers and 760 shares of stock in the company?

31. The Balloon Bow-K company charges a flat rate of $20 for each delivery plus $3 per balloon. How many balloons are there in an order that costs $65?

32. The telegram company charges a flat rate of $1.80 for each message plus 20¢ per word after the first 12 words. How many words are in a message that costs $3.40?

33. The rent-a-car company charged us a flat rate of $25 for the weekend, plus 8¢ a mile for all miles over the first 200. How many miles did we drive if our bill came to $44.76?

34. It costs $1.50 more to make 5 × 7 prints than it costs to make 3 × 5 prints at my local photo store. It cost me $17 to have seven 2 × 5 prints and four 5 × 7 prints made. How much did each kind cost?

35. It costs 55¢ more to have prints made from slides than it does to make them from negatives. If it cost $16.70 to order 6 prints from negatives and 14 from slides, how much did each process cost?

36. The cost for enlarging a photo to 8 × 12 is 5¢ less than twice the cost of making a 5 × 7 enlargement. If it cost $44.75 to get five 8 × 12 and eight 5 × 7 enlargements, how much did each cost?

Check your answers on page 744.

 ### Problems Where We Know the Sum of the Unknowns

Setting up the Expressions

In some problems we do not know how much larger one unknown is than the other, but we are told what their sum is. Although that information is helpful for setting up the problem, it is not enough for us to figure out what they are. For example; there are many pairs of numbers that add up to 20.

EXAMPLE 1   The following pairs of numbers all have the sum of 20.

$$10 + 10; \; 2 + 18; \; 1 + 19; \; 16 + 4; \; 3\frac{1}{2} + 16\frac{1}{2}; \; -5.3 + 25.3$$ ❑

If I am thinking of two numbers and their sum is 20, you can easily figure out what one of them is if I tell you what the other is. If I say that one of them is 5, what is the other?

The answer is 15, but what is more important is how I got 15. Since their sum is 20, the second number must be what is left when we take 5 away from 20. So the second number is $20 - 5$, or 15. You may have found the answer so easily that you were not aware of your method.

EXAMPLE 2   The sum of two numbers is 35. If one is 16, what is the other?
The other is $35 - 16$, or 19.   ❑

If we are told that the sum of two numbers is 20 but do not know the value of either, we can represent one of them by a letter. The other number is then represented as 20 minus that letter.

EXAMPLE 3   The sum of two numbers is 20.
We can call one number x and the other $20 - x$.   ❑

EXAMPLE 4   The sum of two numbers is 46.

We can call one n and the other $46 - n$.   ❑

EXERCISES 9.3A

In each problem I will tell you what the sum of two numbers is and either give you the value of the first number or the letter used to represent it. Show me how to represent the second number. I do not want the answer; I want to see the mathematical computation you do to determine the second number.

Answers

EXAMPLE   The sum is 10, and one number is 2. $10 - 2.$

The other number is _____

The sum is 50, and one number is n. $50 - n.$

The other number is _____

1. The sum is 40, and one number is 18.   1. _____

2. The sum is 50, and one number is 34.   2. _____

3. The sum is 83, and one number is 6.   3. _____

4. The sum is 91, and one number is 25.   4. _____

5. The sum is 67, and one number is 49.   5. _____

6. The sum is 100, and one number is 74.   6. _____

7. The sum is 135, and one number is 88.   7. _____

8. The sum is 280, and one number is 179.   8. _____

9. The sum is 100, and one number is x.   9. _____

10. The sum is 134, and one number is y.   10. _____

11. The sum is 150, and one number is w.   11. _____

12. The sum is 187, and one number is n.   12. _____

13. The sum is 340, and one number is x.   13. _____

14. The sum is 400, and one number is y.   14. _____

Check your answers on page 744.   ✓

Solving Some Problems

In Section 9.2 I suggested that you let the smaller unknown quantity be the one represented by a letter. It is easier to build the expression for the other unknown if it is larger; we get a simpler expression by *adding* or *multiplying* than we do by *subtracting* or *dividing*.

When we are told only the *sum* of two unknowns, we do not know which is larger. Just choose a letter for one of them, and the other will be represented by their *sum minus that letter*.

We can set up an equation using those expressions if we are given more information about the unknowns. We solve it for the letter to find out what the numbers are.

EXAMPLE 5 If we know that the sum of two numbers is **20** and that twice the first one minus the other is 25, we can find the two numbers.

Let's represent the first number as $\underline{x}$ and the other as $\underline{20 - x}$. Then the equation would be as follows.

$$\text{Twice [the \underline{first} number]} - \text{[\underline{other} number]} = 25$$
$$2[\underline{x}] - [\underline{20 - x}] = 25$$
$$2\underline{x} - 20 + \underline{x} = 25$$
$$3x - 20 = 25$$
$$\underline{ + 20 \qquad +20}$$
$$3x = 45$$
$$\frac{3x}{3} = \frac{45}{3}$$
$$x = 15$$

So $x = 15$.

If $x = 15$, then $20 - x$ is $20 - (\mathbf{15}) = 5$.

The two numbers are 5 and 15.   ❏

Check: Is twice the first number minus the other number equal to 25?

Twice 15 is 30, 30 minus 5 is 25.   ✓

It turns out that we let x represent the larger number in Example 5. We could have let n represent the smallest number instead. If we had, this is how it would have been solved:

EXAMPLE 6 The sum of two numbers is **20**, and twice the first one minus the *other* is 25. Find the two numbers.

If we represent the *other* number as $\underline{n}$, then $\underline{20 - n}$ represents the first one. The equation would be as follows.

$$\text{Twice [the \underline{first} number]} - \text{[\underline{other} number]} = 25$$
$$2[\underline{20 - n}] - [\underline{n}] = 25$$
$$40 - 2\underline{n} - \underline{n} = 25$$
$$40 - 3n = 25$$
$$\underline{-40 \qquad\qquad -40}$$
$$-3n = -15$$
$$\frac{-3n}{-3} = \frac{-15}{-3}$$
$$n = 5$$

So $20 - n$ is $20 - (5) = 15$.

In either case, one number is 15 and the other is 5.   ❏

EXAMPLE 7 I went through all the pockets of clothes I was washing at the laundromat. There are certain advantages to that chore if you or someone you live with is absent-minded. I found **70** coins in dimes and pennies.

If the number of dimes minus twice the number of pennies is 10, how many of each coin did I find?

Let's call the number of pennies p.

I can represent the number of dimes as $70 - p$.

Coin	Quantity
Pennies	p
Dimes	$70 - p$

[Number of dimes] − twice [the number of pennies] = 10

$$[70 - p] - 2[p] = 10$$
$$70 - p - 2p = 10$$

$$70 - 3p = 10$$
$$\underline{-70 -70}$$
$$-3p = -60$$
$$\frac{-3p}{-3} = \frac{-60}{-3}$$
$$p = 20$$

So $70 - p$ is $70 - (20) = 50$.

I found 20 pennies and 50 dimes. ❏

Check: Is the number of dimes minus twice the number of pennies equal to 10?

50 minus twice 20 is $50 - 2(20) =$

$$50 - 40 = 10 \quad \checkmark$$

Remember, at present we can only use one letter to represent unknown quantities. That is because we have to be able to set up a linear equation with one letter so we can solve it to answer our questions. Use whatever letter you want.

Once you master the way to set up these problems, you will be ready to handle problems in later sections that otherwise would seem very complicated.

EXERCISES 9.3B

Using only one letter, set up an equation. Solve it to answer each question. Check your solutions.

15. The sum of two numbers is 30. Three times the first number minus the second number is 10.
First number:_____ Second number:_____
Equation:_____

16. The sum of two numbers is 50. The first number minus 4 times the second number is 25.
First number:_____ Second number:_____
Equation:_____

17. The sum of two numbers is 50. The first number minus twice the second number is 2.
First number:_____ Second number:_____
Equation:_____

18. The sum of two numbers is 100. The second one is 20 less than 5 times the first.
First number:_____ Second number :_____
Equation:_____

19. The sum of two numbers is 100. The second is 8 less than 3 times the first.
First number:_____ Second number:_____
Equation:_____

20. The sum of two numbers is 100. The second number is 10 less than 4 times the first.
First number:_____ Second number:_____
Equation:_____

21. The sum of two numbers is 50. One number minus 5 times the other number is 2.
First number:_____ Second number:_____
Equation:_____

22. The sum of two numbers is 43. Twice one number is 24 less than 3 times the other.
First number:_____ Second number:_____
Equation:_____

23. Tim found 38 coins. He multiplied the number of quarters by 10 and added that to twice the number of nickels to get a sum of 100. How many of each coin did he find?

Coin	Quantity
Quarters	
Nickels	

Equation:_____

24. There are 17 pets living on Nick's street that are either cats or dogs. 3 times the number of cats minus the number of dogs is 11. How many of each kind of pet are there?

Pet	Quantity
Dogs	
Cats	

Equation:_____

25. Rebecca found 65 cans at the beach over the last 2 weeks. Twice the number she found the first week plus 3 times the number she found the second week equals 172. How many did she find each week?

Week	Number of cans
first	
second	

Equation:_____

Check your answers on page 744. ✓

9.4 Mixture Problems

Money, Purchases, and Interest Mixture Problems

You are now ready to take on word problems that belong to a category called **mixture problems**. These have to do with mixing together two different kinds of things. Usually the problem consists of figuring out how much (the quantity) of each thing is in the mixture.

In a mixture problem there is some kind of a rate for each item. For example; if the problem involves two different purchases, the rate would be the price or cost of each item. If the problem involves coins, the rate would be the value of each kind of coin. If the problem involves salt solutions, the rate would be the concentration of salt in each solution.

All mixture problems can be easily set up if the data are recorded in a table with one column for quantity, one column for rate, and another column that always represents the product of the other two columns.

We are usually told the rates, which we can record in the table. Using only one letter, represent the quantity of one item by that letter, and write a simple algebraic expression to represent the quantity of each other item. The equation often is set up by adding the rows from the last column of the table.

In a coin *mixture* problem you are told the relationship between the quantities of the different kinds of coins, but instead of the total number of coins, you are given the *total value* of those coins.

EXAMPLE 1   Sharon found $2.05 in coins while walking along the beach. She found 1 more dime than nickels and twice as many quarters as nickels. How many of each kind of coin did she find?

The *rate* of each coin is its *value in cents*. The value of nickels is 5 cents, of dimes is 10 cents, and of quarters is 25 cents.

Let n represent the quantity of nickels. So $n + 1$ represents the quantity of dimes, and $2n$ represents the quantity of quarters.

The last column will represent the **total value of each kind of coin**. The quantity multiplied by the value of each coin equals the total value. (Aren't three nickels worth 3(5 cents) = 15 cents?)

Coin	Quantity	Rate (¢)	Total value (¢)
Nickels	n	5	$5n$
Dimes	$n + 1$	10	$10(n + 1)$
Quarters	$2n$	25	$25(2n) = 50n$
Total			205

Since the value of coins is expressed in cents, $2.05 must be too. So we represent $2.05 as 205 cents.

$$[\text{Value of nickels}] + [\text{value of dimes}] + [\text{value of quarters}] = 205 \text{ cents}$$

$$[5n] + [10(n + 1)] + [50n] = 205$$

$$5n + 10n + 10 + 50n = 205$$

$$65n + 10 = 205$$

$$\underline{ -10 \quad -10}$$

$$65n = 195$$

$$\frac{65n}{65} = \frac{195}{65}$$

$$n = 3$$

n is the quantity of nickels, so there are 3 nickels.

$n + 1$ is the quantity of dimes, so there are (3) + 1 = 4 dimes.

$2n$ is the quantity of quarters, so there are 2(3) = 6 quarters.

Check: Is their total value $2.05?

Value of three nickels is 3(5 cents) = 15 cents.

Value of four dimes is 4(10 cents) = 40 cents.

Value of six quarters is 6(25 cents) = 150 cents.

15 cents + 40 cents + 150 cents = 205 cents.

Another kind of mixture problem deals with combining two things that have different costs. The rate for each item is its cost. The last column will represent the total value of each item, which is obtained by multiplying its quantity by its cost.

EXAMPLE 2 You are trying to get into the business of writing children's books. The annual report of Consolidated Fiction Enterprise says that the company markets two lines of books: children's stories in hardback, which sell for $5.00, and romances in paperback, which sell for $3.00. Altogether they sold 4,100 books for total sales of $13,900. You want to know whether they sell enough children's books to make it worth your while to submit a manuscript to them.

How can you figure this out?

Let C represent the number of children's books sold. Then we can represent the number of romance books sold by

$$[\text{Total books}] - [\text{number of children's books}] = 4{,}100 - C.$$

Kind of book	Quantity	Rate ($)	Total cost ($)
Children's	C	5	$5C$
Romances	$4{,}100 - C$	3	$3(4{,}100 - C)$
Total	4,100		13,900

The equation to solve is

[Value of children's books] plus [value of romance books] equals [total sales].

$$[5C] + [3(4,100 - C)] = [13,900]$$
$$5C + 12,300 - 3C = 13,900$$
$$2C + 12,300 = 13,900$$
$$\underline{-12,300 \quad -12,300}$$
$$2C = 1,600$$
$$\frac{\cancel{2}C}{\cancel{2}} = \frac{1,600}{2}$$
$$C = 800 \text{ and } 4.100 - C =$$
$$4,100 - (800) = 3,300$$

So there were 800 children's book sales and 3,300 romance book sales. ❏

Check: Do the total sales come to $13,900 if there were 800 children's books and 3,300 romances?

Sales from children's books are $5(800) = $4,000, and sales from romances are $3(3,300) = $9,900.

Total sales are $4,000 + $9,900 = $13,900 ✓

EXAMPLE 3   Johanna spent $7.00 at the arcade. She played some games that cost 50 cents each and others that cost only 25 cents each. If she played 23 games altogether, how many of each kind did she play?

If we let x represent the number of 25-cent games played, $23 - x$ represents the number of 50-cent games played.

Kind of game	Quantity	Rate (¢)	Total cost (¢)
25 cent	x	25	$25x$
50 cent	$23 - x$	50	$50(23 - x)$
Total	23	not relevant	700

$$[\text{Cost of 25¢ games}] + [\text{cost of 50¢ games}] = [700]$$
$$[25x] + [50(23 - x)] = 700$$
$$25x + 1150 - 50x = 700$$
$$1150 - 25x = 700$$
$$\underline{-1150 \qquad -1150}$$
$$-25x = -450$$
$$\frac{-25x}{(-25)} = \frac{-450}{(-25)}$$
$$x = 18$$

x represents the number of 25-cent games, so she played 18 of them.

$23 - x$ represents the number of 50-cent games, so she played $23 - (18) = 5$ of them. ❏

Check: Was the total cost $7?

Cost of 18 games at 25 cents each is 18($0.25) = $4.50.

Cost of 5 games at 50 cents each is 5($0.50) = $2.50.

$4.50 + $2.50 = $7 ✓

EXAMPLE 4 Emily won $20,000 in the lottery. She tells the tax auditor that she invested some of it in a regular account at 6% interest and the rest in a special account at 9% interest. The bank reports to the IRS that her total (simple) interest at the end of the year was $1,620. The auditor figures how much she invested in each account.

The rate involved here is the rate of interest, so the last column represents the amount of **interest** in each account.

Account	Quantity ($)	Rate (%)	Interest ($)
Regular	x	6%	6% of x
Special	$20,000 - x$	9%	9% of $(20,000 - x)$
Total	$20,000	Not relevant	1,620

$$[\text{Regular account interest}] + [\text{special account interest}] = \$1,620$$
$$[6\% \text{ of } x] + [9\% \text{ of } (\$20,000 - x)] = \$1,620$$
$$0.06x + 0.09(\$20,000 - x) = \$1,620$$
$$0.06x + \$1,800 - 0.09x = \$1,620$$
$$\$1,800 - 0.03x = \$1,620$$
$$\underline{-1,800 \qquad\qquad -1,800}$$
$$-0.03x = -\$180$$
$$\frac{-0.03x}{(-0.03)} = \frac{-180}{-0.03}$$
$$x = \$6,000$$

x, the amount invested in the regular account, is $6,000. $20,000 − x, the amount invested in the special account, is

$$\$20,000 - (\$6,000) = \$14,000. \qquad \square$$

Check: Does the total interest equal $1,620?

Interest from investing $6,000 at 6% is 6% of $6,000 = $360.

Interest from investing $14,000 at 9% is 9% of $14,000 = $1,260.

$360 + $1,260 = $1,620 ✓

EXAMPLE 5 Frances sells used records every Saturday at the flea market. She meant to keep track of album sales versus sales of single records, but today she forgot to think about that. She knows that she has made 19 sales so far. She sells albums for 75 cents and singles for a quarter. Frances has taken in $8.25. How many albums and how many singles has she sold? To simplify our calculation we can express the costs in cents rather than dollars, thereby avoiding decimals ($8.25 = 825 cents).

Record	Quantity	Rate (¢)	Total cost (¢)
Single	S	25	$25S$
Album	$19 - S$	75	$75(19 - S)$
Total	19	Not relevant	825

$$[\text{Cost of singles sales}] + [\text{cost of album sales}] = 825$$
$$[25S] + [75(19 - S)] = 825$$
$$25\underline{S} + 1425 - 75\underline{S} = 825$$
$$1425 - 50S = 825$$
$$\underline{-1425 \qquad\qquad -1425}$$
$$-50S = -600$$
$$\frac{-50S}{-50} = \frac{-600}{-50}$$
$$S = 12, \quad \text{so } 19 - S =$$
$$19 - (12) = 7$$

So Frances sold 12 single records and 7 albums.

Check: Is the total cost equal to $8.25?

Cost of 7 albums at 75 cents each is 7($0.75) = $5.25.

Cost of 12 singles at 25 cents each is 12($0.25) = $3.00.

Total cost is $5.25 + $3.00 = $8.25. ✓

EXERCISES 9.4A

Use a table to find the expressions you need to set up an equation. Then solve each equation. Check your work.

1. I found $2.35 in change in my pockets before doing the laundry. There were 2 more dimes than nickels and 1 fewer quarter than nickels. How many of each coin did I find?

Coin	Quantity	Rate (¢)	Total value (¢)
Nickels			
Dimes			
Quarters			
Total			

Equation: _____

2. I have 6 more nickels than dimes and 1 fewer quarter than nickels. They total $3.15. How many of each coin do I have?

Coin	Quantity	Rate (¢)	Total value (¢)
Nickels			
Dimes			
Quarters			
Total			

Equation: _____

3. I have 4 more pennies than dimes and 3 fewer quarters than dimes. They total $1.09. How many of each coin do I have?

Coin	Quantity	Rate (¢)	Total value (¢)
Pennies			
Dimes			
Quarters			
Total			

Equation: _____

4. I have twice as many dimes as pennies and 3 fewer nickels than pennies. They total $1.41. How many of each coin do I have?

Coin	Quantity	Rate (¢)	Total value (¢)
Dimes			
Pennies			
Nickels			
Total			

Equation: _____

5. I have 1 more dime than nickels and 4 fewer quarters than nickels. They total $1.90. How many of each coin do I have? (Make your own table.)

6. I have the same number of nickels as dimes and 5 fewer quarters than nickels. They total $3.15. How many of each coin do I have? (Make your own table.)

7. I have 2 fewer nickels than pennies, 5 more dimes than pennies, and 1 more quarter than nickels. They total $1.79. How many of each coin do I have? (Make your own table.)

8. I have twice as many nickels as pennies, 1 more dime than nickels, and 3 fewer quarters than pennies. They total $2.71. How many of each coin do I have? (Make your own table.)

9. Granola costs 50¢ per pound, and coconut costs 65¢ per pound. A 13 lb mixture is worth $7.25. How many pounds of each are used?

Product	Quantity (lb)	Rate per lb (¢)	Total cost (¢)
Granola			
Coconut			
Mixture			

10. Red confetti costs 20¢ an ounce, and white confetti costs 8¢ an ounce. If 12 oz of the mixture cost $1.20, how many ounces of each kind are used?

Product	Quantity (oz)	Rate per oz (¢)	Total cost (¢)
White confetti			
Red confetti			
Mixture			

11. Thin noodles cost 12¢ per pound, and thick noodles cost 30¢ per pound. A 9 lb mixture costs $1.80. How many pounds of each kind are used? (Make your own table.)

12. M & M's cost $2.10 per pound, and sesame seeds cost 40¢ per pound. How many pounds of each are in an 8 lb mixture that costs $8.30? (Make your own table.)

13. If lemonade costs 5¢ per ounce, and orange juice costs 8¢ per ounce, how many ounces of each are in a 10 oz mixture costing 59¢? (Make your own table.)

14. Shrimp are $6.50 per pound, and sardines are $1.25 per pound. A 6 lb mixture is worth $3.00 per pound. How many pounds of each are used?

Product	Quantity (lb)	Rate per lb (¢)	Total cost (¢)
Shrimp			
Sardines			
Mixture			

15. Strawberries cost 15¢ per ounce, and bananas cost 5¢ per ounce. How many ounces of each are used in a 50 oz mixture worth 7¢ per ounce?

Product	Quantity (oz)	Rate per oz (¢)	Total cost (¢)
Strawberries			
Bananas			
Mixture			

16. If the strawberries and bananas in Exercise 15 are made into a 10 oz mixture worth 11¢ per ounce, how many ounces of each are used? (Make your own table.)

17. If white paint costs $6 per gallon, and orange paint costs $14 per gallon, how many gallons of each went into 20 gal of a mixture worth $10 per gallon? (Make your own table.)

18. Beef costs $2.10 per pound, and pork costs $1.60 per pound. How many pounds of each went into a 10 lb mixture worth $2.00 per pound? (Make your own table.)

19. Tracy invested part of the $1,000 prize he won at 2% and the rest at 8% simple interest. Those investments earned $59 in a year. How much was invested at each rate?

20. Annie earned $4,000 and decided to invest some of it at 4% and the rest at 9% simple interest. Those investments earned $200 interest in a year. How much was invested at each rate?

21. Jimmy invested some of his $4,000 inheritance at 3% and the rest at 7% simple interest. Those investments earned $218 in a year. How much was invested at each rate?

22. Ben won $6,000 and invested some at 3%, the same amount at 5%, and the rest at 9%. Those investments earned him $422 in a year. How much was invested at each rate?

Check your answers on page 745.

We will discuss mixture problems again in Chapter 10, where we will learn how to solve systems of equations with two letters.

Concentration-of-Solution Mixture Problems

In Section 4.8 we learned that a 34% salt solution is a mixture in which 34 of every 100 parts is salt. In a lab we might have to change the salinity (saltiness) of a solution from one percentage to another. We can set up problems involving changes in the concentration of a solution by treating them as mixture problems and using tables like the ones in Examples 1–5 of this section.

Lab procedures refer to a solution's weight rather than its quantity. The rate is the percent of salt concentration of the solution, and the last column in the table represents the *amount of salt* in each solution. As in all mixture problems, the amount of salt is the product of the solution's weight multiplied by its rate of salt.

Sometimes we want to dilute a solution. Solutions are diluted when the rate of salt is decreased. Theoretically there are three ways to *dilute* a solution: add some water, add a weaker solution, or remove some salt.

It is difficult to remove dissolved salt from water. The easiest way to dilute is to add pure water. *Pure water has no salt, so its rate of salt is* 0%.

EXAMPLE 6   How much water must we add to 10 grams of a 30% salt solution to dilute it into a 25% solution?

Solution	Weight (g)	Rate of salt	Amount of salt (g)
Original	10	30%	30% of 10 = 3
Added water	x	0%	0% of x = 0
Diluted solution	10 + x	25%	25% of (10 + x)

$$[\text{Salt in original solution}] + [\text{salt added}] = [\text{salt in diluted solution}]$$
$$[3] + [0] = [25\% \,(10 + x)]$$
$$3 = 0.25(10 + x)$$
$$3 = 0.25(10) + 0.25x$$

$$
\begin{array}{r}
3.0 = 2.5 + 0.25x \\
-2.5 -2.5 \\
\hline
0.5 = 0.25x
\end{array}
$$

$$\frac{0.50}{0.25} = \frac{0.25x}{0.25}$$

$$\frac{50}{25} = x$$

$$2 = x$$

We should add 2 grams of water to dilute the solution.

Check: If we add 2 grams of water to 10 grams of a 30% solution, will the resulting concentration of salt be 25%?

Weight of resulting solution is $10 + 2 = 12$ g.

We added no salt, so there are still 3 grams of salt.

Does 3 g = 25% of 12 g? 25% of 12 g = 0.25(12 g) = 3 g ✓

In some situations we want to *strengthen* rather than dilute a solution. Solutions are strengthened when the rate of salt is increased. Theoretically there are three ways to strengthen a solution: Add salt, add a stronger solution, or remove water.

Again it is easiest to add salt. To increase the rate of salt, we must determine *how many grams of salt should be added. Pure salt is considered a 100% salt solution.*

EXAMPLE 7 How much salt must we add to 10 grams of a 20% salt solution to strengthen it into a 50% solution?

Solution	Weight (g)	Rate of salt	Amount of salt (g)
Original	10	20%	20% of 10 = 2
Added salt	x	100%	100% of x = x
Stronger solution	$10 + x$	50%	50% of $(10 + x)$

$$[\text{Salt in original solution}] + [\text{salt added}] = [\text{salt in stronger solution}]$$
$$[2] + [x] = [50\% \,(10 + x)]$$
$$2 + x = (0.50)(10) + 0.50x$$

$$
\begin{array}{r}
2 + x = 5 + 0.5x \\
-0.5x -0.5x \\
\hline
2 + 0.5x = 5 \\
-2 -2 \\
\hline
0.5x = 3
\end{array}
$$

$$\frac{0.5x}{0.5} = \frac{3.0}{0.5}$$

$$x = \frac{30}{5}$$

$$x = 6$$

We should add 6 grams of salt.

Check: If we add 6 grams of salt to 10 grams of a 20 % solution, will its rate of concentration be 50 %?

Weight is $10 + 6 = 16$ g. Salt in solution is $2 + 6 = 8$ g.

Does 8 g $= 50$ % of 16 g? 50 % of 16 g =

$$0.50(16 \text{ g}) = 8 \text{ g} \quad \checkmark$$

Notice that we set up these problems differently when we dilute the solution from when we strengthen it. Look to see whether the *rate of concentration* is being increased or decreased. If it is being *increased*, the solution is to be *strengthened*, and that means adding *x* grams of salt. If it is being *decreased*, the solution is to be *diluted*, and that means adding *x* grams of water.

These problems are not really hard, but you have to think about what you are doing rather than just mechanically setting up a problem.

EXERCISES 9.4B

For each problem determine whether you are diluting the solution or strengthening it. Add water to dilute; add salt to strengthen. Then set up a table like the example for that problem and set up an equation. Solve the equation to determine how much water or salt is to be added. In your answer be sure to identify whether it is water or salt that is to be added, and check your answers.

23. What must be added to 40 g of a 25 % salt solution to convert it into a 20 % solution?

24. What must be added to 10 g of a 40 % salt solution to convert it into a 25 % solution?

25. What must be added to 20 g of an 18 % salt solution to convert it into a 10 % solution?

26. What must be added to 15 g of a 10 % salt solution to convert it into a 25 % solution?

27. What must be added to 80 g of a 25 % salt solution to convert it into a 20 % solution?

28. What must be added to 20 g of a 5 % salt solution to convert it into a 24 % solution?

29. What must be added to 30 g of an 80 % salt solution to convert it into a 88 % solution?

30. What must be added to 100 g of a 35 % salt solution to convert it into a 28 % solution ?

Check your answers on page 745.

Distance Problems

Using the Formula: Distance = (Rate)(Time)

The last category of problems we will deal with in this chapter concerns distance. You probably often do calculations involving distance traveled and are familiar with the following relationship:

Distance traveled is the product of rate multiplied by time ($D = RT$).

(We discussed this relationship earlier in Section 7.2.)

Note that the rate is always expressed as some unit of distance per some unit of time, as in *miles* per *hour*. The ratio of miles to hours is expressed as the fraction $\dfrac{\text{miles}}{\text{hour}}$.

The unit of time in the rate must be the same as the unit in the time. As we saw in Section 3.11, the time unit cancels when time and rate are multiplied, and so the product is the unit for distance.

EXAMPLE 1 If the rate is expressed in miles per hour, the time should be hours, and the product will be expressed in miles.

$$(\text{Miles per hour}) \, (\text{hours}) =$$

$$\left(\frac{\text{miles}}{\text{hour}}\right) (\text{hours}) =$$

$$\left(\frac{\text{miles}}{\cancel{\text{hour}}}\right) \left(\frac{\cancel{\text{hours}}}{1}\right) = \text{miles} \qquad \square$$

EXAMPLE 2 If rate is 10 feet per second, then time must be expressed in seconds, and the unit for distance will be expressed in feet.

$$(\text{Feet per second}) \, (\text{seconds}) =$$

$$\left(\frac{\text{feet}}{\cancel{\text{second}}}\right) (\cancel{\text{seconds}}) = \text{feet} \qquad \square$$

When we know the rate and time traveled, we multiply them to determine the distance traveled: $D = RT$.

EXAMPLE 3 Find the distance you will travel if your rate is 30 miles per hour for 4 hours.

$$\text{Distance} = \left(\frac{30 \text{ mi}}{\cancel{\text{hr}}}\right) (4 \cancel{\text{ hr}})$$

$$= (30) \, (4) \text{ mi}$$

$$= 120 \text{ mi} \qquad \square$$

Before looking at word problems that require setting up an equation to solve, you should do some exercises that will help you feel comfortable with the distance formula.

EXERCISES 9.5A

Find the distance traveled for each situation.

1. How far can you travel in 3 hr if you go at the rate of 48 mph?

2. How far can you travel in 8 hr if you go at the rate of 38 mph?

3. How far have you walked in 20 min if you walk at a rate of 100 ft/min?

4. How far have you traveled if you ride your bike for half an hour and pedal at the rate of 8 mph?

5. How far did Lynn walk if she was out for $3\frac{1}{2}$ hr and covered 4 mi each hour?

6. How far did Wayne ride if he was out for 4.6 hr and traveled 21.2 mi each hour?

7. How far did Benita ride if she was out for three-quarters of an hour and rode at the rate of 12 mph?

8. How far did we carry the piano in 20 min if we were moving 15 ft each minute?

Check your answers on page 745.

Setting up Word Problems

If a word problem involves distance traveled, you can set it up in a table with columns for rate, time, and distance.

⇨ *Strategy*:   Fill the rate and time columns with data given or algebraic expressions representing unknown data, and use the distance formula ($D = RT$) to fill in the distance column.

However, in these problems the information you need to write the equation will not appear in a sentence or phrase for you to translate into an equation as it did in earlier sections of this chapter. You will have to make up an equation by observing relationships in the given information.

Once you have an algebraic expression for each person's distance, set up an equation that expresses whatever you know about the relationship between one distance and the other: Are they equal? Do they have a certain sum? Is one twice the other? Drawing a sketch of the problem will help you to visualize what is happening.

EXAMPLE 4   David drove toward the town where Mary lives at the rate of 50 miles per hour. Mary left an hour before he did and drove toward his town at the rate of 30 miles per hour. If they live 270 miles apart, how many hours did each of them travel before they met?

She left an hour before he did, so she drove 1 hour more than he did. Call David's time x and Mary's time $x + 1$. Remember, the product of the expressions for each one's time multiplied by the rate represents each person's distance.

Person	Rate (mph)	Time (hr)	Distance = RT (mi)
Mary	30	$x + 1$	$30(x + 1)$
David	50	x	$50x$

Now you have the table, but how do you set up the equation?

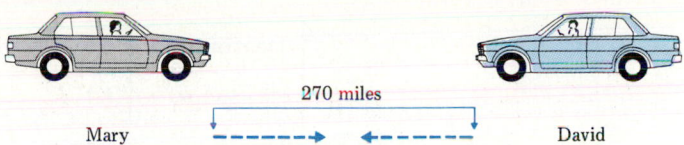

Mary   David

From this sketch we see that the distance Mary traveled plus the distance David traveled must add up to 270. So the equation is as follows.

$$[\text{Mary's distance}] + [\text{David's distance}] = 270 \text{ mi}$$
$$[30(x + 1)] + [50x] = 270$$
$$30x + 30 + 50x = 270$$
$$80x + 30 = 270$$
$$\underline{\quad -30 \quad\quad -30\quad}$$
$$80x = 240$$
$$\frac{80x}{80} = \frac{240}{80}$$
$$x = 3$$

x represents David's time, so he traveled 3 hours.
$x + 1$ represents Mary's time, so she traveled (3) + 1 = 4 hours

Check: Their distances should add up to 270 since that is how far away they were when they began.

Mary's distance =

$$\left(\frac{30 \text{ mi}}{\text{hr}}\right)(4 \text{ hr}) = 120 \text{ mi}$$

David's distance =

$$\left(\frac{50 \text{ mi}}{\text{hr}}\right)(3 \text{ hr}) = 150 \text{ mi}$$

120 mi + 150 mi = 270 mi

In every word problem that involves two people driving toward each other, and where you are told how far they were from each other, you can always find expressions for the distance of each. The sum of the distances each traveled will always be the distance they were apart at the beginning. That kind of distance problem can be set up like Example 4.

The following problems deal with distances that you or I might travel. But they could apply to an astronomer's planning a space probe or a telescope observation of moving bodies.

EXAMPLE 5   On a ski trip Willo drove 60 miles per hour on the way to the lodge, but on the way home she drove only 36 miles per hour.

a. If it took her *2 hours longer to drive home* than it did to drive to the ski lodge, how long did each trip take?

Let's call the time going there x. The time returning is 2 hours longer: $x + 2$

Destination	Rate (mph)	Time (hr)	Distance (mi)
Ski lodge	60	x	$60x$
Home	36	$x + 2$	$36(x + 2)$

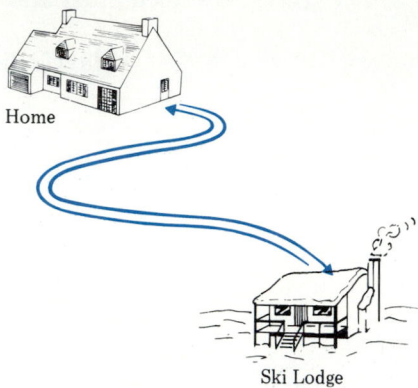

Home

Ski Lodge

It is more noticeable from the sketch than from the words in the problem that, of course, the distances are the same! Both trips are between home and the ski lodge (assume she went the same way each time).

$$[\text{Distance going to ski lodge}] = [\text{distance returning home}]$$
$$[60x] = [36(x + 2)]$$

$$
\begin{array}{rl}
60x = & 36x + 72 \\
-36x & -36x \\
\hline
24x = & 72
\end{array}
$$

$$\frac{24x}{24} = \frac{72}{24}$$

$$x = 3$$

x represents the number of hours it took Willo to drive to the ski lodge, so she drove there in 3 hours.

$x + 2$ represents the number of hours it took Willo to drive home, so she drove home in 5 hours $[(3) + 2 = 5]$.

b. How far away was the ski lodge from her home?

Distance is (60 mph)(**3 hr**) = 180 mi. ❏

Check: Do both trips come to 180 miles?

Distance to ski lodge is (60 mph) (3 hr) = 180 mi.

Distance returning home is (36 mph) (5 hr) = 180 mi. ✓

Every word problem that compares the drive to some place with the drive back always assumes those distances are the *same*. Set the expressions for each distance equal to set up that kind of distance problem. That kind of distance problem can always be set up like Example 5.

Another form of distance-rate-time problem, described below, can be set up as a linear equation.

EXAMPLE 6 My dog Aurie and I began running together. He can run 5 *times as fast* as I can. If I run 4 miles per hour, how long will it take before Aurie is 112 miles ahead of me?

My rate is 4 mph.
Aurie's rate is 5(my rate) =

$$5(4 \text{ mph}) = 20 \text{ mph}.$$

Assume I run x hours. Since we began together and the problem ends when he is 112 miles ahead of me, we stop running at the same time too. So our times are the same. If I run for x hours, so does he.

Runner	Rate (mph)	Time (hr)	Distance (mi)
Me	4	x	$4x$
Aurie	20	x	$20x$

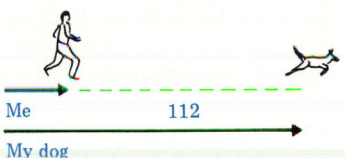

Me   112
My dog

From the sketch we observe that Aurie's distance is 112 miles more than mine So the equation is as follows.

$$[\text{Aurie's distance}] = [\text{my distance}] + 112$$
$$[20x] = [4x] + 112$$
$$\frac{-4x \qquad -4x}{16x = 112}$$
$$\frac{\cancel{16}x}{\cancel{16}} = \frac{112}{16}$$
$$x = 7$$

It would take 7 hours for him to get 112 miles ahead of me. Fortunately that is longer than either of us is capable of running, so he will never get that far ahead of me. ❏

Check. If each of us did run for 7 hours, would Aurie's distance be 112 miles greater than mine?

My distance: Aurie's distance:
(4 mph)(7 hr) = 28 mi   (20 mph)(7 hr) = 140 mi

140 mi − 28 mi = 112 mi ✓

Word problems in which people start and stop at the same time but one person moves faster than the other, and where you are told how much more distance one covers than the other, are always set up like Example 6.

After we learn how to solve equations that are not linear, we will see more examples of problems involving distance, rate, and time that we will be able to represent and solve by equations.

EXERCISES 9.5B

Set up each problem by filling in the table and writing an equation. Draw a simple sketch to help you. Check all solutions.

9. Ben drove toward Becca at 40 mph. Becca started 4 hr after Ben did and drove toward him at 25 mph. If they started out 355 mi apart, how long was Ben driving before they met?

10. On my way to the cabin I drove 55 mph. I returned at 45 mph. If it took me 2 hr longer to get home than it did to get to the cabin, how far away was the cabin?

11. John rode his bike at 9 mph toward Rose. She rode at 4 mph for twice as long as he until they met. How long was each of them riding if they started out 51 mi apart?

12. It took me 12 hr to drive to my cousin's farm and 14 hr to drive home. How far away was the farm if my average speed going there was 5 mph faster than my average speed driving home?

13. On the trip to the cabin we traveled at an average of 32 mph. On the way home we averaged 40 mph. If it took an hour longer to drive there than it did to drive home, how far away was the cabin?

14. Patti left her house and traveled 42 mph toward Robert's house. He left his house an hour after she began her trip, and he traveled 55 mph toward her house. If they live 236 mi apart, how far were they from Patti's house when they passed each other?

15. Kilin can ride his bike at a rate of 3 mph. Ami's rate on her bike is 7 times as fast as Kilin's. If they started at the same time, riding in the same direction, how many hours would they be riding before she was 72 mi ahead of him?

16. I can run at a rate of 5 mph. Yertle can run 3 times as fast as I can. If we began running together in the same direction, how long would it take Yertle to get 60 mi in front of me?

17. Clara and Sarah rode toward each other. Clara rode at 6 mph, and Sarah rode at 4 mph. If Clara began 1 hr before Sarah, and they began 26 mi apart, how long did each ride until they met?

18. If Lisa and Ari start riding their bikes at the same time, how long will it take for Lisa to get 12 mi ahead of Ari if Lisa rides at 10 mph but he rides at only 7 mph?

19. Sheila drove 26.7 mph toward Phyllis's house. Phyllis drove 34.6 mph from her house toward Sheila. If they live 272.28 mi apart, and Phyllis began her trip 2.2 hr before Sheila, how long was Sheila driving before they passed each other?

20. Kathy and Frank started out 64.3 mi apart and rode toward each other. Frank rode for 4 hr at a rate that was 1.5 mph faster than Kathy. How far were they from where Kathy began if she rode for 7 hr before they met each other?

Check your answers on page 745.

Glossary

age problem Usually riddlelike problem that tells the relationship between the ages of certain people and enough information about them so you can figure out how old each person is.

concentration of a solution The percent of a substance's weight in the total weight of a solution.

Example: A 30% salt solution means that 30% of its weight is salt, and 70% of its weight is water.

distance The product of the time and rate traveled: $D = RT$.

Example: If rate is 10 miles/hour and time is two hours:

$$\text{Distance} = (10 \text{ miles/hour})(2 \text{ hours})$$
$$= 20 \text{ miles}$$

interest problem A form of mixture problem that deals with accounts that earn interest at different rates.

mixture problem Problem that involves combining more than one type of thing when each of the things has a different rate that would be its value in a coin problem, its price in a problem about commodities, or its percentage in a solution or interest problem. In each case the sum of the value, cost, weight, or interest of each of the individual items is equal to the value, cost, or weight of their mixture.

money problem The kind of money problem you learn to solve in this book has to do with knowing the total number of coins or the amount they are worth, and being able to figure out how many of each kind of coin you have.

table A method used to record all the information and expressions used in a problem to help you set up an equation to solve it.

Example: Motion problems involving distance, rate, and time are set up in a table like the one below.

Person	Rate	Time	Distance
Henry			
Kate			

Chapter 9 Review

Do each of the problems. After you are confident that you have done them as accurately as possible, compare your answers with those in the back of the book. If any are wrong, go back to the section where the problem came from (indicated to the left of the problem) and review the section.

Once you understand all the sections, you will be sure to learn the skills solidly and remember how to do them if you practice them a little bit more. Turn to the Supplementary Exercises and do all the problems from any section where you had difficulty on these review exercises.

9.1

1. Write an expression for the number that is 7 less than twice as big as x.

2. Mike is 3 years older than Rick, who is 4 years older than Lisa, and Ari is 7 years younger than Rick.

   a. Write an expression involving x for each person's age if you let x represent Lisa's age:

   b. Write an expression involving x for each person's age in 8 years.

   c. Write an expression involving x for each person's age 3 years ago.

   d. What would the equation be if the sum of their ages now is 56?

   e. What would the equation be if the sum of their ages in 8 years will be 88?

   f. What would the equation be if the sum of their ages 3 years ago was 44?

All of the remaining questions must be answered by solving an equation.

9.2

3. If Laura is 7 years older than Bob, who is twice as old as Peter, how old is each of them if in 5 years the sum of their ages will be 37?

4. While doing the laundry I found 22 coins in the family's pockets. There were twice as many dimes as pennies, 2 more nickels than dimes, and 1 fewer quarter than dimes. How many of each coin did I find?

5. The florist charged a flat rate of $5 for the order and 75¢ per flower. How many flowers are in an order that cost $11.00?

9.3

6. How can we represent a number if we know that the sum of it and N is 100?

Answers
1.
2a. Mike
Rick
Ari
2b. Lisa
Mike
Rick
Ari
2c. Lisa
Mike
Rick
Ari
2d.
2e.
2f.
3. Laura
Bob
Peter
4. pennies
nickels
dimes
quarters
5.
6.

7. There are 140 photos in my album. There are 6 times as many in color as there are in black and white. How many of each kind of photo is in the album?

9.4 8. I have $2.69 in change. I have 3 more nickels than pennies, twice as many dimes as pennies, and 1 fewer quarter than nickels. How many of each kind of coin do I have?

9. A new batch of granola was made up from grain that costs 72¢ per pound and fruit that costs $1.13 per pound. If 12 lb of this granola are worth $10.69, how many pounds of grain were used? How many pounds of fruit were used?

10. Dot won $2,000. She invested some at 5% and the rest at 8% (simple interest). Those investments earned $121 in a year. How much was invested at each rate?

11. What must be added (water or salt, and how much) to 10 g of an 80% salt solution to make it into a 50% solution?

9.5 12. How far did we travel after 4 hr if we averaged 38 mph?

13. How far did I drive if I averaged 50 mph going and 60 mph returning, and the trip out took a half-hour longer than the trip back?

Answers

7. color _____

black
and
white _____

8. nickels _____

dimes _____

pennies _____

quarters _____

9. grain _____

fruit _____

10. 5% _____

8% _____

11. _____

12. _____

13. _____

Check your answers on page 745.

Supplementary Exercises

Do all the problems in every section that involves a skill in which you now lack complete mastery. With a little more practice you will achieve that sense of really understanding the topics, and you will remember how to do these problems.

9.1   Write an expression for the number that is:

1. 4 more than 3 times as big as x.

2. 8 less than twice x.

3. Twice as much as 5 more than x.

4. 6 more than half of x.

5. a. Bill is 6 years younger than Ron, who is twice as old as Phil. If Phil's age is represented by x, write expressions involving x to represent the other men's ages.
   b. Write expressions to represent how old Bill, Ron, and Phil will be in 9 years.
   c. Write an equation to represent that in 9 years the sum of Bill's, Ron's, and Phil's ages will be 120.

6. a. If Benita is 5 years older than JoAnn, and Roberta is 9 years younger than JoAnn, write an expression involving x to represent their ages if JoAnn's age is represented by x.
   b. Write expressions to represent how old Benita, JoAnn, and Roberta were 13 years ago.
   c. Write an equation to represent that 13 years ago the sum of Benita's, JoAnn's, and Roberta's ages was 55.

Set up an equation to solve each of the remaining questions.

9.2   7. If one number is 7 larger than another, and a third number is twice as big as the smaller of the first two numbers, what are the numbers if their sum is 91?

8. If George is 5 years younger than Mark, and Susan is twice as old as George, how old is each of them if in 2 years the sum of their ages will be 47?

9. If I have 4 more nickels than quarters, 1 fewer dime than quarters, and a total of 12 coins, how many of each kind do I have?

Answers
1.
2.
3.
4.
5a. Bill
Ron
5b. Phil
Bill
Ron
5c.
6a. Benita
Roberta
6b. JoAnn
Benita
Roberta
6c.
7. 1st
2nd
3rd
8. Susan
George
Mark
9. nickels
dimes
quarters

10. The Chinese restaurant charges $3.50 for an order of Moo Shi pork and 15¢ for each additional pancake. If Julie's order of Moo Shi pork cost $4.10, how many extra pancakes did she order?

9.3 **11.** How can we represent a number if we know that the sum of it and N is 78?

12. There are 84 recipes in my collection. There are 3 more than twice as many recipes for desserts than for entrées. (All my recipes are either for entrées or for desserts.) How many of each do I have?

9.4 **13.** I have twice as many nickels as quarters and 4 more dimes than quarters. The total value of these coins is $1.75. How many of each kind of coin do I have?

14. If solid-colored kerchiefs cost 80¢ each, and patterned kerchiefs cost 96¢ each, how many of each kind of kerchief is in a batch of a dozen kerchiefs worth $10.24?

15. If strawberries cost $1.40 per pound, and blueberries cost $1.23 per pound, how many pounds of each fruit is in a 10 lb mixture worth $12.98?

16. Jack won $1,000. He invested some of it at 4% and the rest at 12% simple interest. Those investments earned $96.80 interest in a year. How much was invested at each rate?

17. What must be added (salt or water, and how much) to 30 g of a 10% salt solution to make it a 25% solution?

18. What must be added (salt or water, and how much) to 20 g of a 30% salt solution to make it an 20% solution?

9.5 **19.** After running for 2 hr at 3.4 mph, how far have we traveled?

20. Two people begin driving toward each other at the same time when they are 300 mi apart. One person travels 40 mph, and the other person travels 20 mph faster than the first. How long have they been driving before they pass each other?

21. If I drive 40 mph going in one direction and 10 mph slower on the way back, how far did I travel in each direction if the slower trip took me 2 hr longer than the faster trip?

Answers

10. _____

11. _____

12. desserts _____

entrées _____

13. nickels _____

dimes _____

quarters _____

14. solid _____

patterned _____

15. strawberries _____

blueberries _____

16. 4% _____

12% _____

17. _____

18. _____

19. _____

20. _____

21. _____

Check your answers on page 745. ✓

Chapter 9 Test

Set up an equation and solve each problem.

1. If one number is 11 less than a second number, and a third number is twice as big as the second number, what are the numbers if the sum of the three numbers is 169?

2. Of Meg is 5 years older than Ken, and Bill is a year younger than Meg, how old is each of them if in 5 years the sum of their ages will be 87?

3. How many of each kind of coin do I have if I have twice as many pennies as quarters, 7 more nickels than quarters, 3 more dimes than pennies, and I have 40 coins altogether?

4. How many of each kind of coin do I have if I have 5 more pennies than quarters, twice as many dimes as quarters, 9 more nickels than dimes, and the total value of the coins is $1.62?

5. It costs a flat rate of $1.50 plus 40¢ for each word after the first 15 words to send a telegram. How many words are in a message that costs $16.30?

6. There are 58 toys in the game room, and 4 more than twice as many of the ones that work are broken. How many work, and how many are broken?

7. If daisies cost 38¢ each, and carnations cost 78¢ each, how many of each flower is in a combination of 15 flowers worth $8.10?

8. Cathy won $5,000 and invested it at simple interest in three different accounts. She put some in an account earning 2%, twice that amount in an account earning 5%, and the rest in an account earning 9%. If those investments earned $270 in a year, how much was invested at each rate?

9. What must be added (salt or water, and how much) to 10 g of a 40% salt solution to make it into a 60% solution?

10. Fran and Steve are driving toward each other. Fran is traveling at 45 mph, and Steve is traveling at 56 mph. How far are they from where Steve started out if when they pass each other Fran has been driving for 1 more hour than Steve, and they started out 348 miles apart?

Answers

1. _____

2. Ken _____

   Meg _____

   Bill _____

3. pennies _____

   nickels _____

   quarters _____

4. pennies _____

   quarters _____

   dimes _____

5. _____

6. work _____

   broken _____

7. _____

8. 2% _____

   5% _____

   9% _____

9. _____

10. _____

Check your answers on page 745. ✓

Cumulative Reviews
CHAPTERS 1–9

Simplify.

1. $\dfrac{1}{5} - \dfrac{3}{5} + \dfrac{1}{5} - \dfrac{4}{5} - \dfrac{1}{5} =$

1. _____

2. _____

2. $\left(\dfrac{20}{35}\right) \div \left(\dfrac{26}{84}\right) =$

3. _____

3. $\left(4\dfrac{1}{3}\right)\left(-5\dfrac{1}{2}\right) =$

4. _____

4. $6.03 - 15.8 - 0.09 - 4.03 + 0.007 - 12.34 =$

5. _____

5. $300:450 =$

6. _____

6. $ab - 3cb - 8ab + bc - 5ca + 7ac - 6a + 9ba =$

7. _____

7. $5.3w - 18.1w^2 + w^3 - 5.8w^2 - 0.6w - 0.34w^3 =$

8. _____

8. $2(mn - 4ck + jp) - 8(4nm - 7kc - 3pj) =$

9. _____

9. $x(a + 7b - 5) - 3a(2x - b + 9) =$

10. _____

Evaluate.

10. 13.7% of $\$5,600$

11. _____

11. $6a - 7b - 5c$, if $a = -1$, $b = -2$, and $c = -3$

12. _____

12. $x^7 - x^6 - x^5 + x^4 - x^3 - x^2 + x - 3$, if $x = -1$

13. _____

Solve for the letter.

13. $8m - 12 = 11m - 3$

14. _____

14. $7 - 3(2y - 1) + 4y = y - 8 + 3y + 5 + 4$

15. _____

15. $13w + 6 > -7$

16. _____

16. $-8 \le 5w + 7 < 47$

17. _____

17. $4.3:p = 0.05:19.5$

18. _____

18. 58% of N is $4,524$

19. _____

19. How much did Rick pay for his stereo if it lists for $\$112$ but he bought it during a 15% discount sale?

20. _____

20. When my nephew came to visit, he calculated that the ratio of the number of corners with street signs to those that are unmarked is 9 to 2. Out of 275 corners, how many have street signs?

Continued

Cumulative Review (*Cont.*)

What is the slope and *y*-intercept of the following.

21. $y = 4x - 9$
22. $f(x) = x + 2$

23. $y = 18$
24. $4y - 20x = 12$

Sketch the graph.

25. $y = 3x - 5$
26. $f(x) = 4x$

27. $x = -7$
28. $y > 2x - 1$

29. If small cans of dog food cost 37¢ per can, and cat food costs 28¢ per can, how many of each kind of can is in a combination box of pet food containing 20 cans if the boxful has a value of $6.14?

30. What must be added (salt or water, and how much) to 10 g of a 60% salt solution to make it an 80% solution?

21. Slope _____

_y_-int. _____

22. Slope _____

_y_-int. _____

23. Slope _____

_y_-int. _____

24. Slope _____

_y_-int. _____

29. _____

30. _____

Check your answers on page 745.

10 Systems of Linear Equations and Inequalities

 **Solutions to Systems of Equations**

Any group of more than one equation is called a **system of equations**. In this chapter we will work with systems of two linear equations or of two or more linear inequalities.

We use systems when we need to find solutions that satisfy criteria from more than one situation. For instance, instead of finding a point that is on a given line, we might want to locate a point that is common to *two different lines*. We would then want to solve the system of the two equations that represent those two lines.

Whenever we have a real-life problem involving two unknown quantities, we can find values for each *if* we can represent different information about them in two different equations. (We will learn how to do this in Section 10.5.)

Every point on each line represents a solution to one equation. A solution to two equations is the point where those two lines intersect.

Some systems do not have solutions. Parallel lines never intersect, and so a system of equations representing parallel lines will not have a solution. You can save yourself a lot of work if you know that there will not be a solution. When the equations are in the slope-intercept form, you can read their slopes. Remember, lines with the same slope and different y-intercepts are parallel and have no solutions in common.

EXAMPLE 1 $y = 3x + 5$ and $y = 3x - 1$ are parallel, so the system of these two equations has no solution. ❑

If two distinct straight lines do intersect, there is only one point that could be on both lines. To visualize this situation, think about two raw pieces of spaghetti. You can cross them to meet in one spot, but they cannot meet again unless you break or cook them. So if a system of different equations has any solution, there can only be *one* solution.

The solution to an equation with two unknowns (two letters) is always a combination of two numbers that represent the coordinates of a point. To be a solution to a system of equations, those numbers must be a solution to every equation in the system.

To determine whether the coordinates of a point represent a solution to an equation, replace each letter by the coordinate that represents it. Those coordinates represent a solution to the equation only if the expressions on each side of the equal sign simplify to the *same* value.

EXAMPLE 2 $(3, 5)$ represents the point where $x = 3$ and $y = 5$

So $(3, 5)$ is the solution to the system

$$y = x + 2 \quad \text{and} \quad y = -3x + 14$$

if $x = 3$ and $y = 5$ are solutions to *both* equations.

Is $(\underline{3}, \underline{5})$ a solution to $\underline{y} = \underline{x} + 2$?   Is $(\underline{3}, \underline{5})$ a solution to $\underline{y} = -3\underline{x} + 14$?

$$(5) \overset{?}{=} (3) + 2 \qquad\qquad\qquad (5) \overset{?}{=} -3(3) + 14$$
$$5 \overset{?}{=} 5 \quad \checkmark \qquad\qquad\qquad\quad 5 \overset{?}{=} -9 + 14$$
$$5 \overset{?}{=} 5 \quad \checkmark$$

$(3, 5)$ is a solution to both equations, so $(3, 5)$ is the solution to this system of equations. ◻

EXAMPLE 3 $(2, -4)$ represents the point where $x = 2$ and $y = -4$.

$(2, -4)$ is *not* the solution to the system

$$y = 3x - 10 \quad \text{and} \quad y = -2x + 1$$

unless $x = 2$ and $y = -4$ is a solution to both equations.

Is $(\underline{2}, \underline{-4})$ a solution to $y = 3x - 10$?

$$-4 \overset{?}{=} 3(\underline{2}) - 10$$
$$-4 \overset{?}{=} 6 - 10$$
$$-4 \overset{?}{=} -4 \quad \checkmark$$

Is $(\underline{2}, \underline{-4})$ a solution to $y = -2x + 1$?

$$\underline{-4} \overset{?}{=} -2(\underline{2}) + 1$$
$$-4 \overset{?}{=} -4 + 1$$
$$-4 \overset{?}{=} -3 \quad \checkmark$$

$(2, -4)$ is *not* a solution to $y = -2x + 1$, so it cannot be a solution to any system involving $y = -2x + 1$. ◻

EXAMPLE 4   Is $(4, -1)$ the solution to $2x + 3y = 5$ and $5y + 6 = 3x - 11$?

$(4, -1)$ represents the point where $x = 4$ and $y = -1$.

Is $(\underline{4}, \underline{-1})$ a solution to $2\underline{x} + 3\underline{y} = 5$?

$$2(\underline{4}) + 3(\underline{-1}) \overset{?}{=} 5$$
$$8 - 3 \overset{?}{=} 5$$
$$5 = 5 \quad \checkmark$$

Is $(4, -1)$ a solution to $5\underline{y} + 6 = 3\underline{x} - 11$?

$$5(\underline{-1}) + 6 \overset{?}{=} 3(\underline{4}) - 11$$
$$-5 + 6 \overset{?}{=} 12 - 11$$
$$+1 = +1 \quad \checkmark$$

So $(4, -1)$ is the solution to the system of equations. ◻

EXERCISES 10.1

For each system of equations, test the proposed point to see whether it is the solution to the system.

1. Is $(1, 3)$ the solution to $y = x + 2$ and $y = 3x$?

2. Is $(2, 0)$ the solution to $y = 4x - 8$ and $y = x + 5$?

3. Is $(5, 4)$ the solution to $y = -x + 9$ and $y = 3x - 1$?

4. Is $(0, -7)$ the solution to $y - 4 = 3x$ and $y = 5x$?

5. Is $(-1, 3)$ the solution to $y = 5x + 8$ and $y = -x + 2$?

6. Is $(-2, 2)$ the solution to $y = 3x + 8$ and $y = 5x + 12$?

7. Is $(2, 1)$ the solution to $y - 4 = 2x$ and $y - 7x = 8$?

8. Is $(-1, 3)$ the solution to $y + 5 = -8x$ and $2y = 6x + 12$?

9. Is $(-4, 0)$ the solution to $3y = 6x + 8$ and $y - 9 = 2x - 1$?

10. Is $(-3, -2)$ the solution to $y - 6 = 2x - 2$ and $y = -x - 5$?

Check your answers on page 746. ✓

10.2 Solving a System of Equations by Graphing

One way to find the point where two lines meet is to graph both lines on the same Cartesian coordinate system. If the coordinates of the points are integers, and if you use graph paper, you should be able to read the coordinates of the point where the lines cross

To avoid errors, graph each line separately. First determine the coordinates of three points on one line, plot them, and sketch the line that goes through all of them. Then do the same for the other line.

EXAMPLE 1   Find the point where $y = x + 1$ and $y = 2x$ meet.

$y = x + 1$

x	0	1	2
y	1	2	3

$y = 2x$

x	0	1	2
y	0	2	4

These lines are on the following graph. They meet at $(1, 2)$.

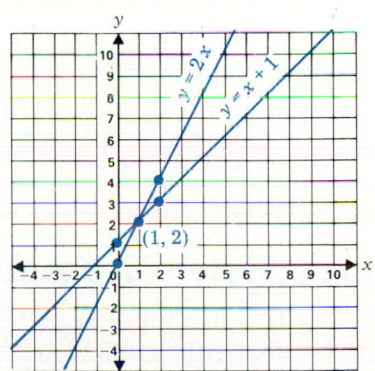

EXERCISES 10.2

Find three solutions for each equation. Using graph paper, graph each line and name the coordinates of the point where the lines of each pair meet.

1. $y = 2x + 1$

x		
y		

$y = x + 4$

x		
y		

Solution: _____

2. $y = 3x + 5$

x		
y		

$y = 2x + 3$

x		
y		

Solution: _____

3. $y = 5x - 2$

x		
y		

$y = 2x + 1$

x		
y		

Solution: _____

4. $y = 4x - 3$

x		
y		

$y = -x - 3$

x		
y		

Solution: _____

5. $y = 7x + 2$

x		
y		

$y = 3x - 6$

x		
y		

Solution: _____

6. $y = -2x + 6$

x		
y		

$y = 2x - 2$

x		
y		

Solution: _____

7. $y = 4x + 1$

x		
y		

$y = -3x - 6$

x		
y		

Solution: _____

Check your answers on page 746. ✓

 **Solving a System of Equations by Substitution**

It is helpful to know how to solve a system of equations by graphing because that process helps reinforce your understanding of what the solution to a system is. But now that you know *what* you are doing, it is important for you to learn other methods for finding the solution. That is because there are two serious drawbacks to graphing:

1. You cannot *precisely* read coordinates that are *not integers* from a Cartesian coordinate system.

2. You cannot *precisely* read *any* coordinates from a Cartesian coordinate system unless you use graph paper.

Fortunately there are algebraic methods that you can use to determine quickly and precisely the exact coordinates of every solution to a system of two linear equations. In this section we will study the algebraic method called **substitution**. I hope the discussion in the next few paragraphs demonstrates a justification for you of the procedure we must follow to solve a system of equations by substitution.

For a point to be on the line $y = x + 1$, the y-coordinate must be $x + 1$ for each value of x. Every point on the line $y = x + 1$ can be written as $(x, x + 1)$.

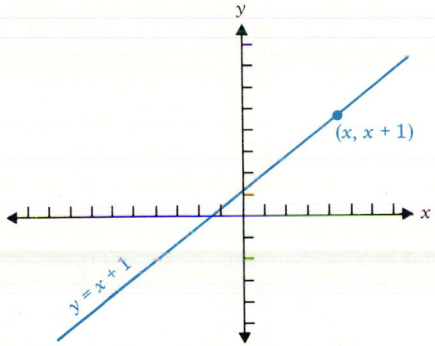

For a point to be on the line $y = 2x$, the y-coordinate must be $2x$ for each value of x. Every point on the line $y = 2x$ can be written as $(x, 2x)$.

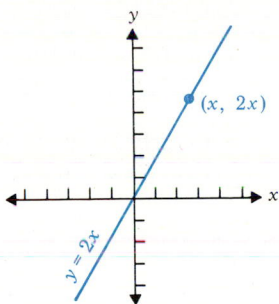

The solution to the system $y = x + 1$ and $y = 2x$ is the point where those two lines meet. We can see that solution by graphing both lines on the same Cartesian coordinate system. The value of x of that point produces the same value of y from the expression $x + 1$ as it does from the expression $2x$. So the solution is the value of x that makes $x + 1 = 2x$.

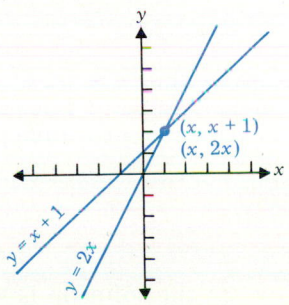

The Substitution Method

If one letter is isolated on one side of either equation in a system, we know that the algebraic expression on the other side of the equation is equal to that letter. Therefore we can *substitute that expression* for the same letter in the other equation. The result will always be an equation with only *one* letter.

Solve that equation and substitute the solution into either equation to solve for the other letter. This method is called *substitution*.

EXAMPLE 1 Use the substitution method to solve the system

$$y = x + 1 \quad \text{and} \quad y = 2x.$$

Since $y = [x + 1]$, we can substitute $[x + 1]$ for y in the other equation, $[y] = 2x$:

$$y = [x + 1] \qquad [y] = 2x$$

So

$$
\begin{aligned}
[y] &= 2x \\
[x + 1] &= 2x \\
-x & \quad\quad -x \\
\hline
1 &= x
\end{aligned}
$$

Use $x = 1$ to solve for y.

$$
\begin{aligned}
y &= x + 1 \\
y &= (1) + 1 \\
y &= 2
\end{aligned}
$$

So the solution to this system is $x = 1$ and $y = 2$. ❑

In Example 1 of Section 10.2, we solved the same system by graphing and found the solution to be that point (1, 2). When asked to find the solutions to a system of equations, simply state the value of each letter. For Example 1 above, for instance, you would say $x = 1$ and $y = 2$.

EXAMPLE 2 Use the substitution method to solve this system:

$$x = 2y - 5 \quad \text{and} \quad 7x - 3y = 9.$$

The only letter that is isolated in this system is x, so substitute $x = [2y - 5]$ in $7[x] - 3y = 9$ for x.

$$x = [2y - 5] \qquad 7[x] - 3y = 9$$

So

$$
\begin{aligned}
7[x] - 3y &= 9 \\
7[2y - 5] - 3y &= 9 \\
14y - 35 - 3y &= 9 \\
11y - 35 &= 9 \\
+ 35 & \quad\quad +35 \\
\hline
11y &= 44 \\
\frac{11y}{11} &= \frac{44}{11} \\
y &= 4
\end{aligned}
$$

Use $y = 4$ to solve for x in $x = 2y - 5$.

$$
\begin{aligned}
x &= 2(4) - 5 \\
&= 8 - 5 \\
&= 3
\end{aligned}
$$

So the solution is $x = 3$ and $y = 4$. ❑

Check: Is $x = 3$ and $y = 4$ a solution to both equations?

If $x = 3$ and $y = 4$, then $\underline{x} = 2\underline{y} - 5$ $7\underline{x} - 3\underline{y} = 9$

$\qquad (3) \stackrel{?}{=} 2(4) - 5 \qquad 7(3) - 3(4) \stackrel{?}{=} 9$

$\qquad\qquad 3 \stackrel{?}{=} 8 - 5 \qquad\qquad 21 - 12 \stackrel{?}{=} 9$

$\qquad\qquad 3 = 3 \quad \checkmark \qquad\qquad\qquad 9 = 9 \quad \checkmark$

Substitution can be used to solve any system of equations if one of the letters is isolated. The letters are not always x and y.

EXAMPLE 3   To solve the system $a = 5b + 4$ and $2a - b = -10$, substitute $[5b + 4]$ for a in the second equation.

$$a = [5b + 4] \qquad 2[a] - b = -10$$

So

$$2[5b + 4] - b = -10$$
$$10\underline{b} + 8 - \underline{b} = -10$$
$$9b + 8 = -10$$
$$\underline{\qquad -8 \qquad -8\qquad}$$
$$9b = -18$$
$$b = -2$$

Now solve for a:

$$a = 5[b] + 4$$
$$= 5[-2] + 4$$
$$= -10 + 4$$
$$= -6$$

The solution is $b = -2$ and $a = -6$.   ❏

Check:

$$a = 5b + 4 \qquad\qquad 2a - b = -10$$
$$(-6) \stackrel{?}{=} 5(-2) + 4 \qquad 2(-6) - (-2) \stackrel{?}{=} -10$$
$$-6 \stackrel{?}{=} -10 + 4 \qquad\qquad -12 + 2 \stackrel{?}{=} -10$$
$$-6 \stackrel{?}{=} -6 \quad \checkmark \qquad\qquad -10 \stackrel{?}{=} -10 \quad \checkmark$$

Substitution can also be used on any system of equations where at least one coefficient is 1, whether or not that term is isolated on one side of the equation. Solve that equation for the letter with the coefficient 1 and then substitute the expression to which it is equal into the other equation.

EXAMPLE 4   Solve the system $3n + m = -1$ and $4m - 3n = 11$.

Solve $3n + m = -1$ for m.

$$3n + m = -1$$
$$\underline{-3n \qquad\qquad -3n}$$
$$m = [-3n - 1],$$

which we can substitute for m in the other equation, $4[m] - 3n = 11$:

$$4[m] - 3n = 11 \longrightarrow 4[-3n - 1] - 3n = 11$$
$$-12n - 4 - 3n = 11$$
$$-15n - 4 = 11$$
$$\underline{+4 \qquad +4}$$
$$-15n = 15$$
$$\frac{-15n}{-15} = \frac{15}{-15}$$
$$n = -1$$

Use $n = -1$ to solve for m:

$$m = -3n - 1$$
$$m = -3(-1) - 1$$
$$m = 3 - 1$$
$$m = 2$$

So the solution to this system is $n = -1$ and $m = 2$.

Always check the solutions in the *original* equations.

Check:

$$3n + m = -1 \quad \text{and} \quad 4m - 3n = 11$$
$$3(-1) + (2) \overset{?}{=} -1 \qquad 4(2) - 3(-1) = 11$$
$$-3 + 2 \overset{?}{=} -1 \qquad \qquad 8 + 3 = 11$$
$$-1 = -1 \;\checkmark \qquad \qquad 11 = 11 \;\checkmark$$

The word **monic** refers to having the coefficient 1. In a monic system of equations, at least one term has the coefficient 1. So we can say that the substitution method works well on monic systems.

EXERCISES 10.3A

Solve each system of equations by using the substitution method, and check your solutions.

1. $y = x + 3$ and
$5y - 2x = 21$

2. $y = x - 5$ and
$6y + 9 = 3x$

3. $x = 3y$ and
$4x = 5y + 14$

4. $x = -5y$ and
$-3x = 9y + 6$

5. $a - 2b = -3$ and
$3b - 4a = -13$

6. $n = 4m + 1$ and
$-2n + 5m = 1$

7. $y = 3x - 4$ and
$4y + 6 = 7x$

8. $y - 7x = -4$ and
$2y - 3 = 3x$

9. $x + y = -4$ and
$y + 2x = 4$

10. $w = -2a + 1$ and
$-3a - w = 2$

11. $x = 5y$ and
$3y - 4x = 17$

12. $x = y + 3$ and
$4x + 3y = -2$

13. $x = 4y - 1$ and
$2x - 5y = 4$

14. $x = 2y - 5$ and
$4y - 3x = 9$

15. $2y - x = 7$ and
$y = -7x - 4$

16. $x = 2y + 3$ and
$6x - 4y = -14$

Check your answers on page 746.

Solving Systems with Fractional Solutions

Now that you are familiar with the substitution method, we can use it on systems where the solutions are fractions. Be careful when you substitute the value of one letter in an equation to find the value of the other letter.

If you are not comfortable working with fractions, review Chapter 3 before proceeding.

EXAMPLE 5   Use the substitution method to solve the system $y = 2x - 5$ and $2y = 8x - 13$.

$$y = [2x - 5] \quad \text{and} \quad 2[\,y\,] = 8x - 13$$

$$
\begin{aligned}
2[2x - 5] &= 8x - 13 \\
4x - 10 &= 8x - 13 \\
-4x &\quad -4x \\
\hline
-10 &= 4x - 13 \\
+13 &\quad\quad +13 \\
\hline
3 &= 4x \\
\frac{3}{4} &= \frac{4x}{4} \\
\frac{3}{4} &= x
\end{aligned}
$$

Now solve for y:

$$
\begin{aligned}
y &= 2x - 5 \\
&= 2\left(\frac{3}{4}\right) - 5 \\
&= \frac{2 \cdot 3}{2 \cdot 2} - 5 \\
&= \frac{3}{2} - 5 \\
&= \frac{3}{2} - \frac{10}{2} \\
&= -\frac{7}{2} \ \text{or} \ -3\frac{1}{2}
\end{aligned}
$$

The solutions are $x = \dfrac{3}{4}$ and $y = -3\dfrac{1}{2}$.   ❑

Check: Use the improper fraction form of y, $\left(-\dfrac{7}{2}\right)$ to check.

$$
\begin{array}{ll}
y = 2x - 5 & 2y = 8x - 13 \\[4pt]
\left(-\dfrac{7}{2}\right) \overset{?}{=} 2\left(\dfrac{3}{4}\right) - 5 \qquad & 2\left(-\dfrac{7}{2}\right) \overset{?}{=} 8\left(\dfrac{3}{4}\right) - 13 \\[8pt]
-\dfrac{7}{2} \overset{?}{=} \dfrac{3}{2} - 5 & \dfrac{2 \cdot (-7)}{2} = \dfrac{4 \cdot 2 \cdot 3}{4} - 13 \\[8pt]
-\dfrac{7}{2} \overset{?}{=} \dfrac{3}{2} - \dfrac{10}{2} & -7 \overset{?}{=} 6 - 13 \\[8pt]
-\dfrac{7}{2} = -\dfrac{7}{2} \quad \checkmark & -7 = -7 \quad \checkmark
\end{array}
$$

Do not become intimidated when working with fractions, but do be careful! If you take your time, do only one calculation per line, and write down each step instead of doing it in your head, the work will not be difficult.

It is especially important to check fractional answers. Do not wait for someone else to find your mistakes. If *you* find them, you can *correct them* and will eventually *stop making them*. Good luck!

EXERCISES 10.3B

Use the substitution method to solve each system of equations, and check your solutions.

17. $y = 2x + 5$ and
$3y - 10x = 13$

18. $x = 3y + 1$ and
$9y + 4x = 11$

19. $y = 4x - 2$ and
$12y + 9x = -5$

20. $x = 6y - 2$ and
$15y - 20x = -2$

21. $x = 9y$ and
$2x - 3y = -10$

22. $y = 3x - 5$ and
$6x - 3y = 14$

Check your answers on page 746.

10.4   Solving a System of Equations by Addition

In theory the substitution method is the only one we need for transforming a system of linear equations into a one-letter equation that we can solve. As we learned in Section 7.2, we can always solve an equation for any letter in it, thus isolating that letter on one side.

But to solve an equation for a letter that has a coefficient other than 1, we must divide the equation by that coefficient. The resulting expression will probably be too complicated to substitute successfully into the other equation. That is why I recommend using the substitution method only when *one letter appears as a monic term* (with the coefficient 1).

EXAMPLE 1   If we tried to solve the system

$$2y + 3x = 11 \quad \text{and} \quad 5y - 3x = 17$$

by substitution, we would have to solve for x or y in one of the equations. I will solve each for y:

$$
\begin{array}{ll}
2y + 3x = 11 & 5y - 3x = 17 \\
\underline{ - 3x \quad\quad -3x} & \underline{ + 3x \quad\quad +3x} \\
2y = 11 - 3x & 5y = 17 + 3x \\
\dfrac{2y}{2} = \dfrac{11 - 3x}{2} & \dfrac{5y}{5} = \dfrac{17 + 3x}{5} \\
y = \dfrac{11}{2} - \dfrac{3x}{2} & y = \dfrac{17}{5} + \dfrac{3x}{5}
\end{array}
$$

I would not want to substitute either of those expressions into the other equation, would you? Fortunately there is another algebraic method that is much easier to use on systems that are not monic. This method is called the **addition method**.

The Addition Method

In the addition method, as its name suggests, we add the equations together. Just as we did in the substitution method, we want to transform the system into *one equation* with only *one letter*. So we want *one of the letters* temporarily to *fall out of the system*.

You can always *add two equations together*, because the two sides to any equation are always equal. When we add equations together, we are adding equal amounts to equal amounts, and both sides continue to be equal.

EXAMPLE 2   Let's add the equations from Example 1 together. The x terms will fall out, since $+3x - 3x = 0x$, leaving an equation that we can solve for y. Once we know y, we can substitute it into either equation and solve for x.

$$\begin{array}{r} 2y + 3x = 11 \\ +5y - 3x = 17 \\ \hline 7y + 0x = 28 \end{array}$$

$$7y = 28$$

$$\frac{\cancel{7}y}{\cancel{7}} = \frac{28}{7}$$

$$y = 4$$

Now, to solve for x substitute $y = 4$ into $2y + 3x = 11$.

$$\begin{array}{r} 2y + 3x = 11 \\ 2(4) + 3x = 11 \\ 8 + 3x = 11 \\ -8 -8 \\ \hline 3x = 3 \end{array}$$

$$\frac{3x}{3} = \frac{3}{3}$$

$$x = 1$$

The solution is $x = 1$ and $y = 4$.   ❏

Check:

$$\begin{array}{cc} 2y + 3x = 11 & \qquad 5y - 3x = 17 \\ 2(4) + 3(1) = 11 & \qquad 5(4) - 3(1) = 17 \\ 8 + 3 = 11 & \qquad 20 - 3 = 17 \\ 11 = 11 \quad \checkmark & \qquad 17 = 17 \quad \checkmark \end{array}$$

The x terms fell out when we added the equations because the coefficients of the x terms were **opposites** (same numbers but different signs). The sum of opposites is always zero.

EXAMPLE 3   To solve the system $5y = 4x - 22$ and $3x - 5y = 19$, add the equations to eliminate the y terms ($5y$ and $-5y$).

$$\begin{array}{r} 5y = 4x - 22 \\ +3x - 5y = 19 \\ \hline 3x + 0y = 4x - 3 \\ 3x = 4x - 3 \\ -4x -4x \\ \hline -x = -3 \end{array}$$

$$\frac{-x}{-1} = \frac{-3}{-1}$$

$$x = 3$$

Now solve for y by substituting $x = 3$ into $5y = 4x - 22$.

$$5y = 4x - 22$$
$$5y = 4(3) - 22$$
$$5y = 12 - 22$$
$$5y = -10$$
$$\frac{5y}{5} = \frac{-10}{5}$$
$$y = -2$$

The solution is $x = 3$ and $y = -2$.

Check:

$$5y = 4x - 22 \qquad\qquad 3x - 5y = 19$$
$$5(-2) = 4(3) - 22 \qquad 3(3) - 5(-2) = 19$$
$$-10 = 12 - 22 \qquad\qquad 9 + 10 = 19$$
$$-10 = -10 \quad\checkmark \qquad\qquad 19 = 19 \quad\checkmark$$

EXERCISES 10.4A

Solve each system of equations by using the addition method, and check your solutions.

1. $2y - 5x = -1$ and
   $3y + 5x = 11$

2. $7y - 4x = -1$ and
   $-7y + 3x = -1$

3. $4y + 3x = -2$ and
   $5y - 3x = 11$

4. $6x - 2y = 2$ and
   $-6x + 3y = 3$

5. $4x - y = 4$ and
   $3x + y = 17$

6. $5y - 2x = 5$ and
   $7y + 2x = 55$

7. $6y = 4x$ and
   $-3y = -4x + 30$

8. $3x = 5y$ and
   $4y - 3x = -3$

9. $2y = 6x$ and
   $2x - 2y = 20$

10. $7x - 3y = 6$ and
    $4x + 3y = 27$

Check your answers on page 746. ✓

Preparing Systems for the Addition Method

Adding the equations in the systems from Examples 2 and 3 eliminated a letter because the coefficients of that letter term in each equation were opposites. If neither letter term has opposites as coefficients, we have to *prepare* the equations so the addition method will work.

Remember, you can always *multiply* both sides of an equation by the same nonzero number without changing the equation's value. (If we multiply by zero, we get $0 = 0$, which is also true but does not help us determine the value of the letters.) Always look over the system first and make the work as easy for yourself as possible.

> **To Prepare a System for the Addition Method:**
> If the coefficients of a letter term are already opposites, just add the equations together.
>
> If not, is the coefficient of one letter term a multiple of that letter's coefficient in the other equation? If so, multiply the other equation by the number that will produce the *opposite* coefficient.

EXAMPLE 4   Solve $2x - 3y = -5$ and $4x - 4y = 8$.

No coefficient is 1, so we should use the addition method.

The coefficients of x are 2 and 4; of y are -3 and -4.

Neither pair of coefficients are opposites, but 4 *is a multiple of* 2, so let's try to *remove the x terms*.

We must multiply both sides of the equation involving the $2x$ term by -2 to produce a coefficient of -4 (the opposite of 4). Then we add that equation to the other equation.

$$-2[2x - 3y] = -2[-5] \longrightarrow \begin{array}{r} -4x + 6y = 10 \\ +4x - 4y = 8 \\ \hline 0x + 2y = 18 \\ 2y = 18 \\ y = 9 \end{array}$$

We're not through solving this system until we find the value of both letters. Once we have solved for one letter, we must solve for the other letter by substituting into one of the *original* equations. I usually use the equation with the smallest coefficients.

Solve for x by substituting $y = 9$ into $2x - 3y = -5$.

$$\begin{array}{r} 2x - 3y = -5 \\ 2x - 3(9) = -5 \\ 2x - 27 = -5 \\ +27 +27 \\ \hline 2x = 22 \\ x = 11 \end{array}$$

The solution is $x = 11$ and $y = 9$.   ❏

Check:

$$\begin{array}{cc} 2x - 3y = -5 & 4x - y = 8 \\ 2(11) - 3(9) \stackrel{?}{=} -5 & 4(11) - 4(9) \stackrel{?}{=} 8 \\ 22 - 27 \stackrel{?}{=} -5 & 44 - 36 \stackrel{?}{=} 8 \\ -5 = -5 \quad ✓ & 8 = 8 \quad ✓ \end{array}$$

EXAMPLE 5   Solve $7x - 9y = -5$ and $2x + 3y = -7$.

The coefficients of y are -9 and 3. -9 is a multiple of 3, so let's remove the y terms.

Multiply both sides of $2x + 3y = -7$ by $+3$, so the y term ($+3y$) will become $+9y$, the opposite of $-9y$ (in the other equation).

$$+3[2x + 3y] = +3[-7] \longrightarrow 6x + 9y = -21$$
$$\underline{+\ 7x - 9y = \quad -5}$$
$$13x + 0y = -26$$
$$13x = -26$$
$$x = -2$$

Solve for y by substituting $x = -2$ into $2x + 3y = -7$.

$$2x + 3y = -7$$
$$2(-2) + 3y = -7$$
$$-4 + 3y = -7$$
$$\underline{+4 \qquad\quad +4}$$
$$3y = -3$$
$$\frac{3y}{3} = \frac{-3}{3}$$
$$y = -1$$

The solution is $x = -2$ and $y = -1$.

Check:

$$2x + 3y = -7 \qquad\qquad 7x - 9y = -5$$
$$2(-2) + 3(-1) = -7 \qquad 7(-2) - 9(-1) = -5$$
$$-4 - 3 = -7 \qquad\qquad -14 + 9 = -5$$
$$-7 = -7 \ \checkmark \qquad\qquad -5 = -5 \ \checkmark$$

EXERCISES 10.4B

Solve each system of equations by multiplying one equation (on both sides) by the number that will result in each equation having the opposite coefficients for one of the letters. Then add the equations together and solve for the resulting equation's one letter. Use that solution to solve for the other letter, and check your solutions.

11. $3y - 4x = 1$ and
$2y + x = 8$

12. $5x + 3y = 5$ and
$2x - y = 13$

13. $4y + 6x = 2$ and
$3y + x = -9$

14. $7x - 4y = 2$ and
$5x - y = 7$

15. $8y - 6x = 6$ and
$3y + 2x = 15$

16. $4y + 10x = 10$ and
$3y - 2x = 17$

17. $7x - 4y = -1$ and
$3x - 2y = -1$

18. $6x + 5y = 3$
$2x + 3y = -3$

19. $5a + 2b = -1$ and
$10a + 3b = -4$

20. $4p - 3m = -1$ and
$-3p + 6m = 12$

21. $2k + 4n = 16$ and
$3k + 2n = 4$

22. $8n - 9m = 6$ and
$2n + 5m = 16$

Check your answers on page 747.

Using Coefficients' LCM to Set up a System for the Addition Method

In some non-monic systems, neither letter's coefficients are multiples of the other.

EXAMPLE 6   In the system $\underline{3x} + \underline{2y} = 11$ and $\underline{5x} - \underline{3y} = 12$, neither the x nor the y coefficients are multiples of each other.

The coefficients of x are 3 and 5. 5 is not a multiple of 3.

The coefficients of y are 2 and -3. -3 is not a multiple of 2.

We cannot just multiply one equation by some integer and get the two coefficients of either letter to be opposites.   ❑

For any pair of numbers, we can always find the lowest common multiple (LCM). The LCM is the smallest number they each divide into. We have used LCMs before—in Section 3.7 to add fractions and in Section 6.4 to clear away denominators before solving equations. If you have forgotten how to find the LCM of numbers, review Section 3.7 before proceeding.

Multiply each equation by the integer that will produce the LCM of one letter's coefficients. Be sure that in one equation the LCM will be positive and in the other equation it will be negative, so that in the resulting system that letter's coefficients will be opposites.

To use the addition method (when no coefficients are multiples):

1. Choose one of the terms to eliminate. (It usually does not matter which letter.)

2. Find the LCM of that term's coefficients.

3. Multiply each equation, on *both sides*, by the integer that will produce the LCM for the coefficients of the term you have chosen to remove. Be sure that the *signs* of the terms end up being *different*.

4. Add the two resulting equations together.

5. Solve the resulting equation for the remaining letter.

6. Substitute that value into one of the *original* equations to solve for the other letter.

7. Check your solutions.

EXAMPLE 7   Solve $\underline{3x} + \underline{2y} = 11$ and $\underline{5x} - 3y = 12$.

1. I'll set this system up for the y terms to fall out since they already have different signs.

2. The coefficients of y are 2 and -3; their LCM is 6 ($2 \cdot 3 = 6$).

3. Multiply both sides of $3x + 2y = 11$ by 3 to convert the y term ($2y$) into $6y$.

   Multiply both sides of $5x - 3y = 12$ by 2 to convert the y term ($-3y$) into $-6y$.

$$
\begin{array}{rcl}
3[3x + 2y] = 3[11] & \longrightarrow & 9x + 6y = 33 \\
2[5x - 3y] = 2[12] & \longrightarrow & +10x - 6y = 24 \\
\end{array}
$$

4. $\qquad\qquad\qquad\qquad\qquad\qquad\qquad\qquad 19x + 0y = 57$

5. $\qquad\qquad\qquad\qquad\qquad\qquad\qquad\qquad\quad 19x = 57$

$$\frac{19x}{19} = \frac{57}{19}$$

$$x = 3$$

6. Solve for y by substituting $x = 3$ into $3x + 2y = 11$.

$$3(3) + 2y = 11$$
$$9 + 2y = 11$$
$$\underline{-9 \qquad -9}$$
$$2y = 2$$
$$y = 1$$

The solution is $x = 3$ and $y = 1$. ❑

7. Check:

$$\begin{array}{cc} 3x + 2y = 11 & 5x - 3y = 12 \\ \underline{3(3) + 2(1) = 11} & \underline{5(3) - 3(1) = 12} \\ 9 + 2 = 11 & 15 - 3 = 12 \\ 11 = 11 \;\checkmark & 12 = 12 \;\checkmark \end{array}$$

EXAMPLE 8   Solve $3x - 2y = 7$ and $5x - 7y = -3$.

It looks like the same amount of work to eliminate either letter. This time let's remove the x terms.

The coefficients of x are 3 and 5. The LCM of 3 and 5 is 15.

Multiply $3x - 2y = 7$ by 5 so the x term ($3x$) will become $15x$.

Multiply $5x - 7y = -3$ by -3 so the x term ($5x$) will become $-15x$.

$$\begin{array}{lll} 5\,[3x - 2y] = 5\,[7] & \longrightarrow & 15x - 10y = 35 \\ -3\,[5x - 7y] = -3\,[-3] & \longrightarrow & \underline{+\; -15x + 21y = \; 9} \\ & & 11y = 44 \\ & & \dfrac{\cancel{11}y}{\cancel{11}} = \dfrac{44}{11} \\ & & y = 4 \end{array}$$

Solve for x by substituting $y = 4$ into $3x - 2y = 7$.

$$3x - 2(4) = \quad 7$$
$$3x - 8 = \quad 7$$
$$\underline{+ 8 \quad +8}$$
$$3x = \quad 15$$
$$x = 5$$

The solution is $x = 5$ and $y = 4$. ❑

Check:

$$\begin{array}{cc} 3x - 2y = 7 & 5x - 7y = -3 \\ 3(5) - 2(4) \stackrel{?}{=} 7 & 5(5) - 7(4) \stackrel{?}{=} -3 \\ 15 - 8 \stackrel{?}{=} 7 & 25 - 28 \stackrel{?}{=} -3 \\ 7 = 7 \;\checkmark & -3 = -3 \;\checkmark \end{array}$$

EXAMPLE 9   Solve $6x - 9y = -3$ and $4x + 6y = 6$. Let's remove the y terms.

The LCM of y's coefficients (-9 and 6) is $3 \cdot 3 \cdot 2 = 18$.

Multiply $6x - 9y = -3$ by 2 so the y term will become $-18y$.

Multiply $4x + 6y = 6$ by 3 so the y term will become $+18y$.

$$2[6x - 9y] = 2[-3] \longrightarrow \quad 12x - 18y = \quad -6$$
$$3[4x + 6y] = 3[6] \quad \longrightarrow \quad + 12x + 18y = \quad 18$$
$$24x = +12$$
$$\frac{24x}{24} = \frac{12}{24}$$
$$x = \frac{1}{2}$$

Solve for y by substituting $x = \dfrac{1}{2}$ into $4x + 6y = 6$.

$$4\left(\frac{1}{2}\right) + 6y = \quad 6$$
$$2 + 6y = \quad 6$$
$$-2 \qquad\qquad -2$$
$$6y = \quad 4$$
$$\frac{6y}{6} = \frac{4}{6}$$
$$y = \frac{2}{3}$$

The solution is $x = \dfrac{1}{2}$ and $y = \dfrac{2}{3}$.

Check:

$$
\begin{array}{cc}
4x + 6y = 6 & 6x - 9y = -3 \\
4\left(\frac{1}{2}\right) + 6\left(\frac{2}{3}\right) \overset{?}{=} 6 & 6\left(\frac{1}{2}\right) - 9\left(\frac{2}{3}\right) \overset{?}{=} -3 \\
2 + 4 \overset{?}{=} 6 & 3 - 6 \overset{?}{=} -3 \\
6 = 6 \ \checkmark & -3 = -3 \ \checkmark
\end{array}
$$

You should now be able to solve any system of two linear equations.

EXERCISES 10.4C

Find the LCM of the coefficients of the letter you want to remove. Multiply each equation (on both sides) by the number that will produce the LCM for that term. Be sure that the resulting coefficients of the term you want to remove are of different signs. Then by adding the equations you can obtain one equation with one letter. Solve it for that letter, then use that solution to solve for the other letter. Check your solutions by substituting them into the *original* equations.

23. $3x + 2y = 16$ and
$2x - 3y = -11$

24. $5x + 3y = 18$ and
$2x + 5y = 11$

25. $7x - 3y = -2$ and
$4x + 4y = 16$

26. $6x + 5y = -3$ and
$4x - 2y = 14$

27. $3y - 8x = 1$ and
$4y + 6x = 18$

28. $10x - 4y = 10$ and
$6x + 5y = 43$

29. $8x + 3y = 1$ and
$10x + 4y = 0$

30. $7x - 4y = -2$ and
$3x - 6y = -18$

31. $5x - 4y = 7$ and
$2x - 4y = -14$

32. $6x + 4y = 10$ and
$4x + 3y = 6$

Check your answers on page 747. ✓

Choosing the Best Method for Solving a System

Now that you know how to use two algebraic methods to solve a system of two linear equations, be sure you use the method that will make the work as simple as possible.

> *Use the substitution method:*
> Whenever the coefficient of one of the letters is 1 in either equation. Solve the equation for that letter, if it is not already isolated, and substitute the resulting expression into the other equation.
>
> *Use the addition method:*
> In all other cases. Look over all the letter terms carefully to see if the coefficients of one letter are already opposites, or if one coefficient is a multiple of the other. If not, find a new coefficient (LCM of old ones) for one of the letters.

Become skilled at both methods. Before you begin working on a problem, think about which will be the easiest way to do it. You can always use addition on any system, but you should not use substitution on a non-monic system.

In the following examples, I will suggest the *method* to use for solving each but will leave the work of solving them up to you.

EXAMPLE 10   I would use substitution to solve $3x + y = 4$ and $2x - 3y = 10$.

Since the y term in $3x + y = 4$ has the coefficient 1, it can be solved for a simple expression for y.

$$\begin{array}{r} 3x + y = 4 \\ -3x \qquad -3x \\ \hline y = [4 - 3x] \end{array}$$

Substitute that expression for y in $2x - 3[y] = 10$.   ❏

EXAMPLE 11   I would use addition to solve $5x + 4y = 7$ and $2x - 4y = -14$.

The y terms will fall out if they are added now since their coefficients, 4 and -4, are already opposites.   ❏

EXAMPLE 12   I would use addition to solve $7y - 2x = 5$ and $4y - 2x = 2$.

The x terms have the *same* coefficients now. Multiply one equation by -1 on both sides to get them to be opposites; then, by adding, the x terms will fall out.   ❏

EXAMPLE 13   I would use addition to solve $3a - 5b = 1$ and $4a - 15b = -7$.

It will be easier to remove the b terms since -15 is a multiple of -5. Do not forget to multiply the first equation by *negative* 3; you want to produce $+15b$ in that equation.   ❏

EXAMPLE 14 I would use addition to solve $2m - 4n = -2$ and $5m - 3n = 2$.
You will need to find an LCM for this one, and it looks like the same amount of work no matter which letter you choose to remove. ❏

EXAMPLE 15 $5k + h = 4$ and $4k + 3h = -10$ can be solved by using either addition or substitution. I would use substitution, as I did in Example 10; I think it is easier. ❏

For some systems of equations, there is no solution.

If the equations represent *parallel* lines, there are *no solutions* (parallel lines never intersect). A system of parallel lines is called an **inconsistent** system and has *no solution*. You can tell if the lines are parallel: They have the same slope (the coefficient of the x term is the same when in slope-intercept form).

EXAMPLE 16 In the system $y = 3x + 5$ and $y = 3x - 4$, both lines have a slope of 3, so they are parallel. This system is inconsistent and has no solution. ❏

If the equations both represent the same line, all points on that line are solutions to the system. A system of equations representing the same line is called a **dependent** system. We cannot list the infinite set of such solutions.

EXAMPLE 17 The system $y + 5x = 9$ and $2y + 10x = 18$ is a dependent system. If you multiply both sides of the first equation by 2, you would see that it is the same as the second equation.

$$2(y + 5x) = 2(9)$$
$$2y + 10x = 18$$ ❏

In this book we only ask you to solve systems of equations that have *one solution*; these systems are called **consistent** or **independent**. There are no inconsistent or dependent systems in any of these exercises, but you will encounter them in more advanced math courses.

EXERCISES 10.4D

First decide which method, substitution or addition, to use; then solve each system and check your solutions.

33. $y = 3x - 5$ and
$y = 4x - 7$

34. $3x + 5y = 13$ and
$2x - y = 0$

35. $5x = 4y + 3$ and
$5x - 3y = 6$

36. $3x - 2y = -1$ and
$y = 2x - 2$

37. $7x - 3y = 2$ and
$3x + 5y = 26$

38. $y = 4$ and
$3x + 7y = 34$

39. $4y = 3x + 1$ and
$8y - 5x = 3$

40. $6x + 15y = -8$ and
$12x + 9y = -2$

41. $3y - 4x = -2$ and
$9y + 8x = 9$

42. $5x - 6y = -2$ and
$10x + 9y = 10$

43. $2x - 6y = -1$ and
$6x + 6y = 5$

44. $7x - 4y = 5$ and
$3x + 2y = 4$

45. $x = 3$ and
$2x - 5y = 16$

46. $5x + y = -5$ and
$9x + 3y = 3$

Check your answers on page 747.

10.5 Solving Real-Life Problems with Systems of Equations

Some real-life problems are easily solved by representing them as systems of two linear equations.

Whenever a problem involves two unknown quantities, you can always represent each of them by a different letter. But to solve for their values, you must be able to translate the information in the problem into **two equations**. You cannot solve for two unknowns unless you have a system of two equations.

Many of the word problems that we set up by a linear equation in Chapter 9 could be set up by systems of two linear equations. If you had difficulty finding expressions for the two unknowns in Chapter 9, you could avoid that problem by representing each unknown as a separate letter.

Learn to set up mixture problems *both* ways, and then you have the flexibility to choose the method you find best. Practice using systems to solve the exercises in this section.

EXAMPLE 1 Millie invested some of her lottery winnings in a NOW account that earned 1.8% simple interest and the rest of her winnings in a savings account that earned 7.5% simple interest. How much was invested in each account if she won $1,000 and earned a total of $57.90 interest from the two accounts after one year?

Account	Quantity Invested ($)	Rate of Interest	Interest ($)
NOW	N	1.8%	$(0.018)N$
Savings	S	7.5%	$(0.075)S$
Total	$1,000		$57.90

The two equations are

1. Quantity in NOW + quantity in savings = $1,000

2. Interest from NOW + interest from savings = $57.90

which translate into

1. $N + S = 1,000$

and

2. $0.018N + 0.075S = 57.90$

The computation will be much easier if we multiply both sides of the second equation by 1,000 to move all decimal points to the right of all digits.

$$(1,000)(0.018N + 0.075S) = (1,000)(57.900) \longrightarrow 18N + 75S = 57,900$$

Let's solve the first equation for N and solve the system by subsitution.

$$N + S = 1,000 \rightarrow N = [1,000 - S]$$

$$18[N] + 75S = 57,900 \rightarrow 18[1000 - S] + 75S = 57,900$$

$$18,000 - 18S + 75S = 57,900$$

$$\begin{aligned} 18,000 + 57S &= 57,900 \\ -18,000 & \quad\quad -18,000 \\ \hline 57S &= 39,900 \end{aligned}$$

$$\frac{57S}{57} = \frac{39,900}{57}$$

$$S = 700$$

Now solve for N by substituting $S = 700$ into $N = 1,000 - S$

$$N = 1,000 - (700)$$

$$N = 300$$

So Millie invested $300 in the NOW account and $700 in the savings account.

❏

Check: Do $300 in the NOW account and $700 in the savings account earn $57.90 in one year?

Interest in NOW =	Interest in savings =
1.8% of 300 =	7.5% of 700 =
(0.018)(300) = 5.40	(0.075)(700) = 52.50

Total interest is $5.40 + $52.50 = $57.90 ✓

There are some word problems that we cannot set up and solve with only one letter and one equation as we did in Chapter 9. That is the case when we do not know any relationship between the two unknowns. For example, in the next problem we want to find the cost of each quahog and each beer. But we are not told that one is a certain amount more than the other, so we cannot represent one by a letter and the other by an expression using that letter.

EXAMPLE 2 One night we went out celebrating after watching the Boston Celtics win the NBA Championship. The bill for seven quahogs and four beers came to $8.85. After discussing Henderson's steal, Parish's dunk, Maxwell's unbelievable baseline turnaround, and Bird's hustle, we ordered again. The second bill for three quahogs and three beers came to $4.95.

Howard paid both bills. Then he wanted to figure out each person's share of the bill so that he could be paid back. But the menu and the bill were both long gone. He needed to determine the cost of each beer and each quahog.

To simplify our calculation we represented everything in terms of cents.

Let Q represent the cost (in cents) of each quahog.
Let B represent the cost (in cents) of each beer.

Our two bills can be translated into the following equations:

7 quahogs + 4 beers = 885¢
3 quahogs + 3 beers = 495¢

These equations translate into

$$7Q + 4B = 885 \quad \text{and} \quad 3Q + 3B = 495,$$

which can be solved by the addition method. Let's remove the B terms. The LCM of 4 and 3 (the coefficients of B) is 12.

$$3[7Q + 4B] = 3[885] \longrightarrow 21Q + 12B = 2655$$
$$-4[3Q + 3B] = -4[495] \qquad + \frac{-12Q - 12B = -1980}{+ \quad 9Q \qquad = \quad 675}$$

$$\frac{9Q}{9} = \frac{675}{9}$$

$$Q = 75$$

Solve for B by substituting $Q = 75$ into $3Q + 3B = 495$.

$$3(75) + 3B = 495$$
$$225 + 3B = 495$$
$$\frac{-225 \qquad\qquad -225}{3B = 270}$$
$$\frac{B}{3} = \frac{270}{3}$$
$$B = 90$$

So each quahog cost 75 cents and each beer cost 90 cents. ❏

Check: At those rates, do seven quahogs and four beers cost $8.85?

$$7 \text{ quahogs cost } 7(\$.75) = \$5.25$$
$$4 \text{ beers cost } 4(\$.90) = \underline{\$3.60}$$
$$\text{Total bill} = \$8.85 \quad \checkmark$$

Do three quahogs and three beers cost $4.95

$$3 \text{ quahogs cost } 3(\$.75) = \$2.25$$
$$3 \text{ beers cost } 3(\$.90) = \underline{\$2.70}$$
$$\text{Total bil} = \$4.95 \quad \checkmark$$

The next example must also be set up with a system of two equations so that we can find the answers. We are not told of any relationship between the commission on used cars and the commission on new cars, so we must represent each unknown commission by a different letter.

EXAMPLE 3 A salesperson at Granny's Best Deal for Your Wheels is paid one commission for each used-car sale and a different commission for each new-car sale.

Her total commissions were $1,275 one month when she sold five used cars and three new cars, and they were $1,950 the next month when she sold two used cars and six new cars.

What kind of commission does she make on each kind of sale?

Let U represent her commission on each used-car sale.

Let N represent her commission on each new-car sale.

We can translate the information from each month's sales into the following equations:

1. Commission on 5 used cars + commission on 3 new cars = $1,275

2. Commission on 2 used cars + commission on 6 new cars = $1,950

These equations become the following systems:

1. $5U + 3N = 1275$

2. $2U + 6N = 1950$

Let's solve this system by the addition method. It will be easiest to eliminate the N terms since 6 is a multiple of 3.

$$-2[5U + 3N] = -2[1275] \longrightarrow \begin{array}{r} -10U - 6N = -2550 \\ +2U + 6N = \underline{1950} \\ -8U = -600 \end{array}$$

$$\frac{-8U}{-8} = \frac{-600}{-8}$$

$$U = 75$$

Solve for N by substituting $U = 75$ into $2U + 6N = 1950$.

$$2(75) + 6N = 1950$$
$$150 + 6N = 1950$$
$$\underline{150 -150}$$
$$6N = 1800$$
$$N = 300$$

So the salesperson makes $75 for each used car and $300 for each new one. ❑

Check: At those rates, does she make $1,275 for selling five used cars and three new cars?

Commission on 5 used cars is 5($75) = $375
Commission on 3 new cars is 3($300) = $900
Total commission = $1,275 ✓

At those rates, does she make $1,950 for selling two used cars and six new cars?

Commission on 2 used cars is 2($75) = $150
Commission on 6 new cars is 6($300) = $1,800
Total commission = $1,950 ✓

EXERCISES 10.5

Set up each problem by using a system of two linear equations with two letters. Solve it to answer each question, and check your solutions.

1. If I find 24 coins, which are all nickels or dimes, how many do I have of each kind of coin if the money has a total value of $2.10?

2. If 10 oz of lemon-pineapple juice is worth 85¢, how much of each juice was used to make the mixture if lemonade costs 7¢ per ounce and organic pineapple juice costs 12¢ per ounce?

3. Rudolph invested $500 in two accounts one earning 5% simple interest and the other 6% simple interest. He earned $26 interest in one year. How much did he put in the 5% account, and how much did he put in the 6% account?

4. If Sandra purchased 10 items on her shopping spree and spent $77.50, how many record albums (at $6.00 each) and how many blouses (at $8.50 each) did she buy?

5. If 60 nickels and quarters are worth $8.00, how many of the coins are nickels and how many are quarters?

6. At the end of the day the merchant figured that he had sold more oldie-but-goodie albums than disco albums. The number of oldies sold was 1 less than twice the discos. If each oldie cost $4 and each disco cost $5, how many of each were sold if the total of album sales was $295?

7. The vendor sold 3 times as many popsicles as ice cream bars. If the total of sales was $8.10, with popsicles selling for 20¢ and ice cream bars selling for 30¢, how many of each treat were sold?

8. The first time Chuck and Peter went to the restaurant, they ordered four hamburgers and three hotdogs. The bill was $15. The next time they ordered two hamburgers and five hotdogs. The bill was $11. How much did each kind of sandwich cost?

9. Mary bought six Cub hats and four Red Sox hats for a total of $38. Ralph bought three Cub hats and three Red Sox hats for a total of $24. How much did each hat cost?

10. Marvin bought 4 box seat tickets, and 10 grandstand tickets for $58. Barbara bought 3 box seat tickets and 6 grandstand tickets for $39. How much did each kind of ticket cost?

11. It cost $5.05 to buy three ice cream cones and two sundaes, and it cost $7.95 to buy five ice cream cones and three sundaes. How much did each cost?

12. Cathy bought seven albums and three singles at the Dusty Disk record shop. Julie bought two singles and five albums. Cathy's purchases cost $38, and Julie's purchases cost $27. How much did the shop charge for each single and each album if all albums cost the same?

13. I bought two pairs of short socks and five pairs of high socks for myself at the Sok Hop Shop and spent $19.85. When I returned to buy seven pairs of short socks and two pairs of high socks for gifts, I spent $19.10. How much did each kind of sock cost?

Check your answers on page 747.

 **Systems of Linear Inequalities**

In Section 8.5 we learned that the solutions to a strict linear inequality are all the points in the coordinate system that are on one side of the line that the linear *equation* represents. If the inequality is not strict, then all the points on that line are solutions too.

EXAMPLE 1   In Example 4 of Section 8.5 we saw that the solution to $y < 2x - 5$ was the following graph.

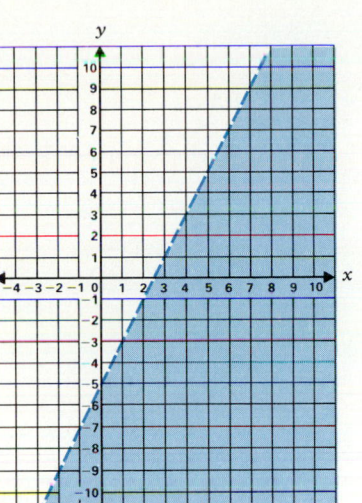

Graph of $y < 2x - 5$

EXAMPLE 2   In Example 6 of that same section we saw that the solution to $x \geq 2$ was the following graph.

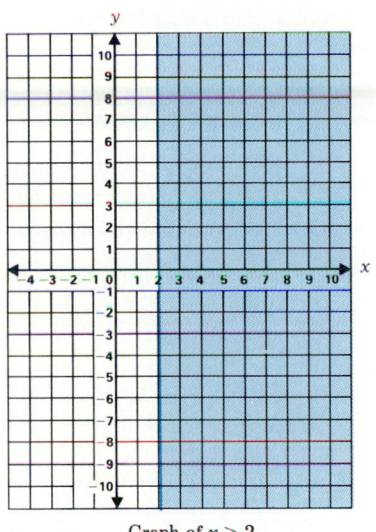

Graph of $x \geq 2$

A **system of inequalities** is made up of two or more linear inequalities. The solution set to a system of inequalities is all the points that are solutions to *every* inequality in the system.

There is an infinite number of points in each solution set, so we cannot list them. Instead we use graphs to represent solution sets. The way to sketch the solution is to graph all the inequalities on the same Cartesian coordinate system. You could shade in the solutions to each inequality by using either a different pattern or a different color. The portion of the graph where all the different shadings overlap is the set of points that are solutions to the system.

EXAMPLE 3   Sketch the solution to this system of inequalities:

$$y < 2x - 5 \quad \text{and} \quad x \geq 2.$$

If we use [   ] shading to represent $y < 2x - 5$ and [   ] shading to represent $x \geq 2$, then the portion with both shadings [   ] is the graph of the solution.

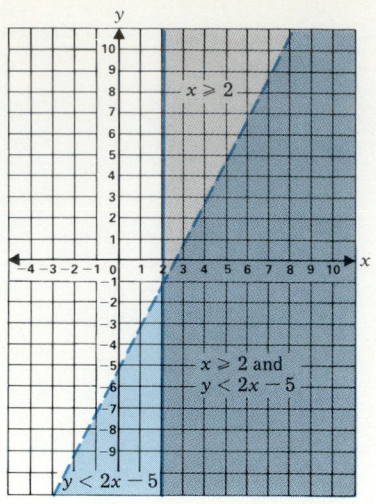

Graph of $y < 2x - 5$ and $x \geqslant 2$

So the solution to the system $y < 2x - 5$ and $x \geq 2$ is

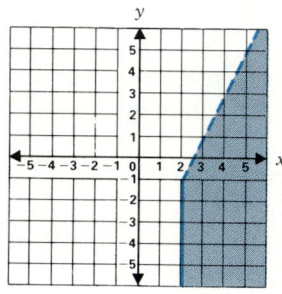

⇒ **Note**: The solution to $y < 2x - 5$ does *not* include the points on the line $y = 2x - 5$, and so that boundary is dotted. The solution to $x \geq 2$ does include the points on the line $x = 2$, and so that boundary is left solid. ❑

The systems of equations that we have learned to solve in this chapter involved only *two linear equations*, but we can sketch the solution to *three linear equalities*.

EXAMPLE 4   Sketch the solution to this system of inequalities:

$$y > 2x - 4; \quad y < 1; \quad \text{and} \quad y > -x - 2.$$

Since all these inequalities are strict, none of the boundaries will be part of the solution, and so all should be dotted.

The solution to each separate inequality follows.

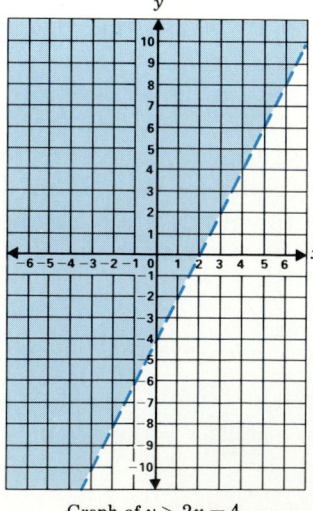

Graph of $y > 2x - 4$

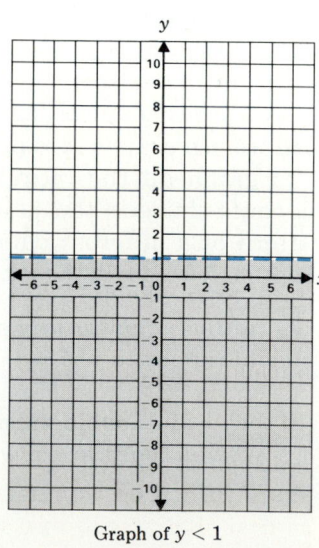

Graph of $y < 1$

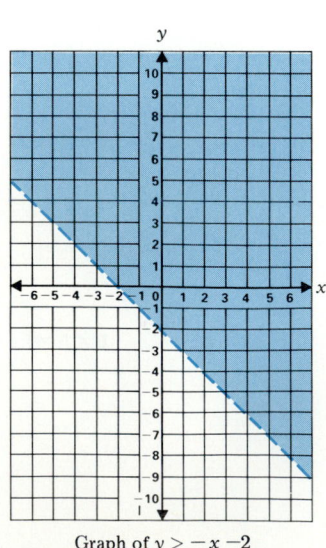

Graph of $y > -x - 2$

To sketch the solution to the *system*, we sketch each of those graphs on the same grid. The portion with all three shadings is our solution.

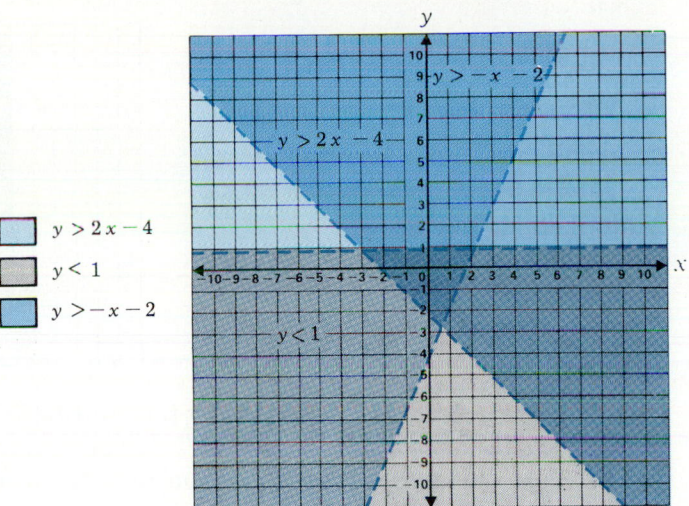

$y > 2x - 4$

$y < 1$

$y > -x - 2$

That solution is graphed below.

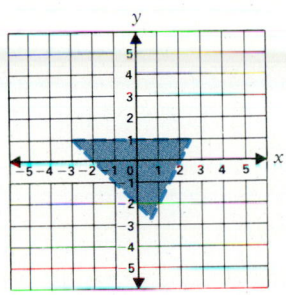

EXAMPLE 5   Sketch the solution to this system:

$$y \geq -3; \quad x > -5; \quad \text{and} \quad y \leq -2x + 1.$$

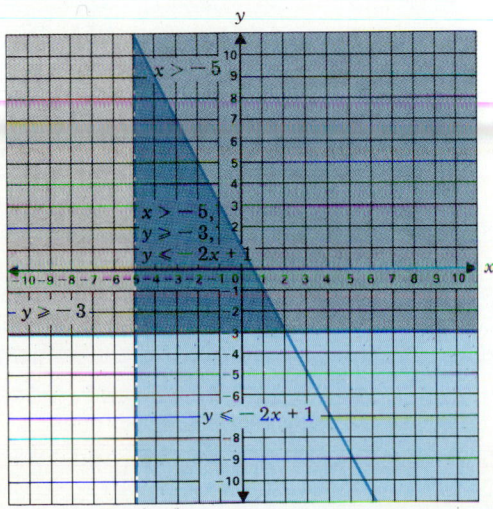

The solution is the portion with all three shadings. The bottom and right boundary of that section is included, and those lines are solid. The left boundary is not included, and that line is dotted. That solution is graphed below:

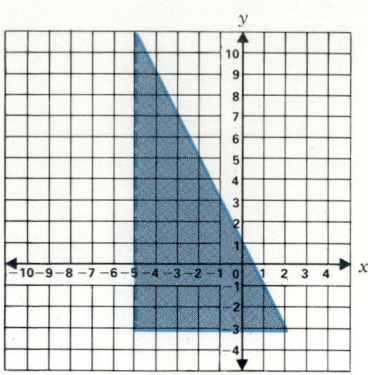

It is possible that the solution set of a system of inequalities could be empty (contain no points). This happens when there is *no overlapped* shaded area. If no point is a solution to every inequality in the system, we say there is no solution to that system.

EXAMPLE 6   We will see that there is no solution to this system:

$$y < -2 \quad y > 3x \quad \text{and} \quad x > 4,$$

There is no portion with all three shadings ▮ .

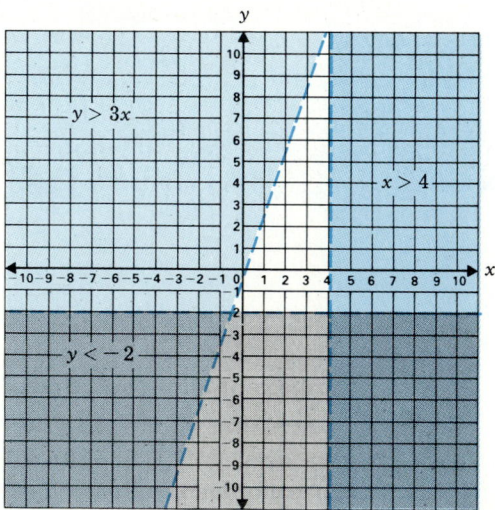

Sketch the solution to each system of inequalities. Be sure to graph boundaries that are part of the solution as solid lines, and boundaries that are not part of the solution as dotted lines.

1. $y < x + 2$ and $y > -3$

2. $y > 3x - 1$ and $y < 5$

3. $y \geq 2x - 3$ and $y < 2$

4. $y \leq -x + 3$ and $y \geq 2x - 5$

5. $y \leq -x + 3$; $y \geq 2x - 5$;   and $x \geq -1$

6. $y < x + 2$; $y > -3$;   and $x < 5$

7. $y > 3x - 1$; $y < 5$;   and $x > -3$

8. $x \geq -4$; $y \geq 2x - 3$;   and $y < 2$

9. $y > 3x - 2$; $y \leq -2x + 2$;   and $x \geq -2$

10. $y > x$; $y < -2x + 5$;   and $y \geq 4x - 5$

Check your answers on page 747.

Glossary

addition method   Method of solving a system of equations by getting them set up so that, when the equations are added, one of the letters will fall out of the resulting equation, leaving an equation with only one letter than can be solved for.

Example: $x + y = 5$
$\underline{+\ x - y = 1}$
$2x \qquad = 6$

$\dfrac{2x}{2} = \dfrac{6}{2}$

$x = 3$ This is used in either equation to solve for y.

$x + y = 5$
$(3) + y = 5$
$y = 2$ Solution is $x = 3$, $y = 2$.

graphing method   Method of finding the coordinates of the point where two lines intersect by graphing both lines on the same Cartesian coordinate system and reading off the coordinates of the point that is on both lines.

Example: The graph of two lines that intersect at $(-4, -2)$ is shown below.

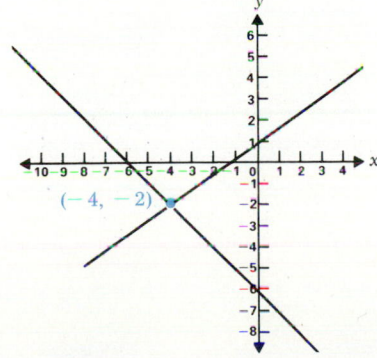

LCM   Lowest common multiple of numbers; the smallest number that each divides into exactly.

Example: The LCM of 6 and 10 is 30.

monic   Refers to a term whose coefficient is 1. A system of equations is monic if it includes at least one monic term.

Example: The system $3x + y = 7$ and $5x - 2y = 8$ is monic because the term y is monic.

opposites   Two numbers that have different signs but equal magnitudes. Their sum is zero.

Example: 7 and -7 are opposites: $(7) + (-7) = 0$.

solution to a system of equations   Coordinates of the point that is a solution to both equations of a system. If the system includes two parallel lines, there is no solution. It called an *inconsistent* system.

Example:

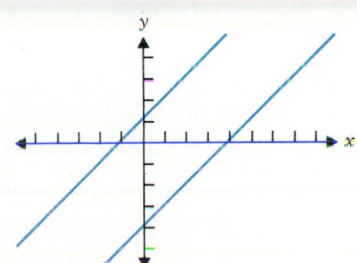

If the system includes two equivalent forms of the same equation, all points on the line are solutions. It is called a *dependent* system.

Example:

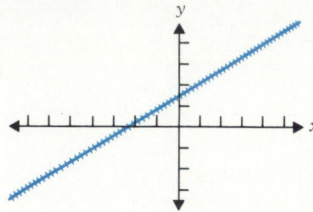

If the system includes two different equations with different slopes, one point with two coordinates will be the solution. It is called a *consistent* or *independent* system.

Example:

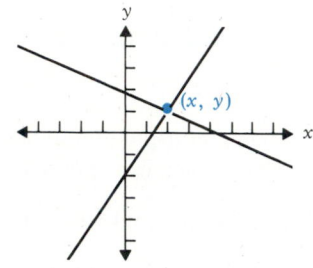

solution to a system of inequalities   All points that are solutions to every inequality in the system.

Example: The solution to $y < 2x - 5$ and $x \geq 2$ is graphed below.

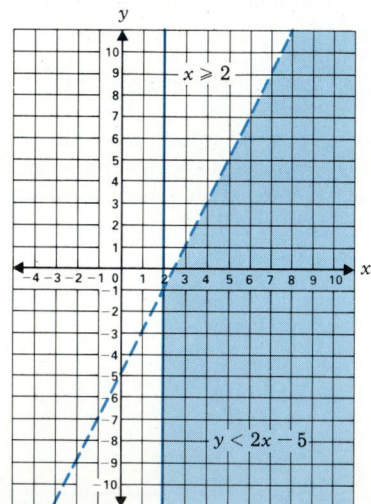

substitution method   Method of solving a system of equations by replacing one letter by an expression to which it is equal and that involves the other letter. (One equation gives you that expression, and it is substituted into the other equation.) Substitution should be used only on monic systems of equations.

Example: Solve $y = x + 2$ and $2y + 3x = 24$ by substitution. Replace y in $2[y] + 3x = 24$ by $[x + 2]$.

$$2[y] + 3x = 24$$
$$2(x + 2) + 3x = 24$$
$$\underline{2x} + 4 + \underline{3x} = 24$$
$$5x + 4 = 24$$
$$\frac{-4 \quad -4}{5x \quad\;\; = 20}$$
$$\frac{\cancel{5}x}{\cancel{5}} = \frac{20}{5}$$
$$x = 4$$
$$y = x + 2$$
$$y = (4) + 2$$
$$y = 6 \qquad \text{The solution is } x = 4 \text{ and } y = 6.$$

system of equations   Any group of more than one equation. In this chapter we have studied systems of two linear equations.

Example: $y = x + 2$ and $2y + 3x = 24$ is a system of two linear equations.

system of inequalities   Any group of more than one inequality.

Example: $y < 2x - 5$ and $x \geq 2$ is a system of two inequalities.

Chapter 10 Review

Do each of the problems. After you are confident that you have done them as accurately as possible, compare your answers with those in the back of the book. If any are wrong, go back to the section where the problem came from (indicated to the left of the problem) and review the section.

Once you understand all the sections, you will be sure to learn the skills solidly and remember how to do them if you practice them a little bit more. Turn to the Supplementary Exercises and do all the problems from any section where you had difficulty on these review exercises.

Answers

1. _____

2. _____

3. _____

4. _____

5. _____

6. _____

7. _____

8. _____

9. _____

10. _____

11. _____

12. _____

13. _____

10.1   1. Is $(4, -5)$ the solution to $y = 3x - 17$ and $2y - 5x = -30$?

10.2   Use graphing to find and state the point where these lines meet.

   2. $y = x + 5$ and $y = -2x - 4$

10.3   Find the solution to these systems by substitution.

   3. $y = x - 3$ and
      $y = -3x + 5$

   4. $x = 3y + 2$ and
      $3x - 4y = -4$

   5. $y - 4x = 0$ and
      $8x = -3y + 3$

10.4   Find the solution to these systems by addition.

   6. $5x - 2y = 3$ and
      $4x + 2y = 6$

   7. $3x - 5y = 11$ and
      $3x + 4y = 2$

   8. $7m + 4n = 10$ and
      $5m - 6n = 16$

   9. $4a + 7b = 5$ and
      $6a - 3b = 21$

10.5   10. $2x + 3y = 2$ and
      $6x - 12y = -1$

   11. $5x = 6y$ and
      $10x - 4y = 40$

Solve each problem by setting up a system of two linear equations.

   12. If part of $3,000 was invested in an account earning 5%, and the rest was invested in an account earning 7%, how much was invested in each account if the total interest in one year was $198?

   13. It cost $11 to buy 4 lb of hamburger and three packages of buns. If it cost $8 to buy 3 lb of hamburger and two packages of buns, how much is each pound of hamburger and each package of buns?

10.6   Find the solutions to these systems of inequalities by graphing them.

   14. $y < 3x$ and $x \le 5$

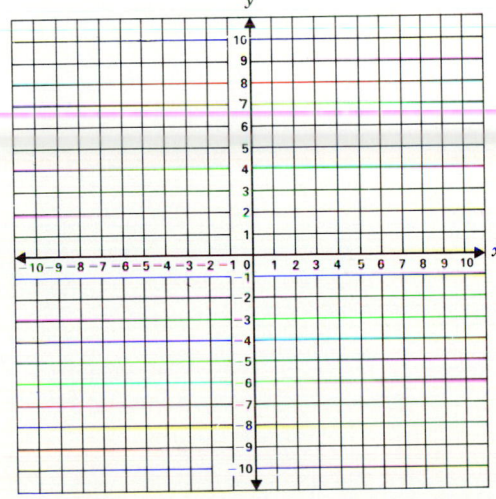

Continued

Chapter 10 Review (*Cont.*)

    15. $y < 4$; $x > -3$; and $y \geq 2x - 1$

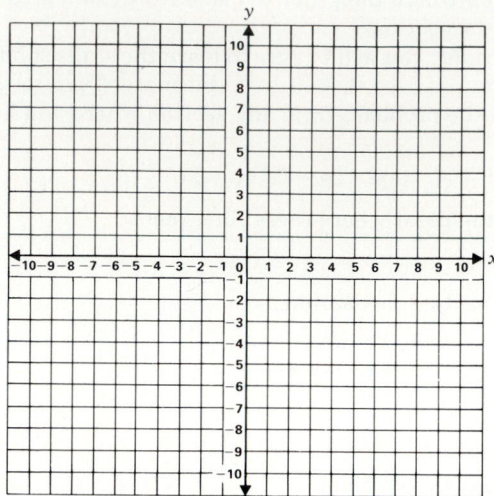

Check your answers on page 748.

Supplementary Exercises

Do all the problems in every section that involves a skill in which you now lack complete mastery. With a little more practice you will achieve that sense of really understanding the topics, and you will remember how to do these problems.

10.1 Determine whether the point given is the solution to the system of equations.

1. $(1, -3)$ $y = 4x - 7$ and
$y = 3x - 4$

2. $(-2, 5)$ $y = 3x + 11$ and
$y + x = 3$

3. $(-4, 0)$ $2y - 5x = 20$ and
$3y + 4x = -16$

4. $(0, 3)$ $y = 2x - 6$
$5y - x = 12$

10.2 Use graphing to find where these lines meet.

5. $y = 3x$
$y = -4x + 7$

6. $y = x + 4$
$y = -3x - 4$

7. $y = -2$
$y = 3x + 1$

8. $x = -3$
$y = x + 5$

10.3 Find the solution to these systems of equations by the substitution method.

9. $y = 4x$ and
$y = 3x - 2$

10. $y = x - 5$ and
$3y - x = -3$

11. $m = n + 3$ and
$4n = -3m - 12$

12. $x = 2y + 5$ and
$3x - 4y = 9$

13. $a - 5b = -1$ and
$10b - 3a = 0$

14. $x - 8y = -5$ and
$3x + 4y = 6$

10.4 Solve these systems of equations by the addition method.

15. $4x - y = -15$ and
$3x + y = 1$

16. $7y - 3x = 1$ and
$5y + 3x = 11$

17. $6p + 4m = -2$ and
$2p + 3m = -4$

18. $5a - 2b = 2$ and
$15a - 7b = 2$

19. $9x - 4y = 1$
$4x + 5y = 14$

20. $6x - 2y = 2$ and
$4x + 5y = 33$

21. $5x + 4y = 1$ and
$10x + 6y = 3$

22. $4x - 6y = 1$ and
$12x - 15y = 4$

Answers

1. _____

2. _____

3. _____

4. _____

5. _____

6. _____

7. _____

8. _____

9. _____

10. _____

11. _____

12. _____

13. _____

14. _____

15. _____

16. _____

17. _____

18. _____

19. _____

20. _____

21. _____

22. _____

Continued

Supplementary Exercises (*Cont.*)

10.5 Solve each problem by setting up a system of two linear equations.

Answers

23. Flo bought three pairs of socks and two pillow cases for $4.05. Frank bought six pairs of socks and three pillow cases for $7.35. How much did each pair of socks cost, and how much did each pillow case cost?

23. socks _____

cases _____

24. Ruth invested $1,500 in two different ways and earned a total of $81 in interest in a year. Find out how much she invested each way if she put some of it in a savings account that earned 7% interest and lent the rest to friends at 5% interest.

24. 7% _____

5% _____

25. The carpenter's bill was twice as much money for labor as it was for supplies. If the total for labor and supplies was $255, how much did the supplies cost, and how much was the labor?

25. labor _____

supplies _____

10.6 Graph the solution to each system of inequalities.

26. $y < 6$ and $y \geq x$

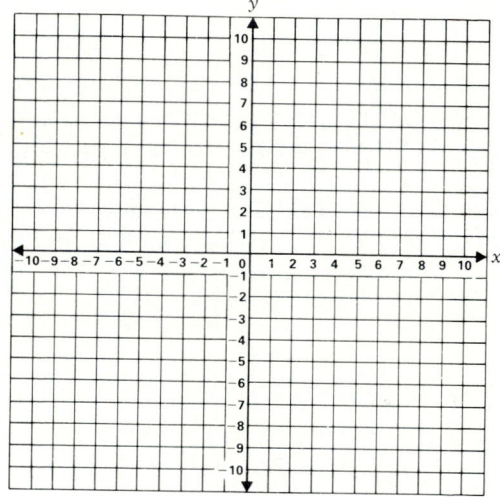

27. $x \leq 4$; $y > -2$;   and $y < x + 5$

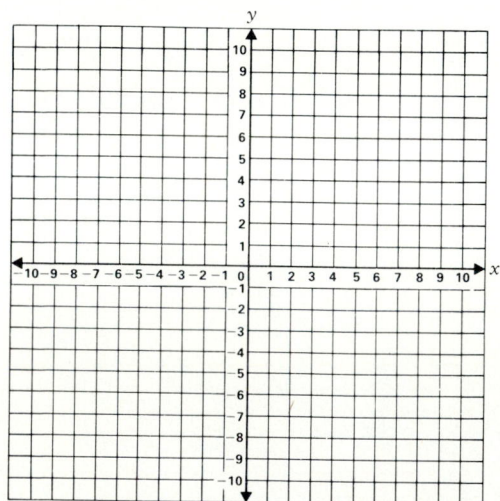

Check your answers on page 748.

Chapter 10 Test

1. Is $(2, -1)$ the solution to $3x + 5y = 1$ and $2x - 3y = 7$?

2. a. Is $(-2, 2)$ a solution to $x > -3$; $y \geq 0$; and $y \leq -2x$?

   b. Is $(3, 5)$ a solution to $x > -3$; $y \geq 0$; and $y \leq -2x$?

Use graphing to determine the point of intersection of the two lines. Indicate the coordinates of the point.

3. $y = 6$
   $y = 3x - 9$

4. $y = 2x - 5$
   $y = 5x - 17$

Find the solution to each system of equations by using whichever algebraic method you prefer; check your solutions.

5. $y = 5x - 3$ and
   $4x - y = 2$

6. $y + x = 4$ and
   $y + 2x = 1$

7. $3x - 4y = 7$ and
   $2x + 5y = 20$

8. $x = 6y - 3$ and
   $2x - 7y = 4$

9. $4a - b = 2$ and
   $12a - 8b = 7$

10. $x = -6$ and
    $3y - 7x = 3$

Solve each problem by setting up a system of two linear equations.

11. Ned bought four rubber worms and two containers of real worms for his fishing trip. If it cost him $4.30, and he noticed that each container of real worms costs 10¢ less than 3 times as much as each rubber worm costs, how much did each of the kinds of bait cost?

12. Each field goal is worth two points, but every free throw is worth only one point. Beth's total of points scored was 37. How many of each did she score if she shot 22 baskets altogether?

13. Graph the solution to $y > 3x - 5$ and $y \leq 4$.

14. Graph the solution to $x > -3$; $y \geq 0$; and $y \leq -2x$.

Answers

1. _____

2a. _____

2b. _____

3. _____

4. _____

5. _____

6. _____

7. _____

8. _____

9. _____

10. _____

11. _____

12. _____

Check your answers on page 749. ✓

Cumulative Review

CHAPTERS 1—10

Combine.

1. $\dfrac{11}{12} - \dfrac{2}{3} + \dfrac{1}{6}$

2. $\dfrac{\dfrac{8}{15}}{\dfrac{12}{30}}$

3. $6.04 - 3.0051$

4. $(12.3)(0.004)$

5. $0.000171 \div 0.003$

6. 11.3% of 96

7. $(-43) + (-21) + (17) - (132)$

8. $(-1)(-1)(2)(-3)(-1)(5)$

9. $\dfrac{44}{-11}$

10. $13 - 5(7 - 4) + (5 - 2)(7 - 8)(3 - 1)$

11. $4(k^2 + 3k - 5) - (k^2 - 4k + 2)$

12. $-2x(3x^2 - 5x + 4)$

13. $8 - 4(9 - 7)^2 + 2^2 + (-3)^2$

14. $(-5 + 3 - 4)^2(-2)^2$

15. $7a(6a - 5b + 1) - 2b(b - 4a - 5)$

16. Solve for p. $4m + 5pn = 7k$

Solve for each unknown.

17. $8x - 4 = 5x + 8$

18. $3(x - 5) + 4x = 3x$

19. $5x - 4y = 6$
    $3x + 4y = 10$

20. $x = 5y - 1$
    $4y - 2x = -10$

What is the slope and y-intercept for each equation?

21. $y = -5x + 7$

22. $7y - 3x = 8$

Find the equation of the line that:

23. has slope -7 and y-intercept 4.

24. goes through points $(1, -4)$ and $3, 8)$.

Answers

1. _____
2. _____
3. _____
4. _____
5. _____
6. _____
7. _____
8. _____
9. _____
10. _____
11. _____
12. _____
13. _____
14. _____
15. _____
16. _____
17. _____
18. _____
19. _____
20. _____
21. _____
22. _____
23. _____
24. _____

25. If 128 out of every 400 of my golf shots are well hit, what is my percentage of good shots?

26. If there were 7 people age 30 or over at the Roy Orbison concert for every 2 people under age 30, out of an audience of 333 people how many were 30 or over?

27. a. How many miles did Ron travel if he drove 45 mph for 6 hr? His car gets 30 mi per gallon of gas, and the gas costs him $1.20 per gallon.
   b. How much did he spend for fuel on that trip?

28. If I have a total of $1.70 in change, and the number of dimes is 1 less than twice the number of quarters, how many of each coin do I have?

Are these lines parallel?

29. $y = -4x + 3$
    $y = 4x + 1$

30. $3y - 15x - 11$
    $4y = 20x + 12$

Answers

25. _____

26. _____

27a. _____

27b. _____

28. _____

29. _____

30. _____

Check your answers on page 749.

Integral Powers and Scientific Notation

 **Integral Exponents**

Positive Exponents

In Section 5.1 we introduced the notation of exponents. Many computations can be greatly simplified through the use of exponents. In this chapter you will learn to perform such computations and to simplify expressions involving exponents.

An exponent that is an integer is called an **integral exponent**. A **positive integral exponent** expresses the number of times that a factor is repeated in a term. We call the factor being repeated the **base**, and the exponent its **power**.

EXAMPLE 1 The term x^3 represents the base x raised to the power 3.

Written without exponents, $x^3 = xxx$. ❏

You will not need to *memorize* any of the rules that we develop for computing with expressions in exponential notation. If you remember that the power in an exponential term represents the number of times its base appears as a factor in that term, you will always be able to recall the rules because they follow logically from that definition of exponential notation.

EXAMPLE 2 Express the term $aaaaabbb$ in exponential notation.

$$aaaaabbb = a^5b^3$$ ❏

Remember that the exponent applies only to the number or letter in front of it (to its left).

EXAMPLE 3 In the term $3ab^4$, the exponent 4 applies only to b.

So $3ab^4 = 3abbbb$ ❏

EXAMPLE 4 In the term $-x^2$ the exponent 2 applies only to x, not to $-x$.

So $-x^2 = -xx$. ❏

If we want an exponent to apply to more than one factor, we must group all the applicable factors within parentheses and write the exponent above the parentheses.

EXAMPLE 5 $(3ab)^4$ represents the fact that the exponent 4 applies to $3ab$.

$$(3ab)^4 = (3ab)(3ab)(3ab)(3ab)$$ ❏

EXAMPLE 6 $(-x)^2 = (-x)(-x)$
$$= +xx$$
$$= x^2$$

So $-x^2 \neq (-x)^2$. These terms represent different expressions.   ❑

Negative Exponents

Recall from our work in Chapter 3 that whenever a factor appears in both the numerator and the denominator of a fraction, we always cancel them to reduce the fraction. If a factor appears *more times in the denominator* than it does in the numerator, that factor will appear in the denominator when the fraction is reduced.

EXAMPLE 7   Simplify $\dfrac{xxx}{xxxx}$.

$$\frac{xxx}{xxxx} =$$

$$\frac{\cancel{x}\cancel{x}\cancel{x}}{\cancel{x}\cancel{x}\cancel{x}x} = \frac{1}{x}$$ ❑

There is an exponential notation that indicates that a factor appears in the denominator and therefore is *dividing* into the numerator. The notation is a **negative exponent**, and it represents the number of times that a factor is divided into the numerator (or appears as a factor in the denominator).

EXAMPLE 8 x^{-1} represents $\dfrac{1}{x}$; x^{-3} represents $\dfrac{1}{x^3}$ or $\dfrac{1}{xxx}$.

a^{-2} represents $\dfrac{1}{a^2}$ or $\dfrac{1}{aa}$; b^{-5} represents $\dfrac{1}{b^5}$ or $\dfrac{1}{bbbbb}$.   ❑

A negative exponent instructs you to *divide* by the base. You must divide by the base as many times as the value of the exponent. *The fact that an exponent is negative has no effect on the sign of the expression.*

To compute with negative exponents, move the base that has a negative exponent into the denominator. This exchange takes care of the instruction from the negative exponent, and the exponent becomes positive.

A term with a negative exponent always becomes a fraction, because the negative exponent represents division by its base.

EXAMPLE 9   Express with all positive exponents: $3ab^{-2}c^3d^{-1}$.

$$3ab^{-2}c^3d^{-1} = \frac{3ac^3}{b^2d}$$ ❑

To simplify a term that includes a negative exponent, you might want to represent it without exponential notation. Always *represent the expression first with positive exponents only*, and then it will be easy to write it without any exponents.

EXAMPLE 10   Represent without any exponents: $2x^2y^{-1}wz^{-4}$.

$$2x^2y^{-1}wz^{-4} =$$

$$\frac{2x^2w}{yz^4} = \frac{2xxw}{yzzzz}$$ ❑

In order **to evaluate** an expression involving exponents, you must express it with *positive* exponents only.

EXAMPLE 11   To evaluate 3^{-2} we must represent it as $\dfrac{1}{3^2}$.

$$3^{-2} =$$

$$\frac{1}{3^2} =$$

$$\frac{1}{3 \cdot 3} = \frac{1}{9} \qquad \Box$$

EXAMPLE 12   Evaluate $(-5)^{-3}$.

$$(-5)^{-3} =$$

$$\frac{1}{(-5)^3} =$$

$$\frac{1}{(-5)(-5)(-5)} =$$

$$\frac{1}{(25)(-5)} = -\frac{1}{125} \qquad \Box$$

EXAMPLE 13   Evaluate -8^{-2}. Remember, the exponent does not apply to the negative sign in front of the 8: $[-8^{-2} = -(8^{-2})]$.

$$-(8^{-2}) =$$

$$-\frac{1}{8^2} =$$

$$-\frac{1}{8 \cdot 8} = -\frac{1}{64} \qquad \Box$$

You have to switch the base into the denominator only if the **exponent** in a product is **negative**. Do not confuse negative *exponents* and negative *bases*.

EXAMPLE 14   Evaluate $(-4)^3$. The exponent (3) is positive, so we do not have to move anything into a denominator. We can evaluate this expression directly.

$$(-4)^3 =$$

$$\underline{(-4)(-4)}(-4) =$$

$$\underline{(16)(-4)} = -64 \qquad \Box$$

Zero Exponents

Any nonzero* base raised to the power zero is always equal to 1.

$$x^0 = 1 \quad \text{if} \quad x \neq 0$$

EXAMPLE 15 $9^0 = 1;\quad (3ab)^0 = 1;\quad 23{,}847^0 = 1;\quad (58x^3y^8)^0 = 1$

as long as a, b, x, and y are not equal to zero. $\qquad \Box$

In Section 11.3 you will learn how to simplify expressions in exponential notation, and you will see *why negative exponents* represent *division* and *why* terms raised to *zero exponents* represent *1*. But now it is important for you to feel comfortable in recognizing what exponential notation represents.

---

* x^0 is not defined when $x = 0$. The term 0^0 does not exist in the real number system.

No Exponent

If no exponent is written above a factor, that always implies that the factor appears only *once*. Remember that, just as with the coefficient 1, generally we never write the exponent 1.

EXAMPLE 16   We do not write ab^4c^2 as $a^1b^4c^2$, but that is what is meant.

$$ab^4c^2 = abbbbcc$$

In summary—this is what exponential notation represents:

$x^n = x \cdot x \cdots x$ if n is a **positive** integer.

  (x appears n times)

  Example: $x^5 = x \cdot x \cdot x \cdot x \cdot x$

$x^{-n} = \dfrac{1}{x \cdot x \cdots x}$ if $-n$ is a **negative** integer.

  (x appears n times)

  Example: $x^{-4} = \dfrac{1}{x \cdot x \cdot x \cdot x}$

$x^0 = 1$ if $x \neq 0$.

  Example: $52^0 = 1$

x represents x^1.

EXERCISES 11A

Express each term without exponents.

1. $2x^4 =$

2. $5y^3 =$

3. $7a^3b^5 =$

4. $5ab^4c^2 =$

5. $(7ab)^2 =$

6. $(2pm)^2 =$

7. $(-4w)^2 =$

8. $(-yz)^4 =$

9. $-2ab^2 =$

10. $-3x^2y^3 =$

11. $y^{-4} =$

12. $w^{-7} =$

13. $x^3y^{-5} =$

14. $5ab^{-3}c^{-2}d =$

15. $6a^{-1}b^2cd^3g^0 =$

16. $2x^3yw^{-2}z^{-3} =$

17. $4xy^{-3}w^0 =$

18. $7a^0bc^{-2}d^3 =$

Evaluate each expression. Replace all negative exponents with expressions involving positive exponents before computing.

19. $6^3 =$

20. $38^0 =$

21. $-7^2 =$

22. $-2^4 =$

23. $(-2)^3 =$

24. $(-7)^2 =$

25. $4^{-2} =$ **26.** $2^{-1} =$

27. $39,506^{0} =$ **28.** $(-1)^{500} =$

29. $(-3)^{-3} =$ **30.** $(-5)^{-2} =$

31. $6^{-3} =$ **32.** $(-2)^{-3} =$

33. $-2^{-3} =$ **34.** $(-1)^{-7} =$

Check your answers on page 749.

Applying Negative Exponents to Fractions

When we apply negative exponents to fractions, the computation can get pretty messy. Remember that every fraction can always be represented as a division problem. To compute a complicated fraction (where the numerator or denominator is itself a fraction) represent it as a division problem (as we learned in Section 3.5).

EXAMPLE 17   Simplify $\left(\dfrac{2}{3}\right)^{-2}$.

$$\left(\frac{2}{3}\right)^{-2} =$$

$$\frac{1}{\left(\dfrac{2}{3}\right)^{2}} =$$

$$1 \div \left(\frac{2}{3}\right)^{2} =$$

$$1 \div \left[\left(\frac{2}{3}\right)\left(\frac{2}{3}\right)\right] =$$

$$1 \div \frac{4}{9} =$$

$$1 \times \left(\frac{9}{4}\right) = \frac{9}{4}$$

Fortunately there is a shortcut that makes such computations much easier.

Since a negative exponent implies that the term is to be divided by the base, when that base is a fraction we can say that a fraction raised to a negative exponent means to divide by that fraction. But as we learned in Section 3.5, to *divide* by a fraction is the same as to *multiply by its reciprocal*.

So when a fraction is raised to a negative power, we can make that exponent positive by replacing the fraction by its reciprocal.

EXAMPLE 17   Using the shortcut, simplify $\left(\dfrac{2}{3}\right)^{-2}$.

$$\left(\frac{2}{3}\right)^{-2} =$$

$$\left(\frac{3}{2}\right)^{2} =$$

$$\left(\frac{3}{2}\right)\left(\frac{3}{2}\right) = \frac{9}{4}$$

EXAMPLE 18   Simplify $\left(-\dfrac{2}{5}\right)^{-3}$.

$$\left(-\dfrac{2}{5}\right)^{-3} =$$

$$\left(-\dfrac{5}{2}\right)^{3} =$$

$$\left(-\dfrac{5}{2}\right)\left(-\dfrac{5}{2}\right)\left(-\dfrac{5}{2}\right) =$$

$$\left(\dfrac{25}{4}\right)\left(-\dfrac{5}{2}\right) =$$

$$-\dfrac{125}{8} = -15\dfrac{5}{8}$$

EXAMPLE 19   Simplify $\left(-\dfrac{1}{4}\right)^{-2}$.

$$\left(-\dfrac{1}{4}\right)^{-2} =$$

$$\left(-\dfrac{4}{1}\right)^{2} =$$

$$(-4)^{2} =$$

$$(-4)(-4) = 16$$

EXERCISES 11.1B

Evaluate each expression.

35. $\left(\dfrac{1}{2}\right)^{-3} =$ 36. $\left(\dfrac{1}{3}\right)^{-2} =$

37. $\left(\dfrac{2}{3}\right)^{-1} =$ 38. $\left(\dfrac{3}{5}\right)^{-2} =$

39. $\left(-\dfrac{2}{5}\right)^{-2} =$ 40. $\left(-\dfrac{2}{3}\right)^{-3} =$

41. $\left(-\dfrac{3}{7}\right)^{-2} =$ 42. $\left(-\dfrac{1}{5}\right)^{-2} =$

43. $\left(\dfrac{3}{4}\right)^{-2} =$ 44. $\left(-\dfrac{3}{5}\right)^{-2} =$

Check your answers on page 749.   ✓

11.2   Multiplication of Expressions with Exponents

Multiplying Terms with the Same Base

In this section we learn how to multiply terms that involve exponents. We always want to leave our answers expressed in the most simplified form.

When multiplying two terms with the same base, we want to combine them under one exponent.

To simplify expressions with small exponents, rewrite each term without any exponents, count the number of times each letter factor appears, and then express the term in exponential notation.

EXAMPLE 1 Simplify $(x^2)(x^3)$.

$$(x^2)(x^3) =$$
$$(x \cdot x)(x \cdot x \cdot x) = x^5 \qquad \square$$

EXAMPLE 2 Simplify $(y^3)(y^4)$.

$$(y^3)(y^4) =$$
$$(y \cdot y \cdot y)(y \cdot y \cdot y \cdot y) = y^7 \qquad \square$$

Since the exponent simply represents the number of times that a factor appears in a term, there is a shortcut we can use: To *multiply* two terms that have the *same base*, you will know the total number of times that base is a factor by *adding its exponents*.

EXAMPLE 3 Use that observation to simplify each expression:

$$(x^2)(x^3) = \qquad (y^3)(y^4) =$$
$$x^{2+3} = x^5 \qquad y^{3+4} = y^7 \qquad \square$$

We can express that observation in general:

To *multiply* expressions with the same base, just *add exponents*.
$$x^m \cdot x^n = x^{n+m} \text{ for all numbers } m \text{ and } n.$$

Now that you are familiar with these relationships, you can apply that general rule when exponents are too large or too complicated to replace by all the factors they represent. The rule makes it easy for us to simplify the expressions in the following examples.

EXAMPLE 4 Simplify $x^{15} \cdot x^{27}$.

We wouldn't want to write out 15 x's and then 27 x's.

$$x^{15} \cdot x^{27} =$$
$$x^{15+27} = x^{42} \qquad \square$$

Don't forget that writing a letter with no exponent implies that the exponent is one.

EXAMPLE 5 Simplify $(x)(x^{53})$.

$$(x)(x^{53}) =$$
$$x^1 x^{53} =$$
$$x^{1+53} = x^{54} \qquad \square$$

EXAMPLE 6 Simplify $w^{218} \cdot w^{341} \cdot w^{806}$.

$$w^{218} \cdot w^{341} \cdot w^{806} =$$
$$w^{218+341+806} = w^{1365} \qquad \square$$

If any of the exponents are negative, you must remember the rules of signed numbers when you are adding them:

If the signs are the same—*add* the magnitudes and keep their sign.

If the signs are different—*subtract* the magnitudes and keep the sign of the number with the larger magnitude.

EXAMPLE 7 Simplify $x^5 x^{-3}$.

$$x^5 x^{-3} =$$
$$x^{5-3} = x^2$$ ❏

We could have simplified that expression by writing it without exponents, as we did in Examples 1 and 2.

$$x^5 x^{-3} =$$
$$(xxxxx)\left(\frac{1}{xxx}\right) =$$
$$\left(\frac{xxxxx}{xxx}\right) =$$
$$\left(\frac{\not x \not x \not x xx}{\not x \not x \not x}\right) =$$
$$xx = x^2$$

It is much easier to simplify such expressions by simply adding the exponents.

EXAMPLE 8 Simplify $x^{23} \cdot x^{-15}$.

$$x^{23} \cdot x^{-15} =$$
$$x^{23-15} = x^8$$ ❏

To simplify expressions as much as possible, I will leave them all expressed with positive exponents.

EXAMPLE 9 Simplify $y^{-37} \cdot y^{-28}$.

$$y^{-37} \cdot y^{-28} =$$
$$y^{-37-28} =$$
$$y^{-65} = \frac{1}{y^{65}}$$ ❏

If terms involve more than one base, add the exponents from the similar bases together. Since exponents represent the total number of occurrences of every factor in a term, each base in the simplified term should be different.

EXAMPLE 10 Simplify $(xy^{40})(x^{25}y^{37})$.

$$(xy^{40})(x^{25}y^{37}) =$$
$$x^{1+25}y^{40+37} = x^{26}y^{77}$$ ❏

EXAMPLE 11 Simplify $(x^{39}y^{20}w^{17})(x^{41}y^{14}w^{54})$.

$$(x^{39}y^{20}w^{17})(x^{41}y^{14}w^{54}) =$$
$$x^{39+41}y^{20+14}w^{17+54} = x^{80}y^{34}w^{71}$$ ❏

EXAMPLE 12 Simplify $(a^{45}b^{73}c^{-18})(a^{-24}b^{-90}c^{-21})$.

$$(a^{45}b^{73}c^{-18})(a^{-24}b^{-90}c^{-21}) =$$
$$a^{45-24}b^{73-90}c^{-18-21} =$$
$$a^{21}b^{-17}c^{-39} = \frac{a^{21}}{b^{17}c^{39}}$$ ❏

We can multiply more than two terms together in the same way.

EXAMPLE 13   Simplify $(pm^{50})(n^{38}m^{41}p^{25})(m^{16}n^{57})$.

$$(pm^{50})(n^{38}m^{41}p^{25})(m^{16}n^{57}) =$$
$$p^{1+25}m^{50+41+16}n^{38+57} = p^{26}m^{107}n^{95} \qquad \square$$

EXERCISES 11.2A

Simplify each expression. Leave your answers expressed in exponential notation.

1. $a^3a^2 =$ **2.** $yy^7 =$

3. $m^{13}m =$ **4.** $m^3m^3 =$

5. $x^8x^5 =$ **6.** $k^5k^{11} =$

7. $x^5x^3 =$ **8.** $c^5c^4 =$

9. $nn^8 =$ **10.** $y^5y =$

11. $j^4j^7 =$ **12.** $pp^2 =$

13. $x^{47}x^{86} =$ **14.** $a^{69}a^{43} =$

15. $w^{32}w^{17}w^{45} =$ **16.** $n^{64}n^{208}n^{147} =$

Simplify. Leave your answers expressed with positive exponents.

17. $a^9a^{-2} =$ **18.** $b^{-4}b^2 =$

19. $a^{-7}a^3 =$ **20.** $b^5b^{-1} =$

21. $p^{38}p^{-27} =$ **22.** $k^{-46}k^{-52} =$

23. $n^{-23}n^{-61}n^{-108} =$ **24.** $x^{340}x^{-246} =$

25. $y^{-718}y^{490} =$ **26.** $a^{375}a^{-237}a^{180} =$

27. $a^3b^7a^{-1}b^{-4} =$ **28.** $a^7b^{-2}a^{-5}b^3 =$

29. $a^3b^{-1}a^{-1}b^6 =$ **30.** $a^{-4}b^3ab^{-7} =$

31. $(a^2bc^5)(a^3b^2c) =$ **32.** $(x^3yw^2)(x^3yw^2) =$

33. $(m^2nw^3)(mnw) =$ **34.** $(a^2bc^3)(abc)(a^3b^2c) =$

35. $(x^5y^3z^2)(xy^2z)(x^3y^3z) =$ **36.** $(mnp^2)(m^2n^2p)(mn^3p^2) =$

37. $(a^{216}b^{740})(a^{328}b^{403}) =$

38. $(x^{126}y^{648})(x^{-460}y^{-538}) =$

39. $(n^{150}m^{-407}p^{683})(n^{625}m^{139}p^{-488})(n^{-400}m^{-261}p^{344}) =$

40. $(a^{63}b^{-270}c^{362})(a^{325}b^{-360}c^{-427})(a^{561}b^{-228}c^{397}) =$

Check your answers on page 750.   ✓

Multiplication of Terms with Coefficients

> *To multiply terms involving coefficients and exponents:*
>
> 1. Determine the sign of the product by using the rules of signed numbers you learned in Section 2.4 for the signs of each *term*.
> 2. Multiply the coefficients to determine the coefficient of the product.
> 3. Indicate the number of times each letter factor appears by adding the exponents of that letter.

EXAMPLE 14 Simplify $(2x^5 y)(-3x^8 y^6)$.

The product of a positive term multiplied by a negative term is *negative*.

$$(2x^5 y)(-3x^8 y^6) =$$
$$-(2)(3)x^{5+8}y^{1+6} = -6x^{13}y^7$$

EXAMPLE 15 Simplify $(-5a^{24}b^{-57}c^{38})(4a^{-20}bc^{31})(-a^8 b^{13}c^{-11})$.

The product of two negative terms is *positive*.

$$(-5a^{24}b^{-57}c^{38})(4a^{-20}bc^{31})(-a^8 b^{13}c^{-11}) =$$
$$+ (5)(4)a^{24-20+8}b^{-57+1+13}c^{38+31-11} =$$
$$20a^{12}b^{-43}c^{58} = \frac{20a^{12}c^{58}}{b^{43}}$$

EXERCISES 11.2B

Simplify each expression. Leave your answers expressed in exponential notation with positive exponents.

41. $(2ab^3 c^3)(5a^2 b^2) =$ **42.** $(3x^2 y^3 w)(2xy)(x^2 w^2) =$

43. $(7ab^2)(2bc^3)(a^3 c^2) =$ **44.** $(4w^3 x)(x^2 yw^5)(3yxw) =$

45. $(abc^3)(b^3 c)(7a^2 c) =$ **46.** $(6m^3 n)(2pn^2)(nmp) =$

47. $(4x^{37}y^{50})(-6x^{75}y^{38}) =$ **48.** $(-5m^{64}n^{43})(-2m^{34}n^{80}) =$

49. $(3p^{57}m^{42})(7p^{-60}m^{-25}) =$ **50.** $(-2a^{41}b^{-27})(6a^{-30}b^{-12}) =$

51. $(-5nm^{30})(-2n^{-28}m^{-38})(3n^{56}m^{-7}) =$

52. $(7w^{14}p^{43})(-4w^{-20}p^{-25})(-w^{41}p^{-20}) =$

Check your answers on page 750. ✓

11.3 Division of Expressions with Exponents

Dividing Terms with the Same Base

In this section we learn how to divide terms that involve exponents. The easiest way to divide one term by another is to represent it as a fraction whose denominator is the term you are dividing by.

Always leave the answer in the most simplified form. Try rewriting all factors without exponents and canceling any pair of similar factors appearing in both the numerator and the denominator of the fraction. Then express the number of times each factor appears with an exponent.

EXAMPLE 1   Simplify (x^7) divided by (x^3).

$$\frac{x^7}{x^3} =$$

$$\frac{x \cdot x \cdot x \cdot x \cdot x \cdot x \cdot x}{x \cdot x \cdot x} =$$

$$\frac{\cancel{x} \cdot \cancel{x} \cdot \cancel{x} \cdot x \cdot x \cdot x \cdot x}{\cancel{x} \cdot \cancel{x} \cdot \cancel{x}} = x^4 \qquad \square$$

EXAMPLE 2   Simplify $(a^3b^2c^4) \div (ab^5c^3)$.

$$\frac{a^3b^2c^4}{ab^5c^3} =$$

$$\frac{a \cdot a \cdot a \cdot b \cdot b \cdot c \cdot c \cdot c \cdot c}{a \cdot b \cdot b \cdot b \cdot b \cdot b \cdot c \cdot c \cdot c} =$$

$$\frac{\cancel{a} \cdot a \cdot a \cdot \cancel{b} \cdot \cancel{b} \cdot \cancel{c} \cdot \cancel{c} \cdot \cancel{c} \cdot c}{\cancel{a} \cdot \cancel{b} \cdot \cancel{b} \cdot b \cdot b \cdot b \cdot \cancel{c} \cdot \cancel{c} \cdot \cancel{c}} =$$

$$\frac{a \cdot a \cdot c}{b \cdot b \cdot b} = \frac{a^2c}{b^3} \qquad \square$$

Fortunately we can use a shortcut that does not require us to write each factor. Since pairs of factors that appear in both the numerator and the denominator will cancel out, we can subtract the number of times a factor appears in one of those places from the number of times it appears in the other place. We will subtract the pairs that cancel.

Each term will always be left with positive exponents (and therefore ready for us to evaluate if we need to) if we take the smaller number of occurrences of a factor from the larger number.

For example, in $\dfrac{a^3b^2c^4}{ab^5c^3}$ from Example 2, there are fewer a's in the denominator than there are in the numerator, so we should take the one a in the denominator from the three a's in the numerator, leaving us two factors of a in the numerator.

But we take the two factors of b in the numerator from the five that appear in the denominator, leaving us three factors of b in the denominator since there were more factors of b in the denominator.

We take the three factors of c in the denominator from the four in the numerator, leaving us one factor of c in the numerator.

EXAMPLE 2   Using that shortcut, simplify $(a^3b^2c^4) \div (ab^5c^3)$.

$$\frac{a^3b^2c^4}{ab^5c^3} =$$

$$\frac{a^{3-1}c^{4-3}}{b^{5-2}} = \frac{a^2c}{b^3} \qquad \square$$

We can generalize the shortcut as follows:

$$\frac{x^m}{x^n} = x^{m-n} \quad \text{if} \quad m > n$$

$$\frac{x^m}{x^n} = \frac{1}{x^{n-m}} \quad \text{if} \quad n > m$$

If a factor appears the same number of times in the numerator as in the denominator, that factor will completely cancel out of the term. When we use our subtraction shortcut, that factor ends up with a zero exponent; and since any factor to the zero power is 1, that factor becomes 1 and cancels out of the term.

EXAMPLE 3 Simplify $x^{23}y^{80}w^{54}$ divided by $x^{38}y^{61}w^{54}$.

$$\frac{x^{23}y^{80}w^{54}}{x^{38}y^{61}w^{54}} =$$

$$\frac{y^{80-61}w^{54-54}}{x^{38-23}} =$$

$$\frac{y^{19}w^0}{x^{15}} =$$

$$\frac{y^{19}(1)}{x^{15}} = \frac{y^{19}}{x^{15}} \qquad \square$$

Now I can show you why *negative exponents* represent *division* and why terms raised to *zero exponents* represent 1.

EXAMPLE 4 Simplify $\dfrac{x^3}{x^3}$.

We can use the subtraction shortcut:

$$\frac{x^3}{x^3} =$$

$$x^{3-3} = x^0$$

But any fraction with the same numerator and denominator is equal to 1! $\left(\text{So } \dfrac{x^3}{x^3} = 1.\right)$

That is why in Section 11.1 we stated that $x^0 = 1$. $\qquad \square$

EXAMPLE 5 In Section 11.1 we defined x^{-2} as $\dfrac{1}{x^2}$.

To see why, let's simplify the expression $\dfrac{x^3}{x^5}$.

$$\frac{x^3}{x^5} =$$

$$\frac{xxx}{xxxxx} =$$

$$\frac{xxx}{xxxxx} = \frac{1}{x^2}$$

Now let's simplify it by using the subtraction shortcut.

$$\frac{x^3}{x^5} = \qquad \text{or we could simplify it as} \qquad \frac{x^3}{x^5} =$$

$$\frac{1}{x^{5-3}} = \frac{1}{x^2} \qquad\qquad\qquad x^{3-5} = x^{-2}$$

To ensure that all our rules are consistent, so there is only one *value to any expression*, we interpret *negative exponents* as representing the *number of times that the base appears as a factor in the denominator*. $\qquad \square$

EXERCISES 11.3A

Divide these terms by subtracting exponents of similar bases.

1. $\dfrac{x^5}{x^2} =$

2. $\dfrac{r^{19}}{r^3} =$

3. $\dfrac{y^7}{y^3} =$

4. $\dfrac{k^{53}}{k^{20}} =$

5. $\dfrac{m^9}{m^{11}} =$

6. $\dfrac{x^{114}}{x^{38}} =$

7. $\dfrac{w}{w^5} =$

8. $\dfrac{y^{513}}{y^{286}} =$

9. $\dfrac{z^{23}}{z^{35}} =$

10. $\dfrac{x^{704}}{x^{479}} =$

11. $\dfrac{a^7}{a} =$

12. $\dfrac{y^{433}}{y^{76}} =$

13. $\dfrac{a^5b^4c^2}{abc} =$

14. $\dfrac{x^5y^7w^3}{x^2y^4w} =$

15. $\dfrac{x^9y^5w^7}{x^4y^2w^5} =$

16. $\dfrac{m^0n^5p^{15}}{mn^3p^9} =$

17. $\dfrac{a^3bc^5}{ab^5c^8} =$

18. $\dfrac{p^5m^3}{p^5m^7} =$

19. $\dfrac{x^7y^3z^2}{x^5y^5z^5} =$

20. $\dfrac{kr^3p^4}{k^2rp^2} =$

21. $\dfrac{a^3b^2c^2}{ab^4c^2} =$

22. $\dfrac{x^4yw^3}{xy^2w} =$

Check your answers on page 750. ✓

Division of Terms with Coefficients

If there are coefficients in the terms you are dividing, be careful. Remember that when you *subtract* the *exponents*, you are really *canceling factors*. There is no similar shortcut for the coefficients.

If the coefficients are not prime, factor them and cancel any pair of factors appearing in both the numerator and the denominator, just as you did with numerical fractions in Chapter 3.

EXAMPLE 6 Simplify $20a^{13}b^{37}c^{59}$ divided as $35a^{28}b^{17}c^{46}$.

$$\frac{20a^{13}b^{37}c^{59}}{35a^{28}b^{17}c^{46}} =$$

$$\frac{4\cdot \cancel{5}\cdot a^{13}b^{37}c^{59}}{\cancel{5}\cdot 7\cdot a^{28}b^{17}c^{46}} =$$

$$\frac{4b^{37-17}c^{59-46}}{7a^{28-13}} = \frac{4b^{20}c^{13}}{7a^{15}}$$

If either *term* is negative, use the rule of signs to determine the sign of their quotient. That is, if both signs are negative, the quotient is *positive*; if only one sign is negative, the quotient is *negative*.

EXAMPLE 7   Simplify $(12x^{40}y^{14})$ divided by $(-10x^{32}y^{51})$.

The quotient of a positive term and a negative term is *negative*.

$$\frac{12x^{40}y^{14}}{-10x^{32}y^{51}} =$$

$$-\frac{\cancel{2} \cdot 6 \cdot x^{40}y^{14}}{\cancel{2} \cdot 5 \cdot x^{32}y^{51}} =$$

$$-\frac{6 \cdot x^{40-32}}{5 \cdot y^{51-14}} = -\frac{6x^8}{5y^{37}}$$

EXAMPLE 8   Simplify $(-14p^{26}m^{33}n^{21})$ divided by $(-21p^{18}m^{50}n^{35})$.

The quotient of two negative terms is *positive*.

$$\frac{-14p^{26}m^{33}n^{21}}{-21p^{18}m^{50}n^{35}} =$$

$$+\frac{2 \cdot \cancel{7} \cdot p^{26}m^{33}n^{21}}{3 \cdot \cancel{7} \cdot p^{18}m^{50}n^{35}} =$$

$$\frac{2 \cdot p^{26-18}}{3 \cdot m^{50-33}n^{35-21}} = \frac{2p^8}{3m^{17}n^{14}}$$

EXERCISES 11.3B

Divide these terms and simplify each quotient.

23. $\dfrac{4m^5n^2p^3}{2m^2n^5p} =$

24. $\dfrac{15x^3y^2w}{6x^3y^4w^4} =$

25. $\dfrac{-8a^{15}b^{12}c^{18}}{20a^{21}b^{11}c^{20}} =$

26. $\dfrac{6w^4yz^2}{-14y^3wz^5} =$

27. $\dfrac{24m^5pn^2}{60m^3p^5n^2} =$

28. $\dfrac{-12x^4yw^3}{-20xy^5w^3} =$

29. $(20a^{40}b^{25})$ divided by $(35a^{18}b^{19}) =$

30. $(-12pm^{13}n^{30})$ divided by $(20p^{15}m^7n^{14}) =$

31. $(-10x^{37}y^{50}w^{86})$ divided by $(-15x^{46}y^{32}w^{40}) =$

32. $(14n^{130}p^{86}m^{50})$ divided by $(-21n^{63}p^{74}m^{83}) =$

Check your answers on page 750.   ✓

Dividing Terms with Coefficients and Negative Exponents

If the single term in the numerator or denominator has any factors with negative exponents, it will be easier to simplify the expression if we remove the negative exponents. Since negative exponents represent division, any factors of the numerator with negative exponents can be put into the denominator with positive exponents.

If any factors with the same base end up in the numerator or denominator, they are being multiplied, and so their exponents should be *added*.

EXAMPLE 9 Simplify $(a^5b^{-2}c)$ divided by $(a^3b^7c^4)$.

$$\frac{a^5b^{-2}c}{a^3b^7c^4} =$$

$$\frac{a^5c}{a^3b^7c^4b^2} =$$

$$\frac{a^5c}{a^3b^{7+2}c^4} =$$

$$\frac{a^{5-3}}{b^9c^{4-1}} = \frac{a^2}{b^9c^3} \qquad \square$$

If a factor of the denominator has a negative exponent, it can be made positive by putting it in the numerator. I'll illustrate why that manipulation is valid in the next example.

EXAMPLE 10 Simplify $\dfrac{3}{x^{-1}}$. Remember, a fraction can represent the numerator divided by the denominator.

$$\frac{3}{x^{-1}} =$$

$$3 \div x^{-1} =$$

$$3 \div \frac{1}{x} =$$

To divide by $\dfrac{1}{x}$ is the same as to multiply by its reciprocal, $\dfrac{x}{1}$.

$$3\left(\frac{x}{1}\right) = 3x. \quad \text{So} \quad \frac{3}{x^{-1}} = 3x. \qquad \square$$

Instead of going through all those steps for each problem, it's okay to generalize as follows:

$$\frac{1}{x^{-1}} = x; \quad \frac{1}{x^{-n}} = x^n; \quad \frac{x^{-a}}{y^{-b}} = \frac{y^b}{x^a}$$

where n, a, and b are positive integers, and x and $y \neq 0$.

EXAMPLE 11 Simplify $\dfrac{a^7bc^{-3}d^{10}}{a^{-2}b^5c^{-1}d^6}$.

$$\frac{a^7bc^{-3}d^{10}}{a^{-2}b^5c^{-1}d^6} =$$

$$\frac{a^7bd^{10}a^2c^1}{b^5d^6c^3} =$$

$$\frac{a^{7+2}bd^{10}c^1}{b^5d^6c^3} =$$

$$\frac{a^9d^{10-6}}{b^{5-1}c^{3-1}} = \frac{a^9d^4}{b^4c^2} \qquad \square$$

Always decide on the sign of your quotient before proceeding with the rest of your computation.

EXAMPLE 12 Simplify $\dfrac{-6x^5y^{-4}wz^{-7}}{14x^4y^3w^{-1}z^{-4}}$.

The quotient of terms with different signs is *negative*.

$$\frac{-6x^5y^{-4}wz^{-7}}{14x^4y^3w^{-1}z^{-4}} =$$

$$-\frac{6x^5ww^1z^4}{14x^4y^3y^4z^7} =$$

$$-\frac{3\cdot\not{2}\cdot x^5wwz^4}{\not{2}\cdot 7\cdot x^4y^3y^4z^7} =$$

$$-\frac{3x^5w^{1+1}z^4}{7x^4y^{3+4}z^7} =$$

$$-\frac{3x^{5-4}w^2}{7y^7z^{7-4}} = -\frac{3xw^2}{7y^7z^3}$$

In Summary—to simplify an expression with one term in the numerator and one term in the denominator whose factors are raised to exponents:

1. Determine the sign of the quotient.

2. Replace all negative exponents by switching the base between the numerator and the denominator to make its exponent positive.

3. Factor coefficients and cancel any factors appearing in both the numerator and the denominator.

4. Add the exponents of any similar bases appearing in the numerator, and do the same for similar bases in the denominator.

5. Subtract the exponents from similar bases appearing both in the numerator and in the denominator.

EXERCISES 11.3C

Simplify each expression. Leave your answers expressed with positive exponents.

33. $\dfrac{a^{-5}b^2}{a^3b} =$

34. $\dfrac{x^3y^2}{xy^{-2}} =$

35. $\dfrac{w^4y^{-2}}{yw^2} =$

36. $\dfrac{ab^{-2}c}{a^5b^4c^{-3}} =$

37. $\dfrac{m^2n^{-1}}{m^{-3}n} =$

38. $\dfrac{p^{-2}r^{-3}m}{p^3r^{-1}m^2} =$

39. $\dfrac{xy^{-5}w^2}{x^{-3}y^2w^{-5}} =$

40. $\dfrac{b^5a^3c^{-2}}{b^{-2}a^{-7}} =$

41. $\dfrac{5a^{14}b^{34}}{7a^{23}b^{16}} =$

42. $\dfrac{10x^{23}y^{48}}{12x^{34}y^{35}} =$

43. $\dfrac{-8x^{24}w^{-13}}{10x^{-17}w^{31}} =$

44. $\dfrac{-14p^{43}m^{58}}{-49p^{58}m^{-12}} =$

45. $\dfrac{20ab^{14}c^{-8}}{28a^8b^{-10}c^{23}} =$

46. $\dfrac{9z^7k^{-12}m^{28}}{-15z^{-12}k^{-40}m^{19}} =$

Check your answers on page 750. ✓

11.4 Raising Terms with Exponents to a Power

Simplifying Expressions Like $(x^n)^m$

To simplify an expression like $(x^3)^2$, we can go back to the definition of exponential notation. The outer exponent (2) tells us how many times (twice) the base (x^3) appears as a factor. We can remove the outer exponent by writing that factor twice. Then we have an expression we can simplify.

EXAMPLE 1 Simplify $(x^3)^2$.

$$(x^3)^2 =$$
$$(x^3)(x^3) =$$
$$x^{3+3} = x^6 \qquad \square$$

EXAMPLE 2 Simplify $(y^7)^4$.

The exponent 4 tells us the base (y^7) appears four times.

$$(y^7)^4 =$$
$$(y^7)(y^7)(y^7)(y^7) =$$
$$y^{7+7+7+7} = y^{28} \qquad \square$$

This method will become tedious if the outer exponent is very large. There is a shortcut we can use. The outer exponent will always represent the number of times we add the inner exponent to itself. Recall from Section 1.3 that an instruction to add a number to itself is really an instruction to multiply.

EXAMPLE 3 Just as

$$8+8+8+8+8 =$$
$$8 \cdot 5 = 40,$$

so

$$(w^8)^5 =$$
$$(w^8)(w^8)(w^8)(w^8)(w^8) =$$
$$w^{8+8+8+8+8} =$$
$$w^{8 \cdot 5} = w^{40}. \qquad \square$$

We can represent that shortcut in general as follows:

$$(x^n)^m = x^{nm}$$

EXAMPLE 1   Using that shortcut, $(x^3)^2$ is $x^{3(2)} = x^6$.   ❑

EXAMPLE 2   Using that shortcut, $(y^7)^4$ is $y^{7(4)} = y^{28}$.   ❑

EXAMPLE 4   Simplify $(a^8)^{10}$.

$$(a^8)^{10} =$$
$$a^{8(10)} = a^{80}$$ ❑

The shortcut also makes it easy to raise terms involving exponents to powers when either exponent is negative.

EXAMPLE 5   Simplify $(x^{15})^{-3}$.

$$(x^{15})^{-3} =$$
$$x^{15(-3)} =$$
$$x^{-45} = \frac{1}{x^{45}}$$ ❑

EXERCISES 11.4A

Simplify each term. Leave your answers expressed with one positive exponent.

1. $(x^5)^2 =$

2. $(y^4)^2 =$

3. $(a^3)^5 =$

4. $(m^7)^6 =$

5. $(b^2)^4 =$

6. $(n^4)^7 =$

7. $(c^3)^{10} =$

8. $(w^6)^5 =$

9. $(x^4)^5 =$

10. $(x^5)^4 =$

11. $(a^{12})^5 =$

12. $(k^{10})^7 =$

13. $(m^4)^{-5} =$

14. $(c^{-7})^{-6} =$

15. $(x^{-8})^{-9} =$

16. $(w^{-6})^5 =$

Check your answers on page 750.   ✓

Simplifying Expressions Like $(xy)^n$

When we want an exponent to apply to more than one base, we enclose in parentheses all the factors it should apply to. We can remove the parentheses from those expressions by utilizing the exponential notation definition. Let's do it this way so you will see what would happen in such problems; then we'll find a shortcut.

EXAMPLE 6   Simplify $(xy)^4$.

$$(xy)^4 =$$
$$(xy)(xy)(xy)(xy) =$$
$$(x \cdot x \cdot x \cdot x \cdot y \cdot y \cdot y \cdot y) = x^4y^4$$ ❑

That method works fine for small exponents but would be troublesome for large or negative exponents. Each factor within the parentheses will appear the number of times indicated by the exponent, so we can adopt the shortcut of simply applying the exponent to each factor within the parentheses.

In general we say:

$$(xy)^n = x^n y^n.$$

EXAMPLE 7   Simplify $(abc)^{12}$.

$$(abc)^{12} = a^{12} b^{12} c^{12} \qquad \square$$

EXAMPLE 8   Simplify $(pkm)^{-5}$.

$$(pkm)^{-5} =$$

$$p^{-5} k^{-5} m^{-5} = \frac{1}{p^5 k^5 m^5} \qquad \square$$

It is not wrong to leave letter factors with negative exponents, but I prefer to leave answers expressed with all *positive* exponents so that the expressions will be ready to be evaluated if I am told the values of those letters.

If there is a coefficent, always convert expressions so that the exponent of a *numerical base is positive*, and *evaluate* the number raised to that power in your answer.

EXAMPLE 9   Simplify $(2xyw)^{-3}$.

$$(2xyw)^{-3} =$$

$$2^{-3} x^{-3} y^{-3} w^{-3} =$$

$$\frac{1}{2^3 x^3 y^3 w^3} = \frac{1}{8x^3 y^3 w^3} \qquad \square$$

Simplifying Expressions Like $(x^a y^b)^n$

Now we can simplify expressions where the factors within parentheses are raised to exponents. We can remove the outer exponent by applying the definition of exponential notation. Let's do it this way first so that you will see what would happen in such problems, then we'll find a shortcut.

EXAMPLE 10   Simplify $(a^3 b^5)^2$.

$$(a^3 b^5)^2 =$$

$$(a^3 b^5)(a^3 b^5) =$$

$$a^{3+3} b^{5+5} = a^6 b^{10} \qquad \square$$

The effect of this calculation is the application of the outer exponent to each factor within the parentheses, so we can simplify our work by doing that step directly.

In general we say:

$$(x^a y^b)^n =$$

$$(x^a)^n (y^b)^n = x^{an} y^{bn} \text{ for any integers } a, b, \text{ and } n.$$

EXAMPLE 10   Using that shortcut, simplify $(a^3 b^5)^2$.

$$(a^3 b^5)^2 =$$

$$a^{3(2)} b^{5(2)} = a^6 b^{10} \qquad \square$$

The shortcut makes it easy for us to simplify expressions involving large or negative exponents.

EXAMPLE 11   Simplify $(a^{15}b^{20})^9$.

$$(a^{15}b^{20})^9 =$$
$$a^{15(9)}b^{20(9)} = a^{135}b^{180}$$ ❑

EXAMPLE 12   Simplify $(pm^7k^{-4})^{13}$.

$$(pm^7k^{-4})^{13} =$$
$$p^{13}m^{7(13)}k^{-4(13)} =$$
$$p^{13}m^{91}k^{-52} = \frac{p^{13}m^{91}}{k^{52}}$$ ❑

EXAMPLE 13   Simplify $(5a^7b^{-8}cd^{-4})^3$.

$$(5a^7b^{-8}cd^{-4})^3 =$$
$$5^3a^{7(3)}b^{-8(3)}c^3d^{-4(3)} =$$
$$5^3a^{21}b^{-24}c^3d^{-12} =$$
$$\frac{5^3a^{21}c^3}{b^{24}d^{12}} = \frac{125a^{21}c^3}{b^{24}d^{12}}$$ ❑

EXAMPLE 14   Simplify $(2xy^{-5}w^8z^{-1})^{-4}$.

$$(2xy^{-5}w^8z^{-1})^{-4} =$$
$$2^{-4}x^{-4}y^{-5(-4)}w^{8(-4)}z^{-1(-4)} =$$
$$2^{-4}x^{-4}y^{20}w^{-32}z^4 =$$
$$\frac{y^{20}z^4}{2^4x^4w^{32}} = \frac{y^{20}z^4}{16x^4w^{32}}$$ ❑

Note that there is no need to replace negative exponents by expressions involving positive ones before calculating when terms are simply multiplied or raised to a power. It is only when we *divide* terms or *evaluate* powers of numbers that we have to make that replacement before calculating.

EXERCISES 11.4B

Simplify each expression. Leave your answers expressed in terms of *positive* exponents.

17. $(mn)^8 =$

18. $(pmn)^7 =$

19. $(abc)^{14} =$

20. $(xyw)^{38} =$

21. $(abc)^{-7} =$

22. $(xyz)^{-13} =$

23. $(3ab)^2 =$

24. $(5pk)^3 =$

25. $(7mp)^{-2} =$

26. $(2abc)^{-3} =$

27. $(2x^3y)^3 =$

28. $(5a^7b^3)^3 =$

29. $(3w^4y)^3 =$

30. $(2m^3n^6)^3 =$

31. $(7x^4w^3)^2 =$

32. $(4pm^4)^2 =$

33. $(6r^5pm^3)^2 =$

34. $(10t^6kw^8)^2 =$

35. $(a^3b^{-2})^2 =$

36. $(x^{-1}y^3)^5 =$

37. $(w^{-3}y^{-1})^3 =$

38. $(ab^{-5}c^{-2})^4 =$

39. $(x^2y^3)^{-4} =$

40. $(w^5y^{-2})^{-3} =$

41. $(r^4p^{-4}l^2)^{-3} =$

42. $(a^2b^{-1}c^{-3})^{-2} =$

43. $(2ab^{-3})^{-2} =$

44. $(3x^{12}y^{-1})^{-2} =$

45. $(5x^{-3}y^7)^{-2} =$

46. $(7wy^3)^{-2} =$

47. $(2a^8b^{-2})^{-3} =$

48. $(3ab^{-4})^{-3} =$

49. $(5w^5y^{-3}x)^{-2} =$

50. $(4x^7y^{-3}w^{-1})^{-2} =$

Check your answers on page 750.

Raising Negative Terms to Powers

When we first discussed the multiplication of signed numbers in Section 2.4, we saw that when multiplying more than two terms, only the negative terms affected the sign of the answer.

1. If an *even* number of negative numbers are multiplied, their product will be *positive*.

2. If an *odd* number of negative numbers are multiplied, their product will be *negative*.

Since the positive integral power to which we raise a term represents the number of times that the term will be multiplied, we can translate those rules into terminology that applies to exponents.

1. When we raise any term to an *even* power, the answer will always be *positive*.

2. When we raise a *negative* term to an *odd* power, the answer will always be *negative*.

We can express this rule in general as follows:

$$(-x)^n = -x^n \quad \text{if } n \text{ is } odd, \text{ and } \quad (-x)^m = +x^m \quad \text{if } m \text{ is } even.$$

EXAMPLE 15   Simplify $(-3ab^4c^{-7})^2$.

This expression will be *positive* because the power it is raised to (2) is *even*.

$$(-3ab^4c^{-7})^2 =$$
$$+3^2a^2b^{4(2)}c^{-7(2)} =$$
$$+3^2a^2b^8c^{-14} =$$
$$\frac{3^2a^2b^8}{c^{14}} = \frac{9a^2b^8}{c^{14}}$$

EXAMPLE 16 Simplify $(-2x^{-4}y^6w^{-5})^3$.

This expression will be *negative* because the power it is raised to (3) is *odd*.

$$(-2x^{-4}y^6w^{-5})^3 =$$
$$-2^3x^{-4(3)}y^{6(3)}w^{-5(3)} =$$
$$-2^3x^{-12}y^{18}w^{-15} =$$
$$-\frac{2^3y^{18}}{x^{12}w^{15}} = -\frac{8y^{18}}{x^{12}w^{15}}$$ ❑

Remember, the sign of the exponent has no effect on the *sign* of the answer. If the coefficient of a term is negative, the sign of the answer depends on whether the exponent of that term is *even* or *odd*.

EXAMPLE 17 Simplify $(-4a^3b^{-1}c^8)^{-1}$.

This expression will be *negative* because the power it is raised to (-1) is *odd*.

$$(-4a^3b^{-1}c^8)^{-1} =$$
$$-4^{-1}a^{3(-1)}b^{-1(-1)}c^{8(-1)} =$$
$$-4^{-1}a^{-3}b^1c^{-8} =$$
$$-\frac{b}{4^1a^3c^8} = -\frac{b}{4a^3c^8}$$ ❑

EXAMPLE 18 Simplify $(-d^{12}f^{-15}g^{23})^{-6}$.

This expression will be *positive* because the power it is raised to (-6) is *even*.

$$(-d^{12}f^{-15}g^{23})^{-6} =$$
$$+d^{12(-6)}f^{-15(-6)}g^{23(-6)} =$$
$$d^{-72}f^{90}g^{-138} = \frac{f^{90}}{d^{72}g^{138}}$$ ❑

EXAMPLE 19 Simplify $(7pr^5m^{-6})^{-9}$.

This expression will be *positive* because the *term being raised to a power is positive*. Remember, the only way a term *can* be negative is when its base is negative!

$$(7pr^5m^{-6})^{-9} =$$
$$7^{-9}p^{-9}r^{5(-9)}m^{-6(-9)} =$$
$$7^{-9}p^{-9}r^{-45}m^{54} = \frac{m^{54}}{7^9p^9r^{45}}$$ ❑

7^9 is too large a number to evaluate. It's fine to leave it expressed in exponential notation.

EXERCISES 11.4C

Simplify each expression. Leave your answers expressed in terms of *positive* exponents.

51. $(-5x^3y)^2 =$

52. $(-3ab^5)^3 =$

53. $(-2w^3y)^3 =$

54. $(-7m^4n^2)^2 =$

55. $(-4y^4w)^2 =$

56. $(-a^3b^7)^5 =$

57. $(-k^3m^4)^4 =$

58. $(-p^6m^6n)^7 =$

59. $(-2ab^{-5})^2 =$

60. $(5xy^{-4}w^6)^2 =$

61. $(-3pm^{-6}n^{-1})^3 =$

62. $(-7a^{23}b^{-35}c^{40})^{-1} =$

63. $(-5w^4y^{-6}z^{-2})^2 =$

64. $(a^{-8}b^6c^{-9})^{-5} =$

65. $(-2ab^3c^{-4})^{-2} =$

66. $(-x^3y^{-4}w^{-5})^{-4} =$

67. $(-2n^2p^{-3}m^5)^{-3} =$

68. $(-5ab^6c^{-7})^{-8} =$

Check your answers on page 750.

Simplifying Complicated Expressions

Multiplying Terms Raised to Powers

The order of operations we use to compute problems involving more than one operation is always our guide for simplifying complicated problems. Remember what that order is.

> ***Order of Operations:***
>
> 1. Simplify the expression within the *parentheses* if possible. (Usually it is already simplified.)
> 2. Apply the *power* outside the parentheses to every factor inside, simplify each factor, and determine what the sign of that term will be.
> 3. *Multiply* factors with the same base by adding their exponents.
> 4. *Switch* all factors with negative exponents into the denominator so the exponents will be positive.
> 5. *Evaluate* and simplify the numerical factors.

EXAMPLE 1 Simplify $(a^6bc^7)^2(a^5b^4)^3$.

$$(a^6bc^7)^2(a^5b^4)^3 =$$
$$(a^{6(2)}b^2c^{7(2)})(a^{5(3)}b^{4(3)}) =$$
$$(a^{12}b^2c^{14})(a^{15}b^{12}) =$$
$$a^{12+15}b^{2+12}c^{14} = a^{27}b^{14}c^{14}$$

EXAMPLE 2 Simplify $(-2x^{-5}y^8w)^{-3}(-x^7y^5w^6)^4$.

$$(-2x^{-5}y^8w)^{-3}(-x^7y^5w^6)^4 =$$
$$[-2^{-3}x^{-5(-3)}y^{8(-3)}w^{-3}][+\,x^{7(4)}y^{5(4)}w^{6(4)}] =$$
$$[-2^{-3}x^{15}y^{-24}w^{-3}][+\,x^{28}y^{20}w^{24}] =$$
$$-\,[2^{-3}x^{15+28}y^{-24+20}w^{-3+24}] =$$
$$-\,(2^{-3}x^{43}y^{-4}w^{21}) =$$
$$-\,\frac{x^{43}w^{21}}{2^3y^4} = -\,\frac{x^{43}w^{21}}{8y^4}$$

EXAMPLE 3 Simplify $(-5mn^3p^{-5})^{-2}(-2m^{-7}n^{-4}p^8)^{-3}$.

$$(-5mn^3p^{-5})^{-2}(-2m^{-7}n^{-4}p^8)^{-3} =$$
$$[+5^{-2}m^{-2}n^{3(-2)}p^{-5(-2)}][-2^{-3}m^{-7(-3)}n^{-4(-3)}p^{8(-3)}] =$$
$$[5^{-2}m^{-2}n^{-6}p^{10}][-2^{-3}m^{21}n^{12}p^{-24}] =$$
$$-(5^{-2})(2^{-3})m^{-2+21}n^{-6+12}p^{10-24} =$$
$$-(5^{-2})(2^{-3})m^{19}n^6p^{-14} =$$
$$-\frac{m^{19}n^6}{5^2 \cdot 2^3 p^{14}} =$$
$$-\frac{m^{19}n^6}{25 \cdot 8 p^{14}} = -\frac{m^{19}n^6}{200p^{14}} \quad \square$$

EXAMPLE 4 Simplify $(-xw^3z^{-7})^{-5}(-2x^5y^2w^{-4})^4(-3x^{-1}y^4z^{-3})^{-2}$.

$$(-xw^3z^{-7})^{-5}(-2x^5y^2w^{-4})^4(-3x^{-1}y^4z^{-3})^{-2} =$$
$$(-x^{-5}w^{3(-5)}z^{-7(-5)})(+2^4x^{5(4)}y^{2(4)}w^{-4(4)})(+3^{-2}x^{-1(-2)}y^{4(-2)}z^{-3(-2)}) =$$
$$(-x^{-5}w^{-15}z^{35})(+2^4x^{20}y^8w^{-16})(+3^{-2}x^2y^{-8}z^6) =$$
$$-[2^4 \cdot 3^{-2} \cdot x^{-5+20+2}w^{-15-16}z^{35+6}y^{8-8}] =$$
$$-[2^4 \cdot 3^{-2}x^{17}w^{-31}z^{41}y^0 =$$
$$-\frac{2^4x^{17}z^{41}(1)}{3^2w^{31}} = -\frac{16x^{17}z^{41}}{9w^{31}} \quad \square$$

EXERCISES 11.5A

Simplify and leave your answers expressed with all exponents positive.

1. $(x^2y)^3(xy^2)^2 =$ **2.** $(x^5y)^3(x^2y^3)^2 =$

3. $(2a^5b^2)^2(a^3b^2)^3 =$ **4.** $(2m^5n^2p)^3(m^2n^2p^2)^3 =$

5. $(5xw^{-3})^2(x^2w^{-4})^5 =$ **6.** $(-m^2p^{-3}n^7)^4(-m^3p^5n^{-6})^7 =$

7. $(-2ab^3c^{-4})^3(-3a^5b^{-4}c^{-1})^2 =$

8. $(3xy^6w^{-3})^{-2}(5x^{-2}y^3w^4)^{-2} =$

9. $(-mp^6)^{-5}(-m^3p^{-4})^{-6} =$

10. $(-ab^5c^{-3})^{-4}(-a^4b^{-3}c^{-2})^{-1} =$

11. $(-3xy^6w^3)^{-2}(-5x^2y^{-3}w^2)^2 =$

12. $(-2pm^{-4}k^7)^{-3}(-p^5m^3k^{-4})^{-5} =$

13. $(-ab^3c^{-4})^{-6}(-2a^2b^{-1}c^3)^{-2}(-5a^{-1}b^5c^{-6})^{-1} =$

14. $(-3w^5yz^{-3})^{-2}(2wy^4z^{-7})^{-3}(-2w^{-3}y^6z^2)^{-1} =$

15. $(-4pm^8)^{-2}(5p^{-3}m^{-2})(-3p^4m^{-5})^{-2} =$

16. $(-a^5b^4c^{-3})^{-5}(-2ab^2c^{-6})^{-2}(-3a^2\,bc^{-7})^{-2} =$

Check your answers on page 750. ✓

Multiplying and Dividing Terms Raised to Powers

The strategy to simplify problems that involve multiplication and division of terms is very similar to the one we use when terms are multiplied.

> **To simplify the multiplication and division of terms:**
>
> 1. Apply the *power* outside the parentheses to every factor inside, simplify each factor, and determine what the sign of that term will be.
>
> 2. *Exchange* all factors with negative exponents so the exponent will be positive. (Factors that were in the numerator go to the denominator and vice versa.)
>
> 3. *Multiply* factors that are in the numerator with the same base, and multiply factors that are in the denominator with the same base by adding their exponents.
>
> 4. *Cancel* factors with the same base that appear in the numerator and denominator—by subtracting the smaller exponent from the larger.
>
> 5. *Evaluate* and simplify the numerical factors.

EXAMPLE 5   Simplify $\dfrac{(2a^3b^7)^3}{(5a^6b^3)^2}$.

$$\frac{(2a^3b^7)^3}{(5a^6b^3)^2} =$$

$$\frac{2^3a^{3(3)}b^{7(3)}}{5^2a^{6(2)}b^{3(2)}} =$$

$$\frac{2^3a^9b^{21}}{5^2a^{12}b^6} =$$

$$\frac{2^3b^{21-6}}{5^2a^{12-9}} = \frac{8b^{15}}{25a^3} \qquad \square$$

EXAMPLE 6   Simplify $\dfrac{(a^2b^3)^2(a^5b^2)^4}{(a^3b)^3}$.

$$\frac{(a^2b^3)^2(a^5b^2)^4}{(a^3b)^3} =$$

$$\frac{(a^{2(2)}b^{3(2)})(a^{5(4)}b^{2(4)})}{(a^{3(3)}b^3)} =$$

$$\frac{a^4b^6a^{20}b^8}{a^9b^3} =$$

$$\frac{a^{4+20}b^{6+8}}{a^9b^3} =$$

$$\frac{a^{24}b^{14}}{a^9b^3} =$$

$$a^{24-9}b^{14-3} = a^{15}b^{11} \qquad \square$$

EXAMPLE 7 Simplify $\dfrac{(-3k^9m^4)^2(-k^3p^{-8})^5}{(-2k^{-1}p^6)^3(-m^2k^7)^4}$.

$$\frac{(-3k^9m^4)^2(-k^3p^{-8})^5}{(-2k^{-1}p^6)^3(-m^2k^7)^4} =$$

$$\frac{[+3^2k^{9(2)}m^{4(2)}][-k^{3(5)}p^{-8(5)}]}{[-2^3k^{-1(3)}p^{6(3)}][+m^{2(4)}k^{7(4)}]} =$$

$$\frac{(+3^2k^{18}m^8)(-k^{15}p^{-40})}{(-2^3k^{-3}p^{18})(+m^8k^{28})} =$$

$$\frac{-3^2k^{18}k^{15}k^3m^8}{-2^3p^{18}p^{40}m^8k^{28}} =$$

$$+\frac{3^2k^{18+15+3}m^8}{2^3p^{18+40}m^8k^{28}} =$$

$$\frac{3^2k^{36}m^8}{2^3p^{58}m^8k^{28}} =$$

$$\frac{3^2k^{36-28}m^{8-8}}{2^3p^{58}} =$$

$$\frac{3^2k^8m^0}{2^3p^{58}} = \frac{9k^8}{8p^{58}} \quad \square$$

EXAMPLE 8 Simplify $\dfrac{(-2xy^{-3})^{-1}(w^5x^{-2})^4}{(5y^2w^{-7})^{-2}(2x^3y^4w^{-1})^2}$.

$$\frac{(-2xy^{-3})^{-1}(w^5x^{-2})^4}{(5y^2w^{-7})^{-2}(2x^3y^4w^{-1})^2} =$$

$$\frac{(-2^{-1}x^{-1}y^3)(w^{20}x^{-8})}{(5^{-2}y^{-4}w^{14})(2^2x^6y^8w^{-2})} =$$

$$\frac{-(5^2)y^3y^4w^{20}w^2}{(2^1)(2^2)x^6x^1x^8w^{14}y^8} =$$

$$-\frac{5^2y^{3+4}w^{20+2}}{2(2)^2x^{6+1+8}y^8w^{14}} =$$

$$-\frac{5^2y^7w^{22}}{2(2)^2x^{15}y^8w^{14}} =$$

$$-\frac{5^2w^{22-14}}{2(2)^2x^{15}y^{8-7}} =$$

$$-\frac{5^2w^8}{2(2)^2x^{15}y^1} =$$

$$-\frac{25w^8}{2(4)x^{15}y} = -\frac{25w^8}{8x^{15}y} \quad \square$$

EXERCISES 11.5B

Simplify and leave your answers expressed with all exponents positive.

17. $\dfrac{(a^2b^3)^2(ab)^2}{(a^2b)^2} =$

18. $\dfrac{(xy^2)^3(x^2y)^3}{(xy^2)^2} =$

19. $\dfrac{(x^5y)^2(xy^3)^4}{(x^5y^4)^3} =$

20. $\dfrac{(x^5y)^3(x^2y^2)^3}{(xy^3)^4} =$

21. $\dfrac{(m^5np^2)^3(mn^2p^3)^2}{(m^7n^2)^3} =$

22. $\dfrac{(a^5b^2)^3(a^7b)^2}{(a^2b)^4} =$

23. $\dfrac{(3a^5b)^2}{(2ab^3)^3} =$

24. $\dfrac{(7x^3y^4)^2}{(5xy^2)^2} =$

25. $\dfrac{(-a^3bc^4)^5(-ab^7c^2)^4}{(-abc^4)^2} =$

26. $\dfrac{(-mp^4n^3)^4(2m^3pn^2)^2}{(-3m^2pn)^3} =$

27. $\dfrac{(-2xw^3z)^2(-5x^4wz^2)^2}{(-3x^4w^5z^3)^2(5x^2wz)^2} =$

28. $\dfrac{(3ab^3c)^{-1}(-2a^2b^2c)^{-2}}{(-2ab^4c^2)^{-3}(-a^4bc^2)^3} =$

29. $\dfrac{(-mp^4n^6)^{-3}(-2m^2p^3n)^{-3}}{(5m^4pn^3)^{-2}} =$

30. $\dfrac{(6wy^5z^2)^{-1}(-3w^4yz^2)^2}{(-2w^3y^2z^2)^{-3}(5wyz^4)^{-1}} =$

Check your answers on page 750.

 ## Scientific Notation

Powers of Ten

Powers of ten are the numbers that result when any integer (positive, negative, or zero) is applied as an exponent to 10. Some powers of ten are:

$$10^0 = 1$$

$$10^1 = 10; \qquad 10^{-1} = \frac{1}{10}$$

$$10^2 = 100; \qquad 10^{-2} = \frac{1}{10^2} \ \text{ or } \ \frac{1}{100}$$

$$10^3 = 1000; \qquad 10^{-3} = \frac{1}{10^3} \ \text{ or } \ \frac{1}{1000}$$

We saw in Sections 4.4 and 4.5 that the easiest number to multiply or divide by is 10.

Every time a number is *multiplied* by 10, the decimal point is moved one place value to the *right*.

Every time a number is *divided* by 10, the decimal point is moved one place value to the *left*.

Exponential notation makes it easy to represent repeated multiplication or division by 10.

Positive exponents represent multiplication.

Negative exponents represent division.

To represent repeated *multiplication* of a number by 10, we can multiply the number by a *positive* power of ten.

EXAMPLE 1 Multiply 528.34 by 10 twice.

$$(528.34)(10)^2 = 52,834$$ ❏

EXAMPLE 2 Multiply 47.000371 by 10 four times.

$$(47.000371)(10)^4 = 470,003.71$$ ❏

To represent repeated *division* of a number by 10, we can multiply the number by a *negative* power of ten.

EXAMPLE 3 Divide 34,896.58 by 10 three times.

$$(34,896.58)(10)^{-3} = 34.89658$$ ❏

If no decimal point is written in a number, it is understood to be to the right of the last digit. You should write in the decimal point, if it is not there, before trying to move it.

If there are not enough digits in the number to move the decimal point past, you should put zeros in to represent those place values.

EXAMPLE 4 Multiply 726 by 10 seven times.

$$(726.0000000)(10)^7 = 7,260,000,000$$ ❏

EXAMPLE 5 Divide 705 by 10 five times.

$$(705)(10)^{-5} =$$
$$(00705.)(10)^{-5} = 0.00705$$ ❏

EXERCISES 11.6A

Compute each repeated multiplication or division by 10 by multiplying by the powers of ten.

1. $(15)(10)^3 =$ **2.** $(430)(10)^5 =$

3. $(5.389)(10)^2 =$ **4.** $(601.58)(10)^3 =$

5. $(68)(10)^{-2} =$ **6.** $(308)(10)^{-4} =$

7. $(13.2)(10)^{-2} =$ **8.** $(538.002)(10)^{-2} =$

9. $(0.05)(10)^{-3} =$ **10.** $(6.1)(10)^{-4} =$

11. Divide 84 by 10 three times.

12. Divide 120 by 10 two times.

13. Divide 2,304 by 10 three times.

14. Divide 5,609,000 by 10 five times.

15. Divide 560 by 10 six times.

16. Divide 0.00056 by 10 three times.

17. Divide 0.0087 by 10 four times.

18. Divide 6,834 by 10 seven times.

19. Divide 5.04 by 10 three times.

20. Divide 14.55 by 10 four times.

Check your answers on page 750.

Scientific Notation

There is a notation, called **scientific notation** that makes it easier to represent and to calculate with very large and very small numbers that have only a few nonzero digits.

> *In scientific notation, numbers are represented by the* **product of**:
>
> 1. Their nonzero digits (this is called the **coefficient**) by
> 2. The power of ten needed to get the decimal point in the right place.
>
> Number expressed in scientific notation = (coefficient) × (power of ten)

Scientific notation is the only situation where × is used to represent multiplication in algebra. The coefficient is written as a number between 1 and 10.

EXAMPLE 6   15,000 is expressed in scientific notation as

$$1.5 \times 10^4.$$

> *To express a number in scientific notation:*
>
> Put a decimal point after the first nonzero digit.
>
> Then determine what power of ten you must multiply or divide by to make their product equal to the original number.

To determine the right exponent for the power of ten, put your pencil to the right of the first nonzero digit in the number and count the number of place values from there to the decimal point.

EXAMPLE 7   Write 280,000 in scientific notation.

The pencil is put between the **2** and **8**, and we must move five places to the right to get to the (understood) decimal point at the end of the number.

$$280,000 \text{ is written as } 2.8 \times 10^5$$

EXAMPLE 8   Write 0.000000035 in scientific notation.

We put the pencil between the **3** and **5** and move eight places to the left to get to the decimal point.

$$0.\underset{\displaystyle\frown}{000000035} \text{ is written as } 3.5 \times 10^{-8}$$ ❑

⇒ *Note*:   If the original number is 10 or larger, the power of ten must *increase* the coefficient (which is a number between 1 and 10), so the exponent will be *positive*.

EXAMPLE 9   Write 3,900,000 in scientific notation.

Since 3,900,000 > 10, we will use a positive power of ten.

$$3,900,000 =$$
$$3,900,000. = 3.9 \times 10^6$$ ❑

If the original number is between 0 and 1, the power of ten must *decrease* the coefficient, so the exponent will be *negative*.

EXAMPLE 10   Write 0.000047 in scientific notation.

Since 0.000047 < 1, we will use a negative power of ten.

$$0.000047 = 4.7 \times 10^{-5}$$ ❑

You don't have to memorize that. Just think about whether the power of ten is making the coefficient larger or smaller.

Here are some examples of numbers used in various scientific fields that are usually expressed in scientific notation:

9.108×10^{-28} g—mass of an electron at rest
1.67×10^{-24} g—mass of a proton at rest
2.8×10^{-13} cm—radius of an electron
5.3×10^{-11} m—radius of a hydrogen atom
1.0×10^{-8} cm—one Angstrom unit (the measure of atomic size and electro-
        magnetic wavelength)
6.02×10^{23}—Avagadro's number (atoms/mole)
9.3×10^7 mi—distance between earth and sun
3.0×10^{10} cm/sec or 1.86×10^5 mi/sec—speed of light
1.0×10^{14}—approximate number of stars in the Milky Way Galaxy
8.65×10^5 mi—diameter of the sun

EXERCISES 11.6B

Write each of these numbers in scientific notation.

21. 236 =   **22.** 53 =

23. 13,908 =   **24.** 405,911 =

25. 5,671,330 =   **26.** 8,300,000 =

27. 0.00047 =   **28.** 0.000009134 =

29. 0.000845 =   **30.** 0.0007 =

31. 0.00000099 =   **32.** 0.000146 =

Check your answers on page 750.

11.7 Computing with Numbers Expressed in Scientific Notation

Multiplying or dividing very large or very small numbers that have many zero digits is tedious. It is easy to make a careless error in such calculations by putting the wrong number of zeros in the answer.

Most calculators display only eight digits, so they do not accurately show the result of calculations where the answer has more than eight place values.

It is faster, easier, and less error prone to do such calculations with numbers expressed in scientific notation.

Multiplication

> **To multiply numbers expressed in scientific notation:**
>
> 1. Group the coefficients together and multiply.
> 2. Group the powers of ten together and multiply by adding their exponents.

EXAMPLE 1   Multiply $[1.5 \times 10^7]$ by $[2.3 \times 10^4]$.

$$[1.5 \times 10^7][2.3 \times 10^4] =$$
$$[(1.5)(2.3)][(10)^7(10)^4] =$$
$$[3.45][10^{7+4}] = 3.45 \times 10^{11} \qquad \square$$

EXAMPLE 2   Multiply $[2.1 \times 10^{-15}]$ by $[3.0 \times 10^{-7}]$.

$$[2.1 \times 10^{-15}][3.0 \times 10^{-7}] =$$
$$[(2.1)(3.0)][(10)^{-15}(10)^{-7}] =$$
$$[6.3][10^{-15-7}] = 6.3 \times 10^{-22} \qquad \square$$

If the numbers are not expressed in scientific notation, you must first translate them into that notation.

EXAMPLE 3   Multiply (0.00000003) by (0.000000000000032).

$$(0.00000003) = 3.0 \times 10^{-8}; (0.000000000000032) = 3.2 \times 10^{-14}$$
$$[0.00000003][0.000000000000032] =$$
$$[3.0 \times 10^{-8}][3.2 \times 10^{-14}] =$$
$$[(3.0)(3.2)][(10)^{-8}(10)^{-14}] =$$
$$[9.6][10^{-8-14}] = 9.6 \times 10^{-22} \qquad \square$$

EXAMPLE 4   Multiply $14,000,000,000$ by 0.00000022.

Convert numbers into scientific notation first.

$$14,000,000,000 = 1.4 \times 10^{10}; \qquad 0.00000022 = 2.2 \times 10^{-7}$$
$$[14,000,000,000][0.00000022] =$$
$$[1.4 \times 10^{10}][2.2 \times 10^{-7}] =$$
$$[(1.4)(2.2)][(10)^{10}(10)^{-7}] =$$
$$[3.08][10^{10-7}] = 3.08 \times 10^3 \qquad \square$$

When computing in scientific notation, leave the answer expressed in scientific notation. Be sure that the coefficient is a number between 1 and 10.

(In other words, the decimal point should be to the right of the first nonzero digit.)

If it is not, the easiest way to convert it correctly is first to rewrite the coefficient in scientific notation and then combine the powers of ten.

EXAMPLE 5 Multiply 7.8×10^{-260} by 9.1×10^{-590}.

$$[7.8 \times 10^{-260}][9.1 \times 10^{-590}] =$$
$$[(7.8)(9.1)][(10^{-260})(10^{-590})] =$$
$$[70.98][10^{-260-590}] = [70.98][10^{-850}]$$

The coefficient (70.98) is larger than 10, so our answer is *not* in scientific notation yet!

$$[70.98][10^{-850}] =$$
$$[7.098 \times 10^1][10^{-850}] =$$
$$[7.098][10^{1-850}] = 7.098 \times 10^{-849}$$

Scientific notation also simplifies our work when we multiply more than two numbers.

EXAMPLE 6 Multiply $[1,300,000][280,000][41,000,000]$.

$$[1,300,000][280,000][41,000,000] =$$
$$[1.3 \times 10^6][2.8 \times 10^5][4.1 \times 10^7] =$$
$$[(1.3)(2.8)(4.1)][(10)^6(10)^5(10)^7] =$$
$$[(3.64)(4.1)][10^{6+5+7}] = [14.924][10^{18}]$$

The coefficient (14.924) is larger than 10, so our answer is *not* in scientific notation yet!

$$[14.924][10^{18}] =$$
$$[1.4924 \times 10][10^{18}] =$$
$$[1.4924][10^{1+18}] = 1.4924 \times 10^{19}$$

If very large or very small numbers have many nonzero digits, we might be instructed to round them off. Often the numbers that we are computing with are expressed more accurately than necessary for our purposes. Remember, the place values get smaller as you read numbers from left to right, so the largest place values are represented by the first digits.

> *If we say, "Round off an integer to two nonzero digits," that means to:*
>
> Replace all nonzero digits (after the first two) by zeros.
> Round down to that number if the third nonzero digit is <5.
> Round up by increasing the second nonzero digit by 1 if the third nonzero digit is ≥5.

EXAMPLE 7 Round off $849,326$ to two nonzero digits.

We'll replace all digits after 4 by zero; and since 9 (the third nonzero digit) is ≥5, we round up. So 84↓9,326 rounds up to 850,000.

When rounding off a decimal, the digits we do not use are dropped rather than replaced by zeros.

EXAMPLE 8   Round off 0.00017359 to two nonzero digits.

We'll drop the digits after the 7, and since 3 (the third nonzero digit) is <5, we round down.

So 0.00017↓359 rounds down to 0.00017. ❑

We discussed rounding off more thoroughly in Section 4.2. Go back and review this topic if you are not yet comfortable with rounding off.

Scientific notation makes calculations easier only with numbers that have few nonzero digits. But if you do round off the numbers in a calculation, you are approximating the answer. You must use the symbol $\approx$ to clarify that your answer is not precise.

EXAMPLE 9   Round off each number and the answer to two nonzero digits, convert into scientific notation, and multiply:

$$(47^{\downarrow}①,329,568)(9,4^{\downarrow}⑦8,465) \approx$$
$$(470,000,000)(9,500,000) \approx$$
$$[4.7 \times 10^8][9.5 \times 10^6] \approx$$
$$[(4.7)(9.5)] \times [(10)^8(10)^6] \approx$$
$$[44.65] \times [10^{8+6}] \approx$$
$$44.65 \times 10^{14} \approx$$
$$(4.4^{\downarrow}65 \times 10^1) \times 10^{14} \approx 4.5 \times 10^{15}$$ ❑

In this book, round off only when told to. (Problems involving money, however, should *never* be expressed more precisely than to the nearest hundredth (cent), since there is no coin for part of a cent.)

EXERCISES 11.7A

Multiply in scientific notation and leave your answers expressed that way.

1. (310)(220,000) =

2. (630,000)(1,200,000) =

3. (0.000047)(0.000016) =

4. (0.00000034)(0.000024) =

5. (510,000)(67,000) =

6. (43,000)(689,000) =

7. (0.000083)(0.000031) =

8. (0.0000056)(0.00034) =

9. (0.000062)(0.0007) =

10. (3,800,000)(55,000) =

11. (520,000)(4,700,000)(1,700) =

12. (7,800,000)(610,000)(8,000,000) =

13. (0.00071)(0.0005)(0.000000011) =

14. (0.00007)(0.0000031)(0.00000072) =

15. (0.00000066)(0.00043)(0.000081) =

16. (0.003)(0.000072)(0.000054) =

17. $(63,000,000)(400)(38,000) =$

18. $(7,300)(45,000)(810) =$

19. $(5,400,000)(0.0006) =$

20. $(0.00013)(7,000) =$

21. $(830,000)(0.0003) =$

22. $(0.000015)(6,200,000) =$

23. $(860,000,000)(0.0005) =$

24. $(0.0000072)(860) =$

25. $(3.1 \times 10^{39})(1.2 \times 10^{41})$

26. $(2.3 \times 10^{71})(2.4 \times 10^{64})$

27. $(4.2 \times 10^{57})(5.3 \times 10^{75})$

28. $(6.1 \times 10^{37})(3.4 \times 10^{56})$

29. $(1.3 \times 10^{-21})(2.6 \times 10^{-35})$

30. $(4.1 \times 10^{-40})(3.2 \times 10^{-61})$

31. $(3.3 \times 10^{-52})(2.4 \times 10^{-85})$

32. $(5.1 \times 10^{-80})(6.1 \times 10^{-57})$

33. $(3.2 \times 10^{35})(1.7 \times 10^{-80})$

34. $(2.6 \times 10^{-25})(3.1 \times 10^{43})$

Round off each number and the answer to two nonzero digits. Express your answers in scientific notation.

35. $(371,425)(63,710) =$

36. $(5,172,869)(713,897) =$

37. $(0.17611154)(0.0008437) =$

38. $(0.0008269)(0.00714) =$

39. $(678,173)(0.00003478) =$

40. $(0.00047681)(973,105) =$

Check your answers on page 750.

Division

Scientific notation makes it easier to divide large and small numbers too.

> **To divide numbers expressed in scientific notation:**
> 1. Group the coefficients together and divide.
> 2. Group the powers of ten together and divide by subtracting their exponents.

Be careful when dividing. Remember that the order of the numbers indicates which is the dividend and which is the divisor.

EXAMPLE 10   Divide $[4.48 \times 10^{12}]$ by $[3.2 \times 10^9]$.

$$[4.48 \times 10^{12}] \div [3.2 \times 10^9] =$$
$$[4.48 \div 3.2][\underline{(10)^{12} \div (10)^9}] =$$
$$[1.4][10^{12-9}] = 1.4 \times 10^3$$

EXAMPLE 11   Divide $[6,630,000,000]$ by $[510,000]$.

$$[6,630,000,000] \div [510,000] =$$
$$[6.63 \times 10^9] \div [5.1 \times 10^5] =$$
$$\left[\frac{6.63}{5.1}\right]\left[\frac{10^9}{10^5}\right] =$$
$$[1.3][10^{9-5}] = 1.3 \times 10^4$$

When we were multiplying, we had no problem working with negative exponents because we were adding them. But to divide powers of ten, we will be subtracting exponents, and it is easy to make careless errors when subtracting negative exponents.

So before dividing the powers of ten, I usually convert all exponents so they are positive.

EXAMPLE 12 Divide $[8.61 \times 10^{-18}]$ by $[4.1 \times 10^{-37}]$.

$$[8.61 \times 10^{18}] \div [4.1 \times 10^{-37}] =$$

$$\left[\frac{8.61}{4.1}\right]\left[\frac{10^{-18}}{10^{-37}}\right] =$$

$$\left[\frac{8.61}{4.1}\right]\left[\frac{10^{+37}}{10^{+18}}\right] =$$

$$[2.1][10^{37-18}] = 2.1 \times 10^{19} \qquad \square$$

EXAMPLE 13 Divide (0.0000000000357) by (0.00000021).

$$(0.0000000000357) \div (0.00000021) =$$

$$[3.57 \times 10^{-11}] \div [2.1 \times 10^{-7}] =$$

$$\left[\frac{3.57}{2.1}\right]\left[\frac{10^{-11}}{10^{-7}}\right] =$$

$$\left[\frac{3.57}{2.1}\right]\left[\frac{10^{+7}}{10^{+11}}\right] =$$

$$[1.7][(10)^{7-11}] = 1.7 \times 10^{-4} \qquad \square$$

EXERCISES 11.7B

Divide each of these numbers in scientific notation.

41. Divide 3.22×10^{17} by 2.3×10^{6} =

42. Divide 8.88×10^{46} by 3.7×10^{35} =

43. Divide 8.68×10^{14} by 2.8×10^{20} =

44. Divide 7.56×10^{26} by 6.3×10^{31} =

45. Divide 7.14×10^{-8} by 3.4×10^{-12} =

46. Divide 7.28×10^{6} by 5.2×10^{-5} =

47. Divide 7.74×10^{-18} by 1.8×10^{-14} =

48. Divide $8,500,000,000$ by $250,000$ =

49. Divide $60,800,000,000,000$ by $3,200,000$ =

50. Divide 0.0000147 by 0.0042 =

51. Divide 0.00000148 by 0.000037 =

52. Divide $3,840,000,000,000$ by 0.0016 =

53. Divide 0.0000575 by 230,000 =

54. Divide 1,230,000,000,000 by 0.00041 =

Check your answers on page 751.   ✓

11.8   Using Scientific Notation to Solve Real-Life Problems

Many calculations of real-life problems are simplified by the use of scientific notation. The goal of this section is for you to become comfortable using scientific notation so you will be successful using it when you *must*. I have used it in the next two examples even though the numbers are small enough to calculate directly.

For real-life purposes it is usually best to convert answers back to normal (decimal) notation.

EXAMPLE 1   If a caterer charges $7.39 for each guest, approximately how much should it cost to pay for a party with 1,823 guests? Each figure can be rounded off to two nonzero digits, and so should the answer.

$$\$7.3{\downarrow}9 \rightarrow 7.4$$
$$18{\downarrow}23 \rightarrow 1800$$

The total cost is the product of price per guest multiplied by the number of guests.

$$\text{Total cost is: } \$(7.4)(1800) =$$
$$\$(7.4)(1.8 \times 10^3) =$$
$$\$(13.32)(10^3) =$$
$$\$13{\downarrow}320 \approx \$13,000 \qquad ❑$$

EXAMPLE 2   If the membership fees in a union are $188.43 per year, approximately how much should a union receive in fees if there are 286,142 members?

We can round off the fees to the nearest dollar: $188{\downarrow}43 \rightarrow \$188.

We can round off the membership to two nonzero digits:

$$28{\downarrow}6,142 \rightarrow 290,000$$

Total income of fees equals product of fee per member multiplied by number of members. Total membership fee income is:

$$\$(188)(290,000) =$$
$$\$(1.88 \times 10^2)(2.9 \times 10^5) =$$
$$\$(1.88)(2.9) \times (10^2)(10^5) =$$
$$\$(5.452) \times (10^7) =$$
$$\$54{\downarrow}520,000 \approx 55,000,000 \qquad ❑$$

EXAMPLE 3   A lump of coal that weighs 12 grams has 6×10^{23} atoms of carbon. How many atoms of carbon are there in a 3,000 gram chunk of coal?

We can set this problem up as a proportion:

Grams of coal: atoms of carbon = grams of coal: atoms of carbon

$$12 : (6 \times 10^{23}) = 3{,}000 : A$$

$$12A = 3{,}000(6 \times 10^{23})$$

$$\frac{12A}{12} = (12 \cdot 250 \cdot 6) \times \frac{10^{23}}{12}$$

$$A = 1500 \times 10^{23}$$

$$A = (1.5 \times 10^{3}) \times 10^{23}$$

$$A = 1.5 \times 10^{26}$$

So there are 1.5×10^{26} atoms of carbon in that big chunk of coal. ❏

EXERCISES 11.8

Round off and use scientific notation where appropriate to compute these problems. Leave only answers to Exercises 13–18 expressed in scientific notation.

1. If the cost of buying and sending invitations comes to 31¢ per person, approximately how much will it cost to invite 893 people to an affair? Round off both figures and the answer to one nonzero digit.

2. If it takes an assembly line worker in an electronics plant 47 sec to handle each transistor, how long will it take her to handle 723,186 of them? Round off each figure and the answer to one nonzero digit.

3. Convert the answer to Exercise 2 into hours (1 hr = 3.6×10^3 sec) by dividing the answer by 3.6×10^3. Round the answer off to the nearest whole hour.

4. If it took 867,462 sec to handle all the transistors, and it took an average of 28 sec per transistor, approximately how many did the worker handle if each figure and the answer are rounded off to one nonzero digit?

5. If the worker in Exercise 4 was paid approximately $900 for working on those transistors, how much was she paid for each one?

6. How many hours is 867,462 sec (divide by 3.6×10^3)? Compute with figures rounded off to two nonzero digits; round off your answer to the nearest whole number of hours.

7. If the worker in Exercise 4 was paid $900 for the work described, how much was she paid per hour? (Round off hours to two nonzero digits.)

8. If the pushcart man makes 21.3¢ profit on each ice cream bar sold at a concert, approximately how much profit does he make if he sells 3,876 bars? Round off each figure and the answer to two nonzero digits.

9. Approximately how much profit would you have figured he made if you had rounded off the figures to one nonzero digit each in Exercise 8?

10. If the cost is $0.00073 per leaflet, approximately how much will it cost the antidraft group to distribute leaflets to 1,479,068 19-year-olds? Round off each figure to one nonzero digit.

11. If the average amount of change each customer got one week was 13¢ per person, and if each person contributed the change to fund a community youth program, approximately how much was contributed if there were 7,018 customers? Round off both figures to one nonzero digit.

12. If the fundraiser brought in $18,316.27 from 471 contributors, approximately how much did each person contribute? Round off both figures to one nonzero digit.

For Exercises 13–18:

If every lump of coal has 6×10^{23} atoms of carbon for every 12 g of weight, how many atoms of carbon are there in each chunk of coal (leave answers in scientific notation):

13. weighing 84,000 g?

14. weighing 240,000 g?

15. weighing 1,800,000 g?

16. weighing 0.006 g?

17. weighing 0.00048 g?

18. weighing 16,800,000 g?

19. Each worker produced 3.7×10^{18} widgets last year. If there were 1.3×10^5 workers, how many widgets were produced?

20. Each worker printed 4.3×10^{13} stickers last year. If there were 1.5×10^4 workers, how many stickers were printed?

21. Each worker processed 3.1×10^6 lb of grain last year. If there were 2.6×10^4 workers, how many pounds of grain were processed?

22. Each worker developed 3.4×10^5 photos last year. If there were 1.2×10^4 workers, how many photos were developed?

Check your answers on page 751.

Glossary

base Factor that is multiplied repeatedly in a term. In exponential notation, the factor or factors written before (and below) the exponent.

Example: In 3^5, 3 is the base.

In x^7, x is the base.

exponent Positive integral number representing the number of times that the base is repeated as a factor in the term. The exponent is written above the base.

Example: In 3^5, 5 is the exponent.

In x^7, 7 is the exponent.

negative exponent Number representing repeated division by a factor. The exponent indicates the number of times the base is a factor in the denominator.

Example: $x^{-5} = \dfrac{1}{x^5}$; $2^{-3} = \dfrac{1}{2^3}$ or $\dfrac{1}{8}$

zero exponent Any term raised to a zero exponent is always equal to 1.

Example: $x^0 = 1$; $3,486^0 = 1$

power Another name for exponent.

Example: x^3 can be read as "x to the third power."

power of ten Concise notation raising the number 10 to an exponent; used to represent repeated multiplication or division by 10.

Example: 10^2 represents multiplication by 10 twice.

10^{-3} represents division by 10 three times.

scientific notation Writing a number as the product of its nonzero digits expressed as a number (between 1 and 10) and the power of ten needed to get the decimal point in the correct place value.

Example: 35,000 in scientific notation is 3.5×10^4.

rules for computing with integral exponents

$x^1 = x \qquad x^0 = 1 \text{ (if } x \neq 0)$

$x^m x^n = x^{m+n}$

$\dfrac{x^m}{x^n} = x^{m-n} \quad \text{if } m \geq n$

$\dfrac{x^m}{x^n} = \dfrac{1}{x^{n-m}} \quad \text{if } n > m$

$x^{-1} = \dfrac{1}{x}; \quad x^{-n} = \dfrac{1}{x^n} \ (x \neq 0); \quad \dfrac{1}{x^{-n}} = x^n; \quad \dfrac{y^{-a}}{w^{-b}} = \dfrac{w^b}{y^a}$

$(xy)^n = x^n y^n$

$(x^m)^n = x^{mn}$

$(x^a y^b)^n = x^{an} y^{bn}$

$(-x)^n = -x^n \quad \text{if } n \text{ is an odd number}$

$(-x)^m = x^m \quad \text{if } m \text{ is an even number}$

Chapter 11 Review

Do each of the problems. After you are confident that you have done them as accurately as possible, compare your answers with those in the back of the book. If any are wrong, go back to the section where the problem came from (indicated to the left of the problem) and review the section.

Once you understand all the sections, you will be sure to learn the skills solidly and remember how to do them if you practice them a little bit more. Turn to the Supplementary Exercises and do all the problems from any section where you had difficulty on these review exercises.

11.1 Express each term without using exponents.

1. $5y^2$

2. $-4xy^3$

3. m^{-1}

4. $2a^3b^{-4}$

Evaluate each expression.

5. 586^0

6. $(4)^{-3}$

7. $(-3)^{-2}$

8. -9^2

9. $\left(\dfrac{4}{5}\right)^{-1}$

10. $\left(\dfrac{3}{4}\right)^2$

Simplify and leave expressed with positive exponents.

11.2 11. a^7a^{18}

12. $x^{24}x^{-38}$

13. $(-xy^3)(-x^4y^{-1})$

14. $(5x^2y^5)(-2xy^{-8})$

11.3 15. $\dfrac{x^{17}}{x^{12}}$

16. $\dfrac{y^8}{y^{13}}$

17. $\dfrac{nm^7p^4}{n^5m^5p}$

18. $\dfrac{-24a^6b^3}{40a^2b^{10}}$

19. $\dfrac{x^3yw^{-4}}{xy^{-7}w^2}$

20. $\dfrac{-10a^5b^{-1}}{-6a^5b}$

11.4 21. $(m^8)^5$

22. $(w^4)^{-8}$

23. $(xy^5)^2$

24. $(-7ab^{-5})^2$

25. $(-nm^{-3})^{-5}$

26. $(-3ab^2c^{-4})^{-2}$

Answers

1. _____

2. _____

3. _____

4. _____

5. _____

6. _____

7. _____

8. _____

9. _____

10. _____

11. _____

12. _____

13. _____

14. _____

15. _____

16. _____

17. _____

18. _____

19. _____

20. _____

21. _____

22. _____

23. _____

24. _____

25. _____

26. _____

11.5   **27.** $(xy^3)^2(x^4y^3)^5$ **28.** $(-2a^3b^{-2})^3(-ab^3)^{-7}$

29. $\dfrac{(3mn^4p)^3}{(2m^3n^2p)^2}$ **30.** $\dfrac{(2ab^5)^3(2a^3b^{-1})^{-2}}{(-a^{-4}b^7)^3}$

11.6   Compute.

31. 40.76×10^4

32. Divide 7,806 by 10 twice.

33. 560×10^{-4}

Express in scientific notation.

34. 560,000,000   **35.** 0.00000000000072

11.7   Compute and leave in scientific notation.

36. $(5.1 \times 10^{23})(1.3 \times 10^{40})$

37. $(4,300,000)(0.00000000000017)$

38. $(8.3 \times 10^{-37})(1.4 \times 10^{-28})(3.0 \times 10^{-53})$

39. Round off each number and the answer to two nonzero digits; express in scientific notation:

$$(9,360,412,773)(1,847,991,586)$$

40. $(8.75 \times 10^{78}) \div (3.5 \times 10^{-46})$

11.8   **41.** Last year each worker in the E-Z-2 Thred factory produced 4.2×10^{80} needles. If there were 7.0×10^4 employees, how many needles were produced?

42. Approximately how much did it cost per vote if the campaign for Alfred E. Newman for governor spent \$813,429.68 that netted 2,056,821 votes? Round off both figures and answer to one nonzero digit.

Answers

27. _____

28. _____

29. _____

30. _____

31. _____

32. _____

33. _____

34. _____

35. _____

36. _____

37. _____

38. _____

39. _____

40. _____

41. _____

42 _____

Check your answers on page 751.

Supplementary Exercises

Do all the problems in every section that involves a skill in which you now lack complete mastery. With a little more practice you will achieve that sense of really understanding the topics, and you will remember how to do these problems.

11.1 Express each term without using exponents.

1. $4a^3$ 2. $-5nm^6$

3. b^{-3} 4. $7n^5p^{-3}$

Evaluate each expression.

5. $1,704^0$ 6. $(6)^{-3}$

7. $(-4)^{-2}$ 8. -7^2

9. $\left(\dfrac{2}{9}\right)^{-1}$ 10. $\left(\dfrac{2}{7}\right)^2$

Simplify and leave your answers expressed with positive exponents.

11.2 11. k^5k^{26} 12. $y^{17}y^{-29}$

13. $(-ab^5)(-a^3b^{-2})$ 14. $(6p^3k^9)(-4pk^{-7})$

11.3 15. $\dfrac{w^{14}}{w^{11}}$ 16. $\dfrac{p^{12}}{p^{25}}$

17. $\dfrac{nm^8p^3}{n^7m^4p}$ 18. $\dfrac{-25a^8b^5}{40a^5b^{13}}$

19. $\dfrac{x^7yw^{-3}}{xy^{-6}w^4}$ 20. $\dfrac{-12a^9b^{-2}}{-6a^2b}$

11.4 21. $(a^7)^8$ 22. $(x^5)^{-6}$

23. $(ab^4)^3$ 24. $(-6wy^{-9})^2$

25. $(-pn^{-4})^{-7}$ 26. $(-2nm^3p^{-5})^{-2}$

Answers

1. _____

2. _____

3. _____

4. _____

5. _____

6. _____

7. _____

8. _____

9. _____

10. _____

11. _____

12. _____

13. _____

14. _____

15. _____

16. _____

17. _____

18. _____

19. _____

20. _____

21. _____

22. _____

23. _____

24. _____

25. _____

26. _____

11.5 **27.** $(xy^4)^6(x^3y^6)^4$

28. $(-2a^5b^{-3})^3(-ab^4)^{-5}$

Answers

29. $\dfrac{(2mn^3p)^3}{(5m^4n^6p)^2}$

30. $\dfrac{(3ab^8)^3(4a^6b^{-3})^{-2}}{(-a^{-5}b^4)^3}$

27. _____

28. _____

11.6 Compute.

31. 640.32×10^5

29. _____

32. Divide 14,375 by 10 four times.

30. _____

33. 37×10^{-5}

31. _____

Express in scientific notation.

34. 382,000,000   **35.** 0.000000000028

32. _____

33. _____

11.7 Compute and leave your answers in scientific notation.

36. $(3.7 \times 10^{34})(1.2 \times 10^{51})$

34. _____

37. $(5,200,000)(0.000000000018)$

35. _____

38. $(2.7 \times 10^{-26})(1.6 \times 10^{-37})(4.0 \times 10^{-49})$

36. _____

39. Round off each number and the answer to two nonzero digits; express in scientific notation:

$$(7,810,459,827)(6,480,672,419)$$

37. _____

36. _____

40. $(6.12 \times 10^{63}) \div (1.8 \times 10^{-35})$

39. _____

11.8 **41.** Find the approximate cost of feeding 1,130,000,000 people if the food costs $2.82 per person. Round each figure and the answer to one nonzero digit.

40. _____

41. _____

42. Approximately how long does it take to address and seal each envelope if 22,743 envelopes can be addressed and sealed in 463,000 sec? Round each figure and the answer to two nonzero digits.

42. _____

Check your answers on page 751.

Chapter 11 Test

Express without exponents.

1. $2c^5$

2. $-3ab^4$

3. k^{-4}

4. $4n^6p^{-2}$

Evaluate each expression.

5. $6,380^0$

6. $(3)^{-4}$

7. $(-8)^{-2}$

8. -6^2

9. $\left(\dfrac{1}{5}\right)^{-1}$

10. $\left(\dfrac{3}{8}\right)^2$

Simplify and leave your answers expressed with positive exponents.

11. $x^{14}x^{65}$

12. $w^{27}w^{-43}$

13. $(np^4)(-n^3p^{-5})$

14. $(-2a^4b^7)(-4ab^{-7})$

15. $\dfrac{m^{28}}{m^{15}}$

16. $\dfrac{w^{16}}{w^{37}}$

17. $\dfrac{ab^6c^4}{a^4bc^5}$

18. $\dfrac{-30p^7n^3}{45p^4n^{18}}$

19. $\dfrac{y^8xw^{-4}}{yx^{-5}w^4}$

20. $\dfrac{-18a^7b^{-3}}{-9a^5b}$

21. $(p^6)^8$

22. $(w^4)^{-7}$

23. $(xz^5)^6$

24. $(-3ab^{-7})^3$

25. $(-ab^{-3})^{-8}$

26. $(-4nm^6p^{-3})^{-2}$

Answers

1. _____

2. _____

3. _____

4. _____

5. _____

6. _____

7. _____

8. _____

9. _____

10. _____

11. _____

12. _____

13. _____

14. _____

15. _____

16. _____

17. _____

18. _____

19. _____

20. _____

21. _____

22. _____

23. _____

24. _____

25. _____

26. _____

27. $(wm^3)^9(w^4m^6)^5$

28. $(3a^6b^{-4})^3(-ab^5)^{-5}$

29. $\dfrac{(3mn^2p)^3}{(4m^3n^7p)^2}$

30. $\dfrac{(2ab^5)^3(3a^4b^{-4})^{-2}}{(-a^{-8}b^2)^3}$

Compute

31. 806.63×10^4

32. Divide 34,028 by 10 three times.

33. 583×10^{-4}

Express in scientific notation.

34. 71,000,000

35. 0.0000000000468

Compute and leave your answers in scientific notation.

36. $(4.8 \times 10^{39})(1.1 \times 10^{46})$

37. $(3,400,000)(0.000000000000013)$

38. $(2.3 \times 10^{-18})(3.1 \times 10^{-42})(5.0 \times 10^{-28})$

39. Round off each number and the answer to two nonzero digits; express in scientific notation:

$$(6,135,459,862)(1,961,236,257)$$

40. $(3.38 \times 10^{82}) \div (1.3 \times 10^{-45})$

41. To come out even, approximately how much should you charge for each picture frame if it costs \$831,789,000 to produce 23,786,650 frames? Round each figure and the answer to one nonzero digit.

Answers

27. _____

28. _____

29. _____

30. _____

31. _____

32. _____

33. _____

34. _____

35. _____

36. _____

37. _____

38. _____

39. _____

40. _____

41. _____

Check your answers on page 751.

Cumulative Review
CHAPTERS 1–11

Add.

1. $\dfrac{3}{10} + \dfrac{2}{10}$

2. $\dfrac{7}{15} + \dfrac{3}{10}$

3. $0.006 + 4.3 + 0.89$

4. $(-5) + (-3) + (6) + (2) + (-8)$

5. $7\dfrac{2}{5} + 3\dfrac{3}{4}$

Subtract.

6. $\dfrac{7}{11} - \dfrac{4}{11}$

7. $\dfrac{7}{20} - \dfrac{1}{12}$

8. $8,006.043 - 789.0008$

9. $(-5) - (-14)$

10. $(-24) - (15)$

11. $3\dfrac{2}{3} - 1\dfrac{1}{4}$

Multiply.

12. $\left(\dfrac{6}{25}\right)\left(\dfrac{45}{180}\right)$

13. $\left(4\dfrac{1}{2}\right)\left(3\dfrac{1}{7}\right)$

14. $(0.0063)(2.4)$

15. $(-5)(3)(-1)(-1)(2)$

16. $(3x^2y)(5xy^4)$

17. $(x^5y^2)^2(xy^3)^3$

18. Multiply and leave in scientific notation:

$$(0.0004)(0.00031)$$

Divide.

19. $\dfrac{8}{15} \div \dfrac{20}{55}$

20. $0.0628 \div 0.002$

21. $(-72) \div (-9)$

22. $(40) \div (-8)$

Simplify.

23. $5ab - 4ac + bc - 8ba + cb - 2ca + 9bc + 6ab - 5ac$

24. $2x(x^2 - 5x - 1) - 4x(x^3 - x^2 + 3x - 5)$

25. $7 + 3(6 - 2) + (5 - 3)^2(7 - 1 - 4)^3$

Answers

1. _____

2. _____

3. _____

4. _____

5. _____

6. _____

7. _____

8. _____

9. _____

10. _____

11. _____

12. _____

13. _____

14. _____

15. _____

16. _____

17. _____

18. _____

19. _____

20. _____

21. _____

22. _____

23. _____

24. _____

25. _____

Evaluate if $a = -1$, $b = 2$, and $c = -3$.

Answers

26. abc **27.** $5a - 4b - 3c$

28. $a^2 + b^2 + c^2$ **29.** $5ab^2 - 3c^2 + (3b)^2$

30. $2(3a + 4b)^2 - c^2$ **31.** $(2ab^2) + 3(5a)^2$

32. Terry figured that since she earned more money than her husband, she should pay 60% of their household expenses. What is Terry's share of their $280.40 oil bill?

33. What is the total bill if I purchase three bowls at $6.45 each, one pitcher for $8.90, four mugs at $2.75 each, and a butter dish for $6.40 at the pottery store where there is a sales tax of 8%?

34. If the ratio of men to women in the union is 14:3, out of 10,880 members how many are women?

35. If only 46% of the teenagers in a community could find summer jobs, and 368 teenagers found jobs, how many teenagers were in the community?

26. _____

27. _____

28. _____

29. _____

30. _____

31. _____

32. _____

33. _____

34. _____

35. _____

Check your answers on page 751.

12 Radicals and Rational (Fractional) Exponents

12.1 Square Roots

Perfect Squares

There's an old joke from the 1950s that goes, "What do 9, 25, 100, and Lawrence Welk all have in common?" Answer: They're all perfect squares.

A number is a **perfect square** if you can produce it by multiplying some integer or fraction by itself.

EXAMPLE 1 25 is a perfect square because $25 = 5 \cdot 5$. ❏

EXAMPLE 2 $\frac{1}{4}$ is a perfect square because $\frac{1}{4} = \left(\frac{1}{2}\right)\left(\frac{1}{2}\right)$. ❏

Perfect squares which are integers are so named because that number of dots can always be arranged in a square. For example 25 is a perfect square and 25 dots are arranged in a square below:

If a number is a perfect square, its prime factors can be separated into two identical lists (each representing the same number). So every prime factor must appear an even number of times in a perfect square.

EXAMPLE 3 $36 = (2 \cdot 3)(2 \cdot 3)$, so 36 is a perfect square. ❏

To determine whether a number is a perfect square, factor it into a product of prime factors and see if they make up two identical lists. You can review how to factor numbers in Section 3.3.

EXAMPLE 4 Is 12 a perfect square?

$12 = 2 \cdot 2 \cdot 3$, which cannot be divided into two identical lists.

So 12 is *not* a perfect square. ❏

EXAMPLE 5 Is 100 a perfect square?

$100 = (2 \cdot 5)(2 \cdot 5)$, so 100 is a perfect square. ❏

What is true for numbers is also true for algebraic expressions. An expression is a perfect square if there exists a term (or terms) that, when multiplied by itself, produces the given expression. In this chapter we will study terms and numbers that are perfect squares, not polynomials.

To determine whether an expression is a perfect square, factor it and see if its factors can be separated into two identical lists.

EXAMPLE 6 $9x^2w^6$ is a perfect square because $9x^2w^6 = (3xwww)(3xwww)$. ❑

EXAMPLE 7 $36xy^4$ is *not* a perfect square because its factors cannot be separated into two identical lists.

$$36xy^4 = 2 \cdot 3 \cdot 2 \cdot 3xyyyy$$

There is only one factor of x. ❑

EXERCISES 12.1A

Determine whether these numbers are perfect squares by factoring them into a product of prime factors.

1. $2 =$ 2. $4 =$

3. $16 =$ 4. $10 =$

5. $7 =$ 6. $36 =$

7. $40 =$ 8. $49 =$

9. $400 =$ 10. $150 =$

Are these expressions perfect squares?

11. $4a^6$ 12. $12a^2b^{10}$

13. $100x^8w^{12}$ 14. $49pm^4k^{14}$

Check your answers on page 751. ✓

Square Roots

A **square root** of an expression is a number or term that, when multiplied by itself, produces that expression.

EXAMPLE 8 A square root of 49 is 7 because $7 \cdot 7 = 49$. ❑

Note that 7 is not the only number that, when multiplied by itself, produces 49; -7 is another such number: $(-7)(-7) = 49$.

There are always two square roots of every positive expression, the positive root and the negative root. That is because every negative number multiplied by itself will always produce the same product as the positive of that number multiplied by itself.

EXAMPLE 9 The square roots of 49 are 7 and -7, which can be expressed as ± 7. (The symbol $\pm$ is read as "plus or minus." ❑

EXAMPLE 10 The square roots of $100 = \pm 10$. ❑

The *positive* square root is called the **principal square root** and is denoted by the symbol $\sqrt{}$.

EXAMPLE 11   The principal square root of 36 is written as $\sqrt{36}$.

$$\sqrt{36} = 6$$ ❏

The principal square root of a perfect square is one of the two identical lists of factors into which it can be separated.

To simplify our discussion, we will assume that when we are simplifying *algebraic* expressions, the letters represent *positive* numbers.

EXAMPLE 12   The principal square root of $9x^2w^6$ is written as $\sqrt{9x^2w^6}$.

$$\sqrt{9x^2w^6} =$$
$$\sqrt{(3xwww)(3xwww)} =$$
$$3xwww = 3xw^3$$ ❏

In this chapter we will learn to find principal roots of expressions. Later, in Section 15.3, we will calculate square roots of numbers to solve equations. It will be important at that point to recall that when you calculate the square root of a given number, you always get *two solutions*, the *positive* and the *negative root*.

Since the square root multiplied by itself provides the given number, we can see why there are square roots only of non-negative numbers:

Remember—the rules for multiplying signed numbers are:

The product of *two negative* numbers is *positive*.

The product of *two positive* numbers is *positive*.

The product of a *positive* and a *negative* number is *negative*.

The only way the product of two numbers can be negative is if the numbers have *different* signs. If two numbers have *different* signs, they are not the same number. So there is no way the product of a real number multiplied by itself can ever be negative.

There cannot be a square root of a negative number
in the real number system.*

EXAMPLE 13 $\sqrt{-3}$ does not exist; $\sqrt{-1}$ does not exist;

$\sqrt{-4}$ does not exist. ❏

Expressions written with a root symbol, $\sqrt{}$, are called **radicals**. "Radical" is a word derived from the Latin word *radix*, which means "root." Another word derived from *radix* is "radish," a vegetable that is a root. The political term "radical" refers to someone committed to going to the root of the problem for change rather than trying to find solutions through reform.

---

* Mathematicians have developed a number system made up of numbers that are the square roots of negative numbers. Since these numbers do not exist in the system of numbers we normally use (called the **real** numbers), this developed system is called the **imaginary** numbers. We will not study them in this book.

EXERCISES 12.1B _____

Express the principal square root of each perfect square. (Assume all letters represent positive numbers.)

15. What is the square root of 4?

16. What is the square root of 36?

17. $\sqrt{49} =$ **18.** $\sqrt{4} =$

19. $\sqrt{64} =$ **20.** $\sqrt{4900} =$

21. $\sqrt{81} =$ **22.** $\sqrt{36} =$

23. $\sqrt{121} =$ **24.** $\sqrt{1} =$

25. $\sqrt{25y^2} =$ **26.** $\sqrt{49a^8b^4} =$

27. $\sqrt{900x^2w^6} =$ **28.** $\sqrt{144p^8m^6n^4} =$

Check your answers on page 752.   ✓

Estimating Irrational Numbers

Square roots of numbers that are not perfect squares are called **irrational numbers.**

EXAMPLE 14 $\sqrt{2}$, $\sqrt{7}$, $\sqrt{6}$, $\sqrt{8}$, $\sqrt{12}$, $\sqrt{18}$, and $\sqrt{200}$ are irrational numbers.   ❑

The word "irrational" means *not rational*. The numbers we call *rational* include all the integers, numerical fractions that have integers as the numerator and denominator, and decimals that end in a repeating pattern of zeros or other integers. Numbers that are called rational can be represented as a *ratio* of two integers (as we saw in Section 4.7). The set of *real* numbers consists of all the rational and irrational numbers.

EXAMPLE 15 $\dfrac{4}{5} = 0.800000\ldots$

$\dfrac{5}{11} = 0.454545\ldots$, which we write as $0.\overline{45}$.   ❑

An *irrational number can be expressed as a decimal*, although it will be an *unending list of digits that never repeat*. Most calculators display irrational numbers rounded off to eight digits, which is more accurate than is usually necessary.

There is a mechanical way to compute approximate decimal expressions of irrational numbers, but it will not be discussed in this book. If you had to express an irrational number as a decimal, the easiest way to compute it would be to press the appropriate keys on a calculator.

Using an inexpensive calculator, I found the approximate decimal representation of the following irrational numbers.

EXAMPLE 16 $\sqrt{2} \approx 1.4142135$; $\sqrt{7} \approx 2.6457513$; $\sqrt{6} \approx 2.4494897$;

$\sqrt{12} \approx 3.4641016$; $\sqrt{18} \approx 4.2426406$; $\sqrt{200} \approx 14.142135$ ❑

Although irrational numbers are hard to visualize, they are real. Many geometric formulas use square roots. You might find that the length of one side of a triangle is $\sqrt{20}$. We cannot express precisely how long $\sqrt{20}$ inches is, but we can come very close.

It is important and easy to estimate (determine approximately) where an irrational number belongs on the number line.

> **To estimate the value of an irrational number:**
>
> Look at the number under the radical symbol (called the radicand).
> Determine the two perfect squares that it lies between.

The square root of that number is between the square root of those perfect squares. Recall that the square root of every perfect square integer is an integer, and it will locate the placement of the irrational number on a number line.

EXAMPLE 17 Estimate where $\sqrt{20}$ is on the number line.
20 is between the perfect squares 16 and 25.

$$\text{Since } 16 < 20 < 25 \longrightarrow \sqrt{16} < \sqrt{20} < \sqrt{25},$$
$$4 < \sqrt{20} < 5.$$

So $\sqrt{20}$ is between 4 and 5 on the number line:

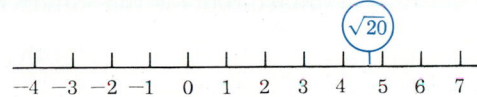

EXAMPLE 18 Estimate where $\sqrt{7}$ is on the number line.
7 is between the perfect squares 4 and 9.

$$\text{Since } 4 < 7 < 9 \longrightarrow \sqrt{4} < \sqrt{7} < \sqrt{9},$$
$$2 < \sqrt{7} < 3.$$

So $\sqrt{7}$ is between 2 and 3 on the number line:

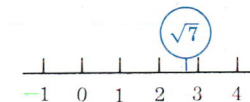

EXERCISES 12.1C

Estimate the value of each irrational number by stating which two integers it is between. Draw a number line and show where the irrational number is located.

29. $\sqrt{3}$ 30. $\sqrt{8}$

31. $\sqrt{30}$ 32. $\sqrt{19}$

33. $\sqrt{23}$ 34. $\sqrt{40}$

35. $\sqrt{10}$ 36. $\sqrt{28}$

Check your answers on page 752.

Simplifying Square Roots

As we saw earlier in Examples 11 and 12, radicals that are perfect squares can be simplified into rational expressions.

Radicals that are not perfect squares are irrational, and so they cannot be simplified into rational expressions.

Square roots of primes are always irrational because primes have no identical pairs of factors.

EXAMPLE 19 $\sqrt{3}$ is irrational and cannot be simplified.   ❏

Radicals of some nonperfect squares can be partially simplified if any of their factors are perfect squares. In other words, a radical can be simplified if any factor appears *twice*. The result will be an irrational expression consisting of the product of a rational term multiplied by an irrational term.

The square root of any repeated factor is that factor, so any factor appearing within the radical symbol twice can be removed and written outside the symbol.

EXAMPLE 20   Simplify $\sqrt{12}$.

$$\sqrt{12} =$$
$$\sqrt{2 \cdot 2 \cdot 3} = 2\sqrt{3}$$ ❏

The reason you can take that factor out of the radical is that the product of square roots is the square root of the products:

$$(\sqrt{a})(\sqrt{b}) = \sqrt{ab}, \quad \text{if } a \geq 0 \text{ and } b \geq 0.$$

So you can consider $\sqrt{2 \cdot 2 \cdot 3}$ as the product of the two radicals: $(\sqrt{2 \cdot 2})(\sqrt{3})$. We know what the square root of a perfect square is

$$\sqrt{2 \cdot 2} = 2.$$

So $$\sqrt{2 \cdot 2 \cdot 3} =$$
$$(\sqrt{2 \cdot 2})(\sqrt{3}) = 2\sqrt{3}.$$

To simplify a radical, factor the expression within the symbol. Replace any factors appearing twice within it by writing those factors once outside the symbol.

EXAMPLE 21   Simplify $\sqrt{50}$.

$$\sqrt{50} =$$
$$\sqrt{2 \cdot 5 \cdot 5} = 5\sqrt{2}$$ ❏

EXAMPLE 22   Simplify $\sqrt{18ab^2c^3}$.

$$\sqrt{18ab^2c^3} =$$
$$\sqrt{2 \cdot 3 \cdot 3\, a\, bb\, c\, cc} = 3bc\sqrt{2ac}$$ ❏

It usually is not obvious when a large number is a perfect square; so until you factor it, you cannot tell whether a radical contains any perfect square factors.

If you recognize a factor as a perfect square, you do not need to factor it down to primes. Remember, the reason for factoring is to discover perfect squares of which you can find the square root.

EXAMPLE 23   Simplify $\sqrt{576}$.

$$\sqrt{576} =$$
$$\sqrt{2 \cdot 288} =$$
$$\sqrt{2 \cdot 2 \cdot 144} = \quad \text{(I noticed } 144 = (12)(12)!)$$
$$\sqrt{2 \cdot 2 \cdot 12 \cdot 12} =$$
$$2 \cdot 12 = 24$$

❑

There is a shortcut for simplifying algebraic expressions with letters raised to even powers. Recall that the exponent represents the number of times that the factor appears, and the square root of a repeated factor is that factor. *So if the exponent is even, half of it is the exponent of that factor's square root.*

EXAMPLE 24 $\sqrt{w^8 z^{14}} = w^4 z^7$, since $\frac{1}{2}$ of 8 is 4 and $\frac{1}{2}$ of 14 is 7.

❑

You can check this out by factoring!

$$\sqrt{w^8 z^{14}} = \sqrt{(ww)(ww)(ww)(ww)(zz)(zz)(zz)(zz)(zz)(zz)(zz)}$$
$$= w^4 z^7$$

> **To simplify an algebraic expression:**
>
> Factor the coefficient and remove any factors from under the radical symbol that appear twice (or that you recognize as perfect squares).
>
> If the exponent of a letter factor is even, that letter will come out of the radical symbol with an exponent half as big as the original one.
>
> If the exponent of a letter factor is odd, rewrite the expression without exponents. Remove from under the radical symbol any factor that appears twice.

EXAMPLE 25   Simplify $\sqrt{98a^4 b^3 c^{20} d^5}$.

$$\sqrt{98a^4 b^3 c^{20} d^5} =$$
$$\sqrt{2 \cdot 7 \cdot 7a^4 \, bbbc^{20} ddddd} = 7a^2 bc^{10} d^2 \sqrt{2bd}$$

❑

You can simplify irrational expressions if the expression within the radical symbol has any repeated factors. Simplify your answer by multiplying any numerical factors that appear outside the radical.

EXAMPLE 26   Simplify $5\sqrt{12}$.

$$5\sqrt{12} =$$
$$5\sqrt{2 \cdot 2 \cdot 3} =$$
$$5(2)\sqrt{3} = 10\sqrt{3}$$

❑

If any additional letter factors can be removed from the radical symbol, use exponents to describe how many times each letter factor appears in the term.

EXAMPLE 27 Simplify $3a^2b\sqrt{20a^3b^5}$.

$$3a^2b\sqrt{20a^3b^5} =$$
$$3a^2b\sqrt{4\cdot5\,aaa\,bbb\,bb} =$$
$$3a^2b(2abb)\sqrt{5ab} =$$
$$(3)(2)a^2ab(bb)\sqrt{5ab} =$$
$$(3)(2)a^{2+1}(bbb)\sqrt{5ab} = 6a^3b^3\sqrt{5ab}$$

EXERCISES 12.1D

Simplify each radical. (Assume all letters represent positive numbers.)

37. $\sqrt{20} =$ | **38.** $\sqrt{45} =$

39. $\sqrt{12} =$ | **40.** $\sqrt{63} =$

41. $\sqrt{48} =$ | **42.** $\sqrt{300} =$

43. $\sqrt{80} =$ | **44.** $\sqrt{60} =$

45. $\sqrt{175} =$ | **46.** $\sqrt{500} =$

47. $\sqrt{600} =$ | **48.** $\sqrt{324} =$

49. $\sqrt{196} =$ | **50.** $\sqrt{900} =$

51. $\sqrt{144} =$ | **52.** $\sqrt{1764} =$

53. $\sqrt{1225} =$ | **54.** $\sqrt{432} =$

55. $\sqrt{n^2m^4} =$ | **56.** $\sqrt{pm^5} =$

57. $\sqrt{4a^2b^8} =$ | **58.** $\sqrt{9ab^3c^6} =$

59. $\sqrt{12mn^4p^7} =$ | **60.** $\sqrt{20nm^5p^{10}} =$

61. $\sqrt{28a^6bc^5} =$ | **62.** $\sqrt{16x^3yw^4} =$

63. $2\sqrt{8ab^3} =$ | **64.** $3\sqrt{18x^3y^2w^4} =$

65. $5ab\sqrt{20a^4b^5} =$ | **66.** $7xy^2\sqrt{12xy^4} =$

67. $4m^3n\sqrt{9m^2n^7} =$ | **68.** $10pk\sqrt{50p^6k^2} =$

Check your answers on page 752.

12.2 Cube Roots and Other Radicals

Cube Roots

There are radicals other than square roots. The **cube root** of an expression is a term that, when used as a factor *three times*, produces that expression.

The symbol for a cube root is $\sqrt[3]{}$.

EXAMPLE 1   The cube root of 8 is 2, since $(2)(2)(2) = 8$.

We express that cube root as $\sqrt[3]{8} = 2$.   ❏

Remember, we say a factor that is raised to the third power is *cubed*. x^3 is read as "*x* cubed."

Unlike square roots, it is possible to have a cube root of a negative number, since the product of three negatives is negative.

The cube root of a negative expression is always negative.

The cube root of a positive expression is always positive.

EXAMPLE 2 $\sqrt[3]{-8} = -2$, since $(-2)(-2)(-2) = -8$.   ❏

EXAMPLE 3 $\sqrt[3]{125} = 5$, since $(5)(5)(5) = 125$.   ❏

EXAMPLE 4 $\sqrt[3]{-1} = -1$, since $(-1)(-1)(-1) = -1$.   ❏

Just like the square roots of nonperfect squares, the cube roots of numbers or expressions that are not cubes are *irrational*.

EXAMPLE 5 $\sqrt[3]{-5}$ cannot be simplified, but it exists as an irrational number.   ❏

Simplifying a cube root is similar to simplifying a square root; but in cube roots, factors are removed from under the radical symbol and written in front of the symbol when they are repeated three times (rather than twice as with square roots).

EXAMPLE 6   Simplify $\sqrt[3]{b^3}$.

$$\sqrt[3]{b^3} =$$
$$\sqrt[3]{bbb} = b$$ ❏

EXAMPLE 7   Simplify $\sqrt[3]{27a^3b^6}$.

$$\sqrt[3]{27a^3b^6} =$$
$$\sqrt[3]{(3\cdot3\cdot3)(aaa)(bbb)(bbb)} =$$
$$3abb = 3ab^2$$ ❏

Remember, the cube root of a negative term exists and is always negative.

EXAMPLE 8   Simplify $\sqrt[3]{-24ab^4c^3d^5}$.

$$\sqrt[3]{-24ab^4c^3d^5} =$$
$$\sqrt[3]{-2\cdot2\cdot2\cdot3abbbbcccddddd} = -2bcd\sqrt[3]{3abd^2}$$ ❏

EXERCISES 12.2A

Simplify each cube root. (Assume all letters represent positive numbers.)

1. $\sqrt[3]{24} =$

2. $\sqrt[3]{27} =$

3. $\sqrt[3]{-125} =$

4. $\sqrt[3]{56} =$

5. $\sqrt[3]{16x^5y^4} =$

6. $\sqrt[3]{135a^4b^9} =$

7. $\sqrt[3]{p^5m^5} =$

8. $\sqrt[3]{nm^4} =$

9. $\sqrt[3]{a^4b^6c^8} =$

10. $\sqrt[3]{xy^4w^9} =$

11. $\sqrt[3]{8a^7b^6} =$

12. $\sqrt[3]{-27x^8} =$

13. $\sqrt[3]{-1000n^{10}m^{11}} =$

14. $\sqrt[3]{16a^3b^6c^9} =$

15. $\sqrt[3]{-54nm^4p^{10}} =$

16. $\sqrt[3]{250k^4p^4m^8} =$

17. $\sqrt[3]{-40a^4b^5c^6} =$

18. $\sqrt[3]{-56w^7y^5z^3} =$

Check your answers on page 752. ✓

Other Radicals

The **index** of a radical indicates the *number* of times that a factor must appear within the radical symbol before that factor can be removed and written in front as a rational expression.

EXAMPLE 9 Square roots are radicals of index 2: $\sqrt{x^2} = x$.

Cube roots are radicals of index 3: $\sqrt[3]{x^3} = x$. ❑

Except for square roots, the index of a radical is always written at the front of the radical symbol. Square roots are the most common radical; so when we represent a square root, we do not write any number on the symbol.

We say a radical of index n is an *nth root*.

EXAMPLE 10 A fourth root is a radical of index 4 and is represented as

$$\sqrt[4]{\quad} ; \qquad \sqrt[4]{x^4} = x.$$ ❑

EXAMPLE 11 A fifth root is a radical of index 5 and is represented as

$$\sqrt[5]{\quad} ; \qquad \sqrt[5]{x^5} = x.$$ ❑

EXAMPLE 12 An *nth* root is a radical of index n and is represented as

$$\sqrt[n]{\quad} ; \qquad \sqrt[n]{x^n} = x.$$ ❑

To simplify a radical, look for factors repeated the same number of times as the number of the index. You can remove these factors from within the radical symbol and write them once outside that symbol.

Radicals of index 4 are simplified by removing factors from the expression under the radical symbol that are repeated 4 times.

EXAMPLE 13 Simplify $\sqrt[4]{16a^2b^3c^4d^5}$.

$$\sqrt[4]{16a^2b^3c^4d^5} =$$

$$\sqrt[4]{2 \cdot 2 \cdot 2 \cdot 2 \cdot aabbbccccddddd} = 2cd\sqrt[4]{a^2b^3d}$$ ❑

Radicals of index 5 are simplified by removing factors from the expression under the radical symbol that are repeated 5 times.

EXAMPLE 14 Simplify $\sqrt[5]{-p^3m^7k^{10}}$.

$$\sqrt[5]{-p^3m^7k^{10}} =$$

$$\sqrt[5]{-pppmmmmmmmkkkkkkkkkk} = -mk^2\sqrt[5]{p^3m^2}$$ ❑

EXAMPLE 15   Simplify $\sqrt[6]{w^8}$.

$$\sqrt[6]{w^8} =$$
$$\sqrt[6]{wwwwwwww} = w\sqrt[6]{w^2} \qquad \square$$

EXAMPLE 16   Simplify $\sqrt[7]{a^{10}}$.

$$\sqrt[7]{a^{10}} =$$
$$\sqrt[7]{aaaaaaaaaa} = a\sqrt[7]{a^3} \qquad \square$$

The argument for why there is no square root of a negative number can be extended to every radical with an even number for an index.

Just as there is a cube root of a negative number, there can always be a radical of a negative number if the index is odd. The radical of an odd-numbered index

of a positive term is always positive;

of a negative term is always negative.*

EXAMPLE 17 $\sqrt[5]{-x}$ exists; $\sqrt[4]{-x}$ does *not* exist (under real numbers).

$\sqrt[7]{-m}$ exists; $\sqrt[10]{-n}$ does *not* exist (under real numbers). $\qquad \square$

We talk about principal roots only when the index is even.

EXAMPLE 18 $\sqrt[8]{x}$ represents the *principal* eighth root of x.

$\sqrt[5]{y}$ represents the fifth root of y. $\qquad \square$

EXERCISES 12.2B

Simplify each radical. (Assume all letters represent positive numbers.)

19. $\sqrt[4]{w^4} =$ 20. $\sqrt[4]{x^6} =$

21. $\sqrt[4]{p^8} =$ 22. $\sqrt[4]{a^{10}} =$

23. $\sqrt[5]{a^5} =$ 24. $\sqrt[5]{w^7} =$

25. $\sqrt[7]{y^7} =$ 26. $\sqrt[9]{r^{10}} =$

27. $\sqrt[5]{x^6 y^{12}} =$ 28. $\sqrt[5]{ab^4 c^8 d^{14}} =$

29. $\sqrt[4]{162} =$ 30. $\sqrt[4]{16} =$

31. $\sqrt[4]{81x^3 y^8 w^7} =$ 32. $\sqrt[4]{16a^2 b^5 c^6} =$

33. $\sqrt[4]{32a^4 b^5 c^6} =$ 34. $\sqrt[4]{81x^3 y^4 w^7} =$

35. $\sqrt[5]{-n^8 p^{12} m^{15}} =$ 36. $\sqrt[5]{-32a^6 bc^8} =$

Check your answers on page 752. ✓

---

* So roots with an odd-numbered index always exist but do not each have a positive and negative root as squared (and other even-numbered index) roots do. For this reason we do not talk about principal odd-numbered roots, but we do represent principal even-numbered roots by the radical symbol.

12.3 Rational (Fractional) Exponents

Evaluating Numbers Raised to Fractional Exponents

In Chapter 11 we learned how to simplify expressions involving exponents that were integers. Now that you know how to simplify radicals, we can work with **fractional exponents**.

When an exponent is a fraction, the numerator and denominator represent different instructions to be applied to the base:

The numerator is the power the expression is raised to.

The denominator represents the index of the radical.

$$x^{m/n} \text{ represents } (\sqrt[n]{x})^m \text{ or } \sqrt[n]{x^m}.$$

We say $x^{m/n}$ is the nth root of x raised to the mth power.
It is more difficult to take a root than it is to raise to a power.

Therefore, to evaluate a number raised to a fractional exponent:

First calculate the radical (with the index from the denominator).

Then raise that result to the power (from the numerator).

By evaluating the root first, you will be simplifying a smaller radical, and the computation will be easier than if you applied the power first and then took the root.

EXAMPLE 1 Evaluate $8^{1/3}$.
$$8^{1/3} =$$
$$(\sqrt[3]{8})^1 =$$
$$2^1 = 2 \qquad \square$$

EXAMPLE 2 Evaluate $(-27)^{2/3}$.
$$(-27)^{2/3} =$$
$$(\sqrt[3]{-27})^2 =$$
$$(-3)^2 = 9 \qquad \square$$

EXAMPLE 3 Evaluate $(25)^{1/2}$.
$$(25)^{1/2} =$$
$$\sqrt{25} = 5 \qquad \square$$

EXERCISES 12.3A

Evaluate these numbers, which are raised to fractional exponents.

1. $4^{1/2} =$　　　　2. $9^{1/2} =$　　　　3. $27^{1/3} =$

4. $(-8)^{2/3} =$　　　　5. $16^{3/4} =$　　　　6. $81^{3/4} =$

7. $(-32)^{3/5} =$　　　　8. $49^{1/2} =$　　　　9. $125^{2/3} =$

10. $(-125)^{1/3} =$　　　　11. $32^{4/5} =$　　　　12. $(-32)^{2/5} =$

Check your answers on page 752. ✓

Negative Fractional Exponents

As we discussed in Section 11.1, a negative exponent is an instruction to divide. If the exponent is negative and the base is an integer, remove the negative exponent by writing the base as the denominator of that term (which represents division).

When evaluating a number raised to a negative exponent, always convert the expression so that its exponent becomes positive *before* you take the root and power indicated by the numerator and denominator of the exponent.

EXAMPLE 4 Evaluate $8^{-1/3}$.

$$8^{-1/3} =$$

$$\frac{1}{8^{1/3}} =$$

$$\frac{1}{\sqrt[3]{8}} = \frac{1}{2} \qquad \square$$

EXAMPLE 5 Evaluate $27^{-2/3}$.

$$27^{-2/3} =$$

$$\frac{1}{27^{2/3}} =$$

$$\frac{1}{(\sqrt[3]{27})^2} =$$

$$\frac{1}{3^2} = \frac{1}{9} \qquad \square$$

When the base is a fraction, recall the shortcut we observed in Section 11.1 when raising the base to a negative exponent: The exponent becomes positive if we replace the fractional base by its reciprocal.

EXAMPLE 6 Evaluate $\left(\dfrac{1}{4}\right)^{-1/2}$.

$$\left(\frac{1}{4}\right)^{-1/2} =$$

$$(4)^{1/2} =$$

$$\sqrt{4} = 2 \qquad \square$$

EXAMPLE 7 Evaluate $\left(\dfrac{8}{27}\right)^{-2/3}$.

$$\left(\frac{8}{27}\right)^{-2/3} =$$

$$\left(\frac{27}{8}\right)^{2/3} =$$

$$\left(\sqrt[3]{\frac{27}{8}}\right)^2 =$$

$$\left(\frac{\sqrt[3]{27}}{\sqrt[3]{8}}\right)^2 =$$

$$\left(\frac{3}{2}\right)^2 = \frac{9}{4} \text{ or } 2\frac{1}{4} \qquad \square$$

Remember, as long as the index (denominator of the exponent) is odd, we can apply it to a negative base.

EXAMPLE 8 Evaluate $\left(-\dfrac{27}{125}\right)^{-1/3}$.

$$\left(-\dfrac{27}{125}\right)^{-1/3} =$$

$$\left(-\dfrac{125}{27}\right)^{1/3} =$$

$$\sqrt[3]{-\dfrac{125}{27}} =$$

$$\dfrac{\sqrt[3]{-125}}{\sqrt[3]{27}} = -\dfrac{5}{3} \text{ or } -1\dfrac{2}{3}$$

EXERCISES 12.3B

Evaluate each number.

13. $9^{-1/2} =$ **14.** $25^{-1/2} =$

15. $125^{-1/3} =$ **16.** $125^{-2/3} =$

17. $1000^{-2/3} =$ **18.** $49^{-1/2} =$

19. $16^{-1/4} =$ **20.** $16^{-3/4} =$

21. $\left(\dfrac{1}{36}\right)^{-1/2} =$ **22.** $\left(\dfrac{1}{100}\right)^{-1/2} =$

23. $\left(\dfrac{8}{125}\right)^{-1/3} =$ **24.** $\left(\dfrac{1}{16}\right)^{-3/4} =$

25. $\left(\dfrac{27}{1000}\right)^{-2/3} =$ **26.** $\left(\dfrac{125}{8}\right)^{-1/3} =$

27. $\left(-\dfrac{1}{8}\right)^{-1/3} =$ **28.** $\left(-\dfrac{1}{125}\right)^{-2/3} =$

Check your answers on page 752. ✓

Algebraic Expressions Raised to Fractional Exponents

To simplify algebraic expressions raised to fractional exponents, use the rules we discussed in Chapter 11:

$$x^m x^n = x^{m+n} \qquad \dfrac{x^m}{x^n} = \begin{cases} x^{m-n}, \text{ if } m \geq n \\ \dfrac{1}{x^{n-m}} \text{ if } n > m \end{cases}$$

$$(xy)^n = x^n y^n$$

$$(x^n)^m = x^{nm} \qquad x^{-n} = \dfrac{1}{x^n} \qquad \dfrac{1}{x^{-n}} = x^n$$

$$x^0 = 1, \text{ if } x \neq 0.$$

EXAMPLE 9 Simplify $(x^{2/5})(x^{1/5})$.

$$(x^{2/5})(x^{1/5}) =$$
$$x^{2/5 + 1/5} = x^{3/5} \qquad \square$$

EXAMPLE 10 Simplify $\dfrac{a^{5/7}}{a^{3/7}}$.

$$\frac{a^{5/7}}{a^{3/7}} =$$
$$a^{5/7 - 3/7} = a^{2/7} \qquad \square$$

EXAMPLE 11 Simplify $\dfrac{p^{1/5}}{p^{4/5}}$.

$$\frac{p^{1/5}}{p^{4/5}} =$$
$$\frac{1}{p^{4/5 - 1/5}} = \frac{1}{p^{3/5}} \qquad \square$$

EXAMPLE 12 Simplify $(b^{2/3})^{3/4}$.

$$(b^{2/3})^{3/4} =$$
$$b^{(2/3)(3/4)} =$$
$$b^{(\cancel{2} \cdot \cancel{3})/(\cancel{3} \cdot 2 \cdot \cancel{2})} = b^{1/2} \qquad \square$$

EXAMPLE 13 Simplify $(nm^3)^{1/2}$.

$$(nm^3)^{1/2} =$$
$$n^{1/2}m^{3(1/2)} = n^{1/2}m^{3/2} \qquad \square$$

EXAMPLE 14 Simplify $p^{-2/3}$.

$$p^{-2/3} = \frac{1}{p^{2/3}} \qquad \square$$

EXAMPLE 15 Simplify $\dfrac{1}{k^{-2/3}}$.

$$\frac{1}{k^{-2/3}} = k^{2/3} \qquad \square$$

EXAMPLE 16 Simplify $\dfrac{ab^{3/5}c^{4/5}}{a^{2/5}b^{1/5}c^3}$.

$$\frac{ab^{3/5}c^{4/5}}{a^{2/5}b^{1/5}c^3} =$$
$$\frac{a^{1 - 2/5}b^{3/5 - 1/5}}{c^{3 - 4/5}} =$$
$$\frac{a^{5/5 - 2/5}b^{3/5 - 1/5}}{c^{15/5 - 4/5}} = \frac{a^{3/5}b^{2/5}}{c^{11/5}} \qquad \square$$

When we worked with fractions in Chapter 3, we pointed out that improper fractions could also be represented as mixed numbers. Since the numerator and denominator of a fractional exponent represent operations that are applied to the base, you must always represent improper fractional exponents as fractions and never as mixed numbers.

EXAMPLE 17 Simplify $\dfrac{(xy^{1/3}w^2)^{1/2}}{(x^{1/2}y^2w^{1/6})^{2/3}}$.

$$\frac{(xy^{1/3}w^2)^{1/2}}{(x^{1/2}y^2w^{1/6})^{2/3}} =$$

$$\frac{x^{1/2}y^{(1/3)(1/2)}w^{2(1/2)}}{x^{(1/2)(2/3)}y^{(2/1)(2/3)}w^{(1/6)(2/3)}} =$$

$$\frac{x^{1/2}y^{1/6}w}{x^{1/3}y^{4/3}w^{1/9}} =$$

$$\frac{x^{1/2-1/3}w^{1-1/9}}{y^{4/3-1/6}} =$$

$$\frac{x^{3/6-2/6}w^{9/9-1/9}}{y^{8/6-1/6}} = \frac{x^{1/6}w^{8/9}}{y^{7/6}}$$

□

EXERCISES 12.3C

Simplify each expression. Express your answers with positive exponents. (Assume all letters represent positive numbers.)

29. $(a^{1/4})(a^{1/4}) =$ **30.** $(x^{3/7})(x^{1/7}) =$

31. $(w^{2/5})(w^{-1/5}) =$ **32.** $(c^{3/8})(c^{5/8}) =$

33. $(w^{-2/9})(w^{-4/9}) =$ **34.** $(y^{1/7})(y^{2/7})(y^{3/7}) =$

35. $\dfrac{a^{4/5}}{a^{1/5}} =$ **36.** $\dfrac{n^{7/8}}{n^{5/8}} =$

37. $\dfrac{p^{7/12}}{p^{5/12}} =$ **38.** $\dfrac{m^{1/4}}{m^{3/4}} =$

39. $\dfrac{k}{k^{2/3}} =$ **40.** $\dfrac{a^{3/4}}{a^2} =$

41. $(n^{1/2})^3 =$ **42.** $(c^{2/3})^{1/3} =$

43. $(m^2)^{-1/5} =$ **44.** $(k^{-1/2})^{-2/3} =$

45. $(ab^2)^{1/2} =$ **46.** $(mn^3)^{-3/4} =$

47. $(pm^{2/3})^{1/2} =$ **48.** $(nm^2p^{1/3})^{3/4} =$

49. $\dfrac{a^2bc^{1/2}}{ab^{2/3}c} =$ **50.** $\dfrac{xy^3w^{3/5}}{x^{1/2}y^2w^{1/5}} =$

51. $\dfrac{n^{2/3}m^{2/3}}{n^{1/3}m} =$ **52.** $\dfrac{ab^{3/4}c^{5/6}}{a^{3/4}bc} =$

53. $\dfrac{k^2m^{3/5}n^{1/3}}{k^{2/3}m^{4/5}n} =$ **54.** $\dfrac{(ab^{1/3})^{1/2}}{(a^2b^{2/3})^{1/2}} =$

55. $\dfrac{(xy^{3/4})^{1/2}}{(x^{2/3}y^{1/2})^{1/4}} =$ **56.** $\dfrac{(pm^2n^{1/4})^2}{(p^2mn^{2/3})^{3/4}} =$

57. $\dfrac{(a^{1/2}b^{5/3})^{1/2}}{(ab^{1/2})^{1/3}} =$ **58.** $\dfrac{(n^{1/3}p^{4/5})^{1/2}}{(np^{3/5})^{1/3}} =$

Check your answers on page 752.

 **Combining Radical Expressions**

Combining Like Radicals

We can combine radical expressions if the irrational parts of the expressions are exactly the same. If you consider the *rational part* of the expression to be the coefficient and the *irrational part* to be the kind of term, then this operation is just like combining like terms, as we did in Section 5.2.

EXAMPLE 1 In the radical expression $4\sqrt{3}$,

$\qquad$ 4 is the coefficient;

$\qquad$ $\sqrt{3}$ is the kind of term. ❏

Radical expressions with the same irrational parts are combined by adding their coefficients.

EXAMPLE 2 Combine these like radicals: $4\sqrt{3} + 6\sqrt{3}$.

$$4\sqrt{3} + 6\sqrt{3} =$$
$$(4 + 6)\sqrt{3} = 10\sqrt{3} \qquad ❏$$

Remember, if no coefficient is written in an expression, it is understood to be 1.

EXAMPLE 3 Simplify $3\sqrt{2} + 6\sqrt{5} + 8\sqrt{2} + \sqrt{5}$ by combining like radicals.

$$3\sqrt{2} + 6\sqrt{5} + 8\sqrt{2} + \sqrt{5} =$$
$$(3\sqrt{2} + 8\sqrt{2}) + (6\sqrt{5} + (1)\sqrt{5}) = 11\sqrt{2} + 7\sqrt{5} \qquad ❏$$

Just as when combining like terms, be careful about the signs. The sign in front of an expression is its sign, and you must add signed numbers by following the rules of signs learned in Section 2.2.

Careless errors can be avoided by combining like radicals with the same signs first, and then combining the positive and negative like radicals that result.

EXAMPLE 4 Simplify $\sqrt{3} - 2\sqrt{7} + 5\sqrt{3} - 2\sqrt{3} - \sqrt{7}$.

$$\sqrt{3} - 2\sqrt{7} + 5\sqrt{3} - 2\sqrt{3} - \sqrt{7} =$$
$$(\sqrt{3} + 5\sqrt{3} - 2\sqrt{3}) + (-2\sqrt{7} - \sqrt{7}) =$$
$$(6\sqrt{3} - 2\sqrt{3}) - 3\sqrt{7} = 4\sqrt{3} - 3\sqrt{7} \qquad ❏$$

When combining algebraic radicals, do not forget that the powers of each letter in the irrational part must be exactly the same in order to combine the radicals.

EXAMPLE 5 Simplify $\sqrt[3]{ab^2} + 5\sqrt[3]{a^2b} - 2\sqrt[3]{ab} + 3\sqrt[3]{ba^2} - 7\sqrt[3]{b^2a}$.

$$\sqrt[3]{ab^2} + 5\sqrt[3]{a^2b} - 2\sqrt[3]{ab} + 3\sqrt[3]{ba^2} - 7\sqrt[3]{b^2a} =$$
$$(\sqrt[3]{ab^2} - 7\sqrt[3]{b^2a}) + (5\sqrt[3]{a^2b} + 3\sqrt[3]{ba^2}) - 2\sqrt[3]{ab} = -6\sqrt[3]{ab^2} + 8\sqrt[3]{a^2b} - 2\sqrt[3]{ab} \qquad ❏$$

Be sure that the *indexes* of two radicals are the *same* before trying to combine the radicals.

EXAMPLE 6 Simplify $\sqrt{w} + \sqrt[3]{w} + \sqrt[4]{w} + \sqrt[3]{w} + \sqrt{w}$.

$$\underline{\sqrt{w}} + \underline{\underline{\sqrt[3]{w}}} + \sqrt[4]{w} + \underline{\underline{\sqrt[3]{w}}} + \underline{\sqrt{w}} =$$

$$(\underline{\sqrt{w}} + \underline{\sqrt{w}}) + (\underline{\underline{\sqrt[3]{w}}} + \underline{\underline{\sqrt[3]{w}}}) + \sqrt[4]{w} = 2\sqrt{w} + 2\sqrt[3]{w} + \sqrt[4]{w} \qquad \square$$

EXAMPLE 7 Simplify $\sqrt{x} - 5\sqrt[3]{x} + 3\sqrt[5]{x} - 4\sqrt{x} + \sqrt[5]{x} - \sqrt[3]{x}$.

$$\underline{\sqrt{x}} - 5\underline{\underline{\sqrt[3]{x}}} + 3\underset{\sim}{\sqrt[5]{x}} - 4\underline{\sqrt{x}} + \underset{\sim}{\sqrt[5]{x}} - \underline{\underline{\sqrt[3]{x}}} =$$

$$(\underline{\sqrt{x}} - 4\underline{\sqrt{x}}) + (-5\underline{\underline{\sqrt[3]{x}}} - \underline{\underline{\sqrt[3]{x}}}) + (3\underset{\sim}{\sqrt[5]{x}} + \underset{\sim}{\sqrt[5]{x}}) = -3\sqrt{x} - 6\sqrt[3]{x} + 4\sqrt[5]{x} \qquad \square$$

EXERCISES 12.4A

Simplify the following radical expressions by combining like radicals. (Assume all letters represent positive numbers.)

1. $4\sqrt{3} + 5\sqrt{3} =$ 　　　　　　　**2.** $8\sqrt{3} - 6\sqrt{3} =$

3. $9\sqrt{7} - \sqrt{7} =$ 　　　　　　　　**4.** $3\sqrt{2} + 5\sqrt{2} + \sqrt{2} =$

5. $4\sqrt{5} + 2\sqrt{3} - \sqrt{5} + 5\sqrt{3} =$ 　　**6.** $7\sqrt{3} + 4\sqrt{2} + \sqrt{2} - 5\sqrt{3} =$

7. $15\sqrt{6} + 11\sqrt{7} - 2\sqrt{7} + \sqrt{7} + 3\sqrt{6} =$

8. $9\sqrt{4} + \sqrt{3} - 2\sqrt{4} + \sqrt{4} - 5\sqrt{3} =$

9. $\sqrt[3]{5} + 4\sqrt[3]{5} =$ 　　　　　　　**10.** $\sqrt[4]{7} + 3\sqrt[4]{7} - 2\sqrt[4]{7} =$

11. $\sqrt[3]{2} + 4\sqrt[3]{3} + \sqrt[3]{3} + 5\sqrt[3]{2} =$

12. $8\sqrt[4]{3} + \sqrt[4]{5} - 3\sqrt[4]{3} + 2\sqrt[4]{3} + 2\sqrt[4]{5} =$

13. $6\sqrt{3} + 3\sqrt[3]{3} + \sqrt{3} + 5\sqrt[3]{3} =$

14. $7\sqrt{2} + 4\sqrt[3]{2} + 3\sqrt[3]{2} - \sqrt{2} + 8\sqrt{2} =$

15. $\sqrt[3]{2} + 5\sqrt{7} + 8\sqrt{2} - 3\sqrt{7} + 5\sqrt[3]{2} =$

16. $7\sqrt{3} + 4\sqrt[3]{2} - 3\sqrt{3} + \sqrt{2} + 5\sqrt[3]{2} =$

17. $\sqrt{a} + 4\sqrt[3]{a} + \sqrt{a} + 5\sqrt{a} - \sqrt[3]{a} =$

18. $\sqrt{x} + 4\sqrt{y} - 2\sqrt{w} + 5\sqrt{x} + \sqrt{y} - \sqrt{w} =$

19. $\sqrt[3]{y} + \sqrt{y} + \sqrt[3]{y} + \sqrt{y} + \sqrt{y} - 5\sqrt{y} =$

20. $\sqrt[3]{xy^2} + 2\sqrt[3]{x^2y} + 5\sqrt[3]{y^2x} + 7\sqrt[3]{yx^2} + \sqrt[3]{x^2y} =$

21. $\sqrt[3]{ab} + \sqrt[3]{ba} + \sqrt[3]{ab^2} + \sqrt[3]{a^2b} + \sqrt[3]{ba^2} + \sqrt[3]{b^2a} =$

22. $\sqrt[3]{n^2m} + \sqrt[3]{nm} + \sqrt[3]{mn} + \sqrt[3]{nm^2} + \sqrt[3]{m^2n} + \sqrt[3]{mn^2} =$

23. $2\sqrt[3]{ab} + 5\sqrt[3]{ba} - \sqrt[3]{ab^2} + 2\sqrt[3]{a^2b} + \sqrt[3]{ba^2} - \sqrt[3]{b^2a} =$

24. $\sqrt[3]{xy^2} - 5\sqrt[3]{x^2y} - 4\sqrt[3]{y^2x} + 2\sqrt[3]{yx^2} - \sqrt[3]{x^2y} =$

25. $\sqrt[3]{n^2m} + 2\sqrt[3]{nm} - \sqrt[3]{mn} + 4\sqrt[3]{nm^2} - \sqrt[3]{m^2n} - \sqrt[3]{mn^2} =$

Check your answers on page 752. ✓

Simplifying and Combining Radicals

There are lists of radical expressions that at first glance look as if they could not be simplified, since the irrational parts are different. But if the irrational part contains enough repeated factors, it can be simplified; and that radical may then be like one of the others in the list.

EXAMPLE 8 Simplify $5\sqrt{3} + \sqrt{12}$.

$\sqrt{12}$ can be simplied to $\sqrt{2 \cdot 2 \cdot 3} = 2\sqrt{3}$, so

$$5\sqrt{3} + \sqrt{12} =$$
$$5\sqrt{3} + 2\sqrt{3} = 7\sqrt{3}. \qquad ❏$$

Always simplify each radical expression first by factoring out any repeated factors, and then multiply its coefficients before attempting to combine like radicals.

EXAMPLE 9 Simplify $2\sqrt{3} + 7\sqrt{2} + 5\sqrt{12} + \sqrt{18}$.

$$2\sqrt{3} + 7\sqrt{2} + 5\sqrt{12} + \sqrt{18} =$$
$$2\sqrt{3} + 7\sqrt{2} + 5\sqrt{2 \cdot 2 \cdot 3} + \sqrt{2 \cdot 3 \cdot 3} =$$
$$2\sqrt{3} + 7\sqrt{2} + 5(2)\sqrt{3} + 3\sqrt{2} =$$
$$2\sqrt{3} + 7\sqrt{2} + 10\sqrt{3} + 3\sqrt{2} = 12\sqrt{3} + 10\sqrt{2} \qquad ❏$$

EXAMPLE 10 Simplify $\sqrt[3]{27} + \sqrt[3]{16} + \sqrt[3]{125} + \sqrt[3]{54}$.

$$\sqrt[3]{27} + \sqrt[3]{16} + \sqrt[3]{125} + \sqrt[3]{54} =$$
$$\sqrt[3]{3 \cdot 3 \cdot 3} + \sqrt[3]{2 \cdot 2 \cdot 2 \cdot 2} + \sqrt[3]{5 \cdot 5 \cdot 5} + \sqrt[3]{2 \cdot 3 \cdot 3 \cdot 3} =$$
$$3 + 2\sqrt[3]{2} + 5 + 3\sqrt[3]{2} = 8 + 5\sqrt[3]{2} \qquad ❏$$

You must keep in mind that whenever you want to combine like algebraic terms, all letter parts must be exactly the same. So to combine two algebraic radical expressions, not only must the irrational parts and the index be the same, but also the letter factors must be the same. Otherwise the coefficients *cannot* be added together.

EXAMPLE 11 Simplify $3a\sqrt{4a^3} + 7\sqrt{20b} + 3\sqrt{36a^5} - 2\sqrt{45b}$.

$$3a\sqrt{4a^3} + 7\sqrt{20b} + 3\sqrt{36a^5} - 2\sqrt{45b} =$$
$$3a\sqrt{4aaa} + 7\sqrt{4 \cdot 5b} + 3\sqrt{6 \cdot 6aaaaa} - 2\sqrt{9 \cdot 5b} =$$
$$3a(2a)\sqrt{a} + 7(2)\sqrt{5b} + 3(6aa)\sqrt{a} - 2(3)\sqrt{5b} =$$
$$6a^2\sqrt{a} + 14\sqrt{5b} + 18a^2\sqrt{a} - 6\sqrt{5b} =$$
$$(6a^2 + 18a^2)\sqrt{a} + (14 - 6)\sqrt{5b} = 24a^2\sqrt{a} + 8\sqrt{5b} \qquad ❏$$

EXERCISES 12.4B

Simplify each radical expression and then combine any like radicals. (Assume all letters represent positive numbers.)

26. $\sqrt{18} + \sqrt{50} =$

27. $\sqrt{12} - \sqrt{75} =$

28. $5\sqrt{12} + 4\sqrt{3} =$

29. $7\sqrt{2} - 3\sqrt{98} =$

30. $4\sqrt{3} + 7\sqrt{50} + 3\sqrt{8} + 5\sqrt{12} =$

31. $6\sqrt{2} - 3\sqrt{20} + 7\sqrt{45} - 3\sqrt{8} =$

32. $3\sqrt{49} - 4\sqrt{5} + 3\sqrt{20} + 2\sqrt{25} =$

33. $\sqrt{27} + 4\sqrt{9} + 7\sqrt{75} - 2\sqrt{4} + 3\sqrt{18} =$

34. $8\sqrt{12} - 3\sqrt{75} + 2\sqrt{20} - 4\sqrt{5} =$

35. $7\sqrt{3} + \sqrt{36} - 2\sqrt{27} + \sqrt{25} + 4\sqrt{12} =$

36. $\sqrt[3]{8} + \sqrt[3]{27} =$

37. $5\sqrt[3]{2} + 7\sqrt[3]{16} =$

38. $7\sqrt[3]{5} + 2\sqrt[3]{40} =$

39. $\sqrt[3]{125} + 4\sqrt[3]{3} - 7\sqrt[3]{8} + 2\sqrt[3]{24} =$

40. $8\sqrt[3]{1000} + 4\sqrt[3]{2} + \sqrt[3]{250} - 6\sqrt[3]{27} =$

41. $7\sqrt[3]{400} - 3\sqrt[3]{2} + 5\sqrt[3]{50} + 6\sqrt[3]{54} =$

42. $\sqrt{ab^2} + 3\sqrt{ab} + 5\sqrt{a^2b} - 2b\sqrt{a} + a\sqrt{b} - 5\sqrt{ab} =$

43. $\sqrt{x^2} - 5\sqrt{x} + 3\sqrt{x^2} - 4x + 2\sqrt{x} + 7x =$

44. $\sqrt{y^3} + 4\sqrt[3]{y^4} + 2y\sqrt{y} - 5y\sqrt[3]{y} =$

45. $\sqrt{ab^2} - 3\sqrt[3]{a^3b^2} + 2b\sqrt{a} + a\sqrt[3]{b^2} =$

46. $2\sqrt{m} + 6\sqrt{n} - 5\sqrt{n^2} + 8n - 4\sqrt{n} + \sqrt{m} - 5n + \sqrt{m} =$

47. $\sqrt{12ab^2} - \sqrt{8a^2b} + b\sqrt{3a} + 5a\sqrt{2b} - a\sqrt{2b} + 2b\sqrt{3a} =$

48. $2\sqrt{20x^2y} - 5x\sqrt{5y} + 3\sqrt{xy} + y\sqrt{5x} - 2\sqrt{xy} + 3y\sqrt{45x} =$

49. $6\sqrt{2y} + 3\sqrt{xy} - 2x\sqrt{xy} - 5\sqrt{8y} + \sqrt{9xy} - \sqrt{x^3y} + \sqrt{2y} =$

Check your answers on page 752. ✓

Multiplication of Radicals

Multiplying Radicals

We saw in Section 12.1 that the product of two square roots is the square root of their products. We can extend that observation to radicals with an index other than 2.

The product of two radicals with the same index is the radical of the product of the terms within the root symbols:

$$\sqrt{a}\sqrt{b} = \sqrt{ab}$$
$$\sqrt[3]{a}\sqrt[3]{b} = \sqrt[3]{ab}$$
$$\sqrt[n]{a}\sqrt[n]{b} = \sqrt[n]{ab}$$

So to multiply two radicals, put them under the same symbol and multiply the terms.

EXAMPLE 1 $(\sqrt{7})(\sqrt{3}) = \sqrt{21}$ ❏

The product of two radicals might be one that can be simplified, even though neither of the given radicals could be simplified.

EXAMPLE 2 Simplify the product of $(\sqrt{6})(\sqrt{10})$.

$$(\sqrt{6})(\sqrt{10}) =$$
$$\sqrt{2 \cdot 3 \cdot 2 \cdot 5} = 2\sqrt{15}$$ ❏

EXAMPLE 3 Simplify the product of $(\sqrt{18})(\sqrt{2})$.

$$(\sqrt{18})(\sqrt{2}) =$$
$$\sqrt{2 \cdot 3 \cdot 3 \cdot 2} =$$
$$2 \cdot 3 = 6$$ ❏

EXAMPLE 4 Simplify $(\sqrt[3]{12})(\sqrt[3]{14})(\sqrt[3]{9})(\sqrt[3]{7})$.

$$(\sqrt[3]{12})(\sqrt[3]{14})(\sqrt[3]{9})(\sqrt[3]{7}) =$$
$$\sqrt[3]{2 \cdot 2 \cdot 3 \cdot 2 \cdot 7 \cdot 3 \cdot 3 \cdot 7} =$$
$$(2)(3)\sqrt[3]{7 \cdot 7} = 6\sqrt[3]{49}$$ ❏

EXAMPLE 5 Simplify $(\sqrt{3a})(\sqrt{21ab})(\sqrt{14bc})$.

$$(\sqrt{3a})(\sqrt{21ab})(\sqrt{14bc}) =$$
$$\sqrt{3 \cdot a \cdot 3 \cdot 7 \cdot a \cdot b \cdot 2 \cdot 7 \cdot b \cdot c} =$$
$$(3)(7)ab\sqrt{2c} = 21ab\sqrt{2c}$$ ❏

EXERCISES 12.5A

Multiply these radicals and leave each product simplified. (Assume all letters represent positive numbers.)

1. $\sqrt{5}\sqrt{7} =$ **2.** $\sqrt{3}\sqrt{11} =$

3. $\sqrt{6}\sqrt{15} =$ **4.** $\sqrt{10}\sqrt{35} =$

5. $\sqrt{14}\sqrt{35} =$ **6.** $\sqrt{30}\sqrt{6} =$

7. $\sqrt{55}\sqrt{10} =$

8. $\sqrt{22}\sqrt{33} =$

9. $\sqrt{6}\sqrt{10}\sqrt{15} =$

10. $\sqrt{26}\sqrt{39}\sqrt{14} =$

11. $\sqrt[3]{4}\sqrt[3]{6} =$

12. $\sqrt[3]{25}\sqrt[3]{15} =$

13. $\sqrt[3]{36}\sqrt[3]{10} =$

14. $\sqrt[3]{18}\sqrt[3]{9} =$

15. $\sqrt[3]{28}\sqrt[3]{14}\sqrt[3]{21} =$

16. $\sqrt[3]{12}\sqrt[3]{18}\sqrt[3]{10} =$

17. $\sqrt{10a}\sqrt{6a} =$

18. $\sqrt{15ab}\sqrt{20b} =$

19. $\sqrt{5x}\sqrt{10y}\sqrt{6xy} =$

20. $\sqrt{12w}\sqrt{6w} =$

21. $\sqrt{6ab}\sqrt{15bc} =$

22. $\sqrt{8p}\sqrt{6p}\sqrt{3p} =$

Check your answers on page 752.

Multiplying Radical Expressions

Multiplying radical expressions is like multiplying algebraic expressions. The rational parts are multiplied just as the numerical coefficients of terms are, and the radical parts are multiplied just as the letter factors of the terms are.

EXAMPLE 6 Simplify $(2\sqrt{3})(5\sqrt{2})$.

$$(2\sqrt{3})(5\sqrt{2}) =$$
$$(2)(\sqrt{3})(5)(\sqrt{2}) =$$
$$(2)(5)\sqrt{3}\sqrt{2} = 10\sqrt{6}$$

Factor the numbers in the radicals and simplify if possible before multiplying them.

EXAMPLE 7 Simplify $(7\sqrt{6})(-3\sqrt{10})$.

$$(7\sqrt{6})(-3\sqrt{10}) =$$
$$(7)(-3)\sqrt{2\cdot3}\sqrt{2\cdot5} =$$
$$-21\sqrt{2\cdot3\cdot2\cdot5} =$$
$$(-21)(2)\sqrt{3\cdot5} = -42\sqrt{15}$$

EXERCISES 12.5B

Multiply and simplify.

23. $(4\sqrt{7})(2\sqrt{3}) =$

24. $(5\sqrt{2})(7\sqrt{5}) =$

25. $(6\sqrt{3})(-4\sqrt{5}) =$

26. $(-10\sqrt{6})(-4\sqrt{5}) =$

27. $(2\sqrt{7})(3\sqrt{14}) =$

28. $(5\sqrt{6})(7\sqrt{15}) =$

29. $(-4\sqrt{10})(7\sqrt{14}) =$

30. $(-3\sqrt{21})(-2\sqrt{7}) =$

31. $(5\sqrt{20})(2\sqrt{15}) =$

32. $(-6\sqrt{8})(3\sqrt{10}) =$

Check your answers on page 753.

Distributive Law for Multiplying Radicals by a Sum

You already know how to multiply a radical expression by the sum of radical expressions. Use the distributive law to remove the parentheses.

EXAMPLE 8   Simplify $\sqrt{5}(\sqrt{3} + \sqrt{7})$.

$$\sqrt{5}(\sqrt{3} + \sqrt{7}) =$$
$$\sqrt{5}\sqrt{3} + \sqrt{5}\sqrt{7} = \sqrt{15} + \sqrt{35}$$ ❏

Always simplify the product radical if possible.

EXAMPLE 9   Simplify $\sqrt{6}(\sqrt{2} + \sqrt{3})$.

$$\sqrt{6}(\sqrt{2} + \sqrt{3}) =$$
$$\sqrt{2\cdot3\cdot2} + \sqrt{2\cdot3\cdot3} = 2\sqrt{3} + 3\sqrt{2}$$ ❏

EXAMPLE 10   Simplify $\sqrt{15}(\sqrt{20} - \sqrt{21})$.

$$\sqrt{15}(\sqrt{20} - \sqrt{21}) =$$
$$\sqrt{3\cdot5\cdot4\cdot5} - \sqrt{3\cdot5\cdot3\cdot7} =$$
$$5(2)\sqrt{3} - 3\sqrt{5\cdot7} = 10\sqrt{3} - 3\sqrt{35}$$ ❏

EXAMPLE 11   Simplify $-5\sqrt{2}(7 - 3\sqrt{6})$.

$$(-5\sqrt{2})(7 - 3\sqrt{6}) =$$
$$(-5\sqrt{2})(7) + (-5\sqrt{2})(-3\sqrt{6}) =$$
$$-5(7)\sqrt{2} - 5(-3)\sqrt{2\cdot2\cdot3} =$$
$$-35\sqrt{2} + 15(2)\sqrt{3} = -35\sqrt{2} + 30\sqrt{3}$$ ❏

EXERCISES 12.5C

Multiply and simplify.

33. $\sqrt{5}(\sqrt{2} + \sqrt{6}) =$ **34.** $\sqrt{7}(\sqrt{6} + \sqrt{2}) =$

35. $\sqrt{2}(\sqrt{5} - \sqrt{7}) =$ **36.** $\sqrt{6}(11 - \sqrt{5}) =$

37. $\sqrt{3}(\sqrt{6} + \sqrt{15}) =$ **38.** $\sqrt{2}(\sqrt{10} - \sqrt{6}) =$

39. $-\sqrt{5}(3 + \sqrt{10}) =$ **40.** $\sqrt{6}(\sqrt{15} - \sqrt{14}) =$

41. $\sqrt{3}(3\sqrt{6} + 2\sqrt{12}) =$ **42.** $\sqrt{2}(3\sqrt{6} - 5\sqrt{10}) =$

43. $-3\sqrt{7}(2\sqrt{14} + 5\sqrt{2}) =$ **44.** $2\sqrt{3}(6\sqrt{6} - 3\sqrt{5}) =$

Check your answers on page 753.   ✓

Glossary

cube root A radical with index 3. The cube root of an expression is the term that, when raised to the third power, produces that expression.

Example: The cube root of x^3 [$\sqrt[3]{x^3}$] is x.

fractional exponent When an exponent is a fraction, the numerator represents the power, and the denominator represents the index of the radical that is applied to the base.

Example: $x^{m/n} = (\sqrt[n]{x})^m$, so

$$8^{2/3} =$$
$$(\sqrt[3]{8})^2 =$$
$$(2)^2 = 4$$

imaginary numbers Number system that mathematicians developed based on numbers that are the square root of -1. These numbers are not real (see **real numbers**).

index Small number written on the radical symbol; indicates the number of times a factor within the symbol must be repeated for it to be removed from the symbol.

Example: The index of square roots ($\sqrt{}$) is 2:

$$\sqrt{abb} = b\sqrt{a}.$$

The index of cube roots ($\sqrt[3]{}$) is 3:

$$\sqrt[3]{n^3m} = n\sqrt[3]{m}.$$

The index of fourth roots ($\sqrt[4]{}$) is 4:

$$\sqrt[4]{pm^4} = m\sqrt[4]{p}.$$

The index of nth roots ($\sqrt[n]{}$) is n:

$$\sqrt[n]{ck^n} = k\sqrt[n]{c}.$$

irrational expression Term that in its simplest form involves a radical symbol.

Example: $\sqrt{3a}$ and $\sqrt[3]{2x^2}$ are irrational expressions.

irrational number Number that in its simplest form must be represented as a radical. It cannot be equal to a repeating decimal.

Example: $\sqrt{7}$ and $\sqrt[3]{4}$ are irrational numbers.

like radical If the index and expression under a radical symbol are identical, they are like radicals and can be added together by combining their coefficients.

Example: $3\sqrt{a}$ and $5\sqrt{a}$ are like radicals; so $3\sqrt{a} + 5\sqrt{a} = 8\sqrt{a}$.

perfect square Number or term that can be factored into two identical lists of factors that are each an integer, a rational fraction, or a letter. The square root of a perfect square is always rational.

Example: $9a^2b^6$ is a perfect square since $9a^2b^6 = (3abbb)(3abbb)$.

$$\sqrt{9a^2b^6} = 3ab^3$$

principal root Positive root when the index of a radical is an even number. The radical symbol always represents the principal root.

Example: $\sqrt{3ab}$ and $\sqrt[6]{x}$ represent principal roots. The square roots of 4 are $+2$ and -2, but the principal root of 4 ($\sqrt{4}$) is $+2$.

radical Expression that involves a number or term within a root symbol.

Example: $\sqrt{2}, \sqrt{3a}$, and $\sqrt[3]{n}$ are radicals.

rational numbers Set of all integers, numerical fractions with integers as numerators and denominators, and decimals ending with zeros or a pattern of repeating digits.

Example: $6 = 6.\bar{0}$, $\frac{4}{5} = 0.8\bar{0}$, and $0.\overline{317}$ are rational numbers.

real numbers All the rational and irrational numbers. They satisfy the rule that only an even amount of negative numbers, when multiplied together, will always produce a positive number.

square root Radical with index 2. The square root of an expression is the term that, when raised to the second power, produces the expression.

Example: The square root of x^2 is $\pm x$. The square root of 9 is ± 3.

Chapter 12 Review

Do each of the problems. After you are confident that you have done them as accurately as possible, compare your answers with those in the back of the book. If any are wrong, go back to the section where the problem came from (indicated to the left of the problem) and review the section.

Once you understand all the sections, you will be sure to learn the skills solidly and remember how to do them if you practice them a little bit more. Turn to the Supplementary Exercises and do all the problems from any section where you had difficulty on these review exercises.

12.1

1. Is 121 a perfect square?

2. Is $50a^2b^8$ a perfect square?

3. What is the square root of 81?

Simplify. (Assume letters represent positive numbers.)

4. $\sqrt{49} =$

5. $\sqrt{75} =$

6. $\sqrt{200} =$

7. $\sqrt{1600} =$

8. $\sqrt{a^4} =$

9. $\sqrt{w^3} =$

10. $\sqrt{9ab^3c^6} =$

11. $\sqrt{50a^2b^8c^7} =$

12. Estimate between which two whole numbers the irrational number $\sqrt{45}$ is, and show it on this number line.

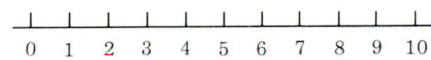

12.2 Simplify each radical. (Assume letters represent positive numbers.)

13. $\sqrt[3]{-40} =$

14. $\sqrt[3]{27x^4y^2} =$

15. $\sqrt[4]{80} =$

16. $\sqrt[4]{16a^6b^2} =$

17. $\sqrt[5]{w^6} =$

18. $\sqrt[8]{n^{10}} =$

12.3 Evaluate.

19. $16^{1/2} =$

20. $(-27)^{1/3} =$

21. $49^{-1/2} =$

22. $\left(\dfrac{1}{8}\right)^{-1/3} =$

Simplify so that exponents are positive.

23. $(x^{1/4})(x^{3/4}) =$

24. $\dfrac{w^{5/9}}{w^{4/9}} =$

25. $(a^{1/4})^{3/5} =$

26. $(ab^2)^{3/4} =$

Continued

Answers
1.
2.
3.
4.
5.
6.
7.
8.
9.
10.
11.
12.
13.
14.
15.
16.
17.
18.
19.
20.
21.
22.
23.
24.
25.
26.

Chapter 12 Review (*Cont.*)

27. $n^{-6/7} =$

28. $\dfrac{1}{m^{-3/4}} =$

29. $\dfrac{a^2 b^{5/6} c^{1/2}}{a^{1/3} b^{1/6} c} =$

30. $\dfrac{(ab^{1/2})^{1/3}}{(a^2 b^{1/3})^{1/2}} =$

12.4 Combine and simplify.

31. $8\sqrt{3} - 5\sqrt{3} =$

32. $6\sqrt{2} + 9\sqrt{5} - \sqrt{5} + 3\sqrt{2} =$

33. $\sqrt[3]{7} + 4\sqrt[3]{7} =$

34. $\sqrt{5} + 3\sqrt[3]{5} - 2\sqrt[3]{5} + 6\sqrt{5} =$

35. $\sqrt{12} - \sqrt{27} =$

36. $6\sqrt{20} + 5\sqrt{45} =$

37. $3\sqrt{28} + 2\sqrt{3} - 5\sqrt{12} + 2\sqrt{63} =$

38. $4\sqrt[3]{24} + 2\sqrt[3]{5} - 4\sqrt[3]{40} - \sqrt[3]{3} =$

39. $\sqrt{m} + 3\sqrt{n} - 5\sqrt{n} + 2\sqrt{m} =$

40. $\sqrt{p} + \sqrt[3]{p} - 4\sqrt{p} + 2\sqrt[3]{p} =$

41. $\sqrt{4a^3} + a\sqrt{25a} + \sqrt{12b} + \sqrt{3b} + \sqrt{16} =$

42. $2\sqrt{12x} + 3\sqrt{45y^3} - y\sqrt{5y} + 2\sqrt{75x} =$

12.5 Multiply and simplify.

43. $\sqrt{6}\sqrt{14} =$

44. $\sqrt{10}\sqrt{18}\sqrt{35} =$

45. $(3\sqrt{5})(4\sqrt{2}) =$

46. $(-6\sqrt{10})(5\sqrt{15}) =$

47. $(\sqrt{2n})(\sqrt{6mn}) =$

48. $(\sqrt[3]{5ab})(\sqrt[3]{10ac^2})(\sqrt[3]{15a^2bc}) =$

49. $(5\sqrt{p})(-3\sqrt{2p})(-2\sqrt{6}) =$

50. $\sqrt{7}(\sqrt{3} - \sqrt{2}) =$

51. $\sqrt{6}(2\sqrt{3} + 5\sqrt{10}) =$

52. $3\sqrt{2}(5 + 2\sqrt{10}) =$

Check your answers on page 753.

Answers

27. _____
28. _____
29. _____
30. _____
31. _____
32. _____
33. _____
34. _____
35. _____
36. _____
37. _____
38. _____
39. _____
40. _____
41. _____
42. _____
43. _____
44. _____
45. _____
46. _____
47. _____
48. _____
49. _____
50. _____
51. _____
52. _____

Supplementary Exercises

Do all the problems in every section that involves a skill in which you now lack complete mastery. With a little more practice you will achieve that sense of really understanding the topics, and you will remember how to do these problems.

12.1

1. Is 225 a perfect square?

2. Is $49a^6b^{10}$ a perfect square?

3. What is the square root of 25?

Simplify.

4. $\sqrt{25} =$

5. $\sqrt{98} =$

6. $\sqrt{300} =$

7. $\sqrt{4000} =$

8. $\sqrt{p^6} =$

9. $\sqrt{m^5} =$

10. $\sqrt{16ab^2c^5} =$

11. $\sqrt{20a^3b^6c^9} =$

Estimate between which two whole numbers these irrational numbers are, and show them on the number lines.

12. $\sqrt{22}$ 0 1 2 3 4 5 6 7 8 9 10

13. $\sqrt{50}$ 0 1 2 3 4 5 6 7 8 9 10

12.2 Simplify each radical. (Assume letters represent positive numbers.)

14. $\sqrt[3]{24} =$

15. $\sqrt[3]{-125x^5y^6} =$

16. $\sqrt[4]{81} =$

17. $\sqrt[4]{32a^7b^3} =$

18. $\sqrt[6]{w^{10}} =$

19. $\sqrt[7]{n^{10}} =$

12.3 Evaluate.

20. $36^{1/2} =$

21. $(-8)^{1/3} =$

22. $121^{-1/2} =$

23. $\left(\dfrac{1}{27}\right)^{-1/3} =$

Simplify so that exponents are positive.

24. $(x^{1/3})(x^{3/5}) =$

25. $\dfrac{w^{1/7}}{w^{3/7}} =$

Answers

1. _____
2. _____
3. _____
4. _____
5. _____
6. _____
7. _____
8. _____
9. _____
10. _____
11. _____
12. _____
13. _____
14. _____
15. _____
16. _____
17. _____
18. _____
19. _____
20. _____
21. _____
22. _____
23. _____
24. _____
25. _____

Continued

Supplementary Exercises (*Cont.*)

26. $(a^{2/5})^{3/4} =$

27. $(ab^3)^{1/6} =$

28. $n^{-4/7} =$

29. $\dfrac{1}{m^{-2/7}} =$

30. $\dfrac{a^3 b^{4/5} c^{1/3}}{a^{1/3} b^{2/5} c} =$

31. $\dfrac{(ab^{1/4})^{2/3}}{(a^2 b^{1/6})^{1/2}} =$

12.4   Combine and simplify.

32. $2\sqrt{10} - 3\sqrt{10} =$

33. $2\sqrt{5} + 4\sqrt{3} - \sqrt{3} + 7\sqrt{5} =$

34. $5\sqrt{11} + 2\sqrt{13} - 4\sqrt{13} - \sqrt{11} + 5\sqrt{11} =$

35. $\sqrt[3]{5} + 3\sqrt[3]{5} =$

36. $4\sqrt[3]{2} + 7\sqrt{2} - 3\sqrt{2} + 8\sqrt[3]{2} - \sqrt[3]{2} =$

37. $2\sqrt[3]{5} + 3\sqrt{6} + 4\sqrt{6} - \sqrt[3]{5} =$

38. $\sqrt{20} - \sqrt{45} =$

39. $\sqrt{28} + \sqrt{63} =$

40. $5\sqrt{20} + 7\sqrt{45} =$

41. $5\sqrt{12} + 3\sqrt{2} - 6\sqrt{8} + 2\sqrt{3} =$

42. $6\sqrt{45} + 3\sqrt{3} - 6\sqrt{5} - 2\sqrt{12} =$

43. $5\sqrt[3]{2} + 3\sqrt[3]{16} - 4\sqrt[3]{5} - \sqrt[3]{40} =$

44. $7\sqrt[3]{81} + 4\sqrt[3]{5} + 7\sqrt[3]{3} - \sqrt[3]{135} =$

45. $\sqrt{p} + 2\sqrt{k} - 4\sqrt{k} + 5\sqrt{p} =$

46. $\sqrt{a} + \sqrt[3]{a} - 7\sqrt{a} + 4\sqrt[3]{a} =$

47. $\sqrt{9a^3} + a\sqrt{16a} + \sqrt{12b} + \sqrt{3b} + \sqrt{36} =$

48. $3\sqrt{18x} + 4\sqrt{50y^3} - y\sqrt{2y} + 2\sqrt{2x} =$

12.5   Multiply and simplify.

49. $\sqrt{10}\sqrt{20} =$

50. $\sqrt{8}\sqrt{6} =$

Answers

26. _____

27. _____

28. _____

29. _____

30. _____

31. _____

32. _____

33. _____

34. _____

35. _____

36. _____

37. _____

38. _____

39. _____

40. _____

41. _____

42. _____

43. _____

44. _____

45. _____

46. _____

47. _____

48. _____

49. _____

50. _____

51. $\sqrt{6}\sqrt{12}\sqrt{14} =$

52. $\sqrt{5}\sqrt{35}\sqrt{14} =$

53. $(2\sqrt{7})(4\sqrt{3}) =$

54. $(5\sqrt{6})(2\sqrt{10}) =$

55. $(-3\sqrt{6})(5\sqrt{2}) =$

56. $(-4\sqrt{10})(-2\sqrt{30}) =$

57. $(\sqrt{2n})(\sqrt{6mn}) =$

58. $(5\sqrt{p})(-3\sqrt{2p})(-2\sqrt{6}) =$

59. $(\sqrt[3]{5ab})(\sqrt[3]{10ac^2})(\sqrt[3]{15a^2bc}) =$

60. $\sqrt{2}(\sqrt{7} + \sqrt{5}) =$

61. $\sqrt{3}(\sqrt{6} - \sqrt{15}) =$

62. $2\sqrt{3}(1 - 3\sqrt{6}) =$

63. $\sqrt{5}(2\sqrt{3} + 4\sqrt{10}) =$

64. $3\sqrt{2}(4\sqrt{6} - 5\sqrt{10}) =$

Answers

51. _____

52. _____

53. _____

54. _____

55. _____

56. _____

57. _____

58. _____

59. _____

60. _____

61. _____

62. _____

63. _____

64. _____

Check your answers on page 753.

Chapter 12 Test

1. What is the square root of 64?

Simplify each expression.

2. $\sqrt{16} =$

3. $\sqrt{40} =$

4. $\sqrt[3]{125} =$

5. $\sqrt[3]{-405} =$

6. $\sqrt[4]{48} =$

7. $\sqrt[4]{324} =$

8. $\sqrt{28x^3y^4} =$

9. $\sqrt[3]{40a^3b^2c^8} =$

10. $4\sqrt{5} - \sqrt{5} =$

11. $4\sqrt{3} + 2\sqrt{2} + 6\sqrt{2} - \sqrt{3} =$

12. $\sqrt[3]{4} + 3\sqrt[3]{4} =$

13. $5\sqrt{2} + 7\sqrt[3]{2} - \sqrt[3]{2} + 3\sqrt{2} =$

14. $\sqrt{54} - \sqrt{24} =$

15. $2\sqrt{75} + 7\sqrt{12} =$

16. $5\sqrt{63} + 4\sqrt{18} - 3\sqrt{3} + 2\sqrt{28} =$

17. $6\sqrt[3]{16} + 3\sqrt[3]{2} - 5\sqrt[3]{2} - \sqrt[3]{54} =$

18. $\sqrt{a} + 5\sqrt{b} - 3\sqrt{c} - \sqrt{b} + 4\sqrt{a} + \sqrt{c} =$

19. $\sqrt{x} + 2\sqrt[3]{x} + 5\sqrt{x} - \sqrt[3]{x} + 2\sqrt{x} =$

20. $\sqrt{20a} + \sqrt{12a} + \sqrt{3a} - 2\sqrt{5a} + 3\sqrt{75a} =$

21. $\sqrt{12}\sqrt{15} =$

22. $\sqrt{6}\sqrt{8}\sqrt{15} =$

23. $(2\sqrt{7})(-3\sqrt{2}) =$

24. $(-4\sqrt{6})(-3\sqrt{21}) =$

1. _____

2. _____

3. _____

4. _____

5. _____

6. _____

7. _____

8. _____

9. _____

10. _____

11. _____

12. _____

13. _____

14. _____

15. _____

16. _____

17. _____

18. _____

19. _____

20. _____

21. _____

22. _____

23. _____

24. _____

Answers

25. $\sqrt{3}(\sqrt{5} - \sqrt{11}) =$

26. $\sqrt{10}(\sqrt{14} - \sqrt{15}) =$

27. $\sqrt{6}(2\sqrt{3} + 5\sqrt{15}) =$

Evaluate.

28. $25^{1/2} =$

29. $25^{-1/2} =$

30. $\left(\dfrac{27}{1000}\right)^{2/3} =$

31. $\left(-\dfrac{27}{1000}\right)^{-1/3} =$

Simplify.

32. $(a^{3/8})(a^{1/8}) =$

33. $(m^{3/5})(m^{-2/5}) =$

34. $\dfrac{p^{7/8}}{p^{3/8}} =$

35. $(k^{2/3})^{1/4} =$

36. $\dfrac{xy^{3/5}w^{2/3}}{x^{1/3}y^{4/5}w^{1/3}} =$

37. $\dfrac{(n^2 m^{3/4})^{1/3}}{(n^{2/3} m^{1/2})^{1/2}} =$

38. Estimate between which two whole numbers $\sqrt{17}$ is, and show it on the number line.

25. _____

26. _____

27. _____

28. _____

29. _____

30. _____

31. _____

32. _____

33. _____

34. _____

35. _____

36. _____

37. _____

38. _____

Check your answers on page 753.

Cumulative Review
CHAPTERS 1–12

Combine and simplify.

Answers

1. $\dfrac{4}{7} + \dfrac{3}{14} - \dfrac{1}{28} =$

2. $\dfrac{\frac{5}{12}}{\frac{10}{9}} =$

3. $54.02 - 6.041 =$

4. $(7.02)(0.31) =$

5. $0.000042 \div 0.003 =$

6. 5.1% of $346 =$

7. $12 - (-4) + (-7) + (5) - (-3) =$

8. $(3)(-2)(-1)(-1)(3)(-1) =$

9. $\dfrac{-52}{-4} =$

10. $-5 - 3(-4) + 2(-3)(-1) + 10 \div (-5) =$

11. $2(y^2 - 6y + 8) + 5(2y^2 + 3y - 4) =$

12. $-3x^2(4x^2 - 5x - 3) =$

13. $\dfrac{6a^3bc^5}{10abc} =$

14. $(-3)^{-2} =$

Solve each equation.

15. $4y - 3 + 2(y - 5) = 4y - 9$

16. $\dfrac{3x}{2} + 5 = -1$

Solve each system of equations.

17. $x = 3y - 1$
    $2x - 3y = 4$

18. $4x - 3y = 7$
    $3x + 2y = -16$

What is the equation of the line that

19. has slope 4 and passes through point $(-2, -3)$?

20. passes through points $(2, -3)$ and $(4, 5)$?

Leave your answers in scientific notation.

21. $(2.3 \times 10^{27})(4.8 \times 10^{56}) =$

22. $(4.0 \times 10^{-18})(3.4 \times 10^{-30})(2.7 \times 10^{21}) =$

1. _____

2 _____

3. _____

4. _____

5. _____

6. _____

7. _____

8. _____

9. _____

10. _____

11. _____

12. _____

13. _____

14. _____

15. _____

16. _____

17. _____

18. _____

19. _____

20. _____

21. _____

22. _____

Sketch the following graphs.

23. $y = 3x - 2$

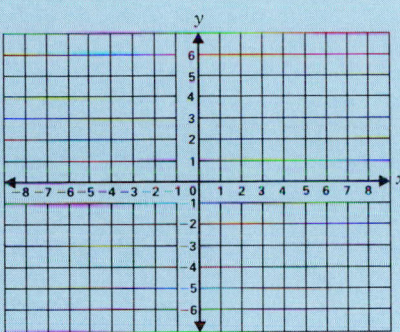

24. $y < 3x - 2$

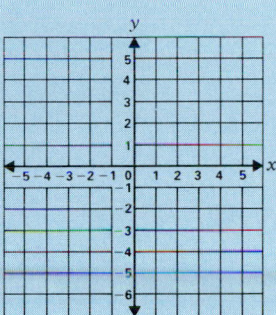

25. $y \geq 3x - 2$

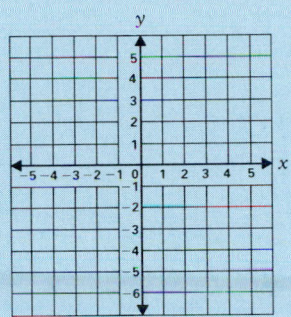

23. _____

24. _____

25. _____

Solve the following problems.

26. How much would it cost to buy three shirts if each costs $8.95, two pairs of socks if each costs $1.50, and a sweater that costs $24, if the store is having a sale of a 15 % discount on everything?

27. If there are three foreign cars for every five American-made cars in a parking lot, out of a total of 280 cars how many are foreign made?

28. Two children were playing a game they made up. It was possible to score points in two ways. They decided that players would get a certain number of points every time they scored by catching a fly ball, and would get a different number of points every time they scored by catching the ball on one bounce. Frank scored seven times by catching on a bounce and three times by catching on a fly. Anne scored four times by catching on a bounce and nine times by catching on a fly. If Frank earned a total of 29 points and Anne earned a total of 53 points, how many points were scored each way?

26. _____

27. _____

28. _____

Check your answers on page 753.

13 Multiplying and Factoring Polynomials

13.1 Multiplication of Polynomials by Polynomials

In Section 5.5 we learned how to use the distributive law to multiply a term by a polynomial. In this chapter we discuss how to multiply polynomials by polynomials.

To begin, recall that the distributive law tells us that to multiply a term by a polynomial, we must multiply that term by *every* term of the polynomial.

EXAMPLE 1   Multiply $(x - 2y + 5w)$ by $3x$.

$$3x(x - 2y + 5w) =$$
$$3x(x) + 3x(-2y) + 3x(5w) = 3x^2 - 6xy + 15xw$$ ❑

We can use the distributive law to multiply polynomials together too. Treat one of the polynomials as the distributing term and multiply it by each term in the other polynomial. Then apply the distributive law again to multiply each term by that polynomial.

EXAMPLE 2   Let's multiply the polynomial $(a + b + c)$ by the term x.

$$x(a + b + c) = xa + xb + xc$$ ❑

EXAMPLE 3   Now let's multiply the polynomial $(a + b + c)$ by the binomial $(y + w)$, treating $(y + w)$ as the distributing term.

$$(y + w)(a + b + c) =$$
$$(y + w)(a) + (y + w)(b) + (y + w)(c) =$$

Now we apply the distributive law again to each product.

$$(y + w)(a) + (y + w)(b) + (y + w)(c) = ya + wa + yb + wb + yc + wc$$ ❑

After multiplying out all the terms, simplify the resulting polynomial by combining any *like* terms.

EXAMPLE 4 Multiply $(x + y + 5)$ by $(x + 3)$.

Let's treat $(x + y + 5)$ as the distributing term.

$$(x + y + 5)(x + 3) =$$

$$(x + y + 5)(x) + (x + y + 5)(3) =$$

$$\underline{x(x)} + \underline{x(y)} + \underline{x(5)} + \underline{3(x)} + \underline{3(y)} + \underline{3(5)} =$$

$$x^2 + xy + 5\underline{x} + 3\underline{x} + 3y + 15 = x^2 + xy + 8x + 3y + 15 \quad \square$$

You may have noticed that by the time you completed all the multiplying that was necessary, you ended up multiplying each term in one polynomial by each term in the other polynomial. So here is a shortcut to avoid applying the distributive law twice, as we did in Examples 3 and 4.

Shortcut for Multiplying Polynomials.

Take the first term in one polynomial and multiply it by every term in the second polynomial.

Then take the second term in the first polynomial and multiply it by every term in the other polynomial, and so on, until all the terms in the first polynomial have been used as the distributing term through the other polynomial.

EXAMPLE 5 Multiply $(x + y + 5)$ by $(x + 3)$ using this shortcut.

$$(\underline{x} + \underline{\underline{y}} + \underline{5})(x + 3) =$$

$$\underline{x}(x) + \underline{x}(3) + \underline{\underline{y}}(x) + \underline{\underline{y}}(3) + \underline{5}(x) + \underline{5}(3) =$$

$$x^2 + 3\underline{x} + xy + 3y + 5\underline{x} + 15 = x^2 + 8x + xy + 3y + 15 \quad \square$$

As you multiply terms, you can write their product beneath any like terms that have already been produced. This step helps to prevent careless errors when simplifying the resulting polynomial.

EXAMPLE 6 Multiply $(2x + 3y - 5w)$ by $(3x - 2w)$.

$$(\underline{2x} + \underline{\underline{3y}} - \underline{5w})(3x - 2w) =$$

$$\underline{2x}(3x) + \underline{2x}(-2w) + \underline{\underline{3y}}(3x) + \underline{\underline{3y}}(-2w) - \underline{5w}(3x) - \underline{5w}(-2w) =$$

$$6x^2 \quad - 4wx + 9xy - 6wy + 10w^2$$
$$\quad\quad - 15wx$$
$$\overline{6x^2 - 19wx + 9xy - 6wy + 10w^2} \quad \square$$

The shortcut simples the job of multiplying polynomials that contain more than two terms.

EXAMPLE 7 Find the product of $(a^2 + 3a - 4)(5a^2 - a - 1)$.

$$(a^2 + 3a - 4)(5a^2 - a - 1) =$$

$$\underline{(a^2)(5a^2)} + \underline{(a^2)(-a)} + \underline{(a^2)(-1)} + \underline{(3a)(5a^2)} + \underline{(3a)(-a)} + \underline{(3a)(-1)}$$

$$+ \underline{(-4)(5a^2)} + \underline{(-4)(-a)} + \underline{(-4)(-1)} = 5a^4 \quad - a^3 \quad - a^2$$
$$+ 15a^3 \quad - 3a^2 - 3a$$
$$- 20a^2 + 4a + 4$$
$$\overline{5a^4 + 14a^3 - 24a^2 + a + 4} \quad \square$$

In Examples 8 and 9 we will see that the way to multiply polynomials is really the same as the way to multiply large numbers.

EXAMPLE 8 To multiply $2x^2 + 3x + 1$ by $2x + 1$, first multiply each term by 1, then multiply each term by $2x$ (from $2x + 1$).

$$
\begin{array}{r}
2x^2 + 3x + 1 \\
2x + 1 \\
\hline
1(2x^2) + 1(3x) + 1(1) \\
2x(2x^2) + 2x(3x) + 2x(1)
\end{array}
\Rightarrow
\begin{array}{r}
2x^2 + 3x + 1 \\
2x + 1 \\
\hline
2x^2 + 3x + 1 \\
4x^3 + 6x^2 + 2x \\
\hline
4x^3 + 8x^2 + 5x + 1
\end{array}
$$

❏

EXAMPLE 9 Multiply 231 by 21.

$$
\begin{array}{r}
2\ 3\ 1 \\
\times\ 2\ 1 \\
\hline
1(2) + 1(3) + 1(1) \\
2(2) + 2(3) + 2(1)
\end{array}
\Rightarrow
\begin{array}{r}
2\ 3\ 1 \\
\times\ 2\ 1 \\
\hline
2\ 3\ 1 \\
4\ 6\ 2 \\
\hline
4\ 8\ 5\ 1
\end{array}
$$

So $(321)(21) = 4,851$

❏

The number we write as 231 is really $2(10)^2 + 3(10) + 1$, and 21 is really $2(10) + 1$. So multiplying 231 by 21 is the same as multiplying $2x^2 + 3x + 1$ by $2x + 1$ when x is equal to 10! $[4,851 = 4(10)^3 + 8(10)^2 + 5(10) + 1]$.

Learning algebra allows us to understand *why* we perform the computations of arithmetic the way we do—and vice versa.

EXERCISES 13.1

Multiply these polynomials. Simplify your answers by combining like terms.

1. $(x + w)(2x + y + w) =$

2. $(2y - 1)(y^2 + 5y - 2) =$

3. $(3w - 2)(2w^2 + 3w + 1) =$

4. $(5n + 3)(3n^2 + 2n - 3) =$

5. $(2n - 3m)(2n + 3m + 1) =$

6. $(a + b + c)(a + b + c) =$

7. $(x + y - w)(x - y + w) =$

8. $(k + j - 3)(k - j + 2) =$

9. $(x + 2)(x^2 - 3x + 5) =$

10. $(y^2 - y + 1)(y + 5) =$

11. $(w - 2)(w^2 + 3w + 1) =$

12. $(3a + 1)(2a^2 - a + 4) =$

13. $(2x^2 + x - 1)(3x^2 - 2x + 7) =$

14. $(5y^2 + 5y - 3)(y^2 + y - 1) =$

Check your answers on page 754.

13.2 Multiplication of Binomials by FOIL

FOIL

When multiplying a binomial by a binomial, we can expect to produce four terms. That is because each of the two terms in one binomial is multiplied by the two terms in the other binomial. The product will have four terms, although after we simplify and combine like terms, our final answer may have fewer terms.

To avoid careless errors and for reasons that will become clear to you in Section 13.4, it is a good idea to *anticipate producing four terms* when two binomials are multiplied.

EXAMPLE 1 $(\underline{a} + \underline{b})(c + d) = \underline{a}c + \underline{a}d + \underline{b}c + \underline{b}d$ ❏

An easy way to remember which terms are to be multiplied is to remember the word "**FOIL**." Referring to the binomials in Example 1:

F represents the product of the first terms from each binomial, a and c, producing the *first* term: ac.

O represents the product of the outside terms from each binomial, a and d, producing the *outside* terms: ad.

I represents the product of the inside terms from each binomial, b and c, producing the *inside* term: bc.

L represents the product of the last terms from each binomial, b and d, producing the *last* term: bd.

The product of two binomials is the sum of the four product terms:

$$ac \;+\; ad \;+\; bc \;+\; bd$$
$$\text{First} + \text{Outside} + \text{Inside} + \text{Last}$$

EXAMPLE 2   Use FOIL to multiply $(4y - 1)(3y + 2)$.

$$(4y - 1)(3y + 2) =$$
$$\underbrace{[(4y)(3y)]}_{F} + \underbrace{[(4y)(2)]}_{O} + \underbrace{[(-1)(3y)]}_{I} + \underbrace{[(-1)(2)]}_{L} =$$
$$12y^2 + 8\underline{y} - 3\underline{y} - 2 = 12y^2 + 5y - 2$$ ❏

EXAMPLE 3   Use FOIL to multiply $(3a - 5b)(2a + b)$.

$$(3a - 5b)(2a + b) =$$
$$[\underline{(3a)(2a)}] + [\underline{(3a)(b)}] + [\underline{(-5b)(2a)}] + [\underline{(-5b)(b)}] =$$
$$6a^2 + 3\underline{ab} - 10\underline{ab} - 5b^2 = 6a^2 - 7ab - 5b^2$$ ❏

Now that you know how to multiply binomials, we can represent the multiplication of mixed numbers (which we first discussed in Section 3.8) as the product of two binomials.

EXAMPLE 4   Use FOIL to multiply $2\dfrac{1}{3}$ by $3\dfrac{1}{2}$.

$$\left(2\dfrac{1}{3}\right)\left(3\dfrac{1}{2}\right) =$$
$$\left(2 + \dfrac{1}{3}\right)\left(3 + \dfrac{1}{2}\right) =$$
$$\underline{(2)(3)} + (2)\left(\dfrac{1}{2}\right) + \left(\dfrac{1}{3}\right)(3) + \left(\dfrac{1}{3}\right)\left(\dfrac{1}{2}\right) =$$
$$\underline{6} + \underline{1} + \underline{1} + \dfrac{1}{6} = 8\dfrac{1}{6}$$ ❏

If we were to multiply mixed numbers in the form of mixed numbers, that is how we would have to do it. In Section 3.8 I recommended that you multiply mixed numbers as improper fractions.

EXAMPLE 5   Multiply $2\frac{1}{3}$ by $3\frac{1}{2}$ as improper fractions.

$$\left(2\frac{1}{3}\right)\left(3\frac{1}{2}\right) =$$

$$\left(\frac{7}{3}\right)\left(\frac{7}{2}\right) =$$

$$\frac{49}{6} = 8\frac{1}{6} \qquad \square$$

Do you understand why I recommended the multiplication of mixed numbers in the form of improper fractions? Isn't it simpler?

EXERCISES 13.2A

Use the FOIL method to multiply these binomials. Simplify your answers.

1. $(x - 3)(x + 1) =$ **2.** $(y + 4)(y + 3) =$

3. $(y + 2)(y + 7) =$ **4.** $(m - 2)(m + 5) =$

5. $(w + 7)(2w - 1) =$ **6.** $(3x + 1)(2x + 5) =$

7. $(4n - 3)(3n + 1) =$ **8.** $(6r - 1)(3r + 2) =$

9. $(x + 3)(x + 3) =$ **10.** $(7y + 1)(7y + 1) =$

11. $(4x - 1)(x + 7) =$ **12.** $(3y - 2)(y + 7) =$

Check your answers on page 754.

Squaring Binomials

To square anything means to multiply it by itself. So **squaring a binomial** means to multiply it by itself.

EXAMPLE 6 $(x + 3)^2$ denotes the squaring of the binomial $(x + 3)$.

$$(x + 3)^2 =$$

$$(x + 3)(x + 3) =$$

$$x^2 + 3\underline{x} + 3\underline{x} + 9 = x^2 + 6x + 9 \qquad \square$$

Too many people make the careless **error** of *simply squaring each term*. $(x + 3)^2 = x^2 + 6x + 9$; $(x + 3)^2$ does **not** equal $x^2 + 3^2$ or $x^2 + 9$. We have come too far for you to make that kind of mistake. You will not do so if you always take a second to rewrite the problem as *the binomial times itself* instead of performing the multiplication in your head from the exponential notation.

EXAMPLE 7   Express $(3x - 2)^2$ as $(3x - 2)(3x - 2)$ before multiplying.

$$(3x - 2)(3x - 2) =$$

$$(3x)(3x) + (3x)(-2) + (-2)(3x) + (-2)(-2) =$$

$$9x^2 - 6\underline{x} - 6\underline{x} + 4 = 9x^2 - 12x + 4 \qquad \square$$

Besides knowing from FOIL that this is the right way to calculate a binomial squared, we can visualize it in terms of geometric shapes.

As we learned in Section 3.12, the area of a square is calculated by multiplying the length of a side by itself. So the area of a square with side $(a + b)$ is

$$(a + b)^2 =$$
$$(a + b)(a + b) =$$
$$a^2 + \underline{ab} + \underline{ab} + b^2 = a^2 + 2ab + b^2.$$

We can visualize this by making a square in which each side is equal to length $a + b$:

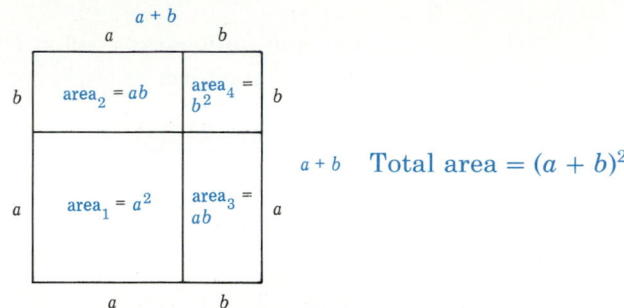

The area of this square is equal to the sum of the areas of the four blocks. So the area of a square with side $(a + b)$ is:

$$\text{Total area} =$$
$$\text{area}_1 + \text{area}_2 + \text{area}_3 + \text{area}_4 =$$
$$a^2 + ab + ab + b^2 = a^2 + 2ab + b^2.$$

But the area of that square with side $(a + b)$ is $(a + b)^2$. So we can see that $(a + b)^2 = a^2 + 2ab + b^2$.

EXERCISES 13.2B

Use FOIL to square each binomial. Combine like terms to leave your answers in the most simplified form.

13. $(x + 9)^2 =$ **14.** $(w - 1)^2 =$

15. $(x + 5)^2 =$ **16.** $(y - 4)^2 =$

17. $(n + 1)^2 =$ **18.** $(2w + 5)^2 =$

19. $(3m - 6)^2 =$ **20.** $(7k - p)^2 =$

21. $(2a - 3b)^2 =$ **22.** $(5x + 2y)^2 =$

23. $(5x - 2y)^2 =$ **24.** $(3n - 2p)^2 =$

Check your answers on page 754. ✓

Producing the Difference of Two Squares

We call a term that results from multiplying a term by itself a **perfect square**. The **difference of squares** is the name given to a binomial in which both terms are *perfect squares* and one of the terms is *negative*.

EXAMPLE 8   Here are some differences of squares:

$$a^2 - b^2; \, x^2 - 49; \, 9p^4 - 36; \, 4w^2 - 25m^4; \text{ and } y^2 - 1.$$ ❑

A difference of squares is always produced by multiplying two binomials that are exactly the same except that the sign of the second term is positive in one binomial and negative in the other.

EXAMPLE 9   The product of $(a - b)(a + b)$ will be a difference of two squares, since the other two terms add to zero.

$$(a - b)(a + b) =$$
$$a^2 + \underline{ab} - \underline{ab} - b^2 = a^2 - b^2 \qquad \square$$

EXAMPLE 10 $(3p^2 + 6)(3p^2 - 6)$ will be a difference of two squares.

$$(3p^2 + 6)(3p^2 - 6) =$$
$$(3p^2)(3p^2) - 6(3p^2) + 6(3p^2) + 6(-6) =$$
$$9p^4 - 18p^2 + 18p^2 - 36 = 9p^4 - 36 \qquad \square$$

EXAMPLE 11 $(y - 1)(y + 1)$ will be a difference of two squares.

$$(y - 1)(y + 1) =$$
$$y^2 + y - y - 1 = y^2 - 1 \qquad \square$$

EXERCISES 13.2C

Multiply each pair of binomials and simplify your answers.

25. $(x + 3)(x - 3) =$ **26.** $(z - 4)(z + 4) =$

27. $(k - 9)(k + 9) =$ **28.** $(y + 6)(y - 6) =$

29. $(3z + 4)(3z - 4) =$ **30.** $(7x - 2)(7x + 2) =$

31. $(5r - 1)(5r + 1) =$ **32.** $(4m - 3)(4m + 3) =$

33. $(6y + 7)(6y - 7) =$ **34.** $(2w + 9)(2w - 9) =$

35. $(x^2 - 5)(x^2 + 5) =$ **36.** $(3n^2 + 2p)(3n^2 - 2p) =$

37. $(2y^2 - x)(2y^2 + x) =$ **38.** $(5r^3 + 2n)(5r^3 - 2n) =$

Check your answers on page 754. ✓

Multiplying and Squaring Radical Binomials

We can also use FOIL to multiply binomials made up of radical terms.

EXAMPLE 12   Multiply and simplify $(\sqrt{3} + 5)(\sqrt{2} + 7)$.

$$(\sqrt{3} + 5)(\sqrt{2} + 7) =$$
$$(\sqrt{3})(\sqrt{2}) + (7)(\sqrt{3}) + (5)(\sqrt{2}) + (5)(7) = \sqrt{6} + 7\sqrt{3} + 5\sqrt{2} + 35 \qquad \square$$

After multiplying the terms, simplify each radical if possible.

EXAMPLE 13   Multiply and simplify $(\sqrt{6} + \sqrt{5})(\sqrt{2} - \sqrt{10})$.

$$(\sqrt{6} + \sqrt{5})(\sqrt{2} - \sqrt{10}) =$$
$$\sqrt{2 \cdot 3 \cdot 2} - \sqrt{2 \cdot 3 \cdot 2 \cdot 5} + \sqrt{5 \cdot 2} - \sqrt{5 \cdot 2 \cdot 5} = 2\sqrt{3} - 2\sqrt{15} + \sqrt{10} - 5\sqrt{2} \qquad \square$$

If any of the resulting radicals are the same, combine them to leave the answer in the simplest form. You may be surprised at how simple a big, intimidating expression can become!

EXAMPLE 14 Multiply and simplify $(2\sqrt{3} + 3\sqrt{5})(7\sqrt{3} - 2\sqrt{5})$.

$$(2\sqrt{3} + 3\sqrt{5})(7\sqrt{3} - 2\sqrt{5}) =$$
$$2(7)\sqrt{3 \cdot 3} - 2(2)\sqrt{3 \cdot 5} + 3(7)\sqrt{5 \cdot 3} - 3(2)\sqrt{5 \cdot 5} =$$
$$14(3) - 4\sqrt{15} + 21\sqrt{15} - 6(5) =$$
$$42 - 4\sqrt{15} + 21\sqrt{15} - 30 = 12 + 17\sqrt{15}$$

To square any polynomial means to multiply the polynomial by itself. You will avoid many careless errors if, before you begin multiplying the terms together, you write the polynomial twice when you are asked to square it.

EXAMPLE 15 Multiply and simplify $(\sqrt{2} + 1)^2$.

$$(\sqrt{2} + 1)^2 =$$
$$(\sqrt{2} + 1)(\sqrt{2} + 1) =$$
$$\sqrt{2 \cdot 2} + \sqrt{2} + \sqrt{2} + 1 =$$
$$2 + \sqrt{2} + \sqrt{2} + 1 = 3 + 2\sqrt{2}$$

EXAMPLE 16 Multiply and simplify $(2\sqrt{3} - \sqrt{5})^2$.

$$(2\sqrt{3} - \sqrt{5})^2 =$$
$$(2\sqrt{3} - \sqrt{5})(2\sqrt{3} - \sqrt{5}) =$$
$$2(2)\sqrt{3 \cdot 3} - 2\sqrt{3 \cdot 5} - 2\sqrt{5 \cdot 3} + \sqrt{5 \cdot 5} =$$
$$4(3) - 2\sqrt{15} - 2\sqrt{15} + 5 =$$
$$12 - 2\sqrt{15} - 2\sqrt{15} + 5 = 17 - 4\sqrt{15}$$

EXERCISES 13.2D

Multiply and simplify.

39. $(\sqrt{2} + \sqrt{3})(\sqrt{5} + \sqrt{7}) =$

40. $(\sqrt{3} + \sqrt{11})(\sqrt{2} - \sqrt{5}) =$

41. $(5 + \sqrt{3})(2 - \sqrt{2}) =$

42. $(7 - \sqrt{2})(3 - \sqrt{3}) =$

43. $(\sqrt{10} + \sqrt{14})(\sqrt{15} + \sqrt{7}) =$

44. $(\sqrt{6} + \sqrt{5})(\sqrt{6} - \sqrt{15}) =$

45. $(\sqrt{12} + \sqrt{6})(\sqrt{6} - \sqrt{3}) =$

46. $(\sqrt{10} - \sqrt{6})(\sqrt{15} - \sqrt{2}) =$

47. $(2\sqrt{7} + 3\sqrt{2})(5\sqrt{2} + 3\sqrt{7}) =$

48. $(3\sqrt{3} + 5\sqrt{2})(3\sqrt{2} - 2\sqrt{3}) =$

49. $(\sqrt{3} + \sqrt{2})^2 =$

50. $(\sqrt{7} + \sqrt{3})^2 =$

51. $(\sqrt{2} - \sqrt{3})^2 =$

52. $(\sqrt{5} - \sqrt{2})^2 =$

53. $(\sqrt{7} - \sqrt{5})^2 =$

54. $(\sqrt{3} + \sqrt{5})^2 =$

55. $(3\sqrt{2} + 2\sqrt{5})^2 =$

56. $(2\sqrt{5} + 5\sqrt{3})^2 =$

57. $(2\sqrt{3} - 5\sqrt{2})^2 =$

58. $(3\sqrt{2} - 2\sqrt{7})^2 =$

59. $(2\sqrt{2} - 3)^2 =$

60. $(5\sqrt{3} + 2)^2 =$

Check your answers on page 754.

13.3 Factoring Binomials That Are Differences of Two Squares

In order to simplify algebraic fractions (as we will do in Chapter 14) or to solve quadratic equations (as we will do in Chapter 15), we must be able to **factor polynomials**.

Factoring a polynomial means to write it as the product of two or more simpler expressions. In Section 5.5 we learned to factor a polynomial when there was some common factor in all terms of that polynomial.

EXAMPLE 1 $3xy - 5xw + x$ contains x as a common factor and so can be factored into $x(3y - 5w + 1)$. ❏

You should always check whether there is a common factor in a polynomial; if so, factor it out by using the distributive law. Remember, a common factor is one that appears in *every* term.

When *no common factor* exists, there are still other ways you might be able to factor a polynomial. One is to note whether the polynomial is a *difference of two squares*. (We will discuss another way in Section 13.4.)

We saw in Section 13.2 that a difference of squares is produced from binomials that are exactly the same except that the second term in one is *positive* while the second term in the other is *negative*.

If you recognize the polynomial to be factored as the difference of two perfect square terms, you can directly state its factors:

The first term of each factor is the *square root* of the *positive term* in the difference of squares.

The second term of each factor is a *square root* of the term that *follows the minus sign*. It is *positive* in one binomial and *negative* in the other.

In general we say: $(a^2 - b^2)$ factors into $(a + b)(a - b)$.

EXAMPLE 2 Factor $x^2 - 49$.

x^2 and 49 are both perfect squares, so $x^2 - 49$ is the difference of two squares.

The first term of each binomial factor is $\sqrt{x^2} = x$.

The second term of each binomial factor is $\pm\sqrt{49} = \pm 7$ (it is 7 in one factor and -7 in the other).

So $x^2 - 49$ factors into $(x + 7)(x - 7)$. ❏

You can always check whether you have correctly factored a polynomial: The factors, when multiplied together, *should produce the original polynomial*.

EXAMPLE 3 Check whether $x^2 - 49 = (x + 7)(x - 7)$.

Multiply by FOIL, we see that:

$$(x + 7)(x - 7) =$$
$$(x)(x) - 7(x) + 7(x) + 7(-7) =$$
$$x^2 - 7x + 7x - 49 =$$
$$x^2 + 0x - 49 = x^2 - 49 \checkmark$$

❏

EXAMPLE 4 Factor $4w^2 - 25m^4$. It is a difference of squares!

$$\sqrt{4w^2} = 2w \quad \text{and} \quad \pm\sqrt{25m^4} = \pm 5m^2$$

So $4w^2 - 25m^4 = (2w + 5m^2)(2w - 5m^2)$.

Check: Multiply $(2w + 5m^2)(2w - 5m^2)$ by FOIL.

Is that product equal to $4w^2 - 25m^4$?

$$(2w + 5m^2)(2w - 5m^2) =$$
$$\underline{(2w)(2w)} + \underline{(2w)(-5m^2)} + \underline{(5m^2)(2w)} + \underline{(5m^2)(-5m^2)} =$$
$$4w^2 - \underline{10m^2w} + \underline{10m^2w} - 25m^4 = 4w^2 - 25m^4 \quad \checkmark \quad \square$$

EXAMPLE 5 Factor $16y^2 - 1$. It is a difference of squares!

$$\sqrt{16y^2} = 4y \text{ and } \pm\sqrt{1} = \pm 1$$

So $16y^2 - 1 = (4y + 1)(4y - 1)$.

Check: Does $(4y + 1)(4y - 1) = 16y^2 - 1$?

$$(4y + 1)(4y - 1) =$$
$$(4y)(4y) - 1(4y) + 1(4y) + 1(-1) =$$
$$16y^2 - 4y + 4y - 1 = 16y^2 - 1 \quad \checkmark \quad \square$$

EXERCISES 13.3

Factor each of these differences of squares and check your answers.

1. $x^2 - 4 =$ **2.** $x^2 - 9 =$

3. $x^2 - 1 =$ **4.** $x^2 - 36 =$

5. $x^2 - 49 =$ **6.** $x^2 - 100 =$

7. $x^2 - 16 =$ **8.** $x^2 - 81 =$

9. $x^2 - 64 =$ **10.** $x^2 - 121 =$

11. $9x^2 - 25 =$ **12.** $4x^2 - 1 =$

13. $16w^2 - 9 =$ **14.** $49m^2 - 16 =$

15. $36n^2 - 25 =$ **16.** $81y^2 - 49 =$

17. $m^2 - p^2 =$ **18.** $x^2 - y^2 =$

19. $4w^2 - z^2 =$ **20.** $9y^2 - w^2 =$

21. $9a^2 - 16b^2 =$ **22.** $25z^2 - 9x^2 =$

23. $49p^2 - 36m^4 =$ **24.** $4k^4 - 9n^6 =$

25. $25m^6 - 16n^4 =$ **26.** $p^8 - 9m^2 =$

Check your answers on page 754.

 **Factoring Trinomials: $ax^2 + bx + c$**

Factoring Monics of Degree 2

In this section we learn how to factor trinomials that involve only one letter and no common factors. They will each include a **constant term**, a term with *one variable (letter) to the power* 1, and a term with *one variable (letter) squared*.

I'll refer to the term with the letter to the power 1 as the **linear term**, and the term with the letter squared as the **squared term**.

We can represent these trinomials in general as:

$$ax^2 + bx + c, \text{ where } a, b, \text{ and } c \text{ are numbers.}$$

In Section 5.1 we learned that the degree of a term is the number of variable (letter) factors in it. In Section 5.2 we defined the degree of a polynomial as the degree of the term with the largest degree. So in this section we learn to factor trinomials of degree 2.

To factor these polynomials, we will try to "unmultiply them by FOIL." Let's multiply two binomial factors by FOIL and see what observations we can make that will help us "undo" it.

EXAMPLE 1   Multiply $(x + 3)(x + 7)$ by FOIL.

$$(x + 3)(x + 7) =$$
$$x^2 + 7\underline{x} + 3\underline{x} + 21 = x^2 + 10x + 21 \qquad \square$$

If the two binomials each have the same kind of terms in them, (as those above *each have an x term and a constant term*), their product will usually be a trinomial with a squared, a linear, and a constant term.

If the terms of the trinomial are written in that order, we can make the following observations from FOIL about the binomial factors:

The *squared term* is the product of the First terms.

The *linear term* is the sum of the products of the Outside and Inside terms.

The *constant term* is the product of the Last terms.

To begin, we will look only at trinomials that are **monic**. If the coefficient of the term with the highest degree (most letter factors) is **1**, the polynomial is called **monic**. So trinomials of the form $ax^2 + bx + c$ are monic if $a = 1$. The squared term in a monic trinomial of degree 2 simply consists of some letter multiplied by itself. These observations lead to the following strategy.

Strategy to Factor ("Unmultiply") a Monic Trinomial [$x^2 + bx + c$] by FOIL:

1. Assume the trinomial will factor into two binomials, and so **set up two sets of parentheses**. (Each will contain a letter as its first term and a constant as its second term.)

2. Put one letter factor from the squared term inside each set of parentheses. This letter will be the first term of each binomial factor.

3. *Look at the constant term* to determine the signs of the constant terms of each binomial factor.

If it is *negative*, the constant terms in the bionomial factors must have *different* signs. Put a plus sign in one and a minus sign in the other.

If it is *positive*, the constant terms in the binomial factors must have the *same* sign. It will always be the sign of the trinomial's linear term.

4. Determine all possible *pairs of factors* of the constant term. Try each pair in the parentheses and multiply them by using FOIL until you find the pair that produces the trinomial's linear term as the sum of the products of the inside and outside terms.

5. Always *check* that your binomial factors, when multiplied together, do produce the original trinomial.

As we do some examples, you will see why this strategy works. Let's look at the different kinds of polynomials that we will factor.

Monics in Which All Terms Are Positive: [$x^2 + bx + c$, Where b and c Are Positive Numbers]

If the constant term is positive, the signs of the constants in the parentheses will be the same. Since their sum is the trinomial's linear term, and it is positive too, the signs in the parentheses must be positive. (Recall from Section 2.2 that when adding numbers of the *same sign*, we keep that sign for their sum. So if the sum is positive, each of the numbers must be positive.)

We begin to factor by placing the letter in each set of parentheses and following each by a $+$ sign.

EXAMPLE 2   Factor $x^2 + 7x + 6$. Begin with $(x + \ )(x + \ )$.

The pairs of factors of 6 are 3 and 2 and 1 and 6.

Try each pair to see which will produce $+7x$ as the linear term.

Try 3 and 2: $(x + 3)(x + 2) = x^2 + \mathbf{5x} + 6$.   No!
Try 1 and 6: $(x + 1)(x + 6) = x^2 + 7x + 6$.   Yes!

So $x^2 + 7x + 6$ factors into $(x + 1)(x + 6)$.   ❏

EXAMPLE 3   Factor $x^2 + 8x + 12$. Again, begin with $(x + \ )(x + \ )$.

The pairs of factors of 12 are 1 and 12, 3 and 4, and 2 and 6.

Which pair will produce $+8x$?

Try 1 and 12: $(x + 1)(x + 12) = x^2 + \mathbf{13x} + 12$.   No!
Try 3 and 4: $(x + 3)(x + 4) = x^2 + \mathbf{7x} + 12$.   No!
Try 2 and 6: $(x + 2)(x + 6) = x^2 + 8x + 12$.   Yes!
So $x^2 + 8x + 12$ factors into $(x + 2)(x + 6)$.   ❏

I was unlucky and tried the right one last in each of those examples. You usually will not have to try *all* the possibilities; once you find the right one, you are finished.

Monics with Positive Constant Terms But Negative Linear Terms: [$x^2 - bx + c$, Where b and c Are Positive Numbers]

If the constant term is positive but the linear term is negative, both signs in parentheses will be negative. (Again, recall from our rules of adding signed numbers that for the sum of two numbers of the same sign to be negative, they must each be negative.)

We begin to factor by placing the letter in each set of parentheses and following each by a $-$ sign.

EXAMPLE 4   Factor $x^2 - 12x + 20$. Begin with $(x - \quad)(x - \quad)$.

The pairs of factors of 20 are 1 and 20, 4 and 5, and 2 and 10.
Which pair will produce $-12x$?

$$(x - 1)(x - 20) = x^2 - 21x + 20. \quad \text{No!}$$
$$(x - 4)(x - 5) = x^2 - 9x + 20. \quad \text{No!}$$
$$(x - 2)(x - 10) = x^2 - 12x + 20. \quad \text{Yes!}$$

So $x^2 - 12x + 20$ factors into $(x - 2)(x - 10)$. ❑

There is a shortcut for finding the constant terms in the factors of *monic* trinomials that have *positive constant terms*. We observed that in FOIL, the linear term is the sum of the inside and outside products. So if the constant terms in the binomial factors have the same sign, their *sum* will always be the coefficient of the linear term.

Instead of trying all pairs of factors and multiplying by FOIL, we can just *add* each pair of factors of the constant term until we find the pair whose sum is equal to the coefficient of the trimonial's linear term. To avoid careless errors, always *check* that their product is the trinomial.

EXAMPLE 5   Factor $y^2 - 10y + 24$. Begin with $(y - \quad)(y - \quad)$.

Which pair of factors of 24 has the sum of 10?

The pairs of factors of 24 are:

1 and 24, with sum $= 25$;

2 and 12, with sum $= 14$;

3 and 8, with sum $= 11$;

4 and 6, with sum 10. ✓   So try this pair.

$$(y - 4)(y - 6) =$$
$$y^2 - 6y - 4y + 24 = y^2 - 10y + 24 \quad ✓$$

So $y^2 - 10y + 24$ factors into $(y - 4)(y - 6)$. ❑

EXAMPLE 6   Factor $w^2 + 17w + 30$. Begin with $(w + \quad)(w + \quad)$.

Which pair of factors of 30 has the sum of 17?

The pairs of factors of 30 are:

1 and 30, with sum $= 31$;

3 and 10, with sum $= 13$;

5 and 6, with sum $= 11$;

2 and 15, with sum $= 17$. ✓   So try this pair.

$$(w + 2)(w + 15) =$$
$$w^2 + 15w + 2w + 30 = w^2 + 17w + 30 \quad ✓$$

So $w^2 + 17w + 30$ factors into $(w + 2)(w + 15)$. ❑

It is easier to factor a trinomial if you list the terms so that the squared term is first, the linear term is second, and the constant term is third.

EXAMPLE 7   Factor $12w + 35 + w^2$.

First we rearrange the trinomial as $w^2 + 12w + 35$.

$$w^2 + 12w + 35 = (w + 5)(w + 7)$$

You must try *all* pairs of the constant term's factors until you get the right linear term in the product. If you try all but one pair, do not assume that the last pair will work; just as there are prime numbers that do not factor (11, 13, 37, and so on), *there are prime polynomials that do not factor either!*

If you systematically try all the possibilities and none works, you must conclude that the trinomial is prime and does not factor.

EXAMPLE 8   Factor $x^2 - 4x + 10$. Begin with $(x - \ )(x - \ )$.

Which pair of factors of 10 has the sum of 4?

The only pairs of factors of 10 are:

  1 and 10, with sum 11;

  2 and 5, with sum 7.

Neither of these pairs works, so $x^2 - 4x + 10$ is a *prime polynomial* that does *not* factor.

EXERCISES 13.4A

Factor each trinomial into a pair of prime binomial factors. Check whether your factors really do produce that trinomial by multiplying them out by FOIL. Check what the signs should be and then factor each polynomial.

1. $x^2 + 3x + 2 =$ **2.** $x^2 + 4x + 4 =$

3. $x^2 + 5x + 4 =$ **4.** $y^2 + 7y + 6 =$

5. $x^2 + x + 1 =$ **6.** $w^2 + 10w + 9 =$

7. $x^2 - 3x + 2 =$ **8.** $y^2 - 6y + 5 =$

9. $w^2 - 7w + 12 =$ **10.** $a^2 - 8a + 12 =$

11. $x^2 + 9x + 20 =$ **12.** $w^2 + 8w + 12 =$

13. $y^2 - 12y + 20 =$ **14.** $x^2 + 8x + 15 =$

15. $x^2 + x - 5 =$ **16.** $y^2 - 10y + 25 =$

Before trying to factor, rearrange the terms so that their order is the squared term, then the linear term, and finally the constant term.

17. $4 + x^2 - 4x =$ **18.** $-10y + y^2 + 9 =$

19. $21 + w^2 + 10w =$ **20.** $8w + 16 + w^2 =$

21. $18 + w^2 + 9w =$ **22.** $16 + w^2 - 8w =$

23. $4x + x^2 + 3 =$ **24.** $8y + y^2 + 7 =$

Check your answers on page 754.

Monics with Negative Constant Terms: [$x^2 + bx - c$, Where c Is a Positive Number]

If the constant term is negative, the signs in the parentheses will be different. Put a $+$ in one and a $-$ in the other. (Recall from the rules of signed numbers, Section 2.4, that if the product of two numbers is negative, the numbers had to have different signs.)

We begin to factor by placing the letter in each set of parentheses and following each by a different sign.

Try each pair of factors of the constant term until you find the one that will produce the coefficient of the trinomial's linear term.

EXAMPLE 9 Factor $x^2 + 2x - 15$. Begin with $(x + \ )(x - \ )$.

Which pair of factors will produce $+2x$?

The pairs of factors of 15 are 1 and 15 and 3 and 5.

$$\text{Try } \underline{1 \text{ and } 15}: (x + 1)(x - 15) =$$
$$x^2 - 15x + x - 15 = x^2 - 14x - 15. \quad \text{No!}$$

$$\text{Try } \underline{3 \text{ and } 5}: (x + 3)(x - 5) =$$
$$x^2 - 5x + 3x - 15 = x^2 - 2x - 15$$

The coefficient of x is the right *number*, 2, but the sign is wrong: It is supposed to be $+2$. This will happen when we use the *right factors* but place them in the set of parentheses with the wrong sign. Try switching each number to the other set of parentheses.

$$\text{Try } \underline{3 \text{ and } 5} \text{ again}: (x + 5)(x - 3) =$$
$$x^2 - 3x + 5x - 15 = x^2 + 2x - 15 \quad \checkmark \quad \text{That worked!}$$

So $x^2 + 2x - 15$ factors into $(x + 5)(x - 3)$. ❏

There is a shortcut for factoring monic trinomials with *negative* constant terms that will avoid the problems of having to switch signs later. Since the signs are different, when we add the products of the inside and outside terms in FOIL, their sum will always be the *difference* of those numbers (just as it was their *sum* when their signs were the same).

We look for the pair of factors whose difference is the coefficient of the linear term. The sign of that difference is always the sign of the number with the larger magnitude, so we place the larger number in the set of parentheses with the sign of the linear term.

It is because of algebraic strategies like this that I encourage you to master computations with signed numbers. Memorizing this strategy would be difficult, but if you know the rules of signed numbers, you will understand the strategy and be able to *remember* it after much practice.

If you need a review of these concepts, go back to Chapter 2 before proceeding.

Remember, if the *constant term is negative*, the signs will be *different*. Find the pair of factors whose difference is equal to the coefficient of the linear term.

EXAMPLE 10 Factor $y^2 - 4y - 21$. Begin with $(y + \ )(y - \ )$.

Which pair of factors of 21 have a difference of 4?

The pairs of factors of 21 are:

1 and 21, with difference **20**;

3 and 7, with difference 4. ✓ So try these.

The coefficient of the linear term is *negative* 4, so put the larger factor (7) in with the **negative** sign.

$$(y + 3)(y - 7) =$$
$$y^2 - 7y + 3y - 21 = y^2 - 4y - 21 \quad \checkmark$$

So $y^2 - 4y - 21$ factors into $(y + 3)(y - 7)$.

EXAMPLE 11   Factor $w^2 + w - 6$. Begin with $(w +)(w -)$.

Which pair of factors has a difference of 1? $(w = 1w)$

The pairs of factors of 6 are:

1 and 6, with difference **5**.

2 and 3, with difference **1**. $\checkmark$ So try these.

The coefficient of the linear term (w) is *positive*, so put the larger factor (3) in with the **positive sign**.

$$(w + 3)(w - 2) =$$
$$w^2 - 2w + 3w - 6 = w^2 + w - 6 \quad \checkmark$$

So $w^2 + w - 6$ factors into $(w + 3)(w - 2)$.

Use the method you are most comfortable with to factor polynomials.

EXERCISES 13.4B

Factor each trinomial and check your answers.

25. $x^2 + x - 6 =$

26. $x^2 - 2x - 8 =$

27. $y^2 - y - 12 =$

28. $w^2 - 6w - 16 =$

29. $a^2 + a - 12 =$

30. $b^2 + 3b - 18 =$

31. $y^2 - y - 30 =$

32. $w^2 + 4w - 5 =$

33. $x^2 + 2x - 35 =$

34. $m^2 - 4m - 12 =$

35. $x^2 - 7x + 12 =$

36. $2y + y^2 - 15 =$

37. $-21 + 4x + x^2 =$

38. $-18 + x^2 - 7x =$

39. $3x - 18 + x^2 =$

40. $5x - 14 + x^2 =$

41. $y^2 + 12y - 13 =$

42. $w^2 - 3w - 88 =$

43. $y^2 + 9y + 18 =$

44. $w^2 - 5w + 4 =$

45. $x^2 - 4x - 32 =$

46. $y^2 + 11y + 30 =$

47. $x^2 + x - 42 =$

48. $10x + x^2 + 25 =$

49. $x^2 + 4x - 5 =$

50. $6x + 9 + x^2 =$

51. $2x + x^2 - 63 =$

52. $y^2 + 12 + 7y =$

Check your answers on page 754.

Factoring Non-Monics: [$ax^2 + bx + c$, Where $a \neq 1$]

If the coefficient of the term with the highest degree in a polynomial is not 1, that polynomial is called **non-monic**. So far all the polynomials we have factored by unmultiplying by FOIL were monics. The strategy we have used applies only to monics.

Now that you have become skilled at factoring monics, you are ready to learn how to factor non-monic trinomials of degree 2.

EXAMPLE 12   Here are some non-monic trinomials of degree 2:

$$-x^2 - 3x - 2; 2y^2 + 7y + 3; -7n^2 + 16n - 4;$$
$$4p^2 + 8p + 3; \text{ and } 6w^2 - 25w - 25. \qquad \square$$

The easiest non-monics to factor are the ones with -1 as the coefficient of the squared term. We can simply factor a -1 out of every term, representing the original trinomial as the product of -1 and a monic polynomial that you can then factor.

This procedure will give you three factors: -1 and the two binomial factors of the resulting monic trinomial. We usually do not write out -1 but represent it simply by a minus sign.

EXAMPLE 13   Factor $-x^2 - 3x - 2$.

$$-x^2 - 3x - 2 =$$
$$-1(x^2 + 3x + 2) = -(x + 2)(x + 1) \qquad \square$$

You can factor -1 out of every term even if the terms are not all negative. Factoring -1 out is the same as dividing each term by -1: It *changes the sign* of each term.

EXAMPLE 14   Factor $-y^2 + 5y - 6$.

$$-y^2 + 5y - 6 =$$
$$-(y^2 - 5y + 6) = -(y - 3)(y - 2) \qquad \square$$

The main difficulty in factoring non-monics occurs when the coefficient of the squared term is not -1 but something like -2 or 2 or 15. For instance, suppose you have to factor $2y^2 + 7y + 3$. One of the letter terms in the binomial factors will have to contain a non-monic coefficient. One binomial will have $2y$ as its first term, while the other will have y.

That non-monic letter term affects the Inside or Outside product, so we cannot use any shortcuts to determine which pair of factors of the constant term will give us the appropriate linear term. We have to try every possible pair of factors and multiply them out, and we have to do it *twice for each pair*. That is because we will get different Inside and Outside terms when we place each candidate in the set of parentheses with $2y$ compared with when we put it in the one with y.

Begin by setting up two sets of parentheses again, and place one linear letter term in each. Choose terms that are factors of the non-monic squared term. For example, if the squared term is $2y^2$, begin with

$$(2y \quad)(y \quad).$$

Use the same guide for determining the *signs* as the one we use for monics.

EXAMPLE 15   Factor $2y^2 + 7y + 3$. Begin with $(2y + \quad)(y + \quad)$.

Fortunately the only pair of factors of 3 is 1 and 3.

First try: $(2y + 3)(y + 1) =$
$$2y^2 + 2y + 3y + 3 = 2y^2 + 5y + 3$$

This produced the wrong linear term ($5y$). We needed $7y$.

Now try again, switching each number to the other set of parentheses:

$$(2y + 1)(y + 3) =$$
$$2y^2 + 6y + y + 3 = 2y^2 + 7y + 3, \text{ which is right.} \quad \checkmark$$

See what a difference the choice of *where* we put each number makes with non-monics?

So $2y^2 + 7y + 3$ factors into $(2y + 1)(y + 3)$. ❏

When there is more than one pair of factors to try, I suggest that you test each combination in each set of parentheses before discarding that possibility. You do not want to decide prematurely that a pair will not work.

EXAMPLE 16 Factor $3x^2 + 17x + 10$. Begin with $(3x + \;)(x + \;)$.

The pairs of factors of 10 are 1 and 10 and 2 and 5.

$$(3x + 1)(x + 10) =$$
$$3x^2 + 30x + x + 10 = 3x^2 + 31x + 10 \quad \text{No, so let's switch them.}$$
$$(3x + 10)(x + 1) =$$
$$3x^2 + 3x + 10x + 10 = 3x^2 + 13x + 10 \quad \text{No, so discard that choice of}$$
$$\text{factors and try 5 and 2.}$$

$$(3x + 5)(x + 2) =$$
$$3x^2 + 6x + 5x + 10 = 3x^2 + 11x + 10 \quad \text{No, so let's switch them.}$$
$$(3x + 2)(x + 5) =$$
$$3x^2 + 15x + 2x + 10 = 3x^2 + 17x + 10 \quad \text{Yes!} \quad \checkmark$$

So $3x^2 + 17x + 10$ factors into $(3x + 2)(x + 5)$. ❏

To use the guide for setting up the signs, the *squared term must be positive.* If it is not, factor out -1, as we did in Examples 13 and 14. To test whether a pair of factors will work, just multiply out the binomial factors and compare the result to the polynomial that resulted after we had factored out the -1.

EXAMPLE 17 Factor $-7n^2 + 16n - 4$. First factor out the -1.

$$-7n^2 + 16n - 4 = -(7n^2 - 16n + 4) \quad \text{Set that polynomial up}$$
$$\text{as } -(7n - \;)(n - \;).$$

The pairs of factors of 4 are 1 and 4 and 2 and 2.

$$-(7n - 1)(n - 4) = -(7n^2 - 29x + 4) \quad \text{No, so try switching them.}$$

$$-(7n - 4)(n - 1) = -(7n^2 - 11n + 4) \quad \text{No, we can discard that pair of}$$
$$\text{factors and try 2 and 2.}$$

$$-(7n - 2)(n - 2) = -(7n^2 - 16n + 4) \quad \text{Yes!} \quad \checkmark$$

So $-7n^2 + 16n - 4$ factors into $-(7n - 2)(n - 2)$. ❏

When the constant term is negative, the signs in the parentheses will be different. Put a plus sign in one and a negative sign in the other. Then proceed as we did in the last four examples until the linear term's coefficient is correct. If the sign of that linear term comes out wrong, just switch the signs from one set of parentheses to the other.

Always check by multiplying your factors out.

EXAMPLE 18   Factor $5x^2 + 13x - 6$. Let's begin with $(5x +)(x -)$.
The pairs of factors of 6 are 1 and 6 and 2 and 3.

$(5x + 6)(x - 1) = 5x^2 + x - 6$ No, so let's switch the 1 and 6.

$(5x + 1)(x - 6) = 5x^2 - 29x - 6$ No, we can discard that choice.

$(5x + 3)(x - 2) = 5x^2 - 7x - 6$ No, so let's switch the 3 and 2.

$(5x + 2)(x - 3) = 5x^2 - 13x - 6$ The number is right, but not the sign, so let's switch the signs.

$(5x - 2)(x + 3) = 5x^2 + 13x - 6$ Yes! ✓

So $5x^2 + 13x - 6$ factors into $(5x - 2)(x + 3)$.   ❑

EXERCISES 13.4C

Factor each trinomial and check your answers.

53. $2x^2 + 3x + 1 =$ **54.** $3x^2 + 4x + 1 =$

55. $5x^2 + 11x + 2 =$ **56.** $7y^2 + 4y - 3 =$

57. $3x^2 + 8x - 3 =$ **58.** $5x^2 - 34x - 7 =$

59. $-2w^2 + 5w + 3 =$ **60.** $2y^2 - 3y - 9 =$

61. $3x^2 + 8x + 4 =$ **62.** $-2x^2 + x + 10 =$

63. $-x^2 + 9x + 5 =$ **64.** $7y^2 + 11y - 6 =$

65. $5w^2 - 6w + 1 =$ **66.** $2x^2 + 13x - 7 =$

Check your answers on page 755. ✓

Non-Monics with Non-prime Coefficients of the Squared Term: [$ax^2 + bx + c$, Where a Is Not a Prime Number]

One more complication in factoring non-monic trinomials may occur. All the trinomials in the previous section contained prime numbers as coefficients for the squared term, and so there was only one pair of possibilities for the first terms of the binomial factors.

If the squared term was $2y^2$, the first terms were $2y$ and y.

If the squared term was $3x^2$, the first terms were $3x$ and x.

But now let's see what happens when the squared term's coefficients are not primes. There will be more than one way to divide them for the first terms.

Unless you have a functioning crystal ball, there is no way to know which way is right at the beginning of the problem.

So try one way and test all the possibilities for the constant terms as we have been doing. But if none of them works, instead of assuming that the polynomial cannot be factored, you will have to begin again with each possible choice for the letter terms and go through the list of possible constants again until you find the right one. You will test each possibility once this way.

In this book, I promise to ask you to factor non-monics with nonprime coefficients of squared terms only if they can be factored. So keep going until you find the combination that works.

EXAMPLE 19 Factor $4p^2 + 7p + 3$.

$4p^2$ factors into $2p$ and $2p$ or $4p$ and p.

Let's begin with $(2p + \ )(2p + \ )$.

The only pair of factors of 3 is 1 and 3.

$$(2p + 1)(2p + 3) = 4p^2 + 8p + 3 \quad \text{No, that wasn't it.}$$

In this case both letter terms are the same, so there is no need to switch the numbers. We would get the same result. So we can discard that choice and try $4p$ and p.

Now let's try $(4p + \ )(p + \ )$.

$$(4p + 1)(p + 3) = 4p^2 + 13p + 3 \quad \text{No, so switch the 1 and 3.}$$
$$(4p + 3)(p + 1) = 4p^2 + 7p + 3 \quad \text{Yes!} \ \checkmark$$

So $4p^2 + 7p + 3$ factors into $(4p + 3)(p + 1)$. ❏

EXAMPLE 20 Factor $6w^2 - 25w - 25$.

$6w^2$ factors into $2w$ and $3w$ and $6w$ and w.

25 factors into 1 and 25 and 5 and 5.

Let's begin with $(3w + \ )(2w - \ )$.

$$(3w + 1)(2w - 25) = 6w^2 - 73w - 25 \quad \text{No, so switch the 1 and 25.}$$
$$(3w + 25)(2w - 1) = 6w^2 + 47w - 25 \quad \text{No, so discard this choice.}$$
$$(3w + 5)(2w - 5) = 6w^2 - 5w - 25 \quad \text{No, and since both constants are the same here, there is no use switching them. Discard this choice too.}$$

Now let's try $(w + \ )(6w - \ )$.

$$(w + 1)(6w - 25) = 6w^2 - 19w - 25 \quad \text{No, so switch the 1 and 25.}$$
$$(w + 25)(6w - 1) = 6w^2 + 149w - 25 \quad \text{No, so discard this choice.}$$
$$(w + 5)(6w - 5) = 6w^2 + 25w - 25 \quad \text{The number is right, but the sign is wrong, so switch the signs.}$$

$$(w - 5)(6w + 5) = 6w^2 - 25w - 25 \quad \checkmark$$

So $6w^2 - 25w - 25$ factors into $(w - 5)(6w + 5)$. ❏

In summary—here is how to factor non-monic trinomials:

1. If the squared term is *negative*, factor out a -1 from every term.
2. Factor the squared term into pairs each containing one letter. Choose one pair of factors for the first terms in each set of parentheses. Be prepared to try the other pair(s) later if necessary.

3. Look at the signs of the constant and linear terms to decide what signs should be put into the parentheses.

4. Try putting each pair of factors of the constant term into the parentheses. If a pair of factors does not produce the trinomial you started with, try switching them to the other set of parentheses before discarding that choice of factors.

5. Multiply out each pair of binomials until the correct trinomial is produced. If no pair of factors of the constant worked with your original choice for the letter terms, go through the list again with the other choice.

6. Be patient and careful. Be sure you are well organized about testing all the choices so that you do not leave any out (it might be the right one) and so you know when you have tried all of them.

EXERCISES 13.4D

Factor each trinomial and check your answers.

67. $10x^2 + x - 3 =$

68. $6x^2 + x - 2 =$

69. $4x^2 - 4x - 3 =$

70. $4x^2 + x - 5 =$

71. $8x^2 - 14x + 3 =$

72. $6x^2 + 13x + 2 =$

73. $6x^2 + 23x - 4 =$

74. $10x^2 + 11x - 6 =$

75. $4x^2 - 4x - 15 =$

76. $12x^2 - x - 6 =$

Check your answers on page 755.

Factoring Trinomials, Difference of Squares, and Polynomials with Common Factors

You now know how to factor polynomials by three methods:

Common factor, difference of squares, and unmultiplying by FOIL.

When factoring a polynomial, you have to look at it carefully and decide which of the three methods of factoring applies to that particular one.

Here's a guide for factoring:

1. Always try the common factor method first. Check whether there is a factor common to all terms. If there is, factor it out by the distributive law:

$$ax + aw + ay = a(x + w + y).$$

2. If there is no common factor, check whether you have a binomial that is a difference of squares. If so, find the square root of each term and write the factors from them:

$$a^2 - b^2 = (a - b)(a + b).$$

3. If neither of those two methods applies, try unmultiplying the polynomial by FOIL:

$$[(\quad)(\quad)].$$

To determine the signs of the constant terms in each binomial factor, look at the trinomial's constant term:

If it is *negative*, the constant terms in the binomial factors must have *different* signs.

Put a $+$ in one and a $-$ in the other.

If it is *positive*, the constant terms in the binomial factors must have the *same* sign. Put in the sign of the trinomial's linear term.

4. Some polynomials are primes and cannot be factored. Be sure that you do not jump too quickly to the conclusion that your polynomial cannot be factored, without first exhausting all the possibilities!

EXAMPLE 21 $x^2 - x - 4$ is prime and does not factor.   ❑

We say a polynomial is *prime* if there are no integers that, when tried as the constants (or coefficients) in the binomial factors, will produce that polynomial.

Once you get enough practice in recognizing the right method and using each method to factor polynomials, you will find it easy to replace polynomials by their factors.

Here is a collection of different polynomials that you can factor. I have recommended the method to use in each example.

EXAMPLE 22 $x^3 + x^2 + x = x(x^2 + x + 1)$; $(x^2 + x + 1)$ is a prime polynomial.
Common factor   ❑

EXAMPLE 23 $x^2 - 4 = (x - 2)(x + 2)$ Difference of squares   ❑

EXAMPLE 24 $x^2 + 6x + 9 = (x + 3)(x + 3)$ or $(x + 3)^2$ Unmultiplying by FOIL   ❑

EXAMPLE 25 $x^2 - 9x + 20 = (x - 4)(x - 5)$ Unmultiplying by FOIL   ❑

EXAMPLE 26 $x^2 + 4x - 45 = (x + 9)(x - 5)$ Unmultiplying by FOIL   ❑

EXAMPLE 27 $x^2 - 3x - 4 = (x - 4)(x + 1)$ Unmultiplying by FOIL   ❑

EXAMPLE 28 $-x^2 + 4x + 45 =$
$-(x^2 - 4x - 45) = -(x - 9)(x + 5)$ Unmultiplying by FOIL   ❑

EXAMPLE 29 $7x^2 + 10x + 3 = (7x + 3)(x + 1)$ Unmultiplying by FOIL   ❑

EXAMPLE 30 $10x^2 + 11x - 6 = (2x + 3)(5x - 2)$ Unmultiplying by FOIL   ❑

EXAMPLE 31 $9a^2 - 25p^4 = (3a + 5p^2)(3a - 5p^2)$ Difference of squares   ❑

EXERCISES 13.4E

Factor each polynomial and check your answers.

77. $x^2 + 5x + 6 =$ **78.** $y^2 - 6y + 8 =$

79. $w^2 - 2w - 15 =$ **80.** $x^2 + 2x - 35 =$

81. $y^2 - 2y - 35 =$ **82.** $x^2 - 16 =$

83. $xy + xz - xw =$ **84.** $x^2 - 4x - 32 =$

85. $ax + ay - aw =$ **86.** $2x^2 - 3x + 1 =$

87. $25w^2 - 9 =$ **88.** $4m^4 - 9w^2 =$

89. $15x^2 - x - 2 =$ **90.** $4w^2 - 7w + 3 =$

91. $y^2 - 2y - 24 =$ **92.** $x^2y^2 + 7xy^2 - 4xyw =$

93. $6x^2 + 7x - 10 =$ **94.** $6x^2 - 5x - 6 =$

Check your answers on page 755.

13.5 Factoring Complicated Polynomials

Factoring Polynomials with More Than One Letter by FOIL

Now that you have mastered the process of factoring by FOIL in polynomials with only one letter, we can generalize the process and apply it to polynomials with more than one letter.

EXAMPLE 1   Factor $x^2 + 2xy + y^2$.

There is no common factor, and it is not a difference of squares, so let's try to unmultiply it by FOIL.

There are two squared terms, x^2 and y^2. One of them is the product of the binomial factors' *First* terms when they are multiplied by FOIL. To begin, let's let x^2 be that term: $(x \quad)(x \quad)$.

Since all three terms are positive, both signs in parentheses will be positive too: $(x + \quad)(x + \quad)$.

There is no constant term, but the third term is y^2. It will be the product of the binomial factors' *Last* terms when they are multiplied by FOIL. So put the factors (y and y) into the parentheses as we did when we had a constant term: $(x + y)(x + y)$.

Check out the product!

$$(x + y)(x + y) =$$
$$x^2 + xy + xy + y^2 = x^2 + 2xy + y^2 \quad \checkmark$$

So $x^2 + 2xy + y^2$ factors into $(x + y)(x + y)$.   ❑

Whether or not the third term is a constant, find its factors for the second terms of the binomial factors. Use the same analysis for the signs that we used in Section 13.4.

EXAMPLE 2 Factor $w^2 - 3wp + 2p^2$. Begin with $(w -\ \ )(w -\ \ )$.

The factors of $2p^2$ are p and $2p$. Put them in and see if those factors work.

$$(w - p)(w - 2p) =$$
$$w^2 - \mathbf{2pw} - \mathbf{pw} + 2p^2 = w^2 - 3wp + 2p^2 \quad \checkmark$$

So $w^2 - 3wp + 2p^2$ factors into $(w - p)(w - 2p)$. ☐

EXAMPLE 3 Factor $m^2 - 3mn - 4n^2$. Begin with $(m +\ \ )(m -\ \ )$.

Divide $4n^2$ into $2n$ and $2n$ or $4n$ and n.

$$(m + 2n)(m - 2n) =$$
$$m^2 - \mathbf{2mn} + \mathbf{2mn} - 4n^2 = m^2 - 4n^2 \quad \text{No, so try } 4n \text{ and } n.$$
$$(m - 4n)(m + n) =$$
$$m^2 + \mathbf{mn} - \mathbf{4mn} - 4n^2 = m^2 - 3mn - 4n^2 \quad \checkmark$$

So $m^2 - 3mn - 4n^2$ factors into $(m - 4n)(m + n)$. ☐

If the first term is monic, we can use the shortcut for factoring monics that we observed in Section 13.4. In Example 3, the trinomial is monic and the signs are different. The coefficients of the terms, $4n$ and n, do have a difference of 3 (which is the coefficient of the middle term). So we should have tried that combination first.

EXAMPLE 4 Factor $x^2 - 8xy + 12y^2$. Begin with $(x -\ \ )(x -\ \ )$.

The pairs of factors of $12y^2$ are:

$$12y \text{ and } y, \quad 4y \text{ and } 3y, \text{ and } 6y \text{ and } 2y.$$

Since the signs are the same for this monic, the pair we want is the one with coefficients whose *sum* is -8. So let's try $6y$ and $2y$.

$$(x - 6y)(x - 2y) =$$
$$x^2 - \mathbf{2xy} - \mathbf{6xy} + 12y^2 = x^2 - 8xy + 12y^2 \quad \checkmark$$

So $x^2 - 8xy + 12y^2$ factors into $(x - 6y)(x - 2y)$. ☐

Whenever you have a trinomial without a common factor, you can always try to factor it by "unmultiplying by FOIL."

EXERCISES 13.5A

Factor each trinomial by unmultiplying by FOIL.

1. $x^2 - 2xy + y^2 =$ 2. $x^2 + 3xw + 2w^2 =$

3. $y^2 + 5yz + 4z^2 =$ 4. $w^2 - 3wx + 2x^2 =$

5. $w^2 + 7wy + 10y^2 =$ 6. $y^2 - 4yw - 21w^2 =$

7. $x^2 - 5xy + 6y^2 =$ 8. $m^2 + 6mn - 7n^2 =$

9. $p^2 - pm - 6m^2 =$ 10. $k^2 + 2kw - 35w^2 =$

Check your answers on page 755. ✓

Polynomials Factored More Than Once to Get Prime Factors

When we factor large numbers (as we did in Section 3.3), we begin by factoring each into two factors and continue to factor them until all factors are prime.

EXAMPLE 5 To factor 40 into a product of prime factors, we might begin by factoring 40 into $4 \cdot 10$ and proceed from there.

$$40 =$$
$$(4) \cdot (10) =$$
$$(2 \cdot 2)(2 \cdot 5) = 2 \cdot 2 \cdot 2 \cdot 5$$

We can illustrate this concept with a factoring tree:

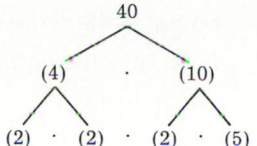

Whether we factor numbers or polynomials, we always want to continue until each factor is a prime.

To factor a polynomial, always apply the three methods of factoring in this order:

1. Factor out any common factor first.
2. If you recognize any binomial factor as a difference of squares, use that method next.
3. Try to factor any trinomial factors that remain by unmultiplying by FOIL.

To check, multiply any pair(s) of factors together, and then multiply those results with any factor(s) not yet multiplied, until one product results. The final product should be the original polynomial.

EXAMPLE 6 Factor $x^3 + 6x^2 + 5x$.

There is a common factor of x in every term. Factor it out.

$$x^3 + 6x^2 + 5x = x(x^2 + 6x + 5)$$

The factor x is a prime, but let's see if we can factor the trinomial $x^2 + 6x + 5$ by FOIL.

$$x(x^2 + 6x + 5) = x(x + 5)(x + 1)$$

So $x^3 + 6x^2 + 5x$ factors into a product of the three prime factors:

$$x, \ x + 5, \text{ and } x + 1.$$

We can illustrate this concept with a factoring tree:

$$x^3 + 6x^2 + 5x$$
$$(x) \ \cdot \ (x^2 + 6x + 5)$$
$$(x) \ \cdot \ (x + 5) \ \cdot \ (x + 1)$$

So $x^3 + 6x^2 + 5x = (x)(x + 5)(x + 1)$.

Check: Does $x(x + 5)(x + 1) = x^3 + 6x^2 + 5x$?

$$x(x + 5)(x + 1) =$$
$$(x^2 + 5x)(x + 1) =$$
$$x^3 + x^2 + 5x^2 + 5x = x^3 + 6x^2 + 5x \quad \checkmark$$

EXAMPLE 7 Factor $5z^3 - 45y^2z$.

There are common factors of 5 and z in each term, so factor them out first.

$$5z^3 - 45y^2z = 5z(z^2 - 9y^2)$$

The third factor, $(z^2 - 9y^2)$, is a difference of squares, so factor it by that method.

$$5z(z^2 - 9y^2) = 5z(z - 3y)(z + 3y)$$

So $5z^3 - 45y^2z$ factors into $5z(z - 3y)(z + 3y)$.

It factors into a product of the four prime factors:

$$5, z, z - 3y, \text{ and } z + 3y.$$

Check: Does $5z(z - 3y)(z + 3y) = 5z^3 - 45y^2z$?

$$5z(z - 3y)(z + 3y) =$$
$$5z(z^2 + 3yz - 3yz - 9y^2) =$$
$$5z(z^2 - 9y^2) = 5z^3 - 45y^2z \quad \checkmark$$

EXAMPLE 8 Factor $5x^3y - 15x^2y + 10xy$.

Each term has common factors of 5, x, and y. Factor all of them out and then factor the remaining trinomial by FOIL.

$$5x^3y - 15x^2y + 10xy =$$
$$5xxxy - 3 \cdot 5xxy + 2 \cdot 5xy =$$
$$5xy(\mathbf{x^2 - 3x + 2}) = 5xy(\mathbf{x - 1})(\mathbf{x - 2})$$

So $5x^3y - 15x^2y + 10xy$ factors into the five prime factors of:

$$5, x, y, x - 1, \text{ and } x - 2.$$

Check: Does $5xy(x - 1)(x - 2) = 5x^3y - 15x^2y + 10xy$?

$$5xy(x - 1)(x - 2) =$$
$$5xy(x^2 - 2x - x + 2) =$$
$$5xy(x^2 - 3x + 2) = 5x^3y - 15x^2y + 10xy \quad \checkmark$$

EXAMPLE 9 Factor $a^5 + 2a^3b + ab^2$.

First factor out the common factor, a.

$$a^5 + 2a^3b + ab^2 = a(a^4 + 2a^2b + b^2)$$

Let's try unmultiplying by FOIL on $a^4 + 2a^2b + b^2$.

$$a(a^4 + 2a^2b + b^2) = a(a^2 + b)(a^2 + b)$$

Check:

$$a(a^2 + b)(a^2 + b) =$$
$$(a^3 + ab)(a^2 + b) =$$
$$(a^3)(a^2) + (a^3)(b) + (ab)(a^2) + (ab)(b) =$$
$$a^5 + a^3b + a^3b + ab^2 = a^5 + 2a^3b + ab^2 \quad \checkmark$$

So $a^5 + 2a^3b + ab^2$ does factor into $a(a^2 + b)(a^2 + b)$.

As these examples show, there are many complicated-looking polynomials that you can factor by using the three methods of common factor, difference of squares, and unmultiplying by FOIL.

Now that you can factor polynomials, you will be able to simplify and compute with *algebraic* fractions (you will learn how to do so in Chapter 14), and you will be able to solve many equations of a degree higher than 1 (you will learn how to do so in Chapter 15).

EXERCISES 13.5B

Factor each polynomial into a product of prime factors. Check by multiplying your factors together.

11. $3x^2 + 12x + 9 =$

12. $2x^2 - 6x - 20 =$

13. $5x^2 - 45 =$

14. $7y^2 - 28 =$

15. $x^3 + 5x^2 + 6x =$

16. $y^4 - 6y^3 + 8y^2 =$

17. $8w^2 - 50 =$

18. $12m^3 - 75m =$

19. $2x^3 + 14x^2 + 20x =$

20. $5y^3 - 35y^2 + 60y =$

21. $8y^3 - 2y =$

22. $50x^2y^3 - 18x^2y =$

23. $2x^2y - 10xy - 12y =$

24. $5m^2n^3 + 10m^2n^2 + 5m^2n =$

25. $12xy^3 - 27xy =$

26. $20w^3m - 45wm^3 =$

27. $3ab^2 - 12ab - 36a =$

28. $4x^3 + 13x^2 + 3x =$

29. $x^4y + 2x^2y^2 + y^3 =$

30. $2yw^3 + 5y^2w^3 - 3y^3x =$

Check your answers on page 755.

Glossary

constant term Term that is only a number, with no variable (letter) factors.

Example: In the polynomial $ax^2 + bx + c$, where a, b, and c are numbers, c is the constant term.

difference of squares Binomial in which both terms are perfect squares and only one of the terms is negative.

Example: $a^2 - b^2$ is a difference of squares.

FOIL Letters representing the four sets of terms that are multiplied when two binomials are multiplied:

First + Outside + Inside + Last

F O I L

Example: $(a + b)(c + d) = ac + ad + bc + bd$

We use FOIL to help factor trinomials like $ax^2 + bx + c$.

linear term Term containing one factor that is a letter with the exponent 1; it is a 1-degree term.

Example: In the polynomial $ax^2 + bx + c$, where a, b, and c are numbers, bx is the linear term.

monic polynomial Polynomial in which the coefficient of the term with the largest degree (most letter factors) is 1.

Example: $x^2 - 5x + 7$ is a monic polynomial.

non-monic polynomial Polynomial in which the coefficient of the term with the largest degree (most letter factors) is *not* 1.

Example: $3x^2 - 5x + 7$ is a non-monic polynomial.

squared term Term containing a letter raised to the power 2; it is a 2nd degree term.

Example: In the polynomial $ax^2 + bx + c$, ax^2 is the squared term.

squaring a binomial Multiplying a binomial by itself.

Example: $(a + b)^2 = (a + b)(a + b)$

Chapter 13 Review

Do each of the problems. After you are confident that you have done them as accurately as possible, compare your answers with those in the back of the book. If any are wrong, go back to the section where the problem came from (indicated to the left of the problem) and review the section.

Once you have mastered all the sections, you will be sure to learn the skills solidly and remember how to do them if you practice them a little more. Turn to the Supplementary Exercises and do all the problems from any section where you had difficulty on these review exercises.

Multiply and simplify.

13.1   **1.** $(x + 4)(2x^2 - 5x + 7) =$

   2. $(4y^2 - 6y + 1)(y^2 + 2y - 5) =$

13.2   **3.** $(w - 5)(w + 8)$ **4.** $(3y - 7)(2y - 3) =$

   5. $(x - 4)^2 =$ **6.** $(3y + 5)^2 =$

   7. $(4y - 3)(4y + 3) =$ **8.** $(\sqrt{5} + 3\sqrt{2})(\sqrt{5} - 4\sqrt{2}) =$

Factor.

13.3   **9.** $w^2 - 144 =$ **10.** $4y^2 - 49x^2 =$

13.4   **11.** $y^2 + 7y + 10 =$ **12.** $w^2 - 9w + 20 =$

   13. $x^2 - 5x - 6 =$ **14.** $k^2 + 4k - 77 =$

   15. $3x^2 + 10x + 8 =$ **16.** $12y^2 + 11y - 15 =$

13.5   **17.** $9x^2 + 6x - 24 =$ **18.** $x^3 - 9x^2 + 14x =$

   19. $18x^2 - 8$ **20.** $100x^4 - 36x^2 =$

   21. $w^2 + 7wn + 12n^2 =$ **22.** $2x^3y + 4x^2y - 30xy =$

Answers

1. _____
2. _____
3. _____
4. _____
5. _____
6. _____
7. _____
8. _____
9. _____
10. _____
11. _____
12. _____
13. _____
14. _____
15. _____
16. _____
17. _____
18. _____
19. _____
20. _____
21. _____
22. _____

Check your answers on page 755.

Supplementary Exercises

Do all the problems in every section that involves a skill in which you now lack complete mastery. With a little more practice you will achieve that sense of really understanding the topics, and you will remember how to do these problems.

Multiply and simplify.

13.1

1. $(y - 2)(y^2 + 3y + 4) =$

2. $(3w - 5)(2w^2 - 3w + 1) =$

3. $(x^2 - 2x + 5)(x^2 + x - 3) =$

4. $(y^2 + 6y - 1)(y^2 - 2y + 4) =$

13.2

5. $(x - 4)(x - 3) =$

6. $(w - 5)(w + 2) =$

7. $(4x - 1)(3x + 6) =$

8. $(7y - 4)(2y - 3) =$

9. $(w - 7)^2 =$

10. $(y + 6)^2 =$

11. $(2x - 5)^2 =$

12. $(3b + 4)^2 =$

13. $(y - 8)(y + 8) =$

14. $(7x - 5)(7x + 5) =$

15. $(2\sqrt{3} - 7)(5\sqrt{3} + 1) =$

16. $(\sqrt{7} + \sqrt{2})(\sqrt{3} + \sqrt{5}) =$

17. $(5\sqrt{2} - 2\sqrt{3})(3\sqrt{2} - 4\sqrt{3}) =$

Factor.

13.3

18. $y^2 - 81 =$

19. $m^2 - 25n^2 =$

13.4

20. $w^2 + 8w + 15 =$

21. $y^2 - 5y - 14 =$

22. $x^2 - 5x - 6 =$

23. $m^2 + 7m + 12 =$

24. $k^2 + 4k - 45 =$

25. $p^2 - 10p + 21 =$

Answers

1. _____
2. _____
3. _____
4. _____
5. _____
6. _____
7. _____
8. _____
9. _____
10. _____
11. _____
12. _____
13. _____
14. _____
15. _____
16. _____
17. _____
18. _____
19. _____
20. _____
21. _____
22. _____
23. _____
24. _____
25. _____

Continued

Supplementary Exercises (*Cont.*)   Answers

26. $5y^2 + 13y - 6 =$ **27.** $2x^2 - x - 10 =$

28. $6y^2 + 5y - 6$ **29.** $8w^2 - 26w + 15 =$

30. $7x^2 + 8x - 12 =$ **31.** $4y^2 - 11y + 6 =$

13.5 **32.** $2y^3 + 10y^2 + 8y =$ **33.** $6x^3 - 150x =$

34. $3x^4 - 27x^2$ **35.** $6x^3 - 15x^2 - 9x =$

36. $y^2 + 4yx - 5x^2 =$ **37.** $w^2 - 3wm - 10m^2 =$

38. $xy^2 - 5xy - 14x =$ **39.** $2wx^2 - 14wx + 24w =$

40. $6x^3y - 9x^2y - 15xy =$ **41.** $12p^3m - 27pm^3 =$

26. _____

27. _____

28. _____

29. _____

30. _____

31. _____

32. _____

33. _____

34. _____

35. _____

36. _____

37. _____

38. _____

39. _____

40. _____

41. _____

Check your answers on page 755.

Chapter 13 Test

Multiply each of these polynomials and combine like terms. Leave your answers in the simplest form.

1. $(x - 5)(x + 6) =$

2. $(3y - 4)(2y - 5) =$

3. $(w + 3)(w^2 + 4w + 7) =$

4. $(4y^2 - 3y + 1)(3y^2 - y - 1) =$

5. $(6n - 1)(6n + 1) =$

6. $(3x + 5y)(3x - 5y) =$

7. $(\sqrt{7} - 2)(3\sqrt{7} + 1) =$

8. $(y - 4)^2 =$

9. $(2x - 5)^2 =$

Factor each polynomial into a product of prime factors.

10. $10y + 35w =$

11. $14x^3y^2 - 28xy^3 + 35x^4y^2 =$

12. $w^2 + 11w + 30 =$

13. $y^2 - 2y - 24 =$

14. $x^2 - 7x + 10 =$

15. $k^2 + 4k - 21 =$

16. $4w^2 - 4w - 24 =$

17. $x^3 - 6x^2 + 8x =$

18. $2x^3 - 8x^2 + 6x =$

19. $p^2 - 1 =$

20. $9x^2 - 25 =$

21. $6x^3 - 3x^2 - 45x =$

22. $6w^2 - 96 =$

23. $75x^5 - 108x^3 =$

24. $p^2 + 4pk + 4k^2 =$

25. $4x^2y - 14xy + 12y =$

Answers

1. _____
2. _____
3. _____
4. _____
5. _____
6. _____
7. _____
8. _____
9. _____
10. _____
11. _____
12. _____
13. _____
14. _____
15. _____
16. _____
17. _____
18. _____
19. _____
20. _____
21. _____
22. _____
23. _____
24. _____
25. _____

Check your answers on page 755.

Cumulative Review
CHAPTERS 1—13

Combine and simplify.

Answers

1. $\dfrac{7}{10} - \dfrac{2}{5} + \dfrac{2}{15} =$

2. $\dfrac{\dfrac{12}{40}}{\dfrac{33}{50}}$

3. $14.013 - 6.0048 =$

4. $(15.3)(0.005) =$

5. $0.0028 \div 0.002 =$

6. 4.3% of $128 =$

7. $(-14) - (3) + (-11) + (-12) =$

8. $(-4)(2)(-1)(-1)(-3) =$

9. $\left(-\dfrac{40}{15}\right) \cdot \left(\dfrac{6}{30}\right) =$

10. $3 - 6(8 - 4 - 1) + 7(3 - 5)(8 - 1 - 4) =$

11. $4(3x^2 + x - 3) + 5(x^2 - 4x - 1) =$

12. $7w^3(3w^2 - 5w + 1) =$

13. $(w - 5)(w - 2) =$

14. $(3y - 2)(y^2 - y + 5) =$

15. $\dfrac{10a^4b^2c}{24ab^3c^4} =$

16. $(-a^3b)^{-2} =$

Solve each equation.

17. $7x - 3 = 2(x + 4) + 4 =$

18. $3.1x - 5.8 = 4.3x + 2.6$

Solve each system of equations.

19. $y = 5x - 3$
    $4x - y = 5$

20. $3x + 5y = 9$
    $2x - 10y = -34$

Find the slope and y-intercept for each line.

21. $4y = 12x - 20$

22. $7y - 3 = 5x$

What is the equation of the line that

23. has slope -5 and goes through point $(-2, 7)$?

24. passes through points $(-3, 5)$ and $(2, 20)$?

Solve the following problems.

25. If the ratio of people with dark eyes to people with light eyes is 7:5, how many people out of a group of 132 have dark eyes?

Answers

1. _____

2. _____

3. _____

4. _____

5. _____

6. _____

7. _____

8. _____

9. _____

10. _____

11. _____

12. _____

13. _____

14. _____

15. _____

16. _____

17. _____

18. _____

19. _____

20. _____

21. _____

22. _____

23. _____

24. _____

25. _____

26. What is the final balance in a checking account that began with $250 after the following transactions: check for $14.78, check for $53.40, check for $125, check for $50, check for $19.83, check for $7.03, deposit $50, check for $15, check for $100, deposit $150, and a service charge of $3.42?

27. (a) How far did I travel if I averaged a speed of 50 mph and drove for 12 hr? (b) If my car gets 30 mi per gallon, how many gallons of gas did the trip require? (c) If each gallon cost $1.25, what was the cost of fuel for the trip?

28. How much did Anne invest in each savings account if she had $1,200 to invest, and invested some at 6% and the rest at 7%, if her interest in one year totaled $76?

29. If after one month 2 goldfish die for every 5 that survive, how many fish should you buy if you hope to end up with 30 live ones at the end of one month?

30. If I win 60% of my squash games, how many games have I played when I have won 12 games?

Answers

26. _____

27a. _____

b. _____

c. _____

28. _____

29. _____

30. _____

Check your answers on page 755.

14 Algebraic Fractions (Rational Expressions)

14.1 Simplifying Algebraic Fractions

Algebraic Factors

In Section 7.3 we studied **ratios**, which can be expressed as fractions. With that in mind, we can say an **algebraic fraction** is the *ratio of two algebraic expressions*. Mathematicians call the ratio of two polynomials a **rational expression**.

EXAMPLE 1
$$\frac{a+3}{5}, \frac{2x-1}{3}, \frac{y}{n}, \frac{3w-2}{n-6}, \text{ and } \frac{5m-1}{3k}$$

are algebraic fractions (rational expressions). ❑

The rules for computing with algebraic fractions are the same as the rules we learned in Chapter 3 for computing with numerical fractions. If you feel at all unsure about working with fractions, go back and review Chapter 3 before proceeding.

Many fractional computations rely on recognizing factors.

EXAMPLE 2 The fraction $\frac{3xy(5w-2)}{2w}$ has

Four *factors* in the numerator: $3, x, y$, and $(5w-2)$;

Two *factors* in the denominator: 2 and w. ❑

When algebraic fractions contain polynomials within their numerators or denominators, people sometimes confuse factors of a term with factors of the numerator or denominator itself. When working with fractions, we are usually concerned with factors of the entire numerator or the entire denominator.

A **factor** of a **numerator** (or **denominator**) is a term or polynomial that, when multiplied by another term(s) or polynomial(s), produces that numerator (or denominator).

When a polynomial is a prime factor of a numerator (or denominator), each of its terms (and the factors of each term) is *not* itself a factor of that numerator (or denominator) but is a *part of a factor*.

EXAMPLE 3 In $\frac{3xy(5w-2)}{2w}$,

w is a factor of the *denominator*;

w is *not* a factor of the *numerator*;

it is only a *part* of the factor $(5w+2)$. ❑

When algebraic fractions contain polynomials, we must try to factor the polynomials. If they cannot be factored, then they are prime factors of the fraction.

EXAMPLE 4 Express the numerator and denominator as a product of prime factors:
$$\frac{x^2 - 4}{x^2 - 3x - 4}.$$

$$\frac{x^2 - 4}{x^2 - 3x - 4} = \frac{(x + 2)(x - 2)}{(x - 4)(x + 1)}$$ ❑

After factoring each polynomial, check whether each factor is prime. If any factors are not prime, then continue to factor them until they are prime.

EXAMPLE 5 Express the numerator and denominator as a product of prime factors:
$$\frac{3x^3w - 12xw^3}{y^4 + 3y^3 + 2y}.$$

$$\frac{3x^3w - 12xw^3}{y^4 + 3y^3 + 2y} =$$

$$\frac{3xw(x^2 - 4w^2)}{y^2(y^2 + 3y + 2)} = \frac{3xw(x - 2w)(x + 2w)}{yy(y + 2)(y + 1)}$$

The numerator has five factors: 3, x, w, $(x - 2w)$, and $(x + 2w)$.

The denominator has four factors: y, y, $(y + 2)$, and $(y + 1)$. ❑

Remember, you know three methods of factoring polynomials (from Chapter 13). Try all three methods, listed below, before assuming that a polynomial is prime. These methods are:

1. Common factor
2. Difference of squares
3. Unmultiplying by FOIL

EXERCISES 14.1A

State how many prime factors there are.

1. $2x(x + 5)$ 2. $3xy$

3. $7x^2y$ 4. $6xy^3$

5. $a^2b(c - 5)$ 6. $x(x + 1)(x - 7)$

7. $7(x + 5)(x - 4)(3x + 1)$ 8. $3x(x - 2)(2x + 5)(x + 7)$

Factor each numerator and denominator into a product of prime factors. State how many prime factors are in each numerator and how many are in each denominator.

9. $\dfrac{7xy}{5yw} =$ 10. $\dfrac{3x^2y(x + 2)}{3wy(x - 2)} =$

11. $\dfrac{2x(y + 5)(x - 3)}{3y(x - 3)(y + 5)} =$ 12. $\dfrac{7(x + 4)(x - 3)}{5x(x - 3)(x + 3)} =$

13. $\dfrac{3x^2(x-2)}{5x(x+1)(x+2)} =$

14. $\dfrac{14y^2x^3(x-1)(x+5)}{21yx^2(x+1)(x-1)} =$

15. $\dfrac{20xy^2(x+2y)}{15x^2y(2x+y)} =$

16. $\dfrac{12x^2yw^3(x+3)}{15xy^5(w+3)} =$

17. $\dfrac{(y-3)(x^2+2x)}{(x-3)(y^2+2y)} =$

18. $\dfrac{(5x+5)(y-4)}{(3y-12)(x-1)} =$

19. $\dfrac{x^2-4}{x^2+3x+2} =$

20. $\dfrac{x^2-4x+4}{x^2-9x+14} =$

Check your answers on page 756. ✓

Simplifying Algebraic Fractions

A **simplified fraction** is one in which no factor of its numerator is also a factor of its denominator.

To simplify a fraction, leaving it in its most reduced form, we factor the numerator and the denominator and cancel all factors common to both.

EXAMPLE 6   Simplify $\dfrac{10}{12}$.

$$\frac{10}{12} =$$

$$\frac{\cancel{2}\cdot 5}{\cancel{2}\cdot 2\cdot 3} = \frac{5}{6} \qquad ❏$$

If a fraction is already factored, like the one below, we can simply cancel any factor common to the numerator and the denominator.

EXAMPLE 7   Simplify $\dfrac{3(x+1)}{(x-3)(x+1)}$.

$(x+1)$ is a common factor that we should cancel.

$$\frac{3\cancel{(x+1)}}{(x-3)\cancel{(x+1)}} = \frac{3}{x-3}$$

The 3s cannot be canceled, because 3 is not a factor of the denominator; it is a part of the factor $(x-3)$. ❏

EXAMPLE 8   Simplify $\dfrac{x^2-9}{x^2+3x}$.

$$\frac{x^2-9}{x^2+3x} =$$

$$\frac{(x-3)(x+3)}{x(x+3)} =$$

$$\frac{(x-3)\cancel{(x+3)}}{x\cancel{(x+3)}} = \frac{x-3}{x} \qquad ❏$$

An algebraic factor is equal to another algebraic factor only if all their parts are exactly alike. This means that the numbers, letters, exponents, and signs must be the same.

Remember, the *order* of terms does not matter.

EXAMPLE 9 $(x + 2)$ is the same as $(2 + x)$.

$(a + 2b - c)$ is the same as $(2b - c + a)$.

$(7y^2 - x)$ is *not* the same as $(7x^2 - y)$.

$(2x^2 - 3)$ is the same as $(-3 + 2x^2)$. ❏

If two algebraic factors look exactly alike, except that the sign of *every* term in one is the opposite of the same term's sign in the other, then one factor is the *negative of the other*. A minus 1 can be factored out of every term of one algebraic factor to represent it more clearly as the negative of the other factor.

EXAMPLE 10 $(a - b + c)$ is the negative of $(-a + b - c)$.

$$(-a + b - c) = -(a - b + c)$$ ❏

To simplify algebraic fractions, never cancel a part of a factor, but always cancel common factors. Answers should *always* be left in factored form except for the coefficients; this way it will be obvious that the fraction has been left simplified.

Exponents can be used to represent repeated factors.

EXAMPLE 11   Simplify $\dfrac{2x(y - 3)^2(5w - 4)}{2y(y + 3)(4 - 5w)}$.

$$\frac{2x(y - 3)^2(5w - 4)}{2y(y + 3)(4 - 5w)} =$$

$$\frac{2x(y - 3)(y - 3)(-1)(-5w + 4)}{2y(y + 3)(4 - 5w)} =$$

$$\frac{(-1)2x(y - 3)(y - 3)(-5w + 4)}{2y(y + 3)(4 - 5w)} = \frac{-x(y - 3)^2}{y(y + 3)}$$ ❏

Remember, if all the factors in either the numerator or the denominator cancel out, that numerator or denominator becomes 1.

EXAMPLE 12   Simplify $\dfrac{7ab(3a + 2b - 5c)}{7abc(2b - 5c + 3a)}$.

$$\frac{7ab(3a + 2b - 5c)}{7abc(2b - 5c + 3a)} = \frac{1}{c}$$ ❏

Fractions with denominators of 1 are no longer fractions and should be expressed simply as their numerators.

EXAMPLE 13   Simplify $\dfrac{(x^2 - 7x + 10)}{(x - 5)}$.

$$\frac{(x^2 - 7x + 10)}{(x - 5)} =$$

$$\frac{(x - 5)(x - 2)}{(x - 5)} =$$

$$\frac{[(x - 5)(x - 2)]}{(x - 5)} =$$

$$\frac{x - 2}{1} = x - 2$$ ❏

Be sure that both the numerator and the denominator are expressed as a product of prime factors before trying to simplify the fraction.

EXAMPLE 14 Simplify $\dfrac{3x^5y - 27x^3y}{x^3y^2 + 5x^2y^2 + 6xy^2}$.

$$\frac{3x^5y - 27x^3y}{x^3y^2 + 5x^2y^2 + 6xy^2} =$$

$$\frac{3x^3y(x^2 - 9)}{xy^2(x^2 + 5x + 6)} =$$

$$\frac{3x^3y(x - 3)(x + 3)}{xy^2(x + 3)(x + 2)} =$$

$$\frac{3xxxyy(x - 3)(x + 3)}{xyy(x + 3)(x + 2)} = \frac{3x^2(x - 3)}{y(x + 2)}$$

EXERCISES 14.1B

State which pairs of factors are equal.

21. $x + 5; \quad 5 + x$ **22.** $y - 4; \quad 4 - y$

23. $3 - m; \quad -m + 3$ **24.** $4x - y; \quad y - 4x$

25. $3n + 2m; \quad 2n + 3m$ **26.** $7a + 2b; \quad 2b + 7a$

27. $2a + 3b - 5; \quad 3b - 5 + 2a$ **28.** $3x + 4y - 7w; \quad 4y - 3x + 7w$

Factor each fraction. Cancel common factors from each numerator and denominator to reduce the fraction. Leave fractions in factored form.

29. $\dfrac{7xy}{5yw} =$ **30.** $\dfrac{3x^2y(x + 2)}{3wy(x - 2)} =$

31. $\dfrac{2x(y + 5)(x - 3)}{3y(x - 3)(y + 5)} =$ **32.** $\dfrac{7(x + 4)(x - 3)}{5x(x - 3)(x + 3)} =$

33. $\dfrac{3x^2(x - 2)}{5x(x + 1)(x + 2)} =$ **34.** $\dfrac{14y^2x^3(x - 1)(x + 5)}{21yx^2(x + 1)(x - 1)} =$

35. $\dfrac{20xy^2(x + 2y)}{15x^2y(2x + y)} =$ **36.** $\dfrac{12x^2yw^3(x + 3)}{15xy^5(w + 3)} =$

37. $\dfrac{x(y - 3)(x + 2)}{y(x + 2)(x - 3)} =$ **38.** $\dfrac{5(x + 1)(y - 4)}{3(y - 4)(x - 1)} =$

39. $\dfrac{x^2 - 4}{x^2 + 3x + 2} =$ **40.** $\dfrac{x^2 - 4x + 4}{x^2 - 9x + 14} =$

41. $\dfrac{x - 5}{x^2 - 8x + 15} =$ **42.** $\dfrac{x^2 - 4x + 3}{x - 1} =$

43. $\dfrac{y^2 - 2y - 15}{y^2 - 25} =$ **44.** $\dfrac{6x^2 - 9x}{x^2 - 7x + 12} =$

45. $\dfrac{x^2y^2 - 4x^2}{xy^2 - 5xy + 6x} =$ **46.** $\dfrac{8y^3 - 4y^2 - 12y}{2y^2 + 4y + 2} =$

47. $\dfrac{3x^2 + 13x - 10}{x^2 y - 25y} =$

48. $\dfrac{x^3 y^2 - 7x^3 y}{y^2 - 49} =$

49. $\dfrac{2x^2 - 5x - 3}{4x^2 + 4x + 1} =$

50. $\dfrac{x^2 y^2 - x^2}{3y^2 + 9y + 6} =$

Check your answers on page 756. ✓

Dividing Polynomials

Now that you know how to factor polynomials, you can use those skills to divide some polynomials.

Division problems can always be set up as fractions. To divide one polynomial by another, you can express the dividend (the one being divided into) as the numerator and the divisor (the one dividing into the other) as the denominator.

If the polynomials can be factored, and if they have any factors in common that can be canceled, the resulting simplified fraction represents their quotient.

EXAMPLE 15 Divide $x^2 - 9x + 20$ by $x^2 - 25$ as $\dfrac{x^2 - 9x + 20}{x^2 - 25}$.

$$\frac{x^2 - 9x + 20}{x^2 - 25} =$$

$$\frac{(x - 5)(x - 4)}{(x - 5)(x + 5)} =$$

$$\frac{\cancel{(x - 5)}(x - 4)}{\cancel{(x - 5)}(x + 5)} = \frac{x - 4}{x + 5}$$

EXAMPLE 16 Divide $y^2 + 2y - 15$ by $y^2 - 6y + 9$.

$$\frac{y^2 + 2y - 15}{y^2 - 6y + 9} =$$

$$\frac{(y + 5)(y - 3)}{(y - 3)(y - 3)} =$$

$$\frac{(y + 5)\cancel{(y - 3)}}{(y - 3)\cancel{(y - 3)}} = \frac{(y + 5)}{(y - 3)}$$

EXAMPLE 17 Divide $w^2 + 6w$ by $w^3 + 7w^2 + 6w$.

$$\frac{w^2 + 6w}{w^3 + 7w^2 + 6w} =$$

$$\frac{w(w + 6)}{w(w^2 + 7w + 6)} =$$

$$\frac{w(w + 6)}{w(w + 6)(w + 1)} =$$

$$\frac{\cancel{w(w + 6)}}{\cancel{w(w + 6)}(w + 1)} = \frac{1}{w + 1}$$

EXERCISES 14.1C

Perform each indicated division.

51. $(x^2 - 1) \div (x^2 + 2x + 1) =$ **52.** $(x^2 + 3x + 2) \div (x^2 + 2x) =$

53. $(y^2 - y - 12) \div (y^3 - 8y^2 + 16y) =$

54. $(w^3 - 4w) \div (wx + 2x) =$

55. $(x^2 + 5x + 6) \div (x^2 - 9) =$ **56.** $(w^3 - 5w^2) \div (xw + 4w) =$

57. $(y^2 - 7y) \div (y^2 - 8y + 7) =$ **58.** $(x^3 - x^2 - 12x) \div (x^2 + 3x) =$

59. $(w^2 + 5w + 6) \div (2w^3 + 10w^2 + 12w) =$

60. $(x^2 + x) \div (x^3 - x) =$

Check your answers on page 756.   ✓

14.2   Multiplication

Multiplying Factored Fractions

To multiply fractions, multiply the numerators to produce the numerator of their product, and multiply the denominators to produce the denominator of their product.

Since fractions must always be left in the most simplified form, factor all numerators and denominators first.

When multiplying, remember that a factor appearing in *any* numerator cancels out if it also appears in *any* denominator. So place all numerator factors in the numerator, place all denominator factors in the denominator, and cancel where you can.

EXAMPLE 1   Multiply $\dfrac{8}{9}$ by $\dfrac{12}{20}$.

$$\left(\frac{8}{9}\right)\left(\frac{12}{20}\right) =$$

$$\frac{(8)(12)}{(9)(20)} =$$

$$\frac{(2 \cdot 2 \cdot 2)(2 \cdot 2 \cdot 3)}{(3 \cdot 3)(2 \cdot 2 \cdot 5)} =$$

$$\frac{\not{2} \cdot \not{2} \cdot 2 \cdot 2 \cdot 2 \cdot \not{3}}{\not{3} \cdot 3 \cdot \not{2} \cdot \not{2} \cdot 5} =$$

$$\frac{2 \cdot 2 \cdot 2}{3 \cdot 5} = \frac{8}{15}$$

Algebraic fractions are multiplied the same way numerical fractions are. But because they should be left in factored form,* there is *no need to multiply together the algebraic factors* that remain after all the canceling is done.

---

* In our answer we usually multiply together any numerical factors rather than leave them in factored form.

Be sure you cancel only *factors* and not *parts* of factors. Before canceling, I usually list numerical factors first, then single-letter factors, and finally any prime polynomial factors.

EXAMPLE 2 Multiply and simplify $\left[\dfrac{3x(y+1)}{(y+2)}\right]\left[\dfrac{5(y+2)}{2xy(y-3)}\right]$.

$$\left[\frac{3x(y+1)}{(y+2)}\right]\left[\frac{5(y+2)}{2xy(y-3)}\right] =$$

$$\frac{[(3x)(y+1)][5(y+2)]}{[(y+2)][2xy(y-3)]} =$$

$$\frac{3\cdot5\cdot x\cdot(y+1)\cdot(y+2)}{2\cdot x\cdot y\cdot(y+2)\cdot(y-3)} =$$

$$\frac{3\cdot5\cdot\cancel{x}\cdot(y+1)\cdot\cancel{(y+2)}}{2\cdot\cancel{x}\cdot y\cdot\cancel{(y+2)}\cdot(y-3)} =$$

$$\frac{3\cdot5\cdot(y+1)}{2\cdot y\cdot(y-3)} = \frac{15(y+1)}{2y(y-3)} \qquad ❑$$

If one factor is the negative of the other, be sure to factor out -1 from one of them so that the similar factors can cancel.

EXAMPLE 3 Multiply and simplify $\left[\dfrac{(3x-5)}{(2x+1)}\right]\left[\dfrac{(2x-1)}{(5-3x)}\right]$.

$$\left[\frac{(3x-5)}{(2x+1)}\right]\left[\frac{(2x-1)}{(5-3x)}\right] =$$

$$\frac{(3x-5)(2x-1)}{(2x+1)(5-3x)} =$$

$$\frac{-(-3x+5)(2x-1)}{(2x+1)(5-3x)} =$$

$$-\frac{\cancel{(-3x+5)}(2x-1)}{(2x+1)\cancel{(5-3x)}} = -\frac{2x-1}{2x+1} \qquad ❑$$

EXERCISES 14.2A

Multiply these algebraic fractions. Cancel common factors and leave your answers in the most simplified factored form.

1. $\dfrac{3xy}{x+2}\cdot\dfrac{x+2}{5y} =$

2. $\dfrac{(x-3)}{(x+4)}\cdot\dfrac{(x-4)}{(x-3)} =$

3. $\dfrac{4xy}{5y(x+1)}\cdot\dfrac{3xy^2}{(x+2)} =$

4. $\dfrac{7x(y+3)}{5(3+y)}\cdot\dfrac{4(x-2)(x+3)}{21(3+x)(x+2)} =$

5. $\dfrac{5(2x-1)}{3(4x+3)}\cdot\dfrac{12y(4x+3)}{20y(1+2x)} =$

6. $\dfrac{(x-3)(x+2)}{4x(x+2)}\cdot\dfrac{7x^2y}{(x+3)(x-3)} =$

7. $\dfrac{4x^2y(x+2)}{10xy^3(x-4)}\cdot\dfrac{(x+2)(x-4)}{2(2+x)} =$

8. $\dfrac{4(x-2)}{y(x+3)}\cdot\dfrac{xy(x+2)(x+3)}{10(x+2)(x-2)} =$

Check your answers on page 756. ✓

Factoring Fractions Before Canceling

Represent the numerator and denominator as a product of prime factors before canceling.

EXAMPLE 4   Multiply and simplify:

$$\left[\frac{(x^2 + 7x)}{(x^2 - 5x)}\right]\left[\frac{(x^2 + 4x + 4)}{(x^2 + 3x + 2)}\right] =$$

$$\frac{(x^2 + 7x)(x^2 + 4x + 4)}{(x^2 - 5x)(x^2 + 3x + 2)} =$$

$$\frac{x(x + 7)(x + 2)(x + 2)}{x(x - 5)(x + 2)(x + 1)} =$$

$$\frac{\cancel{x}(x + 7)\cancel{(x + 2)}(x + 2)}{\cancel{x}(x - 5)\cancel{(x + 2)}(x + 1)} = \frac{(x + 7)(x + 2)}{(x - 5)(x + 1)}$$

EXAMPLE 5   Multiply and simplify:

$$\left[\frac{(x^2 + 5x + 4)}{(x^2 - 1)}\right]\left[\frac{(x^3 - 3x^2 + 2x)}{(x^2 + 4x)}\right] =$$

$$\frac{(x^2 + 5x + 4)(x^3 - 3x^2 + 2x)}{(x^2 - 1)(x^2 + 4x)} =$$

$$\frac{(x + 4)(x + 1)x(x^2 - 3x + 2)}{(x + 1)(x - 1)x(x + 4)} =$$

$$\frac{\cancel{x}\cancel{(x + 4)}\cancel{(x + 1)}(x - 2)\cancel{(x - 1)}}{\cancel{x}\cancel{(x + 1)}\cancel{(x - 1)}\cancel{(x + 4)}} = x - 2$$

The process for multiplying fractions is the same no matter how many fractions are to be multiplied. Remember, for two factors to cancel, one must appear in a numerator and the other in a denominator. In the example below, the $(y - 3)$ factors do not cancel because they both appear in the numerator.

EXAMPLE 6   Multiply and simplify:

$$\frac{y^2 + y}{y^2 + 4y + 3} \cdot \frac{y^2 - y - 6}{y^2 - 7y} \cdot \frac{y^2 - 9}{y^2 + 2y + 1} =$$

$$\frac{(y^2 + y)(y^2 - y - 6)(y^2 - 9)}{(y^2 + 4y + 3)(y^2 - 7y)(y^2 + 2y + 1)} =$$

$$\frac{y(y + 1)(y - 3)(y + 2)(y - 3)(y + 3)}{(y + 3)(y + 1)y(y - 7)(y + 1)(y + 1)} =$$

$$\frac{\cancel{y}\cancel{(y + 1)}(y - 3)(y + 2)(y - 3)\cancel{(y + 3)}}{\cancel{y}\cancel{(y + 3)}\cancel{(y + 1)}(y - 7)(y + 1)(y + 1)} =$$

$$\frac{(y - 3)(y + 2)(y - 3)}{(y - 7)(y + 1)(y + 1)} = \frac{(y - 3)^2(y + 2)}{(y - 7)(y + 1)^2}$$

EXERCISES 14.2B

Find the product of these fractions. Factor and cancel, and leave your answers in factored form.

9. $\dfrac{x^2 - 49}{x^2 - 2x + 1} \cdot \dfrac{x^2 + 4x + 3}{x^2 + 8x + 7} =$

10. $\dfrac{x^2 - 2x - 8}{x^2 + 6x} \cdot \dfrac{x^2 - 36}{x^3 + 2x^2} =$

11. $\dfrac{x^2 - 5x - 6}{x^2y - 3xy} \cdot \dfrac{x^3 - 9x}{x^2 - 8x + 12} =$

12. $\dfrac{y^2 - 7y}{x^2y^2 - 2xy^2} \cdot \dfrac{x^2 - 5x + 6}{y^2 - 6y - 7} =$

13. $\dfrac{x^2 + 6x + 9}{x^2 - 5x} \cdot \dfrac{x^3 - 25x}{x^2 + 8x + 15} =$

14. $\dfrac{x^2 + 9x + 20}{x^2 + 8x + 16} \cdot \dfrac{x^2 + 6x + 8}{x^2 + 12x + 35} =$

15. $\dfrac{x^2 + 9x + 18}{x^2 + 6x - 16} \cdot \dfrac{x^2 - 3x + 2}{x^2 + 12x + 36} =$

16. $\dfrac{4x^3 - 10x^2}{x^2 - 8x + 7} \cdot \dfrac{x^2 - 1}{4x^2 - 25} =$

17. $\dfrac{6x^2 + 5x + 1}{10x^2 - 6x} \cdot \dfrac{2x^2 + 10x}{9x^2 + 6x + 1} =$

18. $\dfrac{4x^2 + x - 3}{4x^2 - 9} \cdot \dfrac{2x^2 - x - 3}{x^2 - 1} =$

19. $\dfrac{x^2 + 5x}{x^2 - 9} \cdot \dfrac{x^2 + 4x + 3}{x^2 + 7x} \cdot \dfrac{x^2 - 49}{x^2 - 2x - 35} =$

20. $\dfrac{x^2 + 4x}{x^2 + 3x + 2} \cdot \dfrac{x^2 + 2x + 1}{x^2 - 16} \cdot \dfrac{x^2 + 4x + 3}{x^2 + x} =$

21. $\dfrac{x^2 - 7x}{x^2 + 4x + 3} \cdot \dfrac{x^2 + 2x + 1}{x^2 - 8x + 7} \cdot \dfrac{x^2 - 2x + 1}{x^2 - 1} =$

22. $\dfrac{x^2 - 5x}{x^2 + 7x} \cdot \dfrac{x^2 + 9x + 14}{x^2 - 6x + 5} \cdot \dfrac{x^2 - 3x + 2}{x^2 + 2x} =$

23. $\dfrac{x^2 + 8x}{x^2 - 5x} \cdot \dfrac{x^2 - 6x + 5}{x^2 + 9x + 8} \cdot \dfrac{x^2 + x}{3x^2} =$

24. $\dfrac{x^2 - 2x}{x^2 - 9} \cdot \dfrac{x^2 + 4x + 3}{x^2 - 4} \cdot \dfrac{x^2 + 2x}{x^2 + 3x} =$

25. $\dfrac{x^2 - 1}{x^2 + 6x} \cdot \dfrac{2x^2}{x + 1} \cdot \dfrac{x^3 + 6x^2}{2x^2 + 4x} =$

26. $\dfrac{x^2 - 5x}{x^2 + 4x} \cdot \dfrac{x^2 + 5x + 4}{x^2 - 25} \cdot \dfrac{x^2 + 7x + 12}{x^2 - 9} =$

Check your answers on page 756. ✓

14.3 Division

Dividing Algebraic Fractions

We learned in Section 3.5 that when numerical fractions are divided, the divisor fraction (the one following the ÷ symbol) is replaced by its reciprocal and the resulting fractions are then multiplied.

The reciprocal of a fraction is the number that fraction must be multiplied by for their product to equal 1. The easiest way to find the reciprocal of any fraction is to invert it (turn it upside down).

EXAMPLE 1 Simplify $\dfrac{3}{4} \div \dfrac{6}{7}$.

$\dfrac{6}{7}$ is the divisor, and its reciprocal is $\dfrac{7}{6}$.

$$\dfrac{3}{4} \div \dfrac{6}{7} =$$

$$\dfrac{3}{4} \cdot \dfrac{7}{6} =$$

$$\dfrac{\cancel{3} \cdot 7}{2 \cdot 2 \cdot 2 \cdot \cancel{3}} = \dfrac{7}{8}$$ ❑

Algebraic fractions are divided exactly as numerical fractions are. The reciprocal of an algebraic fraction is that fraction inverted.

EXAMPLE 2 The reciprocal of $\dfrac{x^2 + 5x}{x^2 + 5x + 6}$ is

$$\dfrac{x^2 + 5x + 6}{x^2 + 5x}.$$

Check: Is their product equal to 1?

$$\left[\dfrac{x^2 + 5x}{x^2 + 5x + 6}\right]\left[\dfrac{x^2 + 5x + 6}{x^2 + 5x}\right] =$$

$$\dfrac{\cancel{(x^2+5x)}\cancel{(x^2+5x+6)}}{\cancel{(x^2+5x+6)}\cancel{(x^2+5x)}} = 1 \quad ✓$$ ❑

To divide algebraic fractions:

1. Replace the divisor fraction by its reciprocal and proceed to multiply those fractions.

2. Replace both numerators by their primes factors, and replace both denominators by their prime factors.

3. Cancel any pair of identical factors appearing in both the numerator and the denominator.

4. Leave the answer in factored form.

EXAMPLE 3 Simplify $\dfrac{x^2 - 25}{x^2 + 3x} \div \dfrac{x^2 + 5x}{x^2 + 5x + 6}$.

$$\dfrac{x^2 - 25}{x^2 + 3x} \div \dfrac{x^2 + 5x}{x^2 + 5x + 6} =$$

$$\left[\dfrac{x^2 - 25}{x^2 + 3x}\right] \cdot \left[\dfrac{x^2 + 5x + 6}{x^2 + 5x}\right] =$$

$$\dfrac{(x^2 - 25)(x^2 + 5x + 6)}{(x^2 + 3x)(x^2 + 5x)} =$$

$$\dfrac{(x - 5)(x + 5)(x + 3)(x + 2)}{x(x + 3)x(x + 5)} =$$

$$\dfrac{(x - 5)\cancel{(x + 5)}\cancel{(x + 3)}(x + 2)}{xx\cancel{(x + 3)}\cancel{(x + 5)}} = \dfrac{(x - 5)(x + 2)}{x^2}$$ ❑

Remember, it is *only in multiplication problems* that factors appearing in the numerator and denominator can be canceled. So always convert division problems into multiplication problems *before* you factor and cancel.

EXERCISES 14.3A

Divide these fractions. Leave your answers in reduced, factored form.

1. $\dfrac{x^2 + 7x}{x^2 - 5x} \div \dfrac{x + 7}{x^2 - 6x + 5} =$

2. $\dfrac{y^2 + 4y + 3}{y^2 - y} \div \dfrac{y^2 - 9}{y^2 - 3y} =$

3. $\dfrac{w^3 - 5w^2}{w + 1} \div \dfrac{w^2 + 2w}{w^2 - 4w - 5} =$

4. $\dfrac{x^2 - 36}{x^3 + 4x^2} \div \dfrac{x^2 + 7x + 6}{x^2 + x} =$

5. $\dfrac{y^2 + 9y}{y - 3} \div \dfrac{y^2 + 7y}{y^2 - 9} =$

6. $\dfrac{m^2 + m}{m^2 - 3m + 2} \div \dfrac{m^2 - 1}{m^2 - 4m + 4} =$

7. $\dfrac{x^3 - x^2}{x^2 - 2x - 35} \div \dfrac{x^2 - x}{x^2 - 49} =$

8. $\dfrac{p^2 + 9p + 20}{p^2 + 7p + 12} \div \dfrac{p^2 + 5p}{p + 3} =$

9. $\dfrac{2p^2 - 7p - 15}{p^2 - 25} \div \dfrac{2p + 3}{2p - 3} =$

10. $\dfrac{2x^2 + 16x + 24}{x^2 + 2x} \div \dfrac{4x^2 + 6x}{x^3 + 7x^2} =$

Check your answers on page 756. ✓

Fractions over Fractions

Another notation used to show that one fraction is to be divided by another is writing the divisor fraction beneath the fraction it is to be divided into.

Whenever you are asked to simplify an expression containing one fraction written beneath another, you should represent the expression as the numerator fraction *divided by* the denominator fraction. Then proceed to convert it into a multiplication problem by replacing the divisor by its reciprocal.

EXAMPLE 4 Simplify $\dfrac{\dfrac{x + 1}{x}}{\dfrac{x - 4}{x + 1}}$.

$$\frac{\dfrac{x + 1}{x}}{\dfrac{x - 4}{x + 1}} =$$

$$\frac{x + 1}{x} \div \frac{x - 4}{x + 1} =$$

$$\left[\frac{x + 1}{x}\right] \cdot \left[\frac{x + 1}{x - 4}\right] =$$

$$\frac{(x + 1)(x + 1)}{x(x - 4)} = \frac{(x + 1)^2}{x(x - 4)}$$

EXAMPLE 5 Simplify $\dfrac{\dfrac{(12xy)}{(5y)}}{\dfrac{(9y^2)}{(20x^2y^2)}}$.

$$\frac{12xy}{5y} \div \frac{9y^2}{20x^2y^2} =$$

$$\left[\frac{12xy}{5y}\right] \cdot \left[\frac{20x^2y^2}{9y^2}\right] =$$

$$\frac{(12xy)(20x^2y^2)}{(5y)(9y^2)} =$$

$$\frac{2 \cdot 2 \cdot 3 \cdot x \cdot y \cdot 2 \cdot 2 \cdot 5 \cdot x \cdot x \cdot y \cdot y}{5 \cdot y \cdot 3 \cdot 3 \cdot y \cdot y} =$$

$$\frac{2 \cdot 2 \cdot \cancel{3} \cdot 2 \cdot 2 \cdot \cancel{5} \cdot x \cdot \cancel{y} \cdot x \cdot x \cdot \cancel{y} \cdot \cancel{y}}{\cancel{5} \cdot \cancel{3} \cdot 3 \cdot \cancel{y} \cdot \cancel{y} \cdot \cancel{y}} =$$

$$\frac{2 \cdot 2 \cdot 2 \cdot 2 \cdot x \cdot x \cdot x}{3} = \frac{16x^3}{3} \qquad \square$$

EXAMPLE 6 Simplify $\dfrac{\left[\dfrac{(y^2-1)}{(y^2+4y+3)}\right]}{\left[\dfrac{(y^2-7y+6)}{(y^2+6y+9)}\right]}$.

$$\left[\frac{y^2-1}{y^2+4y+3}\right] \div \left[\frac{y^2-7y+6}{y^2+6y+9}\right] =$$

$$\left[\frac{y^2-1}{y^2+4y+3}\right] \cdot \left[\frac{y^2+6y+9}{y^2-7y+6}\right] =$$

$$\frac{(y^2-1)(y^2+6y+9)}{(y^2+4y+3)(y^2-7y+6)} =$$

$$\frac{(y-1)(y+1)(y+3)(y+3)}{(y+1)(y+3)(y-6)(y-1)} =$$

$$\frac{\cancel{(y-1)}\cancel{(y+1)}\cancel{(y+3)}(y+3)}{\cancel{(y+1)}\cancel{(y+3)}(y-6)\cancel{(y-1)}} = \frac{y+3}{y-6} \qquad \square$$

EXERCISES 14.3B

Simplify each expression.

11. $\dfrac{\dfrac{x+1}{x-3}}{\dfrac{x-2}{x+7}} =$

12. $\dfrac{\dfrac{35x}{4xy}}{\dfrac{10y}{14xy}} =$

13. $\dfrac{\dfrac{x-4}{x+3}}{\dfrac{x+3}{x+2}} =$

14. $\dfrac{\dfrac{x(x-7)}{(x+1)(x-3)}}{\dfrac{(x+2)(x+1)}{x(x-3)}} =$

15. $\dfrac{\dfrac{x^2+4x}{x-3}}{\dfrac{x^2-16}{x^2-3x}} =$

16. $\dfrac{\dfrac{x^2+4x+4}{x^2-7x}}{\dfrac{x^2+3x+2}{x^2+x}} =$

17. $\dfrac{\dfrac{x^2-4x-5}{x^2-4x+4}}{\dfrac{x^2+x}{x^2+2x}} =$

18. $\dfrac{\dfrac{3x^2-x-10}{x^2-4}}{\dfrac{9x^2+30x+25}{x^2+2x}} =$

19. $\dfrac{\dfrac{x^3+5x^2+6x}{x^2-x-6}}{\dfrac{x^2+3x}{x^2-4}} =$

20. $\dfrac{\dfrac{3x^2+6x+3}{6x-15}}{\dfrac{x^2+8x+7}{4x-10}} =$

Check your answers on page 756. ✓

 **Addition and Subtraction of Fractions with the Same Denominator**

Addition

Just as with numerical fractions, you can add algebraic fractions only if they have the same denominator. Then the numerator of their sum is the sum of the numerators, and the denominator of their sum is the denominator that the fractions share.

EXAMPLE 1

$$\frac{3}{15}+\frac{5}{15}=$$

$$\frac{3+5}{15}=\frac{8}{15} \qquad \square$$

To simplify the sum, combine any like terms in the numerators.

EXAMPLE 2　Combine and simplify $\dfrac{3}{x+2}+\dfrac{5}{x+2}$.

$$\frac{3}{x+2}+\frac{5}{x+2}=$$

$$\frac{3+5}{x+2}=\frac{8}{x+2} \qquad \square$$

EXAMPLE 3　Combine and simplify $\dfrac{x-4}{x+1}+\dfrac{x+7}{x+1}$.

$$\frac{x-4}{x+1}+\frac{x+7}{x+1}=$$

$$\frac{(x-4)+(x+7)}{(x+1)}=\frac{2x+3}{x+1} \qquad \square$$

EXAMPLE 4 Combine and simplify:

$$\frac{2a-b}{3a+b}+\frac{6a+7b}{3a+b}+\frac{2b-a}{3a+b}=$$

$$\frac{(2a-b)+(6a+7b)+(2b-a)}{3a+b}=$$

$$\frac{(2a+6a-a)+(-b+7b+2b)}{3a+b}=\frac{7a+8b}{3a+b} \qquad \square$$

Subtraction

To subtract a second fraction from a first fraction with the same denominator, subtract the numerator of the second fraction from the numerator of the first.

EXAMPLE 5 Combine and simplify $\dfrac{y}{y-1}-\dfrac{5y}{y-1}$.

$$\frac{y}{y-1}-\frac{5y}{y-1}=$$

$$\frac{y-5y}{y-1}=\frac{-4y}{y-1} \qquad \square$$

If the numerator being *subtracted* is a polynomial, just change the signs of each of its terms and then combine like terms to subtract it from the other numerator (as we did in Section 5.3).

EXAMPLE 6 Combine and simplify:

$$\frac{m+2n}{5m-n}-\frac{2m-n}{5m-n}=$$

$$\frac{(m+2n)-(2m-n)}{5m-n}=$$

$$\frac{m+2n-2m+n}{5m-n}=\frac{-m+3n}{5m-n} \qquad \square$$

EXAMPLE 7 Combine and simplify:

$$\frac{a-3b}{a+b}-\frac{5a-b}{a+b}+\frac{-3a}{a+b}=$$

$$\frac{(a-3b)-(5a-b)+(-3a)}{a+b}=$$

$$\frac{a-3b-5a+b-3a}{a+b}=\frac{-7a-2b}{a+b} \qquad \square$$

EXERCISES 14.4A

Combine and simplify.

1. $\dfrac{4}{x-3}+\dfrac{1}{x-3}=$

2. $\dfrac{7}{y+3x}+\dfrac{4}{y+3x}=$

3. $\dfrac{2}{w+5}+\dfrac{3}{w+5}+\dfrac{8}{w+5}=$

4. $\dfrac{x}{x+4}+\dfrac{x-2}{x+4}=$

5. $\dfrac{y}{y-1}+\dfrac{2y}{y-1}+\dfrac{y-5}{y-1}=$

6. $\dfrac{5+x}{3x-4}+\dfrac{2x-1}{3x-4}+\dfrac{4-5x}{3x-4}=$

7. $\dfrac{3a + b}{5a^2b} + \dfrac{7a - 3b}{5a^2b} + \dfrac{a}{5a^2b} =$

8. $\dfrac{8}{w^2 - 4y} + \dfrac{3w + 1}{w^2 - 4y} + \dfrac{3 - 2w}{w^2 - 4y} + \dfrac{4 + w}{w^2 - 4y} =$

9. $\dfrac{6}{x + 3y} + \dfrac{7y - 1}{x + 3y} + \dfrac{4x - y}{x + 3y} =$

10. $\dfrac{a + b}{a - 5b} + \dfrac{2a - 7b}{a - 5b} + \dfrac{2b - a}{a - 5b} =$

11. $\dfrac{5}{x + 3} - \dfrac{2}{x + 3} =$

12. $\dfrac{a}{3a + b} - \dfrac{5b}{3a + b} =$

13. $\dfrac{x + 3}{x - 1} - \dfrac{x - 5}{x - 1} =$

14. $\dfrac{y - 1}{2y + 3} - \dfrac{y + 4}{2y + 3} =$

15. $\dfrac{x + 3y}{x - y} - \dfrac{4x + y}{x - y} =$

16. $\dfrac{7a + b}{a + 2b} - \dfrac{3a - 4b}{a + 2b} =$

17. $\dfrac{3 + x}{2x + 5} - \dfrac{5x + 1}{2x + 5} =$

18. $\dfrac{-5a - 3b}{2a + b} - \dfrac{4a - b}{2a + b} =$

Check your answers on page 756.

Adding, Subtracting, and Simplifying

After fractions have been added or subtracted, always check whether the result can be reduced. Factor the numerator and denominator if possible and cancel any factors appearing in both. The answer to a fractional computation is *correct only if it is left in its most reduced form.*

EXAMPLE 8   Combine and simplify:

$$\left[\frac{(2x + 3)}{(4x - 6)}\right] + \left[\frac{(6x - 5)}{(4x - 6)}\right] =$$

$$\left[\frac{(2x + 3) + (6x - 5)}{(4x - 6)}\right] =$$

$$\frac{[2\underline{x} + \underline{3} + 6\underline{x} - \underline{5}]}{(4x - 6)} =$$

$$\frac{[8x - 2]}{(4x - 6)} =$$

$$\frac{[2(4x - 1)]}{[2(2x - 3)]} =$$

$$\frac{[\cancel{2}(4x - 1)]}{[\cancel{2}(2x - 3)]} = \frac{(4x - 1)}{(2x - 3)}$$

EXAMPLE 9   Combine and simplify:

$$\left[\frac{(3x^2 + 3x + 5)}{(x^2 + 3x + 2)}\right] - \left[\frac{(2x^2 + 5x + 13)}{(x^2 + 3x + 2)}\right] =$$

$$\left[\frac{(3x^2 + 3x + 5) - (2x^2 + 5x + 13)}{(x^2 + 3x + 2)}\right] =$$

$$\left[\frac{3x^2 + 3x + 5 - 2x^2 - 5x - 13}{(x^2 + 3x + 2)}\right] =$$

$$\frac{[x^2 - 2x - 8]}{(x^2 + 3x + 2)} =$$

$$\frac{[(x - 4)(x + 2)]}{[(x + 1)(x + 2)]} =$$

$$\frac{[(x - 4)(x + 2)]}{[(x + 1)(x + 2)]} = \frac{(x - 4)}{(x + 1)}$$

EXAMPLE 10   Combine and simplify:

$$\left[\frac{(3x - 1)}{(x + 5)}\right] + \left[\frac{(x + 4)}{(x + 5)}\right] - \left[\frac{(2x - 7)}{(x + 5)}\right] =$$

$$\frac{[(3x - 1) + (x + 4) - (2x - 7)]}{(x + 5)} =$$

$$\frac{[3x - 1 + x + 4 - 2x + 7]}{(x + 5)} =$$

$$\frac{[2x + 10]}{(x + 5)} =$$

$$\frac{[2(x + 5)]}{(x + 5)} =$$

$$\frac{[2(x + 5)]}{(x + 5)} = \frac{2}{1} = 2$$

EXERCISES 14.4B

Combine, simplify, and leave in the most reduced form.

19. $\dfrac{x^2 + 3x + 5}{x + 4} - \dfrac{x^2 + 2x - 1}{x + 4} =$

20. $\dfrac{x^2 + 3x}{2x^2 + 5} + \dfrac{4 - x}{2x^2 + 5} + \dfrac{3x^2 - 1}{2x^2 + 5} =$

21. $\dfrac{3}{r + 5} - \dfrac{2}{r + 5} + \dfrac{4}{r + 5} =$

22. $\dfrac{x}{r - 3} + \dfrac{3x - 4}{r - 3} - \dfrac{2}{r - 3} =$

23. $\dfrac{3x + 4}{x - 1} + \dfrac{2x - 3}{x - 1} - \dfrac{x + 7}{x - 1} =$

24. $\dfrac{x^2 + 3}{x - 2} + \dfrac{4x - 1}{x - 2} + \dfrac{3 - x^2}{x - 2} =$

25. $\dfrac{3y - 2}{y + 5} - \dfrac{2y - 1}{y + 5} - \dfrac{3y + 4}{y + 5} =$

26. $\dfrac{3x + 1}{x - 2} + \dfrac{5x - 4}{x - 2} - \dfrac{2x + 9}{x - 2} =$

27. $\dfrac{4w - 1}{w + 3} - \dfrac{w}{w + 3} - \dfrac{2w - 2}{w + 3} =$

28. $\dfrac{3x + 2}{x - 1} - \dfrac{x + 5}{x - 1} + \dfrac{1}{x - 1} =$

29. $\dfrac{2x^2 + 3x - 1}{x^2 + 5x + 6} - \dfrac{x^2 - 4x - 13}{x^2 + 5x + 6} =$

30. $\dfrac{4y^2 - 3y + 1}{y^2 + 5y + 4} - \dfrac{3y^2 - y + 7}{y^2 + 5y + 4} =$

Check your answers on page 757.

14.5 Finding the Lowest Common Multiples of Algebraic Expressions

We cannot add or subtract algebraic fractions with different denominators, just as we cannot add or subtract numerical fractions that have different denominators. We have to change the fractions to *equal*, but new, names that have the *same denominator*.

Just as when we worked only with numbers, the new common denominator of the fractions will be a *common multiple* of the old denominators. With algebraic fractions it is important to find the **lowest common multiple** of the old denominators for the new one; otherwise the fractions become very complicated and will be too hard to work with.

Remember, to find the lowest common multiple (LCM) of numbers, factor each one into a product of prime factors and list the factors you need.

First list: all factors of the first denominator.

Then put in: all new factors from the other denominators.

And then put in: any factors appearing more times in later denominators than in the first one.

EXAMPLE 1   Find the lowest common multiple of 10, 15, and 12.

10 is the product of two prime factors: $\underline{2}$ and $\underline{5}$ ($10 = 2 \cdot 5$).

15 is the product of two prime factors: $\underline{3}$ and 5 ($15 = 3 \cdot 5$).

12 is the product of three prime factors: $\underline{2}$, 2, and 3 ($12 = 2 \cdot 2 \cdot 3$).

So we begin with $\underline{2} \cdot \underline{5}$.

Then we expand to $\underline{2} \cdot \underline{5} \cdot \underline{3}$.

And then we expand to $\underline{2} \cdot \underline{5} \cdot \underline{3} \cdot \underline{2}$.

So their LCM is $\underline{2} \cdot \underline{5} \cdot \underline{3} \cdot \underline{2}$, or 60.   ❏

To find the lowest common multiple of algebraic expressions, follow that same procedure. The only difference is that, when listing the LCM, it is better to leave everything in factored form.

EXAMPLE 2   Find the lowest common multiple of $2x + 1$, $x + 1$, and $x + 5$.

Each of those expressions is prime, so their LCM is their product: $(2x + 1)(x + 1)(x + 5)$.   ❏

EXAMPLE 3   Find the lowest common multiple of $2x$, $y(x - 1)$, and $3xy$.

$2x$ is the product of two prime factors: $\underline{2}$ and $\underline{x}$.

$y(x - 1)$ is the product of two prime factors: $\underline{y}$ and $\underline{(x - 1)}$.

$3xy$ is the product of three prime factors: $\underline{3}$, x, and y.

So their LCM is $\underline{2} \cdot \underline{x} \cdot \underline{y} \cdot \underline{(x - 1)} \cdot \underline{3}$, or $6xy(x - 1)$.   ❏

EXAMPLE 4   Find the lowest common multiple of $x^2(x + 1)$, $2x(x - 5)$, and $(x - 1)(x + 1)$.

$x^2(x + 1)$ is the product of three prime factors: $\underline{x}$, $\underline{x}$ and $\underline{x + 1}$.

$2x(x - 5)$ is the product of three prime factors: $\underline{2}$, x, and $\underline{x - 5}$.

$(x - 1)(x + 1)$ is the product of two prime factors: $\underline{x - 1}$ and $x + 1$.

So their LCM is $\underline{x} \cdot \underline{x} \cdot \underline{(x + 1)} \cdot \underline{2} \cdot \underline{(x - 5)} \cdot \underline{(x - 1)}$, or $2x^2(x + 1)(x - 5)(x - 1)$.   ❏

Try to factor each algebraic expression into a product of prime factors before listing the factors you need for the LCM.

EXAMPLE 5 Find the lowest common multiple of $x^2 - 9$, $x^2 + 5x + 6$, and $x^2 - 3x$.

$x^2 - 9$ is the product of two prime factors: $\underline{x - 3}$ and $\underline{x + 3}$.

$x^2 + 5x + 6$ is the product of two prime factors: $\underline{x + 2}$ and $x + 3$.

$x^2 - 3x$ is the product of two prime factors: $\underline{x}$ and $x - 3$.

So their LCM is $(x - 3) \cdot (x + 3) \cdot (x + 2) \cdot x$, or $x(x - 3)(x + 3)(x + 2)$.   ❏

EXAMPLE 6 Find the lowest common multiple of $x^3 - 2x^2$, $x^2 - 4$, $x^2 + 4x + 4$, and x^3y.

$x^3 - 2x^2$ is the product of three prime factors: $\underline{x}$, $\underline{x}$, and $\underline{x - 2}$.

$x^2 - 4$ is the product of two prime factors: $x - 2$ and $\underline{x + 2}$.

$x^2 + 4x + 4$ is the product of two prime factors: $x + 2$ and $\underset{\sim}{x + 2}$.

x^3y is the product of four prime factors: x, x, $\underline{x}$, and $\underline{y}$.

So their LCM is $x \cdot x \cdot (x - 2) \cdot (x + 2) \cdot (x + 2) \cdot x \cdot y$, or $x^3y(x - 2)(x + 2)^2$.   ❏

EXERCISES 14.5

Find the lowest common multiple of these algebraic expressions.

1. x; $x - 5$

2. $y + 3$; $y - 1$

3. x; $5xy$; $3xy^2$

4. $x(x + 1)$; $5xy$

5. $3x^2$; $5xy$; $2x(x - 3y)$

6. $14x^3y$; $10xy^2$; $15xy$

7. $x(x - 1)^2$; $3xy(x - 1)$

8. $6xy^2(x + 2)(x - 3)^2$; $10xy(x - 3)$

9. $x^2 - 25$; $x^2 + 5x$

10. $x^2 + 4x + 3$; $x^2 - 1$

11. $x^3 - 4x^2$; $2x^3 - 4x^2 + 2x$

12. $3x^2 + 2x - 1$; $4x^2 - 4$

13. $6x^2y^3 - 10xy^4$; $5x^4y^2 + 20x^2y$

14. $4a^3b - 2a^2b^2$; $6a^4b^2 - 3a^3b^3$

Check your answers on page 757. ✓

14.6 Addition and Subtraction of Fractions with Different Denominators

To add or subtract fractions with different denominators:

1. Find the lowest common multiple of their denominators.
2. Multiply the numerator and denominator of each fraction by the factor(s) needed so that its denominator will be that LCM.
3. Add or subtract the numerators after all fractions have the same denominators.
4. Factor the resulting numerator and cancel if possible.

Since you want to leave your answers in factored form, do not multiply out the denominator. You must *multiply out the new numerator* of each fraction so that you can recognize like terms which you should combine. Factor the resulting numerator if possible, and cancel any factors common to it and the numerator.

This procedure is the same one we used in Section 3.7 to add numerical fractions. We will look at the different cases for adding algebraic fractions, just as we did in Section 3.7 with numerical fractions. First let's look at fractions with prime denominators.

Prime Denominators

When the denominators of all the fractions are primes, the LCM of their denominators is simply the product of those primes.

EXAMPLE 1   Combine and simplify $\dfrac{2}{3} + \dfrac{1}{5}$.

The denominators 3 and 5 are both primes, so their LCM is their product: $3 \cdot 5 = 15$.

Multiply $\dfrac{2}{3}$ by $\dfrac{5}{5}$, and $\dfrac{1}{5}$ by $\dfrac{3}{3}$ to find the new name of each fraction with denominator 15.

$$\frac{2}{3} = \qquad\qquad \frac{1}{5} =$$

$$\left(\frac{2}{3}\right)\left(\frac{5}{5}\right) = \frac{10}{15} \quad \text{and} \quad \left(\frac{1}{5}\right)\left(\frac{3}{3}\right) = \frac{3}{15}$$

So

$$\frac{2}{3} + \frac{1}{5} =$$

$$\frac{10}{15} + \frac{3}{15} = \frac{13}{15}.$$ ❑

EXAMPLE 2   Combine and simplify $\dfrac{3}{x+5} + \dfrac{2}{x+1}$.

The denominators $(x + 5)$ and $(x + 1)$ are both primes, so their LCM is their product: $(x + 5)(x + 1)$.

To find the new names of the fractions (with the same denominator), multiply $\dfrac{3}{x+5}$ by $\dfrac{x+1}{x+1}$.

$$\frac{3}{x+5} =$$

$$\left(\frac{3}{x+5}\right)\left(\frac{x+1}{x+1}\right) = \frac{3(x+1)}{(x+5)(x+1)}$$

Then multiply $\dfrac{2}{x+1}$ by $\dfrac{x+5}{x+5}$.

$$\frac{2}{x+1} =$$

$$\left(\frac{2}{x+1}\right)\left(\frac{x+5}{x+5}\right) = \frac{2(x+5)}{(x+5)(x+1)}$$

So

$$\frac{3}{x+5} + \frac{2}{x+1} =$$

$$\frac{3(x+1)}{(x+5)(x+1)} + \frac{2(x+5)}{(x+5)(x+1)} =$$

$$\frac{3x+3+2x+10}{(x+5)(x+1)} = \frac{5x+13}{(x+5)(x+1)}.$$

EXAMPLE 3 Combine and simplify $\dfrac{x}{x-3} - \dfrac{x+1}{x}$.

The LCM of $(x-3)$ and x is $x(x-3)$.

$$\frac{x}{x-3} - \frac{x+1}{x} =$$

$$\left[\frac{x}{x-3}\right]\left[\frac{x}{x}\right] - \left[\frac{x+1}{x}\right]\left[\frac{x-3}{x-3}\right] =$$

$$\frac{x(x)}{x(x-3)} - \frac{(x+1)(x-3)}{x(x-3)} =$$

$$\frac{x(x) - (x+1)(x-3)}{x(x-3)} =$$

$$\frac{x^2 - (x^2 - 2x - 3)}{x(x-3)} =$$

$$\frac{x^2 - x^2 + 2x + 3}{x(x-3)} = \frac{2x+3}{x(x-3)}$$

EXAMPLE 4 Combine and simplify $\dfrac{y+1}{y-3} + \dfrac{y+2}{y+1}$.

The LCM of $(y-3)$ and $(y+1)$ is $(y-3)(y+1)$.

$$\frac{y+1}{y-3} + \frac{y+2}{y+1} =$$

$$\left[\frac{y+1}{y-3}\right]\left[\frac{y+1}{y+1}\right] + \left[\frac{y+2}{y+1}\right]\left[\frac{y-3}{y-3}\right] =$$

$$\frac{(y+1)(y+1)}{(y-3)(y+1)} + \frac{(y+2)(y-3)}{(y+1)(y-3)} =$$

$$\frac{(y+1)(y+1) + (y+2)(y-3)}{(y-3)(y+1)} =$$

$$\frac{(y^2 + 2y + 1) + (y^2 - y - 6)}{(y-3)(y+1)} = \frac{2y^2 + y - 5}{(y-3)(y+1)}$$

EXERCISES 14.6A

Combine, simplify, and leave in factored form.

1. $\dfrac{3}{x} + \dfrac{4}{x+2} =$

2. $\dfrac{1}{y+2} + \dfrac{4}{y} =$

3. $\dfrac{7}{x} - \dfrac{2}{x+5} =$

4. $\dfrac{2}{y-4} - \dfrac{3}{y} =$

5. $\dfrac{5}{x+1} + \dfrac{3}{x-2} =$

6. $\dfrac{4x}{x-3} + \dfrac{7}{x-2} =$

7. $\dfrac{8}{y-1} - \dfrac{3}{y+2} =$

8. $\dfrac{2}{w+3} - \dfrac{5}{w-2} =$

9. $\dfrac{x+3}{x} + \dfrac{y-2}{y} =$

10. $\dfrac{4m-1}{m} + \dfrac{n+3}{n} =$

11. $\dfrac{a+3}{a-2} + \dfrac{a-5}{a+1} =$

12. $\dfrac{3b-2}{b+1} + \dfrac{5b+3}{b-2} =$

13. $\dfrac{2a-3}{2a+3} + \dfrac{4a-1}{a-2} =$

14. $\dfrac{x+2}{x+5} - \dfrac{x+3}{x+2} =$

15. $\dfrac{y+4}{y-2} - \dfrac{y+2}{y+1} =$

16. $\dfrac{4x}{3x-1} + \dfrac{2x}{x+4} =$

Check your answers on page 757. ✓

Nonprime Denominators

When the denominators are nonprimes, we must use the methods discussed in Section 14.5 to find their LCM.

If the denominators can be factored, factor them before trying to find their LCM.

Remember, if one denominator is a multiple of a second denominator, that first one is their common multiple!

EXAMPLE 5 Combine and simplify $\dfrac{a}{a^2-9} + \dfrac{5}{a+3}$.

The first denominator, $(a^2 - 9)$, factors to $(a+3)(a-3)$.

The second denominator, $(a+3)$, is prime.

The LCM of $(a+3)(a-3)$ and $(a+3)$ is $(a+3)(a-3)$.

So we do not have to *change* $\dfrac{a}{a^2-9}$; but let's factor it!

We multiply $\dfrac{5}{a+3}$ by $\dfrac{a-3}{a-3}$ to change its name.

$$\frac{a}{a^2-9} + \frac{5}{a+3} =$$

$$\frac{a}{(a+3)(a-3)} + \left[\frac{5}{a+3}\right]\left[\frac{a-3}{a-3}\right] =$$

$$\frac{a}{(a+3)(a-3)} + \frac{5(a-3)}{(a+3)(a-3)} =$$

$$\frac{a+5(a-3)}{(a+3)(a-3)} =$$

$$\frac{a+5a-15}{(a+3)(a-3)} =$$

$$\frac{6a-15}{(a+3)(a-3)} =$$

$$\frac{2\cdot 3 \cdot a - 3 \cdot 5}{(a+3)(a-3)} = \frac{3(2a-5)}{(a+3)(a-3)}$$

EXAMPLE 6 Combine and simplify $\dfrac{2}{x^2 - 1} - \dfrac{5}{x^2 + 2x + 1}$.

$(x^2 - 1)$ factors to $(x - 1)(x + 1)$.

$(x^2 + 2x + 1)$ factors to $(x + 1)(x + 1)$.

So their LCM is $(x - 1)(x + 1)^2$

$$\frac{2}{x^2 - 1} - \frac{5}{x^2 + 2x + 1} =$$

$$\frac{2}{(x - 1)(x + 1)} - \frac{5}{(x + 1)^2} =$$

$$\left[\frac{2}{(x - 1)(x + 1)}\right]\left[\frac{x + 1}{x + 1}\right] - \left[\frac{5}{(x + 1)^2}\right]\left[\frac{x - 1}{x - 1}\right] =$$

$$\frac{2(x + 1)}{(x - 1)(x + 1)(x + 1)} - \frac{5(x - 1)}{(x + 1)(x + 1)(x - 1)} =$$

$$\frac{2(x + 1) - 5(x - 1)}{(x - 1)(x + 1)^2} =$$

$$\frac{2x + 2 - 5x + 5}{(x - 1)(x + 1)^2} = \frac{-3x + 7}{(x - 1)(x + 1)^2} \qquad \square$$

Remember always to check whether the resulting numerator can be factored. If it can be, factor it and cancel any factors appearing in both the numerator and the denominator.

EXAMPLE 7 Combine and simplify $\dfrac{2w + 1}{w^2 - 2w - 8} - \dfrac{w}{w^2 - 4}$.

$(w^2 - 2w - 8)$ factors to $(w - 4)(w + 2)$.

$(w^2 - 4)$ factors to $(w - 2)(w + 2)$.

So their LCM is $(w - 4)(w + 2)(w - 2)$.

$$\frac{2w + 1}{w^2 - 2w - 8} - \frac{w}{w^2 - 4} =$$

$$\frac{2w + 1}{(w - 4)(w + 2)} - \frac{w}{(w - 2)(w + 2)} =$$

$$\left[\frac{2w + 1}{(w - 4)(w + 2)}\right]\left[\frac{w - 2}{w - 2}\right] - \left[\frac{w}{(w - 2)(w + 2)}\right]\left[\frac{w - 4}{w - 4}\right] =$$

$$\frac{(2w + 1)(w - 2)}{(w - 4)(w + 2)(w - 2)} - \frac{w(w - 4)}{(w - 4)(w + 2)(w - 2)} =$$

$$\frac{(2w + 1)(w - 2) - w(w - 4)}{(w - 4)(w + 2)(w - 2)} =$$

$$\frac{2w^2 - 3w - 2 - w^2 + 4w}{(w - 4)(w + 2)(w - 2)} =$$

$$\frac{w^2 + w - 2}{(w - 4)(w + 2)(w - 2)} =$$

$$\frac{(w + 2)(w - 1)}{(w - 4)(w + 2)(w - 2)} =$$

$$\frac{(w + 2)(w - 1)}{(w - 4)(w + 2)(w - 2)} = \frac{w - 1}{(w - 4)(w - 2)}$$

EXAMPLE 8 Combine and simplify $\dfrac{4}{x^2 - 9x + 18} + \dfrac{3}{x^2 - 6x} - \dfrac{1}{x^2 - 3x}$.

$(x^2 - 9x + 18)$ factors to $\underline{(x - 6)(x - 3)}$.

$(x^2 - 6x)$ factors to $\underline{\underline{x}}(x - 6)$.

$(x^2 - 3x)$ factors to $x(x - 3)$.

So their LCM is $x(x - 6)(x - 3)$.

$$\frac{4}{x^2 - 9x + 18} + \frac{3}{x^2 - 6x} - \frac{1}{x^2 - 3x} =$$

$$\frac{4}{(x - 6)(x - 3)} + \frac{3}{x(x - 6)} - \frac{1}{x(x - 3)} =$$

$$\left[\frac{4}{(x - 6)(x - 3)}\right]\left[\frac{x}{x}\right] + \left[\frac{3}{x(x - 6)}\right]\left[\frac{x - 3}{x - 3}\right] - \left[\frac{1}{x(x - 3)}\right]\left[\frac{x - 6}{x - 6}\right] =$$

$$\frac{4x}{x(x - 6)(x - 3)} + \frac{3(x - 3)}{x(x - 6)(x - 3)} - \frac{1(x - 6)}{x(x - 6)(x - 3)} =$$

$$\frac{4x + 3(x - 3) - (x - 6)}{x(x - 6)(x - 3)} =$$

$$\frac{4\underline{x} + 3x - \underline{9} - \underline{x} + \underline{6}}{x(x - 6)(x - 3)} =$$

$$\frac{6x - 3}{x(x - 6)(x - 3)} = \frac{3(2x - 1)}{x(x - 6)(x - 3)}$$

EXERCISES 14.6B

Combine, simplify, and leave in reduced, factored form.

17. $\dfrac{4}{x^2 - 2x} + \dfrac{3}{x - 2} =$

18. $\dfrac{1}{x^2 - 3x} + \dfrac{1}{x^2 + 5x} =$

19. $\dfrac{3}{y^2 - 4} + \dfrac{1}{y^2 - 3y + 2} =$

20. $\dfrac{6}{w^2 - 3w} - \dfrac{2}{w^2 - 9} =$

21. $\dfrac{7}{x^2 + 7x + 12} + \dfrac{1}{x + 4} =$

22. $\dfrac{5}{x^2 - 49} - \dfrac{2}{x + 7} =$

23. $\dfrac{4}{2y^2 - 6y} - \dfrac{4}{y - 3} =$

24. $\dfrac{3}{2x^2 - x - 1} + \dfrac{5}{x - 1} =$

25. $\dfrac{x + 3}{x^2 - 6x + 8} + \dfrac{x - 2}{x^2 - 5x + 4} =$

26. $\dfrac{y + 2}{y^2 - 3y - 4} + \dfrac{y - 1}{y^2 + 2y + 1} =$

27. $\dfrac{w + 6}{w^2 - 5w} - \dfrac{w + 6}{w^2 - 10w + 25} =$

28. $\dfrac{m - 2}{m^3 - 3m^2} - \dfrac{m + 1}{m^2 - 9} =$

29. $\dfrac{3p}{p^2 + 6p + 5} + \dfrac{4p}{p^2 + 7p + 10} =$

30. $\dfrac{2x}{x^2 + 4x} - \dfrac{3x}{x + 4} =$

31. $\dfrac{3}{x^2 - 1} + \dfrac{2}{x^2 + x} + \dfrac{5}{x} =$

32. $\dfrac{4}{x^2 - 6x + 9} + \dfrac{1}{x^2 - 9} + \dfrac{3}{x + 3} =$

33. $\dfrac{7}{y^2 - 6y} + \dfrac{3}{y^2} - \dfrac{4}{y - 6} =$ **34.** $\dfrac{1}{m^2 + 5m} - \dfrac{1}{m^2} - \dfrac{3}{m^2 - 5m} =$

Check your answers on page 757. ✓

Simplifying Fractions That Involve Radical Terms

Fractions that involve radical terms do occur, and it will be easier to learn how to use the quadratic formula in Section 15.6 to solve some equations if you learn to simplify such expressions now.

EXAMPLE 1 $\dfrac{3 + \sqrt{12}}{2}$ is a fraction involving a radical term. ❑

You already know how to perform each step to simplify these expressions:

1. Factor the term within the radical and simplify it if possible.
2. Factor the numerator by the distributive law if the *rational* factors of each term have any common factor(s).
3. Cancel any factors common to both the numerator and the denominator.

It may not be possible to apply all three steps to an expression, but perform as many as possible in the order that they are listed.

EXAMPLE 2   Simplify $\dfrac{5 + \sqrt{18}}{2}$.

$$\frac{5 + \sqrt{18}}{2} =$$

$$\frac{5 + \sqrt{2 \cdot 3 \cdot 3}}{2} = \frac{5 + 3\sqrt{2}}{2} \qquad ❑$$

The answer to example 2 cannot be further simplified because neither the numerator nor the denominator can be factored: They are prime.

Sometimes these complicated-looking fractions simplify into a rational number, which is not obvious until you do the simplification. So always simplify the fractions as much as possible.

EXAMPLE 3   Simplify $\dfrac{7 - 5\sqrt{4}}{3}$.

$$\frac{7 - 5\sqrt{4}}{3} =$$

$$\frac{7 - 5(2)}{3} =$$

$$\frac{7 - 10}{3} =$$

$$\frac{-3}{3} = -1 \qquad ❑$$

EXAMPLE 4   Simplify $\dfrac{2 - \sqrt{20}}{1 - \sqrt{5}}$.

$$\frac{2 - \sqrt{20}}{1 - \sqrt{5}} =$$

$$\frac{2 - \sqrt{4 \cdot 5}}{1 - \sqrt{5}} =$$

$$\frac{2 - 2\sqrt{5}}{1 - \sqrt{5}} =$$

$$\frac{(2)(1 - \sqrt{5})}{1 - \sqrt{5}} =$$

$$\frac{2(1 - \sqrt{5})}{(1 - \sqrt{5})} = 2 \qquad \square$$

EXAMPLE 5   Simplify $\dfrac{5 - \sqrt{45}}{7}$.

$$\frac{5 - \sqrt{45}}{7} =$$

$$\frac{5 - \sqrt{3 \cdot 3 \cdot 5}}{7} = \frac{5 - 3\sqrt{5}}{7} \qquad \square$$

Notice that the numerator in Example 5 could not be factored because the 5 *in the radical is the square root of 5 (not the integer 5)* and so there is no common factor that can be factored out by the distributive law.

EXAMPLE 6   Simplify $\dfrac{6 + \sqrt{24}}{2}$.

$$\frac{6 + \sqrt{24}}{2} =$$

$$\frac{6 + \sqrt{2 \cdot 2 \cdot 2 \cdot 3}}{2} =$$

$$\frac{6 + 2\sqrt{6}}{2} =$$

$$\frac{2 \cdot 3 + 2\sqrt{6}}{2} =$$

$$\frac{2(3 + \sqrt{6})}{2} =$$

$$\frac{2(3 + \sqrt{6})}{2} = 3 + \sqrt{6} \qquad \square$$

EXAMPLE 7 Simplify $\dfrac{10 - \sqrt{12}}{6}$.

$$\frac{10 - \sqrt{12}}{6} =$$

$$\frac{10 - \sqrt{4 \cdot 3}}{6} =$$

$$\frac{10 - 2\sqrt{3}}{(2 \cdot 3)} =$$

$$\frac{2(5) - 2\sqrt{3}}{(2 \cdot 3)} =$$

$$\frac{2(5 - \sqrt{3})}{(2 \cdot 3)} =$$

$$\frac{\cancel{2}(5 - \sqrt{3})}{(\cancel{2} \cdot 3)} = \frac{5 - \sqrt{3}}{3}$$

❑

EXAMPLE 8 Simplify $\dfrac{14 + \sqrt{108}}{10}$.

$$\frac{14 + \sqrt{108}}{10} =$$

$$\frac{14 + \sqrt{2 \cdot 2 \cdot 3 \cdot 3 \cdot 3}}{10} =$$

$$\frac{14 + (2)(3)\sqrt{3}}{10} =$$

$$\frac{2(7) + (2)(3)\sqrt{3}}{(2 \cdot 5)} =$$

$$\frac{2(7 + 3\sqrt{3})}{(2 \cdot 5)} =$$

$$\frac{\cancel{2}(7 + 3\sqrt{3})}{(\cancel{2} \cdot 5)} = \frac{7 + 3\sqrt{3}}{5}$$

❑

EXERCISES 14.7

Simplify each expression.

1. $\dfrac{5 + \sqrt{9}}{2} =$

2. $\dfrac{3 + \sqrt{8}}{2} =$

3. $\dfrac{7 - \sqrt{50}}{3} =$

4. $\dfrac{2 + \sqrt{16}}{10} =$

5. $\dfrac{6 + \sqrt{12}}{5} =$

6. $\dfrac{4 - \sqrt{12}}{2} =$

7. $\dfrac{9 - \sqrt{18}}{3} =$

8. $\dfrac{10 - \sqrt{20}}{6} =$

9. $\dfrac{12 + \sqrt{40}}{4} =$

10. $\dfrac{2 - \sqrt{28}}{4} =$

11. $\dfrac{6 - \sqrt{24}}{10} =$

12. $\dfrac{14 + \sqrt{20}}{10} =$

Check your answers on page 757. ✓

14.8 Solving Rational Equations

In Sections 6.3 and 6.4 we solved equations that involved fractions with numerical denominators. The first thing we did to simplify our work was to clear the equation of the fractions. We multiplied both sides of the equation by the lowest common multiple of the denominators. Then we were able to simplify each side and eventually got the letter isolated on one side of the equation.

EXAMPLE 1 Solve $\dfrac{y}{5} - 7 = \dfrac{3y}{2} - 2y + 7$.

We begin by multiplying both sides by 10, the LCM of 5 and 2.

$$10\left[\frac{y}{5} - 7\right] = 10\left[\frac{3y}{2} - 2y + 7\right]$$

$$10\left(\frac{y}{5}\right) + 10(-7) = 10\left(\frac{3y}{2}\right) + 10(-2y) + 10(7)$$

$$(2 \cdot 5)\left(\frac{y}{5}\right) + 10(-7) = (2 \cdot 5)\left(\frac{3y}{2}\right) + 10(-2y) + 10(7)$$

$$2y - 70 = 15y - 20y + 70$$

Now that we have removed the fractions, this is a linear equation that is easy to solve. I'll leave the rest of the work for you to do. ❏

We removed the fractions by multiplying both sides of the equation by a common multiple of each denominator appearing in the equation. That is exactly what we must do to remove *any* fraction from an equation, whether that fraction is numerical or algebraic. Equations that contain rational expressions are called **rational equations**.

You can simplify rational equations, like any other equations, by multiplying both sides by the same algebraic expression as long as it is *not equal to zero*. Don't forget that *every term* on both sides must be multiplied by the LCM.

If there is only one denominator in the equation, multiply both sides of the equation by that denominator.

EXAMPLE 2 Solve $\dfrac{12}{x} - 1 = 5$.

There is only one denominator in this equation, and that denominator is x. So to remove it, we multiply both sides by x, which we can do as long as $x \neq 0$.

$$x\left[\frac{12}{x} - 1\right] = x(5)$$

$$\cancel{x}\left(\frac{12}{\cancel{x}}\right) - x(1) = x(5)$$

$$12 - x = 5x \qquad \text{Add } x \text{ to both sides.}$$

$$\underline{ + x \quad + x}$$

$$12 = 6x$$

$$\frac{12}{6} = \frac{6x}{6} \qquad \text{Divide both sides by 6.}$$

$$2 = x$$

Always check the solution to an equation by substituting it into the *original* equation.

Check: Does $\dfrac{12}{x} - 1 = 5$ when $x = 2$?

$$\frac{12}{x} - 1 =$$

$$\frac{12}{2} - 1 =$$

$$6 - 1 = 5 \quad \checkmark$$

EXAMPLE 3   Solve $\dfrac{w + 1}{w - 4} + 3 = 9$.

We can multiply both sides by the only denominator, $w - 4$ if $w \neq 4$.

$$(w - 4)\left[\frac{w + 1}{w - 4} + 3\right] = (w - 4)(9)$$

$$\cancel{(w - 4)}\frac{(w + 1)}{\cancel{(w - 4)}} + (w - 4)(3) = (w - 4)(9)$$

$$(w + 1) + (w - 4)(3) = (w - 4)(9)$$

$$w + 1 + 3w - 12 = 9w - 36$$

$$4w - 11 = 9w - 36$$

$$\underline{-4w \qquad\qquad -4w}$$

$$-11 = 5w - 36$$

$$\underline{+36 \qquad\qquad +36}$$

$$25 = 5w$$

$$5 = w$$

Check: Does $\dfrac{w + 1}{w - 4} + 3 = 9$ when $w = 5$?

$$\frac{w + 1}{w - 4} + 3 =$$

$$\frac{(5) + 1}{(5) - 4} + 3 =$$

$$\frac{6}{1} + 3 = 9 \quad \checkmark$$

In Section 14.5 we learned how to find the least common multiple of polynomials. We used the LCM as the new denominator of fractions we were adding. Now we'll use LCMs again, in a different way, to solve equations. If there is more than one denominator in the equation, we'll find their least common multiple and *multiply it on both sides* of the equation.

Write the LCM in factored form so that you can easily recognize what will cancel with each denominator.

EXAMPLE 4 Solve $\dfrac{5}{6} - \dfrac{1}{4} = \dfrac{7}{n}$.

$6 = (2)(3);\quad 4 = (2)(2);\quad$ and $\underline{n}$ is prime.

So the LCM of 6, 4, and n is $(2)(2)(3)n$.

We can multiply both sides by $(2)(2)(3)n$ if $n \neq 0$.

$$(2)(2)(3)n\left[\frac{5}{6} - \frac{1}{4}\right] = (2)(2)(3)n\left[\frac{7}{n}\right]$$

$$(2)(2)(3)n\left[\frac{5}{6}\right] + (2)(2)(3)n\left[\frac{-1}{4}\right] = (2)(2)(3)n\left[\frac{7}{n}\right]$$

$$(2)(\cancel{2})(\cancel{3})n\left[\frac{5}{(\cancel{2})(\cancel{3})}\right] + (\cancel{2})(\cancel{2})(3)n\left[\frac{-1}{(\cancel{2})(\cancel{2})}\right] = (2)(2)(3)\cancel{n}\left[\frac{7}{\cancel{n}}\right]$$

$$2n[5] + 3n[-1] = (2)(2)(3)[7]$$

$$10\underline{n} - 3\underline{n} = (4)(21)$$

$$7n = (4)(21)$$

$$\frac{\cancel{7}n}{\cancel{7}} = \frac{(4)(3)(\cancel{7})}{(\cancel{7})}$$

$$n = 12$$

Check: Does $\dfrac{5}{6} - \dfrac{1}{4} = \dfrac{7}{n}$ when $n = 12$?

$$\frac{5}{6} - \frac{1}{4} = \qquad\qquad \frac{7}{n} =$$

$$\frac{10}{12} - \frac{3}{12} = \frac{7}{12} \;\checkmark \qquad \frac{7}{(12)} = \frac{7}{12} \;\checkmark$$

EXAMPLE 5 Solve $\dfrac{1}{15} + \dfrac{1}{4y} = \dfrac{1}{4} - \dfrac{3}{10y}$.

$15 = (3)(5);\quad 4y = (2)(2)\underline{y};\quad 4 = (2)(2);\quad 10y = (2)(5)y$

So the LCM is $(3)(5)(2)(2)y$.

We can multiply both sides by $(3)(5)(2)(2)y$ if $y \neq 0$.

$$(3)(5)(2)(2)y\left[\frac{1}{15} + \frac{1}{4y}\right] = (3)(5)(2)(2)y\left[\frac{1}{4} - \frac{3}{10y}\right]$$

$$(\cancel{3})(\cancel{5})(2)(2)y\left[\frac{1}{(\cancel{3})(\cancel{5})}\right] + (3)(5)(\cancel{2})(\cancel{2})\cancel{y}\left[\frac{1}{(\cancel{2})(\cancel{2})\cancel{y}}\right] = (3)(5)(\cancel{2})(\cancel{2})y\left[\frac{1}{(\cancel{2})(\cancel{2})}\right]$$

$$+ (3)(\cancel{5})(\cancel{2})(2)\cancel{y}\left[\frac{-3}{(\cancel{2})(\cancel{5})\cancel{y}}\right]$$

$$(2)(2)y + (3)(5) = (3)(5)\underline{y} + (3)(2)(-3)$$

$$4y + 15 = 15y + (6)(-3)$$

$$4y + 15 = 15y - 18$$

$$\underline{-4y \qquad\qquad -4y}$$

$$+ 15 = 11y - 18$$

$$\underline{+ 18 \qquad\qquad + 18}$$

$$33 = 11y$$

$$3 = y$$

Check: Does $\dfrac{1}{15} + \dfrac{1}{4y} = \dfrac{1}{4} - \dfrac{3}{10y}$ when $y = 3$?

$$\dfrac{1}{15} + \dfrac{1}{4y} = \qquad \dfrac{1}{4} - \dfrac{3}{10y} =$$

$$\dfrac{1}{15} + \dfrac{1}{4(3)} = \qquad \dfrac{1}{4} - \dfrac{3}{10(3)} =$$

$$\dfrac{1}{15} + \dfrac{1}{12} = \qquad \dfrac{1}{4} - \dfrac{3}{30} =$$

$$\dfrac{4}{60} + \dfrac{5}{60} = \qquad \dfrac{15}{60} - \dfrac{6}{60} =$$

$$\dfrac{9}{60} = \dfrac{3}{20} \; ✓ \qquad \dfrac{9}{60} = \dfrac{3}{20} \; ✓$$

EXAMPLE 6 Solve $\dfrac{5}{x} - \dfrac{6}{(x+3)} = \dfrac{1}{x}$.

The denominators x and $(x+3)$ are primes. So the LCM is their product: $x(x+3)$

We can multiply both sides by $x(x+3)$ if $x \neq 0$ and $x \neq -3$.

$$x(x+3)\left[\dfrac{5}{x} - \dfrac{6}{(x+3)}\right] = x(x+3)\left[\dfrac{1}{x}\right]$$

$$x(x+3)\left[\dfrac{5}{x}\right] + x(x+3)\left[\dfrac{-6}{(x+3)}\right] = x(x+3)\left[\dfrac{1}{x}\right]$$

$$(x+3)[5] + x[-6] = (x+3)$$

$$5x + 15 - 6x = x + 3$$

$$15 - x = x + 3$$

$$\underline{+x \qquad +x}$$

$$15 = 2x + 3$$

$$\underline{-3 \qquad -3}$$

$$12 = 2x$$

$$6 = x$$

Check: Does $\dfrac{5}{x} - \dfrac{6}{x+3} = \dfrac{1}{x}$ when $x = 6$?

$$\dfrac{5}{x} - \dfrac{6}{x+3} = \qquad \dfrac{1}{x} =$$

$$\dfrac{5}{(6)} - \dfrac{6}{(6)+3} = \qquad \dfrac{1}{(6)} = \dfrac{1}{6} \; ✓$$

$$\dfrac{5}{6} - \dfrac{6}{9} =$$

$$\dfrac{5}{6} - \dfrac{2}{3} =$$

$$\dfrac{5}{6} - \dfrac{4}{6} = \dfrac{1}{6} \; ✓$$

EXAMPLE 7 Solve $\dfrac{1}{a-1} - \dfrac{2a-3}{a^2-1} = \dfrac{2}{a+1}$.

$(a^2 - 1) = (a-1)(a+1)$, so the LCM is $(a-1)(a+1)$

We can multiply both sides by $(a - 1)(a + 1)$ if $a \neq \pm 1$.

$$(a - 1)(a + 1)\left[\frac{1}{a - 1} - \frac{2a - 3}{a^2 - 1}\right] = (a - 1)(a + 1)\left[\frac{2}{a + 1}\right]$$

$$(a - 1)(a + 1)\left[\frac{1}{(a - 1)}\right] + (a - 1)(a + 1)\left[-\frac{2a - 3}{(a - 1)(a + 1)}\right]$$

$$= (a - 1)(a + 1)\left[\frac{2}{(a + 1)}\right]$$

$$(a + 1) - (2a - 3) = (a - 1)[2]$$
$$a + 1 - 2a + 3 = 2a - 2$$
$$4 - a = 2a - 2$$
$$+ a \qquad + a$$
$$\overline{\qquad 4 = 3a - 2}$$
$$+ 2 \qquad + 2$$
$$\overline{\qquad 6 = 3a}$$

$$2 = a$$

Check: Does $\dfrac{1}{a - 1} - \dfrac{2a - 3}{a^2 - 1} = \dfrac{2}{a + 1}$ when $a = 2$?

$$\frac{1}{a - 1} - \frac{2a - 3}{a^2 - 1} = \frac{2}{a + 1} =$$

$$\frac{1}{(2) - 1} - \frac{2(2) - 3}{(2)^2 - 1} = \frac{2}{(2) + 1} =$$

$$\frac{1}{1} - \frac{4 - 3}{4 - 1} = \frac{2}{3} = \frac{2}{3} \checkmark$$

$$1 - \frac{1}{3} = \frac{2}{3} \checkmark$$

We learned *cross multiplication* in Section 3.4—that is, when two fractions are equal, the products are equal when we multiply the numerator of each fraction across the equal sign by the denominator of the other.

If the equation we want to solve consists of just one algebraic fraction equal to another, we can always clear away the fractions by multiplying each side by the product of the two denominators. That step gives us the same result as if we had *cross multiplied* them!

Cross multiplying is the simplest way to remove the denominators when a rational equation consists of only two fractions.

EXAMPLE 8   Solve $\dfrac{6}{n + 4} = \dfrac{20}{n - 3}$ by cross multiplying.

$$\frac{6}{n + 4} \bowtie \frac{20}{n - 3}$$

$$6(n - 3) = 20(n + 4)$$
$$6n - 18 = 20n + 80$$
$$-6n \qquad\quad -6n$$
$$\overline{\qquad -18 = 14n + 80}$$
$$-80 \qquad\quad -80$$
$$\overline{\qquad -98 = 14n}$$
$$-\frac{98}{14} = \frac{14n}{14}$$

$$-7 = n$$

When we cross multiplied the fractions, we were really multiplying both sides by $(n-3)(n+4)$. We could do so only if $n \neq 3$ and $n \neq -4$. Since it turns out that $n = -7$, it was perfectly fine to have done that.

Check: Does $\dfrac{6}{n+4} = \dfrac{20}{n-3}$ when $n = -7$?

$$\dfrac{6}{n+4} = \qquad\qquad \dfrac{20}{n-3} =$$

$$\dfrac{6}{(-7)+4} = \qquad\qquad \dfrac{20}{(-7)-3} =$$

$$\dfrac{6}{-3} = -2 \; \checkmark \qquad\qquad \dfrac{20}{-10} = -2 \; \checkmark \qquad\qquad \square$$

EXAMPLE 9   Solve $\dfrac{p+4}{p+5} = \dfrac{p}{p-3}$.

Since these are two equal fractions, we can use cross multiplication to simplify if $p \neq -5$ and $p \neq 3$.

When we multiply them, there will be a p^2 term on each side. You have not learned how to solve equations with squared terms yet. (You will learn that in Chapter 15.)

But in this example, the *squared term* has the *same coefficient* on *both sides*, so we can eliminate it by adding its opposite to both sides.

$$(p+4)(p-3) = \qquad p(p+5)$$
$$p^2 + p - 12 = \qquad p^2 + 5p$$
$$\underline{-p^2} \qquad\qquad \underline{-p^2}$$
$$p - 12 = \qquad\quad +5p$$
$$\underline{-p} \qquad\qquad\quad \underline{-p}$$
$$-12 = \qquad\quad 4p$$
$$-3 = p$$

Check: Does $\dfrac{p+4}{p+5} = \dfrac{p}{p-3}$ when $p = -3$?

$$\dfrac{p+4}{p+5} = \qquad\qquad \dfrac{p}{p-3} =$$

$$\dfrac{(-3)+4}{(-3)+5} = \dfrac{1}{2} \; \checkmark \qquad\qquad \dfrac{(-3)}{(-3)-3} =$$

$$\dfrac{-3}{-6} = \dfrac{1}{2} \; \checkmark \qquad\qquad \square$$

After a while people sometimes feel so confident that they do not always check their answers (or they decide that checking is too much trouble). Not checking can be disastrous with rational equations, because they may have no solution—even when solving them *appears* to result in a value for the letter.

So be sure to check your solutions.

EXAMPLE 10   Solve $\dfrac{1}{x+1} = \dfrac{1}{x^2+3x+2} + \dfrac{3}{x+2}$.

$x^2 + 3x + 2 = (x+1)(x+2)$, so the LCM is $(x+1)(x+2)$.

We can multiply both sides by $(x + 1)(x + 2)$ if $x \neq -1$ and $x \neq -2$.

$$(x + 1)(x + 2)\left[\frac{1}{x + 1}\right] = (x + 1)(x + 2)\left[\frac{1}{x^2 + 3x + 2} + \frac{3}{x + 2}\right]$$

$$(x + 1)(x + 2)\left[\frac{1}{(x + 1)}\right] = (x + 1)(x + 2)\left[\frac{1}{(x + 1)(x + 2)}\right]$$

$$+ (x + 1)(x + 2)\left[\frac{3}{(x + 2)}\right]$$

$$x + 2 = 1 + (x + 1)3$$

$$x + 2 = \underline{1} + 3x + \underline{3}$$

$$x + 2 = \quad 4 + 3x$$

$$\underline{-x \qquad\qquad - x}$$

$$2 = \quad 4 + 2x$$

$$\underline{-4 \quad -4}$$

$$-2 = \qquad 2x$$

$$-1 = x$$

We have a problem here. If $x = -1$, then $(x + 1) = 0$, so we cannot multiply both sides of the equation by $(x + 1)(x + 2)$ as we did in the first step to cancel out the denominators. Let's see what happens when we check the solution!

Check: Does $\dfrac{1}{x + 1} = \dfrac{1}{x^2 + 3x + 2} + \dfrac{3}{x + 2}$ when $x = -1$?

$$\frac{1}{x + 1} = \qquad\qquad \frac{1}{x^2 + 3x + 2} + \frac{3}{x + 2} =$$

$$\frac{1}{(-1) + 1} = \frac{1}{0} \qquad \frac{1}{(-1)^2 + 3(-1) + 2} + \frac{3}{(-1) + 2} =$$

$$\frac{1}{1 - 3 + 2} + \frac{3}{1} = \frac{1}{0} + \frac{3}{1}$$

Does $\dfrac{1}{0} = \dfrac{1}{0} + 3$? No, because $\dfrac{1}{0}$ is undefined. We cannot divide something by zero! Therefore $x = -1$ is *not* a solution, and so this equation has no solution. ❏

A quick check to tell whether a proposed answer is not really a solution is to see if it will make an *expression in a denominator equal to zero*. We can never divide something by zero, so such values are *not solutions*.

EXAMPLE 11   We can use that shortcut to check whether $x = -1$ is a solution to

$$\frac{1}{x + 1} = \frac{1}{x^2 + 3x + 2} + \frac{3}{x + 2}.$$

If $x = -1$, do any of the denominators become zero?

$$x + 1 =$$

$$(-1) + 1 = 0$$

So $x = -1$ cannot be a solution. ❏

Always check your solutions.

EXERCISES 14.8

Solve each rational equation and check your solutions.

1. $\dfrac{x}{2} - 3 = \dfrac{x}{4} + 3$

2. $\dfrac{w}{3} + 1 = \dfrac{3w}{2} - 6$

3. $\dfrac{10}{x} - 3 = 2$

4. $\dfrac{6}{y} + 5 = 7$

5. $\dfrac{6}{w} + 4 = \dfrac{12}{w} + 3$

6. $\dfrac{(x+1)}{(x-3)} - 2 = 3$

7. $\dfrac{(5x-1)}{(x+3)} + 4 = 1$

8. $\dfrac{3}{4} - \left(\dfrac{1}{2}\right) = \dfrac{3}{x}$

9. $\dfrac{5}{6} - \left(\dfrac{1}{3}\right) = \dfrac{2}{n}$

10. $\dfrac{5}{(2n)} - \left(\dfrac{1}{2}\right) = \dfrac{2}{n} - \dfrac{1}{3}$

11. $\dfrac{3}{x} + \left(\dfrac{1}{3}\right) = \dfrac{7}{(3x)} + \left(\dfrac{1}{2}\right)$

12. $\dfrac{8}{(5y)} - \left(\dfrac{1}{y}\right) = \dfrac{1}{(y+2)}$

13. $\dfrac{7}{(3w)} - \left(\dfrac{1}{w}\right) = \dfrac{1}{(w-1)}$

14. $\dfrac{10}{(x+2)} = \dfrac{6}{x}$

15. $\dfrac{3}{(x+1)} = \dfrac{9}{(4x-1)}$

16. $\dfrac{5}{(x+2)} = \dfrac{10}{(3x-1)}$

17. $\dfrac{(x+2)}{(x+4)} = \dfrac{x}{(x+1)}$

18. $\dfrac{(y-1)}{(y+1)} = \dfrac{(y+2)}{(y+6)}$

19. $\dfrac{(x-5)}{(x-2)} = \dfrac{(x-3)}{(x+6)}$

20. $\dfrac{3}{(x+2)} + \left[\dfrac{2}{(x^2-4)}\right] = \dfrac{1}{(x-2)}$

21. $\dfrac{1}{(x+3)} + \dfrac{10}{(x^2-9)} = \dfrac{3}{(x-3)}$

Check your answers on page 757. ✓

14.9 Applications of Rational Equations

There are two kinds of problems that can easily be represented and solved as rational equations. In this section we discuss some examples of each.

Work Problems

The first kind of problem is called a **work problem**. It involves situations where work is being done by several people or machines that work at different rates. In some cases we know how long it would take each person or machine to complete the job, but we want to find the time needed for them to do the job together. In others we know the length of time it took to complete a job, and we want to know how long it would have taken each person or machine working alone.

All such problems can be solved by setting up an equation that says: *The amount of the job completed in 1 unit of time by one participant + the amount of the job completed in 1 unit of time by the other participant equals the amount of the job completed in that 1 unit of time if they work together.*

Amount done by one + amount done by other = amount done together

To set up this equation, you must find a way to express how much work is done by each in 1 hour (or another appropriate unit of time). Organize the given data in a table where:

One column indicates the time needed to complete the job;

The other column indicates the amount of job completed in 1 hour (or another appropriate unit of time).

EXAMPLE 1 Wendy and Sharon are hired to wallpaper a room. Wendy, an experienced decorator, could do the job alone in 6 hours. Sharon, who is less experienced, needs 30 hours to do the job alone. How long should it take them if they work together?

Let's record the data in the table below. We don't know the time needed to complete the job when they work together, so let's call that x.

Worker	Hours to do job	Amount done in 1 hour
Wendy	6	$\dfrac{1}{6}$
Sharon	30	$\dfrac{1}{30}$
Together	x	$\dfrac{1}{x}$

⇒ *Note*: Amount of job done in 1 hour is the *reciprocal* of hours needed to do job. If you can do a job in 3 hours, haven't you completed $\dfrac{1}{3}$ of it in 1 hour?

Amount done by Wendy + amount done by Sharon = amount done together

Solve:
$$\frac{1}{6} + \frac{1}{30} = \frac{1}{x} \quad \text{(LCM is } 30x)$$

$$30x\left(\frac{1}{6} + \frac{1}{30}\right) = 30x\left(\frac{1}{x}\right)$$

$$30x\left(\frac{1}{6}\right) + 30x\left(\frac{1}{30}\right) = 30x\left(\frac{1}{x}\right)$$

$$5x + x = 30$$

$$6x = 30$$

$$\frac{6x}{6} = \frac{30}{6}$$

$$x = 5$$

So it should take them 5 hours to do the job together.

Check: See if working 5 hours will complete the job. Does the amount of the job done by Sharon plus the amount done by Wendy add up to 1?

Amount of the job done by Sharon is:

(Amount done each hour)(number of hours) =

$$\left(\frac{1}{30}\right)(5) = \frac{1}{6}.$$

Amount of the job done by Wendy is:

(Amount done each hour)(number of hours) =

$$\left(\frac{1}{6}\right)(5) = \frac{5}{6}.$$

If Sharon does $\frac{1}{6}$ and Wendy does $\frac{5}{6}$, they do finish the job, since $\frac{1}{6} + \frac{5}{6} = 1$. ✓

We can also use this method to set up problems that involve more than two workers.

EXAMPLE 2 Rob can replace all the broken clapboards on the house himself in 2 days. Sue can do the same job in 3 days. But Phil needs 6 days to complete the job. How long should it take them if they work together?

The unit of time in this problem is days.

Worker	Days to do job	Amount done in 1 day
Rob	2	$\frac{1}{2}$
Sue	3	$\frac{1}{3}$
Phil	6	$\frac{1}{6}$
Together	x	$\frac{1}{x}$

Amount done by Rob + amount done by Sue + amount done by Phil = amount done together.

Solve:

$$\frac{1}{2} + \frac{1}{3} + \frac{1}{6} = \frac{1}{x} \quad \text{(LCM is } 6x\text{)}$$

$$6x\left(\frac{1}{2} + \frac{1}{3} + \frac{1}{6}\right) = 6x\left(\frac{1}{x}\right)$$

$$6x\left(\frac{1}{2}\right) + 6x\left(\frac{1}{3}\right) + 6x\left(\frac{1}{6}\right) = 6x\left(\frac{1}{x}\right)$$

$$3x + 2x + x = 6$$

$$6x = 6$$

$$x = 1$$

So it should take them 1 day to complete the job if they work together.

Check: If they each work 1 day, will the amount of the job they each do add up to one completed job?

Rob does $\frac{1}{2}$ a job a day and works 1 day, so he completes $\left(\frac{1}{2}\right)(1) = \frac{1}{2}$.

Sue does $\frac{1}{3}$ a job a day and works 1 day, so she completes $\left(\frac{1}{3}\right)(1) = \frac{1}{3}$.

Phil does $\frac{1}{6}$ a job a day and works 1 day, so he completes $\left(\frac{1}{6}\right)(1) = \frac{1}{6}$.

Together they complete $\frac{1}{2} + \frac{1}{3} + \frac{1}{6} =$

$$\frac{3}{6} + \frac{2}{6} + \frac{1}{6} =$$

$$\frac{6}{6} = 1 \text{ job} \quad \checkmark$$

We use the same kind of table (labeled a little differently) if machines, not workers, are involved in the problem.

EXAMPLE 3 Tracy has a large hose that fills the pool in 4 hours and a small hose that takes 6 hours to fill the same pool. How long should it take to fill the pool if both hoses are used?

Hose	Hours to fill pool	Amount filled in 1 hour
Large	4	$\frac{1}{4}$
Small	6	$\frac{1}{6}$
Together	x	$\frac{1}{x}$

Amount done by large hose + amount done by small hose

= amount done together.

Solve:
$$\frac{1}{4} + \frac{1}{6} = \frac{1}{x} \quad \text{(LCM is } 12x\text{)}$$

$$12x\left(\frac{1}{4} + \frac{1}{6}\right) = 12x\left(\frac{1}{x}\right)$$

$$12x\left(\frac{1}{4}\right) + 12x\left(\frac{1}{6}\right) = 12x\left(\frac{1}{x}\right)$$

$$3x + 2x = 12$$

$$5x = 12$$

$$\frac{5x}{5} = \frac{12}{5}$$

$$x = 2\frac{2}{5}$$

So it should take $2\frac{2}{5}$ hours to fill the pool with both hoses.

Check: Does the amount of pool filled by each hose in $2\frac{2}{5}$ hr. add up to 1 pool completely filled?

Large hose fills $\frac{1}{4}$ each hour for $2\frac{2}{5}$ hours:

$$\left(\frac{1}{4}\right)\left(2\frac{2}{5}\right) =$$

$$\left(\frac{1}{4}\right)\left(\frac{12}{5}\right) =$$

$$\left(\frac{4 \cdot 3}{4 \cdot 5}\right) = \frac{3}{5} \text{ of pool}$$

Small hose fills $\frac{1}{6}$ each hour for $2\frac{2}{5}$ hours:

$$\left(\frac{1}{6}\right)\left(2\frac{2}{5}\right) =$$

$$\left(\frac{1}{6}\right)\left(\frac{12}{5}\right) =$$

$$\left(\frac{6 \cdot 2}{6 \cdot 5}\right) = \frac{2}{5} \text{ of pool}$$

Together: $\frac{3}{5} + \frac{2}{5} = 1$ ✓

Now let's look at some examples where we know the time it takes to complete the job when the workers or machines work together. In these problems we'll find how long it will take each when working alone.

EXAMPLE 4 Rosa and Florence tiled the kitchen floor in 6 hours.

Florence noticed that Rosa used up twice as many boxes of tiles, so Rosa must have been working twice as fast. How long would it have taken Florence to do the job alone? How long would it have taken Rosa?

In this case what we don't know is the time each woman needs to do the job. Since Rosa works faster, it takes her less time, so call her time x. Florence takes twice that time, so represent her time as $2x$.

Worker	Hours to do job	Amount done in 1 hour
Rosa	x	$\frac{1}{x}$
Florence	$2x$	$\frac{1}{2x}$
Together	6	$\frac{1}{6}$

Amount done by Rose + amount done by Florence = amount done together.

Solve:
$$\frac{1}{x} + \frac{1}{2x} = \frac{1}{6} \quad \text{(LCM is } 6x\text{)}$$

$$6x\left(\frac{1}{x} + \frac{1}{2x}\right) = 6x\left(\frac{1}{6}\right)$$

$$6x\left(\frac{1}{x}\right) + 6x\left(\frac{1}{2x}\right) = 6x\left(\frac{1}{6}\right)$$

$$6 + 3 = x$$

$$9 = x$$

x represents Rosa's time, so it would take her 9 hours. $2x$ represents Florence's time, so it would take her

$$2x =$$
$$2(9) = 18 \text{ hr.}$$

Check: See if the amount of job each woman does in 1 hour adds up to $\frac{1}{6}$ of the job.

Amount done by Rosa in one hour is $\frac{1}{9}$.

Amount done by Florence in one hour is $\frac{1}{18}$.

$$\frac{1}{9} + \frac{1}{18} =$$

$$\frac{2}{18} + \frac{1}{18} =$$

$$\frac{3}{18} = \frac{1}{6} \checkmark$$

EXAMPLE 5 One company pollutes the air in the neighborhood to an unsafe level 3 times faster than another company that uses emission control devices. It takes 5 hours for the air to get to the unsafe level when both companies are working. How long would it take each company to reach that level when working alone?

Company	Hours to pollute air	Amount done in 1 hour
Without devices	x	$\frac{1}{x}$
With devices	$3x$	$\frac{1}{3x}$
Together	5	$\frac{1}{5}$

Amount polluted by company without devices

+ amount polluted by company with devices

= amount polluted together.

Solve:
$$\frac{1}{x} + \frac{1}{3x} = \frac{1}{5} \quad \text{(LCM is } 15x\text{)}$$

$$15x\left[\frac{1}{x} + \frac{1}{3x}\right] = 15x\left(\frac{1}{5}\right)$$

$$15x\left[\frac{1}{x}\right] + 15x\left[\frac{1}{3x}\right] = 15x\left(\frac{1}{5}\right)$$

$$15 + 5 = 3x$$

$$20 = 3x$$

$$\frac{20}{3} = \frac{3x}{3}$$

$$6\frac{2}{3} = x$$

x represents time for the company without the emission devices to pollute the air; it will do so in $6\frac{2}{3}$ hours.

$$3x =$$

$$3\left(6\frac{2}{3}\right) =$$

$$3\left(\frac{20}{3}\right) = 20$$

$3x$ represents the time for the company with the emission devices to pollute the air; it will do so in 20 hours.

Check: Working together at those rates, would they have reached $\frac{1}{5}$ of the unsafe level of pollution in 1 hour?

Company without devices reached $\left(\dfrac{1}{6\frac{2}{3}}\right) =$

$$\left(\dfrac{1}{\frac{20}{3}}\right) = \frac{3}{20}.$$

Company with devices reached $\left(\dfrac{1}{20}\right)$.

$$\frac{3}{20} + \frac{1}{20} =$$

$$\frac{4}{20} = \frac{1}{5} \quad \checkmark \qquad \square$$

EXAMPLE 6 Nick can clean up the playroom alone in 7 minutes. If it takes 6 minutes when Tim helps, how long should it take Tim to do it alone?

We'll let x represent the time it would take Tim to clean the room alone.

Person	Minutes to clean room	Amount done in 1 minute
Nick	7	$\dfrac{1}{7}$
Tim	x	$\dfrac{1}{x}$
Together	6	$\dfrac{1}{6}$

Amount cleaned by Nick
+ amount cleaned by Tim
= amount cleaned together.

Solve:

$$\frac{1}{7} + \frac{1}{x} = \frac{1}{6} \quad \text{(LCM is } 42x\text{)}$$

$$42x\left(\frac{1}{7} + \frac{1}{x}\right) = 42x\left(\frac{1}{6}\right)$$

$$42x\left(\frac{1}{7}\right) + 42x\left(\frac{1}{x}\right) = 42x\left(\frac{1}{6}\right)$$

$$
\begin{array}{rcl}
6x + 42 &=& 7x \\
-6x & & -6x \\
\hline
42 &=& x
\end{array}
$$

So it should take Tim 42 minutes to clean the room alone.

Check: Together would they complete job in 6 minutes?

Nick would do $\left(\dfrac{1}{7}\right)(6) = \dfrac{6}{7}$ of job.

Tim would do $\left(\dfrac{1}{42}\right)(6) =$

$$\frac{6}{42} = \frac{1}{7} \text{ of job.}$$

So together they would do $\dfrac{6}{7} + \dfrac{1}{7} = 1$ job. ✓

EXERCISES 14.9A

Use a table to record the given data and find the expressions you need to set up an equation. Solve the equation to obtain the answer to the problem.

1. Mary can paint a room in 3 hr by herself. Bill needs 6 hr to complete that job. How long would it take them if they work together?

2. Bob can wrap all the gifts in 4 min, but it will take Tom 12 min. How long would it take them if they work together?

3. It takes 5 min to fill the pool with the large hose, and 10 min with the small hose. How long would it take if both hoses are used at the same time?

4. It takes Judy 6 hr to fix the floor, but Jimmy can do it in 2 hr. How long would it take them if they work together?

5. We printed all the leaflets in 2 hr with the new press, but the old press used to get that job done in 3 hr. How long would it take if we use both presses together?

6. Maury devoured his ice cream sundae in 2 min, and Millie took 5 min to eat hers. How long would it take them to eat a sundae together?

7. Emily wallpapered the room alone in 4 hr. It took Jack 5 hr when he did that job. How long would it take them if they work together?

8. I can fill the tub in 3 min with a new blue hose, 4 min with a new red hose but it takes 6 min with the old hose. How long would it take to fill the tub if I use all 3 hoses?

9. Kate can type the paper 3 times as fast as John can. How long would it take each to type it if it takes 3 hr when they work together?

10. The old factory pollutes the air to an unsafe level 4 times faster than the new factory. How long would it take each to reach that unsafe level if it takes 4 hr when both factories are working?

11. It takes Leah twice as long as Peter to bake all the cookies needed for the party. How long would it take each if they can do it together in 2 hr?

12. The electric ice cream maker works 5 times faster than the old hand-cranking model. Using both of them, we were able to make all the ice cream for the bazaar in 5 hr. How long would it take if we used each alone?

13. The new press can print the job 4 times as fast as the old press. Working them together, it takes 2 hr to run the job. How long would it take using each of them alone?

14. I graded all the papers in 4 hr. When the tutor helped me, it took us 3 hr. How long would it take the tutor to do it herself?

15. Stan replaced the ceiling in 5 hr working alone. It took 3 hr when Janine helped. How long would it take her to do it alone?

Check your answers on page 757.

Motion Problems

The other kind of problem that can be represented and solved by rational equations involves motion.

Since distance = (rate)(time),

$$\text{time} = \frac{\text{distance}}{\text{rate}} \quad \text{and} \quad \text{rate} = \frac{\text{distance}}{\text{time}}.$$

So problems asking us to solve for time or rate often involve those fractional expressions, requiring us to solve rational equations.

Motion problems dealing with boats or planes often involve two speeds that affect the actual speed of the vehicle. With boats, one is the *speed of the boat in still water* (this would be the motor or the rower's speed), and the other is the *speed of the stream*. With planes, one is the *plane's speed on the speedometer*, and the other is the *speed of the wind*.

Canoeing downstream, for example, we go with the current, which pushes us ahead. We would move forward even without paddling. When moving with the stream (or flying with the wind), the vehicle's *actual speed* is the sum of the canoe and stream's (or plane and wind's) speed.

Similarly, canoeing upstream, we go against the current, which pushes us back. We would move backward without paddling. When moving against the current (or flying against the wind), the vehicle's *actual speed* is the difference of the canoe's (plane's) speed minus the stream's (wind's) speed.

If we let r represent the rate of speed of the vehicle (independent of stream or wind effects) and c represent the speed of the current (w for speed of the wind), then

Speed downstream $= r + c$ (with wind speed $= r + w$);

Speed upstream $= r - c$ (against wind speed $= r - w$).

In these motion problems you are either given the vehicle's speed and asked to find the speed of the stream (or wind) or you are given the speed of the stream (or wind) and asked to find the vehicle's speed.

Organize the data in a table with a column each for distance, rate, and time, as we do with every motion problem.

EXAMPLE 7   Pat and Jesse went on a canoe trip last year. It took them the *same time* to go 14 miles downstream as it did to go 8 miles upstream. If they always paddled at the same speed, and the current's speed was measured at 3 miles per hour, how fast did they paddle?

Let r represent the speed (in mph) that they paddled.

We represent time as the expression for distance divided by the expression for speed:

$$\text{time} = \left(\frac{\text{distance}}{\text{speed}}\right).$$

Direction	Distance (in miles)	Speed (in mph)	Time (in hours)
Downstream	14	$r + 3$	$\dfrac{14}{r + 3}$
Upstream	8	$r - 3$	$\dfrac{8}{r - 3}$

Since the time for both excursions was the same,

Time upstream = time downstream.

$$\frac{14}{r+3} \quad \frac{8}{r-3} \qquad \text{(We can cross multiply.)}$$

$$
\begin{aligned}
14(r-3) &= 8(r+3) \\
14r - 14(3) &= 8r + 8(3) \\
14r - 42 &= 8r + 24 \\
+42 &\quad +42 \\
\hline
14r &= 8r + 66 \\
-8r &\quad -8r \\
\hline
6r &= 66 \\
r &= 11
\end{aligned}
$$

They paddled at a rate of 11 miles per hour.

Check: Was the time the same for both excursions?

> *Actual speed* downstream = 14 (11 + 3) mph
> Time to go 14 mi at 14 mph = 1 hr ✓
> *Actual speed* upstream = 8 (11 − 3) mph
> Time to go 8 mi at 8 mph = 1 hr. ✓

Now let's look at an example where we know the boat's speed and want to find the speed of the stream.

EXAMPLE 8 Ben and John have timed their paddling in still lake water and figure they paddle at 8 miles per hour. On a river trip they traveled 30 miles downstream in the *same time* it took them to go 18 miles back upstream. How fast was the river flowing?

Let c represent the speed of the river's current.

Direction	Distance (in miles)	Speed (in mph)	Time (in hours)
Downstream	30	8 + c	$\dfrac{30}{8+c}$
Upstream	18	8 − c	$\dfrac{18}{8-c}$

Time upstream = time downstream

$$\frac{30}{8+c} \quad \frac{18}{8-c} \qquad \text{(Use cross multiplication.)}$$

$$
\begin{aligned}
30(8-c) &= 18(8+c) \\
240 - 30c &= 144 + 18c \\
+30c &\quad\quad +30c \\
\hline
240 &= 144 + 48c \\
-144 &\quad -144 \\
\hline
96 &= 48c \\
2 &= c
\end{aligned}
$$

The river was flowing at 2 miles per hour.

Check: Was the time of both trips the same?

$$\text{Actual speed downstream} = 10\,(8+2)\text{ mph}$$

$$\text{Time} = \frac{30\text{ mi}}{10\text{ mph}} = 3\text{ hr} \quad \checkmark$$

$$\text{Actual speed upstream} = 6\,(8-2)\text{ mph}$$

$$\text{Time} = \frac{18\text{ mi}}{6\text{ mph}} = 3\text{ hr} \quad \checkmark$$

Now let's look at an example involving planes and wind speed.

EXAMPLE 9 A plane travels at an average rate of 180 miles per hour in still air. One day it took the *same time* to fly 832 miles with the wind as it did to fly 608 miles against the wind. How fast was the wind blowing that day? Let w represent the speed at which the wind was blowing.

Direction	Distance (in miles)	Speed (in mph)	Time (in hours)
With wind	832	$180 + w$	$\dfrac{832}{180 + w}$
Against wind	608	$180 - w$	$\dfrac{608}{180 - w}$

Since both trips took the same time,

Time with wind = time against wind.

$$\frac{832}{180 + w} \quad \diagdown \kern-0.9em\diagup \quad \frac{608}{180 - w} \qquad \text{(Cross multiply.)}$$

$$832(180 - w) = 608(180 + w)$$
$$832(180) - 832w = 608(180) + 608w$$
$$149{,}760 - 832w = 109{,}440 + 608w$$
$$\underline{+\,832w \qquad\qquad +\,832w}$$
$$149{,}760 = 109{,}440 + 1{,}440w$$
$$\underline{-109{,}440 \quad -109{,}440}$$
$$40{,}320 = 1{,}440w$$
$$\frac{40{,}320}{1{,}440} = \frac{1{,}440w}{1{,}440}$$
$$28 = w$$

The wind was blowing at 28 miles per hour that day.

Check: Was the time of both trips the same?

$$\text{Actual speed with wind} = 208\,(180 + 28)\text{ mph}$$

$$\text{Time} = \frac{832\text{ mi}}{208\text{ mph}} = 4\text{ hr} \quad \checkmark$$

$$\text{Actual speed against wind} = 152\,(180 - 28)\text{ mph}$$

$$\text{Time} = \frac{608\text{ mi}}{152\text{ mph}} = 4\text{ hr} \quad \checkmark$$

Use a table to record the given data and find the expressions you need to set up an equation. Solve the equation to obtain the answer to the problem.

16. Doree traveled 40 mi with the current in the same time it took her to go 8 mi against the current. If the speed of the river was 4 mph, how fast was her still-water speed?

17. When the speed of the stream was 2 mph, Ami moved 48 mi with the current in the same time it took her to go 32 mi against the current. What was her still-water speed?

18. Mike's boat traveled 14 mph in still water. He went 51 mi with the current in the same time it took to go 33 mi against the current. What was the speed of the stream?

19. Rick's boat traveled 20 mph in still water. He went 96 mi with the current in the same time it took to go 64 mi against the current. What was the speed of the river?

20. A plane went 630 mi with the wind in the same time it took to fly 570 mi against the wind. If the plane's still-air speed was 200 mph, what was the speed of the wind?

21. On a day when the speed of the wind was 8 mph, it took a plane the same time to fly 900 mi with the wind as it took to cover 804 mi against the wind. What was the plane's still-air speed?

22. Ari can row 18 mph in still water. He went 57 mi with the current in the same time it took him to go 51 mi against the current. What was the speed of the current?

23. A jet flew 1,044 mi against the wind in the same time it took to fly 1,116 mi with the wind. What was the speed of the jet if its still-air speed was 360 mph?

24. When the wind was blowing at 6 mph, the plane took the same time to fly 904 mi with the wind as it took to go 856 mi against it. What was the plane's still-air speed?

25. During a 4 mph wind, it takes the same time to go 130 mi against the wind as it takes to go 170 mi with the wind. What is the still-air speed?

Check your answers on page 757.

Glossary

algebraic fraction Fraction with at least one letter in the numerator or denominator; ratio of algebraic terms or polynomials.

Example: $\frac{3x-5}{2x}$ is an algebraic fraction.

LCM (lowest common multiple) Smallest expression into which each expression in a set divides exactly.

Example: The LCM of $2x$, $(x+4)(x-4)$, and $2(x-4)$ is $2x(x+4)(x-4)$.

motion problems   Real-life word problems involving situations where you are given some, but not all, of the information about the rate, time, and distance traveled. You must set up an equation and solve for the unknown information.

Example: (See Section 14.9.)

rational equations   Equations containing rational expressions.

Example: $\dfrac{3}{(x + 2)} + \dfrac{2}{(x^2 - 4)} = \dfrac{1}{(x - 2)}$

rational expressions   Ratio of two polynomials that is expressed as an algebraic fraction.

work problems   Real-life word problems involving situations where you are told something about the rate of time it takes for certain jobs to be done.

If you are given the individuals' rates, you can set up an equation and solve for the rate if they were to work together.

If you are given the rate it takes them working together, you can set up an equation and solve for their individual rates.

Example: (See Section 14.9.)

Guide for Working with Algebraic Fractions and Rational Equations

algebraic fractions   Simplify by factoring the numerator and the denominator; then cancel any factors common to both. Leave answer in factored form.

Fractions must have *equal denominators* before we can *add* or *subtract* them. If they do not have equal denominators, find the LCM of all denominators, and change the name of each fraction by multiplying its numerator and denominator by the factor(s) that will produce that LCM as the denominator.

Add by adding the like terms in numerators; their common denominator is the denominator of their sum.

Subtract by changing the sign of every term in the numerator of the subtracting fraction; then combine like terms from numerators and keep their common denominator.

Fractions do *not* have to have equal denominators in order for us to *multiply* or *divide* them.

Multiply by factoring each numerator and denominator; then cancel any factors common to both.

Divide by replacing the dividing fraction (the one following ÷) by its reciprocal and multiply. The reciprocal is found by reversing a fraction's numerator and denominator.

solve rational equations   Multiply all terms on each side by the LCM of all denominators. That cancels out all denominators so it is no longer rational. In this chapter the equation will be linear, so solve it by isolating the letter on one side.

Always check that your solution does not make any denominator in the original equation equal to zero. Such values are not solutions.

Chapter 14 Review

Do each of the problems. After you are confident that you have done them as accurately as possible, compare your answers with those in the back of the book. If any are wrong, go back to the section where the problem came from (indicated to the left of the problem) and review the section.

Once you understand all sections, you will be sure to learn the skills solidly and remember how to do them if you practice them a little bit more. Turn to the Supplementary Exercises and do all the problems from any section where you had difficulty on these review exercises.

14.1 State the number of prime factors in each expression.

1. $3xy$

2. $4x(x-1)^2$

Determine whether the expressions in each pair are the same or not.

3. $5x - 1$; $5x + 1$

4. $6y + 3$; $3 + 6y$

Reduce each fraction to its simplest form.

5. $\dfrac{15x^2y(x-2)(x+3)}{20xy^3(3+x)(x+2)} =$

6. $\dfrac{3x^2y - 5xy - 2y}{x^2y^2 - 4y^2} =$

7. Divide $y^2 + y - 12$ by $y + 4$.

14.2 Multiply the fractions, leaving your answers in the most simplified factored form.

8. $\dfrac{6m}{m+2} \cdot \dfrac{(m-2)(m+2)}{4m^2} =$

9. $\dfrac{x^2 + 7x + 12}{x^2 - 1} \cdot \dfrac{x^2 - 2x + 1}{x^2 + 4x} =$

10. $\dfrac{3x - 12}{3x^2 - 5x + 2} \cdot \dfrac{x^3 - 25x}{x^2 - 5x + 4} \cdot \dfrac{6x^2 - 4x}{x^2 + 5x} =$

14.3 Divide the fractions, leaving your answers in simplified factored form.

11. $\dfrac{m^2 + 4m + 3}{8m^3 - 6m^2} \div \dfrac{m^2 - 9}{4m^3} =$

12. $\dfrac{\dfrac{x-4}{x^2 + 5x + 6}}{\dfrac{5x^2 - 20x}{x^2 + 3x}} =$

14.4 Combine the fractions.

13. $\dfrac{6x}{3x - 1} + \dfrac{x - 4}{3x - 1} + \dfrac{7}{3x - 1} =$

14. $\dfrac{8x - 1}{4x - 5} - \dfrac{3x - 2}{4x - 5} =$

15. $\dfrac{5}{2x + 1} + \dfrac{3x - 1}{2x + 1} - \dfrac{8}{2x + 1} =$

14.5 Find the least common multiples.

16. $4xy^2$ and $10x^3y$

17. $5x(y - 3)$ and $3y(y - 3)^2$

18. $2x^2 + x - 15$ and $x + 3$

19. $4x^2 + 7x - 2$ and $x^2 + 8x + 12$

14.6 Combine the fractions, leaving your answers in the most simplified factored form.

20. $\dfrac{3x}{x + 4} + \dfrac{2x}{x - 3} =$

21. $\dfrac{a + 2b}{5a + b} + \dfrac{3a - b}{2a - b} =$

22. $\dfrac{x - 1}{x + 3} - \dfrac{x + 2}{x + 4} =$

23. $\dfrac{2}{y^2 + 4y + 4} - \dfrac{3}{y + 2} =$

Continued

Answers

1. _____
2. _____
3. _____
4. _____
5. _____
6. _____
7. _____
8. _____
9. _____
10. _____
11. _____
12. _____
13. _____
14. _____
15. _____
16. _____
17. _____
18. _____
19. _____
20. _____
21. _____
22. _____
23. _____

Chapter 14 Review (*Cont.*)

24. $\dfrac{x+2}{x^2+4x+3} + \dfrac{x-5}{x^2+6x+9} =$

25. $\dfrac{3}{x^2-4x} + \dfrac{1}{x} - \dfrac{2}{x-4} =$

14.7 26. Simplify $\dfrac{10+\sqrt{28}}{6}$.

Solve and check.

14.8 27. $\dfrac{3}{x} + \dfrac{4}{(x-2)} = \dfrac{9}{x}$

28. $\dfrac{7}{(x+4)} + \dfrac{2}{(x^2-16)} = \dfrac{1}{(x-4)}$

14.9 29. Niko can build the model in 5 hr, but when Jani helps him they can do it in 2 hr. How long would it take Jani to do it alone?

30. Fran went 88 mi with the current in the same time it took her to go only 72 mi against the current. If the river was flowing at 2 mph, how fast was her still-water speed?

Answers

24. _____

25. _____

26. _____

27. _____

28. _____

29. _____

30. _____

Check your answers on page 757.

Supplementary Exercises

Do all the problems in every section that involves a skill in which you now lack complete mastery. With a little more practice you will achieve that sense of really understanding the topics, and you will remember how to do these problems.

14.1 How many prime factors are included in each expression?

   1. $3xw$ **2.** $5y(y - 2)$

   3. $15x^2y(x - 4)^2$

Determine whether the expressions in each pair are the same or not.

   4. $3x + 2$ and $2 + 3x$ **5.** $7y - 3$ and $3 - 7y$

Reduce each fraction to its simplest form.

   6. $\dfrac{6xy(x + 1)}{10xw(x - 1)} =$ **7.** $\dfrac{12x^3y(x + 2)(3 - y)}{20xy^2(y + 3)(2 + x)} =$

   8. $\dfrac{y^2 - 7y + 10}{y^3 - 25y} =$ **9.** $\dfrac{3x^2y - 15xy}{6x^2y^2 - 30xy^2} =$

   10. Divide $x^2 - 4x + 3$ by $x - 1$.

14.2 Multiply these fractions, leaving your answers in the most simplified factored form.

   11. $\dfrac{10a^2}{20a^2(a + 3)} \cdot \dfrac{(a - 3)(4ab)}{(2a + 1)(a - 3)} =$

   12. $\dfrac{6ab(b - 3)}{(4 + 5b)(a + b)} \cdot \dfrac{(2b - 3)(5b + 4)}{10a^2(b - 3)} =$

   13. $\dfrac{x^2 + 7x}{x^2 + 10x + 9} \cdot \dfrac{x^2 - 81}{x^2 + 9x} =$ **14.** $\dfrac{x^2 + 11x + 30}{x^2 + 6x} \cdot \dfrac{x^3 - 36x}{x^2 + 10x + 25} =$

   15. $\dfrac{x^2 - 6x}{x^2 + 3x} \cdot \dfrac{x^2 - 9}{x^2 + 2x + 1} \cdot \dfrac{x^2 + 3x + 2}{x^2 - 36} =$ **16.** $\dfrac{x^2 - 3x - 4}{x^2 + 3x} \cdot \dfrac{x^2 - 4x}{20xy^3} \cdot \dfrac{5x^2y}{x^2 - 1} =$

14.3 Divide the fractions.

   17. $\dfrac{3x(x - 4)}{6(x - 1)} \div \dfrac{15(x + 2)}{8(x - 1)} =$ **18.** $\dfrac{x^2 + 4x}{x^2 + 2x} \div \dfrac{x^3 - 16x}{x^2 + 4x + 4} =$

   19. $\dfrac{\dfrac{3xy}{5y^2}}{\dfrac{7y}{6x}}$ **20.** $\dfrac{\dfrac{x^2 - x - 12}{x + 5}}{\dfrac{x^2 - 9}{2x^2 + 10x}} =$

14.4 Combine the fractions.

   21. $\dfrac{4}{x + 1} + \dfrac{5}{x + 1} =$ **22.** $\dfrac{5x}{4x - 3} + \dfrac{2x - 1}{4x - 3} =$

   23. $\dfrac{7}{3x} - \dfrac{9}{3x} =$ **24.** $\dfrac{3x + 2}{5x - 1} - \dfrac{4x - 8}{5x - 1} =$

Answers

1. _____

2. _____

3. _____

4. _____

5. _____

6. _____

7. _____

8. _____

9. _____

10. _____

11. _____

12. _____

13. _____

14. _____

15. _____

16. _____

17. _____

18. _____

19. _____

20. _____

21. _____

22. _____

23. _____

24. _____

Continued

Supplementary Exercises (*Cont.*)

25. $\dfrac{7x}{x+2} + \dfrac{3-2x}{x+2} - \dfrac{4x-1}{x+2} =$

26. $\dfrac{4x-3}{x+8} + \dfrac{8-2x}{x+8} - \dfrac{5x-1}{x+8} =$

14.5 Find the least common multiples.

27. $3ab$ and $5ac$

28. $9xy^2$ and $12x^3y$

29. $8xy(x+3)$ and $6y(x+3)^2$

30. $4x(x-2)(x+3);\quad 10xy(x+2)(3+x)$

31. $x^2 + 3x - 10;\quad x^2 + 6x + 5$

32. $2x+5$ and $6x^2 + 7x - 20$

14.6 Combine the fractions, leaving your answers in the most simplified factored form.

33. $\dfrac{8}{x-1} + \dfrac{3}{x+2} =$

34. $\dfrac{7y}{y-5} + \dfrac{5}{y} =$

35. $\dfrac{7x-1}{x^2+6x} + \dfrac{3x+2}{x^2+3x} =$

36. $\dfrac{3y}{y-4} - \dfrac{8}{y} =$

37. $\dfrac{5x-2}{x^2+4x-5} + \dfrac{3x+1}{x+5} =$

38. $\dfrac{7}{x+4} + \dfrac{3x}{x^2+4x} - \dfrac{5}{x^2} =$

14.7 **39.** Simplify $\dfrac{6-\sqrt{27}}{15}$.

14.8 Solve and check.

40. $\dfrac{3}{y} + \dfrac{1}{(y+2)} = \dfrac{11}{(3y)}$

41. $\dfrac{2}{(x-1)} + \dfrac{2}{(x^2-1)} = \dfrac{4}{(x+1)}$

14.9 **42.** Luke can build a model in 8 hr that Sonya can build in 2 hr. How long should it take them to build it together?

43. The plane flew 795 mi with the wind in the same time it took to fly 765 mi against the wind. How fast was its still-air speed if the wind was blowing at 5 mph?

25. _____

26. _____

27. _____

28. _____

29. _____

30. _____

31. _____

32. _____

33. _____

34. _____

35. _____

36. _____

37. _____

38. _____

39. _____

40. _____

41. _____

42. _____

43. _____

Check your answers on page 757.

Chapter 14 Test

Combine and simplify fractions, leaving your answers in the most simplified factored form.

Answers

1. $\dfrac{8x^2y(2x+3)}{10xy^3(3+2x)} =$

2. $\dfrac{(x^2-x-20)(x^2-7x)}{x^2-12x+35} =$

3. $\dfrac{21x^2(x-3)}{(x-1)(x+1)} \cdot \dfrac{(2x-3)(x+1)}{35x(x+2)} =$

4. $\dfrac{x^2-8x+15}{x^2-49} \cdot \dfrac{x^2-7x}{x^2-5x-6} =$

5. $\dfrac{4(p-1)}{(p-2)^2} \cdot \dfrac{p^2-6p}{p^2-1} \cdot \dfrac{p^2-4p+4}{2p^2-18} =$

6. $\dfrac{k^2+3k-28}{10k^2-6k} \div \dfrac{k^2-16}{k^3-k^2} =$

7. $\dfrac{\dfrac{2x-1}{x^2-5x}}{\dfrac{8x^2-4x}{x^2-25}} =$

8. $\dfrac{4x-3}{3x-5} + \dfrac{7x+1}{3x-5} =$

9. $\dfrac{9y-4}{8y+1} - \dfrac{4y+7}{8y+1} =$

10. $\dfrac{6x-5}{x^2+5x} + \dfrac{3x-1}{x+5} =$

11. $\dfrac{7x}{x^2-4x} - \dfrac{3}{5x} =$

12. $\dfrac{4y+1}{y^2-4y-5} + \dfrac{3y-2}{y^2+3y+2} =$

13. $\dfrac{5}{y} + \dfrac{2y}{y^2-3y} - \dfrac{4}{y-3} =$

14. $\dfrac{15-\sqrt{75}}{10} =$

15. Divide (w^2-2w-8) by $(w+2)$.

Solve and check.

16. $\dfrac{1}{(x+2)} + \dfrac{1}{x} = \dfrac{5}{(3x)}$

17. $\dfrac{14}{(x+3)} - \left[\dfrac{7}{(x^2-9)}\right] = \dfrac{1}{(x-3)}$

18. Lee could paddle his canoe 60 mi with the current in the same time it took him to canoe 40 mi against the current. If the speed of the river was 2 mph, how fast was his still-water speed?

19. Alice can arrange a client's file for trial 3 times as quickly as I can. When we work together, it only takes us 3 hr to prepare a file. How long would it take each of us to do it alone?

20. Betsy can paint the stairs in 5 hr but it takes Dave 10 hr to do the same job. How long should it take them if they work together?

Answers

1. _____

2. _____

3. _____

4. _____

5. _____

6. _____

7. _____

8. _____

9. _____

10. _____

11. _____

12. _____

13. _____

14. _____

15. _____

16. _____

17. _____

18. _____

19. _____

20. _____

Check your answers on page 758.

Cumulative Review
CHAPTERS 1–14

Combine and simplify.

Answers

1. $\dfrac{7}{12} + \dfrac{2}{3} - \dfrac{5}{6} =$

2. $\dfrac{18}{40} \div \dfrac{27}{50} =$

3. $1.008 + 0.34 + 0.06 + 5.1 - 3.0008 =$

4. $\dfrac{0.71352}{0.003} =$

5. 7.2% of $\$198.60 =$

6. $13 - 5(6 - 4) + 3(8 - 5)^2 =$

7. $3 + (-4) - (-8) + (-3) - (7) =$

8. $3 + 4(5 - 2 + 1) + \dfrac{12}{(-6)} =$

9. $4xy^3 - yx + 7y^2x + y^3x - 6xy^2 + 2xy =$

10. $\dfrac{14x^5yw^3}{20xy^4w^2} =$

11. $(-2a^3bc^{-2})^{-1} =$

12. $(6xy^3w^{-2})(-5x^4y^{-1}w^7) =$

13. $(3.1 \times 10^{-4})(2.2 \times 10^7) =$

14. $(3y^2 - 5y + 8) - (y^2 - 5y + 2) =$

15. $-5x(3x^2 - 4x + 1) =$

16. $(8)^{2/3} + (27)^{1/3} + (8)^{-1/3} =$

17. $4(3x^2 - x + 7) - 5(x^2 + 2x - 1) =$

18. $(-x^3yw)(-5x^2yw^5) =$

19. $(8x - 5)(x + 3) =$

20. $(x + 2)(x^2 - 3x + 5) =$

21. $\dfrac{y^2 - 5y}{y(y - 1)} \cdot \dfrac{y^2 - 7y + 6}{y^2 - 25} =$

22. $\sqrt{75a^5b^6} =$

23. $\sqrt{7} - 3\sqrt{5} + \sqrt{28} - \sqrt{20} =$

Solve each equation.

24. $7x - 3(2x + 1) = 2x - 5$

25. $\dfrac{y}{2} + 7 = \dfrac{y}{4} + 6$

1. _____

2. _____

3. _____

4. _____

5. _____

6. _____

7. _____

8. _____

9. _____

10. _____

11. _____

12. _____

13. _____

14. _____

15. _____

16. _____

17. _____

18. _____

19. _____

20. _____

21. _____

22. _____

23. _____

24. _____

25. _____

26. $\dfrac{5}{2x} - \dfrac{1}{x-1} = \dfrac{1}{x}$

Solve each system of equations.

27. $x = 3y - 5$; $4x + 3y = 10$

28. $5x - 2y = 5$; $3x + 3y = 24$

Sketch the graph of each expression.

29. $y = 3x - 1$ **30.** $y < 3x - 1$ **31.** $y \geq 3x - 1$

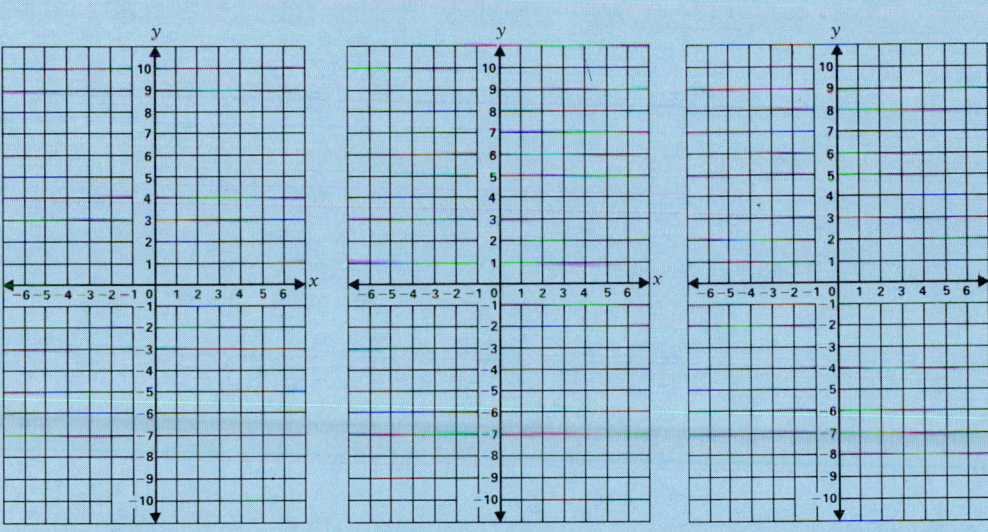

32. Find the slope and y-intercept of $6y - 5x = 7$.

33. Find the equation of the line that is parallel to $y = 3x - 1$ and passes through point $(4, -3)$.

34. Convert $23°F$ into Celsius.

35. What must be added to 20 g of a 60 % salt solution to make it into a 20 % solution?

36. Randy drove at a rate of 40 mph on his trip to the mountains. Fran left 2 hr after Randy did, from the same place as he, to bring him the news that he won the lottery. If she drove at 60 mph, how many hours did it take her to catch up with him?

37. Gregory bought two granola bars and five small boxes of raisins for $1.65. Sharon bought five granola bars and two small boxes of raisins for $1.50. How much did each granola bar and each box of raisins cost?

Answers

26. _____

27. _____

28. _____

32. _____

33. _____

34. _____

35. _____

36. _____

37. _____

Check your answers on page 758.

15 Quadratic Equations, Graphs, and Solutions

15.1 Graphing Quadratic Equations (Parabolas)

The equations we have solved so far have all been linear. As we discussed in Section 5.1, linear equations are also called *equations of degree 1*.

In this chapter we will study **quadratic equations**—equations of *degree 2*. That means that a term in each equation contains two letter factors, but no term has more than two letter factors.

An equation that is quadratic in one variable contains a term with that variable factor repeated, such as $7xx$ (usually written as $7x^2$). We will study equations that are quadratic in only one variable.

Quadratic equations involving one letter can be simplified. We can move all terms to one side of the equation and combine like terms, so the equation simplifies to three terms at most. There must be a term of *degree 2*; there might also be a term of *degree 1*; and there might be a *constant* term.

If the letter involved is x, we can express that quadratic equation in general as

$$ax^2 + bx + c = 0, \text{ where } a, b, \text{ and } c \text{ are numbers.}$$

ax^2 is called the "squared" term. It has a degree of 2. a can be any number other than zero.

bx is called the "linear" term. It has a degree of 1.

c is the constant term. b and c can be any numbers.

EXAMPLE 1 $3x^2 - 5x + 7 = 0$; $y^2 + 4y - 1 = 0$; $2w^2 + 5 = 0$;   and $z^2 - 7z = 0$

are quadratic equations in one letter.   ❑

Quadratic Equations: $f(x) = ax^2 + bx + c$

In Chapter 8 we learned to graph linear equations, $y = mx + b$, where m and b were numbers. In this section we learn to graph quadratic equations in two letters of the form $y = ax^2 + bx + c$, where a, b, and c are numbers.

There is always an infinite number of pairs of solutions to an equation with two letters (linear or quadratic), because for every choice of x we get a value for y. Just as we set up $y = mx + b$ as a function of x [$f(x) = mx + b$] in Section 8.4, we can often represent such equations in two letters as a function of one of them.

We find each value of y in $y = ax^2 + bx + c$ by evaluating the right side of that equation for each choice of x. So we can represent such equations in functional notation too.

$$y = ax^2 + bx + c \text{ can be expressed as } f(x) = ax^2 + bx + c.$$

Functional notation simplifies our work of finding the coordinates of points that are solutions to equations.

In Chapter 8 we graphed equations of degree 1 and found the graphs always to be straight lines. Graphs of equations of degree 2 are curves, not straight lines. One reason is that, in quadratic equations, two different values of x can produce the same value of $f(x)$. For example, if $f(x) = x^2$, every number x, except zero, will produce the same value for $f(x)$ that the negative of x will.

The graphs of quadratic equations, with one squared letter term $f(x) = ax^2 + bx + c$, are symmetric, almost U-shaped curves, called **parabolas**, which either open upward and have a lowest point (called a **vertex**) as in this graph:

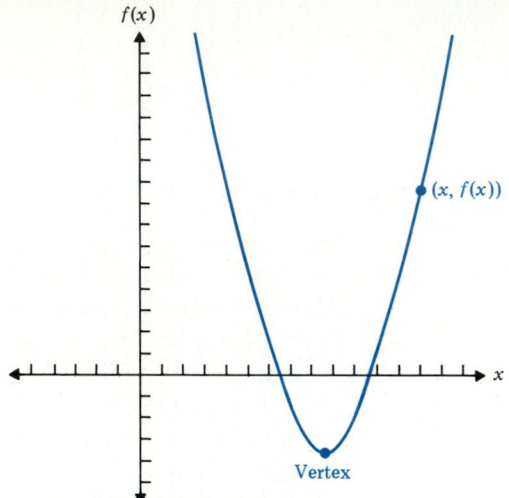

Figure 15.1

or open downward and have a highest point (also called a vertex) as in this graph:

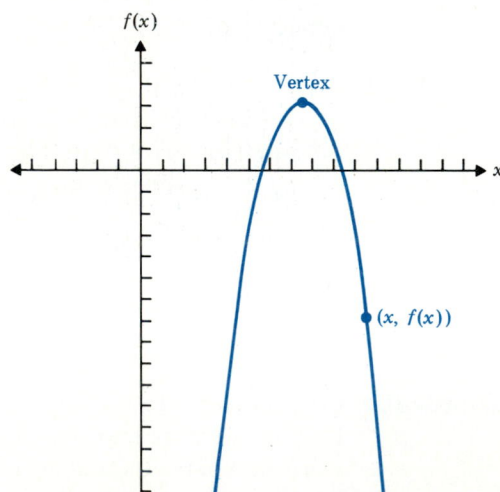

Figure 15.2

Our strategy for graphing linear equations was to find the coordinates of *any three points*. That method worked because the graph of a linear equation will always be a *straight line*. It will not work for graphing quadratic equations. If the three points are on the same side of the vertex, plotting them will not give us enough information about the curve.

Some people graph parabolas by continually choosing numbers for x until enough points are plotted to get a good sense of the curve. If the vertex is far from the origin $(0, 0)$, that method could take a lot of time and much more work than necessary. Let's try graphing a parabola by that method.

EXAMPLE 2 Sketch the graph of $f(x) = x^2 + 2x + 1$.

Let's choose $3, 2, 1, 0, -1, -2,$ and -3 as values for x.

With seven points we should get a good sketch of the parabola.

If $x = 3$, $f(3) = (3)^2 + 2(3) + 1$; so $f(3) = 16$, or point $(3, 16)$.

If $x = 2$, $f(2) = (2)^2 + 2(2) + 1$; so $f(2) = 9$, or point $(2, 9)$.

If $x = 1$, $f(1) = (1)^2 + 2(1) + 1$; so $f(1) = 4$, or point $(1, 4)$.

If $x = 0$, $f(0) = (0)^2 + 2(0) + 1$; so $f(0) = 1$, or point $(0, 1)$.

If $x = -1$, $f(-1) = (-1)^2 + 2(-1) + 1$; so $f(-1) = 0$, or point $(-1, 0)$.

If $x = -2$, $f(-2) = (-2)^2 + 2(-2) + 1$, so $f(-2) = 1$, or point $(-2, 1)$.

If $x = -3$, $f(-3) = (-3)^2 + 2(-3) + 1$, so $f(-3) = 4$, or point $(-3, 4)$.

x	3	2	1	0	-1	-2	-3
$f(x)$	16	9	4	1	0	1	4

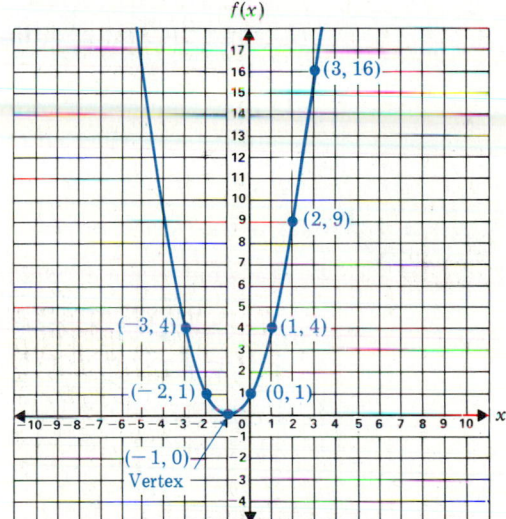

It looks as if the lowest point is $(-1, 0)$, so that point should be the vertex.

We will learn an easier and more exact way of determining the vertex in the next few pages.

You can sketch a parabola by using only three points, as long as the *first* one is the *vertex* and the other two are on either side of the vertex. So the easiest way to graph parabolas is to learn how to identify the vertex.

Vertex: Parabolas That Open Upward

For simplicity, let's first discuss parabolas that open upward like the one illustrated in Fig. 15.1. We'll come back to the ones that open downward later.

The vertex is the lowest point on parabolas that open upward. That means the vertical (second) coordinate of the vertex is the smallest possible value of $ax^2 + bx + c$. There is no easy way to identify the lowest value of $ax^2 + bx + c$. Suppose, however, we could convert the equation into the form of

$$f(x) = \text{(a squared term)} + \text{(a constant term)}, \text{ like } f(x) = x^2 + 3.$$

A *squared* expression is never negative; when you multiply any number by itself (positive or negative), the product is always positive. So the *smallest value of the squared term* will be **zero**, and the *smallest value of f(x)* will therefore be *zero plus the constant (which is just the constant term)*. These observations will make it easy for us to find the coordinates of the vertex.

A quadratic equation, $f(x) = ax^2 + bx + c$, can always be converted into the form of simply a squared term plus a constant, so we can readily find the vertex. We will learn how to perform that conversion on some quadratic equations later in this section. We say the quadratic equation is in **standard form** when expressed that way. (The squared term will not always be as simple as x^2, of course. It might be a binomial squared, like $[x + 2]^2$; or it might be a binomial squared with a coefficient, like $3[x + 2]^2$.) We can write this *standard form* in general as

$$f(x) = a(x + h)^2 + k, \text{ where } a, h, \text{ and } k \text{ are numbers.}$$

Remember that every point on the graph of $f(x) = a(x + h)^2 + k$ has coordinates $[x, f(x)]$.

To find the coordinates of the vertex, set the squared factor $[x + h]$ equal to zero *(its lowest value)*. Then:

1. Find the first coordinate of the vertex by calculating the value of x (when $x + h = 0$).

2. Find the second coordinate of the vertex by calculating the value of $f(x)$ (for that x).

EXAMPLE 3   What is the vertex of $f(x) = (x + 2)^2 + 7$?

1. For what value of x does $(x + 2) = 0$?

$$(x + 2) = 0 \text{ when } x = -2.$$

So that is the first coordinate of the vertex.

2. Evaluate $f(-2)$ to find the second coordinate of the vertex.

$$f(-2) = [(-2) + 2]^2 + 7$$
$$f(-2) = 0 + 7$$
$$f(-2) = 7$$

So the vertex of $f(x) = (x + 2)^2 + 7$ is $(-2, 7)$.   ❑

EXAMPLE 4   What is the vertex of $f(x) = (x - 4)^2$?

1. $(x - 4) = 0$ when $x = 4$.
2. $f(4) = [(4) - 4]^2$
   $f(4) = 0$

So the vertex of $f(x) = (x - 4)^2$ is $(4, 0)$.   ❑

EXAMPLE 5   What is the vertex of $f(x) = 2(x + 1]^2 - 5$?

1. $(x + 1) = 0$ when $x = -1$.
2. $f(-1) = 2[(-1) + 1]^2 - 5$
   $f(-1) = 2[0]^2 - 5$
   $f(-1) = 0 - 5$
   $f(-1) = -5$

So the vertex of $f(x) = 2(x + 1)^2 - 5$ is $(-1, -5)$.   ❑

EXAMPLE 6 What is the vertex of $f(x) = x^2$?

The squared term, x^2, $= 0$ when $x = 0$, and $f(0) = (0)^2$
$$= 0$$

So the vertex of $f(x) = x^2$ is $(0, 0)$. ❑

EXAMPLE 7 What is the vertex of $f(x) = 3x^2$?

The squared term, x^2, $= 0$ when $x = 0$, and $f(0) = 3(0)^2$
$$= 0$$

So the vertex of $f(x) = 3x^2$ is $(0, 0)$. ❑

EXAMPLE 8 What is the vertex of $f(x) = x^2 + 3$?

The squared term, x^2, $= 0$ when $x = 0$, and $f(0) = (0)^2 + 3$
$$= 3$$

So the vertex of $f(x) = x^2 + 3$ is $(0, 3)$. ❑

Vertex: Parabolas That Open Downward

Since squared terms are always positive or zero, the values of $f(x)$ will get bigger for large choices of x if that squared term is multiplied by a positive number. The vertex will be the lowest point on the curve.

If the *squared term is multiplied by a negative number*, the product will always be *negative* or zero, so the values of $f(x)$ will get smaller for large choices of x. These parabolas will open downward (instead of upward), and the vertex will be the *highest* point on the curves.

The parabola will open downward if the coefficient of the squared term is negative.

The parabola will open upward if the coefficient of the squared term is positive.

In other words, when the quadratic equation is in standard form, $f(x) = a(x + h)^2 + k$, it will open upward if a is positive and downward if a is negative. In either event, you find the vertex the same way:

First coordinate: Value of x that makes $(x + h) = 0$.

Second coordinate: Value of $f(x)$ when $(x + h)^2$ is zero.

Now let's look at the parabolas that open downward like the one illustrated in Fig. 15.2. The vertex of such parabolas is the *highest* point, or the *largest* value, that $f(x)$ can be for that equation.

EXAMPLE 9 What is the vertex of $f(x) = -3(x + 4)^2 + 1$?

First coordinate: $(x + 4) = 0$ when $x = -4$.

Second coordinate: $f(-4) = -3(-4 + 4)^2 + 1$
$$f(-4) = -3(0) + 1$$
$$f(-4) = 1.$$

So the vertex is $(-4, 1)$. ❑

EXAMPLE 10 What is the vertex of $f(x) = -7(x - 2)^2 - 3$?

First coordinate: $(x - 2) = 0$ when $x = 2$.

Second cordinate: $f(2) = -7(0) - 3$
$$f(2) = -3.$$

So the vertex is $(2, -3)$. ❑

EXAMPLE 11   What is the vertex of $f(x) = -x^2$?

When $x = 0$, the squared term $(-x^2) = 0$, so $f(0) = 0$.

The vertex is **(0, 0)**.

EXERCISES 15.1A

Find the coordinates of the vertex for each equation.

1. $f(x) = (x - 3)^2 + 4$

2. $f(x) = (x - 8)^2 + 1$

3. $f(x) = (x + 1)^2 - 6$

4. $f(x) = (x + 5)^2 - 2$

5. $f(x) = (x - 2)^2$

6. $f(x) = (x + 7)^2$

7. $f(x) = (x - 4)^2 - 1$

8. $f(x) = (x + 3)^2 + 3$

9. $f(x) = 3(x - 1)^2 + 5$

10. $f(x) = 4(x + 3)^2 - 2$

11. $f(x) = 2x^2$

12. $f(x) = 4x^2$

13. $f(x) = 3x^2 + 1$

14. $f(x) = 5x^2 - 3$

15. $f(x) = x^2 - 7$

16. $f(x) = x^2 + 1$

17. $f(x) = -2(x - 3)^2$

18. $f(x) = -(x + 2)^2$

19. $f(x) = -x^2 + 3$

20. $f(x) = -x^2 + 1$

Check your answers on page 758.

Completing the Square

Now that you can easily find the vertex from any equation of the form $f(x) = a(x + h)^2 + k$, I want to show you how to convert some equations of the form $f(x) = ax^2 + bx + c$ into standard form.

To accomplish that, we must transform the squared and linear terms into a *single* squared term. For instance, in Example 20 we will find that $f(x) = x^2 - 2x - 3$ is the same as $f(x) = (x - 1)^2 - 4$. This process is called **completing the square**.

Recall from Section 13.2 what happens when a binomial of the form $[x + c]$ (for any number c) is squared.

EXAMPLE 12 $[x + 3]^2 = [x + 3][x + 3]$

$= x^2 + 6x + 9$

So $[x + 3]^2 = x^2 + 6x + 9$

EXAMPLE 13 $[x - 5]^2 = x^2 - 10x + 25$

EXAMPLE 14 $[x + c]^2 = [x + c][x + c]$

$= x^2 + 2cx + c^2$

In general, we can observe that when a binomial is squared, it produces a trinomial of degree 2 of the form $(ax^2 + bx + c)$. And certain relationships always hold:

The constant term of the binomial is always *half* the coefficient of the linear term.

What's more, the constant term in the trinomial is always the *square* of the constant term in the binomial.

EXAMPLE 15 $(x + 7)^2 = x^2 + 14x + 49$.

7 is half of 14.   (7 is the constant in the binomial, and 14 is the coefficient of $14x$, the linear term.)

49 is the square of 7.   (49 is the constant in the trinomial, and 7 is the constant in the binomial.)   ❏

We use those observations to complete the square. In this chapter we are going to learn to complete the square only on *monics* (where the coefficient of x^2 is 1), so the quadratics will be $x^2 + bx + c$.

To complete the square, we want to turn the squared term (x^2) and the linear term (bx) into a squared binomial.

We will be given the value of b and must find the value of m and n in $x^2 + bx + n = (x + m)^2$.

First find the constant term in the binomial (m).

$$m \text{ is half of } b \left(\text{or } m = \frac{b}{2} \right)$$

Then find the value of the constant term in the trinomial (n).

$$n \text{ is the square of } m \text{ (or } n = m^2)$$

EXAMPLE 16   Complete the square on $x^2 + 8x$ by finding the values of:

(n) and (m) in the equation $[x + (m)]^2 = x^2 + 8x + (n)$.

(m) should be *one-half* of 8, so $(m) = \dfrac{8}{2}$

$$= 4.$$

(n) should be the *square of* (m), so $(n) = 4^2$

$$= 16.$$

The equation is $[x + 4]^2 = x^2 + 8x + 16$.

You can check this solution by squaring $[x + 4]$ and seeing that it is equal to $x^2 + 8x + 16$.   ❏

EXAMPLE 17   Complete the square on $x^2 - 10x$ by finding the values of (n) and (m) in the equation $[x + (m)]^2 = x^2 - 10x + (n)$.

(m) should be *one-half of* -10, the coefficient of the linear term ($-10x$), so $(m) = \dfrac{-10}{2}$

$$= -5.$$

(n) should be the *square of* (m), so $(n) = (-5)^2$

$$= 25.$$

The equation is $[x - 5]^2 = x^2 - 10x + 25$.

You can check this solution by squaring $[x - 5]$ and seeing that $[x - 5]^2 = x^2 - 10x + 25$.   ❏

EXAMPLE 18   Complete the square on $x^2 - 6x$ by finding the values of m and n in the equation $x^2 - 6x + n = (x + m)^2$.

m is *half of* -6, so $m = -3$.

n is the *square of* -3, so $n = 9$.

$x^2 - 6x + n = (x + m)^2$ is $x^2 - 6x + 9 = (x - 3)^2$ ❏

This simple way of completing the square works only on monics, where a (the coefficient of x^2) is 1. After you learn to complete the square on monics, you will find it easy to learn how to complete the square on non-monics in more advanced algebra books.

We want to complete the square on quadratic equations to convert them into standard form. If we begin with an equation, $f(x) = x^2 + bx + c$, we want to add some number to $(x^2 + bx)$ so it is equal to a binomial squared (just as, in Example 18, we added 9 to $[x^2 - 6x]$ to make it equal to $[x - 3]^2$).

We do not want to change the value of our quadratic equation, though, so whatever number is needed, we *add* it to and *subtract* it from the side of the equation with the squared and linear terms.*

The number we need to complete a square is
half the coefficient of the linear term, squared.

Add that number to the squared and linear terms and replace the resulting trinomial by the squared binomial it is equal to. Remember, the constant term in the binomial will be one-half the coefficient of the linear term.

Set the squared term equal to zero to find the first coordinate of the vertex; then substitute it in and find the second coordinate of the vertex.

EXAMPLE 19   Complete the square to get this equation into standard form and find the coordinates of the vertex:

$$f(x) = x^2 + 12x.$$

The number we need to complete the square is

$$\left(\frac{\mathbf{12}}{2}\right)^2 =$$
$$(6)^2 = 36.$$

So we add 36 to and subtract 36 from the right side of $f(x) = x^2 + 12x$.

$f(x) = (x^2 + 12x + \mathbf{36}) - \mathbf{36}$, which simplifies to

$f(x) = (x + 6)^2 - \mathbf{36}$, which is in standard form.

Now, to find the vertex solve $x + 6 = 0$.

$x = -6$ is the first coordinate of the vertex.

$$f(-6) = [(-6) + 6]^2 - 36$$
$$f(-6) = -36$$

So $(-6, -36)$ is the vertex of $f(x) = (x + 6)^2 - 36$ (which is the completed square form of $f(x) = x^2 + 12x$).   ❑

EXAMPLE 20   Complete the square to get this equation into standard form and find the coordinates of the vertex:

$$f(x) = x^2 - 2x - 3.$$

The number we need is $\left(\dfrac{\mathbf{-2}}{2}\right)^2 =$

$$(-1)^2 = 1.$$

$$f(x) = (x^2 - 2x + 1) - 1 - 3,$$

which simplifies to $f(x) = (x - 1)^2 - 4.$

---

* Adding and subtracting a number on the same side of an equation has the same effect as adding zero (nothing) to it. So it does not alter the equation.

Now we can find the vertex. $(x - 1 = 0$ when $x = 1.)$

$$f(1) = [(1) - 1]^2 - 4$$
$$f(1) = -4$$

So $(1, -4)$ is the vertex. ❑

EXERCISES 15.1B

Complete the square on each equation to get it into standard form, $f(x) = (x + h)^2 + k$, and state the coordinates of the vertex.

21. $f(x) = x^2 + 2x$ **22.** $f(x) = x^2 + 6x$

23. $f(x) = x^2 - 4x$ **24.** $f(x) = x^2 - 8x$

25. $f(x) = x^2 - 2x$ **26.** $f(x) = x^2 - 12x$

27. $f(x) = x^2 + 14x$ **28.** $f(x) = x^2 - 14x$

29. $f(x) = x^2 - 6x + 1$ **30.** $f(x) = x^2 + 4x + 3$

31. $f(x) = x^2 - 2x - 5$ **32.** $f(x) = x^2 + 8x + 20$

33. $f(x) = x^2 + 10x + 30$ **34.** $f(x) = x^2 - 6x - 12$

35. $f(x) = x^2 - 4x - 10$ **36.** $f(x) = x^2 + 2x - 3$

Check your answers on page 758. ✓

Graphing Parabolas

Now that you can recognize the coordinates of the vertex, you are ready to graph parabolas by using only three points. First find the vertex; then find one point to its right and another to its left.

A point to the *right* of the vertex will have a *larger x-coordinate than the vertex* does, so choose any number larger than the *x*-coordinate of the vertex and find the value of $f(x)$ for that number.

A point to the *left* of the vertex will have a *smaller x-coordinate than the vertex* does, so choose any number smaller than the *x*-coordinate of the vertex and find the value of $f(x)$ for that number.

Plot and label those three points and sketch the curve of the parabola.

EXAMPLE 21   Sketch the graph of $f(x) = x^2 + 3$.

As we saw in Example 8, the vertex is $(0, 3)$.

To find a point to the right of it, choose any number >0 for x. For instance, let $x = 1$.

$$f(1) = (1)^2 + 3$$
$$f(1) = 1 + 3$$
$$f(1) = (4) \qquad \text{So } (1, 4) \text{ is a point on the parabola.}$$

To find a point to the left of the vertex, choose any number <0 for x. For instance, let $x = -2$.

$$f(-2) = (-2)^2 + 3$$
$$f(-2) = 4 + 3$$
$$f(-2) = 7 \qquad \text{So } (-2, 7) \text{ is a point on the parabola.}$$

x	0	1	-2
$f(x)$	3	4	7

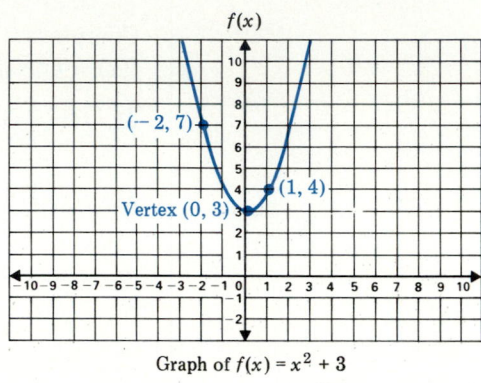

Graph of $f(x) = x^2 + 3$

EXAMPLE 22   Sketch the graph of $f(x) = (x + 2)^2 + 7$.

As we saw in Example 3, the vertex is $(-2, 7)$.

To find a point to the right of it, choose a value of $x > -2$. Let $x = 0$.

$$f(0) = [(0) + 2]^2 + 7$$
$$f(0) = [2]^2 + 7$$
$$f(0) = 4 + 7$$
$$f(0) = 11 \qquad \text{So } (0, 11) \text{ is a point on the parabola.}$$

To find a point to the left of it, choose a value of $x < -2$. Let $x = -3$.

$$f(-3) = [(-3) + 2]^2 + 7$$
$$f(-3) = [-1]^2 + 7$$
$$f(-3) = 1 + 7$$
$$f(-3) = 8 \qquad \text{So } (-3, 8) \text{ is a point on the parabola.}$$

x	-2	0	-3
$f(x)$	7	11	8

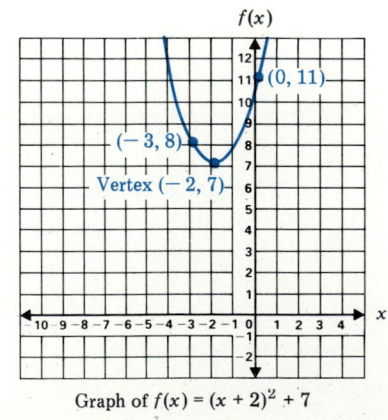

Graph of $f(x) = (x + 2)^2 + 7$

EXAMPLE 23 Sketch the graph of $f(x) = -(x - 4)^2$.

First let's find the vertex:

$$(x - 4) = 0 \text{ when } x = 4 \text{ and } f(4) = 0$$

So the vertex is $(4, 0)$.

To find a point to the right of it, choose a value of $x > 4$. Let $x = 5$.

$$f(5) = -[(5) - 4]^2$$
$$f(5) = -[1]^2 \qquad \text{So } (5, -1) \text{ is a point on the parabola.}$$
$$f(5) = -1$$

To find a point to the left of it, choose a value of $x < 4$. Let $x = 3$.

$$f(3) = -[(3) - 4]^2$$
$$f(3) = -[-1]^2$$
$$f(3) = -1 \qquad \text{So } (3, -1) \text{ is a point on the parabola.}$$

x	4	5	3
$f(x)$	0	-1	-1

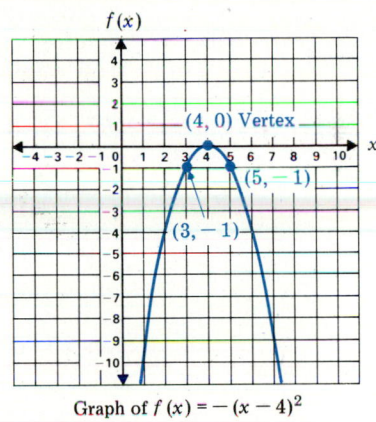

Graph of $f(x) = -(x - 4)^2$

EXAMPLE 24 Sketch the graph of $f(x) = 2(x + 1)^2 - 5$.

As we saw in Example 5, the vertex is $(-1, -5)$.

To find a point to the right of it, choose a value of $x > -1$. Let $x = 0$.

$$f(0) = 2(0 + 1)^2 - 5$$
$$f(0) = 2(1)^2 - 5$$
$$f(0) = 2(1) - 5$$
$$f(0) = 2 - 5$$
$$f(0) = -3$$

So $(0, -3)$ is a point on the parabola.

To find a point to the left of it, choose a value of $x < -1$. Let $x = -2$.

$$f(-2) = 2(-2 + 1)^2 - 5$$
$$f(-2) = 2(-1)^2 - 5$$
$$f(-2) = 2(1) - 5$$
$$f(-2) = 2 - 5$$
$$f(-2) = -3$$

So $(-2, -3)$ is a point on the parabola.

x	-1	0	-2
$f(x)$	-5	-3	-3

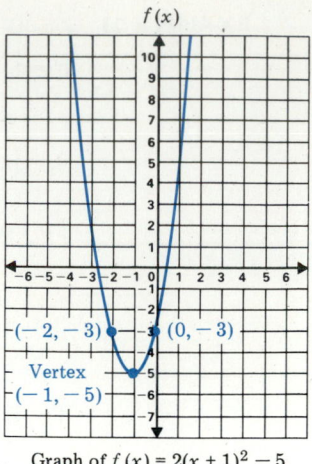

$(-2, -3)$ $(0, -3)$

Vertex
$(-1, -5)$

Graph of $f(x) = 2(x + 1)^2 - 5$

EXAMPLE 25 Complete the square to get this equation into standard form; find the coordinates of the vertex and a point on either side, plot them, and sketch the graph:

$$f(x) = x^2 - 6x + 1$$

The number we need to complete the square is

$$\left(\frac{-6}{2}\right)^2 = 9.$$

$$f(x) = (x^2 - 6x + 9) - 9 + 1$$

$$f(x) = (x - 3)^2 - 8 \text{ is the equation in standard form.}$$

The vertex is $(3, -8)$.

Let $x = 0$. $\begin{aligned} f(0) &= (0 - 3)^2 - 8 \\ &= (-3)^2 - 8 \\ &= 9 - 8 \\ &= 1 \end{aligned}$

So $f(0) = 1$, and $(0, 1)$ is a point on the parabola.

Let $x = 6$. $\begin{aligned} f(6) &= (6 - 3)^2 - 8 \\ &= (3)^2 - 8 \\ &= 9 - 8 \\ &= 1 \end{aligned}$

So $f(6) = 1$, and $(6, 1)$ is another point on the parabola.

x	3	0	6
$f(x)$	-8	1	1

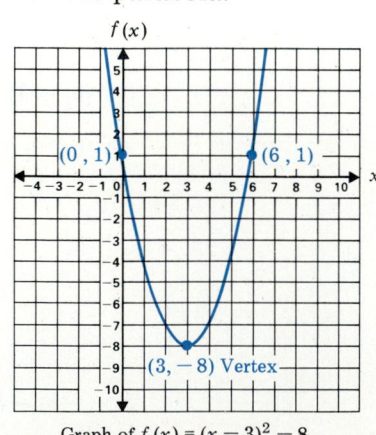

$(0, 1)$ $(6, 1)$

$(3, -8)$ Vertex

Graph of $f(x) = (x - 3)^2 - 8$

Parabolas are symmetric, so the curve is exactly the same on one side of the vertex as it is on the other. You can use that fact to check for careless errors.

After you determine the coordinates of the vertex and the first point, let the x-coordinate of the second point be the *same distance* from the x-coordinate of the vertex, but in the *opposite direction*. Substitute that value into the equation for x to find the second coordinate; it should be the same as the second coordinate of the first point!

In Example 25 the vertex is $(3, -8)$; the first point is $(0, 1)$. The x-coordinate (0) is three steps from the x-coordinate of the vertex (3). The x-coordinate (6) of the second point is also three steps from the x-coordinate of the vertex (3), and both have the same second coordinate (1).

EXERCISES 15.1C

Sketch the graph of each equation.

If the equation is not already in standard form, complete the square to convert it. Find the coordinates of the vertex and a point on either side; plot and label them. Sketch the parabola from those three points.

37. $f(x) = x^2 + 1$ | **38.** $f(x) = x^2 - 2$

39. $f(x) = (x - 3)^2$ | **40.** $f(x) = (x + 3)^2 - 1$

41. $f(x) = (x + 1)^2$ | **42.** $f(x) = (x - 1)^2 + 5$

43. $f(x) = (x - 2)^2 - 3$ | **44.** $f(x) = (x + 1)^2 - 1$

45. $f(x) = (x - 4)^2 - 2$ | **46.** $f(x) = (x + 1)^2 + 2$

47. $f(x) = -(x - 3)^2$ | **48.** $f(x) = -(x + 2)^2$

49. $f(x) = -x^2$ | **50.** $f(x) = -2x^2$

51. $f(x) = -x^2 + 2$ | **52.** $f(x) = -3x^2 - 4$

53. $f(x) = 2(x - 1)^2$ | **54.** $f(x) = 3(x - 2)^2$

55. $f(x) = 2(x + 1)^2 + 1$ | **56.** $f(x) = 3(x - 1)^2 - 2$

57. $f(x) = -2(x - 3)^2$ | **58.** $f(x) = -3(x + 1)^2$

59. $f(x) = -(x - 1)^2 + 3$ | **60.** $f(x) = -2(x - 2)^2 - 4$

61. $f(x) = x^2 - 2x$ | **62.** $f(x) = x^2 + 4x$

63. $f(x) = x^2 + 10x$ | **64.** $f(x) = x^2 - 4x$

65. $f(x) = x^2 - 8x + 3$ | **66.** $f(x) = x^2 + 2x - 5$

67. $f(x) = x^2 - 6x - 1$ | **68.** $f(x) = x^2 + 6x + 7$

69. $f(x) = x^2 - 10x + 8$ | **70.** $f(x) = x^2 + 8x + 12$

Check your answers on page 758.

x- and y-Intercepts

In Section 8.6 we discussed the x- and y-intercepts of straight lines. We can easily find the x- and y-intercepts of any graph, not just a straight line.

Remember, an **x-intercept** is a point where the graph of an equation *crosses the x-axis*. A point is on the x-axis if its second coordinate is 0. So to find an x-intercept, let $f(x) = 0$ and solve for x.

A **y-intercept** is a point where the graph of an equation *crosses the* $y =$ [or $f(x)$] *axis*. A point is on the y-axis if its first coordinate is 0. So to find a y-intercept, let $x = 0$ and solve for $f(0)$.

If the equation does not graph as a straight line, there might be more than one point where it crosses an axis, resulting in more than one x- and y-intercept on the graph of a nonlinear equation.

If we know the value of x^2, we can solve for x by taking the square root. Remember, there is both a positive root and a negative root.

EXAMPLE 26 Find the y- and x-intercepts of $f(x) = x^2 - 9$.

Evaluate $f(0)$ to find the y-intercept.

$$f(0) = (0)^2 - 9$$
$$f(0) = -9$$

So this parabola has one y-intercept, and it is **(0, −9)**.
Let $f(x) = 0$ and solve $0 = x^2 - 9$ for x to find the x-intercept(s).

$$0 = x^2 - 9$$
$$\underline{+9 \qquad\ \ +9}$$
$$9 = x^2 \qquad \text{Solve for } x \text{ by taking the square root of each side.}$$
$$\pm\sqrt{9} = \sqrt{x^2}$$
$$\pm 3 = x$$

So this parabola has two x-intercepts: **(3, 0)** and **(−3, 0)**. ❏

EXAMPLE 27 Find the intercepts of $f(x) = (x + 1)^2 - 4$.

Evaluate $f(0)$ for the y-intercept.

$$f(0) = [(0) + 1]^2 - 4$$
$$= [1]^2 - 4$$
$$= -3$$

So the y-intercept is **(0, −3)**.

If $f(x) = 0$, solve $0 = (x + 1)^2 - 4$ for the x-intercepts.

$$0 = (x + 1)^2 - 4$$
$$\underline{+4 \qquad\qquad\ \ +4}$$
$$4 = (x + 1)^2 \qquad \text{Solve for } x \text{ by taking the square root of each side.}$$
$$\pm\sqrt{4} = \sqrt{(x + 1)^2}$$
$$\pm 2 = x + 1, \text{ which we write as two equations to solve:}$$

$$2 = x + 1 \qquad\qquad -2 = x + 1$$
$$\underline{-1 \quad -1} \qquad\qquad \underline{-1 \qquad -1}$$
$$1 = x \qquad \text{and} \quad -3 = x$$

So the x-intercepts are **(1, 0)** and **(−3, 0)**. ❏

⇒ *Note*: Parabolas from equations like $f(x) = x^2 + bx + c$ will have one
y-intercept and either no, one, or two x-intercepts.

It will have no x-intercept if the vertex is not on the x-axis and the
parabola does not cross the x-axis.

It will have one x-intercept if the vertex is on the x-axis.

It will have two x-intercepts if the parabola crosses the x-axis twice.

EXAMPLE 28   The parabolas below have no x-intercepts

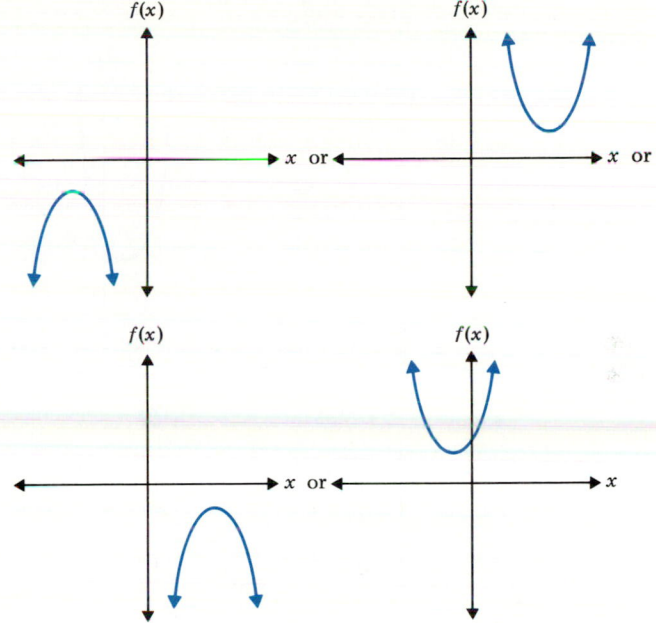

EXAMPLE 29   The parabolas below have one x-intercept

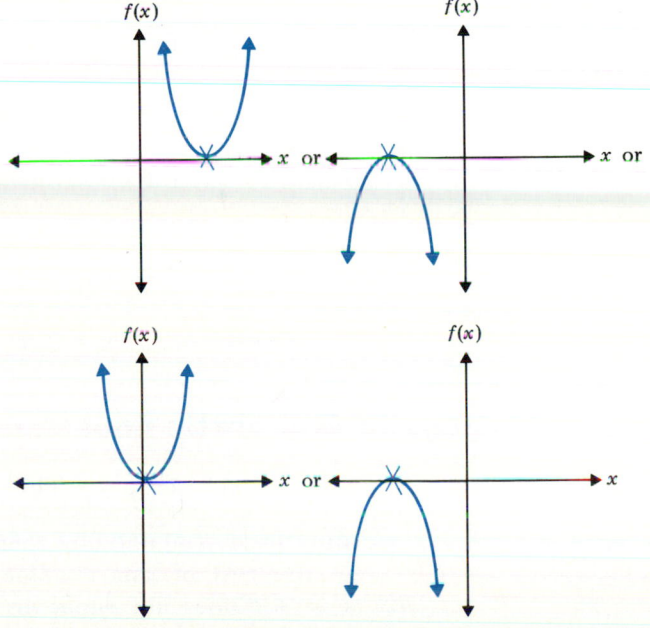

EXAMPLE 30   The parabolas below have *two x-intercepts*

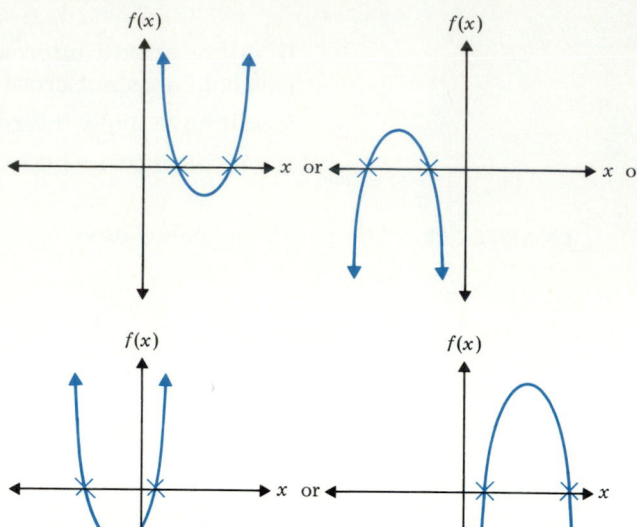

EXERCISES 15.1D

Find the *x*- and *y*-intercepts for each equation.

71. $f(x) = x^2 - 16$ **72.** $f(x) = x^2 - 4$

73. $f(x) = x^2 - 1$ **74.** $f(x) = x^2 - 3$

75. $f(x) = (x - 3)^2$ **76.** $f(x) = (x + 2)^2$

77. $f(x) = (x - 1)^2 - 1$ **78.** $f(x) = (x + 2)^2 - 4$

79. $f(x) = (x - 2)^2 - 9$ **80.** $f(x) = (x + 3)^2 - 1$

81. $f(x) = (x - 4)^2 + 1$ **82.** $f(x) = (x + 2)^2 + 4$

83. $f(x) = -x^2$ **84.** $f(x) = -2x^2$

85. $f(x) = -x^2 + 1$ **86.** $f(x) = -2x^2 + 8$

Check your answers on page 760.

15.2   Solutions to Quadratic Equations

In Section 15.1 we saw that quadratic equations of the form $f(x) = ax^2 + bx + c$ have infinitely many solutions; and if we represent each solution as the coordinates $[x, f(x)]$ and plot them, their graph will be a parabola.

In the rest of this chapter we will focus on finding solutions where $ax^2 + bx + c$ does not equal an unknown value $f(x)$ but equals the known constant 0.

To better understand what our expectations might be regarding the solutions to those equations, it helps to point out that the solutions to $ax^2 + bx + c = 0$ are the points on the parabola $f(x) = ax^2 + bx + c$ where $f(x) = 0$

Remember, if the second coordinate of a point is zero, that point is on the x-axis, and we call such points x-intercepts.

As we discussed in Section 15.1, three different situations could exist:

1. If the *vertex* is *on* the *x-axis*, there will be only one point on the graph of $f(x) = ax^2 + bx + c$ where $f(x)$ will be equal to zero. So $ax^2 + bx + c = 0$ will have *one solution*.

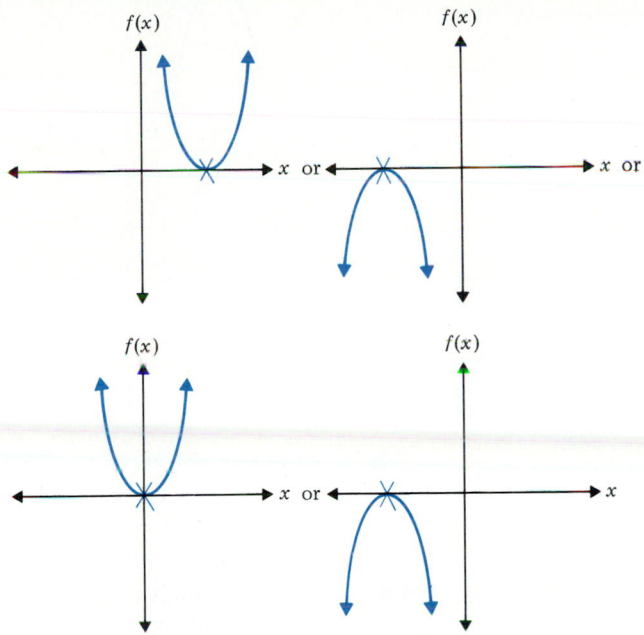

2. If the vertex is not on the x-axis, and the parabola crosses the x-axis, there will be two points on the graph of $f(x) = ax^2 + bx + c$ where $f(x)$ will be equal to zero. So $ax^2 + bx + c = 0$ will have two solutions.

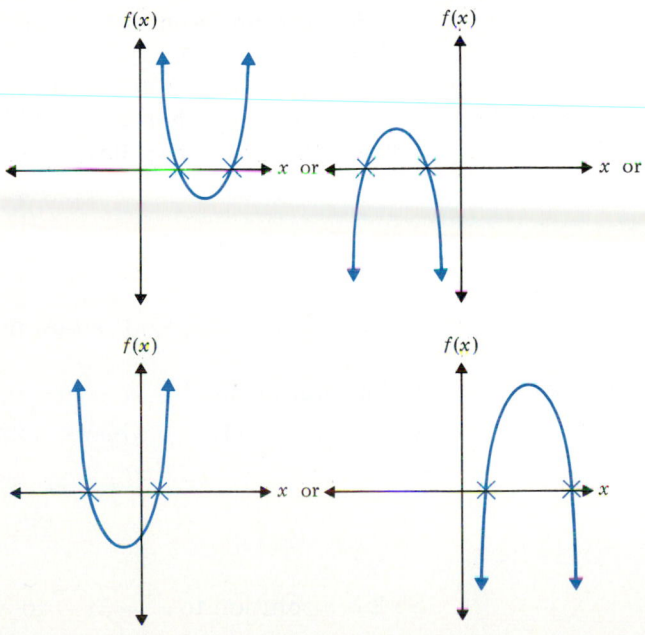

3. If the vertex is not on the x-axis, and the parabola does *not* cross the x-axis, there will be no points on the graph of $f(x) = ax^2 + bx + c$ where $f(x)$ will be equal to zero. So $ax^2 + bx + c = 0$ will have no solutions.

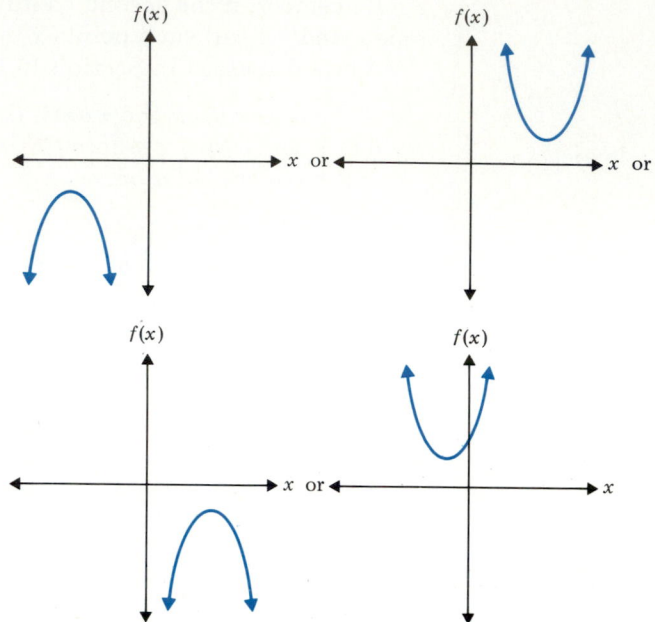

It is not worth the trouble of graphing an equation to determine how many solutions we should expect to find when solving it. Generally we should anticipate that there will be two solutions to most quadratic equations ($ax^2 + bx + c = 0$), but we understand why there might be only one or no solution. Later in this chapter you will learn to recognize, during the solving procedure, when you will not be getting two solutions.

In this section we will learn to check whether a number is actually a solution to a given quadratic equation or not. The procedure is the same one we used earlier to check whether a number is a solution:

Substitute the number for every occurrence of the letter.

See whether both sides of the equation simplify to the *same* numerical value.

EXAMPLE 1   To determine whether 3 is a solution to $x^2 + 3x - 10 = 0$, we must evaluate whether $x^2 + 3x - 10 = 0$ when $x = 3$.

$$\text{If } x = 3, \quad x^2 + 3x - 10 =$$
$$\underline{(3)^2} + \underline{3(3)} - 10 =$$
$$9 + 9 - 10 = 8$$

Since $x^2 + 3x - 10$ equals 8 (not 0) when $x = 3$, 3 is **not** a solution.   ❏

EXAMPLE 2   Is 2 a solution to $x^2 + 3x - 10 = 0$?

Does $x^2 + 3x - 10 = 0$ when $x = 2$?

$$\text{If } x = 2, \quad x^2 + 3x - 10 =$$
$$\underline{(2)^2} + \underline{3(2)} - 10 =$$
$$4 + 6 - 10 = 0 \quad ✓$$

So 2 is a solution to $x^2 + 3x - 10 = 0$.   ❏

EXAMPLE 3   Is -1 a solution to $y^2 - 5y = 6$?

Does $y^2 - 5y = 6$ when $y = -1$?

$$\text{If } y = -1, \quad y^2 - 5y =$$
$$(-1)^2 - 5(-1) =$$
$$+1 + 5 = 6 \quad \checkmark$$

So -1 is a solution to $y^2 - 5y = 6$.

EXAMPLE 4   Is 2 a solution to $w^2 = 5w - 3$?

Does $w^2 = 5w - 3$ when $w = 2$?

$$\text{If } w = 2, \quad w^2 = \qquad 5w - 3 =$$
$$(2)^2 = 4 \qquad 5(2) - 3 =$$
$$10 - 3 = 7$$

The two sides are not equal ($4 \neq 7$), so 2 is **not** a solution to $w^2 = 5w - 3$.

EXERCISES 15.2

Determine whether the given number is a solution to the equation.

1. $x = 5$; $x^2 = 25$

2. $x = 4$; $3x^2 - 23 = 10$

3. $x = -2$; $3x^2 + 5 = 16$

4. $x = 0$; $5x^2 - 7x = 3$

5. $x = -1$; $2x^2 + 2 = 4$

6. $x = -4$; $x^2 + 3x = 28$

7. $x = 5$; $4x^2 - x = 0$

8. $x = 0$; $7x^2 - 4x = 0$

9. $x = 7$; $x^2 - 24 = 25$

10. $x = -5$; $x^2 - 3x = 10$

11. $x = -1$; $x^2 = 5x + 6$

12. $y = 2$; $y^2 = 3y + 2$

13. $w = -2$; $w^2 + 3 = 4w$

14. $y = -1$; $3y - 2 = 2y^2$

15. $w = 3$; $w^2 - 4w = -3$

16. $x = 4$; $x^2 - 7x = 12$

17. $y = -4$; $y^2 - 7y = -12$

18. $w = -1$; $w^2 - 4 = 3w$

19. $x = -2$; $6 - 3x = 3x^2$

20. $y = -3$; $2y^2 - 7 = 1 - 4y$

Check your answers on page 761. $\checkmark$

15.3 Solving Equations by the Square-Root Method

Solving Equations of the Form $ax^2 + c = 0$

Quadratic equations that *do not include a linear term* ($ax^2 + bx + c$, where $b = 0$) can easily be solved by **the square-root method**, which is based on the fact that $\sqrt{x^2} = |x|$ ($|x|$ is the absolute value of x, as introduced in Section 2.1). So if we know the value of a squared term, we can take its square root to solve the equation. We used this principle in Section 15.1 to find x-intercepts.

> *The Square-Root Method:*
>
> Isolate the squared term on one side of the equation by moving all constant terms to the other side.
>
> Then take the *square root* of both sides to solve for the letter.

Don't forget that there are always two solutions when taking square roots of numbers: the positive root and the negative root.

EXAMPLE 1 Solve $x^2 - 9 = 0$ by the square-root method.

$$x^2 - 9 = 0$$
$$\underline{+9 \quad +9}$$
$$x^2 = 9$$
$$\sqrt{x^2} = \pm\sqrt{9}$$
$$x = \pm 3$$

So the two solutions are $x = 3$ and $x = -3$, written as $x = \pm 3$.

Check: Does $x^2 - 9 = 0$ when $x = \pm 3$?

$$\text{If } x = 3, x^2 - 9 = \qquad\qquad \text{If } x = -3, x^2 - 9 =$$
$$(3)^2 - 9 = 0 \quad\checkmark \qquad\qquad (-3)^2 - 9 = 0 \quad\checkmark \qquad \square$$

The square-root method is particularly helpful for solving equations whose solutions are irrational numbers.

EXAMPLE 2 Solve $w^2 - 11 = 0$ by the square-root method.

$$w^2 - 11 = 0$$
$$\underline{+11 \quad +11}$$
$$w^2 = 11$$
$$w = \pm\sqrt{11}$$

So the solutions are $w = \sqrt{11}$ and $w = -\sqrt{11}$.

Check: Does $w^2 - 11 = 0$ when $w = \pm\sqrt{11}$?

$$\text{If } w = \sqrt{11}, \quad w^2 - 11 = \qquad\qquad \text{If } w = -\sqrt{11}, \quad x^2 - 11 =$$
$$(\sqrt{11})^2 - 11 = \qquad\qquad (-\sqrt{11})^2 - 11 =$$
$$11 - 11 = 0 \quad\checkmark \qquad\qquad 11 - 11 = 0 \quad\checkmark \qquad \square$$

EXAMPLE 3 Solve $y^2 - 47 = 3$ by the square-root method.

$$y^2 - 47 = 3$$
$$\underline{+47 \quad +47}$$
$$y^2 = 50$$
$$y = \pm\sqrt{50}$$
$$y = \pm\sqrt{2 \cdot 5 \cdot 5}$$
$$y = \pm 5\sqrt{2}$$

Check: Does $y^2 - 47 = 3$ when $y = \pm 5\sqrt{2}$?

If $y = 5\sqrt{2}$, $y^2 - 47 =$	If $y = -5\sqrt{2}$, $y^2 - 47 =$
$(5\sqrt{2})^2 - 47 =$	$(-5\sqrt{2})^2 - 47 =$
$(5\sqrt{2})(5\sqrt{2}) - 47$	$(-5\sqrt{2})(-5\sqrt{2}) - 47 =$
$(5)(5)(\sqrt{2}\sqrt{2}) - 47 =$	$(-5)(-5)(\sqrt{2}\sqrt{2}) - 47 =$
$25(2) - 47 =$	$25(2) - 47 =$
$50 - 47 = 3$ ✓	$50 - 47 = 3$ ✓ ❑

If the squared term has a non-monic coefficient (not equal to 1), the computation will be easier if you divide both sides of the equation by that coefficient *before* taking the square root of each side. We want the equation to read: "A letter squared is equal to a number." Then when we take the square roots, the resulting equation will read: "The letter is equal to the square roots of that number."

So solve the equation for the squared letter first and then take the square root of each side.

EXAMPLE 4   Solve $2x^2 - 36 = 0$ by the square-root method.

$$2x^2 - 36 = 0$$
$$\underline{+36 \quad +36}$$
$$2x^2 = 36$$
$$\frac{(2x^2)}{2} = \frac{36}{2}$$
$$x^2 = 18$$
$$x = \pm\sqrt{18}$$
$$x = \pm\sqrt{2 \cdot 3 \cdot 3}$$
$$x = \pm 3\sqrt{2}$$

Check: Does $2x^2 - 36 = 0$ when $x = \pm\sqrt{2}$.

If $x = 3\sqrt{2}$, $2x^2 - 36 =$	If $x = -3\sqrt{2}$, $2x^2 - 36 =$
$2(3\sqrt{2})^2 - 36 =$	$2(-3\sqrt{2})^2 - 36 =$
$2(3\sqrt{2})(3\sqrt{2}) - 36 =$	$2(-3\sqrt{2})(-3\sqrt{2}) - 36 =$
$2(3)(3)\sqrt{2}\sqrt{2} - 36 =$	$2(-3)(-3)\sqrt{2}\sqrt{2} - 36 =$
$6(3)(2) - 36 =$	$-6(-3)(2) - 36 =$
$18(2) - 36 =$	$18(2) - 36 =$
$36 - 36 = 0$ ✓	$36 - 36 = 0$ ✓ ❑

We will always get two solutions whenever we use the square-root method: the positive root and the negative root. This method is easy to use when there is no linear term, since we can readily solve the equation for the squared letter.

So a quadratic equation with no linear term, $ax^2 + c = 0$ (if $c \neq 0$), will never have only one solution. That makes sense, since if $c \neq 0$, the vertex of $f(x) = ax^2 + c$ is $(0, c)$, which is *not* on the x-axis (and as we saw in Section 15.2, this is the only situation where we get only one solution).

EXAMPLE 5 Try to solve $x^2 + 7 = 0$ by the square-root method.

$$x^2 + 7 = 0$$
$$\underline{-7 \qquad -7}$$
$$x^2 = -7$$
$$x = \pm\sqrt{-7}$$
$$x = \pm\sqrt{-7}, \text{ which does not exist in the real number system.}$$

So $x^2 + 7 = 0$ is our first example of a quadratic equation that has *no solutions.**

❑

EXERCISES 15.3A

Solve each equation by the square-root method and check both solutions in the original equation.

1. $x^2 = 25$ 2. $y^2 = 36$

3. $w^2 = 100$ 4. $m^2 = 49$

5. $x^2 - 4 = 0$ 6. $y^2 - 81 = 0$

7. $w^2 - 16 = 0$ 8. $n^2 - 64 = 0$

9. $r^2 = 5$ 10. $x^2 = 13$

11. $y^2 = 8$ 12. $k^2 = 10$

13. $a^2 = 20$ 14. $b^2 = 28$

15. $x^2 - 14 = 0$ 16. $y^2 - 45 = 0$

17. $w^2 - 40 = 0$ 18. $r^2 - 12 = 0$

Check your answers on page 761. ✔

Solving Equations of the Form $(ax \pm k)^2 + c = 0$

The square-root method also makes it easy to solve quadratic equations whose *letter term appears only in parentheses that are being squared.*

We can solve these equations for the squared parentheses and then take the square root of both sides. The square root of parentheses squared is the parentheses: $\pm\sqrt{(ax + k)^2} = \pm(ax + k)$.

This will leave us two linear equations that we can solve.

EXAMPLE 6 Solve $(x - 6)^2 - 25 = 0$ by the square-root method.

$$(x - 6)^2 - 25 = 0$$
$$\underline{+ 25 \qquad + 25}$$
$$(x - 6)^2 = 25 \quad \text{Take the square root of each side.}$$
$$(x - 6) = \pm\sqrt{25}$$
$$x - 6 = -\pm 5$$

---

* That is, it has no solutions in the *real number system*, where any number multiplied by itself is always non-negative. In more advanced algebra courses, when imaginary numbers are studied we say the solutions to such equations are *imaginary* instead of nonexistent.

That represents two different linear equations that should each be written out and separately solved.

$x - 6 = \pm 5$ represents **$x - 6 = 5$** and **$x - 6 = -5$**.

$$
\begin{array}{rcl}
x - 6 = & 5 \\
+6 & +6 \\
\hline
x = & 11
\end{array}
\qquad \text{and} \qquad
\begin{array}{rcl}
x - 6 = & -5 \\
+6 & +6 \\
\hline
x = & 1
\end{array}
$$

Check: Does $(x - 6)^2 - 25 = 0$ when $x = 11$ and when $x = 1$?

If $x = 11$, $(x - 6)^2 - 25 =$ 　　　　If $x = 1$, $(x - 6)^2 - 25 =$
　　　　$(11 - 6)^2 - 25 =$ 　　　　　　　　　　$(1 - 6)^2 - 25 =$
　　　　$(\underline{5})^2 - 25 =$ 　　　　　　　　　　　$(\underline{-5})^2 - 25 =$
　　　　$25 - 25 = 0$ ✓ 　　　　　　　　　　$25 - 25 = 0$ ✓ ❑

EXAMPLE 7　Solve $(w - 3)^2 - 7 = 0$ by the square-root method.

$$
\begin{array}{rcl}
(w - 3)^2 - 7 = & 0 \\
+7 & +7 \\
\hline
(w - 3)^2 = & 7
\end{array}
$$

$$(w - 3) = \pm\sqrt{7}$$

$w - 3 = \pm\sqrt{7}$, which we write as the two equations:

$$
\begin{array}{rcl}
w - 3 = & \sqrt{7} \\
+3 & +3 \\
\hline
w = & 3 + \sqrt{7}
\end{array}
\qquad \text{and} \qquad
\begin{array}{rcl}
w - 3 = & -\sqrt{7} \\
+3 & +3 \\
\hline
w = & 3 - \sqrt{7}
\end{array}
$$

Check: Does $(w - 3)^2 - 7 = 0$ when $w = 3 + \sqrt{7}$ and $w = 3 - \sqrt{7}$?

If $w = 3 + \sqrt{7}$, $(w - 3)^2 - 7 =$ 　　　If $w = 3 - \sqrt{7}$, $(w - 3)^2 - 7 =$
　　　$[(\underline{3} + \sqrt{7}) - \underline{3}]^2 - 7 =$ 　　　　　　　$[(\underline{3} - \sqrt{7}) - \underline{3}]^2 - 7 =$
　　　$[\sqrt{7}]^2 - 7 =$ 　　　　　　　　　　　　　$[-\sqrt{7}]^2 - 7 =$
　　　$7 - 7 = 0$ ✓ 　　　　　　　　　　　　　$7 - 7 = 0$ ✓

We can write these solutions as $w = 3 \pm \sqrt{7}$. 　　　　　　　　　　❑

EXAMPLE 8　Solve $(y + 1)^2 = 20$ by the square-root method.

$$(y + 1)^2 = 20$$

$$(y + 1) = \pm\sqrt{20}$$

$$y + 1 = \pm\sqrt{2 \cdot 2 \cdot 5}$$

$y + 1 = \pm 2\sqrt{5}$, which we write as the two equations:

$$
\begin{array}{rcl}
y + 1 = & 2\sqrt{5} \\
-1 & -1 \\
\hline
y = & -1 + 2\sqrt{5}
\end{array}
\qquad \text{and} \qquad
\begin{array}{rcl}
y + 1 = & -2\sqrt{5} \\
-1 & -1 \\
\hline
y = & -1 - 2\sqrt{5}
\end{array}
$$

Check: Does $(y + 1)^2 = 20$ when $y = -1 + 2\sqrt{5}$ and $y = -1 - 2\sqrt{5}$?

$$\text{If } y = -1 + 2\sqrt{5}, \quad (y+1)^2 = \qquad \qquad \text{If } y = -1 - 2\sqrt{5}, \quad (y+1)^2 =$$

$$[(-1 + 2\sqrt{5}) + 1]^2 = \qquad \qquad [(-1 - 2\sqrt{5}) + 1]^2 =$$

$$[2\sqrt{5}]^2 = \qquad \qquad [-2\sqrt{5}]^2 =$$

$$[2\sqrt{5}][2\sqrt{5}] = \qquad \qquad [-2\sqrt{5}][-2\sqrt{5}] =$$

$$[(2)(2)\sqrt{5}\sqrt{5}] = \qquad \qquad [(-2)(-2)\sqrt{5}\sqrt{5}] =$$

$$[4(5)] = 20 \quad \checkmark \qquad \qquad [4(5)] = 20 \quad \checkmark$$

We can write these solutions as $y = -1 \pm 2\sqrt{5}$. $\square$

EXAMPLE 9   Solve $(3z + 2)^2 - 49 = 0$ by the square-root method.

$$(3z + 2)^2 - 49 = \qquad 0$$
$$\underline{ + 49 \qquad +49}$$
$$(3z + 2)^2 = \qquad 49$$

$$(3z + 2) = \pm\sqrt{49}$$

$3z + 2 = \pm 7$, which we write as the two equations:

$$3z + 2 = \quad 7 \qquad\qquad 3z + 2 = -7$$
$$\underline{\quad -2 \quad\; -2} \qquad\qquad \underline{\quad -2 \quad\; -2}$$
$$3z = \quad 5 \qquad\qquad\quad 3z = -9$$

$$\frac{3z}{3} = \frac{5}{3} \qquad\qquad\qquad \frac{3z}{3} = \frac{-9}{3}$$

$$z = \frac{5}{3} \qquad \text{and} \qquad z = -3$$

Check: Does $(3z + 2)^2 - 49 = 0$ when $z = \dfrac{5}{3}$ and $z = -3$?

$$\text{If } z = \frac{5}{3}, \quad (3z + 2)^2 - 49 = \qquad \text{If } z = -3, \quad (3z + 2)^2 - 49 =$$

$$\left[3\left(\frac{5}{3}\right) + 2 \right]^2 - 49 = \qquad\qquad [3(-3) + 2]^2 - 49 =$$

$$[5 + 2]^2 - 49 = \qquad\qquad\qquad [-9 + 2]^2 - 49 =$$

$$[7]^2 - 49 = \qquad\qquad\qquad\qquad [-7]^2 - 49 =$$

$$49 - 49 = 0 \quad \checkmark \qquad\qquad\qquad 49 - 49 = 0 \quad \checkmark \qquad \square$$

EXAMPLE 10   Solve $(5x - 2)^2 = 3$ by the square-root method.

$$(5x - 2)^2 = 3$$

$$(5x - 2) = \pm\sqrt{3}$$

$5x - 2 = \pm\sqrt{3}$, which we write as the two equations:

$$5x - 2 = +\sqrt{3} \qquad\qquad 5x - 2 = -\sqrt{3}$$
$$\underline{\quad +2 \quad\; +2} \qquad\qquad \underline{\quad +2 \quad\; +2}$$
$$5x = 2 + \sqrt{3} \qquad\qquad\quad 5x = 2 - \sqrt{3}$$

$$\frac{5x}{5} = \frac{(2 + \sqrt{3})}{5} \qquad\qquad \frac{5x}{5} = \frac{(2 - \sqrt{3})}{5}$$

$$x = \frac{(2 + \sqrt{3})}{5} \qquad \text{and} \qquad x = \frac{(2 - \sqrt{3})}{5}$$

Check: Does $(5x - 2)^2 = 3$ when $x = \dfrac{(2 + \sqrt{3})}{5}$ and $x = \dfrac{(2 - \sqrt{3})}{5}$?

If $x = \dfrac{(2 + \sqrt{3})}{5}$, $(5x - 2)^2 =$

$$\left[5\left\{\dfrac{(2 + \sqrt{3})}{5}\right\} - 2\right]^2 =$$

$$\left[\not{5}\left\{\dfrac{(2 + \sqrt{3})}{\not{5}}\right\} - 2\right]^2 =$$

$$[2 + \sqrt{3} - 2]^2 =$$

$$[\sqrt{3}]^2 = 3 \quad \checkmark$$

If $x = \dfrac{(2 - \sqrt{3})}{5}$, $(5x - 2)^2 =$

$$\left[5\left\{\dfrac{(2 - \sqrt{3})}{5}\right\} - 2\right]^2 =$$

$$\left[\not{5}\left\{\dfrac{2 - \sqrt{3}}{\not{5}}\right\} - 2\right]^2 =$$

$$[2 - \sqrt{3} - 2]^2 =$$

$$[-\sqrt{3}]^2 = 3 \quad \checkmark$$

We can write these solutions as $x = \dfrac{(2 \pm \sqrt{3})}{5}$. ☐

EXERCISES 15.3B

Solve these equations by the square-root method and check both solutions in the original equation.

19. $(x + 4)^2 = 9$
20. $(y - 2)^2 = 4$

21. $(w - 1)^2 = 25$
22. $(m + 5)^2 = 16$

23. $(x - 1)^2 = 5$
24. $(y + 2)^2 = 3$

25. $(w + 3)^2 = 2$
26. $(x - 2)^2 = 7$

27. $(y - 3)^2 = 8$
28. $(x + 2)^2 = 45$

29. $(m - 5)^2 = 18$
30. $(w + 4)^2 = 50$

31. $(2x - 1)^2 = 4$
32. $(3y + 2)^2 = 9$

33. $(5w + 2)^2 = 49$
34. $(4a - 1)^2 = 25$

35. $(2y + 1)^2 = 3$
36. $(3w - 1)^2 = 2$

Check your answers on page 761. ✓

15.4 The Zero Multiplication Principle

If we were told that the product of two numbers was *zero*, we would know that at least one of the numbers must be zero. That is because the only time a product can be zero is when one or both of the numbers being multiplied is zero; to have a total of zero items, you must have either zero groups of items or zero items in each group!

That statement is called the **zero multiplication principle**:

$$\text{If } (a)(b) = 0, \text{ then } a = 0 \text{ or } b = 0.$$

We use that principle to solve equations in which *one side is a product of algebraic factors* and the other side is equal to *zero*.

According to the zero multiplication principle, the product of algebraic factors can equal zero only when at least one of the factors is equal to zero. So:

Set each factor equal to zero and solve it for
the value that makes it equal to zero.

Every value that makes a factor equal to zero is a solution to the original equation. Thus we might *expect to get the same number of solutions to such equations as there are different algebraic factors*.

EXAMPLE 1   Find all the solutions to $(x + 1)(x - 3) = 0$.

$$
\begin{array}{ll}
\text{If } x + 1 = 0 & \text{If } x - 3 = 0 \\
\underline{-1 \quad -1} & \underline{+3 \quad +3} \\
x = -1 & x = 3
\end{array}
$$

So $x = -1$ and $x = 3$ are the two solutions to $(x + 1)(x - 3) = 0$.

Check: Does $(x + 1)(x - 3) = 0$ when $x = -1$ and when $x = 3$?

$$
\begin{array}{ll}
\text{If } x = -1, \quad (x + 1)(x - 3) = & \text{If } x = 3, \quad (x + 1)(x - 3) = \\
(-1 + 1)(-1 - 3) = & (3 + 1)(3 - 3) = \\
(0)(-4) = 0 \ \checkmark & (4)(0) = 0 \ \checkmark \ \square
\end{array}
$$

If an equation contains two factors equal to zero, but both factors are the same, each factor will produce the same solution, which is called a double solution. So when there are identical algebraic factors, we will get fewer different solutions than the number of factors. (That is, we will get as many solutions as we had algebraic factors, but some of them might be the same.)

EXAMPLE 2   Find all the solutions to $(x + 5)(x + 5) = 0$.

Both factors are the same, so we just set them equal to zero:

$$
\begin{array}{l}
\text{If } x + 5 = 0 \\
\underline{-5 \quad -5} \\
x = -5
\end{array}
$$

So $x = -5$ is the only solution to $(x + 5)(x + 5) = 0$, and it is a double solution.

$\square$

If an equation is the product of three different algebraic factors equal to zero, we expect to find three solutions.

EXAMPLE 3   Find all the solutions to $(x + 1)(x - 3)(x + 2) = 0$.

$$
\begin{array}{lll}
\text{If } x + 1 = 0; & x - 3 = 0; & x + 2 = 0; \\
\underline{-1 \quad -1} & \underline{+3 \quad +3} & \underline{-2 \quad -2} \\
x = -1 & x = 3 & x = -2
\end{array}
$$

So the three solutions are $x = -1, 3, \text{ and } -2$.

$\square$

If an equation is the product of four different algebraic factors equal to zero, we expect to find four solutions.

EXAMPLE 4 Find all the solutions to $y(y - 1)(y + 1)(y - 4) = 0$.

$$
\begin{array}{ccccc}
\text{If } y = 0; & y - 1 = & 0; & y + 1 = & 0; & y - 4 = & 0; \\
 & +1 & +1 & -1 & -1 & +4 & +4 \\
\hline
 & y = & 1 & y = -1 & & y = & 4
\end{array}
$$

So the four solutions are $y = 0, 1, -1,$ and 4.

We could list them as $0, \pm 1,$ and 4.

If the equation contains any factors that are simply numbers, they do not have any effect on the solutions, since no value of any letter can ever make these numbers equal to zero.

EXAMPLE 5 $3y(y - 1)(y + 1)(y - 4) = 0$ has the same solutions as $y(y - 1)(y + 1)(y - 4) = 0$, because that is the equation that results if we divide both sides of the equation $3y(y - 1)(y + 1)(y - 4) = 0$ by 3. Only the four algebraic factors, $y, y - 1, y + 1,$ and $y - 4$, produce solutions.

The factors in those examples were simple enough that you might have solved them in your head. If the factors are more complicated, such as when the letter term has a coefficient, you will make fewer careless errors if you *do not solve them in your head.* They often have *fractional* solutions. Be sure to check them.

You should check the solutions to the simple factors too, to be sure you did not make a careless error with signs.

EXAMPLE 6 Find all the solutions to $(3w - 4)(w + 5) = 0$.

You might try to solve $w + 5 = 0$ in your head, but I would discourage that when solving $3w - 4 = 0$.

$$
\begin{array}{cccc}
\text{If } 3w - 4 = & 0 & w + 5 = & 0 \\
 +4 & +4 & -5 & -5 \\
\hline
3w = & 4 & w = & -5
\end{array}
$$

$$\frac{(3w)}{3} = \frac{4}{3}$$

$$w = \frac{4}{3}$$

So the two solutions are $w = \dfrac{4}{3}$ and -5.

Check: Does $(3w - 4)(w + 5) = 0$ when $w = \dfrac{4}{3}$ and -5?

If $w = \dfrac{4}{3}$, $(3w - 4)(w + 5) =$

$$\left[3\left(\frac{4}{3}\right) - 4\right]\left[\left(\frac{4}{3}\right) + 5\right] =$$

$$[4 - 4]\left[\frac{4}{3} + \frac{15}{3}\right] =$$

$$[0]\left[\frac{19}{3}\right] = 0 \ \checkmark$$

If $w = -5$, $(3w - 4)(w + 5) =$

$$[3(-5) - 4][(-5) + 5] =$$

$$[-15 - 4][-5 + 5] =$$

$$[-19][0] = 0 \ \checkmark$$

EXAMPLE 7   Find all the solutions to $4y(3y - 2) = 0$.

There are only two algebraic factors, y and $3y - 2$, but you will get the same solution whether you use them or $4y$ and $3y - 2$!

$$\text{If} \quad 4y = 0 \qquad\qquad\qquad 3y - 2 = 0$$
$$\qquad\qquad\qquad\qquad\qquad \underline{+2 \quad +2}$$
$$\frac{(4y)}{4} = \frac{0}{4} \qquad\qquad\qquad 3y = 2$$
$$\qquad\qquad\qquad\qquad\qquad \frac{(3y)}{3} = \frac{2}{3}$$
$$y = 0 \qquad\qquad\qquad\qquad y = \frac{2}{3}$$

So the two solutions are $y = 0$ and $\frac{2}{3}$.

Check: Does $4y(3y - 2) = 0$ when $y = 0$ and $\frac{2}{3}$?

$$\text{If} \quad y = 0, \; 4y(3y - 2) = \qquad\qquad \text{If} \quad y = \frac{2}{3}, \; 4y(3y - 2) =$$

$$4(0)[3(0) - 2] = \qquad\qquad 4\left(\frac{2}{3}\right)\left[3\left(\frac{2}{3}\right) - 2\right] =$$

$$\underline{0}[0 - 2] = 0 \;\checkmark \qquad\qquad \left(\frac{8}{3}\right)[2 - 2] =$$

$$\qquad\qquad\qquad\qquad\qquad \left(\frac{8}{3}\right)[\underline{0}] = 0 \;\checkmark \quad \square$$

EXAMPLE 8   Find all the solutions to $(5y - 1)(2y + 3) = 0$.

$$\text{If} \quad 5y - 1 = 0 \qquad\qquad 2y + 3 = 0$$
$$\underline{+1 \quad +1} \qquad\qquad \underline{-3 \quad -3}$$
$$5y = 1 \qquad\qquad\qquad 2y = -3$$
$$\frac{(5y)}{5} = \frac{1}{5} \qquad\qquad \frac{(2y)}{2} = \frac{(-3)}{2}$$
$$y = \frac{1}{5} \qquad\qquad\qquad y = -\frac{3}{2}, \text{ or } -1\frac{1}{2}$$

So the two solutions are $y = \frac{1}{5}$ and $-1\frac{1}{2}$.

Check: Does $(5y - 1)(2y + 3) = 0$ when $y = \frac{1}{5}$ and $-1\frac{1}{2}$?

To check $y = -1\frac{1}{2}$, use the improper fraction form: $-\frac{3}{2}$.

$$\text{If} \quad y = \frac{1}{5}, \; (5y - 1)(2y + 3) = \qquad \text{If} \quad y = -\frac{3}{2}, \; (5y - 1)(2y + 3) =$$

$$\left[5\left(\frac{1}{5}\right) - 1\right]\left[2\left(\frac{1}{5}\right) + 3\right] = \qquad \left[5\left(-\frac{3}{2}\right) - 1\right]\left[2\left(-\frac{3}{2}\right) + 3\right] =$$

$$[1 - 1]\left[\frac{2}{5} + 3\right] = \qquad\qquad \left[-\frac{15}{2} - 1\right][-3 + 3] =$$

$$[\underline{0}]\left[\frac{2}{5} + 3\right] = 0 \;\checkmark \qquad\qquad \left[-\frac{15}{2} - 1\right][\underline{0}] = 0 \;\checkmark \quad \square$$

EXERCISES 15.4

Find all the solutions to these equations and check your solutions.

1. $(x + 3)(x + 5) = 0$

2. $(y - 4)(y - 1) = 0$

3. $w(w + 5) = 0$

4. $(y - 4)(y - 2) = 0$

5. $x(x + 2)(x + 6) = 0$

6. $y(y - 3)(y + 8) = 0$

7. $(x - 2)(x + 2)(x - 6)(x + 1) = 0$

8. $w(w - 3)(w - 2)(w + 3) = 0$

9. $y(y - 4)(y - 1)(y + 1)(y - 2) = 0$

10. $x(x - 1)(x + 1) = 0$

11. $(3x - 1)(x + 4) = 0$

12. $(4y + 3)(y - 6) = 0$

13. $w(2w - 5) = 0$

14. $4x(3x - 7) = 0$

15. $(2y + 6)(5y - 1) = 0$

16. $(3w + 9)(2w - 1) = 0$

17. $(4w - 5)(w - 1) = 0$

18. $(6x + 3)(x - 5) = 0$

19. $x(2x - 1) = 0$

20. $3y(2y - 1) = 0$

Check your answers on page 761.

15.5 Solving Equations by the Factoring Method

Factoring to Solve $ax^2 + bx + c = 0$

In Section 15.3 we learned how to use the square-root method to solve quadratic equations that did not contain a linear term. That is, we can use the square-root method to solve quadratic equations of the form $ax^2 + bx + c = 0$ when $b = 0$.

We can use the zero multiplication principle to solve some quadratic equations of the form $ax^2 + bx + c = 0$ when $b \neq 0$.

To use the zero multiplication principle to solve equations:

First move all the terms to one side, so the other side of the equation is *equal to zero*.

Then factor $ax^2 + bx + c$ into a product of prime algebraic factors, each of degree 1, so the equations will be just like those we solved in Section 15.4.

Set each factor equal to zero and solve it.

Always check solutions in the *original* equations.

EXAMPLE 1 Solve $x^2 - 5x = 0$.

It is already equal to zero, so we just have to factor it.

$$x^2 - 5x = 0$$
$$x(x - 5) = 0$$

Set each factor equal to zero and solve.

$$x = 0 \quad \text{and} \quad x - 5 = 0$$
$$\underline{+5 \quad +5}$$
$$x = 5$$

Check the solutions in the *original* equation.

Check: Does $x^2 - 5x = 0$ when $x = 0$ and $x = 5$?

If $x = 0$, $x^2 - 5x =$ 　　　　If $x = 5$, $x^2 - 5x =$

$$(\underline{0})^2 - 5(\underline{0}) = (\underline{5})^2 - 5(\underline{5}) =$$
$$0 - 5(0) = 25 - 5(5) =$$
$$0 - 0 = 0 \ \checkmark 25 - 25 = 0 \ \checkmark$$

EXAMPLE 2 Solve $2y^2 = -3y$.

We will set the right side equal to zero and factor the equation.

$$2y^2 = -3y$$
$$\underline{+3y \quad +3y}$$
$$2y^2 + 3y = 0$$
$$y(2y + 3) = 0$$

So 　　　　　　　　　$y = 0 \quad \text{and} \quad 2y + 3 = 0$

$$\underline{-3 \quad -3}$$
$$2y = -3$$
$$\frac{(2y)}{2} = -\frac{3}{2}$$
$$y = -\frac{3}{2}$$

The solutions are $y = 0$ and $y = -\dfrac{3}{2}$.

Check: Does $2y^2 = -3y$ when $y = 0$ and $y = -\dfrac{3}{2}$?

If $y = 0$, $2y^2 =$ 　and　 $-3y =$

$$2(0)^2 = -3(0) = 0 \ \checkmark$$
$$2(0) = 0 \ \checkmark$$

If $y = -\dfrac{3}{2}$, $2y^2 =$ 　and　 $-3y =$

$$2\left(-\frac{3}{2}\right)^2 = -3\left(-\frac{3}{2}\right) = \frac{9}{2} \ \checkmark$$
$$2\left(\frac{9}{4}\right) = \frac{9}{2} \ \checkmark$$

EXAMPLE 3 Solve $w^2 + 3w = -2$.

$$w^2 + 3w = -2$$
$$\underline{+2 \quad +2}$$
$$w^2 + 3w + 2 = 0$$
$$(w + 2)(w + 1) = 0$$

So
$$w + 2 = 0 \qquad\qquad w + 1 = 0$$
$$\underline{-2 \quad -2} \qquad\qquad \underline{-1 \quad -1}$$
$$w = -2 \quad \text{and} \quad w = -1 \quad \text{are the solutions.}$$

Check: Does $w^2 + 3w = -2$ when $w = -2$ and $w = -1$?

If $\ w = -2$, $\ w^2 + 3w =$ \qquad\qquad If $\ w = -1$, $\ w^2 + 3w =$
$$(-2)^2 + 3(-2) = \qquad\qquad\qquad (-1)^2 + 3(-1) =$$
$$4 + 3(-2) = \qquad\qquad\qquad\quad 1 + 3(-1) =$$
$$4 - 6 = -2 \ \checkmark \qquad\qquad\qquad 1 - 3 = -2 \ \checkmark \quad \square$$

Notice that quadratic equations factor into two algebraic factors, so we will get two solutions, as we generally expect to find for quadratic equations when we solve them by the factoring method.

A quadratic equation that factors into two identical factors is an example of a quadratic equation with only one solution.

EXAMPLE 4 Solve $x^2 = 2x - 1$.

$$x^2 = 2x - 1$$
$$\underline{-2x + 1 \quad -2x + 1}$$
$$x^2 - 2x + 1 = 0$$
$$x^2 - 2x + 1 = 0$$
$$(x - 1)(x - 1) = 0$$

These factors are identical, so we need only set that factor equal to zero to find the single solution.

If $\ (x - 1) = 0$
$$\underline{+1 \quad +1}$$
$$x = 1 \quad \text{which is a double solution.}$$

Check: Does $x^2 = 2x - 1$ when $x = 1$?

If $\ x = 1$, $\ x^2 =$ \qquad\qquad $2x - 1 =$
$$(1)^2 = 1 \ \checkmark \qquad\qquad 2(1) - 1 =$$
$$\qquad\qquad\qquad\qquad 2 - 1 = 1 \ \checkmark \quad \square$$

Always choose to move terms to the side that will leave the squared term positive, because it is easier to factor a quadratic trinomial by FOIL, as we learned in Section 13.4, if the squared term is positive. So if the squared term is negative, move the terms to the side that does not have the squared term.

EXAMPLE 5 To solve $3y - y^2 = -4$, move all the terms to the right side, since the squared term $-y^2$ is now negative on the left side.

$$3y - y^2 = -4$$
$$\underline{-3y + y^2 \quad -3y + y^2}$$
$$0 = y^2 - 3y - 4$$
$$0 = (y - 4)(y + 1)$$

So

$$y - 4 = 0 \qquad y + 1 = 0$$
$$\underline{+4 \quad +4} \quad \text{and} \quad \underline{-1 \quad -1}$$
$$y = \quad 4 \qquad \qquad y = -1 \quad \text{are the solutions.}$$

Check: Does $3y - y^2 = -4$ when $y = 4$ and $y = -1$?

If $y = 4$, $3y - y^2 =$ If $y = -1$, $3y - y^2 =$

$$3(4) - (\underline{4})^2 = \qquad\qquad 3(\underline{-1}) - (\underline{-1})^2 =$$
$$\underline{3(4)} - 16 = \qquad\qquad \underline{3(-1)} - (1) =$$
$$12 - 16 = -4 \quad \checkmark \qquad -3 - 1 = -4 \quad \checkmark \qquad \square$$

In Section 13.4 we learned that we could always make the squared term in a polynomial positive by factoring out a -1 from every term. That step is not necessary to make the squared term positive *in an equation*. We can move the terms to one side so the squared term becomes positive, as in Example 5. Or we can simply multiply both sides by -1 if all the terms are already on one side of the equation, as in Example 6.

EXAMPLE 6   To solve $-x^2 + 3x - 2 = 0$, begin by multiplying both sides by -1 to make the squared term positive.

$$(-1)(-x^2 + 3x - 2) = (-1)0$$
$$x^2 - 3x + 2 = 0, \quad \text{which is now easy to factor.}$$
$$(x - 2)(x - 1) = 0$$

If $x - 2 = 0$ If $x - 1 = 0$
$$x = 2 \quad \text{and} \quad x = 1$$

Check: Does $-x^2 + 3x - 2 = 0$ when $x = 2$ and $x = 1$?

If $x = 2$, $-x^2 + 3x - 2 =$ If $x = 1$, $-x^2 + 3x - 2 =$

$$-(2)^2 + 3(2) - 2 = \qquad\qquad -(1)^2 + 3(1) - 2 =$$
$$-4 + 6 - 2 = 0 \quad \checkmark \qquad\qquad -1 + 3 - 2 = 0 \quad \checkmark \quad \square$$

EXERCISES 15.5A

Solve each equation and check the solutions in the original equations.

1. $x^2 + 4x = 0$ 2. $y^2 - 7y = 0$

3. $w^2 = -6w$ 4. $3x^2 - 5x = 0$

5. $4x^2 = 3x$ 6. $5y^2 + 3y = 0$

7. $2x^2 = 6x$ 8. $3y^2 = -12y$

9. $10x^2 = 15x$ 10. $6y^2 = -4y$

11. $x^2 = 5x$ 12. $y^2 = y$

13. $x^2 = -4x$ 14. $y^2 + 4y = -4$

15. $w^2 - 3 = 2w$ 16. $x^2 + 5 = 6x$

17. $x^2 = 4x - 3$ **18.** $w^2 = -2w + 15$

19. $y^2 - 6 = y$ **20.** $x^2 = -7x$

21. $-x^2 + 2x - 1 = 0$ **22.** $-x^2 - 5x = 0$

23. $-y^2 + 7y = 0$ **24.** $-w^2 + w + 6 = 0$

25. $-x^2 - 2x + 8 = 0$ **26.** $-y^2 - 3y + 4 = 0$

Check your answers on page 761. ✓

Factoring to Solve Equations of Degree Higher Than 2

You can factor to simplify and solve some equations with degree higher than 2 also.

Always begin by setting the equation equal to zero.

Factor out any common factors first.

Then factor any differences of squares.

And then factor any remaining trinomials by FOIL.

After expressing the equation as a product of prime linear factors, set each factor equal to zero to solve the equation.

EXAMPLE 7 Solve $x^3 + x^2 = 30x$.

$$x^3 + x^2 = 30x$$
$$\underline{- 30x \quad -30x}$$
$$x^3 + x^2 - 30x = 0$$
$$x(x^2 + x - 30) = 0$$
$$x(x + 6)(x - 5) = 0$$

So $\qquad x = 0; \quad x + 6 = 0 \quad$ and $\quad x - 5 = 0$

$$\underline{\qquad -6 \quad -6} \qquad \underline{\quad +5 \quad +5}$$
$$x = -6 \qquad\qquad x = 5$$

Check: Does $x^3 + x^2 = 30x$ when $x = 0$, -6, and 5?

If $x = 0$, $x^3 + x^2 =$ $30x =$

$$(0)^3 + (0)^2 = 0 \;✓ \qquad 30(0) = 0 \;✓$$

If $x = -6$, $x^3 + x^2 =$ $30x =$

$$(-6)^3 + (-6)^2 = \qquad 30(-6) = -180 \;✓$$
$$-216 + 36 = -180 \;✓$$

If $x = 5$, $x^3 + x^2 =$ $30x =$

$$(5)^3 + (5)^2 = \qquad 30(5) = 150 \;✓$$
$$125 + 25 = 150 \;✓$$

EXAMPLE 8 Solve $y^3 - 9y = 0$.

$$y^3 - 9y = 0$$
$$y(y^2 - 9) = 0$$
$$y(y - 3)(y + 3) = 0$$

So $\qquad$ $y = 0;$ $\quad y - 3 = \quad 0;$ $\quad$ and $\quad y + 3 = \quad 0$

$$\begin{array}{rr} +3 & +3 \\ \hline y = & 3 \end{array} \qquad \begin{array}{rr} -3 & -3 \\ \hline y = & -3 \end{array}$$

The solutions are $y = 0, \pm 3$.

Check: Does $y^3 - 9y = 0$ when $y = 0, \pm 3$?

$$\begin{aligned} \text{If} \quad y = 0, \quad y^3 - 9y &= \\ (0)^3 - 9(0) &= \\ 0 - 0 &= 0 \ \checkmark \end{aligned}$$

$$\begin{aligned} \text{If} \quad y = 3, \quad y^3 - 9y &= \\ (3)^3 - 9(3) &= \\ 27 - 27 &= 0 \ \checkmark \end{aligned}$$

$$\begin{aligned} \text{If} \quad y = -3, \quad y^3 - 9y &= \\ (-3)^3 - 9(-3) &= \\ -27 + 27 &= 0 \ \checkmark \end{aligned}$$

We can solve equations that consist of the product of nonlinear factors equal to zero if we can factor them into linear factors.

EXAMPLE 9 Solve $(w^3 - w)(w^3 - 10w^2 + 21w) = 0$.

$$(\underline{w^3} - \underline{w})(\underline{w^3} - 10\underline{w^2} + 21\underline{w}) = 0$$
$$w(w^2 - 1)w(w^2 - 10w + 21) = 0$$
$$w(w - 1)(w + 1)w(w - 3)(w - 7) = 0$$

$w = 0;$ $\quad w - 1 = \quad 0;$ $\quad w + 1 = \quad 0;$ $\quad w - 3 = \quad 0;$ $\quad$ and $\quad w - 7 = \quad 0$

$$\begin{array}{rr} +1 & +1 \\ \hline w = & 1 \end{array} \quad \begin{array}{rr} -1 & -1 \\ \hline w = & -1 \end{array} \quad \begin{array}{rr} +3 & +3 \\ \hline w = & 3 \end{array} \quad \text{and} \quad \begin{array}{rr} +7 & +7 \\ \hline w = & 7 \end{array}$$

There are five solutions: $w = 0, \pm 1, 3,$ and 7.

I'll leave the check for you to do.

Notice that there were two identical factors of w. Do not bother solving identical factors twice, because they will produce the same solution.

EXERCISES 15.5B

Find all the solutions to each equation and check each solution in the original equation.

27. $x^3 - x = 0$ $\qquad$ **28.** $y^2 - 5y = 0$

29. $w^3 - 5w^2 - 6w = 0$ $\qquad$ **30.** $z^3 + 4z^2 + 4z = 0$

31. $y^3 - 4y = 0$ $\qquad$ **32.** $w^3 - 25w = 0$

33. $x^3 - 3x^2 = 18x$ $\qquad$ **34.** $y^3 - 14y = 5y^2$

35. $w^3 + 5w^2 = 14w$ $\qquad$ **36.** $z^3 = 9z$

37. $(x^2 - 4x)(x^3 + 3x^2 + 2x) = 0$

38. $(2w^2 - 6w)(w^3 - 2w^2 + w) = 0$

39. $(y^3 - y^2 - 12y)(y^3 + y^2 - 12y) = 0$

40. $(y^3 + 12y^2)(y^3 - 9y^2 - 10y) = 0$

41. $(x^2 - 4x - 12)(x^2 - 3x + 2) = 0$

42. $(y^2 - 6y + 9)(y^2 - 5y) = 0$

43. $(2w^2 - 6w)(w^2 - 5w + 6) = 0$

44. $(3x - 4)(x^2 - x) = 0$

45. $(y^2 + 4y)(6y^2 + 2y) = 0$

46. $(m^2 - 25)(m^2 + 7m + 12) = 0$

47. $(3r^2 - 2r)(r^2 - r - 20) = 0$

48. $(5h^2 - 2h)(h^2 + 7h) = 0$

49. $(2x^2 - 5x - 3)(2x^2 + 5x + 3) = 0$

50. $(3y^2 - 7y + 2)(5y^2 + 9y - 2) = 0$

Check your answers on page 761.

15.6 The Quadratic Formula

Using $x = \dfrac{-b \pm \sqrt{b^2 - 4ac}}{2a}$ **to Solve** $ax^2 + bx + c = 0$

There are some trinomials, $ax^2 + bx + c$, that are prime and do not factor by FOIL.

Equations consisting of those trinomials equal to zero cannot be solved by either of the methods we have learned so far. They cannot be solved by the factoring method because they do not factor, and they cannot be solved by the square-root method because they include a linear term.

There is another method for solving quadratic equations called the **completing-the-square method**. It involves a modified version of the method we applied to polynomials in Section 15.1 to find the vertex of parabolas.

That method is quite cumbersome to use, so I am not going to show you how it works. When mathematicians applied that method in general to the equation

$ax^2 + bx + c = 0$, they derived the following formula for finding solutions to quadratic equations:

Solutions to $ax^2 + bx + c = 0$ are

$$x = \frac{-b \pm \sqrt{b^2 - 4ac}}{2a}, \quad a \neq 0.$$

We can use this formula on any quadratic equation, whether we can factor it or not. It is an easier method for solving than the completing-the-square method, so we will learn how to use it instead.

When we use this formula to solve equations, it is called the **quadratic formula method**.

To use the quadratic formula method:

First move all the terms to one side so the equation is in the form $ax^2 + bx + c = 0$.

The **a** in the formula is the **coefficient of x^2**.
The **b** in the formula is the **coefficient of x**.
The **c** in the formula is the **constant term**.

Substitute in the values of a, b, and c into the formula and simplify the resulting expression to find the solutions.

Such expressions will be fractional expressions involving radicals, which we learned to simplify in Section 14.7. Go back and review that section if necessary before proceeding.

When the solutions are irrational, you do not have to check them by substituting them back in the given equation, since those expressions would become too cumbersome. In a more advanced algebra course you will learn to simplify such expressions.

Remember, if no coefficient is written in the squared or linear term, the coefficient is 1.

EXAMPLE 1 Solve $x^2 + 5x + 1 = 0$. We can write that as $(1)x^2 + 5x + 1 = 0$.

The numbers to substitute into the formula are $a = \mathbf{1}$, $b = \mathbf{5}$, and $c = \mathbf{1}$.

$$x = \frac{-(\mathbf{5}) \pm \sqrt{(\mathbf{5})^2 - 4(\mathbf{1})(\mathbf{1})}}{2(\mathbf{1})}$$

$$x = \frac{-5 \pm \sqrt{25 - 4}}{2}$$

$$x = \frac{-5 \pm \sqrt{21}}{2}, \quad \text{which cannot be simplified further.}$$

So the two solutions are $x = \dfrac{-5 + \sqrt{21}}{2}$ and $x = \dfrac{-5 - \sqrt{21}}{2}$.

We can write them as $x = \dfrac{-5 \pm \sqrt{21}}{2}$.

EXAMPLE 2 Solve $x^2 - 7x = -1$.

$$x^2 - 7x = -1$$
$$\underline{+1 \quad +1}$$
$$(1)x^2 - 7x + 1 = 0 \qquad \text{So } a = 1;\ b = -7;\ \text{and } c = 1.$$

$$x = \frac{-(-7) \pm \sqrt{(-7)^2 - 4(1)(1)}}{2(1)}$$

$$x = \frac{+7 \pm \sqrt{49 - 4}}{2}$$

$$x = \frac{7 \pm \sqrt{45}}{2}$$

$$x = \frac{7 \pm \sqrt{9 \cdot 5}}{2}$$

$$x = \frac{7 \pm 3\sqrt{5}}{2}, \quad \text{which cannot be simplified further.} \qquad \square$$

EXAMPLE 3 Solve $2w^2 = -4w + 3$.

$$2w^2 = -4w + 3$$
$$\underline{+4w - 3 \quad +4w - 3}$$
$$2w^2 + 4w - 3 = 0 \qquad \text{So } a = 2;\ b = 4;\ \text{and } c = -3.$$

$$w = \frac{-4 \pm \sqrt{4^2 - 4(2)(-3)}}{2(2)}$$

$$w = \frac{-4 \pm \sqrt{16 + 24}}{4}$$

$$w = \frac{-4 \pm \sqrt{40}}{4}$$

$$w = \frac{-4 \pm \sqrt{4 \cdot 2 \cdot 5}}{4}$$

$$w = \frac{-4 \pm 2\sqrt{10}}{4}$$

$$w = \frac{(-2)(2) \pm 2\sqrt{10}}{(2)(2)}$$

$$w = \frac{2(-2 \pm \sqrt{10})}{(2)(2)}$$

$$w = \frac{\cancel{2}(-2 \pm \sqrt{10})}{(\cancel{2})(2)}$$

$$w = \frac{-2 \pm \sqrt{10}}{2}, \quad \text{which cannot be simplified further.} \qquad \square$$

EXAMPLE 4 Solve $2y^2 + 5y + 2 = 0$.

$$a = 2; \quad b = 5; \quad \text{and} \quad c = 2.$$

$$y = \frac{-5 \pm \sqrt{5^2 - 4(2)(2)}}{2(2)}$$

$$y = \frac{-5 \pm \sqrt{25 - 16}}{4}$$

$$y = \frac{-5 \pm \sqrt{9}}{4}$$

$$y = \frac{-5 \pm 3}{4}$$

The radical simplified to a rational number, so the solutions are rational. We should separate them to simplify.

$$y = \frac{-5 + 3}{4} \quad \text{and} \quad y = \frac{-5 - 3}{4}$$

$$= \frac{-2}{4} \qquad\qquad = \frac{-8}{4}$$

$$= \frac{-1}{2} \qquad\qquad = -2$$

So the solutions are $y = \dfrac{-1}{2}$ and -2.

When the solutions simplify into rational numbers, the equation could have been solved by factoring.

$$2y^2 + 5y + 2 = 0 \rightarrow (2y + 1)(y + 2) = 0$$

The computation to solve by factoring is much simpler than the computation involved in simplifying the quadratic formula. So always try to factor the trinomial before turning to the quadratic formula. But if the trinomial does not factor, or you do not notice that it factors, *you can always solve the equation with the quadratic formula.*

Since $y = \dfrac{-5 \pm \sqrt{25 - 16}}{4}$ simplifies to $y = \dfrac{-1}{2}$ and -2, we can appreciate the importance of simplifying as much as possible the expression we get from the quadratic formula.

If the expression under the radical symbol is a negative number, there are no real solutions to the equation, since we cannot take the square root of a negative number in the real number system.

EXAMPLE 5 Solve $x^2 + x + 1 = 0$.

$$a = 1; \quad b = 1; \quad \text{and} \quad c = 1.$$

$$x = \frac{-1 \pm \sqrt{1^2 - 4(1)(1)}}{2(1)}$$

$$x = \frac{-1 \pm \sqrt{1 - 4}}{2}$$

$$x = \frac{-1 \pm \sqrt{-3}}{2}$$

There is *no real solution* to this equation, since $\sqrt{-3}$ does not exist in the real number system.

EXAMPLE 6 Solve $x^2 - 2x + 1 = 0$.

$$a = 1; \quad b = -2; \quad \text{and} \quad c = 1.$$

$$x = \frac{-(-2) \pm \sqrt{(-2)^2 - 4(1)(1)}}{2(1)}$$

$$x = \frac{+2 \pm \sqrt{4 - 4}}{2}$$

$$x = \frac{2 \pm \sqrt{0}}{2}$$

$$x = \frac{2 \pm 0}{2}$$

$$x = \frac{2}{2}$$

$$x = 1$$

There is only one solution to this equation, because the radical simplified to zero. Hence 1 is a double solution. ❏

When the solution is rational (an integer or a fraction), the equation could have been factored, and so we did not have to use the quadratic equation after all!

$x^2 - 2x + 1 = 0$ can be factored into $(x - 1)(x - 1) = 0$.

EXERCISES 15.6A

Use the quadratic formula to solve each equation.

1. $x^2 + x - 1 = 0$ 2. $x^2 + 3x - 1 = 0$

3. $y^2 - 5y - 2 = 0$ 4. $w^2 + 3w - 3 = 0$

5. $m^2 + m = 4$ 6. $n^2 - 3n = 1$

7. $2y^2 - 2y - 3 = 0$ 8. $x^2 - 2x = 2$

9. $w^2 - w = 2$ 10. $k^2 + 2k - 5 = 0$

11. $3x^2 + 3x - 1 = 0$ 12. $5y^2 - 2y - 1 = 0$

Check your answers on page 761. ✓

Discriminant

There is a shortcut you can use to determine whether a trinomial can be factored. It can be factored if the radical term simplifies to a rational number. (In this discussion I assume we are working with equations in the form of $ax^2 + bx + c$, where a, b, and c are all integers.)

> **The Quadratic Formula:**
>
> Solutions to $ax^2 + bx + c = 0$ are $x = \dfrac{-b \pm \sqrt{b^2 - 4ac}}{2a}$.
>
> So the radical becomes rational if $b^2 - 4ac$ is a perfect square.

The expression $b^2 - 4ac$ is called the **discriminant**. By evaluating the discriminant, we can discover whether there will be 1, 2, or no solutions and whether the solutions will be rational or irrational.

If the discriminant is:

Positive
   and a perfect square, there will be *two rational solutions*;
   and not a perfect square, there will be *two irrational solutions*.

Negative, there will be *no real solutions* (only imaginary ones).

Zero, there will be only *one rational solution*.

So we can save a lot of time by evaluating the discriminant.

We do not bother solving by any method if the discriminant is negative.

We try to factor if the discriminant is a positive perfect square (including if it is zero).

We use the quadratic formula if the discriminant is a positive nonperfect square.

In the following examples, evaluate the discriminant for each equation and state what it tells you about the solutions and the best method to use for finding those solutions.

EXAMPLE 7 $x^2 + x + 3 = 0$; $a = 1$, $b = 1$, and $c = 3$,
so $b^2 - 4ac$ is

$$(1)^2 - 4(1)(3) =$$
$$1 - 12 = -11.$$

Since the discriminant is negative, there are *no real* solutions. So we will not bother trying to solve the equation.   ❏

EXAMPLE 8   Analyze the solutions to $2x^2 + 4x + 2 = 0$.

$a = 2$, $b = 4$, and $c = 2$,
so $b^2 - 4ac$ is

$$(4)^2 - 4(2)(2) =$$
$$16 - 16 = 0.$$

Since the discriminant is zero, there will be one rational solution, and we should use the factoring method to solve for it.   ❏

EXAMPLE 9   Analyze the solutions to $8x^2 - 10x + 2 = 0$.

$a = 8$, $b = -10$, and $c = 2$,
so $b^2 - 4ac$ is

$$(-10)^2 - 4(8)(2) =$$
$$100 - 64 = 36.$$

Since the discriminant is a *positive perfect square*, there will be *two rational solutions*. We should try to factor it.   ❏

Do not forget that to use the discriminant, you must convert the equation into the form $ax^2 + bx + c = 0$.

EXAMPLE 10   Analyze the solutions to $x^2 - 3x = 5$.

First we must move 5 from the right side.

$$x^2 - 3x = 5$$
$$\underline{ -5 = -5}$$
$$x^2 - 3x - 5 = 0 \quad ; \text{ so } a = 1, b = -3, \text{ and } c = -5.$$

$b^2 - 4ac$ is

$$(-3)^2 - 4(1)(-5) =$$
$$9 + 20 = 29.$$

Since the discriminant is a *positive nonperfect square*, there will be two irrational solutions. We should use the quadratic formula to solve for them. ❏

EXERCISES 15.6B

Evaluate the discriminant of each equation. State how many real solutions there will be, whether they will be rational or irrational, and which method should be used to solve the equation. Then solve the equations that have real solutions.

13. $x^2 - 2x - 8 = 0$ **14.** $x^2 + x - 3 = 0$

15. $y^2 - 3y + 5 = 0$ **16.** $w^2 - 7w - 13 = 0$

17. $2y^2 - y - 3 = 0$ **18.** $3w^2 + 2w - 2 = 0$

19. $5z^2 - 2z + 1 = 0$ **20.** $-2x^2 + 3x + 2 = 0$

21. $-3y^2 + 2y = 0$ **22.** $7w^2 - 5w + 1 = 0$

23. $x^2 - 1 = -4x$ **24.** $2w^2 - 3w = 1$

25. $-3y^2 + 6y = 3$ **26.** $3y - 2 = y^2$

Check your answers on page 761. ✓

Review of Solutions to Quadratic Equations

You now know how to solve *every* kind of quadratic equation. Here is a guide to help you recognize which method to use:

$$ax^2 + bx + c = 0$$

If $b = 0$, use the square-root method.

> Isolate the squared letter on one side, and take the square root of each side.

If $b \neq 0$, evaluate the discriminant and

> use the quadratic formula method if it is a positive nonperfect square;
> use the factoring method if it is a positive perfect square (including zero);
> do not waste your time trying to solve it if it is negative.

EXERCISES 15.6C

Use the easiest method to solve each equation. Check all rational solutions by substituting them back into the original equations.

27. $x^2 - 3x = 0$

28. $y^2 = 36$

29. $(x - 2)^2 = 4$

30. $x^2 - x - 3 = 0$

31. $w^2 - 14w - 15 = 0$

32. $y^2 + 6y = 7$

33. $(w + 4)^2 = 7$

34. $2x^2 - x - 5 = 0$

35. $x^2 - 8x = 0$

36. $(3y - 1)^2 = 1$

37. $w^2 - 2w = 7$

38. $2y^2 - 14y - 16 = 0$

39. $x^3 - x^2 = 6x$

40. $(2x + 3)^2 = 2$

Check your answers on page 761.

15.7 Solving Rational Quadratic Equations

In Section 14.8 we learned how to solve rational equations that became linear equations. If they became quadratic when they were simplified, however, we would not have been able to solve them. Now that you can solve quadratic equations, we can solve rational equations that simplify into quadratics.

Whenever an equation contains denominators, the first thing you must always do is find the LCM of those denominators and multiply both sides of the equation by the LCM. Each denominator will cancel with a factor(s) of the LCM, and no terms will be left in the denominator. The result will be a linear or quadratic equation that you can then solve.

Do not forget that to multiply the LCM by each side means to multiply it by *every term* on each side.

EXAMPLE 1 Solve $\dfrac{20}{w - 3} + 4 = \dfrac{7}{w} + 6$.

The two denominators, $w - 3$ and w, are both prime, so the LCM is their product: $w(w - 3)$.

Multiply both sides by $w(w - 3)$ to clear the denominators, provided $w \neq 0$ and $w \neq 3$.

$$w(w - 3)\left[\frac{20}{w - 3} + 4\right] = w(w - 3)\left[\frac{7}{w} + 6\right]$$

$$w(w - 3)\frac{20}{w - 3} + w(w - 3)(4) = w(w - 3)\frac{7}{w} + w(w - 3)(6)$$

$$w(20) + w(w - 3)(4) = (w - 3)(7) + w(w - 3)(6)$$

When there are three factors to multiply, as in $w(w - 3)(4)$ and $w(w - 3)(6)$, it is easiest to multiply the monomials first and then multiply their product by the binomial using the distributive law: Multiply $4w$ by $(w - 3)$ and $6w$ by $(w - 3)$.

So the equation becomes

$$20w + 4w(w - 3) = 7(w - 3) + 6w(w - 3)$$
$$20w + 4w^2 - 12w = 7w - 21 + 6w^2 - 18w$$
$$4w^2 + 8w = 6w^2 - 11w - 21$$
$$\underline{-4w^2 - 8w \quad -4w^2 - 8w}$$
$$0 = 2w^2 - 19w - 21, \quad \text{which factors into}$$
$$0 = (2w - 21)(w + 1).$$

To solve, set $(2w - 21)$ and $(w + 1)$ each equal to 0.

$$(2w - 21) = 0 \qquad (w + 1) = 0$$
$$\underline{\quad + 21 \quad +21} \qquad \underline{\quad -1 \quad -1}$$
$$2w = 21 \qquad w = -1$$

$$w = \frac{21}{2}, \text{ or } 10\frac{1}{2}$$

So the solutions are $w = -1$ and $w = 10\frac{1}{2}$.

Neither of them makes the denominator zero, so both really are solutions. ❏

EXAMPLE 2 Solve $\dfrac{x - 1}{x} + \dfrac{2x - 3}{x + 2} = \dfrac{6}{x^2 + 2x}$

The third denominator can be factored:

$$\frac{x - 1}{x} + \frac{2x - 3}{x + 2} = \frac{6}{x(x + 2)} \qquad \text{LCM} = x(x + 2)$$

We can multiply both sides by $x(x + 2)$, provided $x \neq 0$ and $x \neq -2$.

$$x(x + 2)\left[\frac{x - 1}{x}\right] + x(x + 2)\left[\frac{2x - 3}{x + 2}\right] = x(x + 2)\left[\frac{6}{x(x + 2)}\right]$$

$$\cancel{x}(x + 2)\left[\frac{x - 1}{\cancel{x}}\right] + x\cancel{(x + 2)}\left[\frac{2x - 3}{\cancel{x + 2}}\right] = \cancel{x(x + 2)}\left[\frac{6}{\cancel{x(x + 2)}}\right]$$

$$(x + 2)(x - 1) + x(2x - 3) = 6$$
$$x^2 + x - 2 + 2x^2 - 3x = 6$$
$$3x^2 - 2x - 2 = 6$$
$$\underline{\qquad\qquad - 6 = -6}$$
$$3x^2 - 2x - 8 = 0, \quad \text{which factors into:}$$
$$(3x + 4)(x - 2) = 0$$

To solve; set $(3x + 4)$ and $(x - 2)$ each equal to 0.

$$(3x + 4) = 0 \qquad (x - 2) = 0$$
$$\underline{\quad - 4 \quad -4} \qquad \underline{\quad + 2 \quad +2}$$
$$3x = -4 \qquad x = 2$$

$$x = -\frac{4}{3}, \text{ or } -1\frac{1}{3}$$

The solutions are $x = 2$ and $x = -1\frac{1}{3}$. ❏

EXAMPLE 3 Solve $\dfrac{n-2}{n-1} - \dfrac{n-1}{n+3} = \dfrac{5}{n^2 + 2n - 3}$

$$\frac{n-2}{n-1} - \frac{n-1}{n+3} = \frac{5}{(n-1)(n+3)} \qquad \text{LCM} = (n-1)(n+3)$$

We can multiply both sides by $(n-1)(n+3)$, provided $n \neq 1$ and $n \neq -3$.

$$(n-1)(n+3)\left[\frac{n-2}{n-1}\right] + (n-1)(n+3)\left[\frac{-(n-1)}{(n+3)}\right] =$$

$$(n-1)(n+3)\left[\frac{5}{(n-1)(n+3)}\right]$$

$$(n-1)(n+3)\left[\frac{n-2}{n-1}\right] + (n-1)(n+3)\left[\frac{-(n-1)}{(n+3)}\right] =$$

$$(n-1)(n+3)\left[\frac{5}{(n-1)(n+3)}\right]$$

$$(n+3)(n-2) + (n-1)[-(n-1)] = 5$$
$$n^2 + n - 6 - (n-1)(n-1) = 5$$
$$n^2 + n - 6 - [n^2 - 2n + 1] = 5$$
$$n^2 + n - 6 - n^2 + 2n - 1 = 5,$$

which simplifies to the linear equation:

$$
\begin{aligned}
3n - 7 &= 5 \\
+7 & +7 \\
\hline
3n &= 12 \\
n &= 4
\end{aligned}
$$

This equation simplified into a linear equation, so there is only *one* solution.

When the solution is an integer, you should check by substituting it into the original equation.

Check: If $n = 4$, does $\dfrac{(n-2)}{(n-1)} - \dfrac{(n-1)}{(n+3)} = \dfrac{5}{(n^2 + 2n - 3)}$?

$$\frac{(4)-2}{(4)-1} - \frac{(4)-1}{(4)+3} = \qquad \frac{5}{(4)^2 + 2(4) - 3} =$$

$$\frac{2}{3} - \frac{3}{7} = \qquad\qquad \frac{5}{16 + 8 - 3} = \frac{5}{21} \quad \checkmark$$

$$\frac{7}{7} \cdot \frac{2}{3} - \frac{3}{3} \cdot \frac{3}{7} =$$

$$\frac{14}{21} - \frac{9}{21} = \frac{5}{21} \quad \checkmark$$

EXAMPLE 4 Solve $\dfrac{3x+5}{2x+1} + \dfrac{x+4}{x+3} = \dfrac{4x^2 + 17x + 14}{2x^2 + 7x + 3}$.

$$\frac{3x+5}{2x+1} + \frac{x+4}{x+3} = \frac{4x^2 + 17x + 14}{(2x+1)(x+3)}$$

The LCM $= (2x + 1)(x + 3)$. Multiply both sides by $(2x + 1)(x + 3)$, if $x \neq -\dfrac{1}{2}$ and $x \neq -3$.

$$(2x + 1)(x + 3)\left[\frac{3x + 5}{2x + 1}\right] + (2x + 1)(x + 3)\left[\frac{x + 4}{x + 3}\right] =$$

$$(2x + 1)(x + 3)\left[\frac{4x^2 + 17x + 14}{(2x + 1)(x - 3)}\right]$$

$$\cancel{(2x + 1)}(x + 3)\left[\frac{3x + 5}{\cancel{2x + 1}}\right] + (2x + 1)\cancel{(x + 3)}\left[\frac{x + 4}{\cancel{x + 3}}\right] =$$

$$\cancel{(2x + 1)}\cancel{(x + 3)}\left[\frac{4x^2 + 17x + 14}{\cancel{(2x + 1)}\cancel{(x + 3)}}\right]$$

$$(x + 3)(3x + 5) + (2x + 1)(x + 4) = 4x^2 + 17x + 14$$

$$3x^2 + 14x + 15 + 2x^2 + 9x + 4 = 4x^2 + 17x + 14$$

$$5x^2 + 23x + 19 = \quad 4x^2 + 17x + 14$$
$$\underline{-4x^2 - 17x - 14 \quad -4x^2 - 17x - 14}$$
$$x^2 + 6x + 5 = 0$$

$$(x + 1)(x + 5) = 0$$

$$\begin{array}{ccc} x + 1 = & 0 & \qquad x + 5 = \quad 0 \\ \underline{-1 \quad -1} & & \underline{-5 \quad -5} \\ x = -1 & \text{and} & x = -5 \end{array} \quad \text{are the solutions.}$$

Check these solutions. ❏

Remember, if you cannot easily factor a quadratic equation, evaluate the discriminant and use the quadratic formula if necessary.

EXAMPLE 5   Solve $z - \dfrac{12}{z - 4} = \dfrac{z}{2}$.

The LCM is the product $2(z - 4)$. Multiply both sides by $2(z - 4)$ if $z \neq 4$

$$2(z - 4)\left[z - \frac{12}{z - 4}\right] = 2(z - 4)\left(\frac{z}{2}\right)$$

$$2(z - 4)(z) + 2\cancel{(z - 4)}\left(\frac{-12}{\cancel{z - 4}}\right) = \cancel{2}(z - 4)\left(\frac{z}{\cancel{2}}\right)$$

$$2z(z - 4) + 2(-12) = (z - 4)[z]$$

$$2z^2 - 8z - 24 = \quad z^2 - 4z$$
$$\underline{-z^2 + 4z \qquad -z^2 + 4z}$$
$$z^2 - 4z - 24 = 0$$

Solve this using the quadratic formula where $a = 1$, $b = -4$, and $c = -24$.

The discriminant is $16 - 4(1)(-24) =$

$$16 + 96 =$$

$$112 = (4)(4)(7),$$

which is not a perfect square, so we should use the quadratic formula to solve the equation.

$$z = \frac{+4 \pm \sqrt{(4)(4)(7)}}{2(1)}$$

$$= \frac{+4 \pm 4\sqrt{7}}{2}$$

$$= \frac{4(1 \pm \sqrt{7})}{2}$$

$$= \frac{2 \cdot \cancel{2}(1 + \sqrt{7})}{\cancel{2}}$$

$$= 2(1 \pm \sqrt{7}), \text{ or } 2 \pm 2\sqrt{7}.$$

So $2 + 2\sqrt{7}$ and $2 - 2\sqrt{7}$ are both solutions, since neither is equal to $+4$. ◻

EXERCISES 157

Solve each equation. Factor each denominator, find the LCM, multiply every term by the LCM to cancel out each denominator, and solve the resulting linear or quadratic equation.

1. $\dfrac{14}{x - 2} + 3 = \dfrac{9}{x} + 4$

2. $\dfrac{12}{x + 1} = \dfrac{9}{x}$

3. $\dfrac{12}{y - 1} + 2 = \dfrac{10}{y} + 3$

4. $\dfrac{4}{x - 3} + 5 = \dfrac{14}{x} + 4$

5. $\dfrac{x - 1}{x} + \dfrac{x - 2}{x + 1} = \dfrac{11}{x(x + 1)}$

6. $\dfrac{y - 2}{y} + \dfrac{2y - 5}{y - 1} = \dfrac{5}{y(y - 1)}$

7. $\dfrac{w - 1}{w} - \dfrac{2w - 3}{w + 2} = \dfrac{2}{w^2 + 2w}$

8. $\dfrac{x - 3}{x - 2} + \dfrac{x - 4}{x + 3} = \dfrac{19}{(x - 2)(x + 3)}$

9. $\dfrac{2y + 3}{y + 2} - \dfrac{1}{y + 3} = \dfrac{3y + 4}{y^2 + 5y + 6}$

10. $\dfrac{w - 2}{w - 1} + \dfrac{w - 3}{w - 2} = \dfrac{2w - 1}{w^2 - 3w + 2}$

11. $\dfrac{y - 1}{y + 1} - \dfrac{y - 3}{y - 2} = \dfrac{y^2 - 2y - 7}{y^2 - y - 2}$

12. $\dfrac{x + 4}{x - 1} - \dfrac{x + 3}{x + 1} = \dfrac{x^2 + x - 1}{x^2 - 1}$

13. $\dfrac{8 - w}{w - 1} = \dfrac{w}{3}$

14. $x - \dfrac{10}{x - 3} = \dfrac{x + 2}{3}$

15. $\dfrac{2w - 7}{w - 1} - \dfrac{w - 3}{w - 2} = \dfrac{w^2 - 3w - 9}{w^2 - 3w + 2}$

16. $\dfrac{2y - 3}{y + 2} + \dfrac{y - 1}{y} = \dfrac{y^2 + 4y - 2}{y^2 + 2y}$

Check your answers on page 761. ✓

 **Real-Life Quadratic Equations**

Now that you know how to solve quadratic equations and to locate the vertex of parabolas, there are more real-life problems that you can solve. In this section we will use quadratic equations to set up the problems.

With real-life problems, you must check not only that the solutions correctly fit the data of the problem, but also that they *make sense.* Often a solution to the equation is *not* an answer to the real-life problem. For example, if the problem involves finding the dimensions of a painting, negative solutions would not make sense. So even though you may get two solutions to the *equation,* check whether both could be *answers to the question.*

EXAMPLE 1 The area of a rectangular painting is 24 square inches. The length is 2 inches longer than the width. What are the dimensions?

Let's call the width x and the length $x + 2$.

Since the area is the product of the two dimensions, the equation we can set up is (width)(length) $= 24$

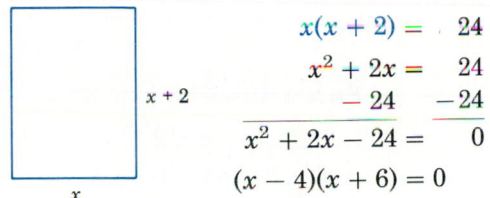

$$x(x + 2) = 24$$
$$x^2 + 2x = 24$$
$$\underline{ -24 \quad -24}$$
$$x^2 + 2x - 24 = 0$$
$$(x - 4)(x + 6) = 0$$

The solution to $x - 4 = 0$ is $x = 4$.
The solution to $x + 6 = 0$ is $x = -6$.

The dimension of a picture cannot be negative, so we must discard the -6 solution.

$$\text{If} \quad x = 4, x + 2 = 6$$

So the width is 4 inches and the length is 6 inches.

Check: Is the area of a 4-inch by 6-inch picture equal to 24 square inches?

$$(4 \text{ in.}) (6 \text{ in.}) = 24 \text{ sq in.} \quad \checkmark \qquad \blacksquare$$

EXAMPLE 2 The photographer wants to mount a photo that measures 8 inches by 6 inches so that the area of the mounted photo is three and a half times the area of the unmounted photo.

Call x the width (in inches) of the matting she should use.

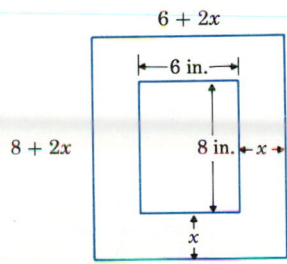

So the width of the mounted photo is $6 + 2x$.

The length of the mounted photo is $8 + 2x$.

The area of the unmounted photo is $6(8) = 48$ sq. in.

The area of the mounted photo is

$$\left(3\frac{1}{2}\right) \text{ times (the area of the unmounted photo)} =$$

$$(3.5)(48 \text{ sq. in.}) = 168 \text{ sq. in.}$$

The equation is

Area of mounted photo = (width of mounted photo) (length of mounted photo)

$$168 = (6 + 2x)(8 + 2x)$$

$$168 = 48 + 12\underline{x} + 16\underline{x} + 4x^2$$

$$168 = 48 + 28x + 4x^2$$

$$\underline{-168 -168}$$

$$0 = -120 + 28x + 4x^2$$

$$0 = 4x^2 + 28x - 120$$

$$0 = 4(x^2 + 7x - 30)$$

$$\frac{0}{4} = \frac{4(x^2 + 7x - 30)}{4}$$

$$0 = x^2 + 7x - 30$$

$$0 = (x + 10)(x - 3)$$

So $x + 10 = 0 $ and $ x - 3 = 0$ are the equations to solve:

$$\underline{- 10 -10} \underline{+3 +3}$$

$x = -10$ and $x = 3$ are the *solutions to the equation.*

Since x is the width of the matting, it does not make sense to have a negative solution to this problem. So $x = -10$ is discarded and the answer is that the matting must be 3 inches wide.

Check: Is the area of the photo mounted with a 3-inch-wide matting equal to three and a half times the area of the unmounted photo?

Width of matted photo would be $(6 + 3 + 3)$ in. = 12 in.

Length of matted photo would be $(8 + 3 + 3)$ in. = 14 in.

$$\text{So mounted area} = (12 \text{ in.})(14 \text{ in.})$$

$$= 168 \text{ sq. in.} \quad \checkmark$$

Area of photo is 48 sq. in.

$$(3.5)[48 \text{ sq. in.}] = 168 \text{ sq. in.} \quad \checkmark \qquad \blacksquare$$

EXAMPLE 3   Buck had 600 feet of fencing to use on a rectangular pasture he was fencing off along the river. He wanted to obtain the largest possible area.

If he ran the sides only 1 foot away from the river, he would use 2 feet of fencing for those sides and have 598 feet for the width. Such a pasture would have an area of

$$(1 \text{ ft})(598 \text{ ft}) = 598 \text{ sq ft.}$$

If he ran the sides 100 feet away from the river, he would use 200 feet of fencing for those sides and have 400 feet for the width. Such a pasture would have an area of

$$(100 \text{ ft})(400 \text{ ft}) = 40,000 \text{ sq ft.}$$

If he ran the sides 250 feet away from the river, he would use 500 feet of fencing for those sides and have 100 feet for the width. Such a pasture would have an area of

(250 ft)(100 ft) = 25,000 sq ft.

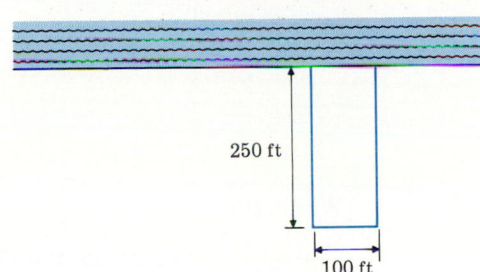

Since the area got bigger and then smaller as he kept increasing those two sides, he decided this problem was too difficult to do in his head, and so he used algebra.

He called the dimension of a side running perpendicular to the river x; the other dimension was $600 - 2x$.

Area $= x(600 - 2x)$

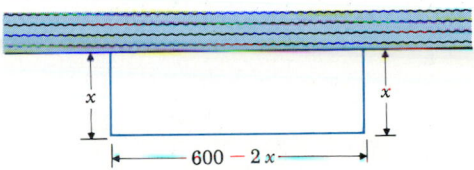

Area $= 600x - 2x^2$, which is the equation of a parabola that opens downward. Since the area is a function of x, the maximum value of the area is the second coordinate of that parabola's vertex.

In a more advanced algebra course, Buck learned to complete the square on that equation and found it to be

Area is $f(x) = -2(x - 150)^2 + 45,000$.

To find the maximum area, find the value of $f(x)$ when the squared term $(x - 150)$ is zero.

$$f(x) = -2(\mathbf{0}) + 45,000 \quad \text{so } f(x) = 45,000.$$

He will get that area when the squared term $(x - 150)$ is zero: $x - 150 = 0$ when $x = 150$.

So Buck should run the fence 150 feet away from the river on the sides. He will use 300 feet of fencing for those two sides, leaving 300 feet for the width of the pasture.

The pasture will be 150 feet by 300 feet.

Check: Will the area be 45,000 square feet with those dimensions?

Area is (150 ft)(300 ft) = 45,000 sq ft. ✓

EXAMPLE 4   The formula $h = 6 + 48t - 16t^2$ represents the distance a ball is from the ground after t seconds when it is thrown upward from 6 feet above the ground at the speed of 48 feet per second. The ball will eventually reach its peak height and come back down, so its path will look like a parabola.

Barb figures she will catch it at a height of 6 feet and wants to calculate how many seconds it should take the ball to go up and come back down to that height.

We want to find the value of t when h is 6 feet.

$$
\begin{array}{rl}
6 = & 6 + 48t - 16t^2 \\
\underline{-6} & \underline{-6} \\
0 = & 48t - 16t^2 \\
0 = & 16t(3 - t)
\end{array}
$$

When $16t = 0$, $t = 0$. When $3 - t = 0$, $t = 3$.

The ball will be at the height of 6 feet when the ball is first thrown ($t = 0$) and after 3 seconds.

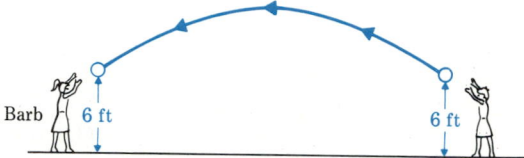

Check: After 3 seconds is the ball's height 6 feet?

$$
\begin{aligned}
\text{Ball's height} &= 6 + 48(3) - \underline{16(3)^2} \\
&= 6 + \underline{48(3)} - \underline{16(9)} \\
&= 6 + 144 - 144 \\
&= 6 \quad ✓
\end{aligned}
$$

EXAMPLE 5   The Angry Arts film collective had been showing films with socially relevant themes each month. They usually sold 200 tickets to each movie and charged $4 a ticket.

Their rent expenses increased $600 a month when they lost their free theater and projector. A financial advisor projected that they would lose 10 patrons for every dollar they increased the ticket price.

How much should they increase the ticket price so they will bring in enough revenue to meet their increased expenses?

Let x represent the number of $1 increases in the ticket price.

Ticket price: $(4 + x)$.

Projected audience:

$$
200 \text{ minus } 10 \text{ times price increase} = 200 - 10x.
$$

New monthly revenue:

$$
\text{(Ticket price) (projected audience)} = (4 + x)(200 - 10x).
$$

In the past their monthly revenue was $4(200) = \$800$. They want the new monthly revenue to be $600 more than that.

So the equation to solve is:

New monthly revenue = $800 + $600, or

$$(4 + x)(200 - 10x) = 1400$$

$$800 - 4(10x) + 200(x) - 10x(x) = 1400$$

$$800 - 40x + 200x - 10x^2 = 1400$$

$$800 + 160x - 10x^2 = 1400$$

$$\frac{-800 - 160x + 10x^2 \qquad -800 - 160x + 10x^2}{0 = \qquad 600 - 160x + 10x^2}$$

$$0 = 10(60 - 16x + x^2)$$

$$\frac{0}{10} = \frac{\cancel{10}(60 - 16x + x^2)}{\cancel{10}}$$

$$0 = x^2 - 16x + 60$$

$$0 = (x - 6)(x - 10)$$

So $x - 6 = 0$ and $x - 10 = 0$, which gives us the solutions: $x = 6$ and 10.

They can increase their tickets by $6 or $10.

Check: Will their monthly revenue be $1,400 if they increase their prices by

$6?   Each ticket would cost $4 + $6 = $10.
Projected audience would be $200 - 10(6) =$

$$200 - 60 = 140.$$

Monthly revenue would be $10(140) = $1,400. ✓

$10?   Each ticket would cost $4 + $10 = $14.
Projected audience would be $200 - 10(10) =$

$$200 - 100 = 100.$$

Monthly revenue would be $14(100) = $1,400. ✓

Either solution would work, but they wanted as many people as possible to see the films, so they chose to increase the price by $6 and did a lot of fund raising in the hope that they would be able to lower the price soon, before the new price drove even more people away. ❑

EXAMPLE 6   A plane traveled 900 miles from Raleigh, N.C. to New Orleans and then 900 miles back. The total flying time for the round trip took 11 hours.

How fast was the plane flying in still air if the wind speed that day was 15 miles per hour?

Let r be the plane's speed in still air.

The speed with the wind is $r + 15$.

The speed against the wind is $r - 15$.

Direction	Distance (in miles)	Rate (in mph)	Time (in hours)
With wind	900	$r + 15$	$\dfrac{900}{(r + 15)}$
Against wind	900	$r - 15$	$\dfrac{900}{(r - 15)}$

The total time of both trips was 11 hours.

Time with wind + time against wind = 11

$$\frac{900}{r+15} + \frac{900}{r-15} = 11 \quad \text{LCM} = (r+15)(r-15)$$

We can multiply both sides by $(r+15)(r-15)$ if $r \neq \pm 15$.

$$(r+15)(r-15)\left[\frac{900}{r+15} + \frac{900}{r-15}\right] = (r+15)(r-15)11$$

$$(r+15)(r-15)\left[\frac{900}{r+15}\right] + (r+15)(r-15)\left[\frac{900}{r-15}\right] = (r+15)(r-15)11$$

$$\cancel{(r+15)}(r-15)\left[\frac{900}{\cancel{r+15}}\right] + (r+15)\cancel{(r-15)}\left[\frac{900}{\cancel{r-15}}\right] = (r+15)(r-15)11$$

$$(r-15)[900] + (r+15)[900] = (r+15)(r-15)11$$

$$900r - 13{,}500 + 900r + 13{,}500 = (r^2 - 225)(11)$$

$$1800r = 11r^2 - 2475$$

$$\underline{-1800r \qquad\qquad\qquad -1800r}$$

$$0 = 11r^2 - 1800r - 2475$$

$$0 = (r - 165)(11r + 15)*$$

So

$$\begin{array}{lll} r - 165 = & 0 & \text{and} \quad 11r + 15 = \quad 0 \\ \underline{+165 \quad +165} & & \underline{\quad -15 \quad -15} \\ r = & 165 & \quad 11r \quad = -15 \\ & & \quad r = -\dfrac{15}{11} \end{array}$$

Again, we discard the negative solution and find that the plane's still-air speed must be 165 miles per hour.

Check: At that rate, would the round trip take 11 hours?

$$\text{Time with wind:} \frac{900 \text{ mi}}{(165+15)\text{mph}} =$$

$$\frac{900 \text{ mi}}{180 \text{ mph}} = 5 \text{ hr}$$

$$\text{Time against wind:} \frac{900 \text{ mi}}{(165-15)\text{mph}} =$$

$$\frac{900 \text{ mi}}{150 \text{ mph}} = 6 \text{ hr}$$

Round trip time: 5 hr + 6 hr = 11 hr ✓

EXAMPLE 7   Edith and Archie each ran 24 miles. She generally runs 2 miles per hour faster than he does. Find the rate and time it took for each of them if the sum of the times of their runs took 7 hours.

Runner	Distance (in miles)	Rate (in mph)	Time (in hours)
Archie	24	x	$\dfrac{24}{x}$
Edith	24	$x + 2$	$\dfrac{24}{(x+2)}$

---

* If you did not notice that this expression factored, you could have solved it by using the quadratic formula!

Archie's time + Edith's time = 7

$$\frac{24}{x} + \frac{24}{x + 2} = 7 \quad \text{LCM} = x(x + 2)$$

We can multiply both sides by $x(x + 2)$ if $x \neq 0$, and $x \neq -2$.

$$x(x + 2)\left[\frac{24}{x} + \frac{24}{x + 2}\right] = x(x + 2)[7]$$

$$\cancel{x}(x + 2)\left[\frac{24}{\cancel{x}}\right] + x\cancel{(x + 2)}\left[\frac{24}{\cancel{x + 2}}\right] = 7x(x + 2)$$

$$24(x + 2) + 24x = 7x(x + 2)$$

$$24x + 48 + 24x = 7x^2 + 14x$$

$$48x + 48 = 7x^2 + 14x$$

$$\underline{-48x - 48 \qquad\qquad -48x - 48}$$

$$0 = 7x^2 - 34x - 48$$

$$0 = (7x + 8)(x - 6)$$

$$
\begin{array}{ll}
7x + 8 = 0 & x - 6 = 6 \\
\underline{-8 \quad -8} & \underline{+6 \quad +6} \\
\dfrac{\cancel{7}x}{\cancel{7}} = -\dfrac{8}{7} & x = 6
\end{array}
$$

$$x = -1\frac{1}{7}$$

Since x represents the rate Archie ran, it cannot be a negative number, so we discard $x = -1\frac{1}{7}$. Archie's rate is $x = 6$.

So Archie ran 6 mph; Edith ran $x + 2 = 8$ mph.

Archie's time: $\dfrac{24}{6} = 4$ hr

Edith's time: $\dfrac{24}{8} = 3$ hr

Check: Is the sum of their times 7 hours?

$$4 \text{ hr} + 3 \text{ hr} = 7 \text{ hr} \quad \checkmark$$

EXAMPLE 8 It takes David 8 hours longer to mow the grass than it takes Alice to do the same job when each works alone. If it takes them 3 hours to do the job together, how long does it take each to do it alone?

Person	Hours to do job	Amount done in 1 hour
Alice	x	$\dfrac{1}{x}$
David	$x + 8$	$\dfrac{1}{x + 8}$
Together	3	$\dfrac{1}{3}$

Amount done by Alice + amount done by David = amount done together

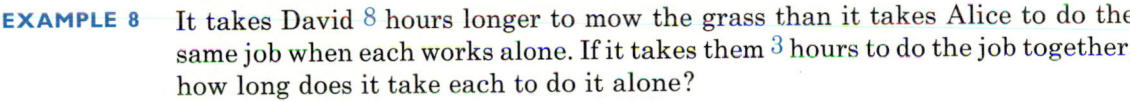

$$\frac{1}{x} + \frac{1}{x + 8} = \frac{1}{3} \quad \text{LCM} = 3x(x + 8)$$

We can multiply both sides by $3x(x + 8)$ if $x \neq 0$, and $x \neq -8$.

$$3x(x + 8)\left[\frac{1}{x}\right] + 3x(x + 8)\left[\frac{1}{x + 8}\right] = 3x(x + 8)\left[\frac{1}{3}\right]$$

$$3\cancel{x}(x + 8)\left[\frac{1}{\cancel{x}}\right] + 3x\cancel{(x + 8)}\left[\frac{1}{\cancel{x+8}}\right] = \cancel{3}x(x + 8)\left[\frac{1}{\cancel{3}}\right]$$

$$3(x + 8) + 3x = x(x + 8)$$

$$3x + 24 + 3x = x^2 + 8x$$

$$6x + 24 = x^2 + 8x$$

$$\underline{-6x - 24 \qquad\quad -6x - 24}$$

$$0 = x^2 + 2x - 24$$

$$0 = (x + 6)(x - 4)$$

So $x + 6 = 0$ and $x - 4 = 0$, or $x = -6$ and 4.

Since x represents the amount of time it takes Alice to do the job, it does not make sense for that value to be a negative number. So $x = -6$ is discarded. The only solution is $x = 4$.

Working alone, Alice can do the job in 4 hours. It takes David $x + 8$, or 12, hours to do the same job.

Check: Does it take them 3 hours to do the job together if one can do it in 4 hours and the other in 12 hours?

Did they complete a third of the job in one hour?

If Alice does the job in 4 hours, she does $\dfrac{1}{4}$ of the job in 1 hour.

If David does the job in 12 hours, he does $\dfrac{1}{12}$ of the job in 1 hour.

Does $\dfrac{1}{4} + \dfrac{1}{12} = \dfrac{1}{3}$?

$$\frac{1}{4} + \frac{1}{12} =$$

$$\left(\frac{3}{3}\right)\left(\frac{1}{4}\right) + \frac{1}{12} =$$

$$\frac{3}{12} + \frac{1}{12} =$$

$$\frac{4}{12} = \frac{1}{3} \checkmark$$

EXERCISES 15.8

Set up a quadratic equation to represent each problem if it is not already given to you. Solve it to answer each question. Check your answers for accuracy and keep only those that make sense in the problem.

1. The length of a rectangular painting is 3 in. more than its width. If the area is 40 sq in., what are the dimensions?

2. A rectangular field has an area of 150 sq yd. What are the dimensions if the length is 5 yd longer than the width?

3. A room is 4 ft longer than it is wide. It would take 140 sq ft of carpet to cover the floor. What are the dimensions of the room?

4. How wide should the matting be around a 5 in. by 7 in. photo if the area of the framed photo is 99 sq in.?

5. How wide should the matting be around a 4 in. by 7 in. photo if the area of the framed photo is 54 sq in.?

6. How wide should the matting be around a 6 in. by 8 in. photo if the area of the framed photo is 120 sq in.?

7. Annie wanted to fence off a rectangular pasture along the side of a river. She had 1,000 yd of fencing material to use on the three sides. If she has the pasture extend x yd from the river, the area of pasture is expressed by the equation $f(x) = -2(x - 250)^2 + 125,000$. To get the maximum area, how long should she make the sides of the pasture that start at the river?

8. If $f(x) = -3(x - 80)^2 + 380$ represents the area of a field, where x ft is the length of one side, what value of x will produce the largest area?

9. In $f(x) = -4(x - 130)^2 + 56,000$ represents the area of a field where x yd is the length of one side, what value of x will produce the largest area?

10. If $h = 6 + 48t - 16t^2$ represents the distance a ball is from the ground after t sec when it is thrown upward from 6 ft at the speed of 48 ft/sec, how long will it take for the ball to reach a height of 38 ft?

11. A theater group charges $5 for tickets and usually sells 300 tickets to each production. They figured they will lose 10 customers for each dollar they increase the ticket price. How much should they increase the price to bring in $240 more from each production?

12. A plane traveled 800 mi between two cities, and the round trip took 9 hr. How fast was the wind blowing if the plane's still-air speed for both trips was 180 mph?

13. We took a boat trip 100 mi down the river and 100 mi back. The round trip took 15 hr. What was the boat's still-water speed if the river was flowing 5 mph?

14. Alice runs 1 mph faster than Betsy. If they each ran 20 mi of a relay race, and the total time of both runs was 9 hr, how fast was each person's speed?

15. Ernie rode his bike 2 mph faster than Cedric did. They each rode 70 mi, and the total time for both trips was 24 hr. How fast did each man ride?

16. It takes Henry 3 hr longer to build the bookcase than it takes Lee. When they work together it takes them 2 hr. How long would each man need to build it alone?

17. It takes Ellen 6 hr longer to do a job than it takes Margaret. Working together they can complete the job in 4 hr. How long would it take each woman if she worked alone to complete that job?

Check your answers on page 761.

Glossary

completing the square   Process of representing the sum of a squared term and a linear term as the sum of a binomial squared and a constant term; used to convert quadratic equations into standard form so coordinates of a parabola's vertex can easily be identified.

Example:
$$x^2 + 6x =$$
$$(x^2 + 6x + 9) - 9 = (x + 3)^2 - 9$$

So $f(x) = x^2 + 6x$ is the same as

$$f(x) = (x + 3)^2 - 9,$$

which is in standard form and has vertex $(-3, -9)$.

constant term   Term consisting of only a number, with no letter factors.

Example: In $3x^2 + 5x - 7$, -7 is the constant term.

discriminant   Expression $b^2 - 4ac$ from the quadratic formula. Evaluating it tells us how many real solutions an equation has and whether they are rational or irrational.

If the discriminant is:
   Positive
      and a perfect square, there will be *two rational solutions*;
      and not a perfect square, there will be *two irrational solutions*.
   Negative, there will be *no real solutions* (only imaginary ones).
   Zero, there will be only *one rational solution*.

factoring method   One way to solve quadratic (and other equations of degree higher than 1) when you can express it as a product of linear factors equal to zero. Using the zero multiplication principle, we find every solution that makes each factor equal to zero.

Example: $x^2 + 5x + 4 = 0$ can be factored into
$(x + 4)(x + 1) = 0$;   so $x + 4 = 0$ or $x + 1 = 0$

The solutions are $x = -1$ and -4.

intercepts   *x*-intercept: Point where a curve crosses the *x*-axis. The second coordinate must be zero.

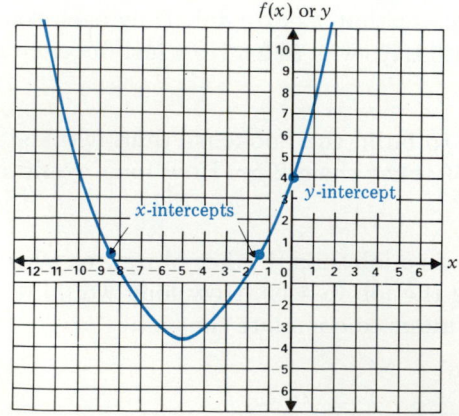

y-intercept: Point where a curve crosses the *y*-axis. The first coordinate must be zero.

linear term   Term of degree 1, with exactly one letter factor.

Example: In $3x^2 + 5x - 7$, **5x** is the linear term.

parabola   Symmetric curve that is the graph of all quadratic equations of the form $f(x) = ax^2 + bx + c$, where a, b, and c are numbers. An example of a parabola is sketched below.

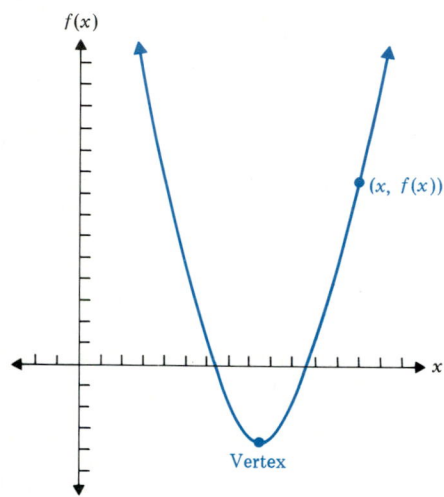

quadratic equation   Equation of degree 2. In this chapter, we study the solutions and graphs of equations which are quadratic in one variable and may also have a linear and/or constant term.

Example: We have solved quadratic equations, $ax^2 + bx + c = 0$, and graphed quadratic functions, $f(x) = ax^2 + bx + c$, where a, b, and c are numbers.

quadratic formula   Formula that gives us the solutions to $ax^2 + bx + c = 0$. These solutions are

$$x = \frac{-b + \sqrt{b^2 - 4ac}}{2a}.$$

square-root method   One way to solve quadratic equations that can be represented as a squared term equal to a constant term. We can solve such equations by taking the square root of both sides.

Example: Solve $(x - 3)^2 = 25$ by the square-root method.

$$(x - 3) = \pm\sqrt{25}$$
$$x - 3 = \pm 5$$

$$\text{So} \quad \begin{array}{r} x - 3 = 5 \\ +3 \quad +3 \\ \hline x = \quad 8 \end{array} \qquad \begin{array}{r} x - 3 = -5 \\ +3 \quad +3 \\ \hline x = -2 \end{array}$$

squared term   Term of degree 2, with two factors of the same letter (and no other letter factors).

Example: In $3x^2 + 5x - 7$, $3x^2$ is the squared term.

standard form   Form of a quadratic equation in two letters   ($y = ax^2 + bx + c$ or $f(x) = ax^2 + bc + c$) when we complete the square and express it as

$f(x) = a(x + h)^2 + k$, where a, h, and k are numbers.

The graph is a parabola with vertex $(-h, k)$.

Example: $f(x) = x^2 + 6x$ is expressed in standard form as

$f(x) = (x + 3)^2 - 9$, which has vertex $(-3, -9)$.

vertex   Point on a parabola where the curve changes direction. If the parabola opens upward, it is the lowest point; if the parabola opens downwards, it is the highest point.

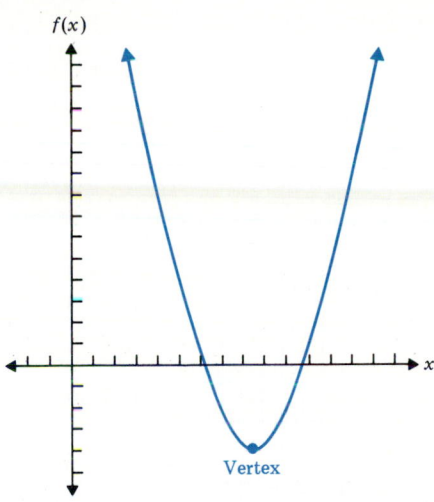

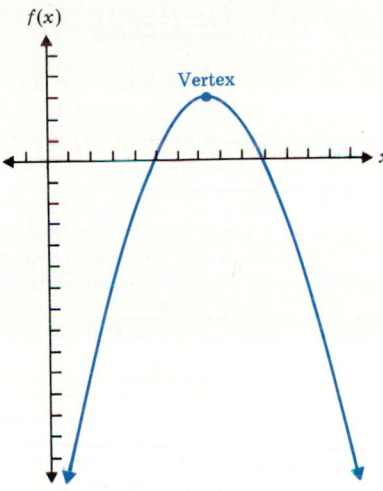

zero multiplication principle   Principle stating that if the product of two or more numbers is zero, at least one of them must be equal to zero.

Example: If $(x + 4)(x + 1) = 0$, then

$$x + 4 = 0 \text{ or } x + 1 = 0.$$

Chapter 15 Review

Do each of the problems. After you are confident that you have done them as accurately as possible, compare your answers with those in the back of the book. If any are wrong, go back to the section where the problem came from (indicated to the left of the problem) and review the section.

Once you have mastered all the sections, you will be sure to learn the skills solidly and remember how to do them if you practice them a little bit more. Turn to the Supplementary Exercises and do all the problems from any section where you had difficulty on these review exercises.

15.1 Find the coordinates of the vertex.

1. $f(x) = (x + 3)^2 - 1$ **2.** $f(x) = -2(x + 5)^2 - 6$

3. $f(x) = x^2 - 9$ **4.** $f(x) = -8x^2$

5. $f(x) = -x^2 + 3$ **6.** $f(x) = x^2 + 10x - 7$

Plot and label the vertex and two other points on either side of the vertex, and sketch the graph of the parabola.

7. $f(x) = (x + 3)^2 - 1$ **8.** $f(x) = -x^2 + 3$

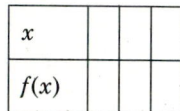

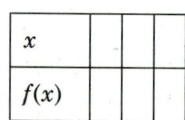

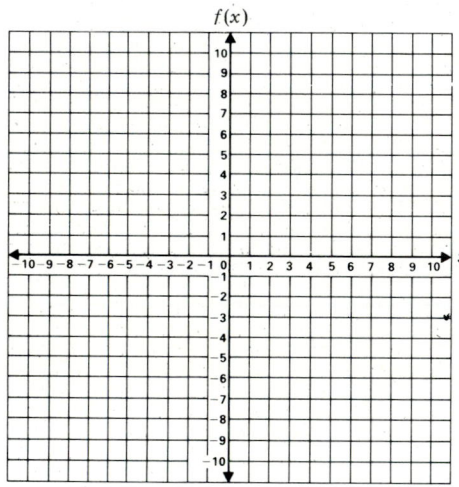

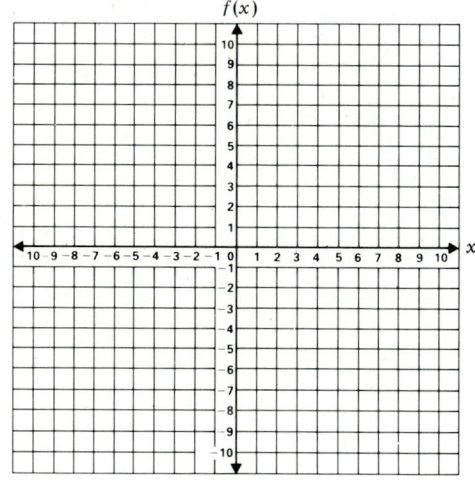

What are the coordinates of the x- and y-intercepts?

9. $f(x) = -5(x - 7)^2$ **10.** $f(x) = x^2 - 16$

11. $f(x) = (x + 5)^2 - 1$

15.2 Determine whether these numbers are solutions to these equations.

12. $x = 3;$ $x^2 + 5x - 24 = 0$ **13.** $y = -1;$ $y^2 + 5y = 4$

15.3 Solve these equations by the square-root method. Check all rational solutions.

14. $x^2 = 49$ **15.** $y^2 = 6$

16. $(w - 2)^2 = 25$ **17.** $(p + 5)^2 = 20$

18. $(3w - 4)^2 = 25$

Answers

1. _____

2. _____

3. _____

4. _____

5. _____

6. _____

9. y-int. _____

    x-int. _____

10. y-int. _____

    x-int. _____

11. y-int. _____

    x-int. _____

12. _____

13. _____

14. _____

15. _____

16. _____

17. _____

18. _____

15.4 Solve these equations by using the zero principle. Check all solutions.

19. $(x - 4)(x + 7) = 0$

20. $(3y - 3)(5y + 4) = 0$

21. $w(w - 1) = 0$

22. $7x(x + 3)(2x - 1) = 0$

15.5 Solve these equations by the factoring method. Check all solutions.

23. $x^2 - 8x = 0$

24. $y^2 - 7y + 12 = 0$

25. $w^2 - 9w = -20$

26. $x^2 + 6x = -8$

27. $y^3 - 4y = 0$

28. $(x^2 + 6x)(x^3 + 7x^2 + 12x) = 0$

15.6 Solve these equations by the quadratic formula. Check all solutions.

29. $2y^2 + y - 5 = 0$

30. $x^2 + 4x - 1 = 0$

31. $5w^2 + w = 3$

32. $x^2 = 3x + 1$

15.7 Solve and check.

33. $\dfrac{y - 6}{y - 3} + \dfrac{2}{y - 4} = \dfrac{11}{(y - 3)(y - 4)}$

34. $\dfrac{x}{x + 1} - \dfrac{x - 3}{x - 2} = \dfrac{3}{10}$

15.8 35. What are the dimensions of a rectangle if the length is 3 in. more than the width, and the area is 54 sq. in.?

36. Molly can do a job 3 times as quickly as John. Together they could do the job in 3 hr. How long should it take each to do it alone?

Answers

19. _____

20. _____

21. _____

22. _____

23. _____

24. _____

25. _____

26. _____

27. _____

28. _____

29. _____

30. _____

31. _____

32. _____

33. _____

34. _____

35. _____

36. _____

Check your answers on page 762.

Supplementary Exercises

Do all the problems in every section that involves a skill in which you now lack complete mastery. With a little more practice you will achieve that sense of really understanding the topics, and you will remember how to do these problems.

15.1 Find the coordinates of the vertex.

1. $f(x) = (x - 4)^2 + 2$ **2.** $f(x) = -3(x + 2)^2 - 7$

3. $f(x) = x^2 - 8$ **4.** $f(x) = -4x^2$

5. $f(x) = -x^2 + 6$ **6.** $f(x) = x^2 + 8x - 5$

Plot and label the vertex and two other points on either side of the vertex, and sketch the graph of the parabola.

7. $f(x) = (x - 4)^2 + 2$ **8.** $f(x) = -x^2 + 6$

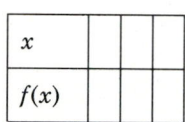

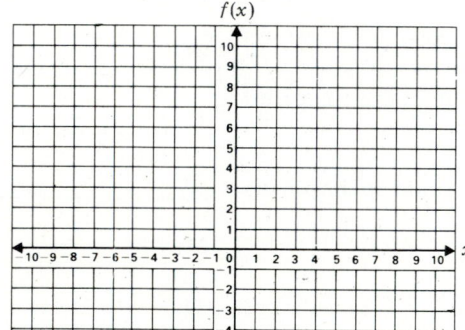

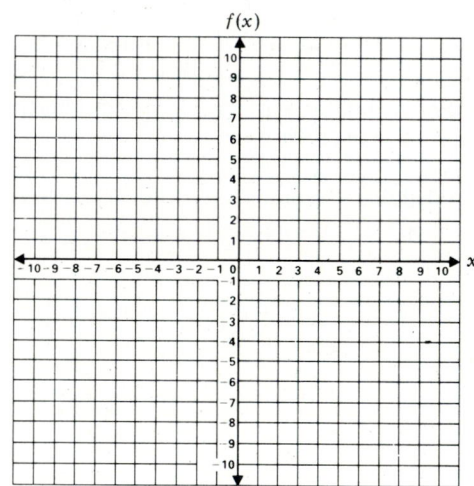

What are the coordinates of the x- and y-intercepts?

9. $f(x) = -7(x - 5)^2$ **10.** $f(x) = x^2 - 36$

11. $f(x) = (x + 2)^2 - 4$

15.2 Determine whether these numbers are solutions to the equations.

12. $x = 1;$ $x^2 - 3x - 2 = 0$ **13.** $y = -2;$ $y^2 - 4y - 12 = 0$

14. $y = -3;$ $2y^2 + 3y = 12$ **15.** $p = 2;$ $5p^2 + 4p = -28$

15.3 Solve these equations by the square-root method.

16. $x^2 = 1$ **17.** $y^2 = 9$

18. $w^2 = 5$ **19.** $m^2 = 45$

20. $(x + 5)^2 = 16$ **21.** $(y - 3)^2 = 36$

22. $(2w - 1)^2 = 4$ **23.** $(3x + 1)^2 = 5$

Answers

1. _____

2. _____

3. _____

4. _____

5. _____

6. _____

9. y-int.: _____

  x-int.: _____

10. y-int.: _____

  x-int.: _____

11. y-int.: _____

  x-int.: _____

12. _____

13. _____

14. _____

15. _____

16. _____

17. _____

18. _____

19. _____

20. _____

21. _____

22. _____

23. _____

15.4 Solve these equations by using the zero principle.

24. $(x + 3)(x - 1) = 0$ **25.** $(y - 4)(y - 5) = 0$

26. $(2w + 3)(5w - 7) = 0$ **27.** $(6p + 5)(2p - 1) = 0$

28. $x(x + 3) = 0$ **29.** $y(y - 6) = 0$

30. $x(x + 8)(x - 3) = 0$ **31.** $5y(3y - 1)(y + 7) = 0$

15.5 Solve these equations by the factoring method.

32. $x^2 + 4x + 4 = 0$ **33.** $y^2 - 8y + 15 = 0$

34. $w^2 - w - 56 = 0$ **35.** $x^2 - 11x = 0$

36. $m^2 + 5m = -6$ **37.** $k^2 - 3 = 2k$

38. $r^2 + 5r = 0$ **39.** $3y^2 - 5y = 0$

40. $x^3 - 16x = 0$ **41.** $y^3 - 8y^2 + 15y = 0$

42. $(w^2 + 5w)(w^3 + w^2 - 20w) = 0$

15.6 Solve these equations by the quadratic formula.

43. $x^2 + x - 5 = 0$ **44.** $y^2 + 2y - 7 = 0$

45. $w^2 - 2w - 1 = 0$ **46.** $m^2 + 3m - 8 = 0$

47. $x^2 + 2x = -7$ **48.** $3x^2 - 1 = 2x$

Answers

24. _____

25. _____

26. _____

27. _____

28. _____

29. _____

30. _____

31. _____

32. _____

33. _____

34. _____

35. _____

36. _____

37. _____

38. _____

39. _____

40. _____

41. _____

42. _____

43. _____

44. _____

45. _____

46. _____

47. _____

48. _____

Continued

Supplementary Exercises *(Cont.)*

15.7 Solve and check.

Answers

49. $\dfrac{y}{y+3} + \dfrac{1}{y+1} = \dfrac{11}{(y+3)(y+1)}$

50. $\dfrac{w-2}{w+3} + \dfrac{1}{w+1} = \dfrac{3w+5}{w^2+4w+3}$

51. $\dfrac{2x+1}{x+4} - \dfrac{x-1}{x+1} = \dfrac{9}{x^2+5x+4}$

15.8 **52.** If the area of a rectangle is 28 sq in. and the length is 3 in. more than the width, what are the dimensions?

53. Jim can do a job twice as fast as Ron can. Together it takes them 4 days to complete the job. How long would it take each working alone?

54. Patti ran her 12 mi leg of the race 1 mph faster than Liz ran her 12 mi leg. What was the rate that each woman ran if the total time for this duo was 7 hr?

49. _____

50. _____

51. _____

52. _____

53. _____

54. _____

Check your answers on page 762.

Chapter 15 Test

State the coordinates of each vertex.

1. $f(x) = x^2 + 7$ **2.** $f(x) = (x - 5)^2$

3. $f(x) = (x + 4)^2 - 5$ **4.** $f(x) = -2(x - 3)^2 + 4$

5. $f(x) = x^2 - 2x$ **6.** $f(x) = x^2 + 10x + 7$

Plot and label the vertex and two other points on either side of the vertex, and sketch the graph of the parabola.

7. $f(x) = (x - 1)^2 - 3$ **8.** $f(x) = -x^2 + 4$

x			
$f(x)$			

x			
$f(x)$			

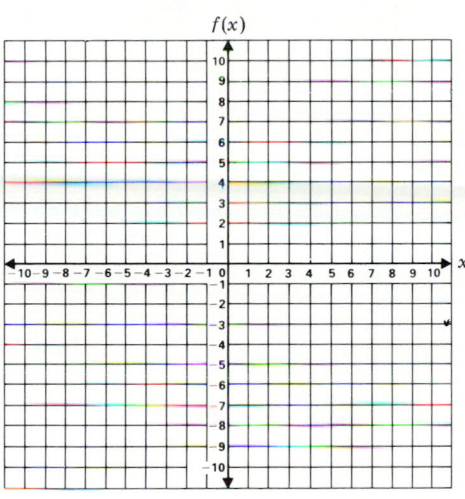

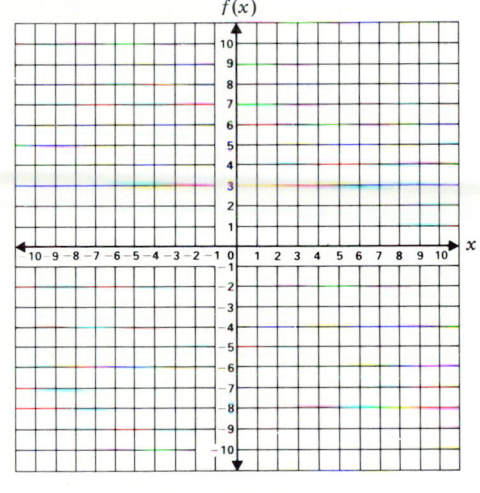

What are the coordinates of the x- and y-intercepts?

9. $f(x) = (x - 6)^2 - 1$ **10.** $f(x) = 5x^2$

Determine whether these numbers are solutions to the equations.

11. $x = -4;$ $x^2 + 3x - 4 = 0$

12. $y = 2;$ $3y^2 - 5y = 26$

Solve these equations by whichever method you prefer. Check all rational solutions.

13. $(y - 3)(y - 7) = 0$ **14.** $x(x + 5)(2x - 1) = 0$

15. $x^2 - 25 = 0$ **16.** $y^2 = 8y$

17. $m^2 + 5m + 4 = 0$ **18.** $r^2 + r - 7 = 0$

19. $w^2 - 11 = 0$ **20.** $r^2 - 3r = 10$

21. $(y - 4)^2 = 81$ **22.** $(x + 1)^2 = 50$

Continued

Answers

1. _____

2. _____

3. _____

4. _____

5. _____

6. _____

9. y-int. _____

    x-int. _____

10. y-int. _____

    x-int. _____

11. _____

12. _____

13. _____

14. _____

15. _____

16. _____

17. _____

18. _____

19. _____

20. _____

21. _____

22. _____

Chapter 15 Test *(Cont.)*

23. $(2w - 3)^2 = 49$

24. $(y^2 + 6y)(y^3 + 8y^2 + 12y) = 0$

25. $\dfrac{x - 1}{x + 3} + \dfrac{2}{x - 1} = \dfrac{5x + 3}{(x + 3)(x - 1)}$

26. $\dfrac{3y - 8}{2y} + \dfrac{y}{y + 1} = \dfrac{7y + 1}{2y^2 + 2y}$

27. $\dfrac{2w}{w + 3} - \dfrac{1}{w} = \dfrac{w + 1}{w^2 + 3w}$

28. What are the dimensions of a rectangle with a length 5 ft greater than its width, if the area is 14 sq ft?

29. Nick can do a job 3 times faster than Jimmy can. When they work together, it takes them 6 min to do the job. How long should it take each to do that job alone?

30. On the first day of our trip we drove 10 mph faster than we did on the second day. We traveled 300 mi each day and drove a total of 11 hr in those 2 days. At what speed did we travel each day?

Answers

23. _____

24. _____

25. _____

26. _____

27. _____

28. _____

29. _____

30. _____

Check your answers on page 762.

Cumulative Review
CHAPTERS 1–15

Combine and simplify.

1. $\dfrac{4}{7} - \dfrac{1}{5} =$

2. $\dfrac{10}{21} \div \dfrac{30}{49} =$

3. $34.073 + 0.056 =$

4. $0.003(1.23) =$

5. 4.3% of $68.95 =$

6. $8 - 3(4 - 2 + 4) + 6 \div 2 =$

7. $-5 + (-3) - (-2) + (9) - (-3) =$

8. $(-6)(2) + (5)(-2)(-1) - 3(-2) =$

9. $2a^3b - 4ab^2 + 6ba^3 + b^2a - 2a^2b^2 + 5a^2b^2 =$

10. $\dfrac{6a^3bc^5}{4abc^3} =$

11. $(-a^3b^{-1})^{-2} =$

12. $(-3ab^2c^5)(8abc^{-3}) =$

13. $(5.2 \times 10^5)(3.1 \times 10^7) =$

14. $(x^2 - 5x + 1) - (3x^2 - 2x - 4) =$

15. $3a^2(a^2 - 4a + 7) =$

16. $\dfrac{10x^2 + 6x - 4}{2x} =$

17. $2(y^2 - 5y + 6) + 5(3y^2 + 4y - 1) =$

18. $(-3a^2b)(5ab) =$

19. $(4w - 3)(2w + 5) =$

20. $(a - 5)(a^2 + 4a - 3) =$

21. $\dfrac{3}{x} + \dfrac{5}{y} =$

22. $\dfrac{(6x^2 - 4x)}{(x + 3)} \cdot \dfrac{(x^2 - 9)}{(8x)} =$

23. $\sqrt{20a^3} =$

24. $\sqrt{3} - \sqrt{12} + 5\sqrt{2} + \sqrt{18} =$

Answers
1.
2.
3.
4.
5.
6.
7.
8.
9.
10.
11.
12.
13.
14.
15.
16.
17.
18.
19.
20.
21.
22.
23.
24.

Continued

Cumulative Review *(Cont.)*

Solve each equation.

25. $3(2x - 5) + 4x - 1 = 5x + 4 + x$ **26.** $\frac{x}{3} + 5 = \frac{x}{6} + 7$

25. _____

Solve each system of equations.

26. _____

27. $y = 3x - 1;$ $2x - 3y = -11$ **28.** $4x + 3y = 5;$ $3x + 2y = 4$

27. _____

Solve each problem.

29. Find the slope and y-intercept of $2y - 12 = 6x$.

28. _____

30. Find the equation of the line that passes through points $(3, -4)$ and $(7, 4)$.

29. Slope _____

31. What must be added to 10 g of a 50% concentration of saltwater to make it into an 80% solution?

    y-int. _____

32. Joy wanted to make a special treat called Truffles with the chocolate that she brought back from California. The recipe called for 9 oz of chocolate and 5 oz of sweet butter. She had only 8 oz of chocolate. How much butter should she use?

30. _____

31. _____

33. How many of each kind of coin do I have if my nickels, dimes, and quarters total $1.45, and I have 3 more nickels than quarters and the dimes are 1 more than 3 times the quarters?

32. _____

33. _____

34. Pat is 4 years older than Mark. Ten years ago Pat's age was 20 less than twice Mark's. How old is each of them now?

34. _____

35. _____

35. Patti borrowed $300 at 11% simple interest per year and plans to pay it back in 2 years. What is the total amount she must pay? (Total means the principal and the interest.)

Check your answers on page 763.

Appendixes

 **Number Bases**

This appendix was developed to help students *develop an understanding about place values*. I believe that it will be easier for them to learn how to compute with decimals and percents in Chapter 4 if they appreciate the significance of place values.

All the numbers used in this book are from the decimal system, which is a number system where all the numbers are represented by combinations of only *ten* different symbols. We call such a system base ten.

Some of the important principles about the decimal system will be easier to understand if we look at some number systems other than base ten.

Our society probably adopted a base ten system because we normally have ten fingers. Early computers (*as in people who compute*) probably counted on their fingers frequently.

When we talk about number base systems, the base tells us how many different digits there are for us to use in representing the various numbers.

Tom Lehrer, a humorous satirist, defines base 8 in his song "New Math" as being just like base 10 if you're missing two fingers. (After studying this lesson, find a copy of his album *That Was the Year That Was* and listen to the song. You'll be one of the few listeners who will understand what he's doing!)

EXAMPLE 1 The number system we usually use is base 10. It's called the decimal system.

Deci is a prefix meaning *ten*.

The ten digits used are 0, 1, 2, 3, 4, 5, 6, 7, 8, and 9. ❏

EXAMPLE 2 In base 8, there are 8 digits. Assume they are 0, 1, 2, 3, 4, 5, 6 and 7. ❏

EXAMPLE 3 In base 2, there are 2 digits. Assume they are 0 and 1. The base 2 system is called the binary number system.

Bi is a prefix meaning *two*.

In a computer, every switch is either on or off. Since there are just these 2 choices, the binary system is used to communicate with computers. ❏

There is a computer language, Hex, that uses a base 16 number system.

EXAMPLE 4 In base 16, there 16 digits. Assume they are 0, 1, 2, 3, 4, 5, 6, 7, 8, 9, a, b, c, d, e, and f. [Since there are only 10 digits in our normal (decimal) number system, we had to use new symbols to represent the 11th–16th digits. The letters a–f work well.) ❏

Below is a listing of how the base-10 numbers one through thirty six would be represented in each of those four number base systems:

Number	Base 10	Base 8	Base 2	Base 16
one	1	1	1	1
two	2	2	10	2
three	3	3	11	3
four	4	4	100	4
five	5	5	101	5
six	6	6	110	6
seven	7	7	111	7
eight	8	10	1000	8
nine	9	11	1001	9
ten	10	12	1010	a
eleven	11	13	1011	b
twelve	12	14	1100	c
thirteen	13	15	1101	d
fourteen	14	16	1110	e
fifteen	15	17	1111	f
sixteen	16	20	10,000	10
seventeen	17	21	10,001	11
eighteen	18	22	10,010	12
nineteen	19	23	10,011	13
twenty	20	24	10,100	14
twenty one	21	25	10,101	15
twenty two	22	26	10,110	16
twenty three	23	27	10,111	17
twenty four	24	30	11,000	18
twenty five	25	31	11,001	19
twenty six	26	32	11,010	1a
twenty seven	27	33	11,011	1b
twenty eight	28	34	11,100	1c
twenty nine	29	35	11,101	1d
thirty	30	36	11,110	1e
thirty one	31	37	11,111	1f
thirty two	32	40	100,000	20
thirty three	33	41	100,001	21
thirty four	34	42	100,010	22
thirty five	35	43	100,011	23
thirty six	36	44	100,100	24

The place value farthest to the right is called the unit place, and it represents the number of ones in all number bases. But the value of all the other places depends on the base of the number system.

In base 10   10 represents one ten.

In base 8   10 represents one eight.

In base 2   10 represents one two.

In base 16   10 represents one sixteen.

There is a notation, called exponents, that will simplify our discussion about place values. Whenever a product of numbers involves the repetition of some number, we can write the amount of its occurrences as an exponent over that number instead of repeating it.

EXAMPLE 5 We could express the product (10)(10)(10) using the exponent 3 (since 10 is repeated 3 times).

$$(10)(10)(10) \text{ is written as } (10)^3 \qquad \square$$

We find the value of each place in a number system by multiplying the value of the place to its right by the number of digits in that system.

EXAMPLE 6 In base 10, we multiply each place value by 10 to find the value of the next place to the left.

The first place has value 1.

The next place (to the left) has value 1(10) or 10.

We call it the ten's place [10].

The next place has value 10(10) or $(10)^2$.

We call it the hundred's place [100].

The next place has value 10(10)(10) or $(10)^3$.

We call it the thousand's place [1,000].

The next place has value 10(10)(10)(10) or $(10)^4$.

We call it the ten thousand's place [10,000].

So, the place values in base 10 are 1, 10, 10^2, 10^3, 10^4, etc. $\square$

EXAMPLE 7 In base 8, we multiply each place value by 8 to find the value of the next place to the left.

The first place has value 1.

The next place (to the left) has value 1(8) or 8.

The next place has value 8(8) or $(8)^2$.

The next place has value 8(8)(8) or $(8)^3$.

The next place has value 8(8)(8)(8) or $(8)^4$.

So, the place values in base 8 are 1, 8, 8^2, 8^3, 8^4, etc. $\square$

EXAMPLE 8 In base 2, we multiply each place value by 2 to find the value of the next place to the left.

The first place has value 1.

The next place (to the left) has value 1(2) or 2.

The next place has value 2(2) or $(2)^2$.

The next place has value 2(2)(2) or $(2)^3$.

The next place has value 2(2)(2)(2) or $(2)^4$.

So, the place values in base 2 are 1, 2, 2^2, 2^3, 2^4, etc. $\square$

EXAMPLE 9 In base 16, we multiply each place value by 16 to find the value of the next place to the left.

The first place has value 1.

The next place (to the left) has value 1(16) or 16.

The next place has value 16(16) or $(16)^2$.

The next place has value 16(16(16) or $(16)^3$.

The next place has value 16(16)(16)(16) or $(16)^4$.

So, the place values in base 16 are 1, 16, 16^2, 16^3, 16^4, etc. $\square$

When we write a number from a base other than 10, we express its base as a subscript.

EXAMPLE 10   A number from base 8 is written with the subscript 8 after it.

234_8 represents a number from the base 8 system. ❑

A number is really the representation of the sum of the products of each digit times the value of the place where it appears.

EXAMPLE 11   573 is in base 10 and represents $3(1) + 7(10) + 5(10)^2$. ❑

EXAMPLE 12 234_8 is in base 8 and represents $4(1) + 3(8) + 2(8)^2$. ❑

To convert a number from some other base to the decimal system, multiply each digit by its place value, and add up those products. *Remember, in all number bases, the place value on the right has value 1.*

EXAMPLE 13   Find the value of 234_8 in the decimal system.

$$234_8 = 4(1) + 3(8) + 2(8)^2$$
$$= 4(1) + 3(8) + 2(64)$$
$$= 4 + 24 + 128$$
$$= 156$$

So in the decimal system, 234_8 represents 156. ❑

EXAMPLE 14   Find the value of 157_{16} in the decimal system.

$$157_{16} = 7(1) + 5(16) + 1(16)^2$$
$$= 7(1) + 5(16) + 1(256)$$
$$= 7 + 80 + 256$$
$$= 343$$

So in the decimal system, 157_{16} represents 343. ❑

Let's look at some computations in base 8.

EXAMPLE 15   Find the sum of $25_8 + 41_8$. Express the answer in base 8.

$$25_8$$
$$+ 41_8$$
$$\overline{66_8}$$

To check: Translate the problem into base 10 and find the sum.
Translate your answer in base 8 into base 10.

If it was correct, they will be the same.

$$25_8 = 5(1) + 2(8) \qquad 41_8 = 1(1) + 4(8)$$
$$= 5 + 16 \qquad\qquad = 1 + 32$$
$$= 21_{10} \qquad\qquad = 33_{10}$$

So, $$25_8 + 41_8 = 21_{10} + 33_{10}$$
$$= 54_{10} \checkmark$$

Our answer was 66_8, which is equal to

$$66_8 =$$
$$6(1) + 6(8) = 6_{10} + 48_{10}$$
$$= 54_{10} \checkmark$$ ❑

Remember we can only use the 8 digits 0, 1, 2, 3, 4, 5, 6, and 7 in base 8, so if a sum is larger than 7 we must carry an 8 to the next place value.

EXAMPLE 16 Find the sum of $364_8 + 135_8$.

$$
\begin{array}{r}
^1 3^1 6\ 4_8 \\
+\ 1\ 3\ 5_8 \\
\hline
5\ 2\ 1_8 \quad \text{Answer is } 521_8.
\end{array}
$$

⟹ ***Note:*** In the units place, $5 + 4 = 9$ but we don't have a 9 in base 8, (7 is the largest digit in base 8). So, I expressed 9 as $8 + 1$, and carried the 8 into the 8's place as a *one* (*one eight*).

In the 8's place, $1 + 6 + 3 = 10$, which is also more than 7. I expressed 10 as $8 + 2$, and carried the 8 into the sixty-four place as a *one* (8^2 is sixty four).

Check:

$$
\begin{aligned}
364_8 &= 4(1) + 6(8) + 3(64) \\
&= 4 + 48 + 192 \\
&= 244_{10} \\[4pt]
135_8 &= 5(1) + 3(8) + 1(64) \\
&= 5 + 24 + 64 \\
&= 93_{10}
\end{aligned}
$$

So,

$$
\begin{aligned}
364_8 + 135_8 &= 244_{10} + 93_{10} \\
&= 337_{10} \quad ✓ \\[6pt]
521_8 &= 1(1) + 2(8) + 5(64) \\
&= 1 + 16 + 320 \\
&= 337_{10} \quad ✓
\end{aligned}
$$

In a number base larger than 10, like base 16, you must remember what number value each of the new symbols (we used letter a–f) represent. If any sum is larger than 15, carry a 16 to the next place value.

EXAMPLE 17 Find the sum in base 16 of $2a7_{16} + c8e_{16}$.

Remember: a is ten, b is eleven, c is twelve, d is thirteen, e is fourteen, f is fifteen.

$$
\begin{array}{r}
2a7_{16} \\
+\ c8e_{16}
\end{array}
$$

In the *units* place, $[7 + e]$ is $7 + fourteen = 21_{10}$. Write 21_{10} as $16 + 5$, and carry the 16 as a 1, to the next place.

In the 16 place, $[a + 8 + 1 \text{ (carried from units place)}]$ is $ten + 8 + 1 = 19_{10}$. Write 19_{10} as $16 + 3$, and carry the 16 to the next place.

In the 256 place, $[2 + c + 1 \text{ (carried from 16 place)}]$ is $2 + twelve + 1 = 15_{10}$. Write 15_{10} as f. So,

$$
\begin{array}{r}
2a7_{16} \\
+\ c8e_{16} \\
\hline
f35_{16}
\end{array}
$$

Another way to get familar with the concept of place value is to add numbers in base 10 using an abacus. You can physically experience how we represent and compute with numbers larger than 10 using only the 10 digits 0, 1, 2, 3, 4, 5, 6, 7, 8, and 9 on an abacus.

EXERCISES

1. Represent the numbers thirty seven through seventy
   a) in base 8; b) in base 2; c) in base 16.

2. Represent the numbers one through forty
   a) in base 6; b) in base 11.

3. Find the value of the following numbers in the decimal system.
   a) 157_8; b) 301_8; c) 436_8; d) 274_8; e) 407_8; f) 1101_2; g) $101,110_2$;
   h) $11,000,111_2$; i) $11,011_2$; j) 157_{16}; k) 301_{16}; l) 436_{16}; m) $2b8_{16}$; n) $13f_{16}$.

4. Find each sum and check your answers as was done in Examples 15 and 16.
   a) $157_8 + 301_8$; b) $436_8 + 274_8$; c) $157_8 + 407_8$; d) $1101_2 + 11,000,111_2$;
   e) $101,110_2 + 11,011_2$; f) $157_{16} + 301_{16}$; g) $436_{16} + 2a8_{16} + 13c_{16} + b01_{16}$.

5. Make up some problems of your own in base 2, 6, 8, 11, or 16, compute them, and check.

Check your answers on page 763.

 Social Science Graphs

"A good picture is worth 1000 words." In math, too, information can be represented more concisely and dramatically by an illustrative graph than it can be through paragraphs of words and numbers. From a glance at a graph we can often recognize significant relationships. That is why newspaper, magazine, and tv reports are frequently accompanied by a graph.

In Chapters 6, 8, 10, and 15, you'll learn to graph algebraic equations and inequalities. But whether or not you've mastered those skills yet, you can learn how to understand pictorial, line, bar, and circle graphs.

Pictorial Graphs

In a pictorial graph, pictures are used to represent data. For example, a graph about the number of women in the work force might represent every 25,000 workers by a picture of one woman in work clothes.

The picture must be labeled so we know exactly what it represents. Each row of the graph must also be labeled so we know what information it represents. For example, if the graph represents the number of women in the work force every 5 years from 1955 to 1980, we would label each row: 1955, 1960, 1965, etc.

Pictorial graphs give us a good, general idea about the data but *do not represent information precisely*. For example, in such a graph, 12,000–15,000 workers would be represented by a picture of approximately one half of a person; 6000–10,000 workers would be represented by a picture of approximately one third of a person, etc.

EXAMPLE 1 I'll represent the numbers of women aged 20–24 in the workforce [with the data below from the Employment Training Report of the President (1973–1981)] with a picture graph. ❏

Year	Number of workers
1955	92,500
1960	94,100
1965	102,200
1970	116,100
1975	121,600
1980	130,600

Represents 25,000 women in workforce

Line Graph

Line graphs are useful in observing how data changes over some period of time. A horizontal line is drawn along the bottom and a vertical line is drawn along the left side of the page. (We call these lines the graph's axes.) Each axis is divided into segments of *equal* size.

For example, the segments of the horizontal axis might represent months, years, or decades; and the vertical axis might represent dollars earned, in segments of a thousand. We must carefully label each axis.

We could graph the profits a company makes each month of the year on such a graph and easily notice the changes and trends. We would do this by placing dots (representing the data) on the graph and connecting them to make a line—the line might be straight or it might zigzag, depending on where the points are. The fluctuations in the line help us notice the changes that have occurred over that time span.

A dot would be placed above each month's segment of the horizontal axis at the level corresponding to the company's profits for that month.

A company with several plants might graph the profits for all the plants on the same graph, each with a different color, to easily compare the profitability of each.

EXAMPLE 2 I'll represent the monthly profits for the Ugotadeel Company from January to June (from the table below) as a line graph. ❏

Month	Profit (thousands of dollars)
January	23
February	34
March	18
April	40
May	33
June	25

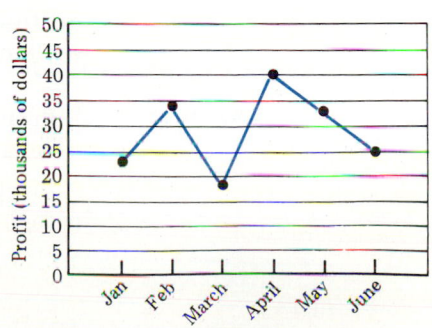

Bar Graph

Bar graphs are often used to demonstrate comparisons between data. They use two axes like we discussed for line graphs, but one axis might represent the data for different individuals instead of for different periods of time.

For example, each month an employer might use a bar graph to demonstrate the sales from each department or salesperson. Each segment on the horizontal axis would be labeled with a different salesperson's name. The vertical axis would represent the amount of sales, in increments of $100. A bar would be drawn up in each person's column to the level representing the amount of sales they made.

The bars can be horizontal instead of vertical, if that would look better. The vertical axis could be labeled with the names of different countries, and the horizontal axis could represent the population in increments of 100,000 people. A bar would be drawn in each row, out to the number representing the population of that country.

EXAMPLE 3   I'll represent the comparison between the sales made by each of the salespeople (from the table below) as a bar graph.   ❏

Salesperson	Total Sales
Bill	360
Millie	450
Judy	780
Jerry	680
Bob	300

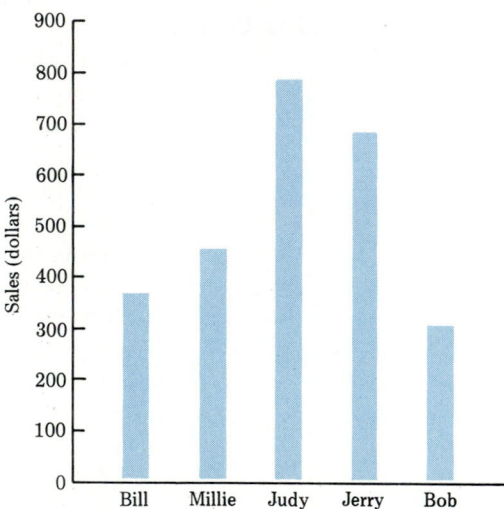

Circle Graphs

(This section should not be studied until the material in Chapter 4 on decimals and percents has been mastered.)

A circle graph is often used to represent how a budget is split up. A circle represents the whole amount, and it is divided into segments (like pieces of a pie) to represent the individual shares.

Begin with the information about what percent of the whole is in each category. For example, a politician may recommend that 30% of tax revenue go toward education, 10% for defense, 40% for health care, and 20% for housing. The percents must add up exactly to 100%!

EXAMPLE 4   I'll represent the portion of the budget proposed for education, defense, health care, and housing with a circle graph.   ❏

A circle contains 360°. To calculate how many degrees would represent each category, we multiply 360 by the percent of taxes going to that category:

$$30\% \text{ of } 360° = 108°, \qquad 10\% \text{ of } 360° = 36°,$$
$$40\% \text{ of } 360° = 144°, \qquad 20\% \text{ of } 360° = 72°.$$

Draw a circle, and mark off an arc (using a protractor) with the correct number of degrees for each category. Label each section with the name and percent of its category:

education by an arc of 108°,
defense by an arc of 36°,
health care by an arc of 144°,
housing by an arc of 72°.

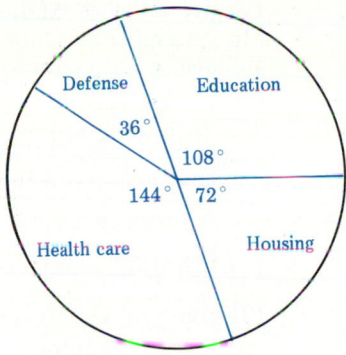

EXERCISE

Make up a graphing project in which you keep track of some data that you are interested in. Record the data, either weekly or daily, over a two-month period of time. List the data neatly in a table, and draw the kind of graph that you think best represents the information.

Metric Measurement System

(Can be done after Chapters 3 and 4 and Section 5.1 are completed.)

Metric Prefixes and Fundamental Units

The **metric system** is the system that most countries use to express units of measurement for length, mass, and volume. It is much easier to compute in the metric system than in the English system; we import so many items from countries using it. In the United States we are seeing more and more measurements expressed in terms of the metric system. Some distance and speed-limit signs on the highway are now expressed in kilometers, the balancing weight on the arm of my record turntable is expressed in grams, and soft drinks now come in liter size bottles. Those are all metric units.

In the English system, conversion from one unit to another may mean multiplying or dividing by 3 (between yards and feet), by 16 (between pounds and ounces), by 64 (between ounces and gallons), and even by 1760 (between yards and miles).

Since we can multiply and divide by 10 simply by moving the decimal point in a number, we can convert from one unit to another in the metric system by merely moving the decimal point. Another advantage is that we don't have to memorize all the different conversion relationships in the metric system as we must in the English system.

Fundamental unit

There is a *fundamental unit* for each of the categories of measurement (length, volume and mass). See Table A.1. All the other units are set up in relation to the fundamental ones.

TABLE A.1 Fundamental units

Category	Unit name	Symbol
Length	Meter	m
Mass (weight)	Gram	g
Volume	Liter	L

The following statements give an idea of what each of those units is.

> One *meter* is a length a few inches longer than one yard.
>
> A *gram* is a very small mass; 450 grams weigh almost 1 pound.
>
> One *liter* is a little larger than one quart.

I'll give you the exact relationships between the metric and English measurement systems later.

Prefixes

In the metric system, prefixes are used to indicate how much larger or smaller a unit is than the fundamental unit. Table A.2 presents the most commonly used metric prefixes, with the symbol and value of each. To read a metric unit, multiply the value of the prefix by the fundamental unit that follows it.

EXAMPLE 1   *Kilo* is the prefix representing 1000, so a *kilometer* is the term used to represent a unit that is 1000 times bigger than one meter.

$$1 \text{ kilometer (km)} = (1000)(\text{one meter}) \qquad \square$$

EXAMPLE 2   *Centi* is the prefix representing $\dfrac{1}{100}$, so a *centimeter* is the term used to represent a unit that is one hundredth of a meter (100 times smaller than 1 meter).

$$1 \text{ centimeter (cm)} = \left(\frac{1}{100}\right)(\text{one meter}) \text{ or}$$

$$100 \text{ centimeters} = 1 \text{ meter} \qquad \square$$

Conversion from one metric unit to another consists of simply multiplying or dividing by ten; the issue becomes *how many times* is 10 multiplied (or divided) in a particular conversion. Remember, every time a number is *multiplied* by 10, the decimal point moves one place to the *right*; and every time a number is *divided* by 10, the decimal point moves one place to the *left*.

Recall (from Sections 5.1 and 11.1) that the notation called **exponent** readily represents the number of times multiplication or division is to be performed.

We represent the number of times we want to multiply by ten by writing that number as a *positive* exponent above ten.

EXAMPLE 3 10^3 represents $(10)(10)(10)$. We say 10^3 is

> "10 raised to the exponent 3."

10^3 instructs us to multiply by 10, 3 *times*. $\qquad \square$

If a number is multiplied by a prefix with value 10^3, its decimal point is moved *three* places to the *right*.

EXAMPLE 4
$$5.0678 \times 10^3 = 5,067.8$$ ❏

We represent that we want to *divide* by ten if we use a *negtive* exponent. Again, the magnitude of the exponent indicates the number of times.

EXAMPLE 5 10^{-2} represents that we divide by ten twice. We say 10^{-2} is

"10 raised to the exponent negative 2." ❏

If a number is multiplied by a prefix with value 10^{-2}, its decimal point is moved *two* places to the *left*.

EXAMPLE 6
$$248.3 \times 10^{-2} = 2.483$$ ❏

Table A.2 presents the most commonly used metric prefixes, with the symbol and value of each.

TABLE A.2 Metric prefixes

Name	Symbol	Value		expressed w/exponents:
mega	M	1,000,000	or	10^6
kilo	k	1,000	or	10^3
deci	d	$\dfrac{1}{10}$	or	10^{-1}
centi	c	$\dfrac{1}{100}$	or	10^{-2}
milli	m	$\dfrac{1}{1000}$	or	10^{-3}
micro	μ	$\dfrac{1}{1,000,000}$	or	10^{-6}
nano	n	$\dfrac{1}{1,000,000,000}$	or	10^{-9}

If one letter appears after a number, that letter always represents the fundamental unit (liter, gram, or meter). If two letters appear after a number, the first represents the prefix, and the second represents the fundamental unit.

EXAMPLE 7
7 m represents 7 meters,

2 mm represents 2 millimeters, and

3 Mg represents 3 megagrams. (Note: But 3 mg represents 3 milligrams.) ❏

Converting To and From Fundamental Units

To convert a number expressed with a metric prefix to the fundamental unit, just move the decimal point according to the exponent in the value of the prefix.

EXAMPLE 8 Convert 73 kg to grams.

$$73\,\text{kg} = (73)(10^3)\ \text{grams}$$
$$= (73.000.)(10^3)\ \text{grams}$$
$$= 73,000.\ \text{grams}$$ ❏

EXAMPLE 9 Convert 234 cm to meters.

$$234\,\text{cm} = (234)(10^{-2})\ \text{meters}$$
$$= (234.)(10^{-2})\ \text{meters}$$
$$= 2.34\,\text{m}$$ ❏

To convert a number expressed in a fundamental unit into one with a metric prefix, just move the decimal point as indicated by the exponent in the value of the prefix. The easiest way to avoid confusion is to remember the following

> If the new unit you convert to is larger than the original unit, move the decimal point to the left. (A distance expressed in *kilometers* is a smaller number than if expressed in *meters*. So to convert from meters to kilometers, the amount becomes a smaller number; thus the decimal point moves to the left.)

> If the new unit you convert to is smaller than the original unit, move the decimal point to the right (A length expressed in *centimeters* is a *larger* number than if expressed in terms of meters. So to convert from meters to centimeters, the amount becomes a larger number, and the decimal point moves to the right.)

EXAMPLE 10 Convert 486 meters to kilometers.

The new unit, kilometers, is larger than meters, so the *number of* kilometers will be less than 486, and the decimal point will move 3 places to the left

$$486. \text{ m} = 486 \text{ km}$$

So, 486 m = 0.486 km.

EXAMPLE 11 Convert 173 grams to centigrams.

The new unit, centigrams, is smaller than grams, so the *number of* centigrams will be more than 173, and the decimal point will move 2 places to the right

$$173. \text{ g} = 173.00 \text{ cg}$$

So, 173 g = 17,300 cg.

EXERCISES

Convert each measure to the fundamental unit.

1. 16 cm to meters

2. 78 kg to grams

3. 280 ML to liters

4. 609 mg to grams

5. 92 mm to meters

6. 280,186,469 ng to grams

7. 0.58 km to meters

8. 928 mL to liters

9. 6.905 kg to grams

10. 3,764 cm to meters.

Convert each of these and be careful about the direction in which you move the decimal point.

11. 800 m to km

12. 760 g to cg

13. 47 m to mm

14. 5,386 g to kg

15. 4,870 nL to L

16. 345 m to cm

17. 6.019 km to meters

18. 26 kg to g

19. 5 mm to m

20. 6.08 L to mL

Check your answers on page 764.

Conversion Between Metric and English Measurement Systems

Table A.3 defines each fundamental metric unit in terms of a unit of measurement from the English system.

TABLE A.3 Metric/English conversion table

Metric	English
1 liter	= 1.06 quarts
454 grams	= 1 pound
1 meter	= 39.37 inches
2.54 cm	= 1 inch

We use a conversion fraction to convert *between* the Metric and English measurement systems.

EXAMPLE 12 Convert 7.42 quarts to liters.

We use the conversion fraction $\left(\dfrac{1 \text{ liter}}{1.06 \text{ quarts}}\right)$ to convert from quarts to liters.

$$7.42 \text{ quarts} =$$

$$(7.42 \text{ quarts})\left(\frac{1 \text{ liter}}{1.06 \text{ quarts}}\right) =$$

$$(7.42 \text{ quarts})\left(\frac{1 \text{ liter}}{1.06 \text{ quarts}}\right) =$$

$$\frac{7.42}{1.06} \text{ liters} = 7 \text{ liters} \qquad \square$$

EXAMPLE 13 Convert 21.59 cm to inches.

$$21.59 \text{ cm} =$$

$$(21.59 \text{ cm})\left(\frac{1 \text{ inch}}{2.54 \text{ cm}}\right) =$$

$$(21.59 \text{ cm})\left(\frac{1 \text{ inch}}{2.54 \text{ cm}}\right) =$$

$$\frac{21.59}{2.54} \text{ in.} = 8.5 \text{ in.} \qquad \square$$

There are more examples of conversions in Section 7.4.

Answers

CHAPTER 1

Section 1.1   **1.** Five thousand, three hundred eighteen   **2.** Twenty-seven thousand, two hundred ninety-four   **3.** One million, three hundred eighty-one thousand, two hundred sixty-seven   **4.** Fifty-seven million, three hundred nine thousand, one hundred seventy-three   **5.** Eight hundred twelve million, ninety-one thousand, four hundred

Section 1.2A   **1.** 1,532   **2.** 2,148   **3.** 26,559   **4.** 722,936   **5.** 44,921   **6.** 16,958

1.2B   **7.** 234   **8.** 151   **9.** 643   **10.** 532   **11.** 117   **12.** 21,646   **13.** 178   **14.** 12,794   **15.** 128,222   **16.** 112,679

Section 1.3A   **1.** 51   **2.** 112   **3.** 159   **4.** 70   **5.** 115   **6.** 164

1.3B   **7.** 65   **8.** 105   **9.** 120   **10.** 92   **11.** 207   **12.** 104

1.3C   **13.** 136   **14.** 160   **15.** 124   **16.** 204   **17.** 266   **18.** 231   **19.** 1,505   **20.** 1,196   **21.** 2,144   **22.** 2,241   **23.** 2,516   **24.** 1,525

Section 1.4A   **1.** 5   **2.** 3   **3.** 4   **4.** 4   **5.** 4   **6.** 3   **7.** 9   **8.** 6

1.4B   **9.** 81   **10.** 321   **11.** 185   **12.** 302   **13.** 108   **14.** 134   **15.** 203   **16.** 124   **17.** 1,023   **18.** 2,101

Section 1.5A   **1.** (1) Multiply 7 by 2;   (2) add that product to 5   **2.** (1) Multiply 3 by 4; (2) add that product to 8   **3.** (1) Multiply 3 by 7; (2) add that product to 4   **4.** (1) Multiply 3 by 4; (2) add that product to 1 and to 2   **5.** (1) Multiply 3 by 4; (2) subtract that product from 16   **6.** (1) Multiply 7 by 3, and multiply 2 by 3; (2) subtract the product of 2 by 3 from the other product

1.5B   **7.** 19   **8.** 14   **9.** 4   **10.** 13   **11.** 39   **12.** 18   **13.** 3   **14.** 5

1.5C   **15.** 33   **16.** 1   **17.** 29   **18.** 90   **19.** 11   **20.** 27   **21.** 2   **22.** 11   **23.** 5   **24.** 1

1.5D   **25.** 22   **26.** 17   **27.** 66   **28.** 30   **29.** 58   **30.** 47   **31.** 27   **32.** 45   **33.** 50   **34.** 42

1.5E   **35.** $P = 16$ **36.** $D = 80$ mi   **37.** $A = 80$ **38.** $A = 78$ **39.** $K = 24$ **40.** $M = 33$ **41.** $R = 33$ **42.** $W = 13$ **43.** $H = 11$ **44.** $J = 14$

Section 1.6   **1.** 170   **2.** 170   **3.** 830   **4.** 1,700   **5.** 4,700   **6.** 47,000   **7.** 370   **8.** 700   **9.** 3,520   **10.** 3,700

Section 1.7A   **1.** 34 ft   **2.** 21 cm   **3.** 20 in.   **4.** 32 ft   **5.** 40 m   **6.** 18 ft   **7.** 33 cm   **8.** 34 in.

1.7B   **9.** 36 sq in.   **10.** 40 sq cm   **11.** 14 sq in.   **12.** 9 sq m   **13.** 36 sq ft   **14.** 18 sq ft   **15.** 81 sq yd   **16.** 18 sq mi   **17.** 35 sq cm

1.7C   **18.** 70 cu ft   **19.** 360 cu cm   **20.** 42 cu in.   **21.** 120 cu in.   **22.** 720 cu cm   **23.** 54 cu yd   **24.** 140 cu ft

Chapter I Review **1.** Two hundred three thousand, one hundred ninety-two **2.** 821 **3.** 1,954 **4.** 16.519 **5.** 1,961 **6.** 307 **7.** 2,032 **8.** 21 **9.** 53 **10.** 29 **11.** 300 **12.** 22 in. **13.** 25 sq ft **14.** 36 cu in.

Supplementary Exercises **1.** Four hundred twenty-one thousand, eighty-four **2.** Three million, nineteen thousand, four hundred thirty-seven **3.** 389 **4.** 876 **5.** 425 **6.** 188 **7.** 555 **8.** 46,376 **9.** 2,812 **10.** 2,204 **11.** 3,486 **12.** 4,550 **13.** 2,034 **14.** 2,105 **15.** 4,201 **16.** 28 **17.** 29 **18.** 36 **19.** 28 **20.** 27 **21.** 24 **22.** 340 **23.** 1,840 **24.** 21 in. **25.** 36 sq m **26.** 26 cu ft

Chapter I Test **1.** Seven hundred thirty-nine thousand, six hundred three **2.** 1,010 **3.** 2,175 **4.** 4,959 **5.** 33,012 **6.** 47 **7.** 930 **8.** 36 in. **9.** 27 sq in. **10.** 48 cu in.

CHAPTER 2

Section 2.1 **1.** Positive 14 **2.** Negative 8 **3.** Negative 17 **4.** Positive 23 **5.** Positive 67 **6.** Negative 84 **7.** 28 **8.** 74 **9.** 308 **10.** 531

2.2A **1.** Positive; +12 **2.** Negative; −7 **3.** Negative; −4 **4.** Negative; −6 **5.** Positive; +10 **6.** Negative −10 **7.** Negative −66 **8.** Negative; −90 **9.** Negative; −435 **10.** Negative; −599

2.2B **11.** −16 **12.** +15 **13.** −16 **14.** −11 **15.** −71 **16.** −130 **17.** −165 **18.** −18 **19.** −12 **20.** −20 **21.** −19 **22.** −136 **23.** −168

2.2C **24.** −6 **25.** +2 **26.** +3 **27.** −2 **28.** +3 **29.** −4 **30.** +12 **31.** −21 **32.** −22 **33.** +13 **34.** +23 **35.** −21

2.2D **36.** +2 **37.** −7 **38.** −1 **39.** +2 **40.** −10 **41.** +6 **42.** −16 **43.** −18 **44.** +2 **45.** +8 **46.** 0 **47.** +26 **48.** +33 **49.** −17 **50.** −28 **51.** −118

2.2E **52.** +20 **53.** −2. **54.** +35 **55.** +52 **56.** −5 **57.** −72 **58.** +89 **59.** −10 **60.** −27

Section 2.3A **1.** −2 **2.** −7 **3.** +14 **4.** −15 **5.** +22 **6.** −25 **7.** −1 **8.** +1 **9.** +25 **10.** −5 **11.** +17 **12.** −46

2.3B **13.** −4 **14.** −9 **15.** +10 **16.** −13 **17.** −13 **18.** +4 **19.** +1 **20.** +46 **21.** −10 **22.** +37

Section 2.4A **1.** Negative; −18 **2.** Positive; +28 **3.** Negative; −5 **4.** Negative; −7 **5.** Positive; +4 **6.** Positive; +3 **7.** Negative; −12 **8.** Negative; −5 **9.** Negative; −12 **10.** Negative; −36 **11.** Positive; +15 **12.** Negative; −16

2.4B **13.** +6 **14.** +24 **15.** −20 **16.** −18 **17.** −12 **18.** +30 **19.** +24 **20.** −28 **21.** +20 **22.** −24

Section 2.5A **1.** +36 **2.** −48 **3.** −80 **4.** −36 **5.** −3 **6.** −21 **7.** +10 **8.** −21 **9.** −8 **10.** +11 **11.** −25 **12.** −15

2.5B **13.** −11 **14.** −11 **15.** −17 **16.** +7 **17.** +26 **18.** −19 **19.** −2 **20.** −8 **21.** +10 **22.** +2 **23.** −14 **24.** +1

Section 2.6 **1.** +$62 **2.** +$185 **3.** +$107 **4.** +18 yd **5.** +29 yd **6.** −2 (down 2) **7.** +12 in. **8.** +62° **9.** +$32 **10.** −14 lb (lost 14 lb) **11.** +$4 **12.** +6 **13.** +44 **14.** +17 **15.** +6

Chapter 2 Review **1.** Negative **2.** 18 **3.** 32 **4.** −21 **5.** −21 **6.** −19 **7.** −13 **8.** −2 **9.** −6 **10.** +5 **11.** −23 **12.** 5 **13.** +27 **14.** −5 **15.** −12 **16.** +1 **17.** +1 **18.** +$140

Supplementary Exercises   1. Negative   2. Magnitude is 137   3. 61   4. 78   5. −21   6. −80
7. −17   8. −18   9. −21   10. −3   11. −1   12. +3   13. +7   14. −29   15. +48   16. +11
17. −1   18. +12   19. −28   20. −3   21. +5   22. +40   23. −21   24. −12   25. +17   26. +1
27. +16   28. $225

Chapter 2 Test   1. Negative   2. 813   3. −11   4. −18   5. −16   6. −8   7. +3   8. −13   9. −17
10. +12   11. +6   12. −28   13. −10   14. +30   15. −47   16. −22   17. −11   18. +$49

Cumulative Review   1. 58,699   2. 46,613   3. −7   4. +68   5. +6   6. +54   7. −4   8. +30
9. 228   10. 817   11. 0   12. +53   13. 1,043   14. 1,591   15. 24 in.   16. 16 sq ft   17. 45 cu in.
18. 12 in.   19. +82 yd   20. +$98

CHAPTER 3

Section 3.1   1. 1/3   2. 2/5   3. 2/6   4. 1/2   5. 3/4   6. 4/8   7. 3/3   8. 2/4   9. 5/5   10. 3/6   11. 7
12. 13   13. Negative   14. Negative   15. Positive   16. Negative   17. Negative   18. Positive
19. Proper   20. Improper   21. Improper   22. Proper   23. 7   24. 1   25. 1   26. 8

Section 3.2A   1. 4/5   2. 9/10   3. 10/11   4. 12/17   5. 4/13   6. 9/14   7. 11/19   8. 7/10   9. −2/17
10. −15/19   11. +1/23   12. −10/31   13. +9/11   14. −6/7   15. −8/13   16. −5/29

3.2B   17. 1/20   18. 4/15   19. 3/10   20. 3/16   21. 5/33   22. 4/21   23. 9/28   24. 6/35   25. 15/4   26. 7/3

Section 3.3   1. 5·7   2. 2·7   3. 2·5·7   4. 2·3·11   5. 2·2·7   6. 2·2·2·5   7. 2·2·5·5   8. 2·2·3·5·5
9. 3·3·3·3   10. 2·2·2·2·3   11. 2·3·7   12. 2·2·3·3·5   13. 2·2·3·3·3·5   14. 2·2·2·5·5·5
15. 2·3·5·5·5   16. 2·2·2·2·2·3·5   17. 3·3·5·7   18. 3·5·7   19. 3·3·3·3·3   20. 2·2·2·5·17

Section 3.4   1. 2/5   2. 5/6   3. 7/10   4. 2/5   5. 5/9   6. 5/7   7. 1/3   8. 2   9. 5/3   10. 2/5
11. 1/3   12. 3/5   13. 1/5   14. 18/25   15. 14/15   16. 1/3   17. 15/49   18. 25/36

Section 3.5A   1. 20/27   2. 3/7   3. −3/5   4. −3/7   5. 14/15   6. 9/16   7. −1/2   8. −21/125
9. −1/9   10. 3/10   11. −16/27   12. 5/3   13. −9/40   14. −2/35

3.5B   15. 8/5   16. 4/21   17. −10/3   18. −21/10   19. 4/3   20. 6/5   21. −45/32   22. 7/5   23. 5/18
24. 42/25   25. 5/12   26. −6/7   27. 125/84   28. 160/189

3.5C   29. 5/6   30. 2/3   31. −16/3   32. −4/5   33. −35/18   34. 30/49

Section 3.6   1. 6/12   2. 5/15   3. 18/24   4. 12/30   5. 15/35   6. 30/48   7. 16/72   8. 35/42   9. 27/45
10. 36/63

Section 3.7A   1. 19/20   2. 13/21   3. 7/15   4. 1/15   5. −6/5   6. 3/5   7. 7/9   8. −2/3   9. −7/6
10. −11/14

3.7B   11. 5/6   12. 9/10   13. −3/14   14. 8/33   15. −22/21   16. −44/65   17. −5/21   18. 12/35
19. 3/22   20. 23/35

3.7C   21. 42   22. 30   23. 42   24. 70   25. 105   26. 110   27. 210   28. 210

3.7D   29. 40   30. 18   31. 36   32. 24   33. 36   34. 120   35. 216   36. 54   37. 80   38. 180   39. 120
40. 120

3.7E **41.** 17/45 **42.** 43/60 **43.** −1/24 **44.** 25/36 **45.** −17/30 **46.** 27/70 **47.** 13/45 **48.** −23/42 **49.** 35/54 **50.** 67/120

3.7F **51.** 46/105 **52.** −1/30 **53.** −11/6 **54.** 7/20 **55.** −1/8 **56.** 3/4 **57.** −4/7 **58.** −11/20 **59.** 11/15 **60.** −3/4 **61.** 43/60 **62.** 7/18 **63.** 21/40 **64.** 83/120

Section 3.8A **1.** 6/5 **2.** 17/5 **3.** 11/2 **4.** 31/7 **5.** 11/4 **6.** 65/9 **7.** 51/8 **8.** 41/5

3.8B **9.** $2\frac{1}{3}$ **10.** $2\frac{1}{2}$ **11.** $3\frac{1}{4}$ **12.** $5\frac{1}{3}$ **13.** $7\frac{1}{2}$ **14.** $4\frac{3}{5}$ **15.** $5\frac{5}{6}$ **16.** $6\frac{1}{7}$

3.8C **17.** $6\frac{3}{5}$ **18.** 7 **19.** $8\frac{2}{3}$ **20.** $-7\frac{14}{15}$ **21.** −32/77 **22.** $2\frac{3}{10}$ **23.** 2/3 **24.** $1\frac{1}{2}$ **25.** $9\frac{9}{10}$ **26.** $1\frac{1}{10}$ **27.** $2\frac{5}{14}$ **28.** $-10\frac{5}{6}$

3.8D **29.** $15\frac{1}{23}$ **30.** $215\frac{2}{31}$ **31.** $203\frac{3}{25}$ **32.** $201\frac{7}{15}$ **33.** $10{,}322\frac{5}{16}$ **34.** $21{,}305\frac{3}{14}$ **35.** $30{,}112\frac{7}{15}$ **36.** $20{,}104\frac{3}{13}$

3.8E **37.** $5\frac{5}{6}$ **38.** $3\frac{4}{5}$ **39.** $7\frac{9}{20}$ **40.** $3\frac{17}{24}$ **41.** $-7\frac{19}{21}$ **42.** $2\frac{7}{10}$ **43.** $2\frac{7}{12}$ **44.** 2/7 **45.** $-5\frac{5}{8}$ **46.** −7 **47.** $6\frac{3}{7}$ **48.** $-9\frac{2}{3}$

Section 3.9A **1.** 0 **2.** 1 **3.** 1 **4.** 0 **5.** 1 **6.** 0 **7.** 0 **8.** 0 **9.** 2 **10.** 3 **11.** 7 **12.** 9 **13.** 0 **14.** 0

3.9B **15.** 3/5 **16.** 3/7 **17.** 1/3 **18.** 4/9 **19.** 4/5 **20.** 3/7 **21.** 6/7 **22.** 7/9 **23.** 11/13 **24.** 8/21 **25.** 2/3 **26.** 1/2 **27.** 3/4 **28.** 3/7 **29.** 4/9 **30.** 4/5 **31.** 2/7 **32.** 5/12 **33.** 5/6 **34.** 1/3 **35.** 11/12 **36.** 3/4

Section 3.10 **1.** 18 lb **2.** $21\frac{2}{5}$ ft **3.** $2\frac{1}{4}$ cups **4.** $1\frac{3}{4}$ cups **5.** $2\frac{7}{12}$ ft, ∼3 ft **6.** $68 **7.** (a) 3/4 (b) 72 hr (c) 96 hr **8.** 8 min **9.** (a) $25\frac{1}{4}$ cm (b) $4\frac{3}{4}$ cm **10.** (a) $51 (b) $17 **11.** $3\frac{1}{4}$ ft **12.** 11 full pairs **13.** 36 full bags **14.** 25° C **15.** 1/3 **16.** (a) $3,600 (b) $14,400 **17.** 165 mi **18.** $7 **19.** Mike, since 7/12 is larger than 5/9 **20.** 53 **21.** $4\frac{1}{2}$ hr

Section 3.11A **1.** 18 tsp **2.** 10 qt **3.** 120 hr **4.** 15,840 ft **5.** 20 tons **6.** 108 in. **7.** 132 months **8.** $7\frac{1}{2}$ pt **9.** $1\frac{2}{3}$ hr **10.** 30 pt **11.** $2\frac{1}{2}$ lb **12.** 14 cups **13.** 8 in. **14.** $13\frac{1}{2}$ tsp **15.** 17,160 ft **16.** 92 oz **17.** 11 pt **18.** $13\frac{1}{3}$ qt **19.** 195 min **20.** 7/18 ft

3.11B **21.** 72 in. **22.** 18,000 sec **23.** 2,880 min **24.** 96,000 oz **25.** 20 cups **26.** 2 gal **27.** 7,040 yd **28.** 432,000 sec **29.** 56 oz **30.** 4,680 min

Section 3.12A **1.** 28 in. **2.** $12\frac{3}{5}$ ft **3.** 18 m **4.** $18\frac{1}{3}$ ft **5.** $14\frac{1}{3}$ in. **6.** $30\frac{2}{3}$ ft **7.** $19\frac{19}{60}$ cm **8.** $13\frac{2}{9}$ sq ft **9.** $162\frac{9}{16}$ sq in. **10.** $12\frac{5}{6}$ sq m **11.** $70\frac{14}{25}$ sq ft **12.** 15/64 sq in. **13.** $1\frac{11}{12}$ in. **14.** 3 cm

3.12B **15.** 92 in. or $7\frac{2}{3}$ ft **16.** 76 in. or $6\frac{1}{3}$ ft **17.** 50 ft or $16\frac{2}{3}$ yd **18.** 4/9 sq ft or 64 sq in. **19.** 156 in. or $4\frac{1}{3}$ yd **20.** $1\frac{1}{6}$ sq ft or 168 sq in. **21.** 74 in. or $6\frac{1}{6}$ ft or $2\frac{1}{18}$ yd **22.** 114 in. or $9\frac{1}{2}$ ft or $3\frac{1}{6}$ yd **23.** 147 in. or $12\frac{1}{4}$ ft or $4\frac{1}{12}$ yd **24.** 1/12 sq yd or 3/4 sq ft.

Chapter 3 Review **1.** 2/5 **2.** Positive **3.** Proper **4.** 20 **5.** 9/13 **6.** 12/23 **7.** 5/8 **8.** 3/35 **9.** 2·2·5·11 **10.** 3/5 **11.** 4/15 **12.** −8/15 **13.** 5/6 **14.** 24/35 **15.** 15/20 **16.** 5/4 or $1\frac{1}{4}$ **17.** 8/21 **18.** 70 **19.** 23/35 **20.** −41/56 **21.** −3/4 **22.** 9/4 **23.** $3\frac{3}{5}$ **24.** $5\frac{7}{10}$ **25.** $1{,}305\frac{2}{23}$ **26.** $3\frac{1}{6}$ **27.** $-8\frac{19}{20}$ **28.** 0 **29.** 6 **30.** 7/12 **31.** 5/6 cup **32.** 2/9 yd **33.** 72 oz **34.** $17\frac{1}{2}$ ft **35.** 17 sq in. or 17/144 sq ft

Supplementary Exercises **1.** 3/7 **2.** Negative **3.** Improper **4.** 13 **5.** −1/17 **6.** 29/31 **7.** 8/9 **8.** −1/13 **9.** 20/63 **10.** 2·2·3·3·5 **11.** 12/17 **12.** 5/28 **13.** −4/45 **14.** −16/25 **15.** 10/21 **16.** 12/42 **17.** 11/10 **18.** −33/35 **19.** 60 **20.** −19/60 **21.** $-3\frac{1}{2}$ **22.** −2/5 **23.** 23/7 **24.** $3\frac{1}{4}$

25. $13\frac{3}{5}$ **26.** $1,085\frac{7}{26}$ **27.** $-12\frac{1}{6}$ **28.** $-2/3$ **29.** 1 **30.** 3 **31.** 9/13 **32.** \$60 **33.** 3/8 qt **34.** 33 in.
35. 204 in. **36.** $16\frac{5}{6}$ ft **37.** $33\frac{3}{4}$ sq in. or 15/64 sq ft

Chapter 3 Test **1.** $2\cdot2\cdot2\cdot3\cdot3\cdot5$ **2.** 9/10 **3.** 20/24 **4.** 7/9 **5.** 5/7 **6.** $-3/10$ **7.** 11/15 **8.** 53/60
9. $-2/5$ **10.** 10/21 **11.** 3/2 or $1\frac{1}{2}$ **12.** $-5\frac{3}{4}$ **13.** $-4/21$ **14.** $-1/4$ **15.** 4 **16.** 1 **17.** 3 **18.** $29\frac{1}{3}$
19. 168 in. **20.** $19\frac{1}{3}$ in. **21.** 7/16 sq ft or 63 sq in.

Cumulative Review **1.** -3 **2.** -5 **3.** -32 **4.** -24 **5.** 15 **6.** -37 **7.** $+6$ **8.** $-4/15$ **9.** 2/9
10. $-1/3$ **11.** $-3\frac{23}{30}$ **12.** $-7/2$ or $-3\frac{1}{2}$ **13.** 39/20 or $1\frac{19}{20}$ **14.** 1 **15.** 7 **16.** 0 **17.** -17 **18.** -20
19. $17\frac{1}{2}$ ft **20.** 45 sq in. or 5/16 sq ft

CHAPTER 4

Section 4.1 **1.** Five hundred seventy-three, thousandths **2.** Three, hundredths **3.** Eight, thousandths
4. Thirty-seven, ten-thousandths **5.** Four, ten-thousandths **6.** Sixty-eight, thousandths **7.** Fifty-six, and
nineteen, hundredths **8.** Twenty-three, and fourteen, thousandths **9.** Six, and thirty-four, hundredths
10. Fifteen, and seven, tenths **11.** Fifteen, and seven, thousandths **12.** Five, and fifty-two, hundredths
13. Eleven, and fifty-two, ten-thousandths **14.** Sixty, and fifty-two, thousandths

Section 4.2 **1.** 5.2 **2.** 14.8 **3.** 20.0 **4.** 0.9 **5.** 5.32 **6.** 638.93 **7.** 0.00 **8.** 70.08 **9.** 5.319
10. 638.927 **11.** 0.004 **12.** 53,280.066 **13.** 16 **14.** 27 **15.** 150 **16.** 67 **17.** 34,700 **18.** 5,800
19. 26,400 **20.** 45,832,100 **21.** \$6, \$23, \$8, \$13, \$25, \$2 **22.** (a) 3.00; 23.76; 408.53; 34.19; 7.09; 186.92 (b) 3.0;
23.8; 408.5; 34.2; 7.1; 186.9 **23.** (a) 9,000 (b) 84 (c) 837 **24.** (a) 64,000 (b) 63,200 (c) 63,180

Section 4.3 **1.** 212.15 **2.** 367.002 **3.** 7.57 **4.** -23.62 **5.** -5.21 **6.** 16.77 **7.** 11.19 **8.** 20.58
9. -24.43 **10.** -24.054 **11.** 11.702 **12.** 41.088 **13.** 12.83 **14.** -2.74 **15.** -15.91 **16.** -53.379
17. 129.306 **18.** -59.293 **19.** 3.86 **20.** 7.9502 **21.** \$76 **22.** (a) 662.493 (b) 662.49 (c) 662.5

Section 4.4A **1.** 34.083 **2.** 5.1 **3.** -5.063 **4.** 7.3062 **5.** 0.0007 **6.** 25.0175 **7.** -8.6912
8. 0.0162 **9.** -0.02601 **10.** -0.0000092 **11.** 0.0192032 **12.** -0.518074 **13.** 0.216405 **14.** 0.280434

4.4B **15.** 3,760 **16.** 670.3 **17.** 50.043 **18.** 0.084 **19.** 4,700 **20.** 16,400 **21.** 307.6 **22.** 71.8
23. 28,062,000 **24.** 462,090 **25.** 38,072,500 **26.** 34,007.6 **27.** 500.83 **28.** 23,547.8 **29.** 6.47 **30.** 3.84
31. 0.07012 **32.** 67.015 **33.** 5,400,238

Section 4.5A **1.** 5.3 **2.** 1.34 **3.** 127.04 **4.** 305.6 **5.** -20.12 **6.** -0.00132 **7.** 0.0007 **8.** 0.0205
9. 0.415 **10.** -0.0325

4.5B **11.** 245.8 **12.** 30.492 **13.** 789,065 **14.** 53.4017 **15.** 50,000.92 **16.** 7,300 **17.** 43.070
18. 0.65008 **19.** 5.08375 **20.** 0.00000052

Section 4.6A **1.** 9.694 **2.** 10.58 **3.** 781.152 **4.** -16.1077 **5.** -8.357 **6.** 337.691 **7.** 18.9196
8. 0.149 **9.** 17.543 **10.** 4.12

4.6B **11.** 3.32 **12.** 6.06 **13.** 6.64 **14.** -6.64 **15.** 12.54 **16.** 9.8 **17.** 6.84 **18.** 12.74 **19.** 6.64
20. -3.46

Section 4.7A **1.** 0.04 **2.** 0.17 **3.** 0.41 **4.** 0.15 **5.** 0.30 **6.** 0.45 **7.** 0.035 **8.** 0.126 **9.** 0.002
10. 0.0005 **11.** 1.25 **12.** 3.40 **13.** 0.000037 **14.** 0.0000014 **15.** 0.00008 **16.** 0.06 **17.** 0.60 **18.** 6.00

4.7B   **19.** 1/25   **20.** 17/100   **21.** 41/100   **22.** 3/20   **23.** 3/10   **24.** 9/20   **25.** 7/200   **26.** 63/500   **27.** 1/500   **28.** 1/2000   **29.** 5/4   **30.** 17/5

4.7C   **31.** 3/10   **32.** 3/5   **33.** 8/25   **34.** 1/100   **35.** 3/1000   **36.** 43/10,000   **37.** 9/200   **38.** 1/8

4.7D   **39.** 0.2   **40.** 0.85   **41.** 0.28   **42.** 0.625   **43.** 0.324   **44.** 0.6   **45.** 0.408   **46.** 0.376   **47.** $0.0\overline{9}$ **48.** $0.\overline{3}$ **49.** $0.58\overline{3}$ **50.** $0.7\overline{3}$ **51.** $0.77\overline{2}$ **52.** $0.\overline{21}$

4.7E   **53.** 41%   **54.** 58%   **55.** 3%   **56.** 15.7%   **57.** 7.2%   **58.** 0.24%   **59.** 4%   **60.** 0.4%   **61.** 0.37%   **62.** 3.7%

Section 4.8A   **1.** 1.52   **2.** 17.46   **3.** 9.75   **4.** 85.0   **5.** 600   **6.** 1,965.6   **7.** 19.995   **8.** 62.5   **9.** 30.6   **10.** 132.44   **11.** 136.8   **12.** 2,160   **13.** 562.5   **14.** 940   **15.** 0.387   **16.** 0.072   **17.** 4.25   **18.** 0.4

4.8B   **19.** $0.34   **20.** $12   **21.** $22.40   **22.** 86 g   **23.** $2.04   **24.** $10.20   **25.** $12   **26.** (a) $104.40 (b) $313.20 (c) $522   **27.** 588 g   **28.** 3,128 g   **29.** $250   **30.** $9.72

Section 4.9   **1.** 217.4 mi   **2.** (a) $37.13 (b) $12.87   **3.** $1.18   **4.** (a) $4.55 (b) $5.45   **5.** $1.33   **6.** $5.25   **7.** $7.05   **8.** (a) 284.8 mi (b) 32 mpg   **9.** $35   **10.** (a) $9.36 (b) $11.51   **11.** 27 mpg   **12.** $4.70   **13.** 0.360   **14.** 0.286   **15.** 326.57   **16.** $172.62   **17.** $18.79 (unless the gas station rounded the gas up, then it would come to $18.80)   **18.** (a) 308.75 mi (b) 43,597   **19.** (a) $295.80 (b) $45.80   **20.** 27.7 mpg   **21.** (a) $68.40 (b) $311.60   **22.** (a) $8.77 (b) $2.19 (c) $54.82   **23.** (a) $4,927.50 (b) $27,922.50   **24.** (a) 38 cents (b) $7.98   **25.** (a) $1.71 (b) $35.91   **26.** $74.80   **27.** $13.65   **28.** $271.50   **29.** $884.94   **30.** $912   **31.** 256   **32.** 30,816   **33.** (a) 4% (b) 210 basketball, 70 hockey, 105 baseball, 85 football, 10 soccer, 20 tennis   **34.** (a) 74 cents (b) $2.22 (c) $17.76   **35.** $1.53   **36.** 36 g   **37.** 0.288   **38.** (a) $211.38 (b) $211   **39.** (a) 7 cents (b) 8 cents

Chapter 4 Review   **1.** Five hundred sixty-eight, ten-thousandths   **2.** 769.1   **3.** 5,300.86   **4.** 54.218   **5.** 309.23   **6.** 91.613   **7.** 203.812   **8.** 57.696   **9.** 0.0000035   **10.** 3400.97   **11.** 400.016   **12.** -50.34 **13.** 0.060005   **14.** 3.9804   **15.** 4.8   **16.** 7.9   **17.** 0.038   **18.** 19/50   **19.** 17/250   **20.** 0.3   **21.** 7.2%   **22.** 62.5%   **23.** 89.1   **24.** $1.24   **25.** $3.80   **26.** (a) $3,880.00 (b) $44,620.00

Supplementary Exercises   **1.** Three hundred forty-one, thousandths   **2.** 28.5   **3.** 618.02   **4.** 161.732   **5.** 1,122.895   **6.** 204.00   **7.** 505.012   **8.** -148.518 **9.** 0.0000024   **10.** 8,800.3   **11.** 166.68   **12.** -20.33 **13.** 0.0128013   **14.** 6.1315   **15.** 22.78   **16.** -3.5 **17.** 0.071   **18.** 21/50   **19.** 7/250   **20.** 0.375   **21.** 30.7%   **22.** 35%   **23.** 70.04   **24.** 97 cents   **25.** 0.560   **26.** 561

Chapter 4 Test   **1.** Two hundred sixty-eight, ten-thousandths   **2.** Seventeen, and ninety-two, thousandths   **3.** (a) 31/500 (b) 9/50 (c) 169/200   **4.** (a) 0.36 (b) $0.\overline{703}$ (c) 0.073   **5.** (a) 5% (b) 37.5% (c) 0.91%   **6.** 6.8   **7.** 421.04   **8.** 0.083   **9.** 42.263   **10.** -23.392 **11.** 0.02046   **12.** -5.03 **13.** 14.82   **14.** 60.3   **15.** $3.73   **16.** (a) $40 (b) $540 (c) $200 (d) $700   **17.** 11.3

Cumulative Review   **1.** -72 **2.** -20 **3.** -24 **4.** 4/5   **5.** 14/25   **6.** $-2/21$ **7.** $1\frac{11}{30}$ **8.** 9/20   **9.** $11\frac{3}{5}$ **10.** $4\frac{7}{12}$ **11.** $-1\frac{11}{14}$ **12.** 3   **13.** -17.095 **14.** -0.0183 **15.** 49.8   **16** (a) 7 (b) 0   **17.** (a) 3.07 (b) 0.04   **18.** $+1$ **19.** $+2$ **20.** -4.0 **21.** 17.2 cm   **22.** 35 sq ft   **23.** $34

CHAPTER 5

Section 5.1A   **1.** $7, xy$ **2.** $3, ac$ **3.** $-2, x$ **4.** $1, yw$ **5.** $16, abc$ **6.** $-4, pwm$

5.1B   **7.** $3a^2b$ **8.** $5x^3y^2$ **9.** $-2ab^3c^2$ **10.** $8k^5m^2$ **11.** 3^4a^6 **12.** $5^2w^3xy^5$ **13.** $7^3x^2y^3w^5$ **14.** $2^5p^2m^3w$

5.1C   **15.** $2xxx$ **16.** $5xyyy$ **17.** $7aabbb$ **18.** $4awwwwwyy$ **19.** $6xxyyyw$ **20.** $3aaaabcc$
21. $8pppmmnnnnn$ **22.** $9abbbcc$

5.1D   **23.** 36   **24.** 54   **25.** 20   **26.** -32 **27.** 7   **28.** -8 **29.** -49 **30.** 49   **31.** 8   **32.** 8   **33.** 72
34. 120   **35.** -18 **36.** 288   **37.** -63 **38.** 36   **39.** 36   **40.** 36   **41.** $-1{,}728$ **42.** -40 **43.** $-1{,}000$
44. 18   **45.** 36

5.1E   **46.** 1   **47.** 3   **48.** 2   **49.** 3   **50.** 8   **51.** 15   **52.** 4   **53.** 7   **54.** 6   **55.** 6

Section 5.2A   **1.** $9a + 10b$ **2.** $17x + 12w$ **3.** $18c + 7m + 13k$ **4.** $12n + 14u + 19f$ **5.** $15k + 14p + 13h$
6. $10a + 12b$ **7.** $7c + 12$ **8.** $17w + 12x + 4$ **9.** $15h + 11k + 12$ **10.** $15b + 22 + 6k$ **11.** $25u + 56 + 36w$
12. $43 + 37x + 20w$ **13.** $22m + 32 + 13n$ **14.** $11x + 6w + 15 + 5y$

5.2B   **15.** $-2a - 6b$ **16.** $-5m - 3k + 7$ **17.** $13 - 11x$ **18.** $-7p - 12w - 12$ **19.** $-10 - 2w$ **20.** $3x$
21. $-5n - 3m$ **22.** $-3k + 8p$ **23.** $-3w - 18$ **24.** $y - 12w$ **25.** $4z + 6w$ **26.** $-6a - 8b - c$
27. $-4k - 7g$ **28.** $12u + 17w$ **29.** $13 - 24r - 9w$ **30.** $24h - 4j - 23$

5.2C   **31.** $8ab + 9ac$ **32.** $9mn + 2pr$ **33.** $6xw + 12 + 14ax$ **34.** $2xw + 4xy + 8yw$
35. $11mn + 6mk + 6 + 3nk$ **36.** $8xy - xw$ **37.** $8ab + 10ac$ **38.** $5xy + 6ab$ **39.** $16xy + 14wx - 7wy$
40. $-2ac$ **41.** $7wx + 11wy$ **42.** $2mn + mp$ **43.** $14wx + 1 + 7wy$ **44.** $9abc + 2bcd$ **45.** $17wxy + 10wxz$
46. $7jkm + 8jmn$

5.2D   **47.** $30xx + 7xxx$ **48.** $12x + 4xx + 12xxx$ **49.** $12yy + 13yyyy$ **50.** $14x + yy - xx + 5y$
51. $5ww - w + 12x + 15xx$ **52.** $16y + 19xx + 10yy + 3x$ **53.** $9w^2 + 10w$ **54.** $12m^3 + 10m + 4m^2$
55. $4x + 11y + x^2 + 11y^2$ **56.** $11m^3 + 6w + m + 3w^3$ **57.** $7x^2 + 8y - 4y^2 + 8x$ **58.** $9x^2 + 18 - 2x$
59. $-2p + 11g - 6 + 7g^2 + p^2$ **60.** $9 - 9k + k^2 + 3m - 3m^2$

5.2E   **61.** $12x^2y^3 + 4x^3y^2 + y^3 + x^2$ **62.** $17xy^3 + 9x^3y$ **63.** $15xy^2 + 6x^2y + 10x^2y^2$
64. $8x + 21x^2 + 8xy^2 + 15x^2y$ **65.** $9x^2 + 9x + 16 + 13x^3y + 4xy^3$ **66.** $13w + 20 + 10wx^2 + x^2$
67. $12x + 13y + 12x^2 + 7 + 5y^2$ **68.** $4x^2w + 5w + 5w^2x + 7x^2 + 4w^2 + 9 + x$ **69.** $w^2x + 12wx^2 + 5wx$
70. $5x^3y^3 - 4xy^3$

Section 5.3A   **1.** $7a + 5b + 3c$ **2.** $10a + 7b + 3c$ **3.** $5x + 8y + 13$ **4.** $2x^2 + 8x + 9$ **5.** $5x^2 + 5x$
6. $2y^2 + 5y + 2$ **7.** $x^5 + 8x^4 + 5x^3 + 12x^2 + x + 11$ **8.** $9x^3 + 7x^2 + x + 4$ **9.** $15x^4 + 7x^3 + 2x^2 - 3x - 8$
10. $8y^3 - 7y^2 - 6y + 11$

5.3B   **11.** $6a + 3b$ **12.** $4x + 3y$ **13.** $4m + 3n$ **14.** $4k$ **15.** $5x - 2w$ **16.** $3m - 2n$ **17.** $13a - 10b$
18. $17r - 4p$ **19.** $5x + 7y - 8z$ **20.** $-a - 5b + 2c$ **21.** $-x^5 + 7x^2 - 5$ **22.** $-3x^2 - 4x$
23. $3x^4 - 7x^3 - 3x^2 - 13x + 1$ **24.** $2y^3 - 12y^2 - 5y - 5$

Section 5.4   **1.** y **2.** $5xw$ **3.** $3a$ **4.** 2   **5.** $12d$ **6.** $10xy$ **7.** $4ax^2$ **8.** $(5bc)/a$ **9.** $(4x)/y$ **10.** $10/b$
11. $w - 2$ **12.** $2a + 6$ **13.** $3a^2 - 4a + 1$ **14.** $2x^2 + 3x - 5$ **15.** $2y + 6/y$ **16.** $(5x)/3 - 5$
17. $(4x)/3 - 2 + 4/x$ **18.** $6w - 15/2 + 2/w$ **19.** $3/x - 4/x^2 + 1/x^3$ **20.** $5/a^2 + 1/a^3 - 5/a^4$
21. $2c + 7ac^2 - 5a^2bc$ **22.** $w/2 - (3x^2w)/2 + 2x^3$

Section 5.5A   **1.** $6abc$ **2.** $20wxy$ **3.** $35klmn$ **4.** $-28abcd$ **5.** $-12mpr$ **6.** $15xy$ **7.** $-2abdk$
8. $-4abc$ **9.** $-14ab^2c$ **10.** $20wxy^2$ **11.** $-15xy^2z$ **12.** $144a^2b^2c^2$ **13.** $-8w^3x^2y^2z^2$ **14.** $-12w^4x^3y$
15. $30k^3lm^5p$ **16.** $36wx^5y^3$ **17.** $-12p^2r^4s^2t$ **18.** $-30w^4x^3y^3$

5.5B   **19.** $6a + 6b$ **20.** $3a + 3b$ **21.** $9x + 63$ **22.** $2r + 2w$ **23.** $7m + 7n$ **24.** $5k + 15$ **25.** $4k + 4m$
26. $8w + 8$ **27.** $6m + 6$ **28.** $4x + 12$ **29.** $2a + 2b + 2c$ **30.** $5x + 5y + 5z$ **31.** $8x + 8y + 8z + 8w$
32. $3m + 3n + 3p + 3r$ **33.** $ab + ac + ad$ **34.** $ax + ay + aw$ **35.** $mx + nx + kx$ **36.** $ax + bx + cx$
37. $an + bn + gn + hn$ **38.** $mr + nr + pr$ **39.** $aw + dw + gw + kw$ **40.** $ay + xy + by + wy$

5.5C   **41.** $5a - 5b$ **42.** $3x - 3y$ **43.** $4x + 4y - 4w$ **44.** $5a - 5b - 5c$ **45.** $ax + bx - cx$ **46.** $ax - ay + az$
47. $6x - 36$ **48.** $8x - 16$ **49.** $7w - 35$ **50.** $6b - 18$ **51.** $5x + 45y - 20z$ **52.** $4a + 12b - 4c$
53. $24a - 32b + 8c$ **54.** $14x - 28y + 21w$ **55.** $33a - 12b - 21c$ **56.** $18m - 24n - 42p$ **57.** $-6x - 10y + 8z$
58. $-6a + 15b + 3c$ **59.** $15d - 20g + 25h - 5k$ **60.** $12x - 4y + 16z - 20w$ **61.** $a^3 + 5a^2 + 7a$
62. $7x^3 - 4x^2 - 3x$ **63.** $2y^3 - 6y^2 - 4y$ **64.** $5m^3 - 5m^2 + 10m$ **65.** $6w^3 - 15w^2 + 3w$
66. $-8x^3 - 6x^2 + 10x$ **67.** $-10ab^2 + 6bc + 6b$ **68.** $-12k^2m + 20kn - 12k$ **69.** $-3x^3 + 15x^2 + 6x$
70. $-18p^4 + 24mp^3 - 12mp^2$

5.5D   **71.** $7(a - b)$ **72.** $3(x - y - w)$ **73.** $-2(a + b + c)$ **74.** $5(p - r + w)$ **75.** $6(y - d - c)$
76. $-5(a + b + c)$ **77.** $-3(a + b + c)$ **78.** $8(v - w + y - x)$ **79.** $y(a + b + c)$
80. $p(k + r - m)$ **81.** $r(3y + 5m - 7c)$ **82.** $n(6p + 5r - 2w)$ **83.** $c(4a - 3d + b)$ **84.** $h(10x - 3j + k - 2)$
85. $b(6a - c - 5d)$ **86.** $w(3a - 5b + 7 + x)$ **87.** $d(2 - 3a + c)$ **88.** $-x(4a + 3b + c + 9w)$

5.5E   **89.** $2a(b + c + d)$ **90.** $5r(a + w + k)$ **91.** $7d(w - h - x)$ **92.** $15q(a - b + c)$
93. $8c(b - d + f - w + g)$ **94.** $ab(3c + 5d)$ **95.** $9s(a - c + d - p)$ **96.** $-7ab(c + d + f)$
97. $pn(6m - 6r - 6k + 5w)$ **98.** $j(4a + 4b - 4c + 3d)$ **99.** $3yw(x + z + 1)$ **100.** $9ac(b - d + 1)$

5.5F   **101.** $10(a + 6b)$ **102.** $15(2x + y - 4w)$ **103.** $4(4f + 5g + 7k)$ **104.** $7(4p + 3k - 6m)$
105. $8(3r - 2p + 2pr)$ **106.** $8k(3h + 2m - 4r)$ **107.** $15b(3a - 2d + p)$ **108.** $6pr(2q + 5k - 7m)$

5.5G   **109.** $y(7y + 8)$ **110.** $w(9 + 5w)$ **111.** $a(7a - 3)$ **112.** $m(5m - 4)$ **113.** $3x(x + 4)$ **114.** $4y(y - 5)$
115. $3w(2w - 5)$ **116.** $5m(2m + 7)$ **117.** $x(x^2 + 4x + 5)$ **118.** $y(2y^2 - 7y + 4)$ **119.** $xy(x^2 + 5x - 3)$
120. $wx(w^2x - 2wx^2 + 4)$ **121.** $2mn(m^2 + 3m + 5)$ **122.** $3ab(4a^3b^2 - 2a^2b + 3)$ **123.** $3w^3y^2(7w^2y - 5 + 4wy^3)$
124. $5xy^2(2y^2 - 8x^2y - 3x)$

Section 5.6   **1.** $9x^2 + 33x + 7$ **2.** $27x^2 + 31x - 33$ **3.** $16x^2 + 28x - 34$ **4.** $7x^2 - 8x + 26$
5. $7x^2 + 28x + 29$ **6.** $8x^2 - 8x - 32$ **7.** $9x^2 + 4x + 13$ **8.** $10x^3 + 24x^2 + 6x + 18$ **9.** $7x^2 + 9x + 20$
10. $2x^3 + 15x^2 - 24x + 19$ **11.** $-16x^2 + 17x - 14$ **12.** $-22x^2 + 6x + 33$ **13.** $9a^2 + 2ab$
14. $9ab + 3bc - 5ac$ **15.** $2x^2 + 3xy - 3x - 7y^2 + 4y$ **16.** $2a^2 + 6ac - 20b^2 + 5bc$
17. $6wx + 6wy - 9wz - 8xy + 2yz$ **18.** $2r^3 - 15r^2 + 27r - 4$ **19.** $-4k^3 - 10k$ **20.** $2m^3 - 10m^2 + 8m - 4$

Section 5.7A   **1.** 5 **2.** -4 **3.** -17 **4.** 17 **5.** 11 **6.** -15 **7.** -4 **8.** -13 **9.** 19 **10.** 32 **11.** 10
12. 31 **13.** 28 **14.** 14 **15.** 2 **16.** 75 **17.** -6 **18.** -28

5.7B   **19.** $5C + 3L + 7S$; $33.80   **20.** $7C + 4P + 8B + 12J$; $59   **21.** $80M + 100W + 25P$; $85
22. $9T + 2Ca + 10S + 3Ch$; $35.27   **23.** $3R + 4S - T + 5Pu - 2Pa - J + 3B$; $-$8.20 (spent $8.20 more than
I took in)   **24.** $8R + 2S - T + 9Pu - 3Pa - J + 7B$; $-$11.65 (spent $11.65 more than I took in)
25. $12C + 8P + 4B - FL - N - FR$; $125   **26.** $9B + 2L + 7S + 5G + 15P - R$; $47.50
27. $-3S - 2P - 2B - J + 3A + C + 4H$; $-$97 (spent $97 more than I took in)

Chapter 5 Review   **1.** -8 **2.** $6xy^3z^2$ **3.** $3mmmnpppp$ **4.** -160 **5.** 6 **6.** $6m + 4k + 13$
7. $7p - 3k + 4m$ **8.** $2ac + 9bc + ab$ **9.** $-8a^3 - 9b + 7b^3 + 5a$ **10.** $2xy - 4xy^2 + 5xy^3$
11. $5y^3 + 4y^2 - 10y + 12$ **12.** $-3x - 4y + 10xy - 13$ **13.** $3b^2/2$ **14.** $5x^2 - 2x + 4y$ **15.** $7x^2 - 35x + 7$
16. $-5wxy^2$ **17.** $-2w^3 + 8w^2 - 14w$ **18.** $6a(b - c + d)$ **19.** $5xy(2x - 5y^2 + 6x^3y)$ **20.** $-6x^2 - 3x + 27$
21. $32a - 20b - 6ab$ **22.** 15 **23.** 17 **24.** $2Ch + H - 3Cu - R + 5Sh - 2Sc$; $12.90 (took in $12.90 more than I
spent)

Supplementary Exercises   **1.** -13 **2.** $5a^2bc^4$ **3.** $7aabbbbccc$ **4.** 12 **5.** 9 **6.** $5x + 8y + 10$
7. $a + 12b - 5c$ **8.** $14mn + 5m^2n + 3m^3n - 2mn^2$ **9.** $-8a^3 - 9b + 7b^3 + 5a$ **10.** $-2wz - 2wz^2 - 2w^3z$
11. $6n^3 - 3n^2 - 3n - 4$ **12.** $-4a - 5b - ab - 13$ **13.** $3k^3$ **14.** $7a^2 - 3a + 6b$ **15.** $4y^2 - 12y - 4$ **16.** $4a^2bc$
17. $-3x^3 + 24x^2 + 15x$ **18.** $5x(y - w - z)$ **19.** $3ab(4a - 5b^2 + 10a^2b^3)$ **20.** $-10x^2 + 34x - 14$

21. $-10x - 12xy + y$ **22.** 13   **23.** 14   **24.** $-H + B + C + Pr - Pz - S + L$; \$10.25 (took in \$10.25 more than I spent)

Chapter 5 Test
1. $7abbccc$ **2.** $6a + 4 + 3b$ **3.** $8wx - 1 + wy$ **4.** $7y^3 + 2y^2 - y$
5. $-5a^3b^2 - 5a^3b - a^2b^2 + ab^3$ **6.** $9a^3 + 2a^2 - 7a - 6$ **7.** $7w^3 - 12w^2 + 10w + 4$ **8.** $12x^3y^6$
9. $12x^3 - 20x + 24$ **10.** $10a^4 + 14a^3b - 10a^3b^2 + 2a^2b^3 - 18ab^2$ **11.** $26c - 31a + 34f + 4$
12. $-3w^3 - 38w^2 + 37w - 1$ **13.** $4x + 2x^2 - 8xy - 10y + 20y^2$ **14.** $4ac^3$ **15.** $4x^2 - 5xy + 3y^2$
16. $m(8w - 4p - k)$ **17.** $7xy^2(2y - x + 3x^3y)$ **18.** 32   **19.** 30   **20.** 1   **21.** $5C + 4P + 7B - Fl - N - Fr$; \$16

Cumulative Review
1. -41 **2.** -9 **3.** $+370$ **4.** $+36$ **5.** 4/15   **6.** 11/10 or $1\frac{1}{10}$ **7.** 13/36
8. $-4\frac{11}{12}$ **9.** $-12/35$ **10.** 7.724   **11.** 30.04   **12.** 843.2   **13.** 0   **14.** 1   **15.** 5   **16.** 7   **17.** 560.09
18. 340.3   **19.** -8 **20.** 4   **21.** -4 **22.** -8 **23.** 8   **24.** -1 **25.** 1   **26.** -5 **27.** 63
28. $-p^3 - 13p^2 + 16p - 2$ **29.** $y^3 - 5y^2 + 2y - 4$ **30.** \$763

CHAPTER 6

Section 6.1
1. Yes   **2.** Yes   **3.** No   **4.** No   **5.** No   **6.** Yes   **7.** No   **8.** Yes   **9.** Yes   **10.** No
11. Yes   **12.** No   **13.** No   **14.** Yes

Section 6.2A
1. 48; $x = 118$ **2.** 22; $x = 440$ **3.** 31; $y = 540$ **4.** -73; $w = 823$ **5.** 108; $k = 1086$
6. -135; $m = 724$ **7.** 213; $r = 1297$ **8.** 347; $p = 1253$ **9.** -815; $f = 2652$ **10.** 391; $a = 5473$
11. $m = -24$ **12.** $k = -768$ **13.** $h = -148$ **14.** $x = -33$ **15.** $y = -54$ **16.** $j = 38$ **17.** $k = -540$
18. $p = 208$ **19.** $x = 4.28$ **20.** $y = 23.95$ **21.** $w = 11.63$ **22.** $a = -22.808$ **23.** $b = 4/5$ **24.** $k = 2/3$
25. $m = 11/15$ **26.** $w = 5/24$ **27.** $r = 1/6$ **28.** $p = 23/60$ **29.** $x = 5$ **30.** $y = 5\frac{1}{6}$

6.2B   **31.** $x = 8$ **32.** $y = 284$ **33.** $x = 15$ **34.** $w = -7$ **35.** $w = 16$ **36.** $y = 28$ **37.** $y = -18$
38. $w = 35$ **39.** $x = 53.7$ **40.** $z = -7$ **41.** $y = 3/5$ **42.** $x = -2/3$ **43.** $m = -4\frac{4}{9}$ **44.** $k = 2\frac{3}{4}$
45. $x = 6\frac{3}{5}$ **46.** $m = -6\frac{1}{4}$ **47.** $w = -6\frac{1}{3}$ **48.** $r = -3\frac{2}{7}$ **49.** $x = 0.02$ **50.** $y = 15,000$
51. $w = 5$ **52.** $a = 150,340$ **53.** $m = 15$ **54.** $b = 5/8$ **55.** $c = 3/4$ **56.** $k = 11/8$ or $1\frac{3}{8}$ **57.** $y = 6$
58. $w = 7/26$ **59.** $a = 21/8$ or $2\frac{5}{8}$ **60.** $w = 20/39$ **61.** $r = -0.007$ **62.** $w = -1,204,400$

6.2C   **63.** $x = 28$ **64.** $y = 55$ **65.** $w = 105$ **66.** $m = -63$ **67.** $p = 38$ **68.** $k = 68$ **69.** $r = -287$
70. $q = 60$ **71.** $t = -560$ **72.** $f = 162$ **73.** $x = 2.8$ **74.** $y = 0.24$ **75.** $m = 0.06$ **76.** $k = 0.602$

Section 6.3A
1. $x = 3$ **2.** $x = 17$ **3.** $x = 15$ **4.** $x = 3$ **5.** $y = -3$ **6.** $x = 7$ **7.** $x = -3$
8. $w = -2$ **9.** $x = 8$ **10.** $x = 10$ **11.** $y = -1$ **12.** $x = 5$ **13.** $y = -3$ **14.** $x = 6$ **15.** $x = 26$
16. $x = 5$ **17.** $x = 11$ **18.** $w = -3$ **19.** $x = 4\frac{1}{3}$ **20.** $x = 4\frac{1}{5}$ **21.** $x = 7\frac{1}{4}$ **22.** $x = 5\frac{2}{7}$
23. $x = 28\frac{1}{2}$ **24.** $x = 24\frac{2}{3}$ **25.** $x = 5\frac{1}{3}$ **26.** $x = 5\frac{1}{6}$ **27.** $x = 5\frac{1}{3}$ **28.** $x = 25\frac{1}{2}$ **29.** $x = 0.6$
30. $y = 0.99$ **31.** $w = 33$ **32.** $a = 1,228$

6.3B   **33.** $x = 6$ **34.** $w = 7$ **35.** $y = 5$ **36.** $x = 3$ **37.** $x = 2$ **38.** $z = 2$ **39.** $m = 5$ **40.** $w = 7$
41. $x = 7$ **42.** $k = 29$ **43.** $m = 9$ **44.** $w = -11$ **45.** $x = 3$ **46.** $y = 5$ **47.** $a = 5$ **48.** $b = 2$
49. $x = 3$ **50.** $x = 5$ **51.** $x = 4$ **52.** $x = 4$ **53.** $x = 2$ **54.** $x = 1$ **55.** $x = 2$ **56.** $x = 5$
57. $x = -2$ **58.** $x = 2$ **59.** $x = 3$ **60.** $x = 3$ **61.** $x = 5$ **62.** $x = 1$ **63.** $x = 5$ **64.** $x = 4$ **65.** $x = 0$
66. $x = 5$ **67.** $x = 12$ **68.** $x = 8$

6.3C   **69.** $w = 2$ **70.** $y = 3$ **71.** $x = 2$ **72.** $x = 3$ **73.** $m = 5/2$ **74.** $b = 3/2$ **75.** $x = 1$ **76.** $y = 2$
77 $p = 12$ **78.** $k = 8$ **79.** $y = 2/3$ **80.** $w = 2/5$

6.3D   **81.** $x = 6$ **82.** $y = -40$ **83.** $w = 1,246$ **84.** $m = -6\frac{2}{3}$ **85.** $a = -5$ **86.** $r = 5$ **87.** $v = 3/4$
88. $p = -3/5$ **89.** $k = -25$ **90.** $j = -7$ **91.** $s = -3$ **92.** $z = -5$ **93.** $x = 2$ **94.** $y = 8$

6

Section 6.4A   **1.** $x = 9$ **2.** $x = 8$ **3.** $x = 9$ **4.** $x = -7$ **5.** $x = -2$ **6.** $x = 4$ **7.** $x = -4$ **8.** $x = 8$
9. $x = -7$ **10.** $x = 4$ **11.** $x = 6$ **12.** $x = 3$ **13.** $x = -2$ **14.** $x = 4$ **15.** $x = -1$ **16.** $x = 3$
17. $x = 2$ **18.** $x = 10$ **19.** $x = -2$ **20.** $x = 7$

6.4B   **21.** $x = 4$ **22.** $x = 7$ **23.** $x = 10$ **24.** $x = 1$ **25.** $x = 5$ **26.** $x = 3$ **27.** $x = 3$ **28.** $x = 2$
29. $x = 3$ **30.** $x = 1$

6.4C   **31.** $x = 6$ **32.** $x = 18$ **33.** $x = 14$ **34.** $x = -8$ **35.** $x = 33$ **36.** $x = 45$ **37.** $x = 20$ **38.** $x = 125$
39. $x = 10$ **40.** $y = 12$ **41.** $x = 6$ **42.** $y = -14$ **43.** $y = 12$ **44.** $w = 6$

6.4D   **45.** $y = 24$ **46.** $k = 21$ **47.** $p = 23$ **48.** $x = 12$ **49.** $y = 12$ **50.** $n = 11$ **51.** $y = -2$
52. $c = -5$ **53.** $b = 2$ **54.** $T = 2$ **55.** $k = -10$ **56.** $h = 1$

Section 6.5A   **1.** m is less than 10.   **2.** y is less than 5.   **3.** n is less than 12.   **4.** p is greater than 14.
5. z is greater than or equal to 8.   **6.** x is greater than or equal to 3.   **7.** h is less than 4.   **8.** x is less than or equal to 23.

6.5B   **9.** (number line 0 1 2 3 4 5)   **10.** (number line 0 1 2 3 4 5)   **11.** (number line $-4\,-3\,-2\,-1\,0\,1\,2$)   **12.** (number line $-4\,-3\,-2\,-1\,0$)
13. (number line $-8\,-7\,-6\,-5\,-4\,-3$)   **14.** (number line $-2\,-1\,0\,1\,2\,3$)   **15.** (number line 4 5 6 7 8 9 10)   **16.** (number line $-2\,-1\,0\,1\,2\,3\,4$)

6.5C   **17.** $x < 10$ (number line 4 5 6 7 8 9 10)   **18.** $x > 6$ (number line 5 6 7 8 9 10)   **19.** $x \geq 7$ (number line 4 5 6 7 8 9)
20. $x \leq 14$ (number line 12 13 14 15 16)   **21.** $x > 5$ (number line 4 5 6 7 8)   **22.** $x \geq 3$ (number line 1 2 3 4 5 6)
23. $x < 10$ (number line 6 7 8 9 10)   **24.** $x \leq 3$ (number line $-1\,0\,1\,2\,3\,4\,5$)   **25.** $x > 10$ (number line 7 8 9 10 11 12)
26. $x < 7$ (number line 5 6 7 8 9)

6.5D   **27.** $x > -5$ (number line $-7\,-6\,-5\,-4\,-3$)   **28.** $x < -4$ (number line $-6\,-5\,-4\,-3\,-2\,-1$)   **29.** $x \geq -6$ (number line $-8\,-7\,-6\,-5\,-4\,-3$)
30. $x \geq -7$ (number line $-9\,-8\,-7\,-6\,-5$)   **31.** $x < -2$ (number line $-5\,-4\,-3\,-2\,-1\,0$)   **32.** $x \leq 6$ (number line 3 4 5 6 7 8)
33. $x \geq -7$ (number line $-9\,-8\,-7\,-6\,-5$)   **34.** $x > -4$ (number line $-5\,-4\,-3\,-2\,-1$)   **35.** $x < -2$ (number line $-4\,-3\,-2\,-1\,0$)   **36.** $x < 5$ (number line 2 3 4 5 6 7)

6.5E   **37.** $6 < x < 13$ (number line 3 4 5 6 7 8 9 10 11 12 13 14 15)   **38.** $9 > x > -1$ (number line $-4\,-3\,-2\,-1\,0\,1\,2\,3\,4\,5\,6\,7\,8\,9\,10$)
39. $-1 < x \leq 6$ (number line $-3\,-2\,-1\,0\,1\,2\,3\,4\,5\,6\,7\,8$)   **40.** $13 > x \geq -1$ (number line $-3\,-2\,-1\,0\,1\,2\,3\,4\,5\,6\,7\,8\,9\,10\,11\,12\,13\,14$)
41. $3 < x \leq 7$ (number line 0 1 2 3 4 5 6 7 8 9 10)   **42.** $1 \leq x < 4$ (number line $-1\,0\,1\,2\,3\,4\,5\,6$)
43. $4 < x < 9$ (number line 0 1 2 3 4 5 6 7 8 9 10)   **44.** $-2 \leq x \leq 5$ (number line $-4\,-3\,-2\,-1\,0\,1\,2\,3\,4\,5\,6\,7$)
45. $3 > x > -2$ (number line $-4\,-3\,-2\,-1\,0\,1\,2\,3\,4\,5$)   **46.** $-2 \geq x > -5$ (number line $-8\,-7\,-6\,-5\,-4\,-3\,-2\,-1\,0$)
47. $-3 \leq x \leq 4$ (number line $-5\,-4\,-3\,-2\,-1\,0\,1\,2\,3\,4\,5\,6$)   **48.** $-2 < x \leq 3$ (number line $-4\,-3\,-2\,-1\,0\,1\,2\,3\,4\,5$)

49. $-3 > x > -7$

50. $2 > x \geq -4$

Chapter 6 Review
1. No **2.** Yes **3.** $m = -27$ **4.** $a = 21.74$ **5.** $m = -5$ **6.** $k = 1/4$
7. $h = 0.03$ **8.** $x = 36$ **9.** $y = -0.155$ **10.** $x = 5$ **11.** $y = -6$ **12.** $m = 57$ **13.** $x = 7$ **14.** $z = -8$
15. $w = 12$ **16.** $x = 20$ **17.** $p = 5$ **18.** $r = 1$ **19.** $f = -1$ **20.** $x = 2$

21. $x < 6$ **22.** $y > 7$ **23.** $k > 4$

24. $w > -2$ **25.** $-7 < z \leq 3$

26. $-1 \leq x \leq 6$

Supplementary Exercises
1. No **2.** Yes **3.** $m = 16$ **4.** $a = 18.47$ **5.** $m = 5$ **6.** $k = 1/10$
7. $h = 0.08$ **8.** $x = 147$ **9.** $y = -0.324$ **10.** $x = 6$ **11.** $y = -9$ **12.** $m = 31$ **13.** $x = 4$ **14.** $z = -1$
15. $w = 12$ **16.** $x = 20$ **17.** $p = 5$ **18.** $r = 1$ **19.** $f = -1$ **20.** $x = 2$

21. $x < 5$ **22.** $y > 8$ **23.** $k \leq 4$

24. $w \leq -4$ **25.** $-7 \leq z < 6$

26. $2 \leq x \leq 7$

Chapter 6 Test
1. No **2.** Yes **3.** Yes **4.** $r = 100$ **5.** $p = -5$ **6.** $m = -8.87$ **7.** $d = 2.1$
8. $y = 40$ **9.** $w = -0.208$ **10.** $j = 8$ **11.** $a = 12$ **12.** $x = -7$ **13.** $y = 3$ **14.** $w = -2$ **15.** $z = 60$
16. $b = 3$ **17.** $c = -2$ **18.** $k = 7$ **19.** $n > 5$ **20.** $p < 3$

21. $a \leq 3$ **23.** $-8 < k < 7$

24. $2 \leq n \leq 6$

Cumulative Review
1. -74 **2.** -30 **3.** -4 **4.** $5/6$ **5.** $13/12$ or $1\frac{1}{12}$ **6.** $10/63$ **7.** $30/49$
8. 53.94 **9.** 235.17 **10.** 51.03 **11.** $10x - 14$ **12.** $-0.3x^2 + 0.3x + 10.4$ **13.** $7x^2 - 4x - 13$
14. $3ab + 21a + 30b$ **15.** $10x^2 - 14xy - 2x + 3y^2 + 12y$ **16.** $x - 5 + 3/x$ **17.** 4 **18.** -8 **19.** -90
20. 6 **21.** -14 **22.** $5a(3bc - 2c + 7bd - 4cd)$ **23.** $3xy(2y - 4x - 5 + 3x^2y)$ **24.** $a = -1/3$ **25.** $n = -2$

26. $0 \leq x < 5$

CHAPTER 7

Section 7.1 **1.** $0.4S = 26; S = 65$ **2.** $0.62B = 186; B = 300$ **3.** $0.12P = 7.200; P = \$60,000$
4. $26,000x = 4,680; x = 18\%$ **5.** $0.03T = 25,020; T = 834,000$ **6.** $25X = 18; X = 72\%$
7. $0.193P = 617.60; P = \$3,200$ **8.** $0.22I = 265; I \approx \$1,205$ **9.** $0.134I = 428.80; I = \$3,200$ **10.** $50X = 37;$
$X = 74\%$ **11.** $56,000X = 6,720; \quad X = 12\%$ **12.** (a) $0.30N = 144; N = 480$ (b) 336 **13.** $0.64T = 179.20;$
$T = \$280$ **14.** $0.875W = 140; \quad W = 160$ lb **15.** $0.125S = 155; S = 1,240$ **16.** $0.025C = 43; C = 1,720$
17. $0.064P = 512; P = 8,000$ **18.** $0.02P = 72; P = 3,600$ **19.** $2.30C = 34,040; C = 14,800$ **20.** $0.042P = 1,260;$

$P = 30,000$ **21.** (a) $P\% = (327,296 - 308,071)/308,071$; $P = 6\%$ (b) Pr Increase = 6% of (409,463); Pr Increase = $24,567.78 or $24,568 (c) Pr Sales = 409,463 + 24,678; Pr Sales = $434,031
22. (a) $P\% = (234,850 - 327,135)327, 135$; $P = -28\%$ (b) Pr Decrease = 28% of $390,250; $PD = \$109,270$ (c) Pr Sales = 390,250 − 109,270; Pr Sales = $280,980

Section 7.2A **1.** 270 mi **2.** 36 mi **3.** 45°C **4.** 77°F **5.** 256 ft **6.** 31.4 mi/sec **7.** 164 ft/sec **8.** 78.4 m **9.** 10°C **10.** 57°F **11.** 164 ft **12.** 135 ft/sec

7.2B **13.** $b = p - a - c$ **14.** $r = C/2\pi$ **15.** $w = a/l$ **16.** $h = 2a/b$ **17.** $m = k/3p$ **18.** $b = R - a + c$ **19.** $r = i/pt$ **20.** $V_0 = V - at$ **21.** $a = (V - V_0)/t$ **22.** $t = (V - V_0)/a$ **23.** $g = (2d)/t^2$ **24.** $c = m - 3ab$ **25.** $a = (k + b)/5$ **26.** $k = (h + 5m)/3$ **27.** $w = x - y - z$ **28.** $b = (a + 3c - 5d)/2$ **29.** $m = (p + 4k)/7$ **30.** $P = (J - 2L)/8$

Section 7.3A **1.** 1:4 **2.** 1:5 **3.** 1:7 **4.** 5:7 **5.** 3:5 **6.** 3:10 **7.** 1:5 **8.** 3:14 **9.** 1:7 **10.** 5:9

7.3B **11.** Yes **12.** No **13.** Yes **14.** No **15.** Yes **16.** No **17.** Yes **18.** Yes **19.** No **20.** Yes

Section 7.4A **1.** $k = 15$ **2.** $y = 8$ **3.** $w = 24$ **4.** $p = 21$ **5.** $x = 7$ **6.** $h = 16$ **7.** $m = 4$ **8.** $r = 17$ **9.** $n = 35$ **10.** $t = 20$ **11.** $b = 25$ **12.** $m = 15$ **13.** $p = 117$ **14.** $r = 85$ **15.** $z = 66$ **16.** $h = 68$ **17.** $j = 6$ **18.** $q = 35$

7.4B **19.** $x = 8/21$ **20.** $y = 3/5$ **21.** $m = 4/15$ **22.** $h = 1.7$ **23.** $k = 0.21$ **24.** $r = 0.5$ **25.** $w = 7\frac{1}{2}$ **26.** $x = 3/4$ **27.** $p = 5\frac{1}{3}$ **28.** $h = 2\frac{2}{5}$

7.4C **29.** 129 women **30.** 16 hr **31.** $1,875 **32.** 175 mi **33.** 14 in. **34.** $3\frac{1}{3}$ cups **35.** $7,600 **36.** 30 adults **37.** 49 prefer lemonade **38.** 33,570 in-state residents **39.** $6,636.60 **40.** 1,200 bu **41.** 35 min **42.** $3.50 **43.** 12 foreign cars **44.** 1,589 g **45.** 4 m **46.** 9 days **47.** $11,200 **48.** 80 days

Section 7.5 **1.** (a) 24 women (b) 18 men **2.** (a) 105 National League fans (b) 30 American League fans **3.** (a) 6 cups milk (b) 9 cups water **4.** (a) 16 drops blue (b) 24 drops red **5.** (a) 6 golf (b) 27 tennis **6.** (a) 20 hot (b) 12 cold **7.** (a) 75 good (b) 50 bad **8.** (a) 20 dog owners (b) 24 cat owners **9.** (a) 2,040 have jobs (b) 85 do not **10.** (a) 1,620 got jobs (b) 240 did not **11.** (a) 32 got it right (b) 12 got it wrong **12.** (a) 6,500 voted for winners (b) 4,500 did not **13.** (a) 238 sang (b) 112 did not **14.** (a) we scored 1,372 (b) opponents scored 1,764 **15.** 80 times **16.** (a) I score 840 points (b) 720 points scored against me **17.** (a) 112 adults (b) 182 children **18.** (a) 80 letters (b) 200 bills **19.** 175 times **20.** 120 leftovers

Chapter 7 Review **1.** $53,750 **2.** 113°F **3.** $M = (K - P)/4$ **4.** 2:5 **5.** Yes **6.** No **7.** $n = 24$ **8.** 4 2/3 **9.** $a = 1.01$ **10.** 52 1/2 mi **11.** 200 are unemployed

Supplementary Exercises **1.** $340 **2.** 10°C **3.** $J = (R - 2A)/5$ **4.** $W = (K - 2N)/5$ **5.** 4:27 **6.** No **7.** Yes **8.** $p = 28$ **9.** $m = 133$ **10.** $n = 3\frac{3}{4}$ **11.** $b = 0.23$ **12.** $8.50 **13.** 69 women **14.** (a) 84 dogs (b) 48 cats

Chapter 7 Test **1.** $66,667 **2.** $P = 17$ **3.** $P = (R - 8 + 3S)/2$ **4.** 10:7 **5.** Yes **6.** $A = 15$ **7.** $N = 0.23$ **8.** $M = 4/9$ **9.** $1.38 **10.** (a) 306 old (b) 68 new

Cumulative Review **1.** 107 **2.** $1\frac{1}{3}$ **3.** 9/100 **4.** 22/29 **5.** -1.298 **6.** $5x^3 - 6x^2 - 4x$ **7.** $5.2a + 4.3b - 5.26c$ **8.** $-2x^2 - x + 2$ **9.** $-3a^3 + 4a^2 + 15a$ **10.** $410 **11.** -23 **12.** $x = 5$ **13.** $y = -3$ **14.** $w = 2$ **15.** $m \geq 8$ **16.** $-3/5 < p \leq 5$ **17.** $a = 2$ **18.** $N = 370$ **19.** 34 checks **20.** (a) 160 men (b) 192 women

CHAPTER 8

Section 8.1A

1.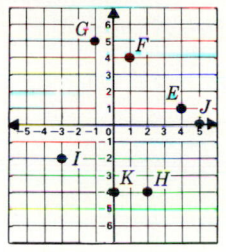

2.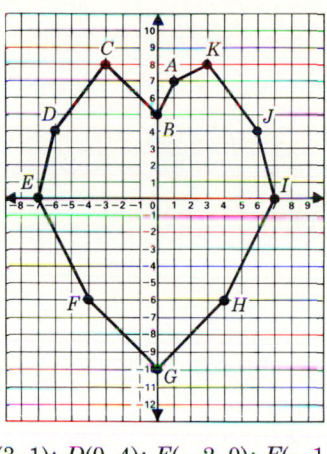

3.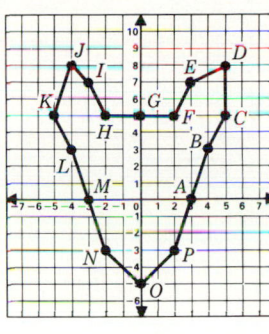

8.1B   **4.** $A(-3, 5)$ $B(1, 3)$; $C(3, 1)$; $D(0, 4)$; $E(-2, 0)$; $F(-1, -3)$; $G(1, -5)$; $H(2, 3)$; $I(0, -2)$; $J(4, 0)$; $K(6, 0)$; $L(6, -3)$; $M(-6, 0)$

8.1C   **(See graph below)**   **5.** Not on line   **6.** On line   **7.** On line   **8.** Not on line   **9.** On line   **10.** Not on line   **11.** Not on line   **12.** On line   **13.** Not on line   **14.** On line   **15.** On line   **16.** Not on line

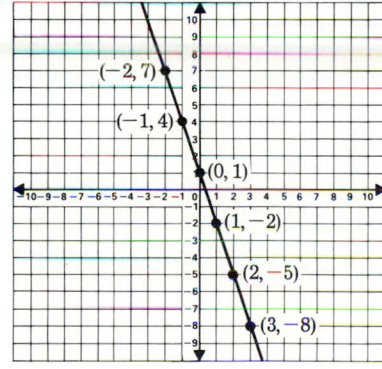

8.1D   **17.** $C(25, -40)$ Quad. 4; $D(-15, 0)$ horizontal axis; $E(-40, 40)$ Quad. 2; $F(-35, -10)$ Quad. 3; $G(30, 0)$ horizontal axis; $H(20, 80)$ Quad 1

18.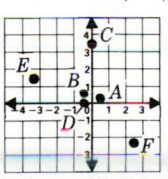

19.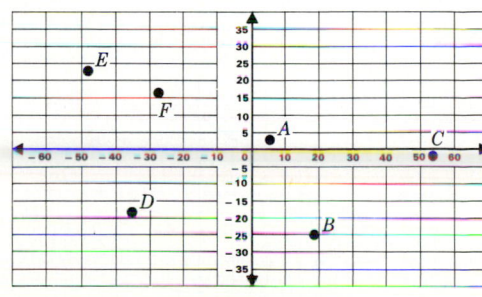

Section 8.2A

1. $y = x + 5$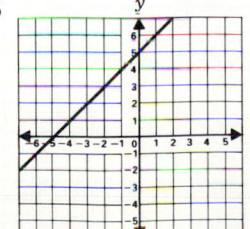

2. $y = x - 3$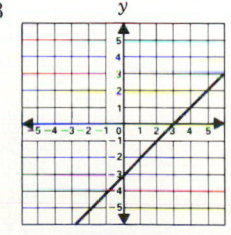

3. $y = 2x$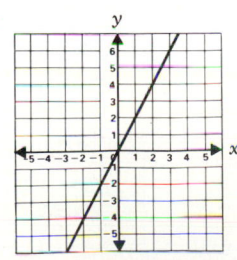

8

4. $y = 2x + 3$

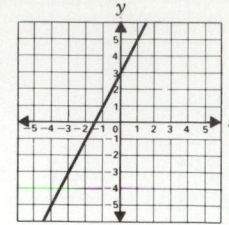

5. $y = 7x - 10$

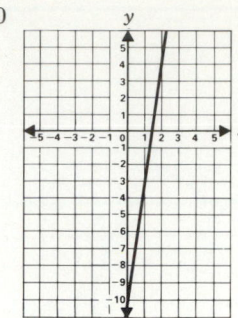

6. $y = x$

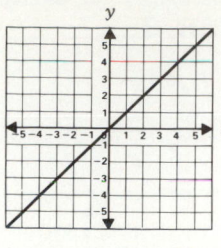

7. $y = -2x + 1$

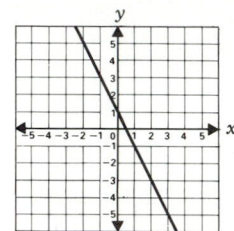

8. $y = 3x$

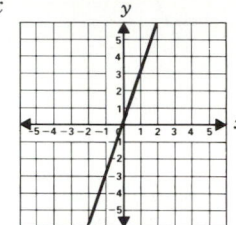

9. $y = -5x$

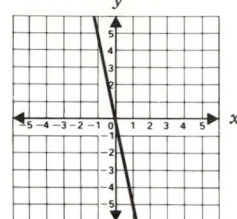

10. $y = 3x - 4$

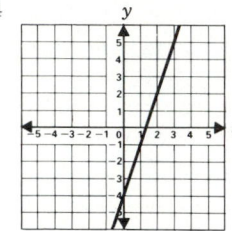

11. $b = -3a - 5$

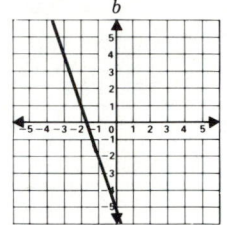

12. $m = 6n - 5$

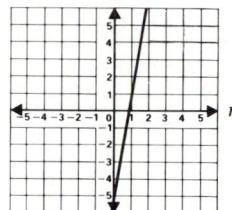

13. $y = 3x + 1$

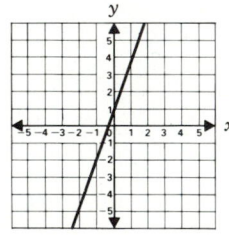

14. $y = 4x - 3$

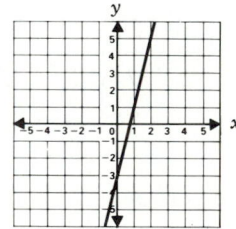

8.2B

15. $y = \frac{3}{4}x$

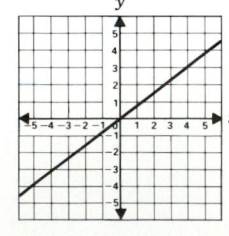

16. $y = \frac{2}{3}x$

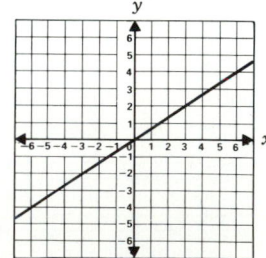

17. $y = \frac{1}{2}x + 3$

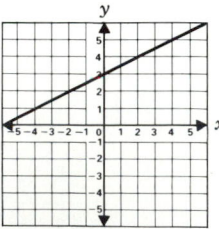

18. $y = \frac{1}{4}x - 2$

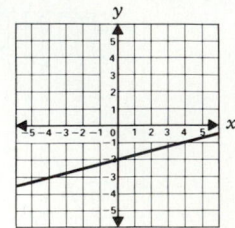

19. $y = \frac{2}{3}x - 1$

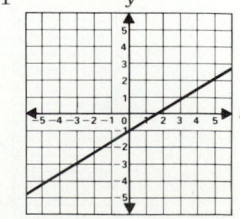

20. $y = \frac{2}{5}x + 3$

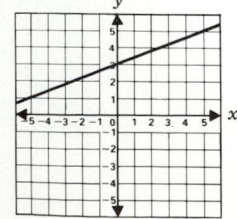

21. $y = \frac{3}{5}x + 2$

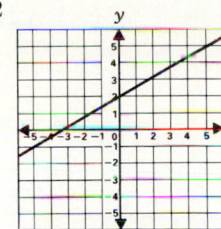

22. $y = \frac{3}{4}x - 2$

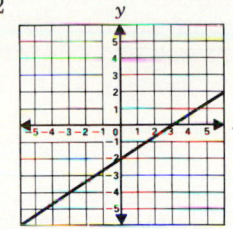

23. $y = \frac{1}{3}x + \frac{2}{3}$

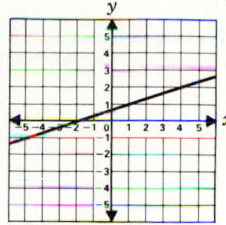

24. $y = \frac{3}{5}x - \frac{1}{5}$

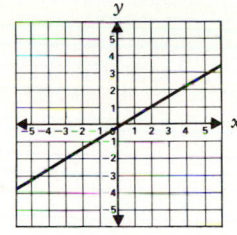

25. $y = \frac{2}{3}x + \frac{1}{2}$

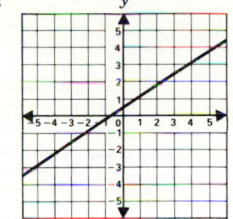

26. $y = \frac{1}{4}x + \frac{1}{3}$

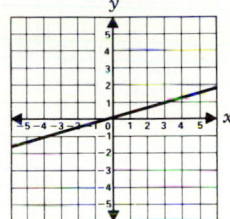

Section 8.3

1. $y = 4$

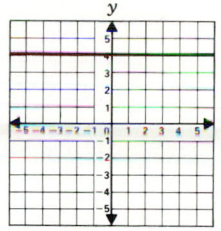

2. $x = 2$

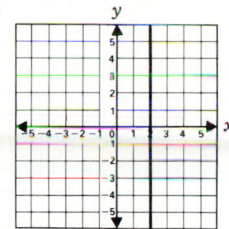

3. $y = -2$

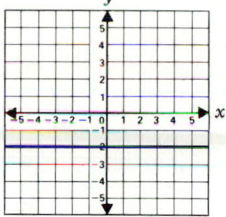

4. $x = -3$

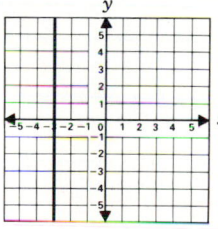

5. $y = 0$

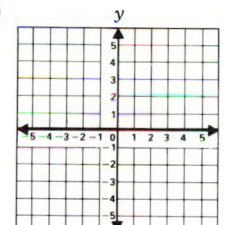

6. $x = 0$

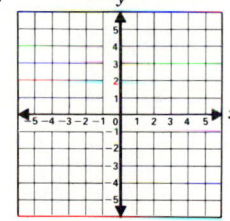

7. $x = -1$

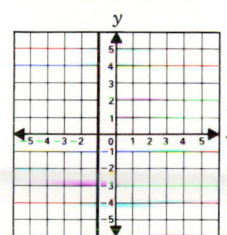

8. $y = -1$

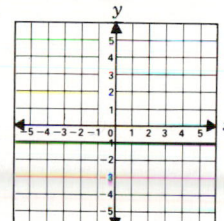

9. $x = -6$

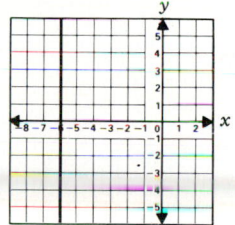

10. $y = 2$

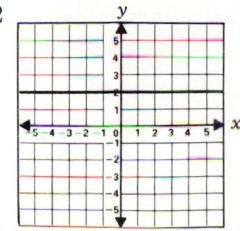

Section 8.4A

1. Solved for y **2.** Not solved for y **3.** Not solved for y **4.** Solved for y **5.** Not solved for y **6.** Solved for y

8.4B

7. $y = 9 - 2x$

$f(x) = 9 - 2x$

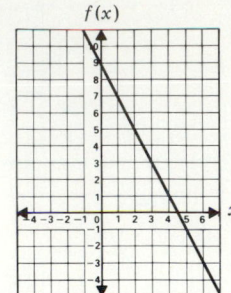

8. $y = 5x - 4$

$f(x) = 5x - 4$

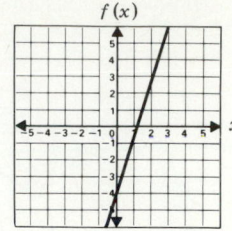

9. $y = x - 3$

$f(x) = x - 3$

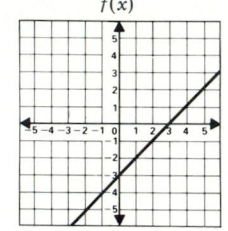

10. $y = 3x + 4$

$f(x) = 3x + 4$

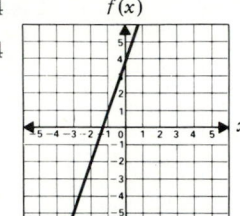

11. $y = 3x - 2$

$f(x) = 3x - 2$

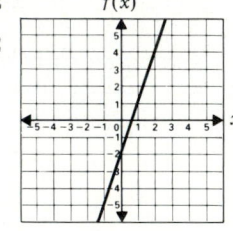

12. $y = 4x - 3$

$f(x) = 4x - 3$

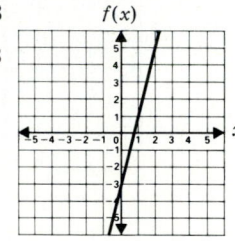

13. $y = 2x - 3$

$f(x) = 2x - 3$

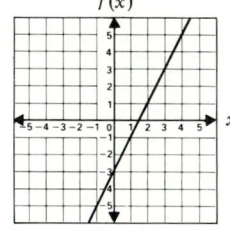

14. $y = 2x + 3$

$f(x) = 2x + 3$

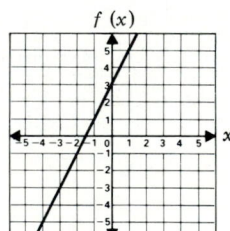

15. $y = -3x - 2$

$f(x) = -3x - 2$

16. $y = 2x + 4$

$f(x) = 2x + 4$

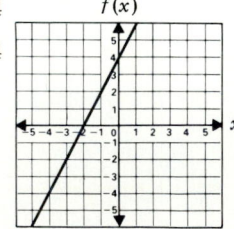

17. $y = \frac{5}{2}x + \frac{3}{2}$

$f(x) = \frac{5}{2}x + \frac{3}{2}$

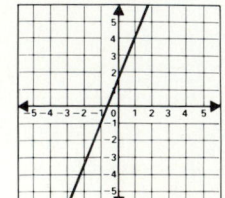

18. $y = 4x - \frac{4}{3}$

$f(x) = 4x - \frac{4}{3}$

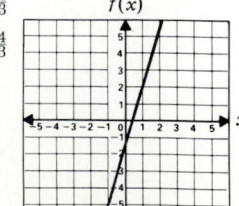

19. $y = \frac{3}{5}x + \frac{1}{5}$

$f(x) = \frac{3}{5}x + \frac{1}{5}$

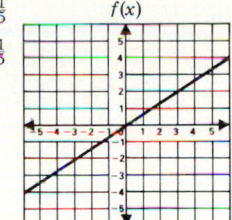

20. $y = \frac{2}{7}x - \frac{4}{7}$

$f(x) = \frac{2}{7}x - \frac{4}{7}$

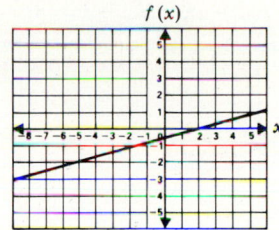

Section 8.5A

1. $y > x - 4$

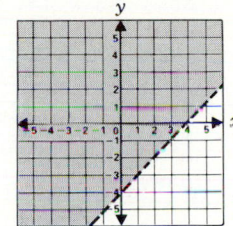

2. $y < x + 5$

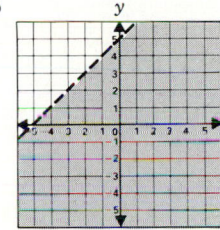

3. $y > 2x$

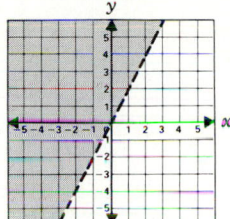

4. $y < -5x$

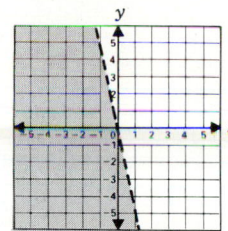

5. $y > 3x - 1$

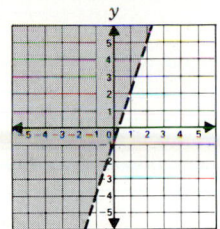

6. $y < 2x + 1$

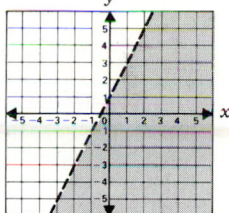

7. $y > -4x + 2$

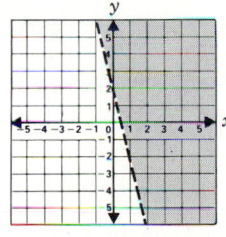

8. $y < -x + 4$

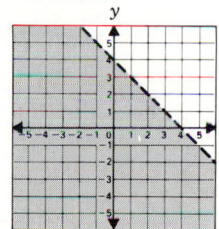

9. $y < -3x - 4$

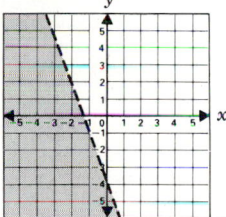

10. $y < -x - 3$

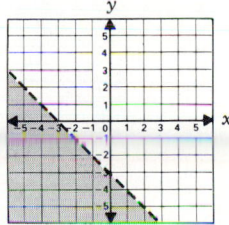

8.5B

11. $y \geq x - 2$

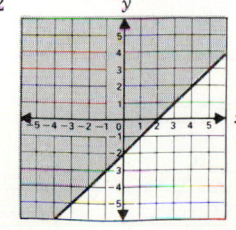

12. $y \leq x + 3$

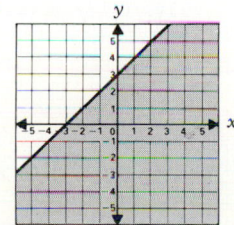

13. $y \geq 3x$

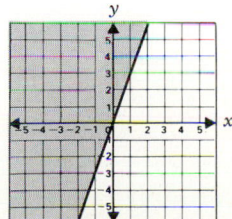

14. $y \leq -2x$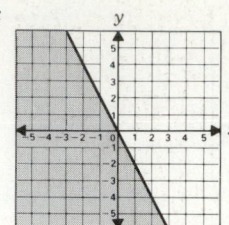

15. $B \leq 5A - 2$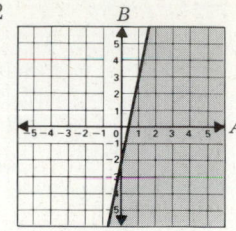

16. $M \geq 3N + 1$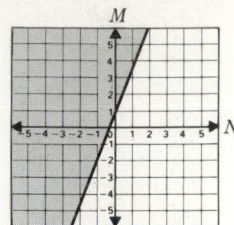

17. $y \leq -2x + 3$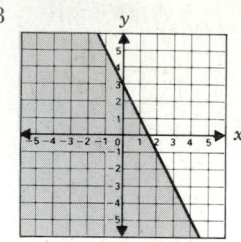

18. $y \geq 2x - 2$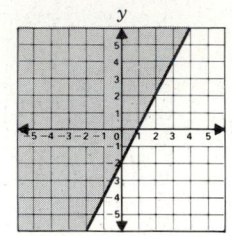

19. $k > 4n - 1$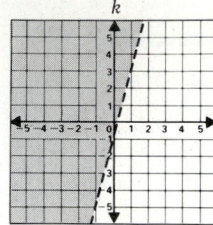

20. $r < 4p + 4$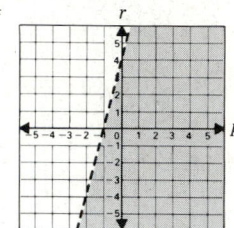

21. $y \leq 3x - 3$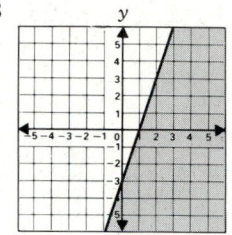

22. $y > -x + 2$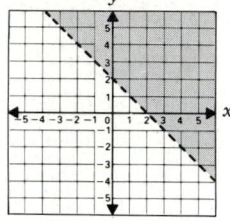

23. $y < 6x - 8$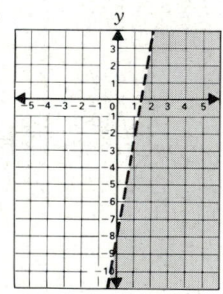

24. $y \geq -4x - 5$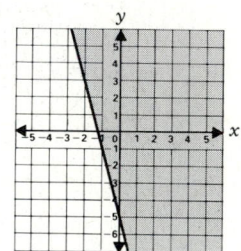

25. $y < 1$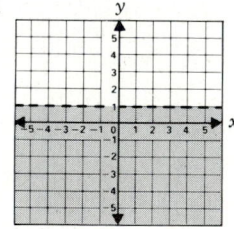

26. $y \geq -2$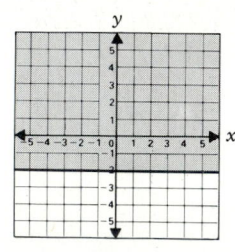

27. $x \geq -6$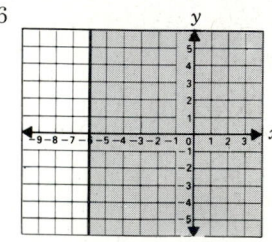

28. $x < 7$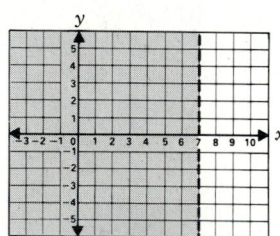

Section 8.6A　**1.** $(0, 2)$　**2.** $(0, 0)$　**3.** $(0. -5)$　**4.** $(0, -1)$　**5.** $(0, 12)$　**6.** $(0, 10)$　**7.** $(0, -11)$
8. $(0, 9)$　**9.** $(0, 4)$　**10.** $(0, -4)$　**11.** $(0, 0)$　**12.** $(0, -8)$

13. $(0, 5)$　$y = 3x + 5$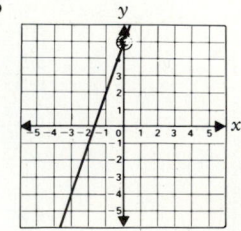

14. $(0, 1)$　$y = -2x + 1$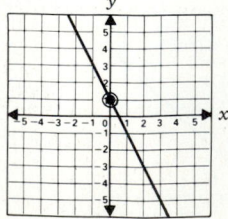

15. $(0, -4)$ $y = 7x - 4$

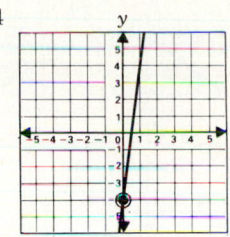

16. $(0, -5)$ $y = x - 5$

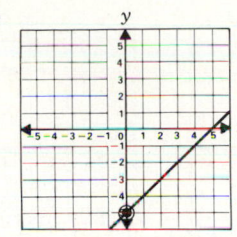

8.6B   **17.** $(0, -1)$ **18.** $(0, 3)$ **19.** $(0, -7)$ **20.** $(0, 0)$ **21.** $(0, 0)$ **22.** $(0, 0)$ **23.** $(0, 0)$ **24.** $(0, 0)$
25. $(0, 9)$ **26.** $(0, -5)$

8.6C   **27.** $(2, 0)$ **28.** $(-4, 0)$ **29.** $(0, 0)$ **30.** $(5, 0)$ **31.** y-int. $(0, 2)$; x-int. $(3, 0)$ **32.** y-int. $(0, -2)$;
x-int. $(2, 0)$ **33.** y-int. $(0, 4)$; x-int. $(-1, 0)$ **34.** y-int. $(0, 4)$; x-int. $(3, 0)$ **35.** y-int. $(0, 0)$;
x-int. $(0, 0)$ **36.** y-int. $(0, -2)$; x-int. $(-2, 0)$

8.6D   **37.** $(0, 0)$ **38.** $(0, 0)$ **39.** $(0, 0)$ **40.** $(0, 0)$ **41.** $(-5, 0)$ **42.** $(3, 0)$ **43.** $(7, 0)$ **44.** $(2, 0)$
45. $(5, 0)$ **46.** $(-3, 0)$ **47.** $(3, 0)$ **48.** $(2/3, 0)$ **49.** $(-5/2, 0)$

Section 8.7A   **1.** 1   **2.** 1   **3.** -4 **4.** $-1/3$ **5.** 4   **6.** 1   **7.** 2   **8.** $-1/2$ **9.** $-1/2$ **10.** $-3/8$

8.7B   **11.** 2   **12.** 3   **13.** 4   **14.** -2 **15.** 1   **16.** 1   **17.** $1/2$ **18.** $5/3$ **19.** 3   **20.** -4 **21.** 1   **22.** 2
23. -5 **24.** -1

8.7C

25.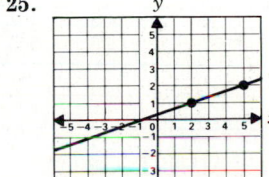

26.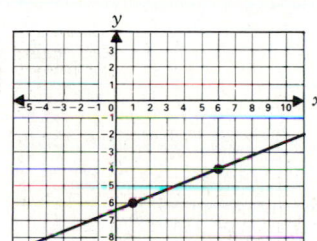

27.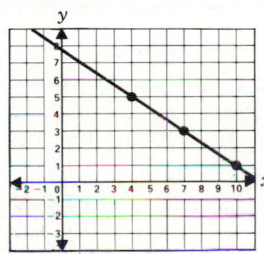

28.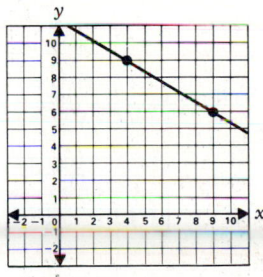

29.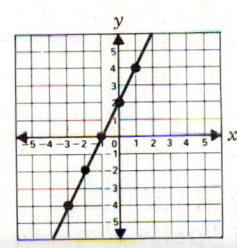

30.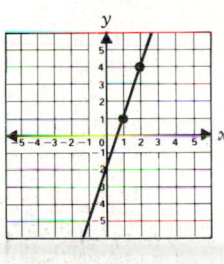

31.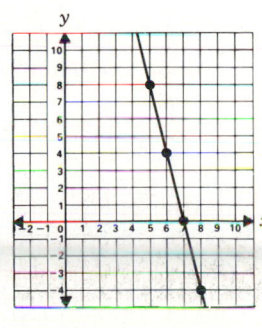

32.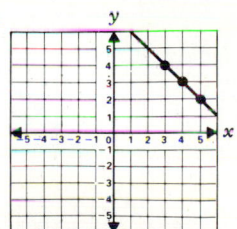

33.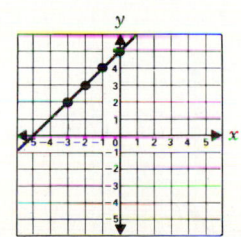

34.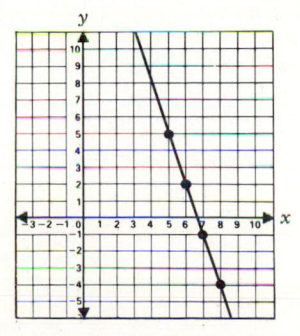

35.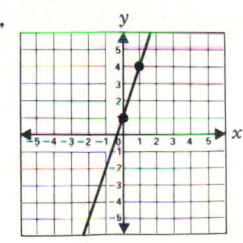

36. 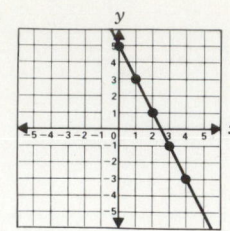 **37.** 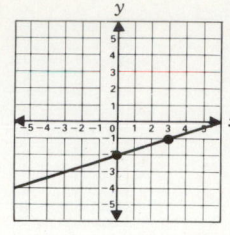 **38.** 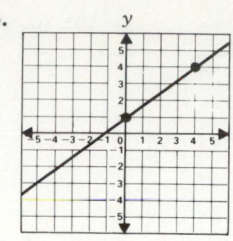 **39.**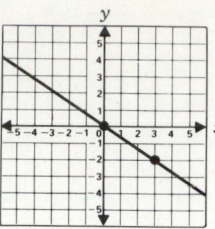

40.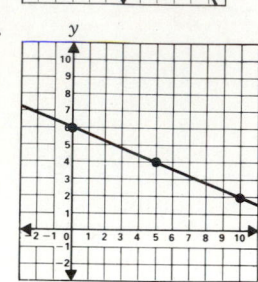

8.7D **41.** 2 **42.** −5 **43.** 0 **44.** No slope **45.** −4 **46.** 3 **47.** −1 **48.** 3 **49.** −1 **50.** 2

Section 8.8A **1.** Slope-int. form **2.** Not slope-int. form **3.** Not slope-int. form **4.** Slope-int. form
5. Not slope-int. form **6.** Slope-int. form

8.8B **7.** $y = 5x + 9$; $(0, 9)$ **8.** $y = 5x − 7$; $(0, −7)$ **9.** $y = −4x + 5$; $(0, 5)$ **10.** $y = 4x + 3$; $(0, 3)$
11. $y = −3x + 4$; $(0, 4)$ **12.** $y = −6x + 3$; $(0, 3)$ **13.** $y = −x − 11$; $(0, −11)$ **14.** $y = (4/5)x + 3/5$;
$(0, 3/5)$ **15.** $y = (3/2)x − 7/2$; $(0, −3\,1/2)$ **16.** $y = −(4/3)x − 1/3$; $(0, −1/3)$

8.8C **17.** $y = 2x − 5$; 2 **18.** $y = −x + 7$; −1 **19.** $y = x − 3$; 1 **20.** $y = 2x + 11$; 2
21. $y = 3x − 2$; 3 **22.** $y = −2x − 5$; −2 **23.** $y = −4x + 2$; −4 **24.** $y = −3x + 5$; −3
25. $y = 7x + 20$; 7 **26.** $y = 5x + 14$; 5

8.8D **27.** $y = 3x − 5$; slope 3; y-int. $(0, −5)$ **28.** $y = 3x − 4$; slope 3; y-int. $(0, −4)$
29. $y = 2x + 7$; slope 2; y-int. $(0, 7)$ **30.** $y = −x + 9$; slope −1; y-int. $(0, 9)$ **31.** $y = −7x + 11$; slope
−7; y-int. $(0, 11)$ **32.** $y = 2x − 5$; slope 2; y-int. $(0, −5)$ **33.** $y = −2x − 3$; slope −2; y-int. $(0, −3)$
34. $y = −3x + 7$; slope −3; y-int. $(0, 7)$ **35.** $y = −3x − 4$; slope −3; y-int. $(0, −4)$
36. $y = 4x − 15$; slope 4; y-int. $(0, −15)$ **37.** $y = (4/3)x + 1/3$; slope 4/3: y-int. $(0, 1/3)$
38. $y = (3/2)x + 7/2$; slope 3/2; y-int. $(0, 3\tfrac{1}{2})$ **39.** $y = (2/5)x + 6/5$; slope 2/5; y-int. $(0, 1\tfrac{1}{5})$
40. $y = (1/2)x + 3/2$; slope 1/2; y-int. $(0, 1\tfrac{1}{2})$

8.8E

41. $y = 3x − 3$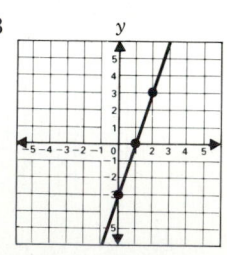

42. $y = 2x + 1$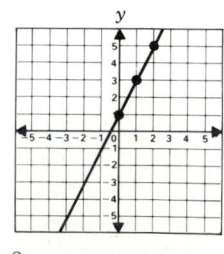

43. $y = −2x + 4$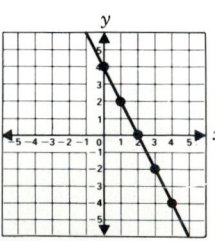

44. $y = −4x + 7$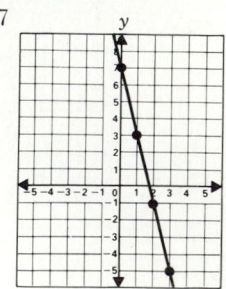

45. $y = x − 3$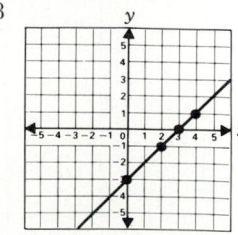

46. $y = x + 1$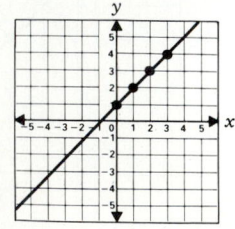

47. $y = -x$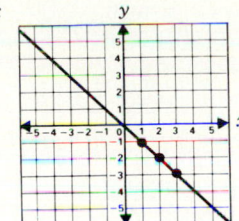

48. $y = 3x$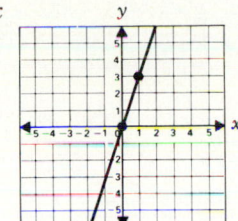

49. $y = \frac{1}{3}x - 2$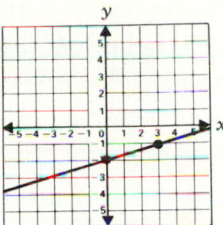

50. $y = \frac{2}{3}x + 1$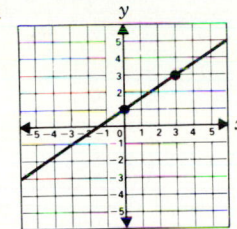

51. $y = -\frac{3}{4}x$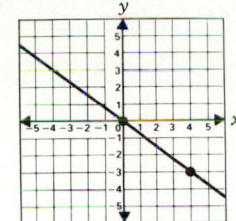

52. $y = -\frac{2}{3}x + 5$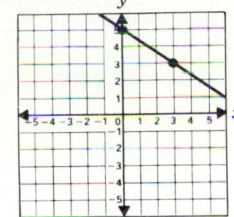

53. $y = \frac{2}{5}x$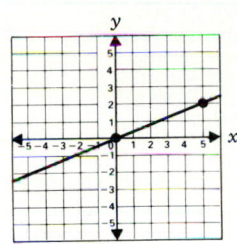

54. $y = \frac{1}{2}x - 3$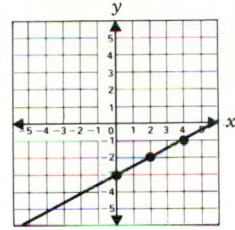

Section 8.9A

1. $y = x - 5$: $y = x + 4$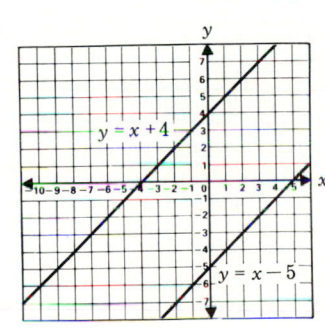

2. $y = -3x + 1$; $y = -3x + 4$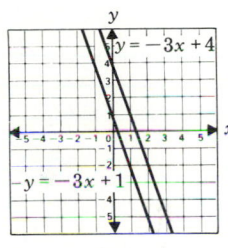

3. $y = 4x - 5$; $y = 4x - 1$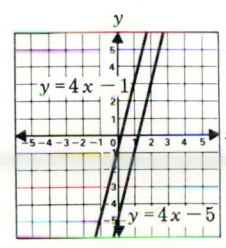

4. $y = 2x - 3$; $y = 2x + 3$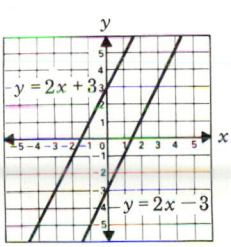

8.9B **5.** Not parallel **6.** Parallel **7.** Not parallel **8.** Parallel **9.** Not parallel **10.** Not parallel

Section 8.10A **1.** $y = 6x + 2$ **2.** $y = -4x$ **3.** $y = -x + 3$ **4.** $y = 5x - 1$ **5.** $y = -3x + 3$
6. $y = 7x - 5$ **7.** $y = 9x + 11$ **8.** $y = (2/3)x + 1/3$ **9.** $y = 0.7x + 0.04$ **10.** $y = (3/5)x - 1/4$

8.10B **11.** $y = 2x - 1$ **12.** $y = -5x + 12$ **13.** $y = 3x + 10$ **14.** $y = -x - 6$ **15.** $y = 7x + 18$
16. $y = x - 8$ **17.** $y = -4x - 19$ **18.** $y = -3x + 6$

8.10C **19.** $y = 2x$ **20.** $y = -3x + 3$ **21.** $y = -x + 2$ **22.** $y = -4x + 5$ **23.** $y = 6x - 2$ **24.** $y = 7x - 4$
25. $y = -4x - 1$ **26.** $y = 3x - 3$ **27.** $y = (4/3)x + 1$ **28.** $y = -2x + 6$ **29.** $y = (7/3)x - 3$

8.10D   **30.** $y = 3x - 5$ **31.** $y = -3x + 16$ **32.** $y = x - 5$ **33.** $y = 3x + 4$ **34.** $y = x + 10$
35. $y = -4x - 13$ **36.** $y = 3x + 22$ **37.** $y = x - 4$

8.10E   **38.** $y = 3x - 11$ **39.** $y = 5x - 1$ **40.** $y = -2x + 14$ **41.** $y = -4x - 2$ **42.** $y = -x - 4$
43. $y = -3x + 8$

8.10F   **44.** $y = x - 4$ **45.** $y = -(2/3)x + 4/5$ **46.** $y = -(1/4)x - 1$ **47.** $y = -5x + 14$ **48.** $y = 4x - 5$
49. $y = -11x - 15$ **50.** $y = -3x$ **51.** $y = x + 6$

Chapter 8 Review

1.
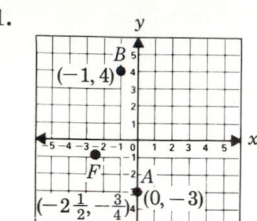

2. $C(4, 0)$; $D(-1, 2)$ **3.** No   **4.** $y = -3x + 1$
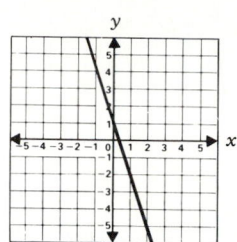

5. $y = \frac{1}{2}x - 5$
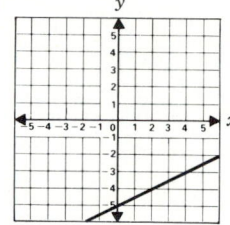

6. $y = \frac{2}{3}x - \frac{1}{4}$
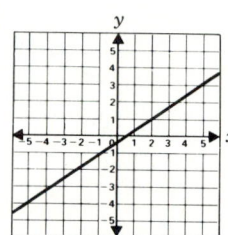

7. $y = -4$
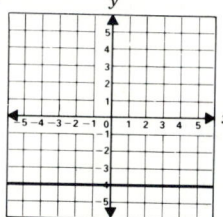

8. $x = 3$
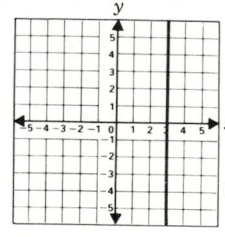

9. $y = 2x + 4$
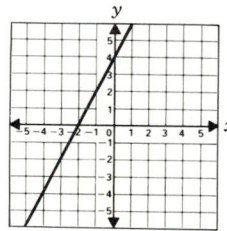

10. $f(x) = 3x - 5$
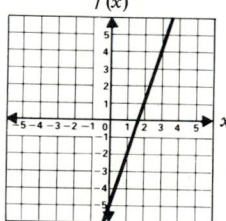

11. $y < x - 4$
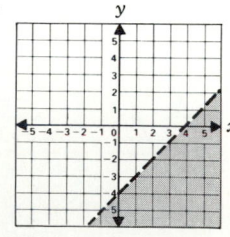

12. $y \geq 2x + 3$
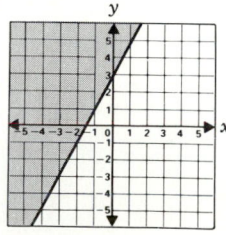

13. $y \geq 3$
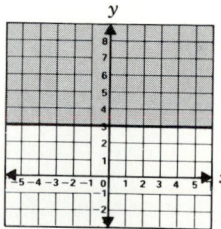

14. $x < 1$
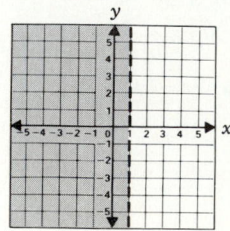

15. y-int. $(0, 4)$; x-int. $(-6, 0)$ **16.** x-int. $(4, 0)$; y-int. $(0, -12)$ **17.** 4   **18.** 3

19. -7 **20.** 0 **21.** No slope   **22.**

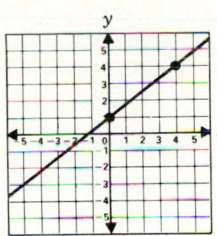

23. Slope 4; y-int. $(0, -3)$ **24.** Slope 5; y-int. $(0, 4)$

25. $y = 4x - 3$

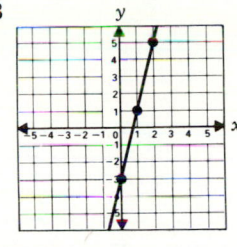

26. Not parallel   **27.** Parallel   **28.** $y = -7x + 1$ **29.** $y = -x - 3$

30. $y = 4x - 7$ **31.** $y = 4x - 3$ **32.** $y = -3x - 2$ **33.** $y = -2x$

Supplementary Exercises

1.

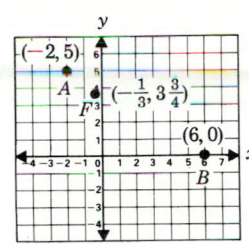

2. $C(2, -5)$; $D(-4, 0)$ **3.** No   **4.** $y = -2x - 1$

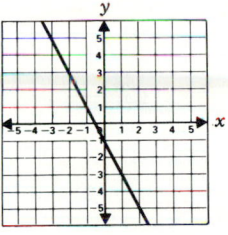

5. $y = \frac{1}{3}x - 2$

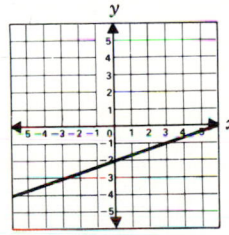

6. $y = \frac{1}{4}x + \frac{2}{3}$

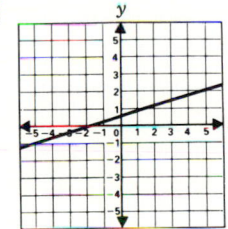

7. $y = 1$

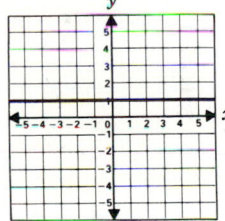

8. $x = -7$

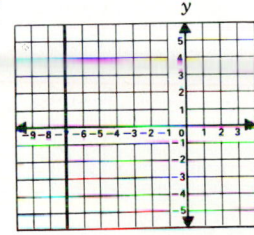

9. $y = 3x + 5$

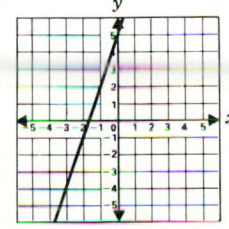

10. $f(x) = -x + 4$

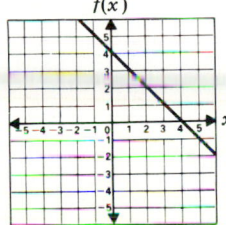

11. $y > x - 2$

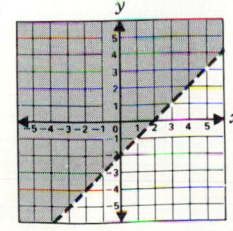

12. $y \leq 3x + 5$

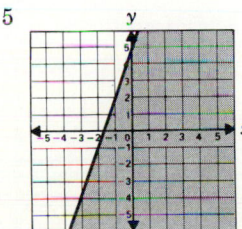

13. $x \geq 6$

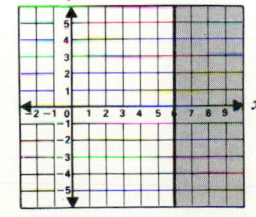

14. $x < -4$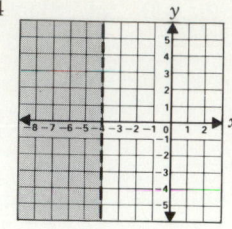

15. y-int. $(0, 2)$; x-int. $(-2, 0)$ **16.** x-int. $(7, 0)$; y-int. $(0, -35)$ **17.** -2

18. 2 **19.** -11 **20.** 0 **21.** No slope **22.** 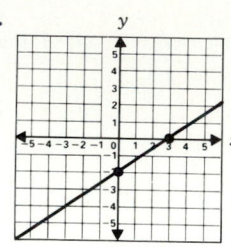 **23.** Slope -7; y-int. $(0, 2)$

24. Slope 4; y-int. $(0, 5)$ **25.** $y = 3x - 5$ 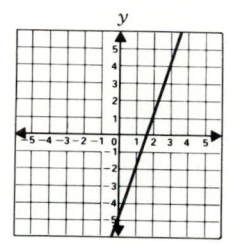 **26.** Not parallel **27.** Parallel

28. $y = -4x + 5$ **29.** $y = x - 4$ **30.** $y = -x + 8$ **31.** $y = -5x + 25$ **32.** $y = -6x - 2$ **33.** $y = -2x + 1$

Chapter 8 Test

1.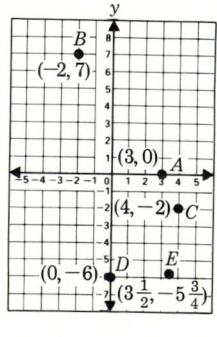

2. $F(8, 3)$; $G(-7, 2)$; $H(5, 0)$; $J(-3, -8)$ **3.** No **4.** Yes

5. $y = 4x - 5$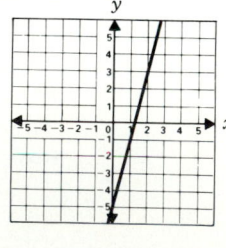

6. $y = -x + 4$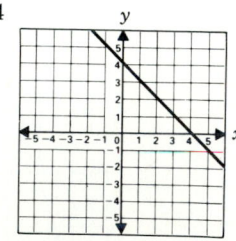

7. $y = \frac{3}{4}x - 1$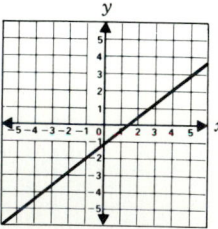

8. $y = -\frac{1}{3}x + \frac{2}{5}$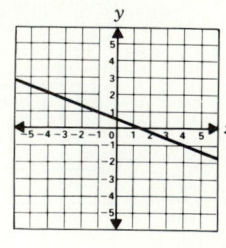

9. $y = -7$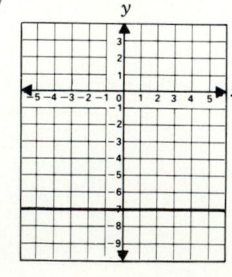

10. $x = 6$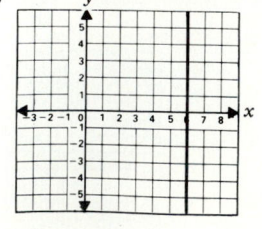

11. $B = -4A + 3$

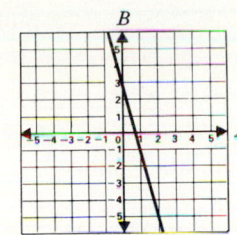

12. $f(x) = -\frac{3}{5}x$

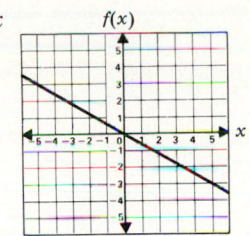

13. $y = 8x + 7$

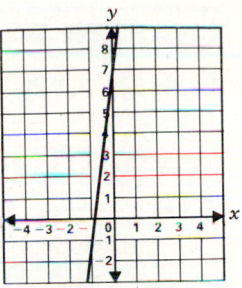

14. $y = (6/7)x - (10/7)$

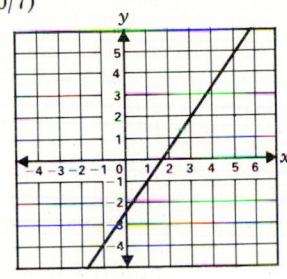

15. $y > 3x - 5$

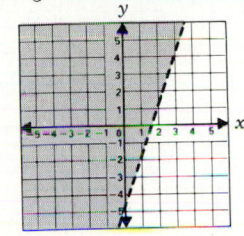

16. $y \le -2x + 3$

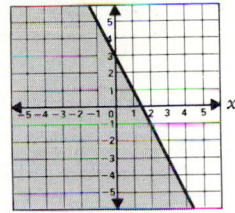

17. $x \ge -5$

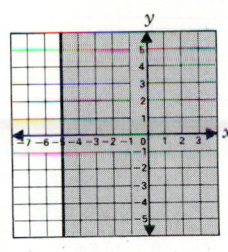

18. $y < 8$

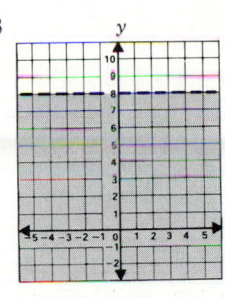

19. y-int. $(0, 1)$; x-int. $(1, 0)$

20. Slope 4; y-int. $(0, -12)$; x-int. $(3, 0)$ **21.** Slope 5; y-int. $(0, 2)$; x-int. $(-2/5, 0)$ **22.** No slope;   no y-int; x-int. $(-3, 0)$ **23.** Slope 0: y-int. $(0, 2)$; no x-int.   **24.** 2   **25.** -1 **26.** 3/4   **27.** $-3/2$

28.

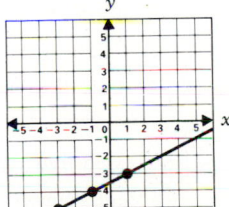

29. No   **30.** $y = 3x - 6$ **31.** $y = -5x$ **32.** $y = 2x - 2$ **33.** $y = 2x + 3$

34. $y = 4x - 1$ **35.** $y = x - 2$

Cumulative Review
1. -22 **2.** 2/3   **3.** 1/4   **4.** 10.791   **5.** $9y^5 - 9y^3 + 5y$ **6.** $-4.2a - 12.7a^2 + 6.1a^3$

7. $5bc - 7ac - 3ab$ **8.** $10x - 8y + 9w$ **9.** $4x^3 - 26x^2 + 15x$ **10.** 499.8   **11.** -18 **12.** 14 sq ft   **13.** $y = 2$

14. $w = -5$ **15.** $a = 1$ **16.** $a = 3$ **17.** $N = \$5,600$ **18.** $a = (d + 5bc)/3$ **19.** $k < 4$

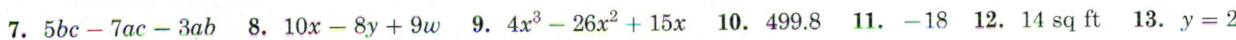

20. $3 \le w < 8$

21. 20 bills   **22.** $3y(x - 5w + 7z)$

23. $y = -2x - 2$

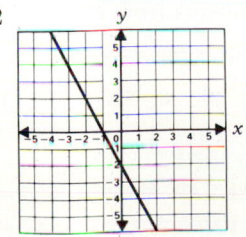

24. $y = 4x + 1$

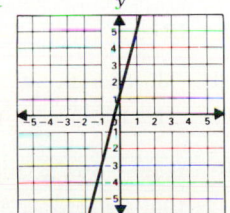

25. $f(x) \le 2x + 1$

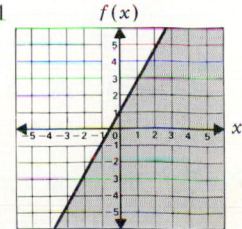

CHAPTER 9

Section 9.1A   **1.** $x + 5$ **2.** $x - 4$ **3.** $x + 7$ **4.** $x - 11$ **5.** $x - 123$ **6.** $x + 86$ **7.** $5x$ **8.** $16x$
9. $(1/2)x$ or $x/2$ **10.** $(1/3)x$ or $x/3$ **11.** $(1/5)x$ or $x/5$ **12.** $(1/15)x$ or $x/15$ **13.** $2x + 3$ **14.** $2x - 5$
15. $5x + 18$ **16.** $4x - 1$ **17.** $(x + 1)/2$ **18.** $(x - 7)/3$ **19.** $(3x + 2)/5$ **20.** $(2x - 5)/8$

9.1B

	Table I		Table II		Table III	
	Fred	**Marge**	**Fred**	**Marge**	**Fred**	**Marge**
21.	$x + 9$	$x - 1$	$x - 6$	$x - 16$	$x + 19$	$x + 9$
22.	$x - 2$	$x + 8$	$x - 17$	$x - 7$	$x + 8$	$x + 18$
23.	$x + 1$	$x + 6$	$x - 14$	$x - 9$	$x + 11$	$x + 16$
24.	$x - 3$	$x + 2$	$x - 18$	$x - 13$	$x + 7$	$x + 12$
25.	$2x$	$2x - 1$	$2x - 15$	$2x - 16$	$2x + 10$	$2x + 9$

9.1C

	Table I Equation	Table II Equation	Table III Equation
21.	$(x) + (x + 9) + (x - 1) = 108$	$(x - 15) + (x - 6) + (x - 16) = 63$	$(x + 10) + (x + 19) + (x + 9) = 138$
22.	$(x) + (x - 2) + (x + 8) = 108$	$(x - 15) + (x - 17) + (x - 7) = 63$	$(x + 10) + (x + 8) + (x + 18) = 138$
23.	$(x) + (x + 1) + (x + 6) = 108$	$(x - 15) + (x - 14) + (x - 9) = 63$	$(x + 10) + (x + 11) + (x + 16) = 138$
24.	$(x) + (x - 3) + (x + 2) = 108$	$(x - 15) + (x - 18) + (x - 13) = 63$	$(x + 10) + (x + 7) + (x + 12) = 138$
25.	$(x) + (2x) + (2x - 1) = 108$	$(x - 15) + (2x - 15) + (2x - 16) = 63$	$(x + 10) + (2x + 10) + (2x + 9) = 138$

Section 9.2A   **1.** Chris (x), Holly $(x + 4)$; $x + (x + 4) = 48$; Chris is 22; Holly is 26.   **2.** Maria (x), Carlos $(x + 8)$; $x + (x + 8) = 90$; Maria is 41; Carlos is 49.   **3.** Harriet (x), Rosa $(x + 3)$; $(x + 5) + (x + 8) = 49$ Harriet is 18; Rosa is 21.   **4.** David (x), Bob $(x + 4)$; $(x - 10) + (x - 6) = 64$ David is 40; Bob is 44.   **5.** Margie (x), Sal $(x + 5)$; $(x + 4) + (x + 9) = 69$ Margie is 28; Sal is 33.   **6.** Ann (x) Phil $(2x)$; $(x + 2) + (2x + 2) = 100$ Ann is 32; Phil is 64.   **7.** Sister (x), Barbara $(x + 2)$; $(x - 1) + (x + 1) = 54$; Sister is 27; Barbara is 29.   **8.** Judy (x), Jane $(x + 3)$; $(x + 12) + (x + 15) = 55$ Judy was 11; Jane was 14.   **9.** Mabel (x), Archie $(x + 7)$; $(x + 27) = 2(x - 3)$ Mabel is 33; Archie is 40.   **10.** Tim (x), Nick $(x + 5)$; $(x + 12) = 2(x - 3)$ Tim is 18; Nick is 23; Nick will be 28 and Tim will be 23 in 5 years.   **11.** Adam (x), Jill $(2x)$, Jesse $(2x + 4)$; $(x + 2) + (2x + 2) + (2x + 6) = 100$; Adam is 18 *now*; he was 13 five *years ago*. Jill is 36 *now*; she was 31 five *years ago*. Jesse is 40 *now*; he was 35 five *years ago*.   **12.** Sonya (x), Luke $(x + 6)$, Rick $(2x - 10)$; $(x + 3) + (x + 9) + (2x - 7) = 105$; Sonya is 25 *now*; she will be 35 *in ten years*. Luke is 31 *now*; he will be 41 *in ten years*. Rick is 40 *now*; he will be 50 *in ten years*.

9.2B   **13.** 12 nickels, 13 dimes, 8 quarters   **14.** 1 nickel, 3 dimes, 2 quarters   **15.** 8 pennies, 12 nickels, 7 quarters, 13 dimes   **16.** 5 pennies, 15 nickels, 5 dimes, 16 quarters   **17.** 7 nickels, 14 dimes, 16 quarters   **18.** 1 ft; 2 ft; 5 ft   **19.** Rachel's 3 min; Zale's 5 min; Jonah's 10 min   **20.** Monday, 52 min; Tuesday, 40 min; Wednesday, 120 min (or 2 hr)   **21.** Monday 46; Tuesday 23; Wednesday 64   **22.** November 60; December 67; January 110   **23.** 80 in Italy; 160 in Africa; 145 in California   **24.** 7 Friday; 10 Saturday; 20 Sunday

9.2C   **25.** 15   **26.** 28   **27.** 13   **28.** 140,000   **29.** $250   **30.** $150 per stock owner; $300 per employee   **31.** 15   **32.** 20   **33.** 447 mi   **34.** $1.00 per 3×5; $2.50 per 5×7 **35.** 45 cents from negatives; $1.00 from slides   **36.** $2.50 per 5×7; $4.95 per 8×12

Section 9.3A   **1.** $40 - 18$ **2.** $50 - 34$ **3.** $83 - 6$ **4.** $91 - 25$ **5.** $67 - 49$ **6.** $100 - 74$ **7.** $135 - 88$
8. $280 - 179$ **9.** $100 - x$ **10.** $134 - y$ **11.** $150 - w$ **12.** $187 - n$ **13.** $340 - x$ **14.** $400 - y$

9.3B   **15.** $x, 30 - x$; $3x - (30 - x) = 10$; 10, 20   **16.** $x, 50 - x$; $x - 4(50 - x) = 25$; 45, 5   **17.** $x, 50 - x$; $x - 2(50 - x) = 2$; 34, 16   **18.** $x, 100 - x$; $100 - x = 5x - 20$; 20, 80   **19.** $x, 100 - x$; $100 - x = 3x - 8$; 27, 73

20. $x, 100 - x$; $100 - x = 4x - 10$; 22, 78   **21.** $x, 50 - x$; $(50 - x) - 5x = 2$; 8, 42   **22.** $x, 43 - x$; $2x = 3(43 - x) - 24$; 21, 22   **23.** $n, 38 - n$; $10(38 - n) + 2n = 100$; 35 nickels, 3 quarters   **24.** $c, 17 - c$; $3c - (17 - c) = 11$; 7 cats, 10 dogs   **25.** $x, 65 - x$; $2x + 3(65 - x) = 172$; 23 first week, 42 second week

Section 9.4A   **1.** 6 nickels, 8 dimes, 5 quarters   **2.** 10 nickels, 4 dimes, 9 quarters   **3.** 9 pennies, 5 dimes, 2 quarters   **4.** 6 pennies, 3 nickels, 12 dimes   **5.** 7 nickels, 8 dimes, 3 quarters   **6.** 11 nickels, 11 dimes, 6 quarters   **7.** 4 pennies, 2 nickels, 9 dimes, 3 quarters   **8.** 6 pennies, 12 nickels, 13 dimes, 3 quarters   **9.** 8 lb granola, 5 lb coconut   **10.** 2 oz red, 10 oz white   **11.** 5 lb thin noodles, 4 lb thick noodles   **12.** 3 lb M & Ms, 5 lb sesame seeds   **13.** 7 oz lemonade, 3 oz orange juice   **14.** 2 lb shrimp, 4 lb sardines   **15.** 10 oz strawberries, 40 oz bananas   **16.** 6 oz strawberries, 4 oz bananas   **17.** 10 gal of each   **18.** 8 lb beef, 2 lb pork   **19.** \$350 at 2%, \$650 at 8%   **20.** \$3,200 at 4%, \$800 at 9%   **21.** \$1,550 at 3%, \$2,450 at 7%   **22.** \$1,180 at 3% and 5%, \$3,640 at 9%

9.4B   **23.** 10 g water   **24.** 6 g water   **25.** 16 g water   **26.** 3 g salt   **27.** 20 g water   **28.** 5 g salt   **29.** 20 g salt   **30.** 25 g water

Section 9.5A   **1.** 144 mi   **2.** 304 mi   **3.** 2,000 ft   **4.** 4 mi   **5.** 14 mi   **6.** 97.52 mi   **7.** 9 mi   **8.** 300 ft

9.5B   **9.** 7 hr   **10.** 495 mi   **11.** John 3 hr, Rose 6 hr   **12.** 420 mi   **13.** 160 mi   **14.** 126 mi   **15.** 4 hr   **16.** 6 hr   **17.** Sarah 2 hr, Clara 3 hr   **18.** 4 hr   **19.** 3.2 hr   **20.** 37.1 mi

Chapter 9 Review   **1.** $2x - 7$ **2.** (a) Mike $x + 7$, Rick $x + 4$, Ari $x - 3$; (b) Lisa $x + 8$, Mike $x + 15$, Rick $x + 12$, Ari $x + 5$; (c) Lisa $x - 3$, Mike $x + 4$, Rick $x + 1$, Ari $x - 6$; (d) $x + (x + 7) + (x + 4) + (x - 3) = 56$ (e) $(x + 8) + (x + 15) + (x + 12) + (x + 5) = 88$ (f) $(x - 3) + (x + 4) + (x + 1) + (x - 6) = 44$ **3.** Peter 3, Bob 6, Laura 13   **4.** 3 pennies, 6 dimes, 8 nickels, 5 quarters   **5.** 8   **6.** $100 - N$ **7.** 20 black and white, 120 color   **8.** 4 pennies, 7 nickels, 8 dimes, 6 quarters   **9.** 7 lb granola, 5 lb fruit   **10.** \$1,300 at 5%, \$700 at 8%   **11.** 6 g water   **12.** 152 mi   **13.** 150 mi

Supplementary Exercises   **1.** $3x + 4$ **2.** $2x - 8$ **3.** $2(x + 5)$ **4.** $x/2 + 6$ **5.** (a) Bill $2x - 6$, Ron $2x$; (b) Phil $x + 9$, Bill $2x + 3$, Ron $2x + 9$; (c) $(x + 9) + (2x + 3) + (2x + 9) = 120$ **6.** (a) Benita $x + 5$, Roberta $x - 9$; (b) JoAnn $x - 13$, Benita $x - 8$, Roberta $x - 22$; (c) $(x - 13) + (x - 8) + (x - 22) = 55$ **7.** 21, 28, 42   **8.** George 9, Mark 14, Susan 18   **9.** 3 quarters, 7 nickels, 2 dimes   **10.** 4   **11.** $78 - N$ **12.** 27 entrées, 57 desserts   **13.** 3 quarters, 6 nickels, 7 dimes   **14.** 4 patterned, 8 solid   **15.** 6 lb blueberries, 4 lb strawberries   **16.** \$290 at 4%, \$710 at 12%   **17.** 6 g salt   **18.** 10 g water   **19.** 6.8 mi   **20.** 3 hr   **21.** 240 mi

Chapter 9 Test   **1.** 34, 45, 90   **2.** Ken 21, Meg 26, Bill 25   **3.** 5 quarters, 10 pennies, 12 nickels, 13 dimes   **4.** 2 quarters, 7 pennies, 4 dimes, 13 nickels   **5.** 52   **6.** 18 work, 40 broken   **7.** 6 carnations, 9 daisies   **8.** \$1,200 at 2%, \$2,400 at 5%, \$1,400 at 9%   **9.** 5 g salt   **10.** 168 mi

Cumulative Review   **1.** $-6/5$ or $-1\frac{1}{5}$ **2.** $1\frac{11}{13}$ **3.** $-23\frac{5}{6}$ **4.** -26.223 **5.** $2:3$ **6.** $2ab - 2bc + 2ac - 6a$ **7.** $4.7w - 23.9w^2 + 0.66w^3$ **8.** $-30mn + 48ck + 26jp$ **9.** $-5ax + 7bx - 5x + 3ab - 27a$ **10.** \$767.20   **11.** 23   **12.** -4 **13.** $m = -3$ **14.** $y = 1\frac{1}{2}$ **15.** $w > -1$ **16.** $-3 \leq w < 8$ **17.** $p = 1,677$ **18.** $N = 7,800$ **19.** \$95.20   **20.** 225   **21.** Slope 4, y-int. $(0, -9)$ **22.** Slope 1, y-int. $(0, 2)$ **23.** Slope 0, y-int. $(0, 18)$ **24.** Slope 5, y-int. $(0, 3)$

25. $y = 3x - 5$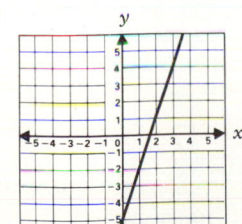

26. $f(x) = 4x$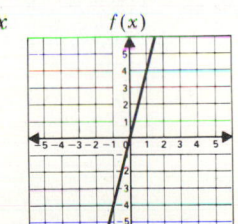

27. $x = -7$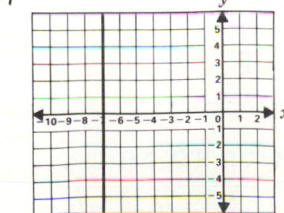

28. $y > 2x - 1$ 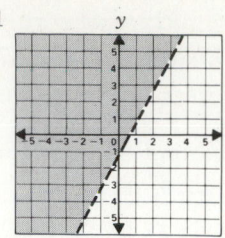 **29.** 14 cat food, 6 dog food **30.** 10 g salt

CHAPTER 10

Section 10.1 **1.** Yes **2.** No **3.** No **4.** No **5.** Yes **6.** Yes **7.** No **8.** Yes **9.** No **10.** Yes

Section 10.2 **1.** $(3, 7)$ 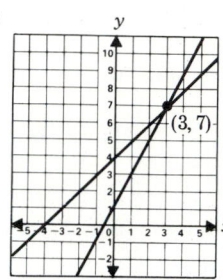 **2.** $(-2, -1)$ 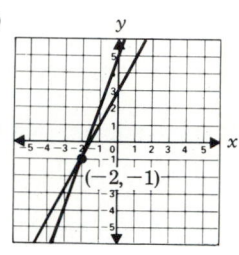 **3.** $(1, 3)$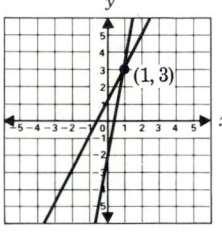

4. $(0, -3)$ 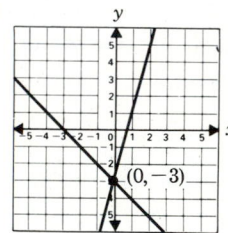 **5.** $(-2, -12)$ 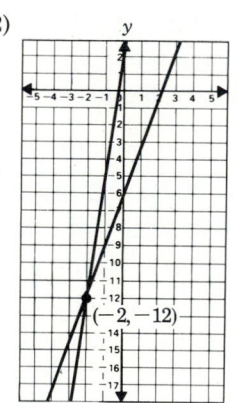 **6.** $(2, 2)$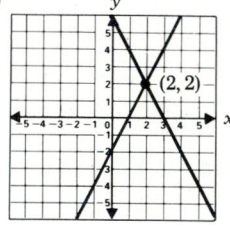

7. $(-1, -3)$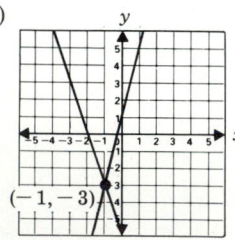

Section 10.3A **1.** $x = 2; y = 5$ **2.** $x = 7; y = 2$ **3.** $x = 6; y = 2$ **4.** $x = -5; y = 1$ **5.** $a = 7; b = 5$ **6.** $n = -3; m = -1$ **7.** $x = 2; y = 2$ **8.** $x = 1; y = 3$ **9.** $x = 8; y = -12$ **10.** $a = -3; w = 7$ **11.** $x = -5;$ $y = -1$ **12.** $x = 1; y = -2$ **13.** $x = 7; y = 2$ **14.** $x = 1; y = 3$ **15.** $x = -1; y = 3$ **16.** $x = -5; y = -4$

10.3B **17.** $x = 1/2; y = 6$ **18.** $x = 2; y = 1/3$ **19.** $x = 1/3; y = -2/3$ **20.** $x = 2/5; y = 2/5$ **21.** $x = -6;$ $y = -2/3$ **22.** $x = 1/3; y = -4$

Section 10.4A **1.** $x = 1; y = 2$ **2.** $x = 2; y = 1$ **3.** $x = -2; y = 1$ **4.** $x = 2; y = 5$ **5.** $x = 3; y = 8$ **6.** $x = 10; y = 5$ **7.** $x = 15; y = 10$ **8.** $x = 5; y = 3$ **9.** $x = -5; y = -15$ **10.** $x = 3; y = 5$

10.4B   **11.** $x = 2; y = 3$ **12.** $x = 4; y = -25$ **13.** $x = 3; y = -4$ **14.** $x = 2; y = 3$ **15.** $x = 3: y = -3$
16. $x = -1; y = 5$ **17.** $x = 1; y = 2$ **18.** $x = 3; y = -3$ **19.** $a = -1; b = 2$ **20.** $p = 2; m = 3$ **21.** $k = -2$;
$n = 5$ **22.** $m = 2; n = 3$

10.4C   **23.** $x = 2; y = 5$ **24.** $x = 3; y = 1$ **25.** $x = 1; y = 3$ **26.** $x = 2; y = -3$ **27.** $x = 1; y = 3$ **28.** $x = 3$;
$y = 5$ **29.** $x = 2; y = -5$ **30.** $x = 2; y = 4$ **31.** $x = 7; y = 7$ **32.** $x = 3; y = -2$

10.4D   **33.** $x = 2; y = 1$ **34.** $x = 1; y = 2$ **35.** $x = 3; y = 3$ **36.** $x = 5; y = 8$ **37.** $x = 2; y = 4$ **38.** $x = 2$;
$y = 4$ **39.** $x = 1; y = 1$ **40.** $x = 1/3; y = -2/3$ **41.** $x = 3/4; y = 1/3$ **42.** $x = 2/5; y = 2/3$ **43.** $x = 1/2$;
$y = 1/3$ **44.** $x = 1; y = 1/2$ **45.** $x = 3; y = -2$ **46.** $x = -3; y = 10$

Section 10.5   **1.** 18 dimes; 6 nickels   **2.** 3 oz pineapple: 7 oz lemonade   **3.** $100 at 6%; $400 at 5%
4. 3 albums; 7 blouses   **5.** 25 quarters; 35 nickels   **6.** 23 discos; 45 oldies   **7.** 9 ice cream bars; 27
popsicles   **8.** $3 hamburger; $1 hotdog   **9.** $5 Red Sox; $3 Cubs   **10.** $3 grandstand; $7 box seats
11. 75 cents ice cream cones; $1.40 sundaes   **12.** $1 singles; $5 albums   **13.** $1.80 short socks; $3.25 high
socks

Section 10.6

1. $y < x + 2$ and
   $y > -3$

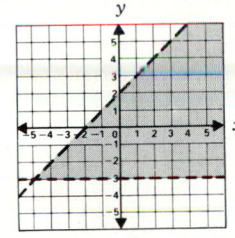

2. $y > 3x - 1$ and
   $y < 5$

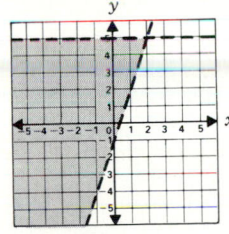

3. $y \geq 2x - 3$ and
   $y < 2$

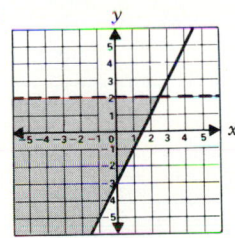

4. $y \leq -x + 3$ and
   $y \geq 2x - 5$

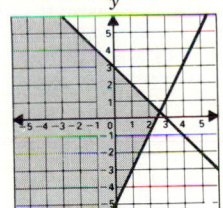

5. $y \leq -x + 3$;
   $y \geq 2x - 5$; and
   $x \geq -1$

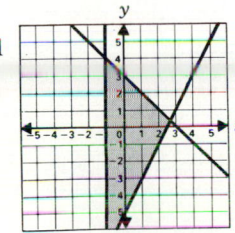

6. $y < x + 2$; $y > -3$;
   and $x < 5$

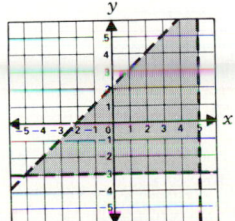

7. $y > 3x - 1$; $y < 5$;
   and $x > -3$

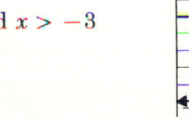

8. $x \geq -4$; $y \geq 2x - 3$;
   and $y < 2$

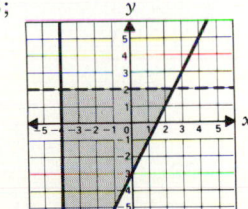

9. $y > 3x - 2; y \le -2x + 2$
and $x \ge -2$

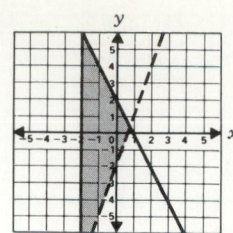

10. $y > x; y < -2x + 5;$
and $y \ge 4x - 5$

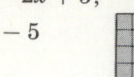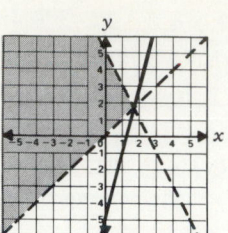

Chapter 10 Review 1. Yes 2. $(-3, 2)$

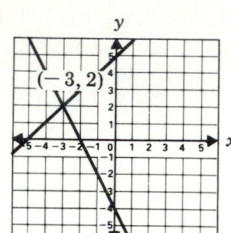

3. $x = 2; y = -1$ 4. $x = -4; y = -2$

5. $x = 3/20; y = 3/5$ 6. $x = 1; y = 1$ 7. $x = 2; y = -1$ 8. $n = -1; m = 2$ 9. $a = 3; b = -1$
10. $x = 1/2; y = 1/3$ 11. $x = 6; y = 5$ 12. $2,400 at 7\%; \$600 at 5\%$ 13. Hamburger $2/lb.; buns \$1/package
14. $y < 3x$ and
$x \le 5$

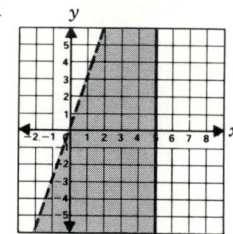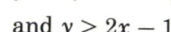

15. $y < 4; x > -3;$
and $y \ge 2x - 1$

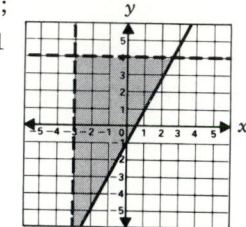

Supplementary Exercises 1. No. 2. Yes 3. Yes 4. No 5. $x = 1; y = 3$

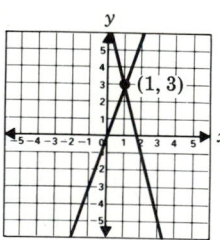

6. $x = -2; y = 2$

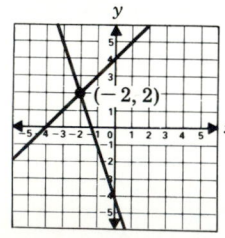

7. $x = -1; y = -2$

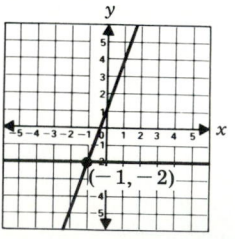

8. $x = -3; y = 2$

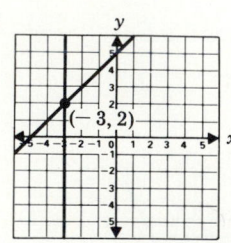

9. $x = -2; y = -8$ 10. $x = 6; y = 1$ 11. $n = -3; m = 0$

12. $x = -1; y = -3$ 13. $a = 2; b = 3/5$ 14. $x = 1; y = 3/4$ 15. $x = -2; y = 7$ 16. $x = 2; y = 1$ 17. $p = 1;$
$m = -2$ 18. $a = 2; b = 4$ 19. $x = 1; y = 2$ 20. $x = 2; y = 5$ 21. $x = 3/5; y = -1/2$ 22. $x = 3/4; y = 1/3$
23. Pillow cases 75 cents; socks 85 cents 24. $300 at 7\%; \$1,200 at 5\%$ 25. Labor $170; supplies \$85

26. $y < 6$ and $y \geq x$

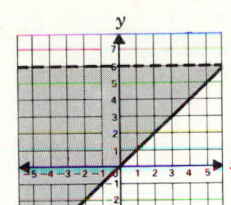

27. $x \leq 4$; $y > -2$; and $y < x + 5$

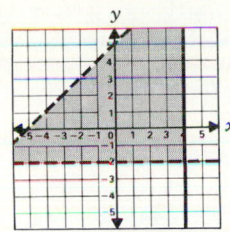

Chapter 10 Test   **1.** Yes   **2.** (a) Yes; (b) No   **3.** (5, 6)

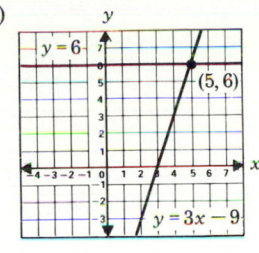

4. (4, 3)

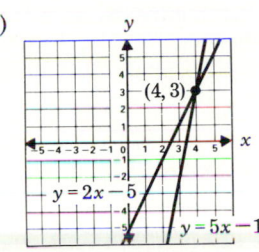

5. $x = 1$; $y = 2$ **6.** $x = -3$; $y = 7$ **7.** $x = 5$; $y = 2$ **8.** $x = 9$; $y = 2$

9. $a = 9/20$; $b = -1/5$ **10.** $x = -6$; $y = -13$ **11.** Rubber worms, 45 cents; real worms can, \$1.25   **12.** 15 field goals; 7 free throws   **13.** $y > 3x - 5$ and $y \leq 4$

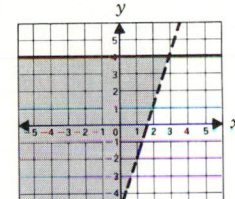

14. $x > -3$; $y \geq 0$; and $y \leq -2x$

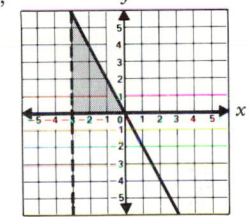

Cumulative Review   **1.** 5/12   **2.** 4/3 or $1\frac{1}{3}$ **3.** 3.0349   **4.** 0.0492   **5.** 0.057   **6.** 10.848   **7.** -179
8. 30   **9.** -4 **10.** -8 **11.** $3k^2 + 16k - 22$ **12.** $-6x^3 + 10x^2 - 8x$ **13.** 5   **14.** 144
15. $42a^2 - 27ab + 7a$ **16.** $p = (7k - 4m)/5n$ **17.** $x = 4$ **18.** $x = 15/4$ or $3\frac{3}{4}$ **19.** $x = 2$; $y = 1$ **20.** $x = 9$; $y = 2$ **21.** Slope -5; y-int. (0, 7)   **22.** Slope 3/7; y-int. $(0, 1\frac{1}{7})$ **23.** $y = -7x + 4$ **24.** $y = 6x - 10$
25. 32%   **26.** 259   **27.** (a) 270 mi; (b) \$10.80   **28.** 4 quarters; 7 dimes   **29.** Not parallel   **30.** Yes; parallel

CHAPTER 11

Section 11.1A   **1.** $2xxxx$ **2.** $5yyy$ **3.** $7aaabbbbb$ **4.** $5abbbbcc$ **5.** $49aabb$ **6.** $4ppmm$ **7.** $16ww$
8. $yyyyzzzz$ **9.** $-2abb$ **10.** $-3xxyyy$ **11.** $1/(yyyy)$ **12.** $1/(wwwwwww)$ **13.** $(xxx)/(yyyyy)$
14. $(5ad)/(bbbcc)$ **15.** $(6bbcddd)/a$ **16.** $(2xxxy)/(wwzzz)$ **17.** $(4x)/(yyy)$ **18.** $(7bddd)/(cc)$ **19.** 216
20. 1   **21.** -49 **22.** -16 **23.** -8 **24.** 49   **25.** 1/(16)   **26.** 1/2   **27.** 1   **28.** 1   **29.** $-1/(27)$
30. 1/(25)   **31.** 1/(216)   **32.** $-1/8$ **33.** $-1/8$ **34.** -1

11.1B   **35.** 8   **36.** 9   **37.** 3/2 or $1\frac{1}{2}$ **38.** 25/9 or $2\frac{7}{9}$ **39.** 25/4 or $6\frac{1}{4}$ **40.** $-(27)/8$ or $-3\frac{3}{8}$ **41.** (49)/9 or $5\frac{4}{9}$
42. 25   **43.** (16)/9 or $1\frac{7}{9}$ **44.** (25)/9 or $2\frac{7}{9}$

Section 11.2A   1. a^5 2. y^8 3. m^{14} 4. m^6 5. x^{13} 6. k^{16} 7. x^8 8. c^9 9. n^9 10. y^6 11. j^{11}
12. p^3 13. x^{133} 14. a^{112} 15. w^{94} 16. n^{419} 17. a^7 18. $1/(b^2)$ 19. $1/(a^4)$ 20. b^4 21. p^{11}
22. $1/(k^{98})$ 23. $1/(n^{192})$ 24. x^{94} 25. $1/(y^{228})$ 26. a^{318} 27. a^2b^3 28. a^2b 29. a^2b^5 30. $1/(a^3b^4)$
31. $a^5b^3c^6$ 32. $x^6y^2w^4$ 33. $m^3n^2w^4$ 34. $a^6b^4c^5$ 35. $x^9y^8z^4$ 36. $m^4n^6p^5$ 37. $a^{544}b^{1143}$ 38. $(y^{110})/(x^{334})$
39. $(n^{375}p^{539})/(m^{529})$ 40. $(a^{949}c^{332})/(b^{858})$

11.2B   41. $10a^3b^5c^3$ 42. $6w^3x^5y^4$ 43. $14a^4b^3c^5$ 44. $12w^9x^4y^2$ 45. $7a^3b^4c^5$ 46. $12m^4n^4p^2$
47. $-24x^{112}y^{88}$ 48. $10m^{98}n^{123}$ 49. $(21m^{17})/(p^3)$ 50. $-(12a^{11})/(b^{39})$ 51. $(30n^{29})/(m^{15})$ 52. $(28w^{35})/(p^2)$

Section 11.3A   1. x^3 2. r^{16} 3. y^4 4. k^{33} 5. $1/(m^2)$ 6. x^{76} 7. $1/(w^4)$ 8. y^{227} 9. $1/(z^{12})$
10. x^{225} 11. a^6 12. y^{357} 13. a^4b^3c 14. $x^3y^3w^2$ 15. x^5y^3w 16. $m^2n^2p^6$ 17. $(a^2)/(b^4c^3)$ 18. $1/(m^4)$
19. $(x^2)/(y^2z^3)$ 20. $(r^2p^2)/k$ 21. $(a^2)/(b^2)$ 22. $(x^3w^2)/y$

11.3B   23. $(2m^3p^2)/(n^3)$ 24. $5/(2y^2w^3)$ 25. $-(2b)/(5a^6c^2)$ 26. $-(3w^3)/(7y^2z^3)$ 27. $(2m^2)/(5p^4)$
28. $(3x^3)/(5y^4)$ 29. $(4a^{22}b^6)/7$ 30. $-(3m^6n^{16})/(5p^{14})$ 31. $(2y^{18}w^{46})/(3x^9)$ 32. $-(2n^{67}p^{12})/(3m^{33})$

11.3C   33. $b/(a^8)$ 34. x^2y^4 35. $(w^2)/(y^3)$ 36. $(c^4)/(a^4b^6)$ 37. $(m^5)/(n^2)$ 38. $1/(p^5r^2m)$ 39. $(x^4w^7)/y^7$
40. $(b^7a^{10})/(c^2)$ 41. $(5b^{18})/(7a^9)$ 42. $(5y^{13})/(6x^{11})$ 43. $-(4x^{41})/(5w^{44})$ 44. $(2m^{70})/(7p^{15})$ 45. $(5b^{24})/(7a^7c^{31})$
46. $-(3z^{19}k^{28}m^9)/5$

Section 11.4A   1. x^{10} 2. y^8 3. a^{15} 4. m^{42} 5. b^8 6. n^{28} 7. c^{30} 8. w^{30} 9. x^{20} 10. x^{20}
11. a^{60} 12. k^{70} 13. $1/(m^{20})$ 14. c^{42} 15. x^{72} 16. $1/(w^{30})$

11.4B   17. m^8n^8 18. $p^7m^7n^7$ 19. $a^{14}b^{14}c^{14}$ 20. $x^{38}y^{38}w^{38}$ 21. $1/(a^7b^7c^7)$ 22. $1/(x^{13}y^{13}z^{13})$ 23. $9a^2b^2$
24. $125p^3k^3$ 25. $1/(49m^2p^2)$ 26. $1/(8a^3b^3c^3)$ 27. $8x^9y^3$ 28. $125a^{21}b^9$ 29. $27w^{12}y^3$ 30. $8m^9n^{18}$
31. $49x^8w^6$ 32. $16p^2m^8$ 33. $36r^{10}p^2m^6$ 34. $100t^{12}k^2w^{16}$ 35. $(a^6)/(b^4)$ 36. $(y^{15})/(x^5)$ 37. $1/(w^9y^3)$
38. $(a^4)/(b^{20}c^8)$ 39. $1/(x^8y^{12})$ 40. $(y^6)/(w^{15})$ 41. $(p^{12})/(r^{12}l^6)$ 42. $(b^2c^6)/(a^4)$ 43. $(b^6)/(4a^2)$
44. $(y^2)/(9x^{24})$ 45. $(x^6)/(25y^{14})$ 46. $1/(49w^2y^6)$ 47. $(b^6)/(8a^{24})$ 48. $(b^{12})/(27a^3)$ 49. $(y^6)/(25w^{10}x^2)$
50. $(y^6w^2)/(16x^{18})$

11.4C   51. $25x^6y^2$ 52. $-27a^3b^{15}$ 53. $-8w^9y^3$ 54. $49m^8n^4$ 55. $16y^8w^2$ 56. $-a^{15}b^{35}$ 57. $k^{12}m^{16}$
58. $-p^{42}m^{42}n^7$ 59. $(4a^2)/(b^{10})$ 60. $(25x^2w^{12})/(y^8)$ 61. $-(27p^3)/(m^{18}n^3)$ 62. $-(b^{35})/(7a^{23}c^{40})$
63. $(25w^8)/(y^{12}z^4)$ 64. $(a^{40}c^{45})/(b^{30})$ 65. $(c^8)/(4a^2b^6)$ 66. $(y^{16}w^{20})/(x^{12})$ 67. $-(p^9)/(8n^6m^{15})$
68. $(c^{56})/(5^8a^8b^{48})$

Section 11.5A   1. x^8y^7 2. $x^{19}y^9$ 3. $4a^{19}b^{10}$ 4. $8m^{21}n^{12}p^9$ 5. $(25x^{13})/(w^{26})$ 6. $-(m^{29}p^{23})/(n^{14})$
7. $-(72a^{13}b)/(c^{14})$ 8. $(x^2)/(225y^{18}w^2)$ 9. $-1/(m^{23}p^6)$ 10. $-(c^{14})/(a^8b^{17})$ 11. $(25x^2)/(9w^2y^{18})$
12. $1/(8m^3kp^{28})$ 13. $-(c^{24})/(20a^9b^{21})$ 14. $-(z^{25})/(144w^{10}y^{20})$ 15. $5/(144p^{13}m^8)$ 16. $-(c^{41})/(36a^{31}b^{26})$

11.5B   17. a^3b^7 18. x^7y^5 19. $(y^2)/x$ 20. $(x^{17})/(y^3)$ 21. $(np^{12})/(m^4)$ 22. $a^{21}b^4$ 23. $(9a^7)/(8b^7)$
24. $(49x^4y^4)/25$ 25. $-a^{17}b^{31}c^{20}$ 26. $-(4m^4p^{15}n^{13})/27$ 27. $4/(9x^2w^4z^2)$ 28. $(2b^2)/(3a^{14}c^3)$
29. $25/(8mp^{19}n^{15})$ 30. $-60w^{17}y^4z^{12}$

Section 11.6A   1. 15,000   2. 43,000,000   3. 538.9   4. 601,580   5. 0.68   6. 0.0308   7. 0.132
8. 5.38002   9. 0.00005   10. 0.00061   11. 0.084   12. 1.2   13. 2.304   14. 56.09   15. 0.00056
16. 0.00000056   17. 0.00000087   18. 0.0006834   19. 0.00504   20. 0.001455

11.6B   21. 2.36×10^2 22. 5.3×10 23. 1.3908×10^4 24. 4.05911×10^5 25. 5.67133×10^6 26. 8.3×10^6
27. 4.7×10^{-4} 28. 9.134×10^{-6} 29. 8.45×10^{-4} 30. 7.0×10^{-4} 31. 9.9×10^{-7} 32. 1.45×10^{-4}

Section 11.7A   1. 6.82×10^7 2. 7.56×10^{11} 3. 7.52×10^{-10} 4. 8.16×10^{-12} 5. 3.417×10^{10}
6. 2.9627×10^{10} 7. 2.573×10^{-9} 8. 1.904×10^{-9} 9. 4.34×10^{-16} 10. 2.09×10^{11} 11. 4.1548×10^{15}
12. 3.8064×10^{19} 13. 3.905×10^{-15} 14. 1.5624×10^{-16} 15. 2.29878×10^{-14} 16. 1.1664×10^{-11}

17. 9.576×10^{14} **18.** 2.66085×10^{11} **19.** 3.24×10^3 **20.** 9.1×10^{-1} **21.** 2.49×10^2 **22.** 9.3×10
23. 4.3×10^5 **24.** 6.192×10^{-6} **25.** 3.72×10^{80} **26.** 5.52×10^{135} **27.** 2.226×10^{133} **28.** 2.074×10^{94}
29. 3.38×10^{-56} **30.** 1.312×10^{-100} **31.** 7.92×10^{-137} **32.** 3.111×10^{-136} **33.** 5.44×10^{-45}
34. 8.06×10^{18} **35.** 2.4×10^{10} **36.** 3.7×10^{12} **37.** 1.5×10^{-4} **38.** 5.9×10^{-6} **39.** 2.4×10
40. 4.7×10^2

11.7B   **41.** 1.4×10^{11} **42.** 2.4×10^{11} **43.** 3.1×10^{-6} **44.** 1.2×10^{-5} **45.** 2.1×10^4 **46.** 1.4×10^{11}
47. 4.3×10^{-4} **48.** 3.4×10^4 **49.** 1.9×10^7 **50.** 3.5×10^{-3} **51.** 4.0×10^{-2} **52.** 2.4×10^5
53. 2.5×10^{-10} **54.** 3.0×10^{15}

Section 11.8   **1.** $300   **2.** 40,000,000 sec   **3.** 11,111 hr   **4.** 30,000 transistors   **5.** 3 cents each
6. 242 hr   **7.** $3.75 per hour   **8.** $820   **9.** $800   **10.** $700   **11.** $700   **12.** $40   **13.** 4.2×10^{27} atoms
14. 1.2×10^{28} atoms   **15.** 9×10^{28} atoms   **16.** 3.0×10^{20} atoms   **17.** 2.4×10^{19} atoms   **18.** 8.4×10^{29} atoms
19. 4.81×10^{23} widgets   **20.** 6.45×10^{17} stickers   **21.** 8.06×10^{10} lb   **22.** 4.08×10^9 photos

Chapter 11 Review   **1.** $5yy$ **2.** $-4xyyy$ **3.** 3. $1/m$ **4.** $(2aaa)/(bbbb)$ **5.** 1   **6.** $1/64$ **7.** $1/9$
8. -81 **9.** $5/4$ **10.** $9/16$ **11.** a^{25} **12.** $1/(x^{14})$ **13.** x^5y^2 **14.** $-(10x^3)/(y^3)$ **15.** x^5 **16.** $1/(y^5)$
17. $(m^2p^3)/n^4$ **18.** $-(3a^4)/(5b^7)$ **19.** $(x^2y^8)/(w^6)$ **20.** $5/(3b^2)$ **21.** m^{40} **22.** $1/(w^{32})$ **23.** x^2y^{10}
24. $(49a^2)/(b^{10})$ **25.** $-(m^{15})/(n^5)$ **26.** $(c^8)/(9a^2b^4)$ **27.** $x^{22}y^{21}$ **28.** $(8a^2)/(b^{27})$ **29.** $(27n^8p)/(4m^3)$
30. $-(2a^9)/(b^4)$ **31.** 407,600   **32.** 78.06   **33.** 0.056   **34.** 5.6×10^8 **35.** 7.2×10^{-13} **36.** 6.63×10^{63}
37. 7.31×10^{-7} **38.** 3.486×10^{-117} **39.** 1.7×10^{19} **40.** 2.5×10^{124} **41.** 2.94×10^{85} **42.** 40 cents per vote

Supplementary Exercises   **1.** $4aaa$ **2.** $-5nmmmmmm$ **3.** $1/(bbb)$ **4.** $(7nnnnn)/(ppp)$ **5.** 1
6. $1/216$ **7.** $1/16$ **8.** -49 **9.** $9/2$ or $4\,1/2$ **10.** $4/49$ **11.** k^{31} **12.** $1/(y^{12})$ **13.** a^4b^3 **14.** $-24p^4k^2$
15. w^3 **16.** $1/(p^{13})$ **17.** $(m^4p^2)/(n^6)$ **18.** $-(5a^3)/(8b^8)$ **19.** $(x^6y^7)/(w^7)$ **20.** $(2a^7)/(b^3)$ **21.** a^{56}
22. $1/(x^{30})$ **23.** a^3b^{12} **24.** $(36w^2)/(y^{18})$ **25.** $-(n^{28})/(p^7)$ **26.** $(p^{10})/(4n^2m^6)$ **27.** $x^{18}y^{48}$ **28.** $(8a^{10})/(b^{29})$
29. $(8p)/(25m^5n^3)$ **30.** $-(27a^6b^{18})/16$ **31.** 64,032,000   **32.** 1.4375   **33.** 0.00037   **34.** 3.82×10^8
35. 2.8×10^{-11} **36.** 4.44×10^{85} **37.** 9.36×10^{-5} **38.** 1.728×10^{-111} **39.** 5.1×10^{19} **40.** 3.4×10^{98}
41. 3×10^9 **42.** 20 sec per envelope

Chapter 11 Test   **1.** $2ccccc$ **2.** $-3abbbb$ **3.** $1/(kkkk)$ **4.** $(4nnnnnn)/(pp)$ **5.** 1   **6.** $1/81$ **7.** $1/64$
8. -36 **9.** 5   **10.** $9/64$ **11.** x^{79} **12.** $1/(w^{16})$ **13.** $-(n^4)/p$ **14.** $8a^5$ **15.** m^{13} **16.** $1/(w^{21})$
17. $(b^5)/(a^3c)$ **18.** $-(2p^3)/(3n^{15})$ **19.** $(y^7x^6)/(w^8)$ **20.** $(2a^2)/(b^4)$ **21.** p^{48} **22.** $1/(w^{28})$ **23.** x^6z^{30}
24. $-(27a^3)/(b^{21})$ **25.** $(b^{24})/(a^8)$ **26.** $(p^6)/(16n^2m^{12})$ **27.** $w^{29}m^{57}$ **28.** $-(27a^{13})/(b^{37})$ **29.** $(27p)/(16m^3n^8)$
30. $-(8a^{19}b^{17})/9$ **31.** 8,066,300   **32.** 34.028   **33.** 0.0583   **34.** 7.1×10^7 **35.** 4.68×10^{-11} **36.** 5.28×10^{85}
37. 4.42×10^{-8} **38.** 3.565×10^{-87} **39.** 1.2×10^{19} **40.** 2.6×10^{127} **41.** $40

Cumulative Review   **1.** $1/2$ **2.** $23/30$ **3.** 5.196   **4.** -8 **5.** $11\frac{3}{20}$ **6.** $3/11$ **7.** $4/15$ **8.** 7217.0422
9. 9   **10.** -39 **11.** $2\frac{5}{12}$ **12.** $3/50$ **13.** $99/7$ or $14\frac{1}{7}$ **14.** 0.01512   **15.** -30 **16.** $15x^3y^5$ **17.** $x^{13}y^{13}$
18. 1.24×10^{-7} **19.** $22/15$ or $1\frac{7}{15}$ **20.** 31.4   **21.** 8   **22.** -5
23. $3ab - 11ac + 11bc$ **24.** $-4x^4 + 6x^3 - 22x^2 + 18x$ **25.** 51   **26.** 6   **27.** -4 **28.** 14   **29.** -11
30. 41   **31.** 67   **32.** $168.24   **33.** $49.30   **34.** 1,920 women   **35.** 800

12

CHAPTER 12

Section 12.1A   **1.** Not perfect square   **2.** Perfect square, $4 = (2)(2)$ **3.** Perfect square, $16 = (4)(4)$
4. Not perfect square   **5.** Not perfect square   **6.** Perfect square, $36 = (6)(6)$ **7.** Not perfect square
8. Perfect square, $49 = (7)(7)$ **9.** Perfect square, $400 = (20)(20)$ **10.** Not perfect square   **11.** Perfect square, $4a^6 = (2a^3)(2a^3)$ **12.** Not perfect square   **13.** Perfect square, $100x^8w^{12} = (10x^4w^6)(10x^4w^6)$ **14.** Not perfect square

12.1B **15.** ± 2 **16.** ± 6 **17.** 7 **18.** 2 **19.** 8 **20.** 70 **21.** 9 **22.** 6 **23.** 11 **24.** 1 **25.** $5y$
26. $7a^4b^2$ **27.** $30xw^3$ **28.** $12p^4m^3n^2$

12.1C

29. $1 < \sqrt{3} < 2$

30. $2 < \sqrt{8} < 3$

31. $5 < \sqrt{30} < 6$

32. $4 < \sqrt{19} < 5$

33. $4 < \sqrt{23} < 5$

34. $6 < \sqrt{40} < 7$

35. $3 < \sqrt{10} < 4$

36. $5 < \sqrt{28} < 6$

12.1D **37.** $2\sqrt{5}$ **38.** $3\sqrt{5}$ **39.** $2\sqrt{3}$ **40.** $3\sqrt{7}$ **41.** $4\sqrt{3}$ **42.** $10\sqrt{3}$ **43.** $4\sqrt{5}$ **44.** $2\sqrt{15}$ **45.** $5\sqrt{7}$
46. $10\sqrt{5}$ **47.** $10\sqrt{6}$ **48.** 18 **49.** 14 **50.** 30 **51.** 12 **52.** 42 **53.** 35 **54.** $12\sqrt{3}$ **55.** nm^2
56. $m^2\sqrt{pm}$ **57.** $2ab^4$ **58.** $3bc^3\sqrt{ab}$ **59.** $2n^2p^3\sqrt{3mp}$ **60.** $2m^2p^5\sqrt{5nm}$ **61.** $2a^3c^2\sqrt{7bc}$ **62.** $4xw^2\sqrt{xy}$
63. $4b\sqrt{2ab}$ **64.** $9xyw^2\sqrt{2x}$ **65.** $10a^3b^3\sqrt{5b}$ **66.** $14xy^4\sqrt{3x}$ **67.** $12m^4n^4\sqrt{n}$ **68.** $50p^4k^2\sqrt{2}$

Section 12.2A **1.** $2\sqrt[3]{3}$ **2.** 3 **3.** -5 **4.** $2\sqrt[3]{7}$ **5.** $2xy\sqrt[3]{2x^2y}$ **6.** $3ab\sqrt[3]{5a}$ **7.** $pm\sqrt[3]{p^2m^2}$
8. $m\sqrt[3]{nm}$ **9.** $ab^2c^2\sqrt[3]{ac^2}$ **10.** $yw\sqrt[3]{3xy}$ **11.** $2a^2b^2\sqrt[3]{a}$ **12.** $-3x^2\sqrt[3]{x^2}$ **13.** $-10n^3m^3\sqrt[3]{nm^2}$
14. $2ab^2c^3\sqrt[3]{2}$ **15.** $-3mp\sqrt[3]{2nmp}$ **16.** $5kpm^2\sqrt[3]{2kpm^2}$ **17.** $-2abc^2\sqrt[3]{5ab^2}$ **18.** $-2w^2yz\sqrt[3]{7wy^2}$

12.2B **19.** w **20.** $x\sqrt[4]{x^2}$ **21.** p^2 **22.** $a^2\sqrt[4]{a^2}$ **23.** a **24.** $w\sqrt[5]{w^2}$ **25.** y **26.** $r\sqrt[9]{r}$ **27.** $xy^2\sqrt[5]{xy^2}$
28. $cd^2\sqrt[5]{ab^4c^3d^4}$ **29.** $3\sqrt[4]{2}$ **30.** 2 **31.** $3y^2w\sqrt[4]{x^3w^3}$ **32.** $2bc\sqrt[4]{a^2bc^2}$ **33.** $2abc\sqrt[4]{2bc^2}$ **34.** $3yw\sqrt[4]{x^3w^3}$
35. $-np^2m^3\sqrt[5]{n^3p^2}$ **36.** $-2ac\sqrt[5]{abc^3}$

Section 12.3A **1.** 2 **2.** 3 **3.** 3 **4.** 4 **5.** 8 **6.** 27 **7.** -8 **8.** 7 **9.** 25 **10.** -5 **11.** 16 **12.** 4

12.3B **13.** $1/3$ **14.** $1/5$ **15.** $1/5$ **16.** $1/25$ **17.** $1/100$ **18.** $1/7$ **19.** $1/2$ **20.** $1/8$ **21.** 6 **22.** 10
23. $5/2$ or $2\frac{1}{2}$ **24.** 8 **25.** $100/9$ or $11\frac{1}{9}$ **26.** $2/5$ **27.** -2 **28.** 25

12.3C **29.** $a^{1/2}$ **30.** $x^{4/7}$ **31.** $w^{1/5}$ **32.** c **33.** $1/w^{2/3}$ **34.** $y^{6/7}$ **35.** $a^{3/5}$ **36.** $n^{1/4}$ **37.** $p^{1/6}$
38. $1/m^{1/2}$ **39.** $k^{1/3}$ **40.** $1/a^{5/4}$ **41.** $n^{3/2}$ **42.** $c^{2/9}$ **43.** $1/m^{2/5}$ **44.** $k^{1/3}$ **45.** $a^{1/2}b$ **46.** $1/(m^{3/4}n^{9/4})$
47. $p^{1/2}m^{1/3}$ **48.** $n^{3/4}m^{3/2}p^4$ **49.** $(ab^{1/3})/c^{1/2}$ **50.** $x^{1/2}yw^{2/5}$ **51.** $(n^{1/3})/(m^{1/3})$ **52.** $(a^{1/4})/(b^{1/4}c^{1/6})$
53. $(k^{4/3})/(m^{1/5}n^{2/3})$ **54.** $1/(a^{1/2}b^{1/6})$ **55.** $x^{1/3}y^{1/4}$ **56.** $p^{1/2}m^{13/4}$ **57.** $(b^{2/3})/(a^{1/12})$ **58.** $(p^{1/5})/(n^{1/6})$

Section 12.4A **1.** $9\sqrt{3}$ **2.** $2\sqrt{3}$ **3.** $8\sqrt{7}$ **4.** $9\sqrt{2}$ **5.** $3\sqrt{5} + 7\sqrt{3}$ **6.** $2\sqrt{3} + 5\sqrt{2}$ **7.** $18\sqrt{6} + 10\sqrt{7}$
8. $16 - 4\sqrt{3}$ **9.** $5\sqrt[3]{5}$ **10.** $2\sqrt[4]{7}$ **11.** $6\sqrt[3]{2} + 5\sqrt[3]{3}$ **12.** $7\sqrt[4]{3} + 3\sqrt[4]{5}$ **13.** $7\sqrt{3} + 8\sqrt[3]{3}$ **14.** $14\sqrt{2} + 7\sqrt[3]{2}$
15. $6\sqrt[3]{2} + 2\sqrt{7} + 8\sqrt{2}$ **16.** $4\sqrt{3} + 9\sqrt[3]{2} + \sqrt{2}$ **17.** $7\sqrt{a} + 3\sqrt[3]{a}$ **18.** $6\sqrt{x} + 5\sqrt{y} - 3\sqrt{w}$
19. $2\sqrt[3]{y} - 2\sqrt{y}$ **20.** $6\sqrt[3]{xy^2} + 10\sqrt[3]{x^2y}$ **21.** $2\sqrt[3]{ab} + 2\sqrt[3]{ab^2} + 2\sqrt[3]{a^2b}$ **22.** $2\sqrt[3]{n^2m} + 2\sqrt[3]{mn} + 2\sqrt[3]{nm^2}\ 2\sqrt[3]{nm^2}$
23. $7\sqrt[3]{ab} - 2\sqrt[3]{ab^2} + 3\sqrt[3]{a^2b}$ **24.** $-3\sqrt[3]{xy^2} - 4\sqrt[3]{x^2y}$ **25.** $\sqrt[3]{mn} + 3\sqrt[3]{nm^2}$

12.4B **26.** $8\sqrt{2}$ **27.** $-3\sqrt{3}$ **28.** $14\sqrt{3}$ **29.** $-14\sqrt{2}$ **30.** $14\sqrt{3} + 41\sqrt{2}$ **31.** $15\sqrt{5}$ **32.** $31 + 2\sqrt{5}$
33. $8 + 38\sqrt{3} + 9\sqrt{2}$ **34.** $\sqrt{3}$ **35.** $11 + 9\sqrt{3}$ **36.** 5 **37.** $19\sqrt[3]{2}$ **38.** $11\sqrt[3]{5}$ **39.** $-9 + 8\sqrt[3]{3}$
40. $62 + 9\sqrt[3]{2}$ **41.** $19\sqrt[3]{50} + 15\sqrt[3]{2}$ **42.** $-b\sqrt{a} - 2\sqrt{ab} + 6a\sqrt{b}$ **43.** $7x - 3\sqrt{x}$ **44.** $3y\sqrt{y} + y\sqrt[3]{y}$
45. $3b\sqrt{a} - 2a\sqrt[3]{b^2}$ **46.** $4\sqrt{m} + 2\sqrt{n} - 2n$ **47.** $5b\sqrt{3a} + 2a\sqrt{2b}$ **48.** $-x\sqrt{5y} + \sqrt{xy} + 10y\sqrt{5x}$
49. $-3\sqrt{2y} + 6\sqrt{xy} - 3x\sqrt{xy}$

Section 12.5A **1.** $\sqrt{35}$ **2.** $\sqrt{33}$ **3.** $3\sqrt{10}$ **4.** $5\sqrt{14}$ **5.** $7\sqrt{10}$ **6.** $6\sqrt{5}$ **7.** $5\sqrt{22}$ **8.** $11\sqrt{6}$
9. 30 **10.** $26\sqrt{21}$ **11.** $2\sqrt[3]{3}$ **12.** $5\sqrt[3]{3}$ **13.** $2\sqrt[3]{45}$ **14.** $3\sqrt[3]{6}$ **15.** $14\sqrt[3]{3}$ **16.** $6\sqrt[3]{10}$ **17.** $2a\sqrt{15}$
18. $10b\sqrt{3a}$ **19.** $10xy\sqrt{3}$ **20.** $6w\sqrt{2}$ **21.** $3b\sqrt{10ac}$ **22.** $12p\sqrt{p}$

12.5B **23.** $8\sqrt{21}$ **24.** $35\sqrt{10}$ **25.** $-24\sqrt{15}$ **26.** $40\sqrt{30}$ **27.** $42\sqrt{2}$ **28.** $105\sqrt{10}$ **29.** $-56\sqrt{35}$ **30.** $42\sqrt{3}$ **31.** $100\sqrt{3}$ **32.** $-72\sqrt{5}$

12.5C **33.** $\sqrt{10}+\sqrt{30}$ **34.** $\sqrt{42}+\sqrt{14}$ **35.** $\sqrt{10}-\sqrt{14}$ **36.** $11\sqrt{6}-\sqrt{30}$ **37.** $3\sqrt{2}+3\sqrt{5}$ **38.** $2\sqrt{5}-2\sqrt{3}$ **39.** $-3\sqrt{5}-5\sqrt{2}$ **40.** $3\sqrt{10}-2\sqrt{21}$ **41.** $9\sqrt{2}+12$ **42.** $6\sqrt{3}-10\sqrt{5}$ **43.** $-42\sqrt{2}-15\sqrt{14}$ **44.** $36\sqrt{2}-6\sqrt{15}$

Chapter 12 Review
1. Yes, perfect square, $121=11^2$ **2.** Not perfect square **3.** ±9 **4.** 7 **5.** $5\sqrt{3}$ **6.** $10\sqrt{2}$ **7.** 40 **8.** a^2 **9.** $w\sqrt{w}$ **10.** $3bc^3\sqrt{ab}$ **11.** $5ab^4c^3\sqrt{2c}$ **12.** $6<\sqrt{45}<7$

13. $-2\sqrt[3]{5}$ **14.** $3x\sqrt[3]{xy^2}$ **15.** $2\sqrt[4]{5}$ **16.** $2a\sqrt[4]{a^2b^2}$ **17.** $w\sqrt[5]{w}$ **18.** $n\sqrt[8]{n^2}$ **19.** 4 **20.** -3 **21.** $1/7$ **22.** 2 **23.** x **24.** $w^{1/9}$ **25.** $a^{3/20}$ **26.** $a^{3/4}b^{3/2}$ **27.** $1/(n^{6/7})$ **28.** $m^{3/4}$ **29.** $(a^{5/3}b^{2/3})/(c^{1/2})$ **30.** $1/(a^{2/3})$ **31.** $3\sqrt{3}$ **32.** $9\sqrt{2}+8\sqrt{5}$ **33.** $5\sqrt[3]{7}$ **34.** $7\sqrt{5}+\sqrt[3]{5}$ **35.** $-\sqrt{3}$ **36.** $27\sqrt{5}$ **37.** $12\sqrt{7}-8\sqrt{3}$ **38.** $7\sqrt[3]{3}-6\sqrt[3]{5}$ **39.** $3\sqrt{m}-2\sqrt{n}$ **40.** $-3\sqrt{p}+3\sqrt[3]{p}$ **41.** $7a\sqrt{a}+3\sqrt{3b}+4$ **42.** $14\sqrt{3x}+8y\sqrt{5y}$ **43.** $2\sqrt{21}$ **44.** $30\sqrt{7}$ **45.** $12\sqrt{10}$ **46.** $-150\sqrt{6}$ **47.** $2n\sqrt{3m}$ **48.** $5ac\sqrt[3]{6ab^2}$ **49.** $60p\sqrt{3}$ **50.** $\sqrt{21}-\sqrt{14}$ **51.** $6\sqrt{2}+10\sqrt{15}$ **52.** $15\sqrt{2}+12\sqrt{5}$

Supplementary Exercises
1. Yes, perfect square, $225=15^2$ **2.** Yes, perfect square, $49a^6b^{10}=(7a^3b^5)^2$ **3.** ±5 **4.** 5 **5.** $7\sqrt{2}$ **6.** $10\sqrt{3}$ **7.** $20\sqrt{10}$ **8.** p^3 **9.** $m^2\sqrt{m}$ **10.** $4bc^2\sqrt{ac}$ **11.** $2ab^3c^4\sqrt{5ac}$ **12.** $4<\sqrt{22}<5$ **13.** $7<\sqrt{50}<8$ **14.** $2\sqrt[3]{3}$ **15.** $-5xy^2\sqrt[3]{x^2}$

16. 3 **17.** $2a\sqrt[4]{2a^3b^3}$ **18.** $w\sqrt[6]{w^4}$ **19.** $n\sqrt[7]{n^3}$ **20.** 6 **21.** -2 **22.** $1/11$ **23.** 3 **24.** $x^{14/15}$ **25.** $1/(w^{2/7})$ **26.** $a^{3/10}$ **27.** $a^{1/6}b^{1/2}$ **28.** $1/(n^{4/7})$ **29.** $m^{2/7}$ **30.** $(a^{8/3}b^{2/5})/(c^{2/3})$ **31.** $(b^{1/12})/(a^{1/3})$ **32.** $-\sqrt{10}$ **33.** $9\sqrt{5}+3\sqrt{3}$ **34.** $9\sqrt{11}-2\sqrt{13}$ **35.** $4\sqrt[3]{5}$ **36.** $11\sqrt[3]{2}+4\sqrt{2}$ **37.** $\sqrt[3]{5}+7\sqrt{6}$ **38.** $-\sqrt{5}$ **39.** $5\sqrt{7}$ **40.** $31\sqrt{5}$ **41.** $12\sqrt{3}-9\sqrt{2}$ **42.** $12\sqrt{5}-\sqrt{3}$ **43.** $11\sqrt[3]{2}-6\sqrt[3]{5}$ **44.** $28\sqrt[3]{3}+\sqrt[3]{5}$ **45.** $6\sqrt{p}-2\sqrt{k}$ **46.** $5\sqrt[3]{a}-6\sqrt{a}$ **47.** $7a\sqrt{a}+3\sqrt{3b}+6$ **48.** $11\sqrt{2x}+19y\sqrt{2y}$ **49.** $10\sqrt{2}$ **50.** $4\sqrt{3}$ **51.** $12\sqrt{7}$ **52.** $35\sqrt{2}$ **53.** $8\sqrt{21}$ **54.** $20\sqrt{15}$ **55.** $-30\sqrt{3}$ **56.** $80\sqrt{3}$ **57.** $2n\sqrt{3m}$ **58.** $60p\sqrt{3}$ **59.** $5ac\sqrt[3]{6ab^2}$ **60.** $\sqrt{14}+\sqrt{10}$ **61.** $3\sqrt{2}-3\sqrt{5}$ **62.** $2\sqrt{3}-18\sqrt{2}$ **63.** $2\sqrt{15}+20\sqrt{2}$ **64.** $24\sqrt{3}-30\sqrt{5}$

Chapter 12 Test
1. ±8 **2.** 4 **3.** $2\sqrt{10}$ **4.** 5 **5.** $-3\sqrt[3]{15}$ **6.** $2\sqrt[4]{3}$ **7.** $3\sqrt[4]{4}$ **8.** $2xy^2\sqrt{7x}$ **9.** $2ac^2\sqrt[3]{5b^2c^2}$ **10.** $3\sqrt{5}$ **11.** $3\sqrt{3}+8\sqrt{2}$ **12.** $4\sqrt[3]{4}$ **13.** $8\sqrt{2}+6\sqrt[3]{2}$ **14.** $\sqrt{6}$ **15.** $24\sqrt{3}$ **16.** $19\sqrt{7}+12\sqrt{2}-3\sqrt{3}$ **17.** $7\sqrt[3]{2}$ **18.** $5\sqrt{a}+4\sqrt{b}-2\sqrt{c}$ **19.** $8\sqrt{x}+\sqrt[3]{x}$ **20.** $18\sqrt{3a}$ **21.** $6\sqrt{5}$ **22.** $12\sqrt{5}$ **23.** $-6\sqrt{14}$ **24.** $36\sqrt{14}$ **25.** $\sqrt{15}-\sqrt{33}$ **26.** $2\sqrt{35}-5\sqrt{6}$ **27.** $6\sqrt{2}+15\sqrt{10}$ **28.** 5 **29.** $1/5$ **30.** $9/100$ **31.** $-10/3$ or $-3\frac{1}{3}$ **32.** $a^{1/2}$ **33.** $m^{1/5}$ **34.** $p^{1/2}$ **35.** $k^{1/6}$ **36.** $(x^{2/3}w^{1/3})/(y^{1/5})$ **37.** $n^{1/3}$ **38.** $4<\sqrt{17}<5$

Cumulative Review
1. $3/4$ **2.** $3/8$ **3.** 47.979 **4.** 2.1762 **5.** 0.014 **6.** 17.646 **7.** 17 **8.** 18 **9.** 13 **10.** 11 **11.** $12y^2+3y-4$ **12.** $-12x^4+15x^3+9x^2$ **13.** $(3a^2c^4)/5$ **14.** $1/9$ **15.** $y=2$ **16.** $x=-4$ **17.** $x=5, y=2$ **18.** $x=-2, y=-5$ **19.** $y=4x+5$ **20.** $y=4x-11$ **21.** 1.104×10^{84} **22.** 3.672×10^{-26} **23.** $y=3x-2$ **24.** $y<3x-2$

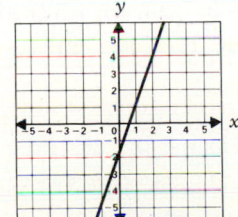

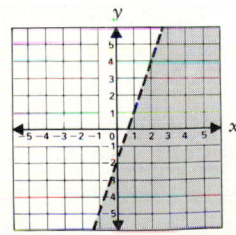

25. $y \geq 3x - 2$

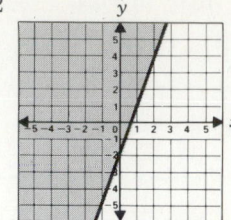

26. $47.77 (or $47.78)   **27.** 105 foreign cars

28. 5 points for fly catch, 2 points for bounce catch

CHAPTER 13

Section 13.1   **1.** $2x^2 + xy + 3xw + wy + w^2$ **2.** $2y^3 + 9y^2 - 9y + 2$ **3.** $6w^3 + 5w^2 - 3w - 2$
4. $15n^3 + 19n^2 - 9n - 9$ **5.** $4n^2 + 2n - 3m - 9m^2$ **6.** $a^2 + 2ab + 2ac + b^2 + 2bc + c^2$
7. $x^2 - y^2 + 2yw - w^2$ **8.** $k^2 - k - j^2 + 5j - 6$ **9.** $x^3 - x^2 - x + 10$ **10.** $y^3 + 4y^2 - 4y + 5$
11. $w^3 + w^2 - 5w - 2$ **12.** $6a^3 - a^2 + 11a + 4$ **13.** $6x^4 - x^3 + 9x^2 + 9x - 7$ **14.** $5y^4 + 10y^3 - 3y^2 - 8y + 3$

Section 13.2A   **1.** $x^2 - 2x - 3$ **2.** $y^2 + 7y + 12$ **3.** $y^2 + 9y + 14$ **4.** $m^2 + 3m - 10$ **5.** $2w^2 + 13w - 7$
6. $6x^2 + 17x + 5$ **7.** $12n^2 - 5n - 3$ **8.** $18r^2 + 9r - 2$ **9.** $x^2 + 6x + 9$ **10.** $49y^2 + 14y + 1$
11. $4x^2 + 27x - 7$ **12.** $3y^2 + 19y - 14$

13.2B   **13.** $x^2 + 18x + 81$ **14.** $w^2 - 2w + 1$ **15.** $x^2 + 10x + 25$ **16.** $y^2 - 8y + 16$ **17.** $n^2 + 2n + 1$
18. $4w^2 + 20w + 25$ **19.** $9m^2 - 36m + 36$ **20.** $49k^2 - 14kp + p^2$ **21.** $4a^2 - 12ab + 9b^2$
22. $25x^2 + 20xy + 4y^2$ **23.** $25x^2 - 20xy + 4y^2$ **24.** $9n^2 - 12np + 4p^2$

13.2C   **25.** $x^2 - 9$ **26.** $z^2 - 16$ **27.** $k^2 - 81$ **28.** $y^2 - 36$ **29.** $9z^2 - 16$ **30.** $49x^2 - 4$ **31.** $25r^2 - 1$
32. $16m^2 - 9$ **33.** $36y^2 - 49$ **34.** $4w^2 - 81$ **35.** $x^4 - 25$ **36.** $9n^4 - 4p^2$ **37.** $4y^4 - x^2$ **38.** $25r^6 - 4n^2$

13.2D   **39.** $\sqrt{10} + \sqrt{14} + \sqrt{15} + \sqrt{21}$ **40.** $\sqrt{6} - \sqrt{15} + \sqrt{22} - \sqrt{55}$ **41.** $10 - 5\sqrt{2} + 2\sqrt{3} - \sqrt{6}$
42. $21 - 7\sqrt{3} - 3\sqrt{2} + \sqrt{6}$ **43.** $5\sqrt{6} + \sqrt{70} + \sqrt{210} + 7\sqrt{2}$ **44.** $6 - 3\sqrt{10} + \sqrt{30} - 5\sqrt{3}$ **45.** $3\sqrt{2}$
46. $5\sqrt{6} - 2\sqrt{5} - 3\sqrt{10} + 2\sqrt{3}$ **47.** $19\sqrt{14} + 72$ **48.** $-\sqrt{6} + 12$ **49.** $5 + 2\sqrt{6}$ **50.** $10 + 2\sqrt{21}$
51. $5 - 2\sqrt{6}$ **52.** $7 - 2\sqrt{10}$ **53.** $12 - 2\sqrt{35}$ **54.** $8 + 2\sqrt{15}$ **55.** $38 + 12\sqrt{10}$ **56.** $95 + 20\sqrt{15}$
57. $62 - 20\sqrt{6}$ **58.** $46 - 12\sqrt{14}$ **59.** $17 - 12\sqrt{2}$ **60.** $79 + 20\sqrt{3}$

Section 13.3   **1.** $(x + 2)(x - 2)$ **2.** $(x + 3)(x - 3)$ **3.** $(x + 1)(x - 1)$ **4.** $(x + 6)(x - 6)$ **5.** $(x + 7)(x - 7)$
6. $(x + 10)(x - 10)$ **7.** $(x + 4)(x - 4)$ **8.** $(x + 9)(x - 9)$ **9.** $(x + 8)(x - 8)$ **10.** $(x + 11)(x - 11)$
11. $(3x + 5)(3x - 5)$ **12.** $(2x + 1)(2x - 1)$ **13.** $(4w + 3)(4w - 3)$ **14.** $(7m + 4)(7m - 4)$
15. $(6n + 5)(6n - 5)$ **16.** $(9y + 7)(9y - 7)$ **17.** $(m + p)(m - p)$ **18.** $(x + y)(x - y)$ **19.** $(2w + z)(2w - z)$
20. $(3y + w)(3y - w)$ **21.** $(3a + 4b)(3a - 4b)$ **22.** $(5z + 3x)(5z - 3x)$ **23.** $(7p + 6m^2)(7p - 6m^2)$
24. $(2k^2 + 3n^3)(2k^2 - 3n^3)$ **25.** $(5m^3 + 4n^2)(5m^3 - 4n^2)$ **26.** $(p^4 + 3m)(p^4 - 3m)$

Section 13.4A   **1.** $(x + 2)(x + 1)$ **2.** $(x + 2)^2$ **3.** $(x + 4)(x + 1)$ **4.** $(y + 6)(y + 1)$ **5.** Prime
6. $(w + 9)(w + 1)$ **7.** $(x - 2)(x - 1)$ **8.** $(y - 5)(y - 1)$ **9.** $(w - 3)(w - 4)$ **10.** $(a - 6)(a - 2)$
11. $(x + 4)(x + 5)$ **12.** $(w + 6)(w + 2)$ **13.** $(y - 10)(y - 2)$ **14.** $(x + 5)(x + 3)$ **15.** Prime   **16.** $(y - 5)^2$
17. $(x - 2)^2$ **18.** $(y - 9)(y - 1)$ **19.** $(w + 7)(w + 3)$ **20.** $(w + 4)^2$ **21.** $(w + 6)(w + 3)$ **22.** $(w - 4)^2$
23. $(x + 3)(x + 1)$ **24.** $(y + 7)(y + 1)$

13.4B   **25.** $(x + 3)(x - 2)$ **26.** $(x - 4)(x + 2)$ **27.** $(y - 4)(y + 3)$ **28.** $(w - 8)(w + 2)$ **29.** $(a + 4)(a - 3)$
30. $(b + 6)(b - 3)$ **31.** $(y - 6)(y + 5)$ **32.** $(w + 5)(w - 1)$ **33.** $(x + 7)(x - 5)$ **34.** $(m - 6)(m + 2)$
35. $(x - 3)(x - 4)$ **36.** $(y + 5)(y - 3)$ **37.** $(x + 7)(x - 3)$ **38.** $(x - 9)(x + 2)$ **39.** $(x + 6)(x - 3)$

40. $(x+7)(x-2)$ **41.** $(y+13)(y-1)$ **42.** $(w-11)(w+8)$ **43.** $(y+6)(y+3)$ **44.** $(w-4)(w-1)$
45. $(x-8)(x+4)$ **46.** $(y+5)(y+6)$ **47.** $(x+7)(x-6)$ **48.** $(x+5)^2$ **49.** $(x+5)(x-1)$ **50.** $(x+3)^2$
51. $(x+9)(x-7)$ **52.** $(y+4)(y+3)$

13.4C   **53.** $(2x+1)(x+1)$ **54.** $(3x+1)(x+1)$ **55.** $(5x+1)(x+2)$ **56.** $(7y-3)(y+1)$ **57.** $(3x-1)(x+3)$
58. $(5x+1)(x-7)$ **59.** $-(2w+1)(w-3)$ **60.** $(2y+3)(y-3)$ **61.** $(3x+2)(x+2)$ **62.** $-(2x-5)(x+2)$
63. $-(x-5)(2x+1)$ **64.** $(7y-3)(y+2)$ **65.** $(5w-1)(w-1)$ **66.** $(2x-1)(x+7)$

13.4D   **67.** $(5x+3)(2x-1)$ **68.** $(3x+2)(2x-1)$ **69.** $(2x-3)(2x+1)$ **70.** $(4x+5)(x-1)$
71. $(4x-1)(2x-3)$ **72.** $(6x+1)(x+2)$ **73.** $(6x-1)(x+4)$ **74.** $(5x-2)(2x+3)$ **75.** $(2x+3)(2x-5)$
76. $(4x-3)(3x+2)$

13.4E   **77.** $(x+3)(x+2)$ **78.** $(y-4)(y-2)$ **79.** $(w-5)(w+3)$ **80.** $(x+7)(x-5)$ **81.** $(y-7)(y+5)$
82. $(x+4)(x-4)$ **83.** $x(y+z-w)$ **84.** $(x-8)(x+4)$ **85.** $a(x+y-w)$ **86.** $(2x-1)(x-1)$
87. $(5w+3)(5w-3)$ **88.** $(2m^2+3w)(2m^2-3w)$ **89.** $(3x+1)(5x-2)$ **90.** $(4w-3)(w-1)$
91. $(y-6)(y+4)$ **92.** $xy(xy+7y-4w)$ **93.** $(6x-5)(x+2)$ **94.** $(3x+2)(2x-3)$

Section 13.5A   **1.** $(x-y)^2$ **2.** $(x+2w)(x+w)$ **3.** $(y+4z)(y+z)$ **4.** $(w-2x)(w-x)$
5. $(w+5y)(w+2y)$ **6.** $(y-7w)(y+3w)$ **7.** $(x-3y)(x-2y)$ **8.** $(m+7n)(m-n)$ **9.** $(p-3m)(p+2m)$
10. $(k+7w)(k-5w)$

13.5B   **11.** $3(x+3)(x+1)$ **12.** $2(x-5)(x+2)$ **13.** $5(x+3)(x-3)$ **14.** $7(y+2)(y-2)$
15. $x(x+3)(x+2)$ **16.** $y^2(y-4)(y-2)$ **17.** $2(2w+5)(2w-5)$ **18.** $3m(2m+5)(2m-5)$
19. $2x(x+5)(x+2)$ **20.** $5y(y-4)(y-3)$ **21.** $2y(2y+1)(2y-1)$ **22.** $2x^2y(5y+3)(5y-3)$
23. $2y(x-6)(x+1)$ **24.** $5m^2n(n+1)^2$ **25.** $3xy(2y+3)(2y-3)$ **26.** $5wm(2w+3m)(2w-3m)$
27. $3a(b-6)(b+2)$ **28.** $x(4x+1)(x+3)$ **29.** $y(x^2+y)^2$ **30.** $y(2w^3+5yw^3-3y^2x)$

Chapter 13 Review   **1.** $2x^3+3x^2-13x+28$ **2.** $4y^4+2y^3-31y^2+32y-5$ **3.** $w^2+3w-40$
4. $6y^2-23y+21$ **5.** $x^2-8x+16$ **6.** $9y^2+30y+25$ **7.** $16y^2-9$ **8.** $-19-\sqrt{10}$ **9.** $(w+12)(w-12)$
10. $(2y+7x)(2y-7x)$ **11.** $(y+5)(y+2)$ **12.** $(w-5)(w-4)$ **13.** $(x-6)(x+1)$ **14.** $(k+11)(k-7)$
15. $(3x+4)(x+2)$ **16.** $(4y-3)(3y+5)$ **17.** $3(x+2)(3x-4)$ **18.** $x(x-7)(x-2)$ **19.** $2(3x+2)(3x-2)$
20. $4x^2(5x+3)(5x-3)$ **21.** $(w+4n)(w+3n)$ **22.** $2xy(x+5)(x-3)$

Supplementary Exercises   **1.** y^3+y^2-2y-8 **2.** $6w^3-19w^2+18w-5$ **3.** $x^4-x^3+11x-15$
4. $y^4+4y^3-9y^2+26y-4$ **5.** $x^2-7x+12$ **6.** $w^2-3w-10$ **7.** $12x^2+21x-6$ **8.** $14y^2-29y+12$
9. $w^2-14w+49$ **10.** $y^2+12y+36$ **11.** $4x^2-20x+25$ **12.** $9b^2+24b+16$ **13.** y^2-64 **14.** $49x^2-25$
15. $23-33\sqrt{3}$ **16.** $\sqrt{21}+\sqrt{35}+\sqrt{6}+\sqrt{10}$ **17.** $54-26\sqrt{6}$ **18.** $(y+9)(y-9)$ **19.** $(m+5n)(m-5n)$
20. $(w+5)(w+3)$ **21.** $(y-7)(y+2)$ **22.** $(x-6)(x+1)$ **23.** $(m+3)(m+4)$ **24.** $(k+9)(k-5)$
25. $(p-7)(p-3)$ **26.** $(5y-2)(y+3)$ **27.** $(2x-5)(x+2)$ **28.** $(3y-2)(2y+3)$ **29.** $(4w-3)(2w-5)$
30. $(7x-6)(x+2)$ **31.** $(4y-3)(y-2)$ **32.** $2y(y+4)(y+1)$ **33.** $6x(x+5)(x-5)$ **34.** $3x^2(x+3)(x-3)$
35. $3x(2x+1)(x-3)$ **36.** $(y+5x)(y-x)$ **37.** $(w-5m)(w+2m)$ **38.** $x(y-7)(y+2)$ **39.** $2w(x-4)(x-3)$
40. $3xy(2x-5)(x+1)$ **41.** $3pm(2p+3m)(2p-3m)$

Chapter 13 Test   **1.** x^2+x-30 **2.** $6y^2-23y+20$ **3.** $w^3+7w^2+19w+21$ **4.** $12y^4-13y^3+2y^2+2y-1$
5. $36n^2-1$ **6.** $9x^2-25y^2$ **7.** $5-5\sqrt{7}$ **8.** $y^2-8y+16$ **9.** $4x^2-20x+25$ **10.** $5(2y+7w)$
11. $7xy^2(2x^2-4y+5x^3)$ **12.** $(w+6)(w+5)$ **13.** $(y-6)(y+4)$ **14.** $(x-2)(x-5)$ **15.** $(k+7)(k-3)$
16. $4(w-3)(w+2)$ **17.** $x(x-4)(x-2)$ **18.** $2x(x-3)(x-1)$ **19.** $(p+1)(p-1)$ **20.** $(3x+5)(3x-5)$
21. $3x(2x+5)(x-3)$ **22.** $6(w+4)(w-4)$ **23.** $3x^3(5x+6)(5x-6)$ **24.** $(p+2k)^2$ **25.** $2y(2x-3)(x-2)$

Cumulative Review   **1.** $13/30$ **2.** $5/11$ **3.** 8.0082 **4.** 0.0765 **5.** 1.4 **6.** 5.504 **7.** -40 **8.** 24
9. $-8/15$ **10.** -57 **11.** $17x^2-16x-17$ **12.** $21w^5-35w^4+7w^3$ **13.** $w^2-7w+10$
14. $3y^3-5y^2+17y-10$ **15.** $(5a^3)/(12bc^3)$ **16.** $1/(a^6b^2)$ **17.** $x=3$ **18.** $x=-7$ **19.** $x=-2, y=-13$

13

20. $x = -2, y = 3$ **21.** Slope 3; y-int. $(0, -5)$ **22.** Slope 5/7; y-int. $(0, 3/7)$ **23.** $y = -5x - 3$
24. $y = 3x + 14$ **25.** 77   **26.** $61.54   **27.** (a) 600 mi (b) 20 gal (c) $25.00   **28.** $400 at 7%; $800 at 6%
29. 42 goldfish   **30.** 20 games

CHAPTER 14

Section 14.1A   **1.** 3   **2.** 3   **3.** 4   **4.** 6   **5.** 4   **6.** 3   **7.** 4   **8.** 5   **9.** Numerator 3; denominator 3
10. $[3xxy(x + 2)]/[3wy(x - 2)]$; numerator 5; denominator 4   **11.** Numerator 4; denominator 4
12. Numerator 3; denominator 4   **13.** $[3xx(x - 2)]/[5x(x + 1)(x + 2)]$; numerator 4; denominator 4
14. $[2 \cdot 7yyxxx(x - 1)(x + 5)]/[3 \cdot 7yxx(x + 1)(x - 1)]$; numerator 9; denominator 7
15. $[2 \cdot 2 \cdot 5xyy(x + 2y)]/[3 \cdot 5xxy(2x + y)]$; numerator 7; denominator 6
16. $[2 \cdot 2 \cdot 3xxywww(x + 3)]/[3 \cdot 5xyyyyy(w + 3)]$; numerator 10; denominator 9
17. $[x(y - 3)(x + 2)]/[y(y + 2)(x - 3)]$; numerator 3; denominator 3   **18.** $[5(x + 1)(y - 4)]/[3(y - 4)(x - 1)]$;
numerator 3; denominator 3   **19.** $[(x - 2)(x + 2)]/[(x + 1)(x + 2)]$; numerator 2: denominator 2
20. $[(x - 2)(x - 2)]/[(x - 2)(x - 7)]$; numerator 2; denominator 2

14.1B   **21.** Same   **22.** Not same   **23.** Same   **24.** Not same   **25.** Not same   **26.** Same   **27.** Same
28. Not same   **29.** $(7x)/(5w)$ **30.** $[x^2(x + 2)]/[w(x - 2)]$ **31.** $(2x)/(3y)$ **32.** $[7(x + 4)]/[5x(x + 3)]$
33. $[3x(x - 2)]/[5(x + 1)(x + 2)]$ **34.** $[2yx(x + 5)]/[3(x + 1)]$ **35.** $[4y(x + 2y)]/[3x(2x + y)]$
36. $[4xw^3(x + 3)]/[5y^4(w + 3)]$ **37.** $[x(y - 3)]/[y(x - 3)]$ **38.** $[5(x + 1)]/[3(x - 1)]$ **39.** $(x - 2)/(x + 1)$
40. $(x - 2)/(x - 7)$ **41.** $1/(x - 3)$ **42.** $(x - 3)$ **43.** $(y + 3)/(y + 5)$ **44.** $[3x(2x - 3)]/[(x - 4)(x - 3)]$
45. $[x(y + 2)]/(y - 3)$ **46.** $[2y(2y - 3)]/(y + 1)$ **47.** $(3x - 2)/[y(x - 5)]$ **48.** $(x^3y)/(y + 7)$
49. $(x - 3)/(2x + 1)$ **50.** $[x^2(y - 1)]/[3(y + 2)]$

14.1C   **51.** $(x - 1)/(x + 1)$ **52.** $(x + 1)/x$ **53.** $(y + 3)/[y(y - 4)]$ **54.** $[w(w - 2)]/x$ **55.** $(x + 2)/(x - 3)$
56. $[w(w - 5)]/(x + 4)$ **57.** $y/(y - 1)$ **58.** $x - 4$ **59.** $1/(2w)$ **60.** $1/(x - 1)$

Section 14.2A   **1.** $(3x)/5$ **2.** $(x - 4)/(x + 4)$ **3.** $(12x^2y^2)/[5(x + 1)(x + 2)]$ **4.** $[4x(x - 2)]/[15(x + 2)]$
5. $(2x - 1)/(1 + 2x)$ **6.** $(7xy)/[4(x + 3)]$ **7.** $[x(x + 2)]/(5y^2)$ **8.** $(2x)/5$

14.2B   **9.** $[(x - 7)(x + 3)]/(x - 1)^2$ **10.** $[(x - 4)(x - 6)]/x^3$ **11.** $[(x + 1)(x + 3)]/[y(x - 2)]$
12. $(x - 3)/[xy(y + 1)]$ **13.** $x + 3$ **14.** $(x + 2)/(x + 7)$ **15.** $[(x + 3)(x - 1)]/[(x + 8)(x + 6)]$
16. $[2x^2(x + 1)]/[(x - 7)(2x + 5)]$ **17.** $[(2x + 1)(x + 5)]/[(5x - 3)(3x + 1)]$
18. $[(4x - 3)(x + 11)]/[(2x + 3)(x - 1)]$ **19.** $(x + 1)/(x - 3)$ **20.** $[(x + 1)(x + 3)]/[(x + 2)(x - 4)]$ **21.** $x/(x + 3)$
22. $(x - 2)/x$ **23.** $(x - 1)/(3x)$ **24.** $[x(x + 1)]/[(x - 3)(x + 3)]$ **25.** $[x^2(x - 1)]/(x + 2)$
26. $[(x + 1)(x + 4)]/[(x + 5)(x - 3)]$

Section 14.3A   **1.** $x - 1$ **2.** $(y + 1)/(y - 1)$ **3.** $(w - 5)^2/(w + 2)$ **4.** $(x - 6)/[x(x + 4)]$
5. $[(y + 9)(y + 3)]/(y + 7)$ **6.** $[m(m - 2)]/(m - 1)^2$ **7.** $[x(x + 7)]/(x + 5)$ **8.** $1/p$ **9.** $(2p - 3)/(p + 5)$
10. $[(x + 6)(x + 7)]/(2x + 3)$

14.3B   **11.** $[(x + 1)(x + 7)]/[(x - 3)(x - 2)]$ **12.** $(49x)/(4y)$ **13.** $[(x - 4)(x + 2)]/(x + 3)^2$
14. $[x^2(x - 7)]/[(x + 1)^2(x + 2)]$ **15.** $x^2/(x - 4)$ **16.** $(x + 2)/(x - 7)$ **17.** $[(x - 5)(x + 2)]/(x - 2)^2$
18. $x/(3x + 5)$ **19.** $[(x + 2)(x - 2)]/(x - 3)$ **20.** $[2(x + 1)]/(x + 7)$

Section 14.4A   **1.** $5/(x - 3)$ **2.** $11/(y + 3x)$ **3.** $13/(w + 5)$ **4.** $[2(x - 1)]/(x + 4)$ **5.** $(4y - 5)/(y - 1)$
6. $[2(4 - x)]/(3x - 4)$ **7.** $(11a - 2b)/(5a^2b)$ **8.** $[2(w + 8)]/(w^2 - 4y)$ **9.** $(6y + 4x + 5)/(x + 3y)$
10. $[2(a - 2b)]/(a - 5b)$ **11.** $3/(x + 3)$ **12.** $(a - 5b)/(3a + b)$ **13.** $8/(x - 1)$ **14.** $-5/(2y + 3)$
15. $(-3x + 2y)/(x - y)$ **16.** $(4a + 5b)/(a + 2b)$ **17.** $[2(1 - 2x)]/(2x + 5)$ **18.** $(-9a - 2b)/(2a + b)$

14.4B **19.** $(x+6)/(x+4)$ **20.** $(4x^2+2x+3)/(2x^2+5)$ **21.** $5/(x+5)$ **22.** $[2(2x-3)]/(x-3)$
23. $[2(2x-3)]/(x-1)$ **24.** $(4x+5)/(x-2)$ **25.** $(-2y-5)/(y+5)$ **26.** 6 **27.** $(w+1)/(w+3)$ **28.** 2
29. $(x+4)/(x+2)$ **30.** $(y^2-2y-6)/[(y+4)(y+1)]$

Section 14.5 **1.** $x(x-5)$ **2.** $(y+3)(y-1)$ **3.** $15xy^2$ **4.** $5xy(x+1)$ **5.** $30x^2y(x-3y)$ **6.** $210x^3y^2$
7. $3xy(x-1)^2$ **8.** $30xy^2(x+2)(x-3)^2$ **9.** $x(x+5)(x-5)$ **10.** $(x+3)(x+1)(x-1)$ **11.** $2x^2(x-4)(x-1)^2$
12. $4(3x-1)(x+1)(x-1)$ **13.** $10x^2y^3(3x-5y)(x^2y+4)$ **14.** $6a^3b^2(2a-b)$

Section 14.6A **1.** $(7x+6)/[x(x+2)]$ **2.** $(5y+8)/[y(y+2)]$ **3.** $[5(x+7)]/[x(x+5)]$ **4.** $(12-y)/[y(y-4)]$
5. $(8x-7)/[(x+1)(x-2)]$ **6.** $(4x^2-x-21)/[(x-2)(x-3)]$ **7.** $(5y+19)/[(y-1)(y+2)]$
8. $(-3w-19)/[(w-2)(w+3)]$ **9.** $(2xy+3y-2x)/(xy)$ **10.** $(5mn-n+3m)/(mn)$
11. $(2a^2-3a+13)/[(a+1)(a-2)]$ **12.** $(8b^2+7)/[(b+1)(b-2)]$ **13.** $(10a^2+3a+3)/[(2a+3)(a-2)]$
14. $(-4x-11)/[(x+5)(x+2)]$ **15.** $(5y+8)/[(y+1)(y-2)]$ **16.** $[2x(5x+7)]/[(3x-1)(x+4)]$

14.6B **17.** $(4+3x)/[x(x-2)]$ **18.** $[2(x+1)]/[x(x-3)(x+5)]$ **19.** $(4y-1)/[(y-2)(y+2)(y-1)]$
20. $[2(2w+9)]/[w(w-3)(w+3)]$ **21.** $(x+10)/[(x+3)(x+4)]$ **22.** $(19-2x)/[(x-7)(x+7)]$
23. $[2(1-2y)]/[y(y-3)]$ **24.** $[2(5x+4)]/[(2x+1)(x-1)]$ **25.** $(2x^2-2x+1)/[(x-4)(x-2)(x-1)]$
26. $[2(y^2-y+3)]/[(y-4)(y+1)^2]$ **27.** $[-5(w+6)]/[w(w-5)^2]$ **28.** $(-m^3+m-6)/[m^2(m-3)(m+3)]$
29. $[p(7p+10)]/[(p+5)(p+1)(p+2)]$ **30.** $(2-3x)/(x+4)$ **31.** $(5x^2+5x-7)/[x(x-1)(x+1)]$
32. $(3x^2-13x+36)/[(x+3)(x-3)^2]$ **33.** $[-2(2y^2-5y+9)]/[y^2(y-6)]$
34. $(-3m^2-20m+25)/[m^2(m+5)(m-5)]$

Section 14.7 **1.** 4 **2.** $(3+2\sqrt{2})/2$ **3.** $(7-5\sqrt{2})/3$ **4.** $3/5$ **5.** $[2(3+\sqrt{3})]/5$ **6.** $2-\sqrt{3}$ **7.** $3-\sqrt{2}$
8. $(5-\sqrt{5})/3$ **9.** $(6+\sqrt{10})/2$ **10.** $(1-\sqrt{7})/2$ **11.** $(3-\sqrt{6})/5$ **12.** $(7+\sqrt{5})/5$

Section 14.8 **1.** $x=24$ **2.** $w=6$ **3.** $x=2$ **4.** $y=3$ **5.** $w=6$ **6.** $x=4$ **7.** $x=-1$ **8.** $x=12$
9. $n=4$ **10.** $n=3$ **11.** $x=4$ **12.** $y=3$ **13.** $w=4$ **14.** $x=3$ **15.** $x=4$ **16.** $x=5$ **17.** $x=2$
18. $y=4$ **19.** $x=6$ **20.** $x=3$ **21.** $x=-1$

Section 14.9A **1.** 2 hr **2.** 3 min **3.** $3\frac{1}{3}$ min **4.** $1\frac{1}{2}$ hr **5.** $1\frac{1}{5}$ hr **6.** $1\frac{3}{7}$ min **7.** $2\frac{2}{9}$ hr
8. $1\frac{1}{3}$ min **9.** Kate 4 hr; John 12 hr **10.** New 20 hr; old 5 hr **11.** Peter 3 hr; Leah 6 hr **12.** Electric 6 hr;
hand 30 hr **13.** New press $2\frac{1}{2}$ hr; old press 10 hr **14.** 12 hr **15.** $7\frac{1}{2}$ hr

14.9B **16.** 6 mph **17.** 10 mph **18.** 3 mph **19.** 4 mph **20.** 10 mph **21.** 142 mph **22.** 1 mph
23. 12 mph **24.** 220 mph **25.** 30 mph

Chapter 14 Review **1.** 3 **2.** $2 \cdot 2x(x-1)(x-1)$; 5 **3.** Not same **4.** Same **5.** $[3x(x-2)]/[4y^2(x+2)]$
6. $(3x+1)/[y(x+2)]$ **7.** $y-3$ **8.** $[3(m-2)]/(2m)$ **9.** $[(x+3)(x-1)]/[x(x+1)]$ **10.** $[6x(x-5)]/(x-1)^2$
11. $[2m(m+1)]/[(m-3)(4m-3)]$ **12.** $1/[5(x+2)]$ **13.** $(7x+3)/(3x-1)$ **14.** $(5x+1)/(4x-5)$
15. $(3x-4)/(2x+1)$ **16.** $20x^3y^2$ **17.** $15xy(y-3)^2$ **18.** $(2x-5)(x+3)$ **19.** $(4x-1)(x+2)(x+6)$
20. $[x(5x-1)]/[(x+4)(x-3)]$ **21.** $(17a^2+ab-3b^2)/[(5a+b)(2a-b)]$ **22.** $[-2(x+5)]/[(x+3)(x+4)]$
23. $(-3y-4)/(y+2)^2$ **24.** $(2x-1)/(x+3)^2$ **25.** $[-(x+1)]/[x(x-4)]$ **26.** $(5+\sqrt{7})/3$ **27.** $x=6$
28. $x=5$ **29.** $3\frac{1}{3}$ hr **30.** 20 mph

Supplementary Exercises **1.** 3 **2.** 3 **3.** $3 \cdot 5xxy(x-4)(x-4)$; 7 **4.** Same **5.** Not same
6. $[3y(x+1)]/[5w(x-1)]$ **7.** $[3x^2(3-y)]/[5y(y+3)]$ **8.** $(y-2)/[y(y+5)]$ **9.** $1/(2y)$
10. $x-3$ **11.** $(2ab)/[(a+3)(2a+1)]$ **12.** $[3b(2b-3)]/[5a(a+b)]$ **13.** $[(x+7)(x-9)]/[(x+1)(x+9)]$
14. $[(x+6)(x-6)]/(x+5)$ **15.** $[(x-3)(x+2)]/[(x+1)(x+6)]$ **16.** $[x(x-4)^2]/[4y^2(x+3)(x-1)]$
17. $[4x(x-4)]/[15(x+2)]$ **18.** $(x+2)/[x(x-4)]$ **19.** $(18x^2)/(35y^2)$ **20.** $[2x(x-4)]/(x-3)$ **21.** $9/(x+1)$
22. $(7x-1)/(4x-3)$ **23.** $-2/(3x)$ **24.** $(-x+10)/(5x-1)$ **25.** $(x+4)/(x+2)$
26. $[-3(x-2)]/(x+8)$ or $[3(2-x)]/(x+8)$ **27.** $15abc$ **28.** $36x^3y^2$ **29.** $24xy(x+3)^2$
30. $20xy(x-2)(x+3)(x+2)$ **31.** $(x+5)(x-2)(x+1)$ **32.** $(2x+5)(3x-4)$ **33.** $(11x+13)/[(x-1)(x+2)]$

14

34. $(7y^2 + 5y - 25)/[y(y - 5)]$ **35.** $(10x^2 + 40x + 9)/[x(x + 6)(x + 3)]$ **36.** $(3y^2 - 8y + 32)/[y(y - 4)]$
37. $[3(x^2 + x - 1)]/[(x + 5)(x - 1)]$ **38.** $[5(2x^2 - x - 4)]/[x^2(x + 4)]$ **39.** $(2 - \sqrt{3})/5$ **40.** $y = 4$
41. $x = 4$ **42.** $1\frac{3}{5}$ hr   **43.** 260 mph

Chapter 14 Test
1. $(4x)/(5y^2)$ **2.** $x(x + 4)$ **3.** $[3x(x - 3)(2x - 3)]/[5(x - 1)(x + 2)]$
4. $[x(x - 5)(x + 3)]/[(x + 7)(x - 6)(x + 1)]$ **5.** $[2p(p - 6)]/[(p + 1)(p + 3)(p - 3)]$
6. $[k(k + 7)(k - 1)]/[2(5k - 3)(k + 4)]$ **7.** $(x + 5)/(4x^2)$ **8.** $(11x - 2)/(3x - 5)$ **9.** $(5y - 11)/(8y + 1)$
10. $(3x^2 + 5x - 5)/[x(x + 5)]$ **11.** $[4(8x - 3)]/[5x(x - 4)]$ **12.** $(7y^2 - 8y + 12)/[(y - 5)(y + 1)(y + 2)]$
13. $[3(y - 5)]/[y(y - 3)]$ **14.** $(3 - \sqrt{3})/2$ **15.** $w - 4$ **16.** $x = 4$ **17.** $x = 4$ **18.** 10 mph   **19.** 4 hr   **20.** $3\frac{1}{3}$ hr

Cumulative Review
1. $5/12$ **2.** $5/6$ **3.** 3.5072 **4.** 237.84 **5.** $\$14.30$ **6.** 30 **7.** -3 **8.** 17
9. $5xy^3 + xy + xy^2$ **10.** $(7x^4w)/(10y^3)$ **11.** $-(c^2)/(2a^3b)$ **12.** $-30x^5y^2w^5$ **13.** 6.82×10^3 **14.** $2y^2 + 6$
15. $-15x^3 + 20x^2 - 5x$ **16.** $7\frac{1}{2}$ **17.** $7x^2 - 14x + 33$ **18.** $5x^5y^2w^6$ **19.** $8x^2 + 19x - 15$
20. $x^3 - x^2 - x + 10$ **21.** $(y - 6)/(y + 5)$ **22.** $5a^2b^3\sqrt{3a}$ **23.** $3\sqrt{7} - 5\sqrt{5}$ **24.** $x = 2$ **25.** $y = -4$
26. $x = 3$ **27.** $x = 1, y = 2$ **28.** $x = 3, y = 5$ **29.** $y = 3x - 1$

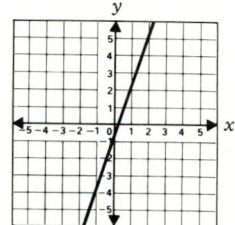

30. $y < 3x - 1$ 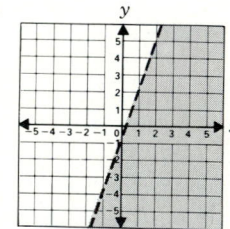   **31.** $y \geq 3x - 1$ 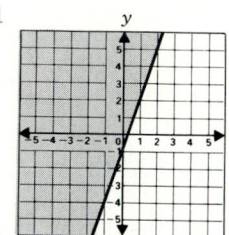   **32.** Slope 5/6; y-int. (0, 1 1/6)

33. $y = 3x - 15$ **34.** $-5°C$ **35.** 40 g water   **36.** 4 hr   **37.** 20 cents per granola bar; 25 cents per box of raisins

CHAPTER 15

Section 15.1A   **1.** $(3, 4)$ **2.** $(8, 1)$ **3.** $(-1, -6)$ **4.** $(-5, -2)$ **5.** $(2, 0)$ **6.** $(-7, 0)$ **7.** $(4, -1)$
8. $(-3, 3)$ **9.** $(1, 5)$ **10.** $(-3, -2)$ **11.** $(0, 0)$ **12.** $(0, 0)$ **13.** $(0, 1)$ **14.** $(0, -3)$ **15.** $(0, -7)$
16. $(0, 1)$ **17.** $(3, 0)$ **18.** $(-2, 0)$ **19.** $(0, 3)$ **20.** $(0, 1)$

15.1B   **21.** $(-1, -1)$ **22.** $(-3, -9)$ **23.** $(2, -4)$ **24.** $(4, -16)$ **25.** $(1, -1)$ **26.** $(6, -36)$
27. $(-7, -49)$ **28.** $(7, -49)$ **29.** $(3, -8)$ **30.** $(-2, -1)$ **31.** $(1, -6)$ **32.** $(-4, 4)$ **33.** $(-5, 5)$
34. $(3, -21)$ **35.** $(2, -14)$ **36.** $(-1, -4)$

15.1C

37. $(0, 1)$ 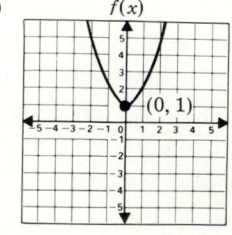   **38.** $(0, -2)$ 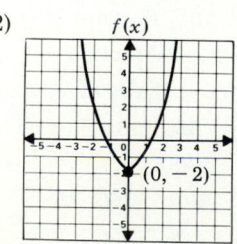   **39.** $(3, 0)$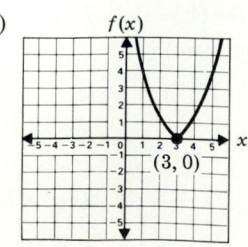

40. $(-3, -1)$

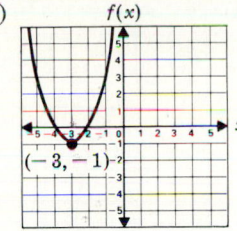

$(-3, -1)$

41. $(-1, 0)$

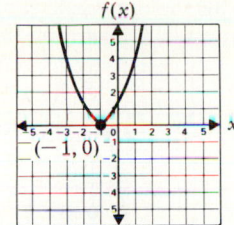

$(-1, 0)$

42. $(1, 5)$

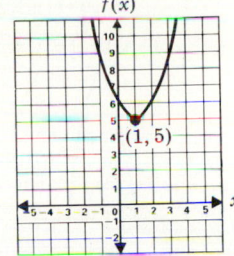

$(1, 5)$

43. $(2, -3)$

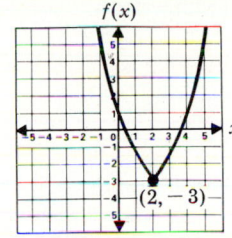

$(2, -3)$

44. $(-1, -1)$

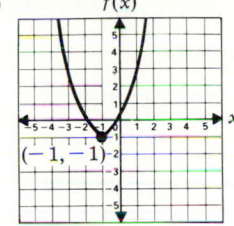

$(-1, -1)$

45. $(4, -2)$

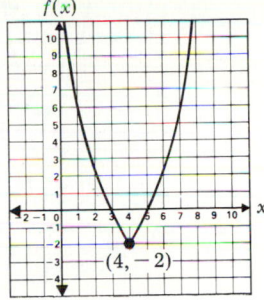

$(4, -2)$

46. $(-1, 2)$

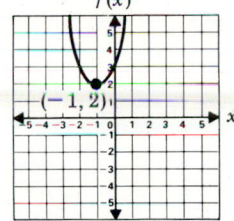

$(-1, 2)$

47. $(3, 0)$

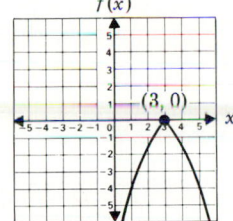

$(3, 0)$

48. $(-2, 0)$

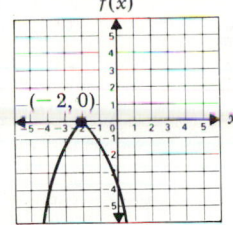

$(-2, 0)$

49. $(0, 0)$

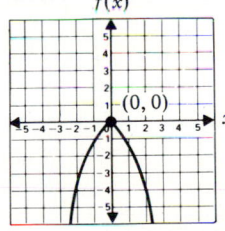

$(0, 0)$

50. $(0, 0)$

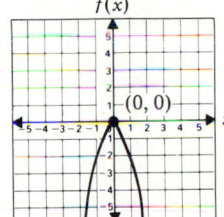

$(0, 0)$

51. $(0, 2)$

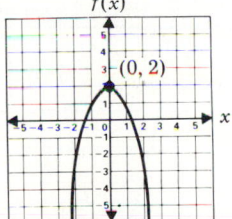

$(0, 2)$

52. $(0, -4)$

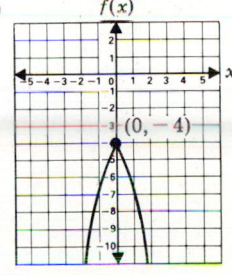

$(0, -4)$

53. $(1, 0)$

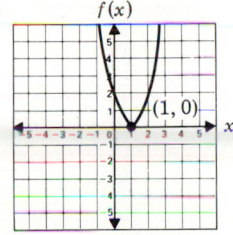

$(1, 0)$

54. $(2, 0)$

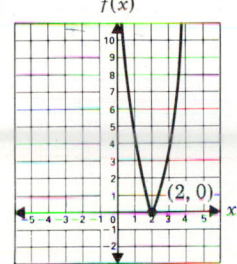

$(2, 0)$

55. $(-1, 1)$

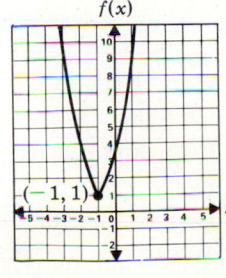

$(-1, 1)$

56. $(1, -2)$

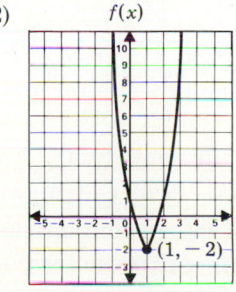

$(1, -2)$

57. $(3, 0)$

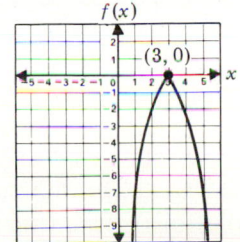

$(3, 0)$

15

58. $(-1, 0)$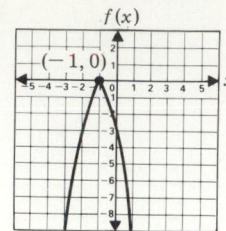

59. $(1, 3)$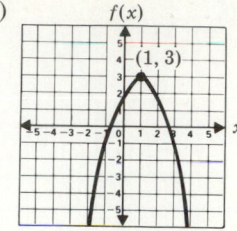

60. $(2, -4)$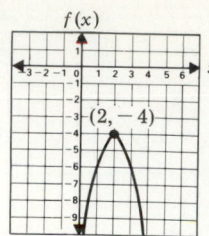

61. $(1, -1)$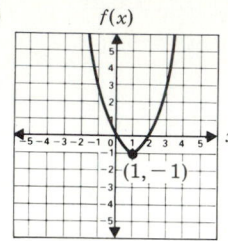

62. $(-2, -4)$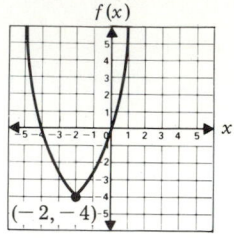

63. $(-5, -25)$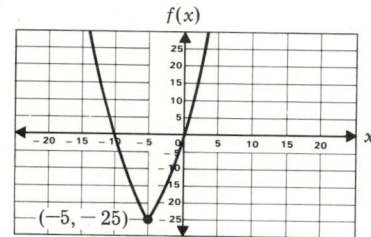

64. $(2, -4)$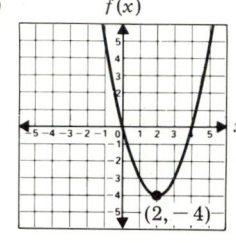

65. $(4, -13)$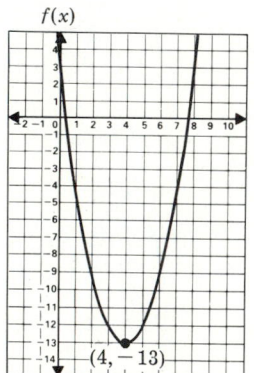

66. $(-1, -6)$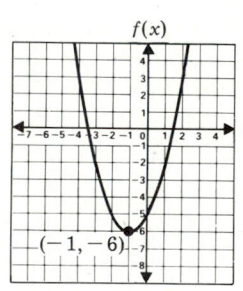

67. $(3, -10)$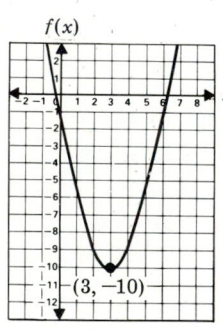

68. $(-3, -2)$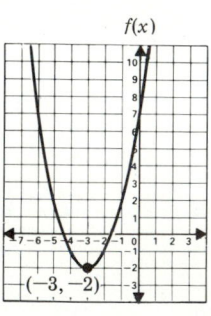

69. $(5, -17)$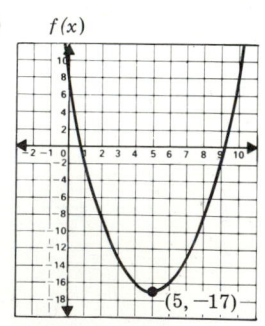

70. $(-4, -4)$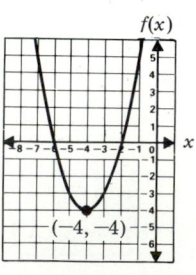

15.1D **71.** x-int.: $(4, 0)$, $(-4, 0)$; y-int.: $(0, -16)$ **72.** x-int.: $(2, 0)$, $(-2, 0)$; y-int.: $(0, -4)$

73. x-int.: $(1, 0)$, $(-1, 0)$; y-int.: $(0, -1)$ **74.** x-int.: $(\sqrt{3}, 0)$, $(-\sqrt{3}, 0)$; y-int.: $(0, -3)$

75. x-int.: $(3, 0)$; y-int.: $(0, 9)$ **76.** x-int.: $(-2, 0)$; y-int.: $(0, 4)$ **77.** x-int.: $(0, 0)$, $(2, 0)$; y-int.: $(0, 0)$

78. x-int.: $(0, 0)$, $(-4, 0)$; y-int.: $(0, 0)$ **79.** x-int.: $(5, 0)$, $(-1, 0)$; y-int.: $(0, -5)$

80. x-int.: $(-2, 0)$, $(-4, 0)$; y-int.: $(0, 8)$ **81.** x-int.: $(0, 0)$; y-int.: $(0, 0)$ **82.** No x-int.; y-int.: $(0, 8)$

83. x-int.: $(0, 0)$; y-int.: $(0, 0)$ **84.** x-int.: $(0, 0)$; y-int.: $(0, 0)$ **85.** x-int.: $(1, 0)$, $(-1, 0)$; y-int.: $(0, 1)$

86. x-int.: $(2, 0)$, $(-2, 0)$; y-int.: $(0, 8)$

Section 15.2 1. Yes 2. No 3. No 4. No 5. Yes 6. No 7. No 8. Yes 9. Yes 10. No
11. Yes 12. No 13. No 14. No 15. Yes 16. No 17. No 18. Yes 19. Yes 20. No

Section 15.3A 1. $x = \pm 5$ 2. $y = \pm 6$ 3. $w = \pm 10$ 4. $m = \pm 7$ 5. $x = \pm 2$ 6. $y = \pm 9$ 7. 7. $w = \pm 4$
8. $n = \pm 8$ 9. $r = \pm\sqrt{5}$ 10. $x = \pm\sqrt{13}$ 11. $y = \pm 2\sqrt{2}$ 12. $k = \pm\sqrt{10}$ 13. $a = \pm 2\sqrt{5}$
14. $b = \pm 2\sqrt{7}$ 15. $x = \pm\sqrt{14}$ 16. $y = \pm 3\sqrt{5}$ 17. $w = \pm 2\sqrt{10}$ 18. $r = \pm 2\sqrt{3}$

15.3B 19. $x = -1; x = -7$ 20. $y = 4; y = 0$ 21. $w = 6; w = -4$ 22. $m = -1; m = -9$ 23. $y = 1 \pm\sqrt{5}$
24. $y = -2 \pm\sqrt{3}$ 25. $w = -3 \pm\sqrt{2}$ 26. $x = 2 \pm\sqrt{7}$ 27. $y = 3 + 2\sqrt{2}$ 28. $x = -2 \pm 3\sqrt{5}$
29. $m = 5 \pm 3\sqrt{2}$ 30. $w = -4 \pm 5\sqrt{2}$ 31. $x = 1\frac{1}{2}; x = -1/2$ 32. $y = 1/3; y = -1\frac{2}{3}$
33. $w = 1; w = -1\frac{4}{5}$ 34. $a = 1\frac{1}{2}; a = -1$ 35. $y = (-1 \pm\sqrt{3})/2$ 36. $w = (1 \pm\sqrt{2})/3$

Section 15.4 1. $x = -3; x = -5$ 2. $y = 4; y = 1$ 3. $w = 0; w = -5$ 4. $y = 4; y = 2$
5. $x = 0; x = -2; x = -6$ 6. $y = 0; y = 3; y = -8$ 7. $x = \pm 2; x = 6; x = -1$ 8. $w = 0; w = \pm 3; w = 2$
9. $y = 0; y = 4; y = \pm 1; y = 2$ 10. $x = 0; x = \pm 1$ 11. $x = -4; x = 1/3$ 12. $y = -3/4; y = 6$
13. $w = 0; w = 2\frac{1}{2}$ 14. $x = 0; x = 2\frac{1}{3}$ 15. $y = -3; y = 1/5$ 16. $w = -3; w = 1/2$
17. $w = 1; w = 1\frac{1}{4}$ 18. $x = -1/2; x = 5$ 19. $x = 0; x = 1/2$ 20. $y = 0; y = 1/2$

Section 15.5A 1. $x = 0; x = -4$ 2. $y = 0; y = 7$ 3. $w = 0; w = -6$ 4. $x = 0; x = 1\frac{2}{3}$
5. $x = 0; x = 3/4$ 6. $y = 0; y = -3/5$ 7. $x = 0; x = 3$ 8. $y = 0; y = -4$ 9. $x = 0; x = 1\frac{1}{2}$
10. $y = 0; y = -2/3$ 11. $x = 0; x = 5$ 12. $y = 0; y = 1$ 13. $x = 0; x = -4$ 14. $y = -2$ 15. $w = 3; w = -1$
16. $x = 5; x = 1$ 17. $x = 3; x = 1$ 18. $w = -5; w = 3$ 19. $y = 3; y = -2$ 20. $x = 0; x = -7$ 21. $x = 1$
22. $x = 0; x = -5$ 23. $y = 0; y = 7$ 24. $w = 3; w = -2$ 25. $x = -4; x = 2$ 26. $y = 1; y = -4$

15.5B 27. $z = 0; x = \pm 1$ 28. $y = 0; y = 5$ 29. $w = 0; w = 6; w = -1$ 30. $z = 0; z = -2$ 31. $y = 0; y = \pm 2$
32. $w = 0; w = \pm 5$ 33. $x = 0; x = 6; x = -3$ 34. $y = 0; y = 7; y = -2$ 35. $w = 0; w = -7; w = 2$
36. $z = 0; z = \pm 3$ 37. $x = 0; x = 4; x = -2; x = -1$ 38. $w = 0; w = 3; w = 1$ 39. $y = 0; y = \pm 4; y = \pm 3$
40. $y = 0; y = -12; y = 10; y = -1$ 41. $x = 6; x = \pm 2; x = 1$ 42. $y = 0; y = 3; y = 5$ 43. $w = 0; w = 3; w = 2$
44. $x = 0; x = 1; x = 1\frac{1}{3}$ 45. $y = 0; y = -4; y = -1/3$ 46. $m = \pm 5; m = -3; m = -4$
47. $r = 0; r = 2/3; r = 5; r = 4$ 48. $h = 0; h = 2/5; h = -7$ 49. $x = 3; x = -1/2; x = -1; x = -1\frac{1}{2}$
50. $y = 1/3; y = \pm 2; y = 1/5$

Section 15.6A 1. $x = (-1 \pm\sqrt{5})/2$ 2. $x = (-3 \pm\sqrt{13})/2$ 3. $y = (5 \pm\sqrt{33})/2$ 4. $w = (-3 + \sqrt{21})/2$
5. $m = (-1 \pm\sqrt{17})/2$ 6. $n = (3 \pm\sqrt{13})/2$ 7. $y = (1 \pm\sqrt{7})/2$ 8. $x = 1 \pm\sqrt{3}$ 9. $w = 2; w = -1$
10. $k = -1 \pm\sqrt{6}$ 11. $x = (-3 \pm\sqrt{21})/6$ 12. $y = (1 \pm\sqrt{6})/5$

15.6B 13. 36; 2 rational; factor; $x = 4, x = -2$ 14. 13; 2 irrational; quadratic formula; $x = (-1 \pm\sqrt{13})/2$
15. -11; no real solutions 16. 101; 2 irrational; quadratic formula; $w = (7 \pm\sqrt{101})/2$ 17. 25; 2 rational; factor;
$y = 1\frac{1}{2}, y = -1$ 18. 28; 2 irrational; quadratic formula; $w = (-1 \pm\sqrt{7})/3$ 19. -16; no real solutions
20. 25; 2 rational; factor; $x = 2, x = -1/2$ 21. 4; rational, factor; $y = 0, y = 2/3$ 22. -3; no real solutions
23. 20; 2 irrational; quadratic formula; $x = -2 \pm\sqrt{5}$ 24. 17; 2 irrational; quadratic formula; $w = (3 \pm\sqrt{17})/4$
25. 0: 1 real, factor; $y = 1$ 26. 1; 2 rational; factor; $y = 2, y = 1$

15.6C 27. $x = 0; x = 3$ 28. $y = \pm 6$ 29. $x = 4; x = 0$ 30. $x = (1 \pm\sqrt{13})/2$ 31. $w = 15; w = -1$
32. $y = 1; y = -7$ 33. $w = -4 \pm\sqrt{7}$ 34. $x = (1 \pm\sqrt{41})/4$ 35. $x = 0; x = 8$ 36. $y = 2/3; y = 0$
37. $w = 1 \pm 2\sqrt{2}$ 38. $y = 8; y = -1$ 39. $x = 0; x = 3; x = -2$ 40. $x = (-3 \pm\sqrt{2})/2$

Section 15.7 1. $x = 9; x = -2$ 2. $x = 3$ 3. $y = 5; y = -2$ 4. $x = 6; x = 7$ 5. $x = 3; x = -2$
6. $y = 3; y = -1/3$ 7. $w = 2$ 8. $x = 5; x = -2$ 9. $y = -1; y = -1\frac{1}{2}$ 10. $w = 4$ 11. $y = 4; y = -3$
12. $x = 4; x = -2$ 13. $w = -6; w = 4$ 14. $x = -2; x = 6$ 15. $w = 5$ 16. $y = 3$

Section 15.8 1. 5 in.: 8 in. 2. 10 yd; 15 yd 3. 10 ft; 14 ft 4. 2 in. 5. 1 in. 6. 2 in. 7. 250 yd
8. 80 ft 9. 130 yd 10. 1 sec: and 2 sec on return 11. $1 ($24 is correct also, but unrealistic)

12. 20 mph   **13.** 15 mph   **14.** Betsy 4 mph; Alice 5 mph   **15.** Cedric 5 mph; Ernie 7 mph
16. Lee 3 hr; Henry 6 hr   **17.** Margaret 6 hr; Ellen 12 hr

Chapter 15 Review
1. $(-3, -1)$ **2.** $(-5, -6)$ **3.** $(0, -9)$ **4.** $(0, 0)$ **5.** $(0, 3)$ **6.** $(-5, -32)$
7. Vertex $(-3, -1)$

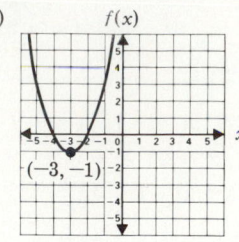

8. Vertex $(0, 3)$

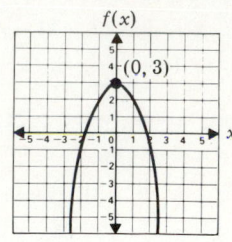

9. x-int.: $(7, 0)$; y-int.: $(0, -245)$ **10.** x-int.: $(4, 0)$, $(-4, 0)$; y-int.: $(0, -16)$ **11.** x-int.: $(-6, 0)$, $(-4, 0)$;
y-int.: $(0, 24)$ **12.** Yes   **13.** No   **14.** $x = \pm 7$ **15.** $y = \pm\sqrt{6}$ **16.** $w = 7$; $w = -3$ **17.** $p = -5 \pm 2\sqrt{5}$
18. $w = 3$; $w = -1/3$ **19.** $x = 4$; $x = -7$ **20.** $y = 1$; $y = -4/5$ **21.** $w = 0$; $w = 1$ **22.** $x = 0$; $x = -3$; $x = 1/2$
23. $x = 0$; $x = 8$ **24.** $y = 4$; $y = 3$ **25.** $w = 5$; $w = 4$ **26.** $x = -2$; $x = -4$ **27.** $y = 0$; $y = \pm 2$
28. $x = 0$; $x = -6$; $x = -3$; $x = -4$ **29.** $y = (-1 \pm \sqrt{41})/4$ **30.** $x = -2 \pm \sqrt{5}$ **31.** $w = (-1 \pm \sqrt{61})/10$
32. $x = (3 \pm \sqrt{13})/2$ **33.** $y = 7$; $y = 1$ **34.** $x = 4$; $x = -3$ **35.** 6 in.; 9 in.   **36.** Molly 4 hr; John 12 hr

Supplementary Exercises
1. $(4, 2)$ **2.** $(-2, -7)$ **3.** $(0, -8)$ **4.** $(0, 0)$ **5.** $(0, 6)$ **6.** $(-4, -21)$
7. Vertex $(4, 2)$

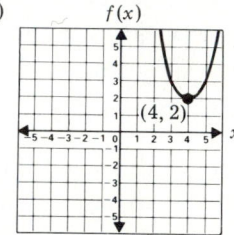

8. Vertex $(0, 6)$

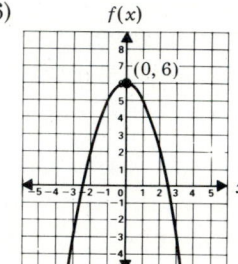

9. x-int.: $(5, 0)$; y-int.: $(0, -175)$

10. x-int.: $(6, 0)$, $(-6, 0)$; y-int.: $(0, -36)$ **11.** x-int.: $(0, 0)$, $(-4, 0)$; y-int.: $(0, 0)$ **12.** No   **13.** Yes   **14.** No
15. No   **16.** $x = \pm 1$ **17.** $y = \pm 3$ **18.** $w = \pm\sqrt{5}$ **19.** $m = \pm 3\sqrt{5}$ **20.** $x = -9$; $x = -1$ **21.** $y = 9$; $y = -3$
22. $w = 1\frac{1}{2}$; $w = -1/2$ **23.** $x = (-1 \pm \sqrt{5})/3$ **24.** $x = -3$; $x = 1$ **25.** $y = 4$; $y = 5$ **26.** $w = -1\frac{1}{2}$; $w = 1\frac{2}{5}$
27. $p = 1/2$; $p = -5/6$ **28.** $x = 0$; $x = -3$ **29.** $y = 0$; $y = 6$ **30.** $x = 0$; $x = -8$; $x = 3$
31. $y = 0$; $y = 1/3$; $y = -7$ **32.** $x = -2$ **33.** $y = 5$; $y = 3$ **34.** $w = 8$; $w = -7$ **35.** $x = 0$; $x = 11$
36. $m = -3$; $m = -2$ **37.** $k = 3$; $k = -1$ **38.** $r = 0$; $r = -5$ **39.** $y = 0$; $y = 1\,2/3$ **40.** $x = 0$; $x = \pm 4$
41. $y = 0$; $y = 3$; $y = 5$ **42.** $w = 0$; $w = -5$; $w = 4$ **43.** $x = (-1 \pm \sqrt{21})/2$ **44.** $y = -1 \pm \sqrt{2}$
45. $w = 1 \pm \sqrt{2}$ **46.** $m = (-3 \pm \sqrt{41})/2$ **47.** No real solutions   **48.** $x = 1$; $x = -1/3$
49. $y = 2$: $y = -4$ **50.** $w = 4$ **51.** $x = \pm 2$ **52.** 4 in.: 7 in.   **53.** Jim 6 days; Ron 12 days   **54.** Liz 3 mph;
Patti 4 mph

Chapter 15 Test
1. $(0, 7)$ **2.** $(5, 0)$ **3.** $(-4, -5)$ **4.** $(3, 4)$ **5.** $(1, -1)$ **6.** $(-5, -18)$
7. Vertex $(1, -3)$

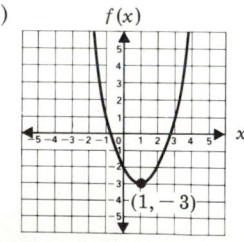

8. Vertex $(0, 4)$

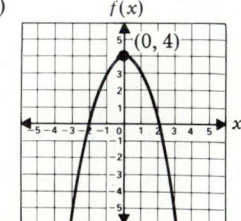

9. x-int.: $(7, 0)$, $(5, 0)$; y-int.: $(0, 35)$

10. x-int.: $(0, 0)$; y-int.: $(0, 0)$ **11.** Yes   **12.** No   **13.** $y = 3$; $y = 7$ **14.** $x = 0$; $x = -5$; $x = 1/2$ **15.** $x = \pm 5$
16. $y = 0$; $y = 8$ **17.** $m = -4$; $m = -1$ **18.** $r = (-1 \pm \sqrt{29})/2$ **19.** $w = \pm\sqrt{11}$ **20.** $r = 5$; $r = -2$
21. $y = 13$; $y = -5$ **22.** $x = -1 \pm 5\sqrt{2}$ **23.** $w = 5$; $w = -2$ **24.** $y = 0$; $y = -6$; $y = -2$ **25.** $x = 4$
26. $y = 3$; $y = -3/5$ **27.** $w = 2$; $w = -1$ **28.** 2 ft; 7 ft   **29.** Nick 8 min; Jim 24 min   **30.** First day 60 mph;
second day 50 mph

Cumulative Review   **1.** 13/35   **2.** 7/9   **3.** 34.129   **4.** 0.00369   **5.** 2.96485   **6.** -7 **7.** 6   **8.** 4

9. $8a^3b - 3ab^2 + 3a^2b^2$ **10.** $(3a^3c^2)/2$ **11.** $(b^2)/(a^6)$ **12.** $-24a^2b^3c^2$ **13.** 1.612×10^{13} **14.** $-2x^2 - 3x + 5$

15. $3a^4 - 12a^3 + 21a^2$ **16.** $5x + 3 - 2/x$ **17.** $17y^2 + 10y + 7$ **18.** $-15a^3b^2$ **19.** $8w^2 + 14w - 15$

20. $a^3 - a^2 - 23a + 15$ **21.** $(3y + 5x)/(xy)$ **22.** $[(3x - 2)(x - 3)]/4$ **23.** $2a\sqrt{5a}$ **24.** $8\sqrt{2} - \sqrt{3}$ **25.** $x = 5$

26. $x = 12$ **27.** $x = 2; y = 5$ **28.** $x = 2; y = -1$ **29.** Slope 3; y-int.: (0, 6)   **30.** $y = 2x - 10$ **31.** 15 g salt

32. 4 4/9 or 4.4 oz   **33.** 5 nickels, 7 dimes, 2 quarters   **34.** Pat 38; Mark 34   **35.** \$366

APPENDIX A

1.

Number	(a) Base 8	(b) Base 2	(c) Base 16
thirty seven	45	100,101	25
thirty eight	46	100,110	26
thirty nine	47	100,111	27
forty	50	101,000	28
forty one	51	101,001	29
forty two	52	101,010	2a
forty three	53	101,011	2b
forty four	54	101,100	2c
forty five	55	101,101	2d
forty six	56	101,110	2e
forty seven	57	111,111	2f
forty eight	60	1,000,000	30
forty nine	61	1,000,001	31
fifty	62	1,000,010	32
fifty one	63	1,000,011	33
fifty two	64	1,000,100	34
fifty three	65	1,000,101	35
fifty four	66	1,000,110	36
fifty five	67	1,000,111	37
fifty six	70	1,001,000	38
fifty seven	71	1,001,001	39
fifty eight	72	1,001,010	3a
fifty nine	73	1,001,011	3b
sixty	74	1,001,100	3c
sixty one	75	1,001,101	3d
sixty two	76	1,001,110	3e
sixty three	77	1,001,111	3f
sixty four	100	1,010,000	40
sixty five	101	1,010,001	41
sixty six	102	1,010,010	42
sixty seven	103	1,010,011	43
sixty eight	104	1,010,100	44
sixty nine	105	1,010,101	45
seventy	106	1,010,110	46

2.

Number	(a) Base 6	(b) Base 11
one	1	1
two	2	2
three	3	3
four	4	4
five	5	5
six	10	6
seven	11	7
eight	12	8
nine	13	9
ten	14	a
eleven	15	10
twelve	20	11
thirteen	21	12
fourteen	22	13
fifteen	23	14
sixteen	24	15
seventeen	25	16
eighteen	30	17
nineteen	31	18
twenty	32	19
twenty one	33	1a
twenty two	34	20
twenty three	35	21
twenty four	40	22
twenty five	41	23
twenty six	42	24
twenty seven	43	25
twenty eight	44	26
twenty nine	45	27
thirty	50	28
thirty one	51	29
thirty two	52	2a
thirty three	53	30
thirty four	54	31
thirty five	55	32
thirty six	100	33
thirty seven	101	34
thirty eight	102	35
thirty nine	103	36
forty	104	37

3. (a) 111 (b) 200 (c) 286 (d) 188 (e) 263 (f) 13 (g) 46 (h) 199 (i) 27 (j) 343 (k) 769 (l) 1078 (m) 696 (n) 319   **4.** (a) 460_8 (b) 732_8 (c) 566_8 (d) $11,010,100_2$ (e) $1,001,001_2$ (f) 458_{16} (g) $1,31b_{16}$

APPENDIX C

1. 0.16 m   **2.** 78,000 g   **3.** 280,000,000 L   **4.** 0.609 g   **5.** 0.092 m   **6.** 0.280186469 g   **7.** 580 m   **8.** 0.928 L

9. 6,905 g   **10.** 37.64 m   **11.** 0.8 km   **12.** 76,000 cg   **13.** 47,000 mm   **14.** 5.386 kg   **15.** 0.00000487 L

16. 34,500 cm   **17.** 6,019 m   **18.** 26,000 g   **19.** 0.005 m   **20.** 6,080 mL

Index